AF389621

Satellite Data Application, Validation and Calibration for Atmospheric Observation

Satellite Data Application, Validation and Calibration for Atmospheric Observation

Editors

Nicholas Nalli
Quanhua Liu
Lori A. Borg

MDPI • Basel • Beijing • Wuhan • Barcelona • Belgrade • Manchester • Tokyo • Cluj • Tianjin

Editors
Nicholas Nalli
I.M. Systems Group, Inc.
NOAA/NESDIS Center
for Satellite Applications
and Research
College Park
USA

Quanhua Liu
Center for Satellite Applications
and Research (STAR)
NOAA/NESDIS
College Park
USA

Lori A. Borg
Space Science Engineering
Center / Cooperative Institute
for Meteorological
Satellite Studies
University of
Wisconsin-Madison
Madison
USA

Editorial Office
MDPI
St. Alban-Anlage 66
4052 Basel, Switzerland

This is a reprint of articles from the Special Issue published online in the open access journal *Remote Sensing* (ISSN 2072-4292) (available at: www.mdpi.com/journal/remotesensing/special_issues/rs_app_val_cal_atmos).

For citation purposes, cite each article independently as indicated on the article page online and as indicated below:

LastName, A.A.; LastName, B.B.; LastName, C.C. Article Title. *Journal Name* **Year**, *Volume Number*, Page Range.

ISBN 978-3-0365-2138-1 (Hbk)
ISBN 978-3-0365-2137-4 (PDF)

Contents

About the Editors . . . . **ix**

Preface to "Satellite Data Application, Validation and Calibration for Atmospheric
Observation" . . . . **xi**

**Bomin Sun, Xavier Calbet, Anthony Reale, Steven Schroeder, Manik Bali, Ryan Smith and
Michael Pettey**
Accuracy of Vaisala RS41 and RS92 Upper Tropospheric Humidity Compared to Satellite
Hyperspectral Infrared Measurements
Reprinted from: *Remote Sens.* **2021**, *13*, 173, doi:10.3390/rs13020173 . . . . **1**

Xingming Liang and Quanhua (Mark) Liu
Applying Deep Learning to Clear-Sky Radiance Simulation for VIIRS with Community
Radiative Transfer Model—Part 2: Model Architecture and Assessment
Reprinted from: *Remote Sens.* **2020**, *12*, 3825, doi:10.3390/rs12223825 . . . . **27**

Song Yang, Richard Bankert and Joshua Cossuth
Tropical Cyclone Climatology from Satellite Passive Microwave Measurements
Reprinted from: *Remote Sens.* **2020**, *12*, 3610, doi:10.3390/rs12213610 . . . . **47**

**Emily Berndt, Nadia Smith, Jason Burks, Kris White, Rebekah Esmaili, Arunas Kuciauskas,
Erika Duran, Roger Allen, Frank LaFontaine and Jeff Szkodzinski**
Gridded Satellite Sounding Retrievals in Operational Weather Forecasting: Product Description
and Emerging Applications
Reprinted from: *Remote Sens.* **2020**, *12*, 3311, doi:10.3390/rs12203311 . . . . **67**

Bingkun Luo and Peter J. Minnett
Comparison of SLSTR Thermal Emissive Bands Clear-Sky Measurements with Those of
Geostationary Imagers
Reprinted from: *Remote Sens.* **2020**, *12*, 3279, doi:10.3390/rs12203279 . . . . **97**

Benjamin Scarino, David R. Doelling, Rajendra Bhatt, Arun Gopalan and Conor Haney
Evaluating the Magnitude of VIIRS Out-of-Band Response for Varying Earth Spectra
Reprinted from: *Remote Sens.* **2020**, *12*, 3267, doi:10.3390/rs12193267 . . . . **117**

**Nicholas R. Nalli, Changyi Tan, Juying Warner, Murty Divakarla, Antonia Gambacorta,
Michael Wilson, Tong Zhu, Tianyuan Wang, Zigang Wei, Ken Pryor, Satya Kalluri, Lihang
Zhou, Colm Sweeney, Bianca C. Baier, Kathryn McKain, Debra Wunch, Nicholas M.
Deutscher, Frank Hase, Laura T. Iraci, Rigel Kivi, Isamu Morino, Justus Notholt, Hirofumi
Ohyama, David F. Pollard, Yao Té, Voltaire A. Velazco, Thorsten Warneke, Ralf Sussmann
and Markus Rettinger**
Validation of Carbon Trace Gas Profile Retrievals from the NOAA-Unique Combined
Atmospheric Processing System for the Cross-Track Infrared Sounder
Reprinted from: *Remote Sens.* **2020**, *12*, 3245, doi:10.3390/rs12193245 . . . . **141**

Yuanyuan Yang and Yong Wang
Using the BFAST Algorithm and Multitemporal AIRS Data to Investigate Variation of
Atmospheric Methane Concentration over Zoige Wetland of China
Reprinted from: *Remote Sens.* **2020**, *12*, 3199, doi:10.3390/rs12193199 . . . . **171**

Yan Zhou and Christopher Grassotti
Development of a Machine Learning-Based Radiometric Bias Correction for NOAA's
Microwave Integrated Retrieval System (MiRS)
Reprinted from: *Remote Sens.* **2020**, *12*, 3160, doi:10.3390/rs12193160 . . . . **189**

Haruki Oshio, Yukio Yoshida, Tsuneo Matsunaga, Nicholas M. Deutscher, Manvendra
Dubey, David W. T. Griffith, Frank Hase, Laura T. Iraci, Rigel Kivi, Cheng Liu, Isamu Morino,
Justus Notholt, Young-Suk Oh, Hirofumi Ohyama, Christof Petri, David F. Pollard, Coleen
Roehl, Kei Shiomi, Ralf Sussmann, Yao Té, Voltaire A. Velazco, Thorsten Warneke and Debra
Wunch
Bias Correction of the Ratio of Total Column CH_4 to CO_2 Retrieved from GOSAT Spectra
Reprinted from: *Remote Sens.* **2020**, *12*, 3155, doi:10.3390/rs12193155 . . . . **205**

Banghua Yan, Junye Chen, Cheng-Zhi Zou, Khalil Ahmad, Haifeng Qian, Kevin Garrett,
Tong Zhu, Dejiang Han and Joseph Green
Calibration and Validation of Antenna and Brightness Temperatures from Metop-C Advanced
Microwave Sounding Unit-A (AMSU-A)
Reprinted from: *Remote Sens.* **2020**, *12*, 2978, doi:10.3390/rs12182978 . . . . **235**

Wenze Yang, Huan Meng, Ralph R. Ferraro and Yong Chen
Inter-Calibration of AMSU-A Window Channels
Reprinted from: *Remote Sens.* **2020**, *12*, 2988, doi:10.3390/rs12182988 . . . . **257**

Cheng-Zhi Zou, Lihang Zhou, Lin Lin, Ninghai Sun, Yong Chen, Lawrence E. Flynn, Bin
Zhang, Changyong Cao, Flavio Iturbide-Sanchez, Trevor Beck, Banghua Yan, Satya Kalluri,
Yan Bai, Slawomir Blonski, Taeyoung Choi, Murty Divakarla, Yalong Gu, Xianjun Hao, Wei
Li, Ding Liang, Jianguo Niu, Xi Shao, Larrabee Strow, David C. Tobin, Denis Tremblay, Sirish
Uprety, Wenhui Wang, Hui Xu, Hu Yang and Mitchell D. Goldberg
The Reprocessed Suomi NPP Satellite Observations
Reprinted from: *Remote Sens.* **2020**, *12*, 2891, doi:10.3390/rs12182891 . . . . **279**

Hartmut H. Aumann, Steven E. Broberg, Evan M. Manning, Thomas S. Pagano and Robert
C. Wilson
Evaluating the Absolute Calibration Accuracy and Stability of AIRS Using the CMC SST
Reprinted from: *Remote Sens.* **2020**, *12*, 2743, doi:10.3390/rs12172743 . . . . **311**

Sang Seo Park, Sang-Woo Kim, Chang-Keun Song, Jong-Uk Park and Kang-Ho Bae
Spatio-Temporal Variability of Aerosol Optical Depth, Total Ozone and NO_2 Over East Asia:
Strategy for the Validation to the GEMS Scientific Products
Reprinted from: *Remote Sens.* **2020**, *12*, 2256, doi:10.3390/rs12142256 . . . . **321**

Bingkun Luo and Peter J. Minnett
Evaluation of the ERA5 Sea Surface Skin Temperature with Remotely-Sensed Shipborne
Marine-Atmospheric Emitted Radiance Interferometer Data
Reprinted from: *Remote Sens.* **2020**, *12*, 1873, doi:10.3390/rs12111873 . . . . **339**

Yunheng Xue, Jun Li, Zhenglong Li, Mathew M. Gunshor and Timothy J. Schmit
Evaluation of the Diurnal Variation of Upper Tropospheric Humidity in Reanalysis Using
Homogenized Observed Radiances from International Geostationary Weather Satellites
Reprinted from: *Remote Sens.* **2020**, *12*, 1628, doi:10.3390/rs12101628 . . . . **359**

Martin J. Burgdorf, Thomas G. Müller, Stefan A. Buehler, Marc Prange and Manfred Brath
Characterization of the High-Resolution Infrared Radiation Sounder Using Lunar Observations
Reprinted from: *Remote Sens.* **2020**, *12*, 1488, doi:10.3390/rs12091488 . . . . **373**

Hou Jiang, Yaping Yang, Hongzhi Wang, Yongqing Bai and Yan Bai
Surface Diffuse Solar Radiation Determined by Reanalysis and Satellite over East Asia:
Evaluation and Comparison
Reprinted from: *Remote Sens.* **2020**, *12*, 1387, doi:10.3390/rs12091387 . . . . **393**

Robbie Iacovazzi, Lin Lin, Ninghai Sun and Quanhua Liu
NOAA Operational Microwave Sounding Radiometer Data Quality Monitoring and Anomaly
Assessment Using COSMIC GNSS Radio-Occultation Soundings
Reprinted from: *Remote Sens.* **2020**, *12*, 828, doi:10.3390/rs12050828 . . . . **413**

Yeeun Lee, Myoung-Hwan Ahn and Mina Kang
The New Potential of Deep Convective Clouds as a Calibration Target for a Geostationary
UV/VIS Hyperspectral Spectrometer
Reprinted from: *Remote Sens.* **2020**, *12*, 446, doi:10.3390/rs12030446 . . . . **437**

About the Editors

Nicholas Nalli

Nicholas R. Nalli received his B.Sc and M.Sc. degrees in education from the State University of New York, College at Oneonta, NY, and his M.Sc. and Ph.D. degrees in atmospheric and oceanic sciences from the University of Wisconsin—Madison, USA. He was awarded a four-year Postdoctoral Fellowship with Colorado State University (CSU), while working onsite at the U.S. National Oceanic and Atmospheric Administration, National Environmental Satellite, Data and Information Service (NOAA/NESDIS) as a Visiting Scientist. He is currently a Senior Research Scientist with the Center for Satellite Applications and Research (STAR), where he performs applied and basic research. His primary research specialty is in environmental infrared remote sensing, radiative transfer and satellite validation, with a focus on oceanic and atmospheric applications. His other research interests include atmospheric aerosols, clouds, air–sea interactions, boundary layer and marine meteorology, forensic science, and climate applications.

Quanhua Liu

Quanhua Liu received a B.Sc. degree from Nanjing University of Information Science and Technology, Nanjing, China, in 1982, a master's degree in physics from the Chinese Academy of Science, Be, China, in 1984, and a Ph.D. degree in meteorology and remote sensing from the University of Kiel, Kiel, Germany, in 1991. He is currently a Physical Scientist with the U.S. National Oceanic and Atmospheric Administration (NOAA), National Environmental Satellite, Data and Information Service (NESDIS), Center for Satellite Applications and Research (STAR), in College Park, Maryland, USA, where he works on microwave sensor data calibration for the Advanced Technology Microwave Sounder (ATMS), and profile retrievals from the Microwave Integrated Retrieval System (MIRS).

Lori A. Borg

Lori Borg received a B.Sc. degree in Mechanical Engineering from the University of Massachusetts at Amherst, Amherst, MA, USA, in 1996, the M.Sc. degree in Mechanical Engineering from the Virginia Polytechnic Institute and State University, Blacksburg, VA, USA, in 2002, and her M.Sc. degree in atmospheric and oceanic sciences from the University of Wisconsin (UW)–Madison, Madison, WI, USA, in 2006. Since 2006, she has been with the Space Science and Engineering Center/Cooperative Institute for Meteorological Satellite Studies, UW–Madison. Her research interests include infrared satellite remote sensing, radiative transfer, and the validation of satellite atmospheric temperature and moisture products. She is part of the Cross-Track Infrared Sounder (CrIS) Sensor Data Records Science Team focusing on CrIS spectral and radiometric calibration, and the CrIS Environmental Data Records Science Team focusing on the assessment of temperature and moisture retrievals.

Preface to "Satellite Data Application, Validation and Calibration for Atmospheric Observation"

Well-calibrated, remotely sensed spectral observations acquired from the growing constellation of environmental satellites flown in low-Earth orbit (LEO) and geosynchronous orbit (GEO) provide the vast majority of data for the purpose of observing the global atmosphere and oceans over varying space and timescales. While environmental satellite data have been critical in the improvement of numerical weather forecasts via data assimilation in recent years, a large complement of derived geophysical products and state parameters (e.g., environmental data records, climate data records) retrieved from sensor data records (i.e., spectral radiances) are used for Earth system observation at microscale, mesoscale, synoptic, and global climate scales. Because multiple independent passive and active sensors are sensitive to different portions of the EM spectrum and deployed onboard different satellite platforms, high absolute calibration accuracy is crucial for synergistic observations and data continuity, as well as for specifying reliable uncertainty estimates. Climate change detection, in particular, requires the capability to resolve small global signals over decadal timescales (approximately 0.1 K per decade), which fundamentally requires stable sensor data records (SDRs) with high calibration accuracy. Routine monitoring of sensor calibration stability is facilitated via the validation of retrieved geophysical state parameters (i.e., SDRs, environmental (EDRs) and climate data records (CDRs)), which includes assessments of both absolute accuracy and precision with respect to independent reference measurements.

We are pleased to bring you this *Remote Sensing* Special Issue volume "Satellite Data Application, Validation, and Calibration for Atmospheric Observation", which features 21 papers covering current topics on the calibration/validation (cal/val) of advanced passive sensors (IR and/or MW) essential for Earth (atmospheric/oceanic) observation onboard operational, experimental, and next-generation environmental satellites. Featured topics range from sensor (SDR) calibration (Luo and Minnett, Scarino et al., W. Yang et al., Yan et al., Burgdorf et al., Iacovazzi et al., Lee et al., Aumann et al.) and algorithm/retrieval (EDR) validation (Sun et al., Nalli et al., Oshio et al., Park et al., Luo and Minnett, Xue et al.), to the subsequent improvements, impacts and applications of derived products (Liang and Liu, S. Yang et al., Berndt et al., Zou et al., Jiang et al., Zhou and Grassotti, Y. Yang and Wang).

Nicholas Nalli, Quanhua Liu, Lori A. Borg
Editors

remote sensing

Article

Accuracy of Vaisala RS41 and RS92 Upper Tropospheric Humidity Compared to Satellite Hyperspectral Infrared Measurements

Bomin Sun [1],*, Xavier Calbet [2], Anthony Reale [3], Steven Schroeder [4], Manik Bali [5], Ryan Smith [1] and Michael Pettey [1]

1   IMSG at NOAA NESDIS Center for Satellite Applications and Research (STAR),
    College Park, MD 20740, USA; ryan.c.smith@noaa.gov (R.S.); michael.pettey@noaa.gov (M.P.)
2   AMET, C/Leonardo Prieto Castro, 8, Ciudad Universitaria, 28071 Madrid, Spain; xcalbeta@aemet.es
3   NOAA NESDIS Center for Satellite Applications and Research (STAR), College Park, MD 20740, USA;
    tony.reale@noaa.gov
4   Department of Atmospheric Sciences, Texas A&M University, College Station, TX 77843, USA;
    s-schroeder@geos.tamu.edu
5   Cooperative Institute for Satellite Earth System Studies, University of Maryland,
    College Park, MD 20740, USA; manik.bali@noaa.gov
*   Correspondence: Bomin.Sun@noaa.gov

Abstract: Radiosondes are important for calibrating satellite sensors and assessing sounding retrievals. Vaisala RS41 radiosondes have mostly replaced RS92 in the Global Climate Observing System (GCOS) Reference Upper Air Network (GRUAN) and the conventional network. This study assesses RS41 and RS92 upper tropospheric humidity (UTH) accuracy by comparing with Infrared Atmospheric Sounding Interferometer (IASI) upper tropospheric water vapor absorption spectrum measurements. Using single RS41 and RS92 soundings at three GRUAN and DOE Atmospheric Radiation Measurement (ARM) sites and dual RS92/RS41 launches at three additional GRUAN sites, collocated with cloud-free IASI radiances (OBS), we compute Line-by-Line Radiative Transfer Model radiances for radiosonde profiles (CAL). We analyze OBS-CAL differences from 2015 to 2020, for daytime, nighttime, and dusk/dawn separately if data is available, for standard (STD) RS92 and RS41 processing, and RS92 GRUAN Data Processing (GDP; RS41 GDP is in development). We find that daytime RS41 (even without GDP) has ~1% smaller UTH errors than GDP RS92. RS41 may still have a dry bias of 1–1.5% for both daytime and nighttime, and a similar error for nighttime RS92 GDP, while standard RS92 may have a dry bias of 3–4%. These sonde humidity biases are probably upper limits since "cloud-free" scenes could still be cloud contaminated. Radiances computed from European Centre for Medium-Range Weather Forecasts (ECMWF) analyses match better than radiosondes with IASI measurements, perhaps because ECMWF assimilates IASI measurements. Relative differences between RS41 STD and RS92 GDP, or between radiosondes and ECMWF humidity profiles obtained from the radiance analysis, are consistent with their differences obtained directly from the RH measurements.

Keywords: radiosondes; satellite; upper tropospheric humidity; infrared radiances; radiative transfer

Citation: Sun, B.; Calbet, X.; Reale, A.; Schroeder, S.; Bali, M.; Smith, R.; Pettey, M. Accuracy of Vaisala RS41 and RS92 Upper Tropospheric Humidity Compared to Satellite Hyperspectral Infrared Measurements. *Remote Sens.* **2021**, 13, 173. https://doi.org/10.3390/rs13020173

Received: 15 October 2020
Accepted: 23 December 2020
Published: 6 January 2021

Publisher's Note: MDPI stays neutral with regard to jurisdictional claims in published maps and institutional affiliations.

1. Introduction

Balloon-borne radiosonde (or "sonde") observations (RAOBs) are critical in numerical weather prediction (NWP), data assimilation and forecasting, satellite data calibration/validation (cal/val), and upper air climate change detection. Vaisala RS92 was a major sonde type in the global operational upper air network and a reference sonde in the Global Climate Observing System (GCOS) Reference Upper Air Network (GRUAN) [1]. However, RS92 has gradually been replaced by Vaisala RS41 starting in late 2013. RS92 production ended in 2017, and all stations analyzed in this study stopped using RS92 for operational flights by early 2019. Vaisala RS41 includes new sensor technologies aimed at improving measurement accuracy for temperature, humidity and other variables throughout the atmosphere. These include a heated humidity sensor to prevent dew or frost formation in clouds and a separate temperature sensor attached to the humidity sensor. When the humidity sensor temperature differs from the free-air temperature sensor (whether the humidity sensor is heated intentionally or by erroneous solar heating), it is simple to express the relative humidity (RH) reading as RH at the free-air temperature. Characterizing the RS41 measurement improvement and accuracy is key to the GRUAN RS92-to-RS41 transition management program.

This study assesses the accuracy of atmospheric humidity observations of Vaisala RS92 and RS41. The first and most-used approach to estimate radiosonde accuracy is to conduct assessments in RH or specific humidity, primarily through comparing the data measured simultaneously by different radiosonde instruments from field experiments e.g., [2–7]. Vömel et al. [7] identify RS92 dry biases in the upper troposphere through comparing with cryogenic frost point hygrometer (CFH) measurements, and they propose a correction method to remove the mean bias.

A second assessment method is conducted in satellite radiance space. "Radiance space" refers to the fact that satellite remote sensing instruments measure the received radiant energy or radiance, which is emitted at each spectral frequency according to temperature and concentration of atmospheric gases, aerosols, and cloud particles. Desired meteorological variables are derived using radiances in carefully selected spectral bands. This study compares observed (OBS) atmospheric satellite radiances in spectral bands sensitive to moisture, with radiances calculated (CAL) from radiosonde temperature and humidity profiles via a forward radiative transfer model (RTM) [8–13]. For example, Moradi et al. [11,12] use microwave radiance values at 183 GHz as the base to analyze humidity characteristics of different radiosondes.

In this paper, OBS satellite radiances are hyperspectral infrared radiances measured in the upper tropospheric water vapor absorption spectral band (1400–1900 cm^{-1}) by the Infrared Atmospheric Sounding Interferometer (IASI). The instrument is on the European Organisation for the Exploitation of Meteorological Satellites (EUMETSAT) MetOp-A satellite. IASI is a Fourier Transform Spectrometer that provides 8461 channels covering the IR spectrum from 3.62–15.5 μm (2762–645 cm^{-1}). IASI is a well-characterized IR instrument and has been considered as the in-orbit reference sensor in the Global Space Inter-Calibration System (GSICS) [Tim Hewison at EUMETSAT, personal communication; http://gsics.atmos.umd.edu/pub/Development/AnnualMeeting2019/GRWG_GDWG_2019_Meeting_Minutes]. The instrument radiometric uncertainty is stable with time, its noise-equivalent delta temperature in the upper tropospheric water vapor absorption band is ~0.1–0.3 K at 280 K, and the corresponding noise in radiance units is ~0.06 mW m^{-2} sr cm^{-1} [14].

This paper calculates radiances (CAL) from radiosonde (or model) temperature and humidity profiles using the Line-by-Line RTM (LBLRTM) [15] over the 1400–1900 cm^{-1} spectral band, which covers practically all atmospheric levels from ~700 hPa and above. LBLRTM is considered a standard in computing radiances by the IR RTM community. It is a highly accurate radiation code that describes the interaction between atmospheric matter and radiation with a very high wavenumber resolution [13]. Calbet et al. [13] similarly use LBLRTM radiances to estimate Vaisala RS92 accuracy at Nauru, the former (1998–2013) Tropical Western Pacific (TWP) ARM site, by comparing with observed IASI radiances. We adopt their approach to understand upper tropospheric humidity (UTH) accuracy, and we extend their study to several GRUAN and ARM sites and compare Vaisala RS41 with RS92. As in Calbet et al. [13], we analyze only cases where the IASI pixel is cloud-free, as discussed in Section 2.2.2, because clouds lead to contaminated radiances.

This study uses the established formula in this field to compare instruments (e.g., [13]), which is the OBS-CAL difference such as IASI-RS92, where IASI (OBS) is considered the reference. Note that the sign of OBS-CAL is opposite from the usual bias formula, which would be CAL-OBS. For example, if RS92 radiances have a positive (high) bias relative to IASI, OBS-CAL is negative.

Operational soundings, including from GRUAN stations, use standard Vaisala procedures and corrections for rapid processing and transmission, but some biases remain and ongoing efforts to reduce these errors are documented at https://www.vaisala.com/en/sounding-data-continuity. Special GRUAN data processing (GDP, using GRUAN software version 2), aims to remove systematic data biases and provide uncertainty estimates [16] so GRUAN soundings can meet climate data record requirements. The NOAA Products Validation System (NPROVS) [17] routinely collects radiosondes and collocates them with satellite sounding data, but when a GRUAN processed sounding and operational sounding are both available, the GRUAN sounding is retained in NPROVS. A test version of GDP is being developed for RS41, so all RS41 soundings collected in NPROVS have standard Vaisala processing (referred to as "RS41 STD"). Depending on site and time period, RS92 soundings collected in NPROVS were processed either through GRUAN software ("RS92 GDP") or standard Vaisala procedures ("RS92 STD").

Validation of retrieved vertical profiles of temperature and humidity obtained from satellite sounding instruments, by comparing them with radiosondes is subject to diverse uncertainties. Among these reasons are significant radiosonde biases, actual profile differences due to the collocation time and distance and separation, and biases in radiative transfer modelling [10]. A detailed comparison exercise, such as the one presented in this paper, is therefore very necessary to properly validate satellite sounder retrievals. In particular, these results are applicable to the validation of satellite-derived products, such as those generated by the NOAA Unique Combined Atmospheric Processing System (NUCAPS, [18,19]) algorithm and several EUMETSAT Nowcasting Satellite Application Facility (SAF) products.

Section 2 describes data and methods. Section 2.1 lists the sites with either single launches (RS92 or RS41) or dual launches (RS92 and RS41 are suspended under the same balloon) collocated with IASI data. Section 2.2 lists methods or procedures used to process radiosonde data as the input to the LBLRTM radiance calculation, select IASI pixels for cloud-free scenes, assess the consistency of radiosonde data with IASI data (in radiance space), and convert the radiance differences (OBS-CAL) into RH differences to compute bias statistics. In Section 3, we first present the OBS-CAL analysis for RS92 GDP vs. RS41 STD, using the dual launch data from three GRUAN sites. Those dual launches allow us to understand the humidity difference of the two sondes in radiance and RH, and verify their consistency using both approaches. We then assess, through analyzing the OBS-CAL difference, the accuracy of RS41 STD, RS92 GDP and RS92 STD based on single launches closely matched with an IASI overpass. Radiosonde and NWP model sounding profiles are the major datasets used as the references for satellite sounding data validation and calibration [20]. Model analysis soundings closely collocated to radiosondes and IASI measurements from European Centre for Medium-Range Weather Forecasts (ECMWF) are also analyzed at those single launch sites with the aim to find out model accuracy in comparison with radiosondes and IASI. Section 4 summarizes specific uncertainties involved in this analysis.

2. Data and Methods

2.1. Collocated Radiosonde Launches and Their Collocations with IASI Data

The target data for the radiosonde data assessment is the radiance measurements of IASI onboard the MetOp-B satellite, with local equator crossing times being 0930 and 2130. Radiosondes at GRUAN and ARM sites are launched at nominal synoptic times (0000, 0600, 1200, and 1800 UTC, with actual launches usually ~1 h earlier, so the radiosonde is in the stratosphere at the stated time. Some stations may not have four launches per day). In addition to synoptic launches, dedicated radiosondes are launched from time to time at ARM sites targeting NOAA polar satellites, including SNPP and NOAA20 [18,19], with local equator crossing times at 0130 and 1330.

Collocation time and distance mismatch errors are the biggest uncertainties in th
assessment using IASI measurements [10,13]. We selected only sondes launched betwee
30 min before and 15 min after satellite overpass and within 50 km of the IASI pixel locatic

The Eastern North Atlantic (ENA) ARM site at Graciosa Airport, Azores, often mee
the criteria with synoptic launches approximately in coincidence with IASI overpasse
In high latitudes, MetOp-B swaths view the same location on several consecutive orbi
about 100 min apart (but not necessarily synchronized with synoptic radiosondes). Th
North Slope of Alaska (NSA) ARM site at Barrow (Utqiaġvik), Alaska, and the Ny Ål
sund, Norway, GRUAN site are used in this study because their radiosondes have high
chances to be close enough in time to IASI overpasses. While these high-latitude sites a
very frequently cloudy, that is not a major concern since our assessment focuses on th
upper troposphere.

Additionally, to support the RS92-to-RS41 transition, some GRUAN sites made RS9
and RS41 dual launches starting 2014. These provide the most rigorous radiosonde com
parisons because both radiosondes sample the same air column, but the comparisons a
still relative because neither RS41 nor RS92 provides absolute accuracy. For dual launche
collected in NPROVS, RS92 soundings are mostly GDP while RS41 soundings are STD. A
the Lauder, New Zealand GRUAN site, synoptic soundings are closely matched with IA
overpasses (within ~1 h before satellite overpass), but synoptic launches at Lindenbe
(11Z) and Payerne (11Z and 23Z) are mostly 1–3 h after the overpass, and prevent dire
determination of radiosonde accuracy using IASI as the reference, as will be discussed f
individual stations in Section 3.1. Nevertheless, dual launches allow us to verify that th
radiance difference of the dual sondes is consistent with their difference in RH observatio
That would give us the confidence to estimate radiosonde RH biases from the radiane
analysis (Sections 3.2–3.4). Of course, the close match of Lauder dual launches with IA
also provides the opportunity to infer the absolute accuracy of RS92 and RS41. The upp
portion of Table 1 lists information about the three dual launch sites and their respectiv
launch numbers.

The lower portion of Table 1 lists three sites with single launches at synoptic time
that often coincide within 30 min before and 15 min after IASI overpasses. The soundir
processing is a mixture of RS41 STD, RS92 GDP, and RS92 STD. Analysis of those da
via OBS-CAL differences is designed to address the absolute accuracy of the radiosonc
humidity data. Table 1 lists the numbers of those sondes along with the respective numbe
of collocated soundings with corresponding cloud-free and all-sky IASI scenes. Soun
ing numbers for nighttime, daytime, or dusk/dawn are stated in Section 3, where th
radiosonde accuracies are analyzed for those diurnal times if they have enough sampl
available for analysis.

As mentioned, all of the radiosonde profiles are collected in NPROVS [17,18], su
ported by the NOAA Joint Polar Satellite System (JPSS) program and operated at NOA
NESDIS office of Satellite Applications and Research (STAR) starting 2008. NPROVS pr
vides routine data access, collocation, and intercomparison of multiple satellite temperatu
and water vapor sounding product suites and NWP model profiles respectively matche
with a) global operational radiosondes and b) GRUAN including dedicated radiosonc
observations. The collocation approach is to select the "single closest" sounding from eac
product suite for each radiosonde.

The EUMETSAT MetOp-B IASI L2 sounding product [21,22] is one of the retriev
products routinely ingested in NPROVS for collocations with radiosonde data. The L
are physical retrievals generated using an optimal estimation method (OEM) by using th
all-sky retrievals as the first guess. The all-sky retrievals are generated using piecewis
regression methods and infrared and microwave channel data. OEM is attempted fo
clear-sky only as identified using strict cloud screening and other testing procedures (se
Appendix A for more information). The L2 physical retrievals are generated at each IA
field-of-view (FOV). IASI level 1c apodized measurements (smoothed to remove artifici
diffractive effects that distort the spectra) are appended to the selected collocations

radiosonde with IASI retrieval profile for use in the study. The level 1c datasets are accessed from the NOAA Comprehensive Large Array-Data Stewardship System (CLASS) (https://www.avl.class.noaa.gov/saa/products/welcome).

Table 1. Data from the radiosonde sites that are used for the analysis. In the last column, the number of soundings given first is those collocated with clear-sky Infrared Atmospheric Sounding Interferometer (IASI) scenes, followed by soundings associated with all sky scenes in parentheses. In that column, the time collocation limits are given in brackets for IASI minus radiosonde observation (RAOB) time.

Launch Types	Station Name (WMO ID)	Radiosonde	Latitude, Longitude (Launch Elevation)	Starting and Ending Date	Number of Soundings [Time Collocation Limits]
RS92 and RS41 dual launches	Lauder, New Zealand (93817)	RS92 GDP, RS41 STD	45.0376°S 169.6826°E (370 m)	Nov 2015 to Nov 2016	14 (26) [0~+1 h]
	Lindenberg, Germany (10393)	RS92 GDP, RS41 STD	52.2094°N 14.1203°E (115 m)	Dec 2014 to Oct 2017	29 (152) [+1~+3 h]
	Payerne, Switzerland (06610)	RS92 GDP, RS41 STD	46.8131°N 6.9437°E (490.5 m)	Aug 2014 to Oct 2017	19 (86) [+1~+3 h]
RS92 or RS41 single launches	Eastern North Atlantic (ENA), Graciosa, Azores (08507)	RS41 STD RS92 GDP RS92 STD	39.0912°N 28.0263°W (31 m)	Jan 2015 to May 2020	225 (1297) [−30~+15 min]
	North Slope Alaska (NSA), Barrow (Utqiaġvik) (70027)	RS41 STD RS92 GDP	71.3226°N 156.6180°W (8 m)	Jan 2015 to May 2020	122 (978) [−30~+15 min]
	Ny Alesund, Norway (01004)	RS41 STD RS92 GDP	78.9230°N 11.9225°E (15.5 m)	May 2015 to May 2020	45 (900) [−30~+15 min]

The selected IASI-RAOB collocations need to go through cloud screening to make sure the IASI FOV scene is cloud-free (see Section 2.2.2 and Appendix A for details) before CAL radiances are computed from radiosonde profiles.

ECMWF operational analysis profiles [23] are also collocated to RAOBs at all IASI-RAOB collocations analyzed. The ECMWF analyses are available at 0000, 0600, 1200, and 1800UTC, with 91 vertical pressure levels thinned from the 137 model sigma levels and horizontal resolution of $0.25° \times 0.25°$ [24]. The collocated ECMWF profiles are over 1 h from IASI overpasses in most of the dual launch cases, while ~1 hr or less from overpasses in most of the single launch cases.

2.2. Methodology

2.2.1. Radiosonde Profile Data Processing

GRUAN RS92 and RS41 soundings report data values at 1-s intervals, or usually ~7000 vertical levels (accessed from gruan.org/data). Those high-density profiles are converted into 100 vertical levels for the rapid transmittance algorithm used in radiative transfer models [20]. The GRUAN sounding objective is to aim for an altitude of 5 hPa, but only about 50% reach 15 hPa and less than 5% reach 5 hPa due to the use of smaller balloons that burst sooner. To apply the radiative transfer equations to the radiosonde profiles, they must be extended above the burst altitude to the top of the atmosphere (TOA) by appending the collocated ECMWF operational analysis to the top of the radiosonde profile. RS92 and RS41 sensors measure the RH of the ambient air, whereas the RTM requires water vapor concentration, typically specific humidity. We convert from RH to specific humidit using the Hyland and Wexler formula [25].

To verify the humidity difference estimated from the radiance difference, for exampl between two radiosondes or between radiosondes and ECMWF as discussed in Section we compute the humidity (and temperature) difference from their sounding profiles. T minimize the impact of different vertical resolutions on the assessment, the 100-leve radiosonde profiles and 91-level ECMWF profiles are averaged to ~1-km coarse layers fc temperature and ~2-km coarse layers for humidity. Statistics are then computed in thos layers with the mid-point coarse layer pressures shown in vertical profile figures (e.g Figure 1d,e). This approach is standard in validating satellite retrieval soundings usin radiosonde or NWP data [18–20,26].

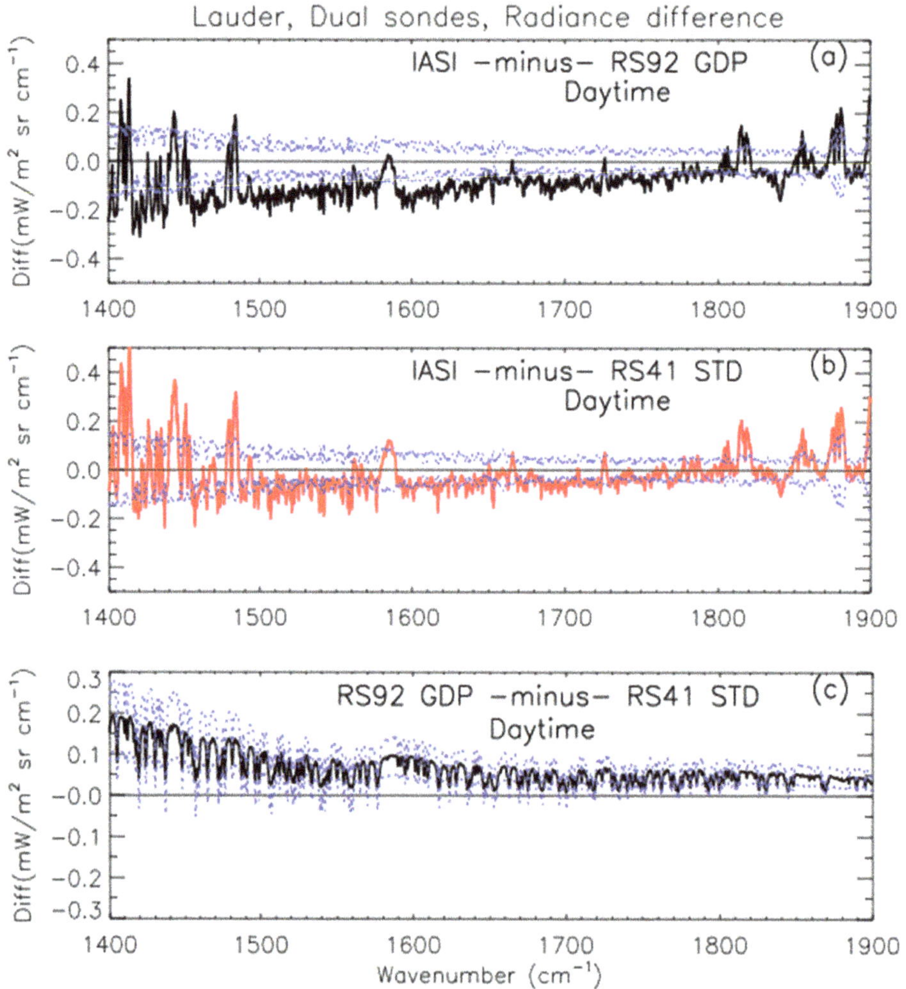

Figure 1. *Cont.*

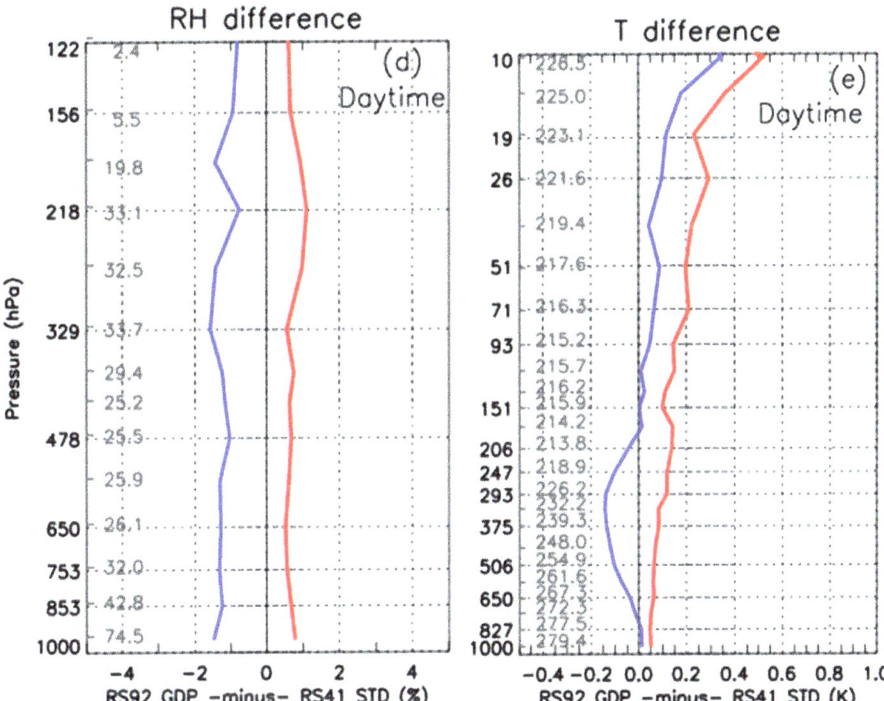

Figure 1. Lauder, New Zealand. (**a–c**) Mean radiance differences based on 14 dual launches, daytime within 1 h of IASI overpass. (**a**) The solid line is the mean difference between IASI observed radiances and calculated radiosonde radiances (OBS–CAL, IASI minus RS92 GDP). Dotted lines show ±2 standard errors (from zero) of the combined uncertainties (see text for more information). (**b**) As in (**a**) except for IASI minus RS41 STD based on the same dual flights. (**c**) The solid line is RS92 GDP minus RS41 STD. The dotted lines show ± one standard deviation of the RS92 GDP minus RS41 STD differences from the solid line. (**d–e**) Mean differences and standard deviations, RS92 GDP minus RS41 STD, at specified pressure levels (hPa), based on same dual launches as in (**a–c**). (**d**) Blue line is the mean atmospheric relative humidity (RH) difference, averaged at specified pressure levels, and the red line is its standard deviation. Gray numbers toward the left of the plot are mean RS41 STD RH values (%) at marked pressure levels. (**e**) As in (**d**) except for mean atmospheric temperature difference, and gray numbers are RS41 STD mean temperature (K).

2.2.2. Cloud Screening for IASI Pixels

A key to this sonde humidity data assessment is that IASI pixels collocated with radiosondes should not be cloud-contaminated. Undetected clouds, primarily high clouds, in the "cloud-free" scenes would bias the assessment. Cloud screening flag information included in the EUMETSAT IASI L2 product is used to find the cloud-free IASI pixels (see Appendix A), and their collocations with RAOBs are then used in the study. Table 1 shows the number of accepted cases after IASI cloud screening. On the average, cloud screening rejects ~87% of the soundings with IASI data within the collocation limits.

2.2.3. Consistency of Radiosonde Data with IASI Measurements

Collocated IASI measurements are compared with the computed radiosonde radiance to find out if the two types of measurements are consistent with each other. Following the proposed rationale [27] for statistical consistency of collocated measurements, IASI and radiosonde data are considered to be consistent with each other if their difference in radiance is within 2 times the k value,

$$m_1 - m_2| < k\sqrt{\sigma^2 + u_1^2 + u_2^2}$$

where "m_1" and "m_2" are OBS and CAL radiances to be compared, "u_1" and "u_2" the associated uncertainties, "σ" the uncertainty due to mismatch and "k" the agreement parameter. The uncertainty in LBLRTM should also be listed as one of the uncertainty components inside the square root but is included in the σ term here to keep the formula general. For this study, the unit for radiance variables is mW m^{-2} sr cm^{-1}.

Ideally, the radiosonde and IASI consistency is assessed for individual collocations by utilizing Equation (1), and based on that, the consistency for the whole collocation sample then statistically determined. At individual collocations, the IASI instrument uncertainty generally available (see Introduction), and uncertainty in computed radiance from GRUAN soundings can be estimated via radiative transfer modelling [13], whether assuming the uncertainty is either fully vertically correlated or not. The spatial and temporal collocation error, however, is unknown. The collocation error is suggested to be much bigger than other uncertainty components [13]. Equation (1) is therefore, not directly used in the assessment, and that could be a limitation of the analysis.

As described by Immler et al. [27], with normally distributed variables and independent uncertainty factors, the standard error (*ste*) of the OBS-CAL difference for an ensemble (for example, all of the collocations of RS41 STD with IASI from a site) is equal to the square root of the ($\sigma^2 + u_1^2 + u_2^2$) term. The *ste* value for a specific wavenumber is calculated from the standard deviation (*std*) of the OBS-CAL difference by dividing *std* by the square root of the number of samples (i.e., collocations).

The uncertainty derived from *ste* based on the ensemble-average is named as the overall or total uncertainty. In this study, this total uncertainty term is used to assess the consistency of ensemble-averaged radiosonde and IASI data in radiance space. RS92 GDP is considered to be consistent with IASI if the mean OBS-CAL difference is less than 2 times *ste* ([13], their Figures 6 and 8), so this paper uses the same definition. This is a 2-sided test of consistency at approximately the 95% statistical significance level. Note that they estimate the "average" collocation uncertainty from the *std* of the OBS-CAL difference.

2.2.4. Converting the Radiance Difference to RH Difference

Radiosonde biases estimated from OBS-CAL differences are stated in terms of radiances (or brightness temperatures). The corresponding biases in RH percentage points can be estimated by simply adding various RH values to the corresponding radiosonde profiles and recomputing the radiances until the OBS-CAL difference for RS92 or RS41 becomes negligible.

Calbet et al. [13] conclude, based on their Figures 7 and 8, that their RS92 OBS-CAL difference of −0.11267 mW m^{-2} sr cm^{-1} (averaged for the spectral band 1500–1570 cm^{-1}) is equivalent to a 2.5% RH dry bias relative to IASI radiances, and our radiance biases infer a daytime RS92 GDP dry bias of 2.58% and a nighttime dry bias of 0.69%. We use this conversion to estimate the radiosonde (or ECMWF) RH biases from their OBS-CAL differences. Note that channels with wavenumber in the range of 1500–1570 cm^{-1} are highly water vapor absorptive with their peak absorption in the middle to upper troposphere. They are not affected by low-level clouds or the underlying surface.

The mean OBS-CAL difference, DIFF, and standard deviation (STD), are formulated as follows:

$$C_i = \frac{\sum_{j=w1}^{w2} (OBS_{i,j} - CAL_{i,j})}{Nw} \quad (2)$$

$$DIFF = \frac{\sum_{i=1}^{N_c} C_i}{Nc} \quad (3)$$

$$STD = \sqrt{\frac{\sum_{i=1}^{N_c} (C_i - DIFF)^2}{Nc - 1}} \quad (4)$$

where $OBS_{i,j}$ is the IASI-observed radiance in wavenumber i for collocation j, $CAL_{i,j}$ is the corresponding LBLRTM-simulated radiance, and N_w and N_c are the number of wavenumbers (or spectral channels) and collocations included in the average, respectively. For the 1500–1570 cm^{-1} spectral region, the wavenumber $w1$ and $w2$, and N_w in Equation (2), are 3421 and 3701, and 281.

Equations (2)–(4) can be applied to any spectral region to compute the radiance bias statistics. Bias statistics in the water vapor absorption band (1615–1800 cm^{-1}) are also computed, another spectral region that is not sensitive to low-level features. For this region, $w1$, $w2$, and N_w in Equation (2) are 3881, 4621, and 741. An equivalence of the OBS-CAL difference of -0.07239 mW m^{-2} sr cm^{-1} averaged for 1615–1800 cm^{-1} to a 2.5% RH dry bias [13] is applied to estimate the RH bias from the radiance difference averaged for this spectral region.

The radiance difference statistics computed using Equations (2)–(4) and the RH bias statistics estimated from the radiance differences are listed in all tables except Table 1 to calculate the radiosonde or ECMWF data accuracy. Unless a spectral region is specified, the RH bias estimated from radiance analysis stated in the text is the average of the values computed from those two regions to better represent the upper tropospheric water vapor absorption across the spectra.

Direct humidity observations from radiosonde and ECMWF profiles are used to verify their consistency with the UTH characteristics estimated from the radiances, as discussed in the next section. The IASI channels at 1400–1900 cm^{-1} are actually sensitive to the water vapor content accumulated through an upper tropospheric layer, rather than a single level. At any wavelength, radiation detected by the satellite originates from the atmospheric layer where there is appreciable water vapor. Above the layer, there is negligible absorption, nor is there enough emission of infrared radiation to be detected. Any radiation emitted below that layer is simply absorbed by the water vapor above it. The layer emitting enough radiation to be detected does not have sharp boundaries. This poses a challenge to define the upper-tropospheric layer in the radiosonde or ECMWF humidity profile that best matches the layer defined in radiance space.

In this study, the 200.9–407.4 hPa pressure interval is used to represent that upper tropospheric layer for all sites and time periods analyzed. RH differences between two dual sondes or between radiosondes and ECMWF (third line of each row in the last column of Tables 2–5) are computed from that pressure interval.

Note that atmospheric structure, including the tropopause and the height of the upper troposphere, varies with location, season or even time of day. Uncertainty can be introduced by the factors discussed in this and preceding paragraphs when we compare the RH characteristics computed from radiances with radiosonde humidity observations. We therefore include figures with RH difference statistics depicted from the lower to the upper troposphere (e.g., Figure 1d) as examples to better understand the consistency of the RH difference between radiance space and humidity observations.

Table 2. Dual launch sounding comparisons. (Col. 1) Station, and in parentheses, period of day (according to category of solar elevation angle, SEA) and number of dual soundings analyzed with this SEA category. For each station and SEA category, there are 3 pairs of rows showing a set of mean difference comparisons. The first row in each pair is a header (shown in Col. 2 only) that summarizes the two instruments in that difference comparison, with the differences shown in the second row, Cols. 2–5 or 2–6 as applicable. All radiance differences have units of mW m^{-2} sr cm^{-1}, RH differences are % (percentage points out of 100), and each number inside parentheses is one standard deviation of the corresponding variable. The third set of comparisons (in *italics*) is based on differences of differences, specifically (IASI-RS41 STD)-(IASI-RS92 GDP) = RS92 GDP-RS41 STD. (Cols. 2–5) OBS-CAL differences are averaged over spectral regions of (Cols. 2–3) 1500–1570 cm^{-1} and (Cols. 4–5) 1615–1800 cm^{-1}. (Cols. 2 and 4) Mean OBS–CAL radiance differences. (Cols. 3 and 5) Corresponding RH differences estimated from OBS–CAL differences. (Col. 6, applicable only to bottom set of comparisons) RH differences calculated from direct observations (radiosonde RH, specifically RS92 GDP-RS41 STD) averaged from 200.9–407.4 hPa. Radiosonde "GDP" or "STD" denotes radiosonde soundings processed either through GRUAN software or standard Vaisala procedures.

Station (Dual Launches)	OBS-CAL Radiance Diff 1500–1570 cm^{-1}	OBS-CAL RH Diff 1500–1570 cm^{-1}	OBS-CAL Radiance Diff 1615–1800 cm^{-1}	OBS-CAL RH Diff 1615–1800 cm^{-1}	Direct Obs RH Diff 200.9–407.4 hPa
Lauder (day, 14)	IASI-RS92 GDP −0.1291 (0.085)	2.86 (1.9)	−0.0818 (0.059)	2.82 (2.0)	
	IASI-RS41 STD −0.0705 (0.081)	1.56 (1.8)	−0.0386 (0.057)	1.33 (2.0)	
	RS92-RS41 0.0585 (0.029)	*−1.30 (0.6)*	*0.0431 (0.019)*	*−1.49 (0.7)*	*−1.33 (0.8)*
Lindenberg (day, 19)	IASI-RS92 GDP −0.2204 (0.107)	4.89 (2.4)	−0.1447 (0.082)	5.00 (2.8)	
	IASI-RS41 STD −0.1307 (0.103)	2.90 (2.3)	−0.0841 (0.080)	2.90 (2.8)	
	RS92-RS41 0.0897 (0.044)	*−1.99 (1.0)*	*0.0606 (0.031)*	*−2.09 (1.1)*	*−1.91 (1.2)*
Payerne (night, 10)	IASI-RS92 GDP −0.1160 (0.196)	2.57 (4.3)	−0.0847 (0.135)	2.92 (4.7)	
	IASI-RS41 STD −0.1318 (0.182)	2.93 (4.0)	−0.0894 (0.126)	3.09 (4.4)	
	RS92-RS41 −0.0158 (0.055)	*0.35 (1.2)*	*−0.0047 (0.033)*	*0.16 (1.1)*	*1.13 (1.9)*
Payerne (day, 9)	IASI-RS92 GDP −0.1578 (0.131)	3.50 (2.9)	−0.0741 (0.108)	2.56 (3.7)	
	IASI-RS41 STD −0.1073 (0.130)	2.38 (2.9)	−0.0423 (0.108)	1.46 (3.7)	
	RS92-RS41 0.0505 (0.042)	*−1.12 (0.9)*	*0.0319 (0.027)*	*−1.10 (0.9)*	*−0.73 (1.0)*

Table 3. Same as Table 2, except that soundings are single launches of RS41 STD, OBS-CAL comparisons are IASI-RS41 STD or IASI-ECMWF, where European Centre for Medium-Range Weather Forecasts (ECMWF) is the collocated reanalysis sounding, and the third set of comparisons for each station and period of day is (IASI-ECMWF)-(IASI-RS41 STD) = RS41 STD-ECMWF. Note that the Arctic stations (NSA and Ny Alesund) have comparisons based on dusk/dawn soundings with SEA between −7.5° and +7.5°.

Station (RS41 STD Single Launches)	OBS-CAL Radiance Diff 1500–1570 cm^{-1}	OBS-CAL RH Diff 1500–1570 cm^{-1}	OBS-CAL Radiance Diff 1615–1800 cm^{-1}	OBS-CAL RH Diff 1615–1800 cm^{-1}	Direct Obs RH Diff 200.9–407.4 hPa
ENA (night, 12)	IASI-RS41 STD −0.0528 (0.054)	1.18 (1.2)	−0.0334 (0.039)	1.15 (1.3)	
	IASI-ECMWF 0.0131 (0.057)	−0.29 (1.3)	−0.0081 (0.042)	0.28 (1.4)	
	RS41-ECMWF 0.0659 (0.053)	*−1.46 (1.2)*	*0.0253 (0.043)*	*−0.87 (1.5)*	*−0.42 (3.5)*

Table 3. *Cont.*

Station (RS41 STD Single Launches)	OBS-CAL Radiance Diff 1500–1570 cm^{-1}	OBS-CAL RH Diff 1500–1570 cm^{-1}	OBS-CAL Radiance Diff 1615–1800 cm^{-1}	OBS-CAL RH Diff 1615–1800 cm^{-1}	Direct Obs RH Diff 200.9–407.4 hPa
ENA (day, 27)	IASI-RS41 STD −0.0633 (0.042)	1.41 (0.9)	−0.0370 (0.028)	1.28 (1.0)	
	IASI-ECMWF 0.0238 (0.063)	−0.53 (1.4)	0.0070 (0.054)	−0.24 (1.9)	
	RS41-ECMWF 0.0871 (0.065)	*−1.93 (1.4)*	*0.0440 (0.053)*	*−1.52 (1.8)*	*−2.38 (7.2)*
NSA (night, 29)	IASI-RS41 STD −0.0610 (0.057)	1.35 (1.3)	−0.0356 (0.032)	1.23 (1.1)	
	IASI-ECMWF −0.0052 (0.065)	0.11 (1.4)	−0.0117 (0.038)	0.41 (1.3)	
	RS41-ECMWF 0.0558 (0.046)	*−1.23 (1.0)*	*0.0239 (0.027)*	*−0.83 (0.9)*	*−2.34 (3.4)*
NSA (dusk/dawn, 36)	IASI-RS41 STD −0.0662 (0.027)	1.47 (0.6)	−0.0416 (0.022)	1.44 (0.8)	
	IASI-ECMWF −0.0338 (0.037)	0.75 (0.8)	−0.0261 (0.030)	0.90 (1.0)	
	RS41-ECMWF 0.0324 (0.043)	*−0.72 (1.0)*	*0.0155 (0.031)*	*−0.54 (1.1)*	*−1.00 (3.8)*
Ny Alesund (dusk/dawn, 12)	IASI-RS41 STD −0.0593 (0.047)	1.31 (1.0)	−0.0372 (0.026)	1.28 (0.9)	
	IASI-ECMWF −0.0012 (0.082)	0.03 (1.9)	−0.0047 (0.066)	0.16 (2.3)	
	RS41-ECMWF 0.0580 (0.096)	*−1.29 (2.1)*	*0.0324 (0.068)*	*−1.12 (2.4)*	*−2.41 (4.0)*
Ny Alesund (day, 15)	IASI-RS41 STD −0.0977 (0.061)	2.17 (1.4)	−0.0422 (0.038)	1.46 (1.3)	
	IASI-ECMWF −0.0065 (0.057)	0.14 (1.3)	−0.0008 (0.041)	0.28 (1.4)	
	RS41-ECMWF 0.0912 (0.067)	*−2.02 (1.5)*	*0.0414 (0.045)*	*−1.43 (1.5)*	*−0.66 (5.8)*

Table 4. Same as Table 3, except that soundings are single launches of RS92 GDP, OBS-CAL comparisons are IASI - RS92 GDP or IASI - ECMWF, and the third set of comparisons for each station and period of day is (IASI-ECMWF) - (IASI-RS92 GDP) = RS92 GDP-ECMWF.

Station (RS92 GDP Single Launches)	OBS-CAL Radiance Diff 1500–1570 cm^{-1}	OBS-CAL RH Diff 1500–1570 cm^{-1}	OBS-CAL Radiance Diff 1615–1800 cm^{-1}	OBS-CAL RH Diff 1615–1800 cm^{-1}	Direct obs RH Diff 200.9–407.4 hPa
ENA (night, 43)	IASI-RS92 GDP −0.0527 (0.088)	1.17 (1.9)	−0.0316 (0.057)	1.09 (2.0)	
	IASI-ECMWF 0.0471 (0.115)	−1.04 (2.6)	0.0208 (0.082)	−0.72 (2.8)	
	RS92-ECMWF 0.0998 (0.134)	*−2.21 (3.0)*	*0.0524 (0.093)*	*−1.81 (3.2)*	*−0.43 (8.4)*
ENA (day, 50)	IASI-RS92 GDP −0.1196 (0.090)	2.65 (2.0)	−0.0722 (0.063)	2.49 (2.2)	
	IASI-ECMWF 0.0097 (0.096)	−0.22 (2.2)	−0.011 (0.078)	0.36 (2.7)	
	RS92-ECMWF 0.1293 (0.127)	*−2.87 (2.8)*	*0.0617 (0.092)*	*−2.13 (3.2)*	*−1.90 (6.1)*

Table 4. *Cont.*

Station (RS92 GDP Single Launches)	OBS-CAL Radiance Diff 1500–1570 cm⁻¹	OBS-CAL RH Diff 1500–1570 cm⁻¹	OBS-CAL Radiance Diff 1615–1800 cm⁻¹	OBS-CAL RH Diff 1615–1800 cm⁻¹	Direct obs RH Diff 200.9–407.4 hPa
NSA (night, 10)	IASI-RS92 GDP -0.0596 (0.063)	1.32 (1.4)	-0.0396 (0.036)	1.37 (1.2)	
	IASI-ECMWF -0.0043 (0.095)	0.10 (2.1)	-0.0135 (0.055)	0.47 (1.9)	
	RS92-ECMWF 0.0553 (0.086)	*-1.23 (1.9)*	*0.0260 (0.055)*	*-0.89 (1.9)*	*-3.65 (5.4)*
NSA (day, 5)	IASI-RS92 GDP -0.0924 (0.039)	2.05 (0.9)	-0.0554 (0.019)	1.91 (0.7)	
	IASI-ECMWF -0.0521 (0.029)	1.16 (0.6)	-0.0413 (0.029)	1.42 (1.0)	
	RS92-ECMWF 0.0403 (0.043)	*-0.90 (1.0)*	*0.0141 (0.037)*	*-0.49 (1.3)*	*-1.70 (4.4)*
Ny Alesund (day, 6)	IASI-RS92 GDP -0.1190 (0.046)	2.64 (1.0)	-0.0558 (0.031)	1.93 (1.1)	
	IASI-ECMWF 0.0030 (0.107)	-0.065 (2.4)	0.0002 (0.086)	-0.01 (3.0)	
	RS92-ECMWF 0.1220 (0.084)	*-2.71 (1.9)*	*0.0643 (0.026)*	*-1.94 (2.2)*	*-3.88 (4.5)*

Table 5. Same as Table 3, except that soundings are single launches of RS92 STD, OBS-CAL comparisons are IASI-RS92 STD or IASI-ECMWF, and the third set of comparisons for each station and period of day is (IASI-ECMWF)-(IASI-RS92 STD) = RS92 STD-ECMWF.

Station (RS92 STD Single Launches)	OBS-CAL Radiance Diff 1500–1570 cm⁻¹	OBS-CAL RH Diff 1500–1570 cm⁻¹	OBS-CAL Radiance Diff 1615–1800 cm⁻¹	OBS-CAL RH Diff 1615–1800 cm⁻¹	Direct Obs RH Diff 200.9–407.4 hPa
ENA (night, 43)	IASI-RS92 STD -0.1758 (0.075)	3.90 (1.7)	-0.1132 (0.05)	3.91 (1.7)	
	IASI-ECMWF 0.0096 (0.080)	-0.21 (1.8)	-0.0133 (0.056)	0.46 (1.9)	
	RS92-ECMWF 0.1854 (0.112)	*-4.11 (2.5)*	*0.1000 (0.075)*	*-3.4 (2.6)*	*-3.08 (6.0)*
ENA (day, 50)	IASI-RS92 STD -0.1448 (0.118)	3.21 (2.6)	-0.0952 (0.075)	3.29 (2.6)	
	IASI-ECMWF 0.0082 (0.105)	-0.18 (2.3)	-0.0681 (0.078)	0.24 (2.7)	
	RS92-ECMWF 0.1529 (0.144)	*-3.39 (3.2)*	*0.0088 (0.094)*	*-3.05 (3.3)*	*-2.32 (7.8)*

3. Results

3.1. Dual Launches of RS92 GDP and RS41 STD

Dual launches of RS92 and RS41 radiosondes at Lauder, Lindenberg, and Payerne were made at synoptic times. Table 2 shows the number of analyzed clear-sky collocations at each station. As in Sun et al. [28], solar elevation angles (SEAs) computed at the radiosonde launch location and time are used to group soundings into three categories for analysis: Nighttime (SEA < $-7.5°$), daytime (SEA $\geq$ +7.5°), and dusk/dawn (any other SEA). While the time of a collocated IASI observation or ECMWF profile may be in a different SEA category from the sounding, the SEA at the IASI or ECMWF location is not different enough from the radiosonde SEA category to reject that case. Fewer cases are analyzed at Lauder and Lindenberg than totals shown in Table 1 due to insufficient night or dusk/dawn collocations for reasonable statistical analysis.

Lauder, New Zealand. Figure 1 shows the average OBS-CAL differences from 14 daytime collocations for (a) RS92 GDP and (b) RS41 STD. These soundings were launched at ~0900 local time, within ~1 h prior to an IASI overpass. The negative OBS-CAL radiance differences shown for both RS92 and RS41 indicate that both sonde types are dry-biased in the upper troposphere. The positive RS92 GDP-RS41 STD radiance difference (Figure 1c) computed using IASI radiance as the transfer standard indicates that RS92 GDP appears to be more dry-biased than RS41 STD.

The dotted lines in Figure 1a,b show ± 2 *ste* (from zero) of the combined uncertainties (as stated in Section 2.2.3), indicating that the CAL radiances for RS92 GDP (solid black line) are statistically inconsistent with IASI measurements while the CAL radiances for RS41 STD (solid red line) are mostly consistent with IASI.

Spikes in the OBS-CAL difference in the spectral regions of 1400–1500 cm^{-1} and 1800–1900 cm^{-1} reflect the sensitivity of narrow spectral lines to the lower troposphere (usually below 700 hPa). Those features are common to all sites analyzed in the study.

As listed in Table 2, the OBS-CAL mean difference for RS92 GDP averaged for 1500–1570 cm^{-1} is -0.1291 (± 0.085) mW m^{-2} sr cm^{-1} and for RS41 STD is -0.0705 (± 0.081) mW m^{-2} sr cm^{-1} for RS41 STD, where the values inside the parentheses are one standard deviation of the difference. Throughout the paper, values inside the parentheses following mean biases or differences are one standard deviation. The RH dry biases in the upper troposphere computed from the radiance differences are 2.86% for RS92 GDP and 1.56% for RS41 STD. Similarly, as indicated in Table 2, the RH dry biases converted from the radiance differences at 1615–1800 cm^{-1} are 2.82% and 1.33%, respectively for RS92 GDP and RS41 STD. The daytime dry bias in RS92 GDP humidity data obtained from Lauder is slightly higher (by 0.30% in the absolute RH value) than found from the former TWP Nauru site [13].

The RH differences between RS92 GDP and RS41 STD estimated from their radiance differences are 1.40% (0.65%), basically consistent with the RH difference of 1.33% (0.8%) based on the measured data (Table 2). Figure 1d indicates that RS92 GDP is systematically drier than R41 STD by 1–1.5% from the lower troposphere to the upper troposphere during daytime (based on only clear sky data). However, the RH (for RS41) STD averages 25.5% at 478 hPa and 5.5% at 156 hPa, and in terms of specific humidity, this means that RS92 GDP (compared to RS41 STD) averages 3.9% drier at 478 hPa, and 18.2% drier at 156 hPa; specific humidity is more fundamental (than RH) to atmospheric radiative transfer.

In the humidity-sensitive channels, atmospheric temperature may also affect the radiation the satellite receives. In Figure 1e, the RS92 GDP temperature appears to be slightly warmer (by <0.2 K except at the highest level) than RS41 STD above 150 hPa, suggesting the existence of a radiation-related warm bias in GRUAN processed data [28]. However, RS92 GDP appears to be colder than RS41 STD in the troposphere with a maximum cold difference of ~0.2 K around 300 hPa.

Appendix B further investigates the impact of atmospheric temperature differences between RS92 GDP and RS41 STD in their CAL radiance differences. It appears the colder temperature in the upper troposphere in RS92 GDP (minus RS41 GDP) leads to slightly more negative radiance differences in the spectrum range of 1400–1900 cm^{-1} (Figure A1), interpreted as being slightly moister in the upper troposphere. However, the radiance difference contributed by the temperature difference is small. For example, -0.0125 mW m^{-2} sr cm^{-1} averaged for 1500–1570 cm^{-1}, is equivalent to 0.28% in RH. Note that the warm temperature difference in the lower stratosphere does not seem to have an impact on the radiance differences in the 1400–1900 cm^{-1} band.

Lindenberg, Germany. Given the longitude of this station, most of the dual sondes (11Z) launched at this site are 1 to 3 h after the MetOp-B overpass. Because it generally takes ~30 min for the balloon to reach ~300 hPa [29], the actual time difference is over 1.5 to 3.5 h in the upper troposphere and lower stratosphere, and systematic (always after overpass).

In Table 2, the OBS-CAL radiance differences averaged for 1500–1570 cm^{-1} for RS92 GDP and RS41 STD are -0.2204 and -0.1307 mW m^{-2} sr cm^{-1}, respectively, and the

radiance derived dry RH biases are 4.89% and 2.90%, respectively, again compared to IAS
Similar values are obtained from the spectral region 1618–1800 cm^{-1}. Those numbers ar
statistically different from zero and are much bigger than the values obtained from Laude
and other single launch sites (see Sections 3.2 and 3.3) where the time differences are withi
0.5 h.

The σ term in Equation (1) may increase with the increase in time difference, bu
the mean difference is not affected much as long as the time differences in the ensembl
are random in sign [30]. We suspect the big difference values in radiance and hence i
RH values estimated from the radiances at Lindenberg are related to the systematic tim
difference between the radiosonde launch and IASI overpass.

Consistent with Lauder, the positive radiance difference between RS92 GDP and RS4
STD obtained by using IASI as the transfer standard (Figure 2a) indicates that RS92 GDP i
drier than RS41 STD. In Table 2, the RH difference estimated from the OBS-CAL difference
averages −2.04% over the 1500–1570 and 1615–1800 cm^{-1} bands, which is close to th
directly observed difference of −1.91%, as also shown in the RH difference vertical profil
in Figure 2b (i.e., ~−2% in RH at ~330 hPa).

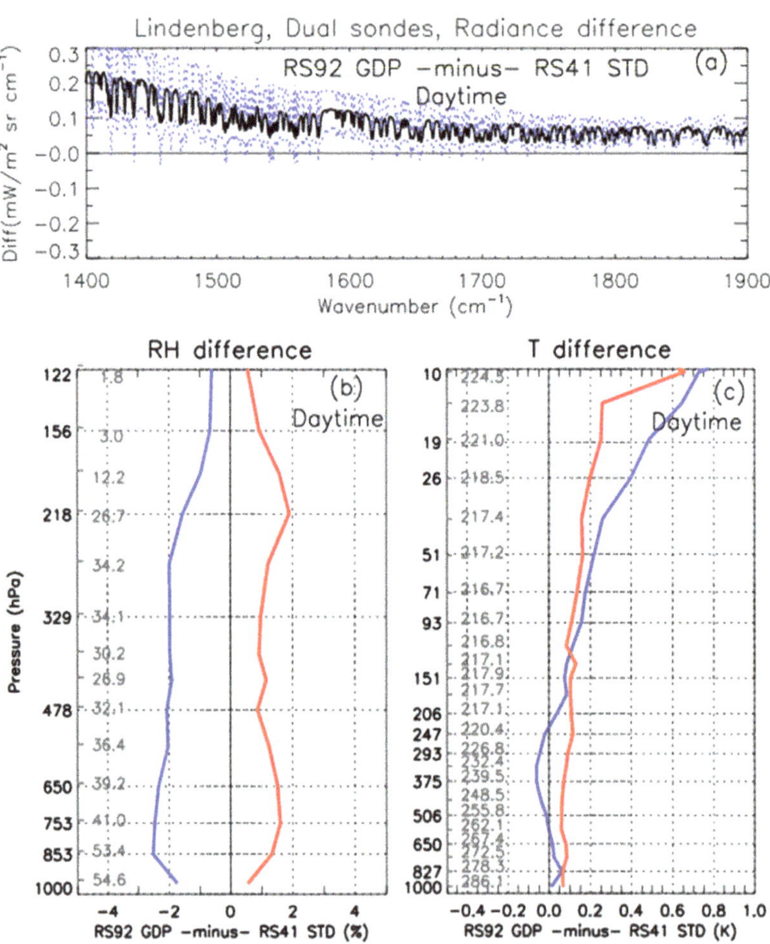

Figure 2. Lindenberg, Germany, differences based on 19 pairs of daytime dual launches. (**a**) Mean radiance differences for
RS92 GDP minus RS41 STD, as in Figure 1c. (**b,c**) RH and T differences for RS92 GDP minus RS41 STD, as in Figure 1d,e.
Gray numbers are the same as in Figure 1d,e.

Appendix B indicates that a small radiance difference is contributed by the small temperature difference at Lindenberg (Figure 2c). Both Lauder and Lindenberg analyses indicate that lower stratospheric temperatures do not have impacts on upper tropospheric humidity-sensitive radiances (1400–1900 cm^{-1}), and mid-upper tropospheric temperatures can have an impact, but impacts of a temperature difference <0.2 K on radiance in the context of RH are negligible.

Payerne, Switzerland. Similar to Lindenberg, dual launches at Payerne are mostly 1–3 h after IASI overpasses. As shown in Table 2, the daytime RH dry bias converted from the OBS-CAL difference averaged for 1500–1570 cm^{-1} is 3.50% for RS92 GDP and 2.38% for RS41 STD; and the corresponding night dry biases are 2.57% and 2.93%. As at Lindenberg, those big RH bias values may be "inflated" by a systematic time difference. In the 1500–1570 cm^{-1} band, the daytime RS92 GDP minus RS41 STD RH difference estimated from radiance differences is −1.12% (0.9%), and the nighttime RH difference is +0.35% (1.2%). Blue lines in Figure 3a,b show similar radiosonde RH differences at ~300 hPa.

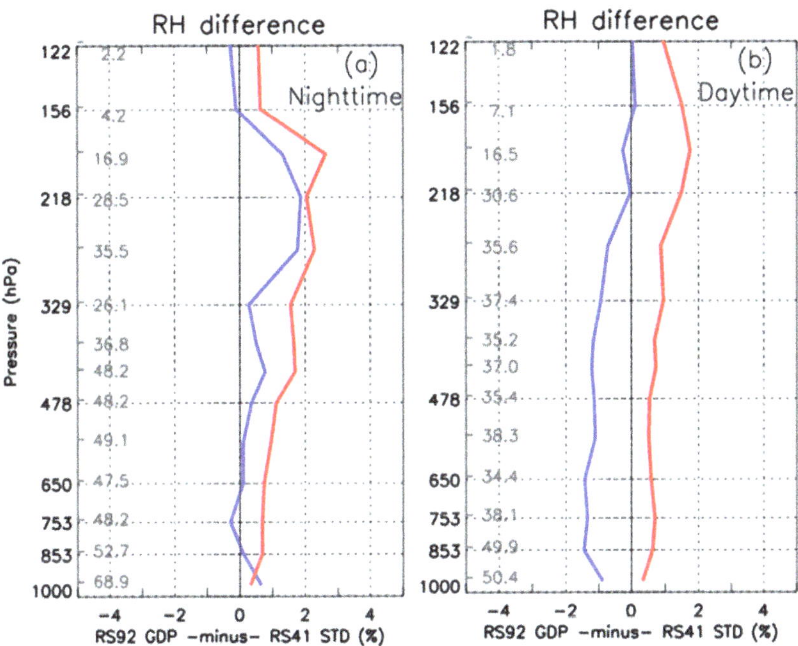

Figure 3. Payerne, Switzerland. Similar to Figure 1d, mean RH differences and standard deviations (Gray numbers are mean RS41 STD RH at marked pressure levels), RS92 GDP minus RS41 STD, at specified pressure levels (hPa). Averaged from (**a**) 10 nighttime collocations and (**b**) 9 daytime collocations.

Analysis of dual sonde data in this subsection indicates that the RH differences estimated from radiance space are basically consistent with the measured upper tropospheric RH differences in the radiosonde observations. This lends confidence in using radiance differences to analyze the sonde accuracy for single launched sondes to be presented in Sections 3.2–3.4.

3.2. Single Launches of RS41 STD

As mentioned in the Introduction, all single launches of radiosondes (including RS41 STD, RS92 GDP, and RS92 STD) analyzed in the study are within 50 km and between 0.5 h before and 0.25 h after IASI MetOp-B overpasses. ECMWF analyses are typically ~1 h or less from the satellite overpasses in those single launch cases.

ENA. A small UTH dry bias for RS41 STD for both nighttime and daytime is suggested by the slightly negative OBS-CAL radiance differences (Figure 4a,b). The RH dry biases estimated from OBS-CAL are 1.17% (1.25%) and 1.34% (0.95%) for nighttime and daytime respectively. The ECMWF analyses collocated with radiosondes are ~0.5 h after overpass in this location. OBS-CAL for ECMWF is close to zero (Figure 4c,d), and the RH biases in Table 3 estimated from the radiance differences are 0.00% (1.35%) and 0.39% (1.65%) for nighttime and daytime.

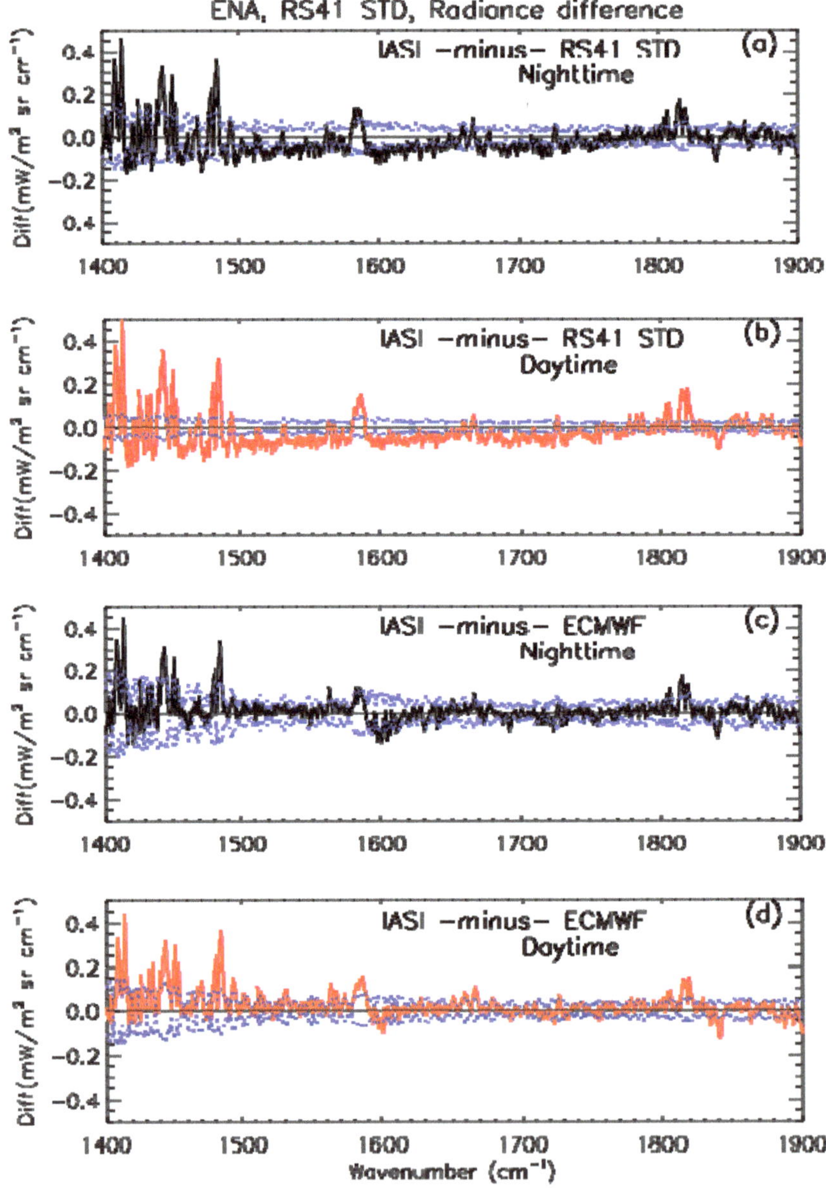

Figure 4. *Cont.*

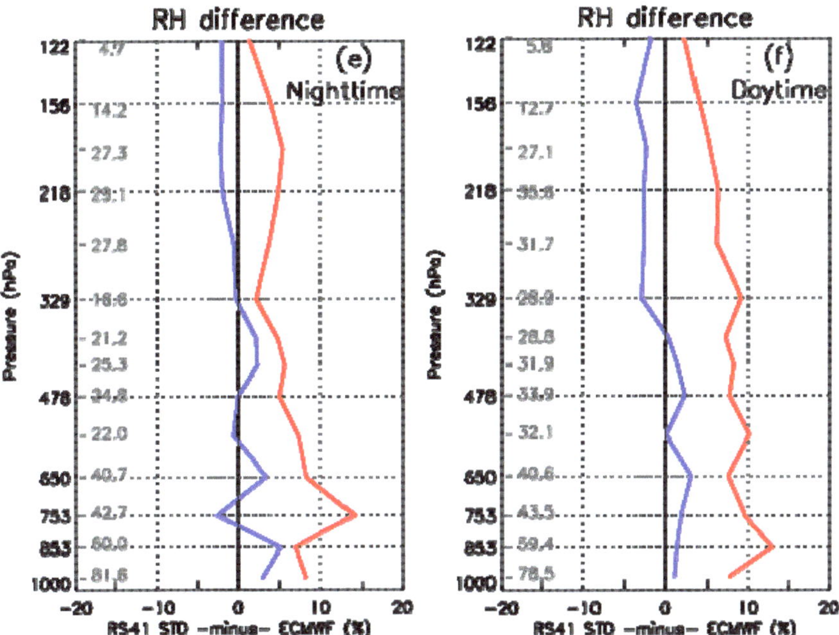

Figure 4. Eastern North Atlantic (ENA) station at Graciosa, Azores, Portugal. (**a–d**) Mean OBS-CAL radiance differences and standard deviations as in Figure 1a. (**a,c**) based on 12 night collocations, (**b,d**) based on 27 daytime collocations. (**a,b**) IASI minus RS41 STD. (**c,d**) IASI minus ECMWF. (**e,f**) As in Figure 3, but for RS41 STD minus ECMWF (plotted gray numbers are RS41 STD mean RH percentages), and based on (**e**) 12 night collocations and (**f**) 27 daytime collocations.

The UTH dry biases of RS41 STD relative to ECMWF estimated from the radiance analysis (1.17% for nighttime and 1.73% for daytime) are basically consistent with those directly computed from the RH profiles (Table 3 and Figure 4e,f). Interestingly, RS41 STD appears to be <1% moister than ECMWF for both nighttime and daytime in the low-middle troposphere (Figure 4e,f).

The standard deviations of the RH differences computed from the RH profile data are, however, much bigger than the ones estimated from the radiance differences (e.g., 7.2% vs. 1.6% for daytime, Table 3). This contrast occurs with all single launches (Tables 3–5), but does not occur with the dual launches, where they are comparable to each other (Table 2). The primary reason is that the standard deviations in column 6 of Tables 3–5 are based on radiosonde RH compared with ECMWF RH that may differ up to 1 h and 10 km from the radiosonde, while those in Table 2 are computed from dual sondes with no collocation time or distance error.

The OBS-CAL differences (solid curves of Figure 4c,d) across 1400–1900 cm^{-1} for ECMWF fall within 2 × *ste*, suggesting that ECMWF and IASI are consistent with each other in the radiance space after taking into account the uncertainty terms discussed in Section 2.2. This consistency happens at other single launch collocations analyzed in this subsection and the following two subsections too (figures not shown).

For RS41 STD (Figure 4a,b), OBS-CAL differences for nighttime marginally fall within 2 × *ste*, while OBS-CAL differences for daytime are far beyond 2 × *ste*. This nighttime vs. daytime contrast is partly related to the mean OBS-CAL differences, which are slightly bigger during daytime (Table 2). However, the major factor is that the daytime collocation sample (27) is much bigger than the night sample (12), so the ensemble-averaged σ (and hence *ste*) is much smaller in the daytime through better averaging out the random collocation noise. Therefore, the consistency evaluation methods discussed in Section 2.2.3 should be exercised cautiously. The collocation sample size and hence the ensemble-averaged uncertainty could play an important role in determining if two variables are consistent with each other.

NSA. The RS41 and IASI cloud-free collocations occur mostly at night and dusk/dawn. At this site, ECMWF is within ~0.5 h after each MetOp-B overpass. Similar to the ENA OBS-CAL radiance patterns, the OBS-CAL differences for RS41 STD for both nighttime and dusk/dawn (Figure 5b,d) are slightly negative, equivalent to a small dry bias in RS41 STD (1.29% and 1.46% respectively, Table 3).

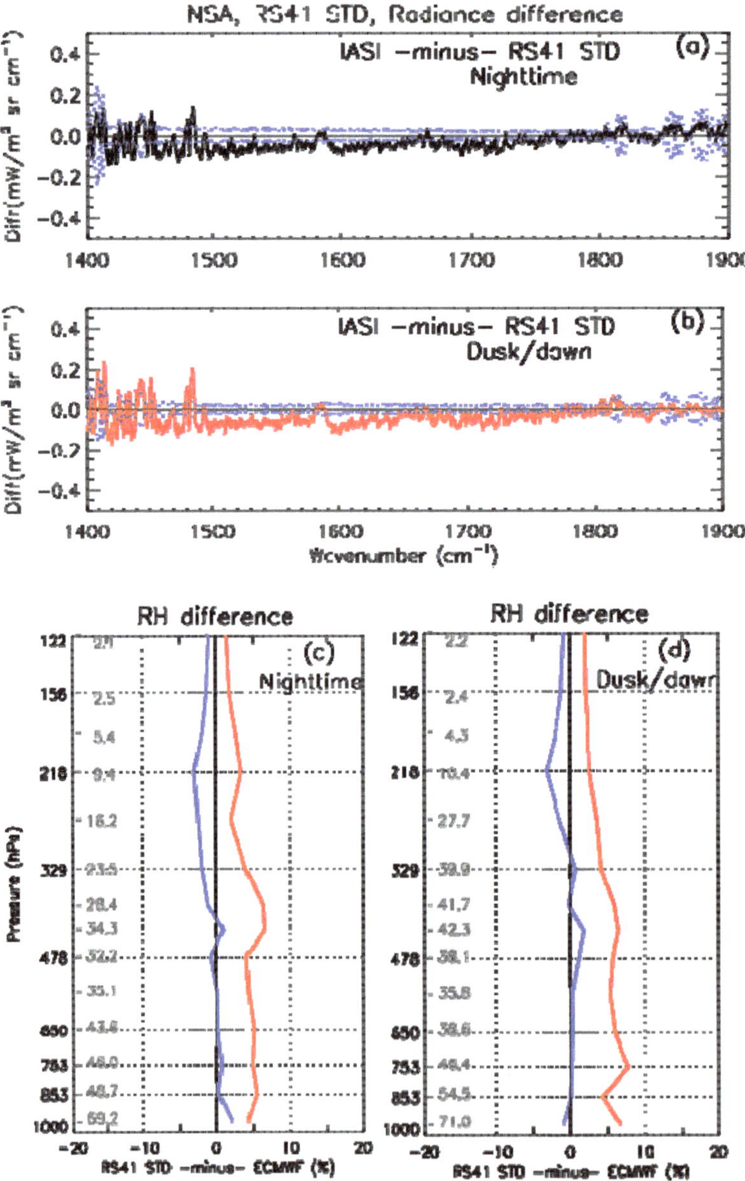

Figure 5. North Slope Alaska (NSA) at Barrow (Utqiagvik). As in Figure 4a,b,e,f), but for (**a**,**c**) 29 night collocations and (**b**,**d**) 36 dusk/dawn collocations.

Relative to the RS41 STD minus ECMWF RH differences estimated from the radiance analysis, 1.02% and 0.63% for nighttime and dusk/dawn, the radiosonde RH differences directly computed over 200.9–407.4 hPa are greater (Table 3). The reason is that the pressure interval does not accurately represent the upper troposphere at the site (see discussion in Section 2.2.4), where the tropopause altitude is generally lower. As a matter of fact, by raising the pressure by ~50 hPa, the UTH dryness of RS41 STD relative to ECMWF obtained from the RH profiles (Figure 5c,d) matches well with RH from the radiance analysis.

Note that in Figure 5a,b, the *ste* values of OBS-CAL for RS41 STD show fluctuations in the channel from 1800 to 1900 cm^{-1} for both nighttime and dusk/dawn, but not in other spectral ranges. This feature is not seen at ENA (Figure 4a–d) or the three dual launch sites (e.g., Figure 1a,b) while it is also observed at Ny Alesund (figures not shown). Atmospheric water vapor content over polar regions tends to be low and channels in 1800–1900 cm^{-1} could be sensitive to surface snow/ice which often occurs there.

Ny Alesund. Most of the radiosonde-satellite collocations for cloud-free scenes are for dusk/dawn and daytime. As listed in Table 3, dry biases of 1.46% and 1.82% are estimated for RS41 STD from the OBS-CAL differences for dusk/dawn and daytime respectively. ECMWF shows a smaller dry bias (<0.7%) estimated from the radiance analysis for both dusk/dawn and daytime, but the bias at this site is slightly greater than that at ENA or NSA. The reason for that could be that ECMWF is ~1 h after satellite overpass at Ny Alesund while the time difference in other two sites is ~0.5 h.

3.3. Single Launches of RS92 GDP

ENA. The negative OBS-CAL radiance differences for RS92 GDP (Figure 6a,b) indicate that the GRUAN processed RS92 has a small upper tropospheric dry bias in both nighttime and daytime, with the daytime dry bias being larger. RH dry biases estimated from the OBS-CAL difference are 1.13% and 2.57% for nighttime and daytime, respectively. The nighttime biases for RS92 GDP and RS41 STD at the same site are comparable, but the daytime RS92 GDP bias is greater (by ~1% in RH) than for RS41 STD.

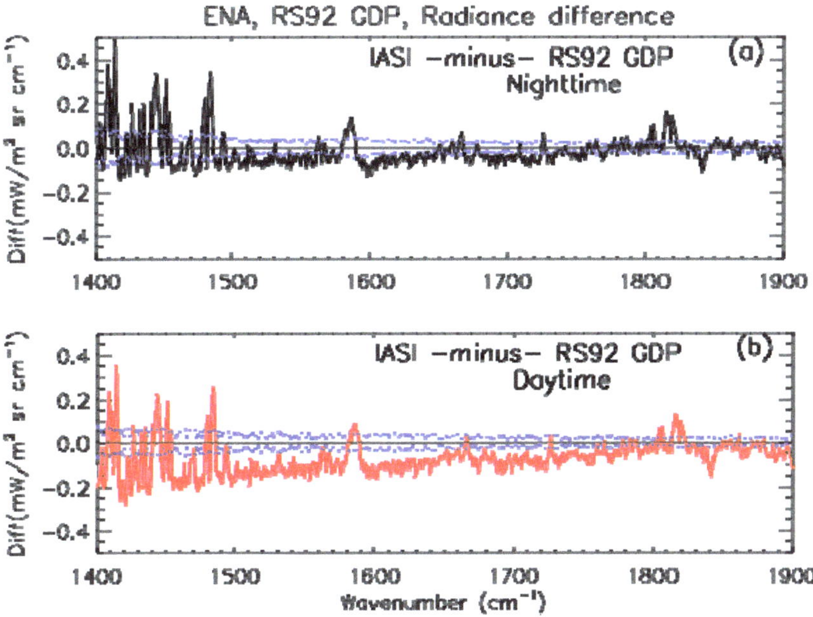

Figure 6. Eastern North Atlantic (ENA) at Graciosa, Azores. As in Figure 4a,b except for RS92 GDP instead of RS41 STD, based on (**a**) 43 night profiles and (**b**) 50 daytime profiles.

The nighttime OBS-CAL differences for RS92 GDP (averaged from 43 collocation) marginally fall within $2 \times ste$, while daytime OBS-CAL differences (averaged from ? collocations) are far beyond $2 \times ste$ (Figure 6a,b). That contrast is primarily related to th mean OBS-CAL differences, which are bigger during daytime than nighttime.

NSA. The sample of collocations with cloud-free IASI is small. A dry bias of 1.35 during nighttime and 1.98% during daytime is obtained from the radiance analysis (Table ? ECMWF is collocated within 0.5 h after the satellite overpass at nighttime and 2–3 h after th overpass for daytime. The OBS-CAL difference for ECMWF averaged over 1500–1570 cm for nighttime is only -0.0043 (0.095) mW m^{-2} sr cm^{-1}, equivalent to a RH dry bias of 0.1 For daytime, the value is -0.0521 (0.029) mW m^{-2} sr cm^{-1}, equivalent to a dry bias 1.16%. We suspect the contrast is related to the difference in ECMWF-IASI collocation tin separation, as discussed in Section 3.1 for data at Lindenberg and Payerne.

Ny Alesund. There are only daytime collocations available for a statistical analys of RS92 GDP launches. A dry bias in RS92 GDP of 2.29% estimated from OBS-CAL shown (Table 4). ECMWF is ~1 h after the satellite overpass. Again, the dryness in RS9 GDP relative to ECMWF estimated from the radiance analysis is verified in radiosonde R observations (Table 4).

3.4. Single Launches of RS92 STD

This study has RS92 STD launches and IASI collocations only at station ENA. striking feature in the OBS-CAL differences for RS92 STD (Figure 7a,b) is their difference are greater than for RS41 STD and RS92 GDP for both nighttime and daytime, suggestir that UTH dry biases of RS92 STD are larger. RH biases estimated from the radiance analys are 3.90% (1.7%) and 3.25% (2.6%) respectively for nighttime and daytime. Those big bias exaggerate the statistical inconsistency between RS92 STD and IASI, compared to betwee RS92 GDP or RS41 STD and IASI.

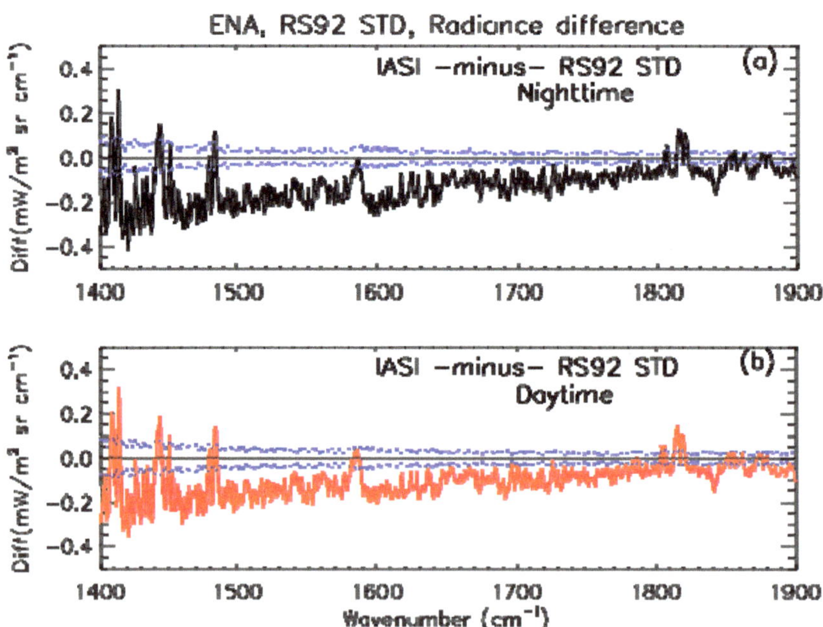

Figure 7. Eastern North Atlantic (ENA) at Graciosa, Azores. As in Figure 5a,b except for RS92 STD instead of RS41 STD, based on (**a**) 43 night profiles and (**b**) 50 daytime profiles.

The RS92 STD dry biases (Figure 7a,b and Table 5) we obtained from the radiance analysis appear to be smaller than those reported by Miloshevich et al. [5]. They notice a dry bias of 4% and 5% for nighttime and daytime respectively by comparing with cryogenic frost point hygrometer measurements. A possible explanation of the discrepancy between the two studies is that radiosonde biases estimated from the radiance analysis are for the whole layer with water vapor detected by IASI, while the biases in Miloshevich et al. [5] are for specific levels of the upper troposphere. Also, https://www.vaisala.com/en/sounding-data-continuity documents a change in Vaisala RS92 operational corrections after 2010.

4. Summary and Discussion

This paper assesses accuracies of upper tropospheric humidity observations for daytime and night separately for Vaisala RS41 STD, RS92 GDP, and RS92 STD, respectively. This is achieved by comparing the humidity sensitive infrared radiances (the 1400–1900 cm^{-1} spectral band) computed using LBLRTM from radiosonde profiles with collocated cloud-free IASI radiance measurements and with radiances similarly computed from collocated ECMWF model profiles. We primarily use single radiosondes from three GRUAN and ARM sites, with launches (primarily at synoptic times) mostly coincident within 30 min before and 15 min after IASI overpasses. We also compare dual launches (RS92 and RS41) at three other GRUAN sites, with radiosondes within 1 h of IASI overpasses at one station and 1–3 h before overpasses at the other two stations. Dual launches provide a direct comparison of RS92 vs. RS41 and with IASI, and are used as a cross-validation of the results obtained from single launches of RS92 or RS41. Accuracy of ECMWF humidity data is assessed in radiance space utilizing the collocations from single launch sites where ECMWF data is mostly at or within ~1 h of IASI. All comparisons of ECMWF vs. IASI radiances show very small systematic ECMWF biases.

Relative to IASI as a practical reference, daytime RS41 (even without GDP) has ~1% (percentage points of RH) smaller UTH errors than RS92 GDP. RS41 may still have a dry bias of 1–1.5% in both daytime and nighttime, and RS92 GDP may have a similar dry bias at night, while standard RS92 may have a dry bias of 3–4%. Those characteristics are obtained independently from 1500–1570 cm^{-1} and 1615–1800 cm^{-1}, indicating the consistency of water vapor spectroscopy between the two bands. The relative differences between RS41 STD and RS92 GDP or between radiosonde and ECMWF obtained from the radiance analysis are consistent with their differences in RH measurements. The small biases of RS41 STD indicate that RS41 at operational stations is probably almost an "absolute" standard. Note also that RS92 GDP improves accuracy to nearly the level of RS41 STD.

Radiosonde-satellite collocation uncertainty plays a big role in assessing their consistency, but collocation uncertainty generally remains unknown for individual collocations. A method was used to investigate the consistency between ensemble-averaged radiosonde (ECMWF) and IASI by computing an overall or total uncertainty term, including noise from radiosonde and satellite instruments, collocation uncertainty, and uncertainty in the LBLRTM (Section 2.2.3). Results show that RS92 STD for both daytime and nighttime and RS92 GDP for daytime are not statistically consistent with IASI. RS92 GDP for nighttime and RS41 STD for both nighttime and daytime are consistent with IASI for some cases while not for some other cases. Interpretation of the biases and consistency results presented in the study requires caution since the size of the collocation sample can directly affect the standard deviation of the overall uncertainty term, and thus the consistency (and confidence) of the assessment. It is interesting to notice, however, that ECMWF analyses are statistically consistent with IASI in almost all of the cases analyzed. We are uncertain about the reason for high model consistency, but ECMWF assimilation of both radiosonde data and IASI radiances may play a role.

The sonde humidity biases obtained from the radiance analysis are likely to be upper limits since the "cloud-free" scenes selected could still be cloud contaminated (Appendix A). The IASI channels used as the target for the analysis sense the water vapor content of an atmospheric layer in the upper troposphere, and caution is needed to compare the sonde accuracy obtained from the radiance analysis with other studies focusing on measurements made at other atmospheric levels.

Author Contributions: Conceptualization contributed by B.S., X.C., A.R., and M.B.; methodology by B.S., X.C., and A.R.; software by X.C., B.S., M.B.; computation, formal analysis, and visualization by B.S. and X.C.; data collection and processing by R.S., B.S., and M.P.; original draft preparation and writing by B.S. and S.S. All authors have read and agreed to the published version of the manuscript.

Funding: This work is supported by NOAA Joint Polar Satellite System (JPSS) NOAA Product Validation System (NPROVS). We thank EUMETSAT's Nowcasting SAF for partially supporting this study.

Data Availability Statement: All data sources are downloaded from the web addresses stated either where each dataset is mentioned in the text or is cited as a reference.

Acknowledgments: We thank Thomas August and Marc Crapeau for providing information about the operational cloud detection algorithm used to produce EUMETSAT IASI products, and Hoger Vömel for commenting on the RS41 humidity sensor measurement mechanism. We thank the GRUAN Lead Centre at Lindenberg, Germany, for access to the radiosonde data and the agencies and individuals who made the radiosonde launches. Disclaimer: The scientific results and conclusions as well as any views or opinions expressed herein, are those of the author(s) and do not necessarily reflect those of NOAA or the Department of Commerce.

Conflicts of Interest: The authors declare no conflict of interest.

Appendix A

Cloud tests used to find cloud-free scenes for the IASI L2 physical retrieval [31] include the following procedures. The IASI window channel test compares the measured radiance in window channels against clear-sky simulated radiance using a collocated NWP profile ("NWP test"). The Advanced Very High Resolution Radiometer (AVHRR) test relies on evaluation of the presence of clouds within each instantaneous field-of-view using collocated AVHRR imager data only ("AVHRR test"). The ANN cloud detection test uses both IASI and collocated AVHRR measurements in combination and implements artificial neural networks to classify the scenes ("ANN test").

Cloud test flag "FLG_CLDNES" for each IASI retrieval generated from those testing procedures has one of the following values: "1" denotes cloud free with high confidence and no clouds detected with the NWP, AVHRR and ANN cloud tests. "2" denotes presumably clear, or potential small cloud contamination (at least one cloud test detected a cloud but a cloud could not be characterized with confidence. "3" denotes cloud detected and characterized and the retrieved cloud amount is $\leq 80\%$. "4" denotes cloud detected and characterized and the retrieved cloud amount is $> 80\%$.

The cloud amount retrieved in the EUMETSAT L2 product is zero when FLG_CLDNES is "1" or "2". We assume that pixels with FLG_CLDNES = "1" have more confidence than "2" to be "clear". We, however, notice that the OBS-CAL radiance differences for the two cases do not show a significant difference from each other. As shown in Table A1, OBS-CAL differences for RS92 GDP with FLG_CLDNES assigned "1" or "2" are nearly identical in both nighttime and daytime. The data used for Table A1 are from ENA because this site has many more samples than other sites. It remains unclear why there is no difference for these two cloud test flags. This could be because both cases have a comparable degree of confidence that the pixel scene is cloud-free, or because the cloud screening method may still be ambiguous. Regardless of the reason, to have more samples for statistical analysis, RAOB-IASI collocations with FLG_CLDNES assigned 1 or 2 in the IASI retrieval are combined to conduct the OBS-CAL analysis.

Table A1. Mean differences of OBS-CAL (in all cases, IASI-RS92 GDP) calculated for IASI scenes with cloud flag being clear with high confidence ("CLD1") and presumably clear ("CLD2") using collocations of IASI-RS92 GDP data at ENA. The RH biases estimated from OBS-CAL are also listed. Numbers of IASI-RAOB collocations are in the parentheses after CLD1 and CLD2 in the second column. The CLD1 or CLD2 header in Col. 2 applies also to Cols. 3–5. Each value in parentheses in the difference lines (lines 2 and 4 in each station and SEA category) is one standard deviation of the difference to its left.

Station (RS92 GDP, Single Launches)	OBS-CAL (IASI-RS92 GDP) Radiance Diff 1500–1570 cm^{-1}	OBS-CAL RH Diff 150–1570 cm^{-1}	OBS-CAL (IASI-RS92 GDP) Radiance Diff 161–1800 cm^{-1}	OBS-CAL RH Diff 161–1800 cm^{-1}
ENA (Night)	CLD1 (14) −0.0529 (0.057)	1.17 (1.3)	−0.0301 (0.045)	1.04 (1.6)
	CLD2 (29) −0.0526 (0.099)	1.17 (2.2)	−0.0323 (0.062)	1.12 (2.1)
ENA (Day)	CLD1 (21) −0.1225 (0.095)	2.72 (2.1)	−0.0702 (0.074)	2.43 (2.5)
	−0.1174 (0.086)	2.61 (1.9)	−0.0736 (0.053)	2.54 (1.8)

Appendix B

The impact of atmospheric temperature differences on the calculated radiance differences at 1400–1900 cm^{-1} is quantified using RS92 and RS41 dual launch radiosonde data, where both sondes sample the same surface and atmosphere. We recalculate the radiances for RS41 STD, but use the RS92 GDP temperature and RS41 STD humidity profiles, and keep other variables needed in the LBLRTM calculation the same. This new RS41 STD is called RS41 STDv. We then compare RS41 STDv radiances with the radiances calculated using RS41 STD temperature and humidity. The difference between the two radiances, CAL (RS41 STDv)–CAL (RS41 STD), if any, is expected to come from their temperature difference.

Figure A1 shows the mean difference and ±2 standard deviations of CAL(RS41 STDv)–CAL(RS41 STD), based on the same Lauder dual launch data used in Figure 1. The radiance differences are negative across 1400–1900 cm^{-1}, indicating that the colder temperature in RS92 GDP (relative to RS41 STD, see Figure 1e) around the upper troposphere tends to "cause" a more "wet" RH. However, the radiance difference is rather small, averaging −0.0125 mW m^{-2} sr cm^{-1} for 1500–1570 cm^{-1}, equivalent to 0.278 % in RH.

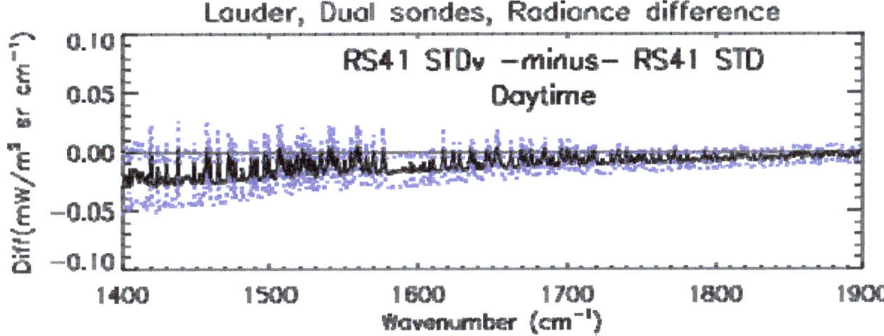

Figure A1. Lauder, New Zealand. CAL radiance differences, RS41 STDv minus RS41 STD, based on 14 daytime launches. Dotted lines show ± one standard deviation from the solid line, as in Figure 1c. RS41 STDv includes temperature profile from RS92 GDP and humidity profile from RS41 STD. See text for discussion on the impact of temperature difference (between RS92 GDP and RS41 STD) on the RH difference estimated from the radiance analysis.

Figure A2 is based on the dual launch data at Lindenberg (also used for Figure 2). The radiance difference between RS42 STDv and RS41 STD is -0.0010 mW m^{-2} sr cm^{-1} for 1500–1570 cm^{-1}, equivalent to 0.023% in RH. The temperature difference between RS92 GDP and RS41 STD in the upper troposphere is much smaller at Lindenberg than at Lauder (for example, -0.06 K vs. -0.14 K at 328.6 hPa); the radiance difference between RS92 GDP and RS41 STD is also smaller at Lindenberg than at Lauder. Since the temperature difference is very small in these analyses, the temperature contribution to the CAL radiances and hence the humidity computed from the radiance is negligible.

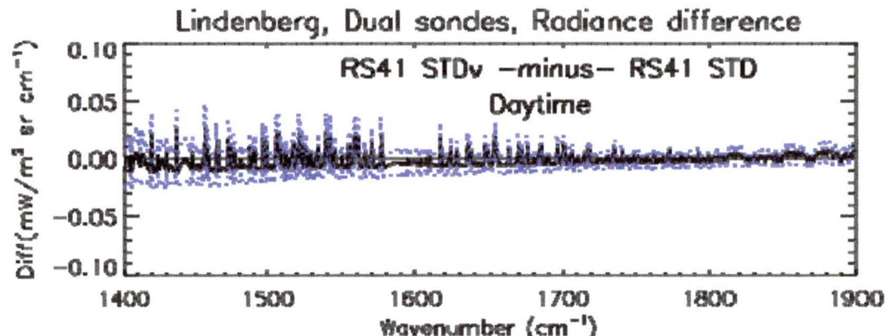

Figure A2. Lindenberg, Germany. Same as Figure A1 but for 19 daytime dual launches.

References

1. Bodeker, G.E.; Bojinski, S.; Cimini, D.; Dirksen, R.J.; Haeffelin, M.; Hannigan, J.W.; Hurst, D.F.; Leblanc, T.; Madonna, Maturilli, M.; et al. Reference upper-air observations for climate: From concept to reality. *Bull. Am. Meteorol. Soc.* **2016**, *97*, 123–135. [CrossRef]
2. Miloshevich, L.M.; Vömel, H.; Paukkunen, A.; Heymsfield, A.J.; Oltmans, S.J. Characterization and correction of relative humidity measurements from Vaisala RS80-A radiosondes at cold temperatures. *J. Atmos. Ocean. Technol.* **2001**, *18*, 135–156. [CrossRef]
3. Miloshevich, L.M.; Paukkunen, A.; Vömel, H.; Oltmans, S.J. Development and validation of a time-lag correction for Vaisala radiosonde humidity measurements. *J. Atmos. Ocean. Technol.* **2004**, *21*, 1305–1327. [CrossRef]
4. Miloshevich, L.M.; Vömel, H.; Whiteman, D.N.; Lesht, B.M.; Schmidlin, F.J.; Russo, F. Absolute accuracy of water vapor measurements from six operational radiosonde types launched during AWEX-G and implications for AIRS validation. *J. Geophys. Res.* **2006**, *111*, D09D10. [CrossRef]
5. Miloshevich, L.M.; Vömel, H.; Whiteman, D.N.; Leblanc, T. Accuracy assessment and correction of Vaisala RS92 radiosonde water vapor measurements. *J. Geophys. Res.* **2009**, *114*, D11305. [CrossRef]
6. Nash, J.; Oakley, T.; Vömel, H.; Wei, L. WMO Intercomparison of High-Quality Radiosonde Systems, Yangjiang, China, 12 July–August 2010. Instruments and Observing Methods Rep. 107 (IOM-107, WMO/TD-No. 1580), 238 pp. World Meteorological Organization: Geneva, 2011. Available online: https://www.wmo.int/pages/prog/www/IMOP/publications/IOM-107_Yangjiang.pdf (accessed on 4 January 2021).
7. Vömel, H.; Selkirk, H.; Miloshevich, L.; Valverde-Canossa, J.; Valdés, J.; Kyrö, E.; Kivi, R.; Stolz, W.; Peng, G.; Diaz, J.A. Radiation dry bias of the Vaisala RS92 humidity sensor. *J. Atmos. Ocean. Technol.* **2007**, *24*, 953–963. [CrossRef]
8. Soden, B.J.; Lanzante, J.R. An assessment of satellite and radiosonde climatologies of upper-tropospheric water vapor. *J. Clim.* **1996**, *9*, 1235–1250. [CrossRef]
9. Kottayil, A.; Buehler, S.A.; John, V.O.; Miloshevich, A.M.; Milz, M.; Holl, G. On the importance of Vaisala RS92 radiosonde humidity corrections for a better agreement between measured and modeled satellite radiances. *J. Atmos. Ocean. Technol.* **2012**, *29*, 248–259. [CrossRef]
10. Calbet, X.; Kivi, R.; Tjemkes, S.; Montagner, F.; Stuhlmann, R. Matching radiative transfer models and radiosonde data from the EPS/Metop Sodankylä campaign to IASI measurements. *Atmos. Meas. Tech.* **2011**, *4*, 1177–1189. [CrossRef]
11. Moradi, I.; Buehler, S.A.; John, V.O.; Reale, A.; Ferro, R.R. Evaluating instrumental inhomogeneities in global radiosonde upper tropospheric humidity data using microwave satellite data. *IEEE Trans. Geosci. Remote Sens.* **2013**, *51*, 3615–3624. [CrossRef]
12. Moradi, I.; Soden, B.; Ferro, R.R.; Arkin, P.; Vömel, H. Assessing the quality of humidity measurements from global operational radiosonde sensors. *J. Geophys. Res. Atmos.* **2013**, *118*, 8040–8053. [CrossRef]
13. Calbet, X.; Peinado-Galan, N.; Ripodas, P.; Trent, T.; Dirksen, R.; Sommer, M. Consistency between GRUAN sondes, LBLRTM and IASI. *Atmos. Meas. Tech.* **2017**, *10*, 2323–2335. [CrossRef]

4. Serio, C.; Masiello, G.; Camy-Peyret, C.; Jacquette, E.; Vandermarcq, O.; Bermudo, F.; Coppens, D.; Tobin, D. PCA determination of the radiometric noise of high spectral resolution infrared observations from spectral residuals: Application to IASI. *J. Quant. Spectrosc. Radiat. Transf.* **2018**, *206*, 8–21. [CrossRef]

5. Clough, S.A.; Shephard, M.W.; Mlawer, E.J.; Delamere, J.S.; Iacono, M.J.; Cady-Pereira, K.; Boukabara, S.; Brown, R.D. Atmospheric radiative transfer modeling: A summary of the AER codes. *J. Quant. Spectrosc. Radiat. Transf.* **2005**, *91*, 233–244. [CrossRef]

6. Dirksen, R.J.; Sommer, M.; Immler, F.J.; Hurst, D.F.; Kivi, R.; Vömel, H. Reference quality upper-air measurements, GRUAN data processing for the Vaisala RS92 radiosonde. *Atmos. Meas. Tech.* **2014**, *7*, 4463–4490. [CrossRef]

7. Reale, A.; Sun, B.; Tilley, F.H.; Pettey, M. The NOAA Products Validation System (NPROVS). *J. Atmos. Ocean. Tech.* **2012**, *29*, 629–645. [CrossRef]

8. Sun, B.; Reale, A.; Tilley, F.H.; Pettey, M.; Nalli, N.R.; Barnet, C.D. Assessment of NUCAPS S-NPP CrIS/ATMS sounding products using reference and conventional radiosonde observations. *IEEE J. Sel. Top. Appl. Earth Obs. Remote Sens.* **2017**, *10*, 1–18. [CrossRef]

9. Nalli, N.R.; Gambacorta, A.; Liu, Q.; Barnet, C.D.; Tan, C.; Iturbide-Sanchez, F.; Reale, T.; Sun, B.; Wilson, M.; Borg, L.; et al. Validation of atmospheric profile retrievals from the SNPP NOAA-Unique Combined Atmospheric Processing System. Part I: Temperature and moisture. *IEEE Trans. Geosci. Remote Sens.* **2018**, *56*, 180–190. [CrossRef]

10. Nalli, N.R.; Barnet, C.D.; Reale, A.; Tobin, D.; Gambacorta, A.; Maddy, E.S.; Joseph, E.; Sun, B.; Borg, L.; Mollner, A.K.; et al. Validation of satellite sounder environmental data records: Application to the Cross-track Infrared Microwave Sounder Suite. *J. Geophys. Res. Atmos.* **2013**, *118*, 13628–13643. [CrossRef]

11. August, T.; Klaes, D.; Schlüssel, P.; Hultberg, T.; Crapeau, M.; Arriaga, A.; 'Carroll, A.O.; Coppens, D.; Munro, R.; Calbet, X. IASI on Metop-A: Operational level 2 retrievals after five years in orbit. *J. Quant. Spectrosc. Radiat. Transf.* **2012**, *113*, 1340–1371. [CrossRef]

12. Sun, B.; Reale, A.; Pettey, M.; Smith, R.; Nalli, N.R.; Zhou, L. Leveraging the strength of dedicated, GRUAN and conventional radiosondes for EUMETSAT IASI atmospheric sounding assessment. In Proceedings of the 2017 EUMETSAT Meteorological Satellite Conference, Rome, Italy, 2–6 October 2017; pp. 1–8.

13. ECMWF. IFS Documentation CY45r1. ECMWF Rep.; 2018; p. 103. Available online: https://www.ecmwf.int/en/publications/ifs-documentation (accessed on 21 December 2018).

14. Eresmaa, R.; McNally, A.P. Diverse Profile Datasets from the ECMWF 137-Level Short-Range Forecasts. NWP Satellite Application Facilities Rep. NWPSAF-EC-TR-017. 2014, p. 12. Available online: https://nwpsaf.eu/oldsite/reports/nwpsaf-ec-tr-017.pdf (accessed on 21 December 2018).

15. Hyland, R.; Wexler, A. Formulations for the thermodynamic properties of the saturated phases of H_2O from 173.15 K to 473.15 K. *ASHRAE Trans.* **1983**, *89*, 500–519.

16. Tobin, D.C.; Revercomb, H.E.; Knuteson, R.O.; Lesht, B.M.; Strow, L.L.; Hannon, S.E.; Feltz, W.F.; Moy, L.A.; Fetzer, E.J.; Cress, T.S. Atmospheric radiation measurement site atmospheric state best estimates for atmospheric infrared sounder temperature and water vapor retrieval validation. *J. Geophys. Res.* **2006**, *111*, D09S14. [CrossRef]

17. Immler, F.J.; Dykema, J.; Gardiner, T.; Whiteman, D.N.; Thorne, P.W.; Vömel, H. Reference Quality Upper Air Measurements: Guidance for developing GRUAN data products. *Atmos. Meas. Tech.* **2010**, *3*, 1217–1231. [CrossRef]

18. Sun, B.; Reale, A.; Schroeder, S.; Pettey, M.; Smith, R. On the accuracy of Vaisala RS41 versus RS92 upper-air temperature observations. *J. Atmos. Ocean. Technol.* **2019**, *36*, 635–653. [CrossRef]

19. Seidel, D.J.; Sun, B.; Pettey, M.; Reale, A. Global radiosonde balloon drift statistics. *J. Geophys. Res.* **2011**, *116*, D07102. [CrossRef]

20. Sun, B.; Reale, A.; Seidel, D.J.; Hunt, D.C. Comparing radiosonde and COSMIC atmospheric profile data to quantify differences among radiosonde types and the effects of imperfect collocations on comparison statistics. *J. Geophys. Res* **2010**, *115*. [CrossRef]

21. EUMETSAT 2014: IASI Level 2: Product Format Specification. Doc. No.: EPS.SYS.SPE.990013, Date: 28 November 2014. Eumetsat-Allee 1, D-64295 Darmstadt, Germany. Available online: https://www.eumetsat.int (accessed on 9 March 2015).

remote sensing

Article

Applying Deep Learning to Clear-Sky Radiance Simulation for VIIRS with Community Radiative Transfer Model—Part 2: Model Architecture and Assessment

Xingming Liang [1,*] and **Quanhua (Mark) Liu** [2]

[1] Earth System Science Interdisciplinary Center, University of Maryland, College Park, MD 20740, USA

[2] Center for Satellite Applications and Research, National Environmental Satellite (STAR), Data, and Information Service (NESDIS), National Oceanic and Atmospheric Administration (NOAA), College Park, MD 20740, USA; quanhua.liu@noaa.gov

* Correspondence: xingming.liang@noaa.gov; Tel.: +1-301-683-3362

Received: 2 November 2020; Accepted: 16 November 2020; Published: 21 November 2020

Abstract: A fully connected "deep" neural network algorithm with the Community Radiative Transfer Model (FCDN_CRTM) is proposed to explore the efficiency and accuracy of reproducing the Visible Infrared Imaging Radiometer Suite (VIIRS) radiances in five thermal emission M (TEB/M) bands. The model was trained and tested in the nighttime global ocean clear-sky domain, in which the VIIRS observation minus CRTM (O-M) biases have been well validated in recent years. The atmosphere profile from the European Centre for Medium-Range Weather Forecasts (ECMWF) and sea surface temperature (SST) from the Canadian Meteorology Centre (CMC) were used as FCDN_CRTM input, and the CRTM-simulated brightness temperatures (BTs) were defined as labels. Six dispersion days' data from 2019 to 2020 were selected to train the FCDN_CRTM, and the clear-sky pixels were identified by an enhanced FCDN clear-sky mask (FCDN_CSM) model, which was demonstrated in Part 1. The trained model was then employed to predict CRTM BTs, which were further validated with the CRTM BTs and the VIIRS sensor data record (SDR) for both efficiency and accuracy. With iterative refinement of the model design and careful treatment of the input data, the agreement between the FCDN_CRTM and the CRTM was generally good, including the satellite zenith angle and column water vapor dependencies. The mean biases of the FCDN_CRTM minus CRTM (F-C) were typically ~0.01 K for all five bands, and the high accuracy persisted during the whole analysis period. Moreover, the standard deviations (STDs) were generally less than 0.1 K and were consistent for approximately half a year, before they significantly degraded. The validation with VIIRS SDR data revealed that both the predicted mean biases and the STD of the VIIRS observation minus FCDN_CRTM (V-F) were comparable with the VIIRS minus direct CRTM simulation (V-C). Meanwhile, both V-F and V-C exhibited consistent global geophysical and statistical distribution, as well as stable long-term performance. Furthermore, the FCDN_CRTM processing time was more than 40 times faster than CRTM simulation. The highly efficient, accurate, and stable performances indicate that the FCDN_CRTM is a potential solution for global and real-time monitoring of sensor observation minus model simulation, particularly for high-resolution sensors.

Keywords: community radiative transfer model (CRTM); deep learning; fully connected "deep" neural network (FCDN); radiative transfer; artificial neural network (ANN); batch normalization (BN); real time; the visible infrared imaging radiometer suite (VIIRS)

1. Introduction

The Community Radiative Transfer Model (CRTM) was developed at the Joint Center for Satellite Data Assimilation (JCSDA). This fast radiative transfer model is used at the *National Oceanic and Atmospheric Administration* (NOAA) and in many institutes and universities, both nationally and internationally [1–7]. The model simulates satellite measurements from visible, infrared, or microwave bands and calculates corresponding tangent-linear, adjoint, and Jacobian values for various geophysical and atmospheric parameters to support radiance assimilation and the retrieval of atmosphere and surface states [4,6]. Trained transmittance coefficients are used in the CRTM instead of the convolution of sensor response function with line-by-line calculations. This approach renders the CRTM highly efficient for application in operational numerical weather prediction, sensor validation and long-term monitoring, development of the environment data record (EDR), and climate research for most polar orbiting and geostationary meteorological satellite sensors [3,8–10]. For instance, at the National Centers for Environmental Prediction (NCEP), the CRTM is a key component of the core of the data assimilation system, called Gridpoint Statistical Interpolation (GSI), to simulate various satellite data [11]. Since the NOAA sea surface temperature (SST) system—the Advanced Clear-sky Processor over Ocean (ACSPO) system—was developed [3], the CRTM has been used to real-time monitor the sensor radiometric bias performance of infrared (IR) window bands for more than a decade on the website of Monitoring of IR Clear-sky Radiances over Ocean for SST (MICROS; https://www.star.nesdis.noaa.gov/sod/sst/micros) [8,9]. Moreover, the monitoring of sensor observations against CRTM simulation (O-M) is a key component of the integrated calibration/validation system (*ICVS*) established by *the* NOAA Center for Satellite Applications and Research (STAR; https://www.star.nesdis.noaa.gov/icvs) [12].

Although the simplified transmittance coefficients have been adopted in the CRTM, with the development of high spatial and temporal resolution sensors, the efficiency of CRTM simulation is still a key issue for global data monitoring of the O-M biases, such as the Visible Infrared Imaging Radiometer Suite (VIIRS) onboard the satellites in the Joint Polar Satellite System (JPSS) and the advanced baseline imager (ABI) onboard the geostationary operational environmental satellite-R (GOES-R). Based on an offline experiment, for the CRTM to reproduce 1440×720 clear-sky radiances for VIIRS five thermal emission M (TEB/M) bands, which are equivalent to approximately 30-s sensor scans, more than two minutes are required on a STAR internal Linux box with a 2.2 G CPU and 200 G memory. It is thus impossible to timeously simulate global VIIRS data with more than 1 billion pixels for real-time monitoring of the sensor radiometric biases using the CRTM as a reference.

To improve CRTM efficiency for high-resolution sensors, MICROS conducted CRTM simulation at the grid level of the NCEP global forecast system (GFS) and then interpolated the model BTs to the sensor pixel. The method renders the global O-M calculation highly efficient even for high-resolution sensors, such as VIIRS and ABI. However, the O-M mean bias and the standard deviation (STD) remain somewhat large [8,9]. For the ICVS, the model data were simulated in selected pixels from a four-by-four moving window, which reduced the solution by one-sixteenth, making real-time O-M monitoring possible for high-resolution sensors [12]. Although reducing the space resolution may speed up CRTM simulation, missing information and dispersed global coverage are problems for some EDR users, such as the SST. Moreover, for simulation in visible bands, the efficiency of atmospheric scattering is a known issue in the remote sensing community.

In recent years, the method of an artificial neural network (ANN) has gradually become a popular algorithm and is applied in most science and technical fields, including atmosphere and ocean remote sensing and climate research [13–18]. Using simple, statistical, nonlinear approximation instead of a complicated physical-based model in ANNs renders a more computationally efficient method to achieve a similar job to that of the physical-based model without significant accuracy loss [13–18]. These advantages have attracted an increasing number of remote sensing scientists to explore the possible replacement of the radiative transfer (RT) forward model or inversion with the ANN model in recent years [19–26]. Given the complicated nature of the RT model and its input, emulating a full RT model using only one ANN architecture is currently impossible. Each study of ANN emulation generally focuses on one specific purpose and limit in some spectrum range, such as visible, long-IR, short-IR, or micro waves. Some ANN emulators have been combined with additional statistics analysis, such as principal component analysis, to reduce the dimensionality of the input features [26].

To explore the efficiency and accuracy of ANN application in the CRTM and in the real-time monitoring of sensor radiometric biases in global, we designed and developed a fully connected deep neural network (FCDN) algorithm and applied it to CRTM simulation for the Suomi-National Polar-orbiting Partnership (*SNPP*) VIIRS in five TEB/M bands. Together with the earlier-developed FCDN clear-sky mask (FCDN_CSM) [27,28], the objectives in this study are (1) to predict global clear-sky BTs using a well-trained FCDN_CRTM for high spatial resolution VIIRS in near real time and (2) to validate the FCDN_CRTM prediction accuracy, efficiency, and long-term stability. Section 2 discusses the methodology of this research. A detailed description of the FCDN_CRTM and data preprocessing is provided. This section also includes a brief summary of the CRTM and its inputs, FCDN_CSM, and batch normalization (BN), which are all used in this study. Section 3 then demonstrates model training, testing, and predicting, along with the model's validation with CRTM BTs and VIIRS SDR data. Thereafter, Section 4 discusses several scientific insights regarding the model and Section 5 provides the conclusion.

2. Methodology

In this section, we first summarize CRTM simulation applied to VIIRS TEB/M bands in the ocean clear-sky domain, in conjunction with upper air profiles from the European Centre for Medium-Range Weather Forecasts (ECMWF) and SST from the Canadian Meteorology Centre (CMC). Then, we discuss the FCDN_CRTM architecture and data preprocessing in detail. In parallel, a summary of the FCDN_CSM is provided, which is used in this study to identify the clear-sky domain efficiently. Finally, we demonstrate the BN algorithm in the FCDN_CRTM to speed up the model convergence.

2.1. The CRTM and Input Data

By excluding the effect of the daytime solar reflection for the mid-IR bands [29] and focusing on more uniformly distributed ocean, CRTM simulation for VIIRS thermal emission bands has been well validated with sensor measurements for over a decade in the nighttime clear-sky ocean domain [3,8–10]. The condition is, thus, used in this study to evaluate the FCDN_CRTM accuracy and stability with mature CRTM simulation.

As the CRTM was used for the VIIRS thermal emission bands, the effects of scattering in the atmosphere were omitted in this study. When excluding the quantitative analyses of the effect of solar reflection and the effect of cloud for all bands, and focusing only on the nighttime ocean clear-sky domain, the radiative transfer equation used for the VIIRS TEB bands is written as follows:

$$\overline{R}(\theta) = \varepsilon(\theta)\overline{B}(Ts)\overline{\tau}(\theta) + \overline{L^{\uparrow}}(\theta) + (1 - \varepsilon(\theta))\overline{L^{\uparrow}}(\theta)\overline{\tau}(\theta) \tag{1}$$

where $\overline{R}(\theta)$ refers to TOA radiance for the VIIRS TEB band; θ is the satellite zenith angle (SZA); and $\varepsilon(\theta)$ depicts the surface emissivity. The diversity and complexity of a land surface can cause unexpected bias and noise in the CRTM simulation; hence, we first selected the more uniform ocean surface in this study. The surface emissivity was defined in line with the wind-speed-dependent emissivity of Wu and Smith [30]. Moreover, Ts denotes surface temperature, and $\overline{B}(Ts)$ is its Planck radiance. Atmospheric transmittance $\overline{\tau}(\theta)$ and both upwelling and downwelling radiances $\overline{L^{\uparrow}}(\theta)$ and $\overline{L^{\downarrow}}(\theta)$ were calculated within the CRTM. The three terms on the right-hand side of the equation are surface emission, upwelling atmospheric emission, and reflected downwelling atmospheric emission, respectively. Trained atmospheric transmittance coefficients were derived against the line-by-line radiative transfer model (LBLRTM) transmittances, and they were then used to calculate $\overline{\tau}(\theta)$, $\overline{L^{\uparrow}}(\theta)$, and $\overline{L^{\downarrow}}(\theta)$ for most sensors onboard NOAA-related polar orbiting and geostationary satellites, such as VIIRS. Resulting errors in TOA BTs were found to be small [31].

Inputs to the CRTM mainly consist of the atmospheric profiles of pressure, temperature, moisture, and ozone; surface temperature; wind speed; solar zenith angle; and satellite view zenith angle; among others. An earlier documentation [3] described in detail the CRTM inputs from the *atmosphere profiles of the National Centers for Environmental Prediction* (*NCEP*) Global Forecast System (*GFS*) and the Reynolds SST, and the CRTM inputs were then further updated using higher resolution ECMWF (https://www.ecmwf.int) and CMC SST to improve simulation accuracy [12]. The ECMWF data are accumulated in the STAR server by the NOAA soundings team and are refreshed daily. This ECMWF product has a 0.25° horizontal resolution with 91 vertical layers in the early release and later updated to 137. The profiles are available up to 0.02 mb (http://www.ecmwf.int/en/forecasts/datasets); therefore, no vertical extrapolation is needed for CRTM calculation. Eight files per day are acquired at 00, 06, 12, and 18 UTC, including four analyses (i.e., 0-h forecast) and four forecasts (3-h and 9-h forecasts at 00 UTC, and 15-h and 21-h forecasts at 12 UTC).

The main difference between the GFS and ECMWF profiles is that the former defines the profile in level but the latter defines the profile in layers. This makes ECMWF atmosphere profiles easier to input into the CRTM, as the complex conversion from levels to layers does not need to be considered [3]. In addition, the ECMWF's reported u and v components of wind vector were used in this study to calculate the near-surface wind speed and direction, and they were then input into the CRTM to determine the sea surface emissivity. In this study, we performed the model simulation in VIIRS pixels, as the simulation results are more accurate than those performed in-grid [12]. The ECMWF fields were, thus, first linearly interpolated in time to match the VIIRS SDR observation times, using two 0-h forecasts separated by 6 h. Since the two 0-h ECMWF forecasts are close to the analysis data, they are more accurate for CRTM simulation than the other forecasts. These time-interpolated fields were further bilinearly interpolated in space to match the VIIRS pixels before simulating CRTM BTs. A 0.1° daily CMC SST analysis (https://podaac.jpl.nasa.gov/dataset/CMC0.1deg-CMC-L4-GLOB-v3.0) was selected as the surface temperature input into the CRTM. It was interpolated in the same way as the ECMWF to match the VIIRS pixels in space. In addition, we did not include the aerosol model in the CRTM simulation in this study. As previously discussed, a missing aerosol in the CRTM simulation may result in a slight overestimation (~0.1 K), particularly for longwave IR window bands [3].

2.2. The FCDN_CRTM Architecture

Due to the issue with efficiency in CRTM simulation for high-resolution sensors, an FCDN was proposed to explore model efficiency. An FCDN is a multilayered artificial neural architecture, which is widely used among deep-learning models to solve problems of function fitting, classification, clustering, and pattern recognition. Liang et al. [27] summarized the details of the FCDN, which was successfully applied to the classification problem of the VIIRS clear-sky mask for efficient and accurate O-M validation in global.

In that early study, we constructed an FCDN including two hidden layers with 40×90 neurons and 11 features as the model input into classify four CSM types. We demonstrated that the FCDN could learn complex nonlinear functional mappings with highly accurate predicted results, given sufficient computational resources and training data. Moreover, the FCDN black-box system reduces the manual work needed for setting up in the traditional methods, including many empirical thresholds in the physics-based CSM retrieval. Furthermore, it offers efficiency and migration advantages.

In the current study, we applied the FCDN to simulate the BTs of five VIIRS TEB/M bands using ECMWF data and CMC SST as input. This model is hereafter referred to as the FCDN_CRTM, as the CRTM simulation was used as the model reference. Furthermore, different from the classification application in the FCDN_CSM, the FCDN_CRTM is a regression problem: to predict a continuous quantity output for an example. We, thus, made several critical updates to the FCDN_CRTM architecture. First, the number of input features for BT calculation included 91-layer profiles for atmospheric temperature, water vapor, and O3, as well as surface and satellite geophysical parameters—which greatly outnumber those of the FCDN_CSM. We discuss the input data further in the next subsection.

Therefore, the design of the FCDN_CRTM architecture was more complex to ensure rapid convergence and to attain a global optimum for the cost function (also called the loss function) during the model training. As discussed in [27], there is no mathematical or physical rule to determine the best hyperparameters, other than early ANN references and fine-tuning by repeated experiments. By effort in extensive experiments and model fine-tuning, we finally designed three hidden layers with 512, 384, and 64 neurons in the layers, respectively. Second, using the mean squared error (MSE) as a cost function for a regression problem was more intuitive than using the cross-entry loss, as in the FCDN_CSM. Furthermore, a regularization term, *known as the L2 norm,* was added in the cost function to avoid possible overfitting when the model was used to predict CRTM BTs [32]. The following equation (Equation (2)) shows the final cost function used in the FCDN_CRTM:

$$J(w,b) = \sum_{i=0}^{n} (y_i - \hat{y}_i)^2 + \lambda \sum_{k=0}^{m} \|W_k^2\| \tag{2}$$

where *w and b* are the weight and bias, respectively, while *n* represents the batch size, and *m* is the total number of weights. As described in part 1, the symbol λ refers to the regularization coefficient, which is a hyper-parameter in the FCDN_CRTM to *decide how much to penalize the flexibility of our model.* In this study, we selected λ to be 0.001.

2.3. Summary of the FCDN_CSM

A new algorithm of the VIIRS clear-sky mask using the FCDN (FCDN_CSM) [27] was developed to replace the traditional physical-based model. The aim is to identify clear-sky domain efficiently for the real-time monitoring of VIIRS O-M biases in the ICVS system. The model was further enhanced recently to include the FCDN_CSM prediction and validation in daytime and improve its long-term stability [28]. Although a slight residual cloud may remain by using the FCDN-CSM, the O-M mean biases are comparable and the maximum degradation of the STDs is only several hundredths of a Kelvin in M16, in comparison to using the ACSPO CSM. On the other hand, the model required less than one minute to generate a day's worth of CSM, at approximately 0.6 billion pixels, in comparison to computationally consuming in the traditional model. Furthermore, the model did not obviously degrade in a half-year analysis period, and it was, thus, used in this study to efficiently identify clear-sky pixels for VIIRS.

2.4. The FCDN_CRTM Input and Preprocessing

As discussed in Section 1, CRTM simulation for VIIRS thermal emission bands in the nighttime clear-sky ocean domain has been well validated for over a decade [3,8–10]. Under the selected condition, which excluded solar contamination in M12, daytime diurnal cycle effects, cloud effects, and complicated land surfaces, the O-M mean biases and STDs are only 0.1 ± 0.3 K for the atmosphere transparency band (M12) and 0.3 ± 0.5 K for the atmosphere opacity band (M16). Achieving these accuracies under the same atmospheric and geographic conditions is, thus, most challenging for the first proposed FCDN_CRTM. Careful treatment of the input data is critical for model accuracy.

All training and testing data were limited to more than 90° of the solar zenith angle and ocean pixels. Similar to the CRTM input, the FCDN_CRTM input features were obtained from ECMWF and CMC SST, including 91-layer atmosphere temperatures; water vapor contents; O3; and each value of surface wind speed, surface temperature, and surface pressure. The ECMWF pressure profiles were calculated by the surface pressure, with the same scales applied to 92 vertical levels for all space grids. Thus, theoretically, surface pressure was adequate to represent a 92-level pressure profile input for the FCDN_CRTM. The result in the next section further verifies this selection.

All ECMWF and CMC gridding data were interpolated with time and space to match the VIIRS SDR pixels. Furthermore, the SZA in VIIRS SDR GEO granules was extracted as a model feature and was roughly separated into positive and negative values by the half-scan swath for model validation.

Although the SZA was directly used as an input feature for the FCDN_CSM to conduct clear-sky classification, a secant of SZA selected in this study is more effective. We further discuss this issue in the next section. While only ocean-type data were selected in this study, the land or sea mask was nonetheless used as a feature in the FCDN_CRTM to allow for an extension of the functionality to include a land analysis in the future. Overall, 278 features were prepared as FCDN_CRTM input, and CRTM Version 2.3.0 was used to generate BT references. Note that some researchers [26] have suggested reducing the dimensionality for input features using principle component analysis (PCA) or other methods to simplify the model and speed up the model training; however, in our case, we kept all 91-layer data as model inputs to include extensive and detailed atmosphere states without any energy loss. Table 1 lists all input features and output BTs in the FCDN_CRTM.

Table 1. Summary of input features and output brightness temperatures (BTs) in the fully connected "deep" neural network algorithm with the Community Radiative Transfer Model (FCDN_CRTM). SZA: satellite zenith angle.

Input Features		Output BTs	
Names	**Number**	**Names**	**Number**
land/sea mask	1	M12 BT	1
Secant of SZA	1	M13 BT	1
Wind Speed	1	M14 BT	1
Surface Temperature	1	M15 BT	1
Surface Pressure	1	M16 BT	1
Air Temperatures	91		
water vapor contents	91		
O3 contents	91		
Total	278	Total	5

2.5. Batch Normalization and Output Mode

Two processing phases were conducted during the FCDN training: forward propagation and backward propagation. Forward propagation enabled the cost function calculation from the left layer of the FCDN architecture to the right, while backward propagation updated the weights and biases by calculating the gradient of the cost function from the right layer to the left. The gradients ideally become steadily smaller from the right layer to the left. However, the weights in the deeper layers are sometimes not updated, and the training of the network is, thus, not highly effective. This is known as the **vanishing gradient** problem, which occurs frequently for complex and deep neuronal networks. The root cause of **vanishing gradients is that** the input distribution that maps to the nonlinear function gradually moves closer to the limit saturation zone as backward propagation progresses to deep layers [33]. To avoid this problem, BN was introduced in the FCDN_CRTM as in Equations (3) to (7) [33]:

$$B = \{x_1, \dots, x_m\} \tag{3}$$

$$\mu_B = \frac{1}{m} \sum_{i=1}^{m} x_i \tag{4}$$

$$\sigma_B^2 = \frac{1}{m} \sum_{i=1}^{m} (x_i - \mu_B)^2 \tag{5}$$

$$\hat{x}_i = \frac{x_i - \mu_B}{\sqrt{\sigma_B^2 + \varepsilon}} \tag{6}$$

$$y_i = \gamma \hat{x}_i + \beta \tag{7}$$

where $\mathcal{B}$ represents x values over a mini batch that was fed into the model in each training iteration, μ and σ^2 refer to the mini batch mean and variance, respectively, and γ and β are two hyper-parameters used to move the original input into a region in which the model is more sensitive to the input. For each hidden layer, the input distribution that moves closer to the limit saturation zone is forced to a relatively normal distribution ($\hat{x}_i$), with a mean of 0 and variance of 1. The input value of the nonlinear transformation function is, hence, in a region that is highly sensitive to the input, thus avoiding the problem of gradient disappearance and dramatically accelerating the training of the deep neural network. Batch normalization also reduces gradients or their initial values' dependence on the scale of the parameters. This enables the use of highly flexible learning rates. Furthermore, BN regularizes the model and reduces the risk of overfitting.

In addition, the prediction BT can be trained together or separately by individual bands. As the possible band-by-band correlation, individual band training and multi-band training may cause different accuracies. To verify the advantage of BN and select the best output mode, in the FCDN_CRTM, we tested the sensitivity of training performance for the following four cases: (1) single-band training with BN and (2) without BN, and (3) multi-band training with BN and (4) without BN. The single-band training, in which the output layer included only one band BT, required five training sessions to obtain all TEB/M BTs for VIIRS. In contrast, the multi-band training trained all five band BTs simultaneously.

ECMWF data on one day (12 October 2019), and the corresponding simulated CRTM data, were separated into training (90%) and testing (10%) data sets to use as model input. The SZA was randomly selected between 0° and 60°, and the solar zenith angle was set to be larger than 90° (nighttime). Figure 1 illustrates the cost function convergence during the training for the four cases. It was clear that all cases converged and reached their optimal results after 400,000 iterations each. For both single-band and multi-band training, the cost functions for the cases with BN converged faster and reached smaller values than those without BN. This finding implies that the predicted BTs from the FCDN model with BN were the most accurate. Furthermore, despite a 0.05 difference for the cost-function convergence between the single and multi-bands, the results were comparable after introducing BN to the model.

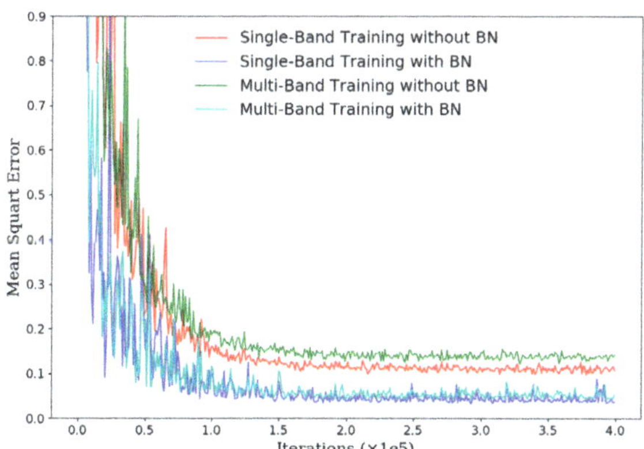

Figure 1. The convergences of the cost function during the training for the four cases: single-band training with batch normalization (BN) and without BN, and multi-band training with BN and without BN.

Table 2 lists the means and STDs of the BT differences between the FCDN_CRTM and the CRTM (F-C) for the testing data set. The mean values for all cases were close to 0, whereas the STDs for the cases with BN were ~0.2 K smaller than those without BN. Furthermore, the STDs for the case of multi-band training with BN were slightly larger than for single-band training with BN, indicating that the latter training was more accurate than the former. The smaller STDs for single-band training might imply that this method avoided interaction among the bands through potential band-band correlation. However, the training and testing accuracies for the multi-band training with BN still remained close to those of single-band training with BN. Furthermore, the multi-band training was more efficient than the single-band model, as all bands were trained at once. It was, thus, reasonable to continue using only the multi-band training with BN thereafter.

Table 2. Global F-C mean and STD for SNPP Visible Infrared Imaging Radiometer Suite (VIIRS) on 21 February 2020 from bands M12-M16 (F-C: difference between FCDN_CRTM BT and CRTM BT; μ: F-C mean bias; σ: corresponding STD; SB: single-band training; MB: multi-bands training).

	μ (F-C, K)				σ (F-C, K)			
	SB	SB and BN	MB	MB and BN	SB	SB and BN	MB	MB and BN
M12	0.0069	0.0096	−0.0076	0.0486	0.3162	0.1127	0.3534	0.1551
M13	0.0000	0.0336	−0.0130	0.0328	0.3142	0.1236	0.3599	0.1644
M14	−0.0159	−0.0479	−0.0088	0.0260	0.3166	0.1253	0.3534	0.1546
M15	0.0131	0.0683	−0.0064	0.0357	0.3196	0.1307	0.3469	0.1561
M16	0.0022	−0.0363	−0.0050	0.0389	0.3285	0.1183	0.3560	0.1633

3. FCDN_CRTM Training, Testing, Predicting, and Validating

In this section, we first demonstrate detailed FCDN_CRTM training and testing. We then employ the trained model to predict CRTM BTs and validate the model with CRTM simulation and VIIRS SDR data.

3.1. FCDN_CRTM Training and Testing

To take account the seasonal cycle effects and to build a robust FCDN_CRTM that can predict BTs accurately and stably, the input data should include most spatial and temporal conditions in global. In this section, six dispersion data points from 2019 to 2020, including 10 March, 5 May, 1 August, 12 October, and 6 November in 2019 and 15 January in 2020, which nearly cover all seasons, were utilized as FCDN_CRTM input data. These data were selected side by side with the CRTM BTs for five VIIRS TEB/M bands. Roughly 40 million samples were accumulated after data preprocessing.

The samples were further separated into training, validation, or testing data sets at a ratio of 90:5:5. The sample data were randomly shuffled and normalized before being fed into the FCDN_CRTM, and the number of iterations was extended to 2.4 million to make the cost function converge adequately. The algorithm was developed by using Tensorflow version 1.4 and Python version 3.7 with parallel processing capability. In total, 6–20 CPUs were used in parallel during the model training, testing, and predicting on a NOAA STAR Linux server that had 200 G of memory and 2.2 G multi-core CPUs, but without GPU support. The whole model training took approximately 8–10 h.

Figure 2 depicts changes in the cost function and the corresponding mean and STD of the testing data for M15 during the training. We recorded the values of the cost function after every 1000 iterations, but we tested the model every 6000 iterations. The value of the cost function began at ~80,000, which is cut from the figure to emphasize the latest convergence. However, it can be estimated by calculating the MSE for the typical BTs of five TEB/M bands. For instance, for a typical BT with 280 K after the first iteration of training, the MSE calculated from the forward propagation should be close to the square of 280, which was close to our expected value. The cost function rapidly reduced from ~80,000 to 0.1 during the first several 10,000 iterations and then gradually became smaller as the iterations increased. During the entire training, the cost function oscillated up and down, but persisted in decreasing, although at increasingly slow speeds, and remained nearly constant at the end of the training. The persistent decreasing of the cost function implies that the BN introduced in the model might mitigate the **vanishing gradient** problem for a long-iteration training, as the change in the cost function became extremely small in later iterations. In the meantime, the massive amount of input data, which covered all seasons, provided a larger data extent to optimize the model more adequately for a long training time.

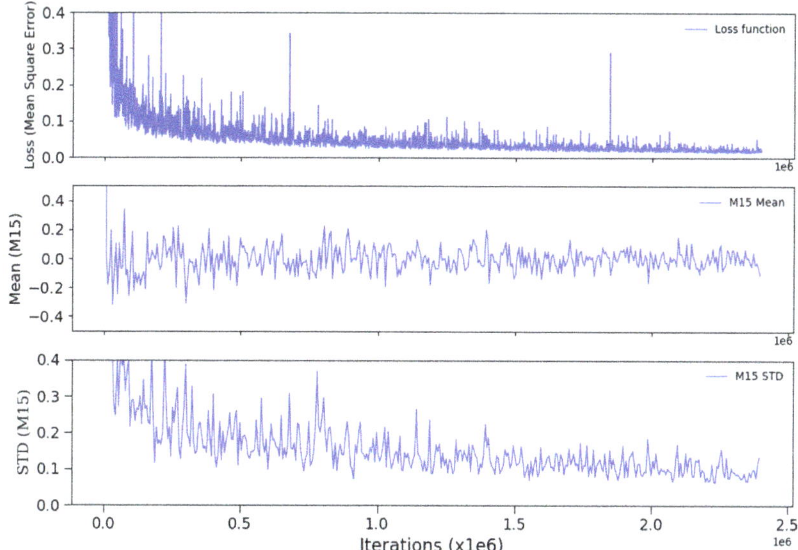

Figure 2. The changes of the loss function (**upper panel**), mean (**middle panel**), and standard deviation (STD) (**bottom panel**) of the testing data for M15 during the FCDN_CRTM training.

Similar to the cost function, the mean and STD were quickly reduced at the beginning of the training and gradually converged as the iterations increased. The mean quickly dropped to its global minimum midway through training, while the STD continued to decrease and was finally stable at the end of the training. We present only the trends for the mean and STD for M15, but the performances of other bands were similar. As discussed in [27], the cost function, mean, and STD oscillated up and down during the training; this was due to using small batch sizes instead of a single sample in each iteration.

Table 3 compares the F-C mean biases and STDs between the training, testing, and prediction data. The prediction data are discussed in the next subsection. For all bands, the F-C means and the STDs were within several thousandths of a Kelvin and several hundredths of a Kelvin, respectively, and are comparable between the training and testing data sets, suggesting that no significant overfitting occurred in the model. Finally, including BN and regularization, together with the substantial all-seasons data fed into the model, resulted in a well-trained model and a significant avoidance of the overfitting effect.

Table 3. The F-C mean and STD of the train and test data, and predicted data for 02/21/2020 (F-C: difference between FCDN_CRTM BT and CRTM BT; μ: F-C mean bias; σ: corresponding STD).

	Train Data		Test Data		Prediction Data	
	μ	σ	μ	σ	μ	σ
M12	−0.0013	0.0313	−0.0013	0.0320	−0.0011	0.0405
M13	−0.0018	0.0329	−0.0018	0.0336	0.0019	0.0408
M14	0.0	0.0444	0.0	0.0454	0.0009	0.0585
M15	−0.0006	0.0505	−0.0005	0.0516	−0.0002	0.0682
M16	−0.0006	0.0620	−0.0006	0.0633	−0.0058	0.0860

3.2. FCDN_CRTM Prediction and Validation with the CRTM

The trained FCDN_CRTM was first used to predict five CRTM BTs for February 21, 2020, which is about one month after the nearest training data. We defined these data as prediction data to distinguish between the training and testing data. As CRTM simulation is quite time consuming, VIIRS data were down sampled by a four-by-four window [12] to speed up CRTM simulation for the model validation. To comprehensively validate the model performance, we did not perform any other quality control for all data, except for the FCDN_CSM clear-sky identification. All ocean clear-sky pixels were selected, including full satellite scan swath and high latitude. As a result, 6.5 million pixels were used for the model validation after the FCDN_CSM clear-sky identification.

The initial experiment used the direct SZA as an input feature to train the FCDN_CRTM and predict the CSM for 02/21/2020, which was similar to its use in the FCDN_CSM. However, a distinct stratification structure persisted in the global distribution of the F-C, regardless of how we tuned the model. Figure 3 (upper panel) depicts this specific texture in the east Pacific Ocean for the M16 band, which is the most pronounced among the five TEB/M bands. As the forward radiance is more related to the cosine of SZA than the SZA itself, by using a secant of SZA as the input feature instead of SZA in the same training condition, the stratification structure was removed completely in the prediction data, as illustrated in the bottom panel of Figure 3. This slight change to the input feature resulted in a significant improvement in the model, strongly indicating that feature selection is important for the ANN model. Hereafter, the secant of SZA was selected as the input feature in this study.

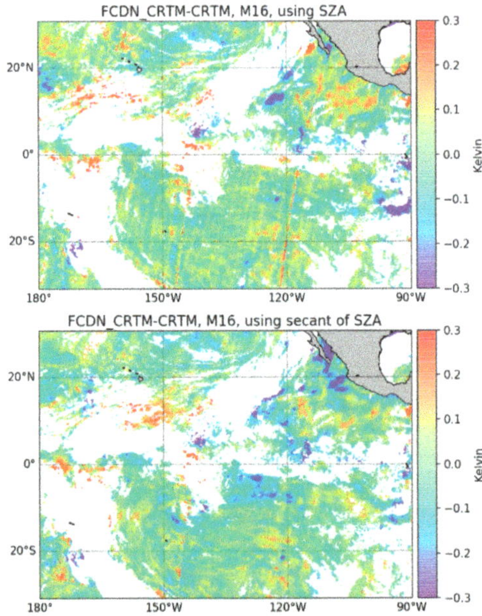

Figure 3. The distribution of the F-C mean biases in M16 with the direct SZA (upper) or secant of SZA (bottom) as an input feature in FCDN_CRTM

Figure 4 portrays the global distribution (left panel) and histograms (right panel) of the F-C mean biases in M12, M15, and M16. A summary of the corresponding F-C statistics for all five bands is listed in the right two columns of Table 3. Note that the train and test data sets were generated using the ACSPO CSM as clear-sky identification, whereas the prediction data used the FCDN_CSM, which was trained with the ACSPO CSM. Therefore, in Table 3, the STDs were slightly reduced for prediction data mainly due to possible residual clouds and outliers, rather than significant overfitting existence. This saying was further verified by the later analyses of the long-term stability. Furthermore, the global distributions were generally uniform, particularly for the most atmosphere-transparent band—M12—followed by M15 and M16. The F-C means were Gaussian distributed, and the global means were typically ±0.002 K, with uncertainties of several hundredths of a Kelvin for all five bands. Further analysis showed that the correlation coefficients between FCDN_CRTM prediction and CRTM simulation are typically 0.9999 for all five bands. All statistics analyses indicated that the model is quite accurate for BT prediction with most atmospheric and geographic conditions. In addition, some outliers had slightly larger biases for M15 and M16 in the high SZA, which may be due to the low accuracy related to a long atmosphere path.

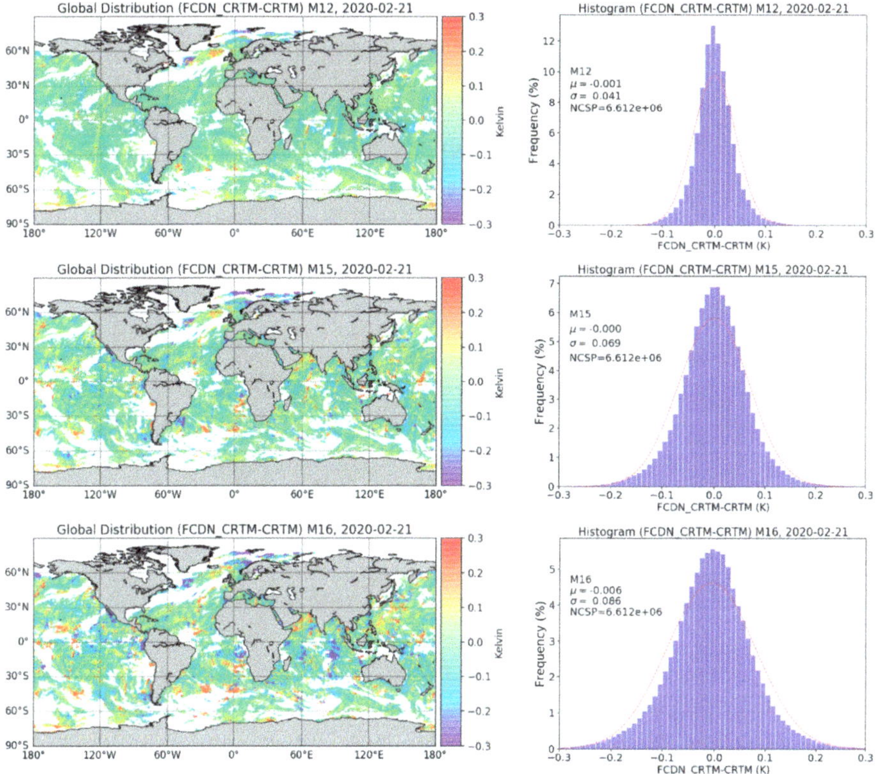

Figure 4. Global distributions (**left panel**) and histograms (**right panel**) of the differences between FCDN_CRTM predicted and CRTM in M12 (**upper**), M15 (**middle**) and M16 (**bottom**) on 02/21/2020.

Figure 5 further validates the model performance in the SZA and total column water vapor (CWV) content dependencies of the F-C differences. Both parameters are the key factors to evaluate radiative transfer model performance. The left panel shows the SZA dependence of the F-C mean, STD, and corresponding histograms. The right panel is the same as the left, but for CWV. For both SZA and CWV, no significant dependencies of the F-C mean biases were observed. All curves of these F-C biases are within a small amplitude range from −0.05 to 0 for all SZA and CWV bins, which suggests that the FCDN_CRTM can reproduce CRTM BTs accurately for different SZAs and CWVs. In addition, there was slight noise at the high CWV, due to the small data portion in the corresponding bin. Moreover, the uniform distribution performance even existed in the SZA dependencies of the STD (e.f. L2 moment) when the SZA ranged from −55° to 55°. The dependencies gradually increased after an SZA larger than 55°, but the maximum increasing amplitude was still ~0.05 K for M16. The amplitudes of the CWV dependencies were [0.02, 0.08] for M12 and M13, [0.04, 0.09] for M14 and M15, and [0.08, 0.12] for M16. Although the CWV dependencies of the STD were slightly larger than those of the SZA, the amplitudes were still within several hundredths of a Kelvin.

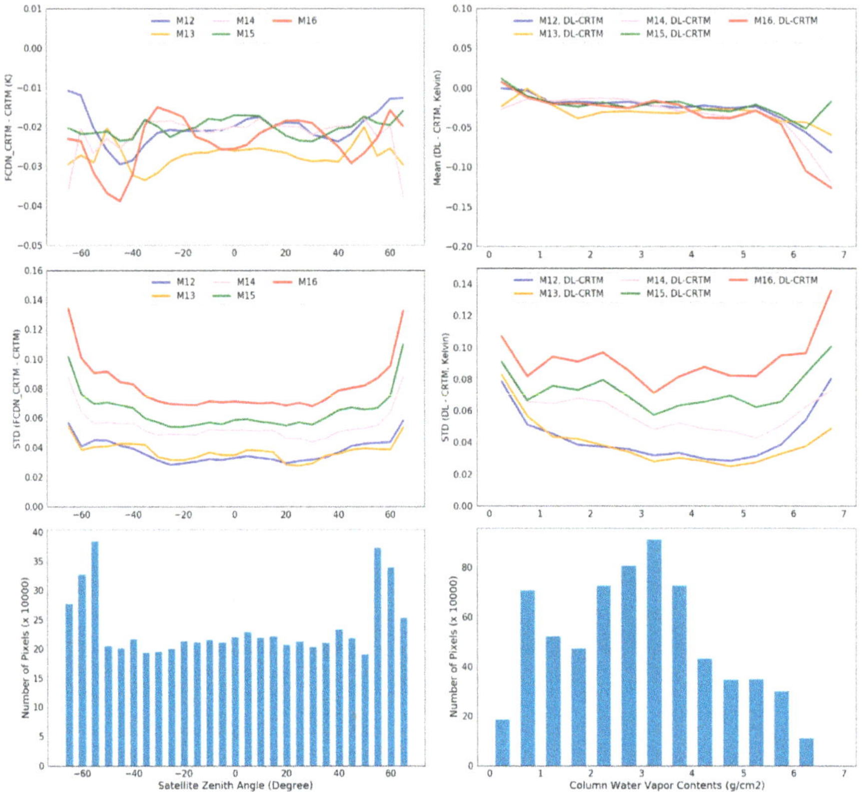

Figure 5. The F-C biases as functions of satellite zenith angle (**left panel**) and column water vapor (**right panel**) in VIIRS M12-M16 for 02/21/2020. (**upper panel**) F-C mean; (**middle panel**) STD; and (**bottom panel**) corresponding histogram.

Overall, for all TEB/M bands, the FCDN_CRTM-predicted BTs are generally consistent with the CRTM for different SZAs and CWVs, suggesting that the model is robust for BT prediction under most atmosphere and geographical conditions. However, slightly large STDs were found with a high SZA and a large CWV, particularly for M16, indicating that the FCDN_CRTM can still be fine-tuned to improve accuracy and spatial stability.

3.3. FCDN_CRTM Validation with VIIRS SDR Data

Similar to the CRTM applications, one ultimate goal of the FCDN_CRTM is to evaluate and monitor the accuracy, stability, and cross-sensor consistency of the VIIRS radiometric biases. Hence, FCDN_CRTM model validation with VIIRS SDR data is necessary to check the model performances in extensive atmosphere and geographical conditions. As the VIIRS O-M biases have been successfully used in the past decade to validate CRTM performance under a global ocean clear-sky condition for infrared atmosphere window bands [3,8–10], in this section, we use a similar method and focus on the consistency between the VIIRS observation minus FCDN_CRTM prediction (V-F) and the VIIRS minus CRTM simulation (V-C).

Figure 6 presents the global distributions of the V-C (left panel) and V-F (right panel) for M12, M15, and M16, and the corresponding histograms are shown in Figure 7. The global distributions were quite consistent between V-F and V-C, and both mean biases for M12 were only negative several hundredths of a Kelvin. Both V-F and V-C exhibited negative biases in long-window IR (LWIR) bands – M15 and M16 (the root sources of the negative V-C biases for LWIR have been discussed in [3,8], wherein one of the key factor is possible residual clouds). However, the negative mean biases for V-F (−0.02 K, −0.27 K, and −0.35 K for M12, M15, and M16, respectively) were all slightly smaller than for V-F, suggesting that the FCDN_CRTM prediction is closer to VIIRS observations. Additionally, the STDs of 0.32, 0.44, and 0.53 K for V-F are extremely comparable to those for V-C, and the largest difference was only 0.005 K in M16. The summary of global statistics of V-C and V-F, including M13 and M14, are listed in Table 4, which shows that the means and STDs for M13 and M14 are also similar to those of M12, M15, and M16.

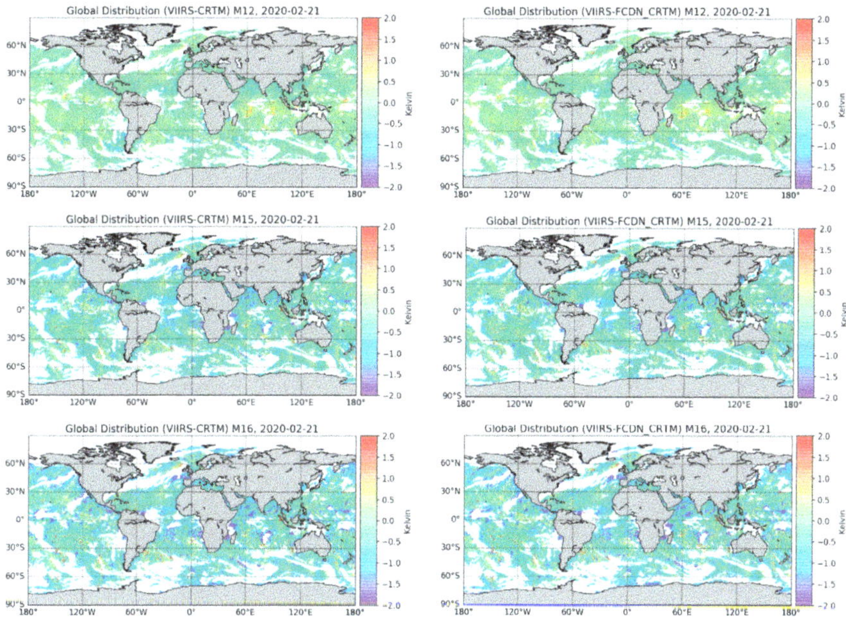

Figure 6. Global distributions of the V-C (**left panels**) and V-F (**right panels**) for M12 (**upper**), M15 (**middle**) and M16 (**bottom**) on 21 February 2020.

Table 4. Global mean and STD for SNPP VIIRS on 21 February 2020 between V-C and V-F (V-C: difference between VIIRS BT and CRTM BT; V-F: difference between VIIRS BT and FCDN_CRTM BT; μ: V-C or V-F mean bias; σ: standard deviation).

	V-C		V-F	
	μ	σ	μ	Σ
M12	−0.0405	0.3160	−0.0383	0.3185
M13	−0.5894	0.2650	−0.5894	0.2680
M14	−0.5212	0.3850	−0.5168	0.3895
M15	−0.2932	0.4390	−0.2885	0.4440
M16	−0.3811	0.5255	−0.3690	0.5318

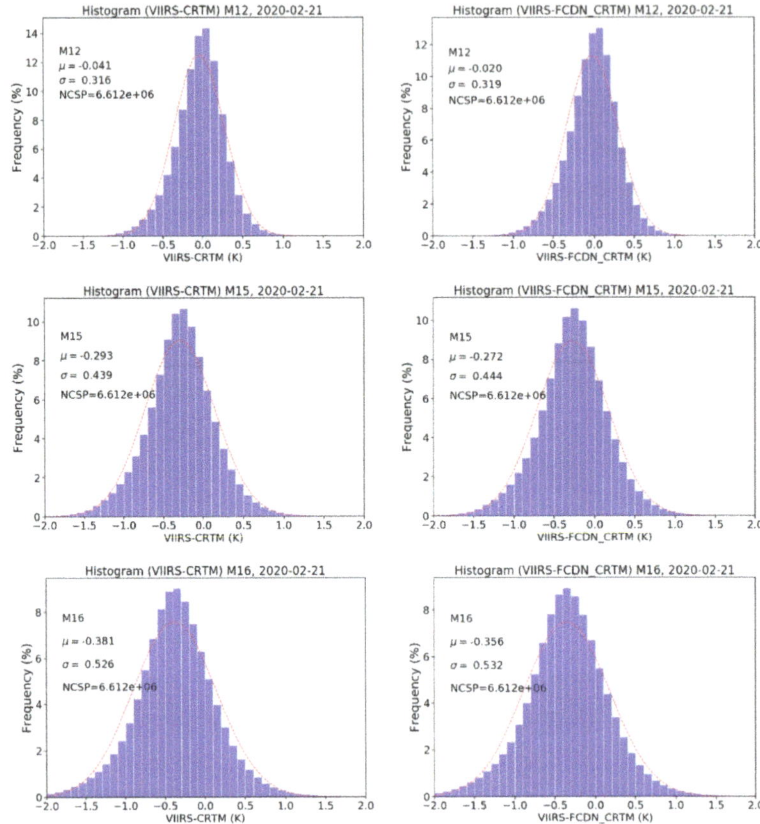

Figure 7. Global histograms of the V-C (**left panels**) and V-F (**right panels**) for M12 (**upper**), M15 (**middle**) and M16 (**bottom**) on February 21, 2020.

Overall, the global distribution, histograms, and statistics data provide strong evidence that V-F is consistent with V-C under most atmosphere and geographical conditions, and the BTs predicted by the FCDN_CRTM were reasonable and accurate in the global ocean clear-sky domain for VIIRS TEB/M bands.

3.4. Long-Term Stability of the FCDN_CRTM

In this study, the stability of the FCDN_CRTM is not only key to the performance for the long-term monitoring of sensor radiometric biases, but also a way to check whether there is any overfitting in the model. For this purpose, we used the trained model to additionally predict BTs for five dispersion days—03/16/2020, 04/15/2020, 05/16/2020, 06/10/2020, 07/01/2020, and 07/30/2020—where we selected one day in each month from March to July 2020. Including 02/21/2020, seven days' data were used to evaluate the stability of the FCDN_CRTM. Note that the day selection was random, and as with data from 02/21/2020, we did not perform any quality control for the data, except for the clear-sky identification by the FCDN_CSM.

Figure 8 illustrates the time series of the F-C error bars from M12 to M16 for the seven days. The VIIRS clear-sky pixels were identified by the FCDN_CSM. The blue dashed line represents the mean for all seven-day data and all bands, and together with two blue dashed lines (y = mean-0.1 and y = mean + 0.1), the three dashed lines help to be more intuitive in checking day-to-day changes of the F-C mean and STD. A corresponding comparison between V-C and V-F is presented in Figure 9 for M12,

M15, and M16. The F-C mean biases persisted for several thousandths of a Kelvin for all analyzed days and all bands. The average of the F-C means were −0.008, −0.006, −0.008, −0.011, and −0.013 K for M12 to M16, respectively, and the change was not significant over the time period. As expected, the STDs on 02/21/2020, listed in Table 3 were the smallest among the seven days for all bands, as this day is closest to the training data period. However, the STD changes were minimal in the first three days, and the amplitude of the change was between 0.001 K and 0.009 K for all bands. Even on 5/15/2020 and 6/10/2020, the STDs only increased by a maximum 0.039 K in M12 in comparison to the most accurate on 02/21/2020. After 06/10/2020, the STDs significantly worsened, and on 07/30/2020, they were 3–4 times more than on other days. Recall that the regularization and BN were introduced in the model, and all season data were included in model training. All efforts were intended to avoid overfitting of the deep learning model. However, 278 input features and a complicated model architecture may result in overfitting not being fully eliminated. Moreover, the seasonal cycle and extreme climate events [34] could cause possible noise during the model prediction. Interestingly, both means and STDs between V-C and V-F persisted consistently longer in Figure 9, wherein the changes in mean and STD from 02/21/2020 to 07/01/2020 are typically only between 0.01 K and 0.038 K for all bands. Then, the V-F STD increased by ~0.055 K on 07/30/2020. Overall, the stable means and STDs of F-C and the consistency between V-F and V-C from 02/21/2020 to 06/10/2020 provide strong evidence that the robust performance of the FCDN_CRTM can be extended from 5 months to half a year. However, model retraining is needed to maintain a high accuracy of the FCDN_CRTM prediction after that period.

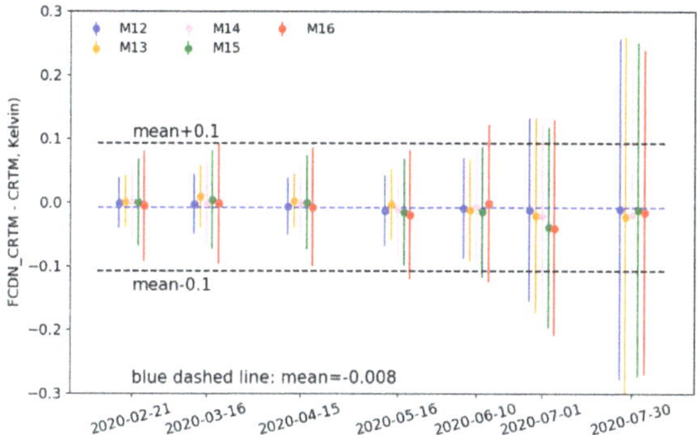

Figure 8. The error bars of FCDN_CRTM minus CRTM from M12 to M16 for seven dispersion days from February to July, 2020. The VIIRS Clear-sky pixels were identified by FCDN_CSM. The blue dashed line represents the mean for the data of all seven days.

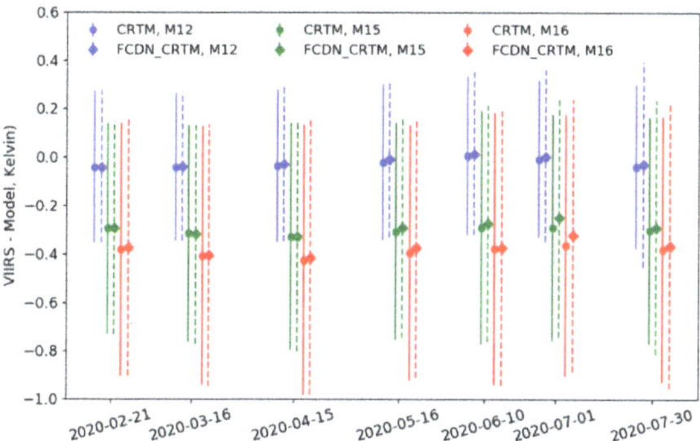

Figure 9. Same as Figure 8, but for VIIRS observations minus model (CRTM or FCDN_CRTM)) in M12, M15, and M16.

4. Discussion

4.1. Efficiency of FCDN_CRTM

As discussed in the last section, one advantage of the FCDN_CRTM is that the model reproduced similar accurate BTs as CRTM simulation, without using a complicated radiative transfer equation. Furthermore, with the same NOAA STAR Linux server (without GPU support), the CRTM simulation for 6 million clear-sky points required approximately 12 min. In contrast, the total processing time for the FCDN_CRTM with multi CPUs is only 17 s, which was about 42 times faster than the CRTM. On the other hand, even we set one CPU to conduct FCDN_CRTM prediction, which is the same condition with that for CRTM simulation, the total processing time to predict the same amount data is no more than 40 s, suggesting that the high efficiency of FCDN_CRTM is mainly due to its inherent high-efficient calculation, rather than just because it utilizes as many as possible CPU resources. This further implies that the model has a strong capability to efficiently simulate high-resolution spatial and temporal sensors, even for insufficient CPU resources. Certainly, the more data are processed, the more memory is needed.

4.2. End-to-End System

In this study, the whole algorithm included data collection and preprocessing, clear-sky mask prediction, and VIIRS BT prediction and validation. In addition, model training of the FCDN_CSM and FCDN_CRTM was separate from the system. Thus, combining all components, we have built an end-to-end AI framework to predict VIIRS BTs. It first inputs VIIRS SDR, ECMWF data, and CMC SST to the data preprocessing module. This module then collocates atmosphere and surface gridding data in space and time to the VIIRS pixel level and generates both the FCDN_CSM input data with 11 features and the FCDN_CRTM with 278 features. Thereafter, the FCDN_CSM input data are fed into the FCDN_CSM model to produce the VIIRS clear-sky mask. The predicted VIIRS CSM are further input into the clear-sky identification module to identify clear-sky pixels for the FCDN_CRTM input data. Finally, the FCDN_CRTM input data with clear-sky mask are fed into the FCDN_CRTM to predict five TEB/M BTs, and the results are input into the validation module to validate prediction data with CRTM simulation or VIIRS SDR data. The whole system is illustrated in Figure 10.

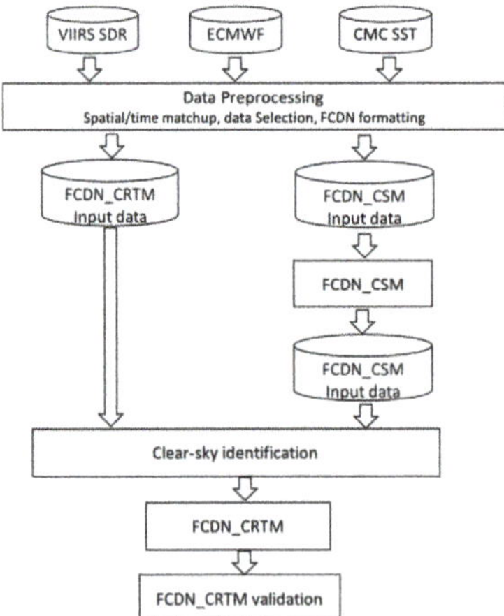

Figure 10. An End-to-End processing chart of FCDN_CRTM System.

This framework has the potential to build a system for real-time monitoring of VIIRS BTs against AI predictions. It can input VIIRS SDR data for granules, orbits, or an entire day, in conjunction with the ECMWF and the CMC, to predict corresponding clear-sky BTs and to evaluate VIIRS data simultaneously. As discussed in the previous section, the FCDN_CRTM is much more efficient than CRTM simulation and has a better design for real-time monitoring of VIIRS radiometric biases. Furthermore, the framework makes it easy to extend our research in the future to include land, cloud, and other conditions.

5. Conclusions

An FCDN algorithm, namely, the FCDN_CRTM, was proposed to explore the efficiency and accuracy for reproducing VIIRS BTs in five TEB/M bands. The model was trained and tested in the nighttime global ocean clear-sky domain, in which the CRTM simulation has been well validated in recent years. The ECMWF atmosphere profile and the CMC SST were used as FCDN_CRTM input, and the CRTM BTs were defined as labels.

Efforts were made to improve model performance by iteratively refining the model design and carefully treating the input data. The FCDN_CRTM was designed with three hidden layers, with 512, 384, and 64 neurons in each layer, respectively. We used 278 features as input and five VIIRS TEB/M BTs as output, and the six dispersed days of data from 2019 and 2020, which constituted approximately 40 million samples and covered all seasons, were selected to train the FCDN_CRTM. The trained model was employed to predict CRTM BTs on seven randomly selected days from 21 February to 30 July 2020—nearly one day per month. The predicted BTs were validated with the CRTM BTs and VIIRS SDR data for both accuracy and stability. Moreover, the earlier published FCDN_CSM was used to quickly identify clear-sky pixels for the FCDN_CRTM prediction, and BN, which was introduced in the FCDN_CRTM, sped up the model convergence and further reduced the STD by ~0.2 K. Furthermore, both BN and regularization used in the model, together with the all-season data fed into the model training, aided in avoiding overfitting and made the model more robust. In addition, a secant of the SZA used as FCDN_CRTM input instead of the SZA itself significantly improved the model prediction performance.

Using a line-by-line RTM (LBLRTM) simulated BT as the FCDN model reference could be more reasonable and accurate than CRTM, as the LBLRTM provides spectral radiance calculations with accuracies most consistent with the sensor measurements [35]. However, its computational inefficiency prevents the possibility of large data sample collection for FCDN_CRTM training, testing, prediction, and validation. In contrast, the CRTM's accuracies have been well validated, although the model is an approximate RTM that uses trained transmittance coefficients. Especially for the TEB bands, the root MSE between the CRTM and the LBLRTM is only ~0.016 K [31], and using CRTM BT as the FCDN_CRTM reference is, thus, adequate for high accuracy and efficiency in this initial study.

As a result, the F-C means were within several thousandths of a Kelvin, and the STDs were within several hundredths of a Kelvin for all bands, and they are comparable between the training and testing data sets. The high accuracies could persist for about half a year before the STDs degrade significantly. In addition, the FCDN_CRTM-predicted BTs are generally consistent with those of the CRTM with different SZAs and CWVs for all TEB/M bands under most atmosphere and geographical conditions. By validation with VIIRS SDR in global distribution and corresponding histograms, V-F was consistent with V-C in most atmosphere and geographical conditions, and the consistencies lasted even longer than the stable F-C period. Furthermore, the FCDN_CRTM processing time was at least one order of magnitude faster than the CRTM simulation. The highly efficient and accurate FCDN_CRTM is, thus, a potential solution to real-time monitoring of global O-M biases for high-resolution VIIRS. We plan to continue to monitor the model's result periodically under the framework to check for any anomalies and find possible physical explanations. Our future work will extend the FCDN_CRTM functionalities to include land, cloud, and other conditions in the FCDN_CRTM end-to-end framework.

Author Contributions: X.L. developed methods, created the evaluation data set, analyzed the results, and wrote the manuscript. Q.L. supervised the research and was the technique guide. All authors have read and agreed to the published version of the manuscript.

Remote Sens. **2020**, *12*, 3825

Funding: This research was funded by NOAA grant NA14NES4320003 (Cooperative Institute for Satellite Earth System Studies-CISESS) at the *Earth System Science Interdisciplinary Center* of University of Maryland.

Acknowledgments: The CRTM are provided by the CRTM team of NOAA STAR. The authors would like to thank Christopher Grassotti and the STAR MiRS team for technical discussions and suggestions. We also thank Changyong Cao, Banghua Yan, Alexander Ignatov, and Sid Boukabara from NOAA STAR for creative advices and helps. The views, opinions, and findings contained in this report are those of the authors and should not be construed as an official NOAA or U.S. Government position, policy, or decision.

Conflicts of Interest: The authors declare no conflict of interest.

References

1. Cheng, X.; Liu, Y.; Xu, X.; You, W.; Zang, Z.; Gao, L.; Chen, Y.; Su, D.; Yan, P. Lidar data assimilation method based on CRTM and WRF-Chem models and its application in PM2.5 forecasts in Beijing. *Sci. Total Environ.* **2019**, *682*, 541–552. [CrossRef] [PubMed]
2. Han, Y.; Delst, P.V.; Liu, Q.; Weng, F.; Yan, B.; Treadon, R.; Derber, J. *Community Radiative Transfer Model. (CRTM)—Version 1*; NOAA Technical Report NESDIS 122; NOAA: Silver Spring, MD, USA, 2006.
3. Liang, X.; Ignatov, A.; Kihai, Y. Implementation of the Community Radiative Transfer Model (CRTM) in Advanced Clear-Sky Processor for Oceans (ACSPO) and validation against nighttime AVHRR radiances. *J. Geophys. Res.* **2009**, *114*, D06112. [CrossRef]
4. Liu, Q.; Boukabara, S. Community Radiation Transfer Model (CRTM) Applications in Supporting the Suomi National Polar-Orbiting Partnership (SNPP) Mission validation and Verification. *Remote Sen. Environ.* **2014**, *140*, 744–754. [CrossRef]
5. Liu, Q.; Cao, C. Analytic expressions of the Transmission, Reflection, and source function for the community radiative transfer model. *J. Quant. Spectrosc. Radiat. Transf.* **2019**, *226*, 115–126. [CrossRef]
6. Liu, Q.; Delst, P.V.; Chen, Y.; Groff, D.; Han, Y.; Collard, A.; Weng, F. Community radiative transfer model for radiance assimilation and applications. In Proceedings of the 2012 IEEE International Geoscience and Remote Sensing Symposium IGARSS, Munich, Germany, 22–27 July 2012; pp. 3700–3703.
7. Stegmann, P.G.; Tang, G.; Yang, P.; Johnson, B.T. A stochastic model for density-dependent microwave Snow- and Graupel scattering coefficients of the NOAA JCSDA community radiative transfer model. *J. Quant. Spectrosc. Radiat. Transf.* **2018**, *211*, 9–24. [CrossRef]
8. Liang, X.; Ignatov, A. Monitoring of IR Clear-sky Radiances over Oceans for SST (MICROS). *J. Atmos. Oceanic Technol.* **2011**, *28*. [CrossRef]
9. Liang, X.; Ignatov, A. AVHRR, MODIS, and VIIRS radiometric stability and consistency in SST bands. *J. Geophys. Res.* **2013**, *118*. [CrossRef]
10. Liang, X.; Ignatov, A. Preliminary Inter-Comparison between AHI, VIIRS and MODIS Clear-Sky Ocean Radiances for Accurate SST Retrievals. *Remote Sens.* **2016**, *8*, 203. [CrossRef]
11. Wang, H.; Fuller-Rowell, T.J.; Akmaev, R.A.; Hu, M.; Kleist, D.T.; Iredell, M.D. First simulations with a whole atmosphere data assimilation and forecast system: The January 2009 major sudden stratospheric warming. *J. Geophys. Res.* **2011**, *116*, A12321. [CrossRef]
12. Liang, X.; Sun, N.; Ignatov, A.; Liu, Q.; Wang, W.; Zhang, B.; Weng, F.; Cao, C. Monitoring of VIIRS ocean clear-sky brightness temperatures against CRTM simulation in ICVS for TEB/M bands. In Proceedings of the SPIE 10402, Earth Observing Systems XXII, San Diego, CA, USA, 6–10 August 2017; p. 104021. [CrossRef]
13. Ball, J.E.; Anderson, D.T.; Chan, C.S. Comprehensive survey of deep learning in remote sensing: Theories, tools, and challenges for the community. *J. Appl. Remote Sens.* **2017**, *11*, 042609. [CrossRef]
14. Feng, X.; Li, Q.; Zhu, Y.; Hou, J.; Jin, L.; Wang, J. Artificial neural networks forecasting of PM2.5 pollution using air mass trajectory based geographic model and wavelet transformation. *Atmos. Environ.* **2015**, *107*, 118–128. [CrossRef]
15. Ma, L.; Liu, Y.; Zhang, X.; Ye, Y.; Yin, G.; Johnson, B.A. Deep learning in remote sensing applications: A meta-analysis and review. *ISPRS J. Photogramm. Remote Sens.* **2019**, *152*, 166–177. [CrossRef]
16. Yuan, Q.; Shen, H.; Li, T.; Li, Z.; Li, S.; Jiang, Y.; Xu, H.; Tan, W.; Yang, Q.; Wang, J.; et al. Deep learning in environmental remote sensing: Achievements and challenges. *Remote Sen. Environ.* **2020**, *241*, 111716. [CrossRef]
17. Zhang, L.; Zhang, L.; Du, B. Deep Learning for Remote Sensing Data: A Technical Tutorial on the State of the Art. *IEEE Geosci. Remote Sens. Mag.* **2016**, *4*, 22–40. [CrossRef]

18. Zhu, X.; Tuia, D.; Mou, L.; Xia, G.; Zhang, L.; Xu, F.; Fraundorfer, F. Deep Learning in Remote Sensing: A Comprehensive Review and List of Resources. *IEEE Geosci. Remote Sens. Mag.* **2017**, *5*, 8–36. [CrossRef]

19. Bue, B.D.; Thompson, D.R.; Deshpande, S.; Eastwood, M.; Green, R.O.; Natraj, V.; Mullen, T.; Parente, M. Neural network radiative transfer for imaging spectroscopy. *Atmos. Meas. Tech.* **2019**, *12*, 2567–2578. [CrossRef]

20. Chevallier, F.; Chéruy, F.; Scott, N.A.; Chédin, A. A Neural Network Approach for a Fast and Accurate Computation of a Longwave Radiative Budget. *J. Appl. Meteor.* **1998**, *37*, 1385–1397. [CrossRef]

21. Liu, Y.; Caballero, R.; Monteiro, J.M. RadNet 1.0: Exploring deep learning architectures for longwave radiative transfer. *Geosci. Model. Dev.* **2020**, *13*, 4399–4412. [CrossRef]

22. Krasnopolsky, V.M.; Fox-Rabinovitz, M.S.; Chalikov, D.V. New Approach to Calculation of Atmospheric Model Physics: Accurate and Fast Neural Network Emulation of Longwave Radiation in a Climate Model. *Mon. Weather Rev.* **2005**, *133*, 1370–1383. [CrossRef]

23. Krasnopolsky, V.M.; Fox-Rabinovitz, M.S.; Belochitski, A.A. Decadal Climate Simulations Using Accurate and Fast Neural Network Emulation of Full, Longwave and Shortwave, Radiation. *Mon. Weather Rev.* **2008**, *136*, 3683–3695. [CrossRef]

24. Krasnopolsky, V.M.; Fox-Rabinovitz, M.S.; Hou, Y.T.; Lord, S.J.; Belochitski, A.A. Accurate and Fast Neural Network Emulations of Model Radiation for the NCEP Coupled Climate Forecast System: Climate Simulations and Seasonal Predictions. *Mon. Weather Rev.* **2009**, *138*, 1822–1842. [CrossRef]

25. Rivera, J.P.; Verrelst, J.; Gomez-Dans, J.; Muoz-Mar, J.; Moreno, J.; Camps-Valls, G. An emulator toolbox to approximate radiative transfer models with statitistical learning. *Remote Sens.* **2015**, *7*, 9347–9370. [CrossRef]

26. Westing, N.; Borghetti, B.; Gross, K.C. Fast and Effective Techniques for LWIR Radiative Transfer Modeling: A Dimension-Reduction Approach. *Remote Sens.* **2019**, *11*, 1866. [CrossRef]

27. Liang, X.; Liu, Q.; Yan, B.; Sun, N. A Deep Learning Trained Clear-Sky Mask Algorithm for VIIRS Radiometric Bias Assessment. *Remote Sens.* **2020**, *12*, 78. [CrossRef]

28. Liang, X.; Liu, Q. Applying Deep Learning to Clear-Sky Radiance Simulation for VIIRS with Community Radiative Transfer Model—Part 1: Develop AI-Based Clear-Sky Mask. *Remote Sens.* **2021**, *13*, 222. [CrossRef]

29. Liang, X.; Ignatov, A. Validation and Improvements of Daytime CRTM Performance Using AVHRR IR 3.7 um Band. In Proceedings of the 13th AMS Conf. Atm. Radiation, Portland, OR, USA, 28 June–2 July 2010; Available online: https://ams.confex.com/ams/pdfpapers/170593.pdf (accessed on 2 November 2020).

30. Wu, X.; Smith, W.L. Emissivity of rough sea surface for 8–13 um: Modeling and verification. *Appl. Opt.* **1997**, *36*, 2609–2618. [CrossRef]

31. Chen, Y.; Weng, F.; Han, Y.; Liu, Q. Planck-Weighted Transmittance and Correction of Solar Reflection for Broadband Infrared Satellite Channels. *J. Atmos. Oceanic Technol.* **2012**, *29*, 382–396. [CrossRef]

32. Ng, A. Feature selection, L1 vs. L2 regularization, and rotational invariance. In Proceedings of the 21st International Conference on Machine Learning (ICML), Banff, AB, Canada, 4–8 July 2004; pp. 78–85.

33. Ioffe, S.; Szegedy, C. Batch normalization: Accelerating deep network training by reducing internal covariate shift. In Proceedings of the 32nd International Conference on Machine Learning (ICML), Lille, France, 6–11 July 2015; Volume 37, pp. 448–456.

34. Planton, S.; Déqué, M.; Chauvin, F.; Terray, L. Expected impacts of climate change on extreme climate events. *Comptes. Rendus. Geosci.* **2008**, *340*, 564–574. [CrossRef]

35. Clough, S.A.; Shephard, M.W.; Mlawer, E.J.; Delamere, J.S.; Iacono, M.J.; Cady-Pereira, K.; Boukabara, S.; Brown, P.D. Atmospheric radiative transfer modeling: A summary of the AER codes, Short Communication. *J. Quant. Spectrosc. Radiat. Transfer* **2005**, *91*, 233–244. [CrossRef]

Publisher's Note: MDPI stays neutral with regard to jurisdictional claims in published maps and institutional affiliations.

Article

Tropical Cyclone Climatology from Satellite Passive Microwave Measurements

Song Yang [1,*], **Richard Bankert** [1] **and Joshua Cossuth** [2]

[1] Naval Research Laboratory, Monterey, CA 93943, USA; richard.bankert@nrlmry.navy.mil
[2] Naval Research Laboratory, Washington, DC 20375, USA; joshua.cossuth@navy.mil
* Correspondence: song.yang@nrlmry.navy.mil

Received: 11 September 2020; Accepted: 29 October 2020; Published: 3 November 2020

Abstract: The satellite passive microwave (PMW) sensor brightness temperatures (TBs) of all tropical cyclones (TCs) from 1987–2012 have been carefully calibrated for inter-sensor frequency differences, center position fixing using the Automated Rotational Center Hurricane Eye Retrieval (ARCHER) scheme, and application of the Backus–Gilbert interpolation scheme for better presentation of the TC horizontal structure. With additional storm motion direction and the 200–850 hPa wind shear direction, a unique and comprehensive TC database is created for this study. A reliable and detailed climatology for each TC category is analyzed and discussed. There is significant annual variability of the number of storms at hurricane intensity, but the annual number of all storms is relatively stable. Results based on the analysis of the 89 GHz horizontal polarization TBs over oceans are presented in this study. An eyewall contraction is clearly displayed with an increase in TC intensity. Three composition schemes are applied to present a reliable and detailed TC climatology at each intensity category and its geographic characteristics. The global composition relative to the North direction is not able to lead a realistic structure for an individual TC. Enhanced convection in the down-motion quadrants relative to direction of TC motion is obvious for Cat 1–3 TCs, while Cat 4–5 TCs still have a concentric pattern of convection within 200 km radius. Regional differences are evident for weak storms. Results indicate the direction of TC movement has more impact on weak storms than on Cat 4–5 TCs. A striking feature is that all TCs have a consistent pattern of minimum TBs at 89 GHz in the downshear left quadrant (DSLQ) for the northern hemisphere basins and in the downshear right quadrant (DSRQ) for the southern hemisphere basin, regarding the direction of the 200–850 hPa wind shear. Tropical depression and tropical storm have the minimum TBs in the downshear quadrants. The axis of the minimum TBs is slightly shifted toward the vertical shear direction. There is no geographic variation of storm structure relative to the vertical wind shear direction except over the southern hemisphere which shows a mirror image of the storm structure over the northern hemisphere. This study indicates that regional variation of storm structure relative to storm motion direction is mainly due to differences of the vertical wind shear direction among these basins. Results demonstrate the direction of the 200–850 hPa wind shear plays a critical role in TC structure.

Keywords: tropical cyclone; climatology; wind shear; storm motion; satellite measurement; brightness temperature

---

1. Introduction

A tropical cyclone (TC) can be one of the most impactful weather systems, causing catastrophic damages to human lives, society, transportation, properties, etc. [1,2]. For example, hurricane Katrina in 2005, with a maximum wind speed of 280 km hr^{-1}, impacted most of the southeast United States (US), making landfall in the greater New Orleans area. It is the costliest hurricane in US history, killing an estimated 1245–1836 people and causing damages of $149 billion US dollars [3]. TCs can attain very strong wind speeds, greater than 260 km hr^{-1}, and bring heavy precipitation. Most TC damage is caused by the force of its strong wind, storm surge, and flash flooding. Flooding from US landfalling TCs is the leading cause of death related to severe storms [4]. TC rainfall can contribute up to 15% of the total precipitation over a hurricane season in the Carolinas of the United States [5]. To mitigate the potential impact of TCs, appropriate preparations should be taken based on accurate monitoring and predictions of TC intensity, structure and precipitation.

The low earth orbit (LEO) satellite passive microwave (PMW) sensor-based measurements are extremely important in TC monitoring and forecasts because of the PMW sensor's capability in penetrating clouds to observe TC temperature and humidity profiles and the horizontal structure [6–9]. The unique horizontal structures of the inner eyewall, outer eyewall, principal convective band and secondary convective band as well as the moat areas are clearly captured by PMW sensors at the 85–91 GHz channels [10,11]. Three types of spiral bands, based on movement, are possible: stationary (non-propagating), apparent propagation (stationary with respect to the TC center), and intrinsic propagation [1]. These special characteristics have been utilized for estimation and prediction of TC intensity and evolution using various approaches, such as Dvorak technique [12], Advanced Dvorak Technique (ADT) [13,14], and satellite consensus [15,16].

TC structure characteristics and evolution are associated with intensification processes [17–25]. Both TC intensifying and weakening periods have maximum precipitation in the downshear left quadrant (DSLQ) and up-shear left quadrant (USLQ). The minimum rain area, especially the greater areal coverage of stratiform rainfall, located in the up-shear quadrants, is associated with a TC rapid intensification process. Another indication of a rapid intensifying vortex is a cyclonic rotation of shallow-to-moderate-to-deep precipitation from the downshear right to downshear left to up-shear left quadrants. Lightning activity is often evident in DSLQ in the TC inner core (0–100 km) area and in the downshear right quadrant (DSRQ) in the outer rainband (100–300 km) region [20]. The inner core lightning burst (ICLB) is also linked to TC intensity change. Results show that TCs with ICLB inside the radius of maximum wind (RMW) lead to intensification and weakening with ICLB outside the RMW [21]. Thus, the vertical wind shear direction is critical to TC horizontal structure patterns. These results highlight importance of the azimuthal coverage of precipitation and the radial location of deep convection for TC intensification. In addition, the environmental impacts on TC structures shows regional differences, but are not well investigated using satellite observations [26–29]. Therefore, a thorough analysis using a long term PMW sensor brightness temperature (TB) TC database will be important to evaluate these results and lead to new insights of TC structure and their geographic variations.

A comprehensive TC database is created with an improved calibration scheme and an advanced interpolation scheme as well as an accurate TC center position from all PMW sensors during 1987–2012 and with TC motion and the 200–850 hPa vertical wind shear information from National Hurricane Center's Hurricane WRF forecasts used in the statistical hurricane intensity prediction scheme (SHIPS) outputs [30]. Detailed structural features of TCs revealed in this study can provide an improved and accurate climatology of TC structure and how it varies with intensity and geographic basins. Results should benefit evaluations of numerical weather prediction (NWP) model TC simulations and the decision-making efforts to mitigate incoming TC impacts. An accurate TC climatology can also be applied to guide improvements of NWP model skills for better predictions of TC intensity and distribution and to improve understanding of TC intensification processes.

2. Methodology and Datasets

The maximum TB difference due to frequency differences among PMW sensors at 85, 89, and 91 GHz is approximately 15 K. These TB differences have to be corrected in order to have consistent TBs from all PMW sensors for TC applications. Yang et al. [31] developed a physical-based inter-sensor calibration scheme to calibrate Special Sensor Microwave Imager (SSM/I) and Tropical Rainfall Measurement Mission (TRMM) Microwave Imager (TMI) TBs at 85 GHz and Special Sensor Microwave Imager and Sounder (SSMIS) TBs at 91 GHZ into 89 GHz so that all PMW sensors will have consistent TBs at 89 GHz to monitor TC activities. The calibrated TBs are applied for the Naval Research Laboratory (NRL) TC webpage products utilized for near real-time monitoring of global TC activities. Accurate center position is very important to present TC structure. The Automated Rotational Center Hurricane Eye Retrieval (ARCHER) scheme [32,33] is used to fix the TC center positions initiated with the 6-hr TC best track locations. The Backus–Gilbert interpolation scheme [34] is implemented for SSM/I and SSMIS to have better presentations of TC structure. The 200–850 hPa vertical wind shear, defined as the difference of mean wind vector over a concentric zone of 200–800 km radius around a TC center between 200 and 850 hPa, is applied to represent the ambient large scale environmental conditions [30,35,36]. The 6-hr best track information of global TCs from Joint Typhoon Warning Center (JTWC), Central Pacific Hurricane Center (CPHC), and National Hurricane Center (NHC) such as the maximum wind speed, minimum center surface pressure, center positions, and direction of TC movement and 200–850 hPa vertical wind shear are collected with the NRL PMW TBs to create a unique comprehensive TC database during 1987–2012. Since the horizontal polarization is better for presentation of storm horizontal structure features, only 89 GHz at horizontal polarization (H) is utilized for this study. Storms over oceans only captured by PMW sensors are included in the dataset.

PMW sensors and their life cycles during 1987–2012 are displayed in Figure 1. SSM/I was onboard the Defense Meteorological Satellite Program (DMSP) F08, F10, F11, F13, F14, and F15, while SSMIS is on DMSP F16-F18. The Advanced Microwave Scanning Radiometer for Earth Observing System (EOS) (AMSR-E) was onboard the EOS Aqua satellite. TMI was on the NASA TRMM satellite. The overlaps of these satellites provide multiple chances to observe a TC per day during its lifecycle. The TC intensity is classified as five categories based on Saffir–Simpson hurricane wind scale, in addition to tropical depression (TD) and tropical storm (TS) [37,38]. Cat 3–5 TCs are classified as major hurricanes. There are six TC basins: Atlantic (AL), Central Pacific (CP), East Pacific (EP), Indian Ocean (IO), Southern Hemisphere (SH) and West Pacific (WP) [6,36]. A TC is classified into one of the basins depending on its center position. A distribution list of all categorized storms observed by PMW sensors during 1987–2012 can be found in Table 1. The thousands of observations for each storm category, except Cat 5 TCs with 615 samples, provide a solid foundation for a robust analysis of global storm climatology. However, there is a significantly uneven distribution of the observed TCs over the six basins, especially for major hurricanes. The number of observed TCs over CP, EP, and IO is much smaller than other basins, especially over IO where only 17 samples of Cat 5 TCs are available.

Satellite Passive Microwave Sensors Applied in NRL Tropical Cyclone Database (1987-2012)

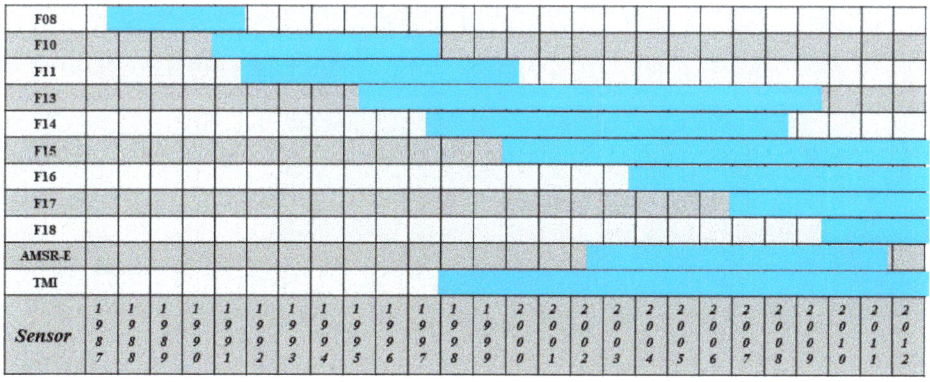

Figure 1. Chart of all satellite passive microwave (PMW) sensors during 1987–2012 used in the Naval Research Laboratory (NRL) Tropical Cyclone (TC) database.

Table 1. Total observations from satellite passive microwave (PMW) sensors in 1987–2012 for every tropical cyclone (TC) basin and category.

	TD	TS	Cat 1	Cat 2	Cat 3	Cat 4	Cat 5
AL	4909	7832	2195	749	548	461	74
CP	785	332	129	68	85	94	47
EP	6016	4633	1178	574	497	355	44
IO	2025	1353	159	48	81	61	17
SH	10,225	8685	2174	1026	1025	726	79
WP	11,718	8794	3101	1699	1418	1520	354
Total	35,678	31,629	8936	4164	3654	3217	615

The consistent TBs at 89 GHz are interpolated at $0.01° \times 0.01°$ spatial resolution and extracted over a $12° \times 12°$ area over the TC center. A polar coordinate system is then adapted to better represent the TC intensity and structure. This system has a radius of 500 km with resolution of 1 km and azimuthal angles of 360° with a resolution of 0.5°. The polar system is ideal for TC climatology when a composite analysis has to be utilized. In order to study impacts of the TC motion and vertical wind shear on TC structures, TBs in the polar coordinate system are rotated accordingly with directions of the TC movement and the 200–850 hPa wind shear before a composition process is conducted for observations at each storm intensity category, respectively.

3. Results

3.1. TC Structure Climatology

3.1.1. Global TC Annual Variability

Figure 2 shows comparison of annual variation of global storm activities from the PMW sensor measurements and the JTWC best track data during 1987–2012. A close agreement is obvious after 1992. The significantly less observed storms from PMW sensors before 1991 is due to the fact that only F08 was available during that time period. Some storms were missed in 1991 because only F08 and F10 sensors were available. The evidence is clear that there is a significantly annual variability of TCs, especially for major hurricanes. Twelve Cat 5 TCs were observed in 1997 while one observed Cat 5 TC in 1993 and 2008. However, the total number of annual storms is relatively stable around 100 with a small annual variation. It is also worth noting there is no obvious trend on the number of annual storms.

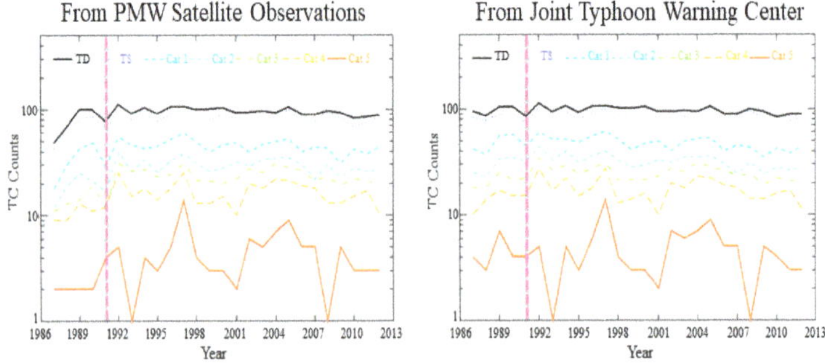

Figure 2. Annual variability of TC activities at each category. Left panel is from the PMW sensor observations while right panel is from Joint Typhoon Warning Center (JTWC). The vertical dashed line is a mark of 1991 which was for start of at least two PMW sensor observations.

3.1.2. Global TC Structure

Radial patterns of the azimuthally-averaged PMW TBs at 89H GHz for a polar coordinate system are shown in Figure 3 for global composite storms in 1987–2012 at each intensity category. The depressed values of TBs at 89 GHz is due to the scattering effect of ice particles from deep convections; therefore, the minimum TB position is an indication of the TC eyewall location. The heavy dashed line is a connection of these minimum TBs for Cat 1–5 TCs. The eyewall of Cat 5 TCs has a radius of 30 km while 50 km for Cat 1 TCs. A slight tilting of the heavy dashed line indicates an eyewall contraction with increase of TC intensity. A TB depression is not obvious near the TD center position because its convection is weak which has a small scattering effect. Another reason is that the TD convection is not well organized so that its center position is not accurately identified. A TB depression around radius of 50 km is evident with a very small amplitude for TS due to less convection intensity and less well-defined eyewall than seen in stronger TCs.

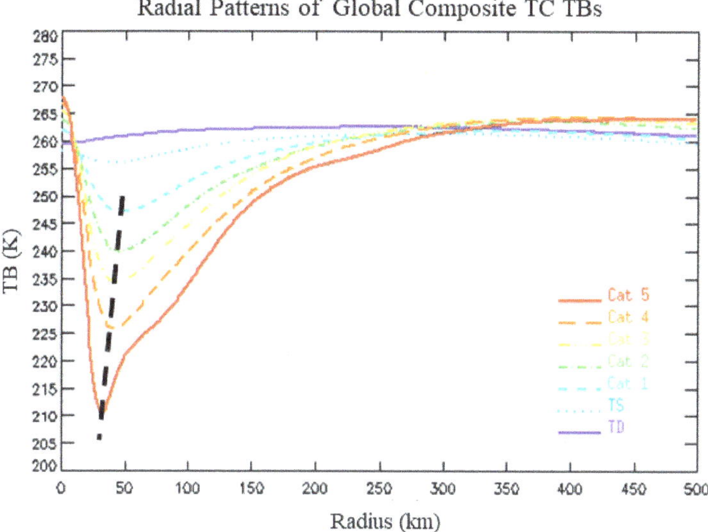

Figure 3. Radial patterns of PMW sensor brightness temperatures (TBs) at 89H GHz for each category of the global composite TCs in 1987–2012. The heavy dashed line is connection of minimum TBs of Cat 1–5 TCs.

Composition of a large sample of observed TCs from PMW sensors is the best way to present an accurate climatology of TC structure. Figure 4 displays distributions of the composite TBs at 89H GHz relative to the North direction for each storm intensity category during 1987–2012. The prominent feature for Cat 1–5 TCs is the consistent concentric pattern of TBs, which is easily understandable because of the TC consistent eyewall and spiral convections. A decrease of TB amplitudes near eyewall associated with increase of TC intensity is evident. The concentric pattern is not clear for TS and TD.

It is well-known that a TC presents a unique asymmetric distribution of spiral convections. The composition analysis for global storms in Figure 4 will not lead to a realistic structure for an individual TC, but results demonstrate the composition process is correctly conducted. Published literature indicates that distribution of strong convection is impacted by TC motion, intensity variability, and especially the vertical wind shear of a TC's ambient condition [28,29,39–43]. A recent study of TC precipitation climatology shows that the maximum TC rainfall is located in the down-motion quadrants with direction of the TC movement and in DSLQ with direction of the 200–850 hPa vertical wind shear [44]. Therefore, we will analyze impacts of the large scale environmental conditions on TC structure.

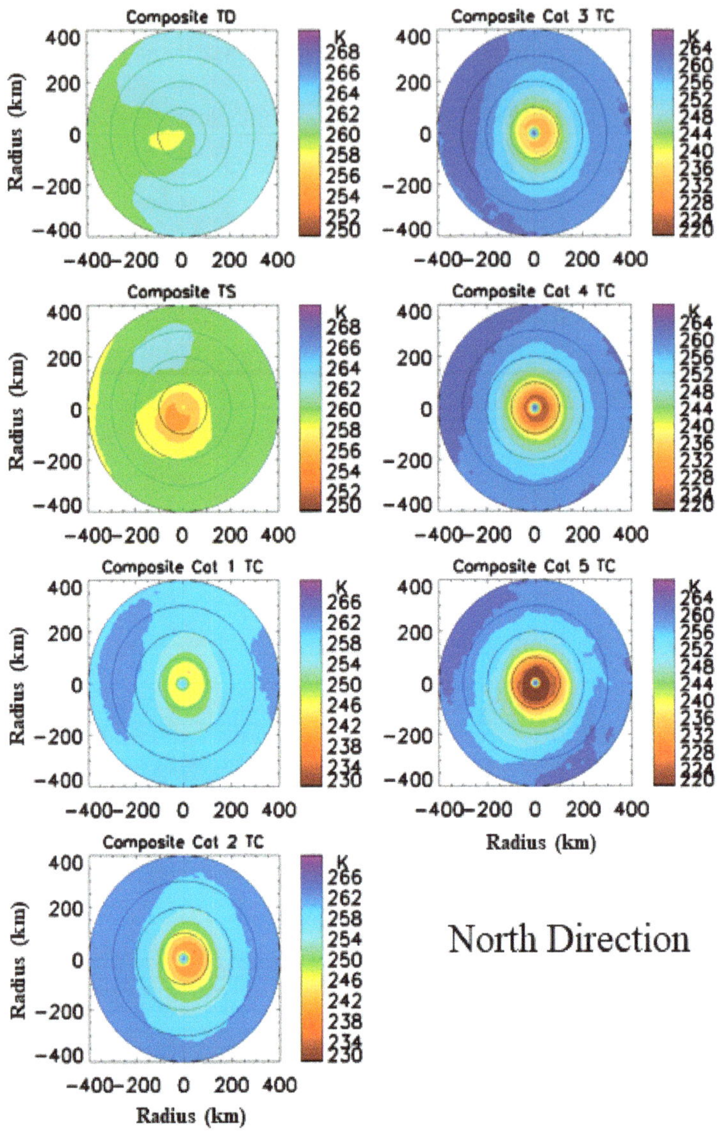

Figure 4. Climatology of the composited global TC structure relative to the North direction at each storm intensity category from PMW sensors at 89H GHz in 1987–2012.

3.1.3. Environmental Impact on TC Structure

Distributions of the composite TBs at 89H GHz for all global storm categories with regards to direction of TC movement are displayed in Figure 5a. The relative low TBs are associated with strong convection and mostly located in the down-motion left quadrant (DMLQ) for TD and TS and in the down-motion quadrants for Cat 1–3 TCs. The strong convection in the down-motion quadrants indicates convergence caused by the storm movement plays an important role on convection enhancement ahead of its motion. However, the concentric pattern is still a dominant feature for storms of higher intensity (Cat 4–5 TCs). It indicates that effect of the TC movement is not able to overtake impact on structure of the thermodynamic processes which generate a strong vortex for Cat 4–5 TCs.

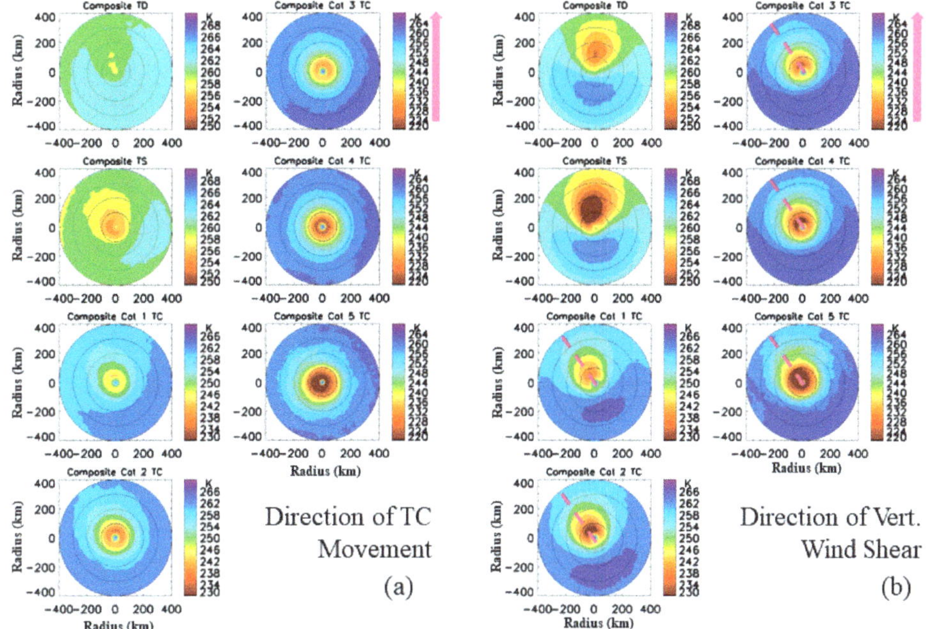

Figure 5. Same Figure 4, except for composition with regard to (**a**) direction of TC movement and (**b**) direction of the 200–850 hPa wind shear. The pink vertical arrow shows direction of the TC movement or 200–850 hPa wind shear. The red dashed line in (**b**) is for axis of the minimum TBs.

By same token, distributions of TBs at 89H GHz with regard to direction of the 200–850 hPa wind shear for all storm intensity categories are given by Figure 5b. The consistent distribution pattern for Cat 1–5 TCs is a distinct feature, i.e., the minimum TBs are always located in DSLQ. However, the minimum TBs are located in the down-motion quadrants for TD and TS. The minimum TBs decrease with an increase of TC intensity is expected due to strong convection which is always correlated with magnitude of higher intensity TCs. Results demonstrate that the direction of the 200–850 hPa wind shear plays a critical role in the distribution of TC convection.

To minimize potential uncertainties associated with limited samples in each category of the Cat 3–5 TCs, Figure 6 presents a distribution comparison of the global composite TBs for major hurricanes between three different composition methods. The combination of major hurricanes in 1987–2012 leads to a more reliable distribution of TC structure than each category of the Cat 3–5 TCs because of more samples involved in the composition process. The classic composition with the North direction shown in the left panel displays an expected concentric pattern of the minimum TBs near the eyewall. Regarding direction of the TC movement, although the concentric pattern is still a dominant feature, a slight forward shift of the minimum TBs is visible. This feature is due to the mixed results from the down-motion quadrants for Cat 3 TC convections and a concentric pattern for Cat 4–5 TC convection. The right panel presents a distinct distribution pattern from the others. The minimum TBs are clearly concentrated in DSLQ with a minimum axis slightly shifted to direction of the 200–850 hPa wind shear. Results from analysis of the major hurricane structure further confirms the findings from Figure 5, except for a clear and smooth distribution pattern. Direction of the 200–850 hPa wind shear is the key factor in affecting distributions of the TC convection, i.e., convection is climatologically favored in DSLQ with a maximum axis slightly shifted to the shear direction, and in downshear quadrants for TD and TS. Direction of the TC movement also plays a more important role in weak storms

(TD, TS, and Cat 1–2 TCs) than in major hurricanes. These features are consistent with locations of TC heavy precipitation [25,28,29,44].

Global Composite Cat 3-5 TC TBs at 89H GHz

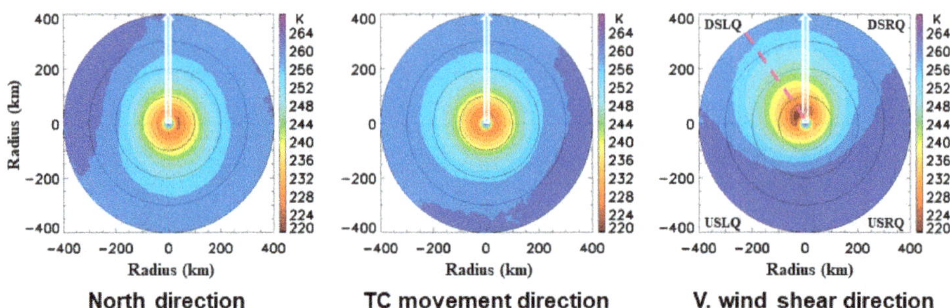

Figure 6. Climatology of major hurricanes at 89H GHz composited from global TCs in 1987–2012. Left, middle and right panel is for composition with direction of northward, TC movement and 200–850 hPa wind shear, respectively. White arrow is shown as direction of the composition method for each panel. The red dashed line in the right panel is for axis of the minimum TBs.

More studies are still needed to explain why direction of the vertical wind shear plays a critical role in distribution of strong TC convection. Black et al. [45] suggested that the rapidly rotating tangential winds near the TC strong convections create a maximum vertical motion in DSLQ which lead to the minimum TBs at 89 GHz and the maximum rainfall there. Nevertheless, a consistent pattern of the minimum TB distributions at 89H GHz with regard to direction of the 200–850 hPa wind shear from the reliable TC database provide valuable information on storm structure to mitigate impacts of approaching storms when the large scale environmental condition is available.

3.2. Structure Differences Among TC Basins

3.2.1. Geographic Features of Environmental Impacts

The environmental impacts on TC structure are clearly shown in Section 3.1.3 and the published literature [22,25,26,28,29]. However, discussion of regional differences on the environmental impacts was not thoroughly verified because of the limited TC observations used in these published studies. With the large TC observations in 1982–2012, a reliable and detailed analysis on geographic features of environmental impacts can be conducted. The mean motion speed of each storm intensity category over all TC basins is listed in Table 2. It is obvious that the averaged storm motion speed is relatively slower over CP, EP, IO, and SH than AL and WP basin. The mean motion speeds (scalar average) of 5.4–6.9, 3.9–6.0, 4.2–6.3, 3.6–5.9, 4.0–4.7, and 4.9–6.0 m s^{-1} are for AL, CP, EP, IO, SH, and WP, respectively. In general, the motion speed increases from TD to Cat 1 TCs, then decreases from Cat 2 to Cat 5 TCs. Table 3 displays the mean 200–850 hPa wind shear amplitude (scalar average) for each storm intensity category over all TC basins. A range of the shear magnitudes are 5.8–9.9, 5.6–7.5, 4.4–7.3, 5.6–9.1, 5.4–8.9, and 4.7–8.0 m s^{-1} for AL, CP, EP, IO, SH, and WP basin, respectively. The prominent feature is a consistent decrease of the vertical wind shear magnitude with increase of the TC intensity category over all basins, except the Cat 4–5 TCs over IO which has a very limited samples. It is well established a relatively weak vertical wind shear is a favorable condition for TC intensification process. Overall, AL basin has a relatively large vertical wind shear, while CP and EP have a relatively small vertical wind shear.

Table 2. Averaged motion speed at each intensity category over each TC basin (m s^{-1}).

	TD	TS	Cat 1	Cat 2	Cat 3	Cat 4	Cat 5
AL	5.4	6.7	6.9	5.8	6.1	5.7	5.6
CP	4.9	5.5	6.0	3.9	4.6	5.0	4.9
EP	4.2	4.5	4.6	4.8	5.0	5.2	6.3
IO	3.6	3.8	3.8	4.4	4.0	4.1	5.9
SH	4.0	4.7	4.4	4.5	4.5	4.1	4.1
WP	4.9	6.0	6.0	5.5	5.4	5.2	5.3

Table 3. Averaged 200–850 hPa wind shear at each intensity category over each TC basin (m s^{-1}).

	TD	TS	Cat 1	Cat 2	Cat 3	Cat 4	Cat 5
AL	8.5	9.4	9.9	8.4	8.0	6.7	5.8
CP	6.0	7.4	7.5	7.3	5.6	5.6	5.7
EP	7.3	6.4	5.5	5.4	4.8	4.5	4.4
IO	9.0	9.1	6.8	6.5	5.6	6.6	6.8
SH	8.3	8.9	8.2	7.4	7.6	6.0	5.4
WP	7.9	8.0	7.8	7.5	7.4	5.6	4.7

Figure 7 displays the composite structures with regard to direction of TC movement at each storm intensity category over AL, SH, and WP. Over the AL basin, the minimum TBs are in DMRQ for Cat 2–3 TCs while an apparent concentric pattern within 200 km radius is for Cat 4–5 TCs. The Cat 1 TC has the minimum TBs in DMLQ. TD and TS have no clear patterns but with relative low TBs in the left quadrants. Over the SH basin, there is a clear concentration of convections in the down-motion quadrants for TD, TS, and Cat 1–3 TCs, while Cat 4–5 TCs still have a dominant concentric pattern. Over the WP basin, TD, TS, and Cat 1–3 TCs have the minimum TBs located in DMLQ, and Cat 4–5 TCs have a concentric pattern. The mean motion vector of each storm category over all basins (average of motion vectors) is listed in Table 4. It is evident that the mean storm motion vector is headed to the northwest direction over CP, EP, and WP basin. The AL basin storms have the mean motion vector directions of 289°–15° from the North and the IO basin storms have mean motion vector directions of 303°–10°, while the SH storms have different mean vector directions of 185°–225°. The relatively smaller amplitudes of the motion vectors compared with the corresponding scalar averages of motion speeds shown in Table 2 are resultant of the varying directions of motion vectors. In addition, percentage of the amplitude decrease of averaged motion vectors is relatively larger for weak storms than for higher intensity TCs, indicating that there are more variations of the motion vector directions for weak storms than for strong TCs. Therefore, there are clear regional structure differences regarding direction of TC movement for TD, TS and Cat 1–3 TCs. It also indicates that direction of TC movement has less impact on higher intensity Cat 4–5 TCs, which demonstrates the TC strong vortex is resilient to impact of the TC motion.

A similar analysis of TC structure with regard to direction of the 200–850 hPa wind shear is displayed in Figure 8. The most striking feature is that AL and WP storms have a consistent structure pattern with the minimum TBs within a radius of 300 km in DSLQ for Cat 1–5 TCs and in downshear quadrants for TD and TS. For SH storms, the minimum TBs is in downshear quadrants for TD and TS, while in DSRQ for Cat 1–5 TCs. The difference in location of the minimum TBs for Cat 1–5 TCs between AL/WP and SH actually reflects the opposite circulation patterns of storms between the Northern and the Southern Hemisphere, which further demonstrates a critical role from direction of the 200–850 hPa wind shear. In addition, a more concentric pattern in the inner core area (radius < 100 km) for Cat 5 TCs demonstrates a more resilience to the environmental impact for strong vortexes than for weak storms. A close review also indicates that the axis of the minimum TBs marked by the red dashed line is slightly shifted toward the vertical wind shear direction. In addition, similar characteristics as the AL/WP storms relative to the vertical wind shear direction are found for CP, EP, and IO basins (Figures are omitted).

The mean 200–850 hPa wind shear vectors (average of the shear vectors) of storms for each intensity category over all basins are shown in Table 5. The overall much smaller mean amplitudes of the vertical wind shear vector compared to the corresponding averaged scalar wind shear shows environmental conditions for different storms. The large geographic variations of the mean vertical wind shear directions among these basins clearly display their different environmental conditions. The mean vertical wind shear vector presents directions of 75°–95°, 54°–157°, 252°–95°, 260°–17°, 90°–135°, and 81°–227° for storms over AL, CP, EP, IO, SH, and WP, respectively.

Table 4. Averaged motion vector at each intensity category over each TC basin. (Direction: clockwise from the North in degree; Speed unit: m s^{-1}).

Basin.	Motion Vector	TD	TS	Cat 1	Cat 2	Cat 3	Cat 4	Cat 5
AL	Speed	2.4	3.0	4.1	3.4	4.2	4.9	5.2
	Direction	323.7	9.7	15.0	344.8	328.7	302.3	289.2
CP	Speed	3.7	2.6	5.1	3.4	4.0	4.5	4.7
	Direction	277.5	317.7	306.0	315.5	316.0	296.7	274.1
EP	Speed	3.0	3.7	4.0	4.0	4.4	4.8	5.0
	Direction	287.0	295.7	296.7	297.6	294.2	291.1	296.4
IO	Speed	2.2	2.2	2.1	3.4	3.0	3.2	5.1
	Direction	303.5	324.2	348.2	1.5	10.5	356.2	342.9
SH	Speed	1.8	2.3	2.4	2.6	2.5	2.4	2.5
	Direction	225.2	185.1	191.4	186.0	191.5	212.5	206.5
WP	Speed	2.9	3.1	3.4	3.3	3.5	3.8	4.5
	Direction	308.5	349.7	345.8	340.5	339.2	323.4	309.4

Table 5. Averaged 200–850 hPa wind shear vector at each intensity category over each TC basin (Direction: clockwise from the North in degree; Wind shear unit: m s^{-1}).

Basin	Wind Shear Vector	TD	TS	Cat 1	Cat 2	Cat 3	Cat 4	Cat 5
AL	WindShear	4.7	6.1	7.1	5.5	5.4	4.5	4.0
	Direction	93.5	91.5	75.7	79.7	79.2	77.3	94.8
CP	WindShear	2.0	3.7	5.4	5.6	4.5	2.4	5.6
	Direction	77.8	68.5	54.0	74.8	59.6	109.6	157.2
EP	WindShear	0.3	0.6	0.8	0.6	0.7	1.3	2.2
	Direction	345.8	255.7	252.9	277.0	288.2	258.4	94.7
IO	WindShear	6.1	6.2	4.0	3.8	3.6	3.1	4.5
	Direction	300.1	299.6	297.8	281.7	260.1	324.2	17.3
SH	WindShear	8.3	8.9	8.2	7.4	7.6	6.0	5.4
	Direction	135.2	123.1	119.6	121.1	117.8	126.9	89.9
WP	WindShear	2.5	1.5	0.9	1.0	2.0	0.8	1.1
	Direction	226.9	197.3	130.4	122.9	81.5	156.0	186.4

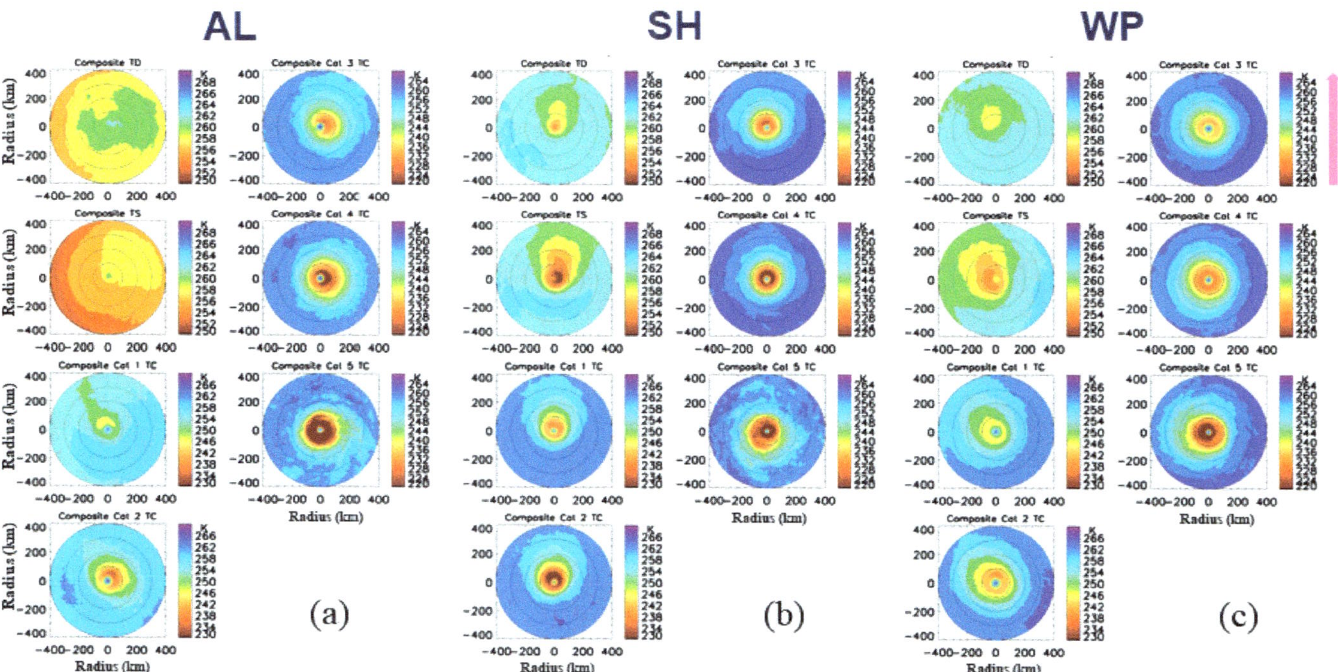

Figure 7. Distributions of TBs at 89H GHz of the composite storms relative to the storm motion direction at tropical depression (TD), tropical storm (TS), and Cat 1–5 TCs during 1987–2012 for (**a**) Atlantic (AL), (**b**) Southern Hemisphere (SH) and (**c**) West Pacific (WP) basin. The pink vertical arrow indicates direction of TC movement.

Remote Sens. **2020**, 12, 3610

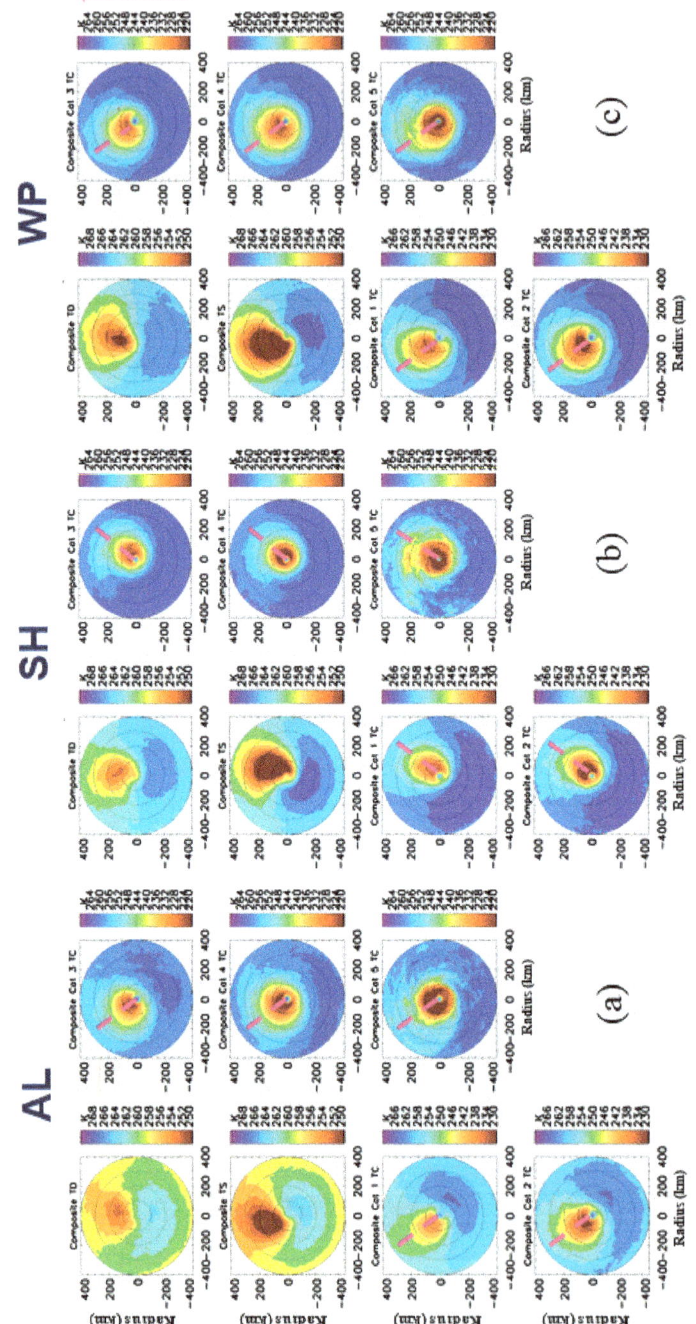

Figure 8. Same as Figure 7, except for relative to direction of the 200–850 hPa wind shear. The red dashed line is for axis of the minimum TBs.

TC structure of the composite major hurricanes with regard to direction of TC movement over the six basins is given in Figure 9a. The prominent feature is a common concentric pattern of TBs at 89H GHz within 200 km radius over all basins except IO, indicating a resilience of the strong vortex to the motion impact. However, a close review still reveals differences on locations of the minimum TBs near the eyewall. It is in the down-motion right quadrant (DMRQ) for AL, right quadrants for CP, left quadrants for EP, down-motion quadrants for SH, and DMLQ for WP. The minimum TBs are obviously shifted into the left quadrants for IO. Similar analysis with regard to direction of the 200–850 hPa wind shear demonstrates a consistent TC structural pattern (Figure 9b). A striking feature is a consistent location of the minimum TBs at 89H GHz in DSLQ for all basins except SH where it is in DSRQ. Since the TC circulation pattern in the southern hemisphere is opposite to what is seen in the northern hemisphere, the TC minimum TBs at 89 HGHz in DSRQ over SH is expected. A close review of Figure 9b indicates a consistent axis of the minimum TBs marked by the red dashed line is slightly shifted toward the vertical shear direction.

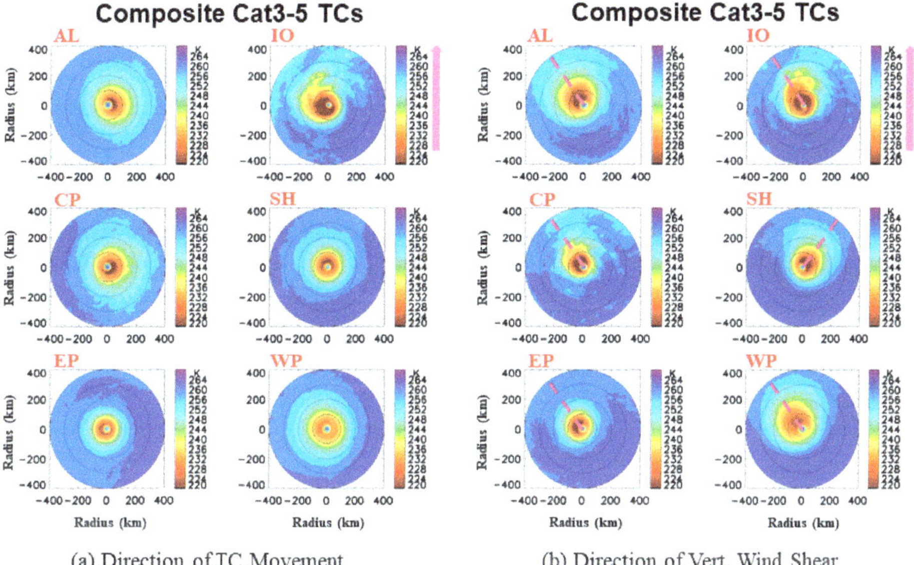

(a) Direction of TC Movement   (b) Direction of Vert. Wind Shear

Figure 9. Distributions of the 89H GHz TBs over six basins in 1987–2012 for the composite Cat 3–5 TCs relative to (**a**) direction of TC movement and (**b**) direction the 200–850 hPa wind shear. The red dashed line in (**b**) is for axis of the minimum TBs.

3.2.2. Regional Characteristics of TC Structure

Distribution patterns relative to the North direction at each storm intensity category over AL, SH and WP are displayed in Figure 10. These kinds of patterns actually reflect the combined impacts of the storm motion and the 200–850 hPa wind shear on storm structures. The similar concentric pattern for Cat 4–5 TCs among these basins are expected because a strong vortex and consistent eyewall appearance associated with intense TCs that are more resilient to the external forcing than the weak storms. The relative minimum TBs at 89H GHz within radius of 300 km are in the northeast quadrant for AL Cat 1–4 TCs, where relatively low TBs are also evident for Cat 5 TCs (Figure 10a). The approximate eastward direction of the vertical wind shear vector indicates that location of the TC minimum TBs is corresponding to DSLQ. Due to the mostly northwest direction of the TC motion, the motion impact is actually in a competitive role of the vertical wind shear. However, results demonstrate the vertical wind shear has a dominant role on the AL TC structure.

The relative minimum 89H GHz TBs are located in the southeast quadrant for the SH Cat 1–4 TCs (Figure 10b). The associated vertical wind shear vectors, generally, have an East direction, which shows that convections are in DSRQ. The associated TC motions are mainly in the south direction. Thus, impacts from motion and the vertical wind shear are supportive to each other which lead to the SH TC convections located in the southeast quadrant. For the WP Cat 1–4 TCs, the minimum TBs at 89H GHz in the southeast quadrant are resultant of the combined impact of the motion and the vertical wind shear (Figure 10c); however, it seems more impacts are linked with the wind shear. The mean vertical wind shear vectors have direction in the southeast quadrant while the mean motion vectors have direction in the northwest quadrant. Differences are obvious for TD and TS among these basins. Both motion directions and the vertical wind shear directions given by Tables 4 and 5 show the combined impacts lead to variations of TD and TS structures.

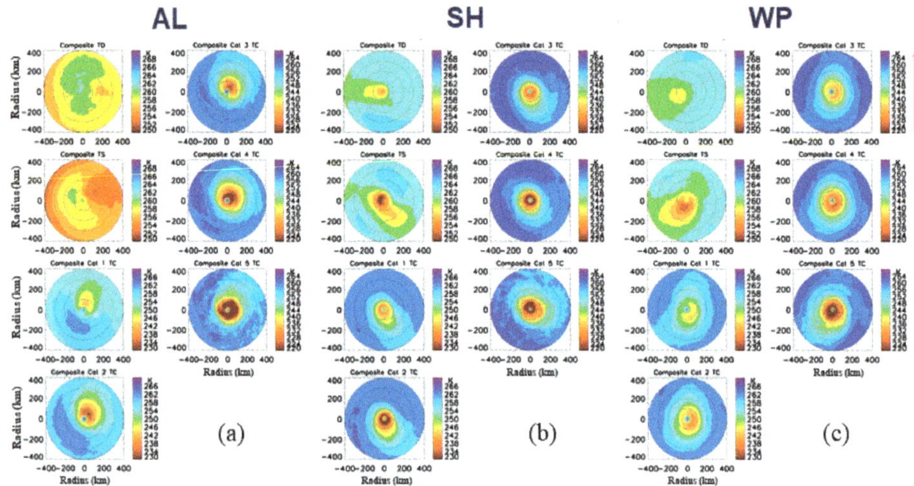

Figure 10. Same as Figure 7, except for relative to the North direction.

Regional differences of TC structure are visible from comparison of 89H GHz TB distributions among the six basins. Figure 11 shows distribution of TBs at 89H GHz for the composite Cat 3–5 TCs over the six basins regarding the North direction. Although the strong TC vortex is resilient to the environmental impacts, a geographic variation is obvious. The concentric patterns of TBs within a 200 km radius are clearly shown in EP, SH, and WP. Both AL and CP basin has the minimum TBs concentrated in the northeast quadrant, while in the west quadrants over the IO basin.

For better display of impacts of storm movement and the 200–850 hPa wind shear on storm structure, a comparison of major hurricanes during 1987–2012 over AL, SH, and WP regarding direction of the North, TC motion and the 200–850 hPa wind shear is shown in Figure 12. For composition relative to direction of TC movement (middle panel), the AL major hurricanes have the minimum TBs located in DMRQ, while both SH and WP major hurricanes have a concentric pattern with a slight shift of minimum TBs in the down-motion quadrants. A consistent pattern for AL and WP major hurricane is displayed with minimum TBs in DSLQ, while SH major hurricanes have the minimum TBs in DSRQ which is opposite to what shown in AL/WP, regarding to direction of the 200–850 hPa wind shear (right panel). The red dashed lines indicate the axis of the minimum TBs is slightly shifted toward the vertical shear direction. The major hurricane structure relative to the North direction shows the minimum TBs are located in the northeast quadrant for AL, the southeast quadrant for WP, and in concentric zones for SH (left panel).

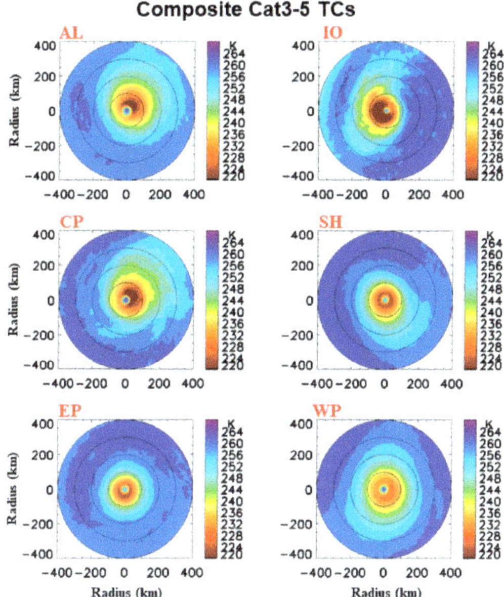

Figure 11. Distributions of 89H GHz TBs for the composite Cat3–5 TCs relative to the North direction over six basins in 1987–2012.

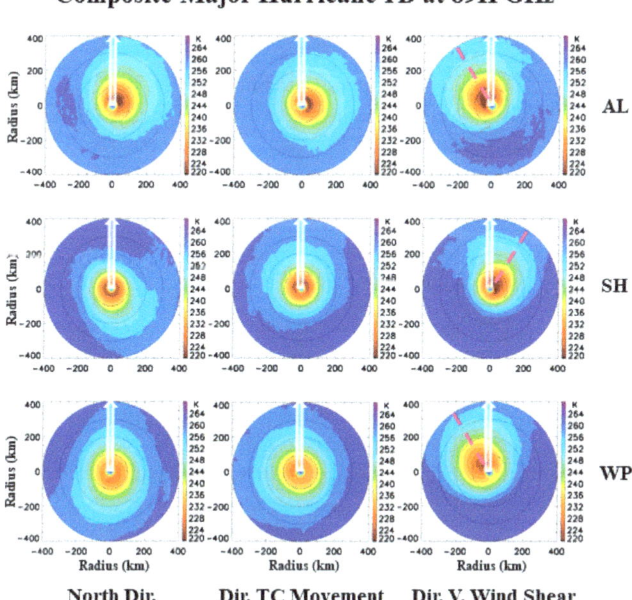

Figure 12. Comparison of the composite TB distributions at 89H GHz for major hurricanes over AL, SH and WP. Left, middle and right panel is for composition with regard to northward, direction of TC movement, and direction of the 200–850 hPa wind shear, respectively. White arrow is direction of the composition method for each panel. The red dashed line in the right panel is for axis of the minimum TBs.

4. Discussion

This analysis demonstrates the obvious regional differences of storm structure relative to storm motion among the six basins, especially for weak storms. TC structure relative to the 200–850 hPa wind shear direction has a consistent pattern with convection located in DSLQ among these basins except the SH basin where convection is located in DSRQ. Results demonstrated in this study are consistent with the published literature [20,21,24,28,46,47]. Exploring mechanisms to explain different impacts on structure from TC motion and the vertical shear are beyond scope of this study. However, evidence revealed here does support the conclusion that the convergence caused by storm motion is most likely responsible for strong convection ahead of storm motion, especially for a weak storm [47,48]. The maximum vertical motion in DSLQ caused by the rapid rotating tangential winds near TC strong convection proposed by Black et al. [45] is verified by observational evidence. Because of the relatively small differences of the TC motion direction among the northern hemispheric basins and large differences of the vertical wind shear direction, the regional variability of TC structure relative to the storm motion is actually caused by differences of the environmental forcing among these basins. The combined impact of storm motion and the vertical wind shear leads to asymmetric TC convection. The vertical wind shear plays a dominant role if the storm motion impact is not lined up with the wind shear. Mctaggart-Cowan et al. [49] demonstrated there are significant differences of large scale systems affecting tropical cyclogenesis among these basins. Results from Wu et al. [50] showed 19.8% TC formations over west Pacific are associated with the Monsoon Gyres, indicating more complex large scale environmental conditions for WP TC formation. Results from this study provide observational evidence from prospective of satellite measurements to confirm that the large scale 200–850 hPa wind shear is critical to storm development and structure.

Systematic analysis not only confirms results from published literature but also leads to an improved and detailed climatology of TCs associated with each storm category and their geographic variability based on long and reliable satellite PMW observations. The satellite-observed TCs from recent years are not included, although they will increase TC samples for a potentially better climatology, especially for major hurricanes. Recent updates on PMW sensor calibrations [51] are also not included. Although these updates could reduce potential uncertainties on TC structure especially for Cat 5 TCs over CP, EP, and IO, they will not, in general, change results from this study. Reprocessing of the new PMW TB datasets for an updated TC database will address these issues, but it is beyond scope of this study. In addition, impacts from speed of TC movement and magnitude of the 200–850 hPa wind shear and their combined impacts are not investigated in details in this study and should be topics of future studies.

5. Conclusions

The historical storms (TD, TS and TCs) observed by all satellite PMW sensors during 1987–2012 are analyzed in a polar coordinate system with different composition methods regarding directions of the North, TC movement and the 200–850 hPa wind shear. The primary goal of this study is to provide a reliable and detailed climatology on global TC activities, structure and geographic characteristics. The TMI and SSM/I TBs at 85 GHz and SSMIS TBs at 91 GHz TBs are calibrated to 89 GHz so that SSM/I, SSMIS, TMI and AMSR-E used in this study have a consistent high frequency channel at 89 GHz. ARCHER is used to accurately fix the TC center positions. The suppression of TBs at 89 GHz due to ice particle scattering effects is strongly associated with TC convection, i.e., the minimum TBs display locations of strong convection.

Analysis shows that having a large number of samples of observed storms in this study leads to robust results on climatology of TC structure and their regional differences. There is significant annual variability of global TC activities, especially for major hurricanes; however, the total number of storms is relatively stable around 100. There is no evidence indicating any trend of TC activities. The radius of the TC eyewall increases with decrease of intensity with a radius of 30 km and 50 km for Cat 5 and Cat 1TCs, respectively.

The composition method regarding the North direction for global storms presents a concentric pattern of TBs within 200 km radius because of the large number of samples of the observed storms used in the composition process. Thus, this method for global storms will not lead to a realistic storm structure for an individual storm due to the fact that TCs have a unique spiral convection structure. The impact of direction of TC movement and the 200–850 hPa wind shear on TC structure are systematically analyzed in this study. The composite structures of global storms at different intensity categories regarding direction of TC movement show the minimum TBs located in DMLQ for TD, TS, and Cat 1–2 TCs, while major hurricanes have an apparent concentric pattern. The climatology of storm structure with regard to direction of the 200–850 hPa wind shear presents a distinctive feature, i.e., the minimum TBs located in DSLQ for Cat 1–5 TCs and in forward quadrants for TD and TS. Results demonstrate direction of TC movement has obvious impacts on the structure of the relatively weak storms; however, direction of the 200–850 hPa wind shear has a critical role in distribution of the TC convection.

The detailed geographic characteristics of TC structure are clearly demonstrated by comparison of three different composition schemes. Regarding direction of TC movement, the composite pattern of major hurricanes has the minimum TBs at 89H GHz located in DMRQ over AL basin, left quadrants over IO basin, and a concentric pattern within 200 km radius over CP, SH, EP, and WP basins. The direction of TC movement has more significant impacts on less intense TCs than strong TCs. The concentric pattern within 200 km radius is always similar for Cat 4–5 TCs among these basins; however, differences are obvious for TD, TS, and Cat 1–3 TCs. The AL Cat 2–3 TCs have the minimum TBs in DMRQ while Cat 1 TCs have a minimum in DMLQ. TD, TS and Cat 1–3 TCs over SH basin have the minimum TBs in the down-motion quadrants. Over the WP basin, TD, TS, and Cat 1–2 TCs have minimum TBs in DMLQ while Cat 3 TCs have minimum TBs in the down-motion quadrants. Results demonstrate that there is a significant regional variation of storm structure relative to the motion direction and that convergence induced by TC movement play an import role on the structure of TD, TS, and Cat 1–2 TCs and less role for Cat 3 TCs and almost no impacts on Cat 4–5 TCs. This study also reveals that the geographic variation of the structure relative to storm motion direction is mainly due to differences of the vertical wind shear direction among these basins.

Regarding direction of the 200–850 hPa wind shear, TD and TS have a consistent minimum TBs in the downshear quadrants, while Cat 1–5 TCs have consistent minimum TBs in DSLQ for all basins except in DSRQ for SH basin, i.e., TC structure has no geographic variations regarding direction of the vertical wind shear. Since the suppressed TBs at 89 GHz are closely linked to deep convection, the consistent patterns shown in TBs and surface precipitation for all storm categories solidify results from this study. This study provides robust observational evidence to confirm the large scale environmental forcing has a critical and consistent impact on TC structure. This feature is important because it could be utilized in TC forecasting and preparation to mitigate impacts from an approaching TC.

The storm structures relative to the North direction are resultant from the combined impact of storm motion and the 200–850 hPa wind shear. Results demonstrate the geographic variation of storm structures relative to motion direction is actually due to regional differences of the vertical wind shear direction. The impact from vertical wind shear dominates the impact from storm motion when their roles are not collaborated each other, indicating the critical role to storm structure by the vertical wind shear direction. Zhang and Tao [52] showed the vertical wind shear has a significant effect on the TC predictability, especially during storm formation and rapid intensification. More studies are needed to investigate the potential mechanisms through carefully designed cloud model simulations.

Author Contributions: S.Y. designed and performed research and wrote the paper. R.B. reviewed and edited this paper. J.C. analyzed and created the satellite PMW tropical cyclone TB dataset. R.B. and S.Y. obtained funding for this study. All authors have read and agreed to the published version of the manuscript.

Funding: This research received financial support from the Office of Naval Research (ONR) program under a 6.4 project "Environmental and Tropical Cyclone Characterization via Sensor Data Exploitation " and a NRL base project "River Influence at Multi-scales" (PE 61153N).

Remote Sens. **2020**, *12*, 3610

Acknowledgments: The authors appreciate the Colorado State University PMW sensors fundamental climate data records (http://rain.atmos.colostate.edu/FCDR) and the TC best track datasets from NOAA and Joint Typhoon Warning Center. Constructive comments from three reviewers are also appreciated for improving quality of this study.

Conflicts of Interest: The authors declare no conflict of interest.

References

1. Anthes, R.A. Tropical Cyclones: Their Evolution, Structure, and Effects. In *Meteorological Monographs*; American Meteorological Society: Boston, MA, USA, 1982; 208p. [CrossRef]
2. Gray, W.M. The Formation of Tropical Cyclones. *Meteorol. Atmos. Phys.* **1998**, *67*, 37–69. [CrossRef]
3. Blake, E.S.; Landsea, C.W.; Gibney, E.J. *The Deadliest, Costliest, and Most Intense United States Tropical Cyclones from 1851 to 2010 (And Other Frequently Requested Hurricane Facts) (PDF) (NOAA Technical Memorandum NWS NHC-6)*; United States National Oceanic and Atmospheric Administration's National Weather Service: Washington, DC, USA, 2011.
4. Rappaort, E.N. Loss of life in the United States associated with recent Atlantic tropical cyclones. *Bull. Am. Meteor. Soc.* **2000**, *81*, 2065–2074. [CrossRef]
5. Knight, D.B.; Davis, R.E. Climatology of tropical cyclone rainfall in the southeastern United States. *Phys. Geogr.* **2007**, *28*, 126–147. [CrossRef]
6. Yang, S.; Cossuth, J. Satellite remote sensing of tropical cyclones. Chapter 7 of Recent Developments. *Trop. Cyclone Dyn. Predict. Detect. Intech* **2016**, 138–170. [CrossRef]
7. Wimmers, A.J.; Velden, C.S. MIMIC: A New Approach to Visualizing Satellite Microwave Imagery of Tropical Cyclones. *Bull. Amer. Meteor. Soc.* **2007**, *88*, 1187–1196. [CrossRef]
8. Hawkins, J.D.; Lee, T.F.; Turk, J.; Sampson, C.; Kent, J.; Richardson, K. Real-time internet distribution of satellite products for tropical cyclone reconnaissance. *Bull. Amer. Meteor. Soc.* **2001**, *82*, 567–578. [CrossRef]
9. Kidder, S.Q.; Goldberg, M.D.; Zehr, R.M.; DeMaria, M.; Purdom, J.F.; Velden, C.S.; Grody, N.C.; Kusselson, S.J. Satellite analysis of tropical cyclones using the Advanced Microwave Sounding Uint (AMSU). *Bull. Am. Meteor. Soc.* **2000**, *81*, 1241–1259. [CrossRef]
10. Hawkins, J.D.; Turk, F.J.; Lee, T.F.; Richardson, K. Observations of tropical cyclones with the SSMIS. *IEEE Trans. Geosci. Remote Sens.* **2008**, *46*, 901–912. [CrossRef]
11. Willoughby, H.E. Mature Structure and Evolution. In *Chapter 2, Global Perspectives on Tropical Cyclones, WMO/TD-No. 693, Report No. TCP-38*; World Meteorological Organization: Geneva, Switzerland, 1995.
12. Dvorak, V.F. Tropical cyclone intensity analysis using satellite data. In *NOAA Technical Report NESDIS*; US Department of Commerce, National Oceanic and Atmospheric Administration: Silver Spring, MD, USA; Montgomery, AL, USA, 1984; Volume 11, pp. 1–47.
13. Velden, C.S.; Olander, T.L.; Zehr, R.M. Development of an Objective Scheme to Estimate Tropical Cyclone Intensity from Digital Geostationary Satellite Infrared Imagery. *Weather. Forecast.* **1998**, *13*, 172–186. [CrossRef]
14. Olander, T.L.; Velden, C.S. The advanced Dvorak technique: Continued development of an objective scheme to estimate tropical cyclone intensity using geostationary infrared satellite imagery. *Weather Forecast.* **2007**, *22*, 287–298. [CrossRef]
15. Hawkins, J.; Velden, C. Supporting Meteorological Field Experiment Missions and Postmission Analysis with Satellite Digital Data and Products. *Bull. Amer. Meteor. Soc.* **2011**, *92*, 1009–1022. [CrossRef]
16. Herndon, D.C.; Velden, C.S.; Hawkins, J.; Olander, T.; Wimmers, A. The CIMSS Satellite Consensus (SATCON) tropical cyclone intensity algorithm. In Proceedings of the 29th Conference on Hurricanes and Tropical Meteorology, Tucson, AZ, USA, 10–14 May 2010.
17. Rogers, R.; Zhang, J.; Zawislak, J.; Jiang, H.; Alvey, G.R., III; Zipser, E.J.; Stevenson, S.N. Observations of the structure and evolution of Hurricane Edouard (2014) during intensity change. Part II: Kinematic structure and the distribution of deep convection. *Mon. Weather Rev.* **2016**, *144*, 3355–3376. [CrossRef]
18. Rogers, R.; Reasor, P.D.; Zhang, J.A. Multiscale structure and evolution of Hurricane Earl (2010) during rapid intensification. *Mon. Weather Rev.* **2015**, *143*, 536–562. [CrossRef]
19. Stevenson, S.N.; Corbosiero, K.L.; Molinari, J. The convective evolution and rapid intensification of Hurricane Earl (2010). *Mon. Weather Rev.* **2014**, *142*, 4364–4380. [CrossRef]
20. Stevenson, S.N.; Corbosiero, K.L.; Abarca, S.F. Lightning in eastern North Pacific tropical cyclones: A comparison to the North Atlantic. *Mon. Weather Rev.* **2016**, *144*, 225–239. [CrossRef]

21. Stevenson, S.N.; Corbosiero, K.L.; DeMaria, M.; Vigh, J.L. A 10-Year Survey of Tropical Cyclone Inner-Core Lightning Bursts and Their Relationship to Intensity Change. *Weather Forecast.* **2018**, *60*, 23–36. [CrossRef]

22. Jiang, H.; Ramirez, E.M. Necessary conditions for tropical cyclone rapid intensification as derived from 11 years of TRMM data. *J. Clim.* **2013**, *26*, 6459–6470. [CrossRef]

23. Zagrodnik, J.P.; Jiang, H. Rainfall, convection, and latent heating distributions in rapidly intensifying tropical cyclones. *J. Atmos. Sci.* **2014**, *71*, 2789–2809. [CrossRef]

24. Tao, C.; Jiang, H. Distributions of shallow to very deep precipitation–convection in rapidly intensifying tropical cyclones. *J. Clim.* **2015**, *28*, 8791–8824. [CrossRef]

25. Tao, C.; Jiang, H.; Zawislak, J. The relative importance of stratiform and convective rainfall in rapidly intensifying tropical cyclones. *Mon. Weather Rev.* **2017**, *145*, 795–809. [CrossRef]

26. Fischer, S.F.; Tang, B.H.; Corbosiero, K.L. Normalized convective characteristics of tropical cyclone rapid intensification events in the North Atlantic and Eastern North Pacific. *Mon. Weather Rev.* **2018**, *146*, 1133–1155. [CrossRef]

27. Fischer, S.F.; Tang, B.H.; Corbosiero, K.L. A climatological analysis of tropical cyclone rapid intensification in environments of upper-tropospheric troughs. *Mon. Weather Rev.* **2019**, *147*, 3693–3719. [CrossRef]

28. Chen, S.S.; Knaff, J.A.; Marks, F.D. Effects of vertical wind shear and storm motion on tropical cyclone rainfall asymmetries deduced from TRMM. *Mon. Weather Rev.* **2006**, *134*, 3190–3208. [CrossRef]

29. Lonfat, M.; Marks, F.D.; Chen, S. Precipitation distribution in tropical cyclones using the Tropical Rainfall Measuring Mission (TRMM) microwave imager: A global perspective. *Mon. Weather Rev.* **2004**, *132*, 1645–1660. [CrossRef]

30. SHIPS. Available online: http://rammb.cira.colostate.edu/research/tropical_cyclones/ships/index.asp. (accessed on 31 October 2020).

31. Yang, S.; Hawkins, J.; Richardson, K. The improved NRL tropical cyclone monitoring system with a unified microwave brightness temperature calibration scheme. *Remote Sens.* **2014**, *6*, 4563–4581. [CrossRef]

32. Wimmers, A.J.; Velden, C.S. Advancements in objective multisatellite tropical cyclone center fixing. *J. Appl. Meteor. Climatol.* **2016**, *55*, 197–212. [CrossRef]

33. Wimmers, A.J.; Velden, C.S. Objectively determining the rotational center of tropical cyclones in passive microwave satellite imagery. *J. Appl. Meteor. Climatol.* **2010**, *49*, 2013–2034. [CrossRef]

34. Poe, G.A. Optimum interpolation of imaging microwave radiance data. *IEEE Trans. Geosci. Remote Sens.* **1990**, *28*, 800–810. [CrossRef]

35. DeMaria, M.; Mainelli, M.; Shay, L.K.; Knaff, J.A.; Kaplan, J. Further Improvements in the Statistical Hurricane Intensity Prediction Scheme (SHIPS). *Weather Forecast.* **2005**, *20*, 531–543. [CrossRef]

36. Knaff, J.; Sampson, C.; DeMaria, M. An Operational Statistical Typhoon Intensity Prediction Scheme for the Western North Pacific. *Weather Forecast.* **2005**, *20*, 688–699. [CrossRef]

37. NOAA. Saffir-Simpson Hurricane Scale Information, National Hurricane Center. Available online: http://www.nhc.noaa.gov/aboutsshws.php (accessed on 5 December 2016).

38. WMO. Regional Association IV Hurricane Operational Plan 2015, World Meteorological Organization Technical Document, 109p. Available online: https://www.wmo.int/pages/prog/www/tcp/documents/OPERATIONALPLAN2015_en_final.pdf. (accessed on 2 November 2020).

39. Leighton, H.; Gopalakrishnan, S.; Zhang, J.; Rogers, R.F.; Zhang, Z.; Tallapragada, V. Azimuthal Distribution of Deep Convection, Environmental Factors, and Tropical Cyclone Rapid Intensification: A Perspective from HWRF Ensemble Forecasts of Hurricane Edouard (2014). *J. Atmos. Sci.* **2018**, *75*, 275–295. [CrossRef]

40. Nguyen, L.T.; Rogers, R.F.; Reasor, P.D. Thermodynamic and Kinematic Influences on Precipitation Symmetry in Sheared Tropical Cyclones: Bertha and Cristobal (2014). *Mon. Weather Rev.* **2017**, *145*, 4423–4446. [CrossRef]

41. Yu, Z.; Wang, Y.; Xu, H. Observed rainfall asymmetry in tropical cyclone making landfall over China. *J. Appl. Meteor. Climatol.* **2015**, *54*, 117–136. [CrossRef]

42. Alvey, R.G., III; Zawislak, J.; Zipser, E. Precipitation properties observed during tropical cyclone intensity change. *Mon. Weather Rev.* **2015**, *143*, 4476–4492. [CrossRef]

43. Rogers, R.F.; Marks, F.D.; Marchok, T. Tropical cyclone rainfall. In *Encyclopedia of Hydrological Sciences*; Anderson, M.G., Ed.; John Wiley and Sons: Chicester, UK, 2009.

44. Yang, S.; Lao, V.; Bankert, R.; Whitcomb, T.R.; Cossuth, J. Improved Climatology of Tropical Cyclone Precipitation from Satellite Passive Microwave Measurements. *J. Clim.* **2020**. (In Revision)

45. Black, M.L.; Gamache, J.F.; Marks, F.D.; Samsury, C.E.; Willoughby, H.E. Eastern Pacific Hurricanes Jimana of 1991 and Olivia of 1994: The effect of vertical shear on structure and intensity. *Mon. Weather Rev.* **2002**, *130*, 2291–2312. [CrossRef]

46. Corbosiero, K.L.; Molinari, J. The effects of vertical wind shear on the distribution of convection in tropical cyclones. *Mon. Weather Rev.* **2002**, *130*, 2110–2123. [CrossRef]
47. Corbosiero, K.L.; Molinari, J. The relationship between storm motion, vertical wind shear, and convective asymmetries in tropical cyclones. *J. Atmos. Sci.* **2003**, *60*, 366–376. [CrossRef]
48. Shapiro, L.J. The asymmetric boundary layer flow under a translating hurricane. *J. Atmos. Sci.* **1983**, *40*, 1984–1998. [CrossRef]
49. McTaggart-Cowan, R.; Galarneau, T.J.; Bosart, L.F.; Moore, R.W.; Martius, O. A global climatology of baroclinically influenced tropical cyclogenesis. *Mon. Weather Rev.* **2013**, *141*, 1963–1989. [CrossRef]
50. Wu, L.; Zong, H.; Liang, J. Observational analysis of tropical cyclone formation associated with monsoon gyres. *J. Atmos. Sci.* **2013**, *70*, 1023–1034. [CrossRef]
51. Berg, W.; Bilanow, S.; Chen, R.; Datta, S.; Draper, D.; Ebrahimi, H.; Farrar, S.; Jones, W.L.; Kroodsma, R.; McKague, D.; et al. Intercalibration of the GPM microwave radiometer constellation. *J. Atmos. Ocean. Technol.* **2016**, *33*, 2639–2654. [CrossRef]
52. Zhang, F.; Tao, D. Effects of vertical wind shear on the predictability of tropical cyclones. *J. Atmos. Sci.* **2013**, *70*, 975–983. [CrossRef]

Publisher's Note: MDPI stays neutral with regard to jurisdictional claims in published maps and institutional affiliations.

 *remote sensing*

Article

Gridded Satellite Sounding Retrievals in Operational Weather Forecasting: Product Description and Emerging Applications

Emily Berndt [1,*], Nadia Smith [2], Jason Burks [3], Kris White [4], Rebekah Esmaili [2], Arunas Kuciauskas [5], Erika Duran [6], Roger Allen [7], Frank LaFontaine [7] and Jeff Szkodzinski [2]

[1] NASA Marshall Space Flight Center/Short-Term Prediction Research and Transition Center, Huntsville, AL 35805, USA

[2] Science and Technology Corporation/National Oceanic and Atmospheric Administration Joint Polar Satellite System Proving Ground and Risk Reduction Program, Columbia, MD 21046, USA; nadias@stcnet.com (N.S.); rebekah@stcnet.com (R.E.); jeff.szkodzinski@stcnet.com (J.S.)

[3] National Weather Service/Cooperative Institute for Research of the Atmosphere/Meteorological Development Lab, Silver Spring, MD 20910, USA; jason.burks@noaa.gov

[4] National Weather Service/Short-Term Prediction Research and Transition Center, Huntsville, AL 35805, USA; kris.white@noaa.gov

[5] United States Naval Research Laboratory, Marine Meteorology Division, Monterey, CA 93943, USA; arunas.kuciauskas@nrlmry.navy.mil

[6] Atmospheric Science Department, Earth System Science Center/Short-Term Prediction Research and Transition Center, University of Alabama in Huntsville, Huntsville, AL 35805, USA; erika.l.duran@nasa.gov

[7] Jacobs Space Exploration Group/Engineering Services and Science Capability Augmentation/Short-Term Prediction Research and Transition Center, Huntsville, AL 35805, USA; roger.e.allen@nasa.gov (R.A.); frank.j.lafontaine@nasa.gov (F.L.)

* Correspondence: emily.b.berndt@nasa.gov

Received: 27 August 2020; Accepted: 9 October 2020; Published: 12 October 2020

Abstract: The National Aeronautics and Space Administration (NASA) Short-term Prediction Research and Transition Center (SPoRT) has been part of a collaborative effort within the National Oceanic and Atmospheric Administration (NOAA) Joint Polar Satellite System (JPSS) Proving Ground and Risk Reduction (PGRR) Program to develop gridded satellite sounding retrievals for the operational weather forecasting community. The NOAA Unique Combined Atmospheric Processing System (NUCAPS) retrieves vertical profiles of temperature, water vapor, trace gases, and cloud properties derived from infrared and microwave sounder measurements. A new, optimized method for deriving NUCAPS level 2 horizontally and vertically gridded products is described here. This work represents the development of approaches to better synthesize remote sensing observations that ultimately increase the availability and usability of NUCAPS observations. This approach, known as "Gridded NUCAPS", was developed to more effectively visualize NUCAPS observations to aid in the quick identification of thermodynamic spatial gradients. Gridded NUCAPS development was based on operations-to-research feedback and is now part of the operational National Weather Service display system. In this paper, we discuss how Gridded NUCAPS was designed, how relevant atmospheric fields are derived, its operational application in pre-convective weather forecasting, and several emerging applications that expand the utility of NUCAPS for monitoring phenomena such as fire weather, the Saharan Air Layer, and stratospheric air intrusions.

Keywords: NUCAPS; satellite soundings; weather forecasting; operational applications; retrievals; infrared; CrIS; severe weather; fire weather; tropical weather; stratospheric intrusions

1. Introduction

The National Oceanic and Atmospheric Administration (NOAA) Joint Polar Satellite System (JPSS) Proving Ground and Risk Reduction (PGRR) Program has fostered the development and application of satellite sounding retrievals for the benefit of end users though a "Sounding Initiative" and competitively funded projects. The National Aeronautics and Space Administration (NASA) Short-term Prediction Research and Transition Center (SPoRT; [1]) has been part of this effort since 2014, contributing expertise associated with their research-to-operations/operations-to-research paradigm. As a result of multi-organizational/multi-agency collaborations within the JPSS PGRR Sounding Initiative, hyperspectral infrared satellite sounding retrievals are contributing to operational weather forecasting in novel ways that were not anticipated two decades ago when the first hyperspectral infrared sounder was launched on Aqua in 2002. The implementation of the NOAA Unique Combined Atmospheric Processing System (NUCAPS; [2–4]) soundings in the United States NOAA National Weather Service (NWS) operational environment inspired much of the work within the JPSS PGRR Sounding Initiative, including the product design and applications that we discuss in this paper. The structure of the level 2 environmental data records, as arrays of vertical soundings, has limited the availability and accessibility of NUCAPS-derived products to assess these observations in plan-view for spatial context and has limited their widespread application for short-term weather forecasting. While a few previous studies have developed and demonstrated the feasibility of level 2, plan-view hyperspectral infrared sounder products [5,6] for convective forecasting, these capabilities have not been widely adopted into operational NUCAPS algorithm processing. Although level 3 gridded products are routinely produced and available as standard NUCAPS products, there has been a gap in the development of gridded, level 2 products or standardized approaches to support short-term forecasting/analysis. In addition, the derivation of more specialized fields beyond basic temperature, moisture, and trace gases has traditionally not been produced due to the lack of standard approaches to easily process and derive level 2 products. A new method and concept for the processing and representation of NUCAPS level 2-derived products is presented here. This work represents the development of approaches to better synthesize remote sensing observations that ultimately increase the availability and usability of NUCAPS observations to benefit scientific analysis and applications. The optimization of basic gridding and interpolation methodologies as appropriately applied to NUCAPS data retains their observational characteristics and enables state-of-the-art product development to further support their application in weather analysis and forecasting, allowing the capability to add or develop new derived products easily. The derived products presented herein, represent the novel development of fields not traditionally derived from hyperspectral infrared sounder observations and new concepts/methods to support applications related to short-term weather forecasting and analysis.

The NUCAPS retrieval system is based on version 5.9 of the NASA Atmospheric Infrared Sounder (AIRS) science team method [7] and runs operationally at NOAA with global coverage in near real-time (~180 min latency) and via direct broadcast sites with regional coverage in real-time (<60 min latency). By "operational", we mean that the system runs continually on every measurement made from space. While NUCAPS has the capability to retrieve soundings from AIRS measurements, it runs operationally at NOAA on measurements made by the Cross-track Infrared Sounder (CrIS), in orbit since 2011 on two different platforms, as well as the Infrared Atmospheric Sounding Interferometer (IASI), in orbit since 2006 on a series of European Meteorological Operational (MetOp) satellite platforms. On any given day at a target scene, there are thus multiple NUCAPS soundings available throughout the diurnal cycle to support any number of applications. Here, we introduce the novel NUCAPS product, known as "Gridded NUCAPS", and the applications it supports. We distinguish between operational applications with a known user-base in weather forecasting and emerging applications with demonstrated relevance to weather forecasting.

NUCAPS sounding products are operationally available to the NOAA weather forecasting community through the Advanced Weather Interactive Processing System (AWIPS) that ingests and displays data products from a wide array of sources to support weather analysis and forecasting.

In 2014, NUCAPS soundings were officially delivered to the NWS AWIPS system operationally, and for the first time forecasters could visualize hyperspectral infrared sounding observations as "Skew-T" diagrams, or thermodynamic plots of temperature and dewpoint profiles. This is also how forecasters view radiosondes, so comparisons between these two sources are easy and intuitive. With thousands of satellite soundings supplementing the sparse radiosonde network, forecasters suddenly had ready access to wide swaths of satellite soundings that helped them characterize the pre-convective environment, when radiosondes are typically sparse or absent [6,8–12]. To date, NUCAPS remains the only NOAA operational sounding product from hyperspectral infrared measurements and the only product of its kind in AWIPS. With active partnerships in the JPSS PGRR Sounding Initiative and this new data source available to NWS Weather Forecast Offices (WFOs) within the United States (including Alaska, Hawaii, and Puerto Rico), forecasters started applying NUCAPS soundings to different forecasting scenarios, such as the cold air aloft aviation hazard described by Weaver et al. [13]. It was this novel application in aviation weather forecasting that inspired the design of Gridded NUCAPS, which allowed forecasters to visualize incoming swaths of NUCAPS soundings as horizontal or vertical cross-sections, instead of individual soundings one Skew-T diagram at a time. With Gridded NUCAPS, Alaskan forecasters could readily determine the spatial and vertical extent of cold-air aloft features and thus speed up their issuing of warnings to the aviation community. The methodology to ingest and display satellite soundings as a series of values at different pressure levels, instead of vertical profiles, was first developed by [5] and later refined by this team through an iterative process involving end user assessment and feedback [11,13–15] with the current method described below. As a result of operations-to-research feedback and collaborative efforts within the JPSS PGRR Sounding Initiative, AWIPS now has the operational ability to display NUCAPS soundings not only as Skew-T diagrams, but also as plan-views and cross-sections of the three-dimensional atmosphere through the Gridded NUCAPS capability.

Gridded NUCAPS has operational applications in severe weather forecasting because with overpasses from CrIS at 01:30 pm local time, it characterizes the summertime, peak afternoon pre-convective environment with observations between typical radiosonde launches that forecasters can use to evaluate forecast models ahead of afternoon thunderstorms. With Gridded NUCAPS, weather forecasters can visualize horizontal swaths of the retrieved sounding observations at different heights and quickly identify areas of convective instability. Gridded NUCAPS has been evaluated within AWIPS by operational forecasters at the Hazardous Weather Testbed (HWT) annually since 2016 to determine its relevance and applicability and refine its quality [8–11,15–17]. The HWT is one of a number of NOAA test beds [18] designed to facilitate a link between researchers and operational forecasters. Esmaili et al. [11] discussed how our partnership with the NWS through the JPSS PGRR program ensures an effective flow of information between the research and operational communities. There is the "research-to-operations" flow that helps to make science operationally available to decision-makers and the "operations-to-research" flow that inspires operationally relevant research and products tailored to operational applications. The Gridded NUCAPS capability, with its emerging applications that we discuss in this paper, is a testimony to the value of this partnership and flow of information.

The main aim of this work is to highlight the Gridded NUCAPS product design and discuss several emerging applications within the NOAA operational environment and beyond. These new applications are an opportunity for research to have value in operations and, in turn, for operations to inform research and product improvement. Section 2 describes the datasets and methodology we implemented to project the NUCAPS soundings from their instrument grid to a standard latitude/longitude grid (0.5° resolution at a fixed set of vertical levels). Section 3 highlights one operational application—namely, surveilling the pre-convective environment—and three emerging applications, including fire weather analysis, monitoring the Saharan Air Layer (SAL), and identifying stratospheric air influence and tropopause folding. The latter was first conceptualized and demonstrated by [19–21] for the AIRS version 6 suite of products. Section 4 is a discussion of the significance of this work, while the manuscript is concluded in Section 5.

2. Materials and Methods

2.1. Datasets: NUCAPS Satellite Soundings

We focus here on the NUCAPS retrieved profiles of temperature, moisture, and ozone from CrIS and the Advanced Technology Microwave Sounder (ATMS) on the Suomi-National Polar-orbiting Partnership (S-NPP) and NOAA-20 platforms. NUCAPS is based on the AIRS version 5.9 algorithm [7]. S-NPP NUCAPS soundings were made available to the NWS in 2014 through the satellite broadcast network. Today, only NOAA-20 soundings are made available to the NWS, since the S-NPP CrIS side-b electronics anomaly during 2019 impacted the availability of S-NPP NUCAPS for a short period of time, and its feed into AWIPS was shut off as a result. Although much of this work depends on the real-time delivery of NUCAPS soundings to the NWS, the examples in this work utilize both NOAA-20 and S-NPP NUCAPS data obtained from the NOAA Comprehensive Large Array Stewardship System (CLASS), either reprocessed for AWIPS display or processed and displayed with the Gridded NUCAPS stand-alone python code base.

For use in real-time forecasting, the NUCAPS algorithm was designed to achieve high-quality profiles across the globe, generate traceable error estimates, and maintain a high computational efficiency. NUCAPS is an optimal estimation retrieval system [22]. Optimal estimation is a method employed in many other retrieval systems also [23–27] that adds information from the radiance measurements to an estimate of the atmospheric state (known as the a-priori, or first guess), while propagating error estimates from both sources to the final solution. This technique has been widely adopted because the space-based radiance measurements do not contain enough information to fully resolve the vertical atmospheric state at every retrieval footprint, and an a-priori estimate helps stabilize the solution. An optimal estimation temperature retrieval can, thus, be interpreted as an improvement in prior assumptions about atmospheric temperature based on measurements from space. One can use any number of data sources to function as an a-priori, as seen in these systems [23–27]. NUCAPS calculates an a-priori for temperature, moisture, and ozone by applying regression coefficients to the CrIS/ATMS measurements. These coefficients are calculated off-line as the correlation between four global days of CrIS/ATMS measurements and co-located atmospheric state variables from the European Centre for Medium-range Weather Forecasts (ECMWF) reanalysis model. Even though these regression coefficients have a model dependence, one can regard the regression retrievals from radiance measurements to have a minimal dependence on forecast models because most of the information about the instantaneous atmospheric state is derived from the radiances themselves. NUCAPS uses a linear regression approach, as described by [28], though other approaches exist here [29–33]. Operational meteorologists value the fact that NUCAPS soundings are largely model-independent, because this allows them to verify forecast models in real-time.

NUCAPS is a multi-step retrieval system that we will not describe in-depth, but it is worth highlighting how NUCAPS retrieves soundings in cloudy atmospheres, because this has direct relevance to discussions here. NUCAPS uses a technique called "cloud clearing" [7,24,34] to derive a cloud-free radiance estimate from each cluster of 9 CrIS radiances (3 × 3 fields of view). This technique is a simple, robust means with which to remove the effects of clouds from the measured radiances without prior knowledge of clouds or the requirement for complex radiative transfer calculations through clouds. With cloud clearing, NUCAPS retrievals in partly cloudy scenes can be interpreted as the state of the atmosphere around or past the clouds, not through the clouds. Cloud clearing does reduce the spatial resolution of NUCAPS retrievals, since a sounding is retrieved for every aggregate of 9 CrIS fields of view (~50 km at nadir and ~150 km at the edge of the scan), but it significantly increases the retrieval yield to a ~75% success rate from a global set of measurements and allows sounding observations in complex, partly cloudy scenes to characterize the environment within storms. Another aspect of NUCAPS is that it retrieves temperature and moisture twice—first from a microwave-only (MW-only) set of ATMS channels [25,35,36], and second from a set of infrared plus microwave (IR + MW) channels [2,4]. Both MW-only and IR + MW retrievals are part of the NUCAPS product, but only the latter is available in AWIPS. The MW-only retrievals contribute to evaluating whether the IR + MW retrievals failed or succeeded.

2.2. Methods: Gridded NUCAPS Product Design

The current Gridded NUCAPS capability was released in the AWIPS baseline distribution in 2019. AWIPS is the primary visualization and decision-support platform for the NWS WFOs. "Baseline" means that the same configuration and software capability is distributed to all WFOs within the United States. This gives each WFO the ability to generate the same Gridded NUCAPS products from the real-time flow of satellite data into AWIPS. The Gridded NUCAPS capability is still under active development to refine the initial AWIPS capability and to create a robust code base for processing gridded sounding products for real-time web-based visualizations for non-AWIPS users and to support applied research and validation studies.

Before the horizontal grids are created, a vertical interpolation is independently applied to each sounding to interpolate the data to standard pressure levels (P_{std}). In Gridded NUCAPS, the 100 native NUCAPS levels are transformed to standard meteorological levels in the AWIPS operational environment for inter-comparison with other data sets, such as models and radiosondes. The set of standard gridded levels we defined are 41 levels from 1100 to 100 hPa every 25 hPa to match the NWP models and enable easier comparison in AWIPS. The observations are interpolated from the Earth's surface to 100 hPa with linear interpolation. In NUCAPS, trace gases are retrieved on pressure layers and temperature on pressure levels. Trace gas quantities, such as water vapor and ozone, must first be converted from a layer quantity to a level quantity. The conversion takes the midpoint between two layer quantities to calculate the level quantity, where $V_{lev,i}$ in Equation (1) represents the trace gas variable such as water vapor or ozone and i represents the index of the native 100 NUCAPS pressure levels, where $i = 1$ and $i = 100$ are at the top and bottom of the atmosphere, respectively:

$$V_{lev, \, i} = \begin{cases} \frac{V_i + V_{i-1}}{2} & i > 1 \\ V_1 & i = 1 \end{cases}. \tag{1}$$

Note that index i may be less than 100 for soundings where the topography is higher and surface pressure is lower than 1100 hPa. This methodology is described below. Separate functions (Equations (2) and (3)) are used to interpolate temperature to standard levels ($T_{std,j}$) compared to water vapor and ozone ($V_{std,j}$). To preserve the mass, water vapor and ozone are linearized by interpolating the standard logarithm of the column density. Below, j represents the index of the 41 standard pressure levels where $P_{i-1} \leq P_{std,j} \leq P_i$. Like the index i: $j = 1$ and $j = 41$ are at the top and bottom of the atmospheric column, respectively:

$$T_{std, \, j} = T_{i-1} + \left(T_{std, \, j} - T_{i-1}\right) \times \frac{P_{std,j} - P_{i-1}}{P_i - P_{i-1}}, \tag{2}$$

$$V_{std,j} = \left[\log_{10}(V_{i-1}) + \left(\log_{10}(V_{std, \, j}) - \log_{10}(V_{i-1})\right) \times \frac{\log_{10}(P_{std,j}) - \log_{10}(P_{i-1})}{\log_{10}(P_i) - \log_{10}(P_{i-1})} \right]^{10}. \tag{3}$$

Figure 1a is an example of the impact of the vertical interpolation on the sounding. There are only slight differences between the resampled profile (blue) and the original sounding (pink). With the cold air aloft aviation hazard in mind, critical temperatures < −65 °C (gray shading in Figure 1b,c) are still identified in the vertically interpolated sounding, with only a 10 hPa difference between the bottom of the cold air aloft layer when comparing the interpolated and native NUCAPS temperature. Based on forecaster feedback, the slight differences in the 250–200 hPa layer are not significant enough to impact the integrity of the sounding or drastically change decisions related to forecasting applications.

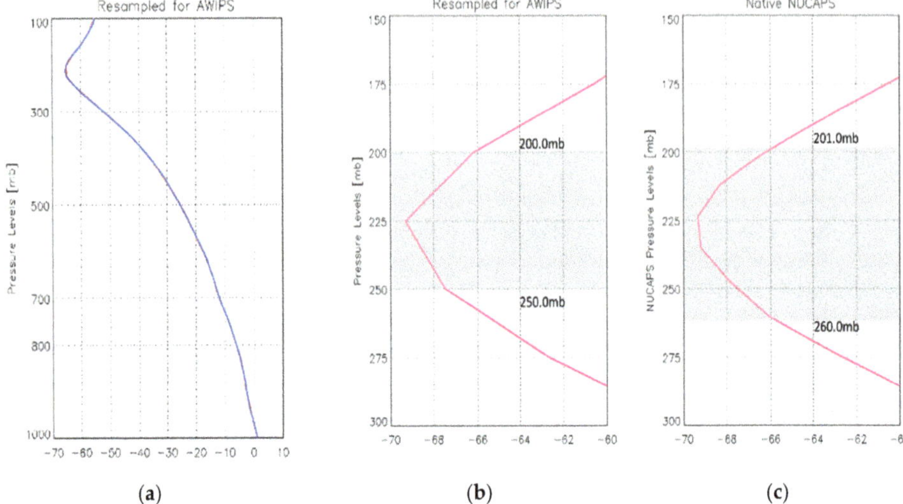

Figure 1. Example of the vertical interpolation compared to the native National Oceanic and Atmospheric Administration (NOAA)'s NOAA Unique Combined Atmospheric Processing System (NUCAPS) resolution. (**a**) NUCAPS vertical temperature sounding plotted on native 100 levels in blue overlaid on NUCAPS temperature interpolated to 41 standard levels (pink). Comparison of the 300–150 hPa upper-level region between (**b**) temperature interpolated to 41 vertical levels, and (**c**) native NUCAPS 100 levels. The gray region represents the region of the sounding $<-65\,^\circ$C, the criteria for identifying the potential for the cold air aloft aviation hazard.

The horizontal gridding is performed on temperature, relative humidity, and additional derived fields. Each array of aggregated soundings is added to a 0.5° latitude/longitude grid over a global domain using nearest neighbor and minimal interpolation. Regions outside the swath are masked where data are unavailable before the gridding takes place. Horizontal fields are created for temperature and relative humidity on 41 standard levels and at the surface (e.g., 2-m), quality flags, and derived single layer products: total precipitable water (TPW) and layer precipitable water (LPW), total ozone, ozone anomaly, and tropopause level are also gridded. In AWIPS, the data are output as a grid record and made available for display. The derived parameters in AWIPS are leveraged to calculate and display additional fields such as lapse rates, theta-e/theta-e lapse rates, Haines Index, and other stability parameters derived from temperature and moisture. The derived parameters are baseline python functions in AWIPS that perform calculations on model and even satellite data to "derive" fields for display. Given that Gridded NUCAPS is ingested as a grid record, akin to model data, any derived parameter that uses temperature or moisture fields for its derivation can be calculated and displayed. Therefore, a wide array of display fields are available through AWIPS-derived parameters. Some fields derived by AWIPS such as stability indices that rely on levels within the boundary layer still need further evaluation for accuracy and efficacy. The specific variables and levels/layers presented here were chosen based on operations-to-research feedback gathered during annual participation in the HWT spring experiment [11,16,17].

Ideally, prior to vertical interpolation and horizontal gridding, the bottom of each sounding should be found based on comparing the surface pressure to the NUCAPS pressure levels. Correctly adjusting the surface and boundary layer conditions according to local changes in topography and surface pressure benefits the interpretation of satellite soundings and prevents the propagation of systematic uncertainty in derived geophysical variables (i.e., lapse rate, stability indices). Note that NUCAPS sounding files include 100 levels from 1100 hPa (P_{100}) up to top of the atmosphere (0.0016 hPa; P_1), and 1100 hPa is often below the Earth's surface and unrepresentative of actual conditions. In the event that the Earth's surface is higher than P_{100}, the remainder of the pressure grid is filled in with values identical to surface temperature, thus creating an isothermal profile below the surface. The technique outlined in Figure 2 removes any isothermal layer from the NUCAPS sounding and correctly assigns the bottom level. This technique to find the bottom portion of the sounding is implemented in the current AWIPS capability. This technique can be taken one step further to adjust the boundary layer temperature and moisture values in the sounding. The boundary layer multiplier (BLMULT; Equation (4)), can be calculated to either narrow or broaden the boundary layer to within 0.2 to 1.2 hPa. Then, a representative fraction of the temperature or moisture can be added or removed from the bottom of the sounding. Since the NUCAPS level closest to the surface pressure will never be an exact match, BLMULT can account for this discrepancy. BLMULT is calculated by:

$$BLMULT = \frac{P_{surf} - P_{botlev-1}}{P_{botlev} - P_{botlev-1}}, \tag{4}$$

where P is the array of 100 NUCAPS pressure levels, surface pressure (P_{surf}) is obtained from the Global Forecast System as part of the NUCAPS algorithm, and botlev is the bottom-level pressure index found using one of the three conditions in Figure 2. Then, the surface temperature (T_{surf}) is calculated by Equation (5) as follows:

$$T_{surf} = T_{botlev-1} + BLMULT \times [T_{botlev} - T_{botlev-1}]. \tag{5}$$

Note that Equation (5) is modified to calculate the surface relative humidity in the same manner. For the total column fields such as ozone and total precipitable water (V_{tot}), BLMULT is applied to the concentration density at the bottom level (V_{botlev}) and added to the total column:

$$V_{tot} = BLMULT \times V_{botlev} + \sum_{i=1}^{botlev-1} V_i. \tag{6}$$

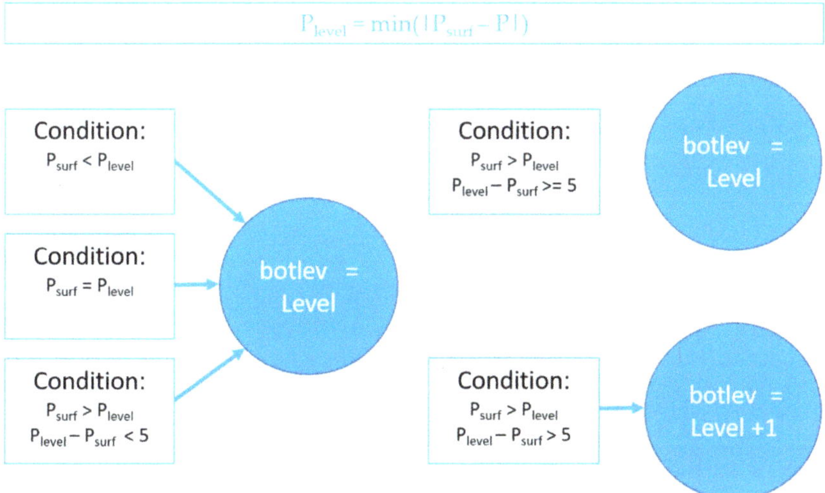

Figure 2. The conditions for finding the index of the bottom level (botlev) in a NUCAPS sounding. Level is the index of the pressure level satisfying $min(|P_{surf}-P|)$. The index botlev is required for accurately calculating the temperature and trace gas surface values.

Figure 3 is an example application of the surface adjustment and BLMULT to a NUCAPS sounding. With a surface pressure of 1029 hPa, the bottom of the sounding is the 1042 hPa NUCAPS level and BLMULT represents an expansion of the boundary layer by 0.5576 hPa. Note that the isothermal layer starting at 1042 hPa and downward is below the topography. The BLMULT can then be applied to temperature and moisture to adjust the surface value. In this case, the fraction of temperature within the bottom layer is added to the temperature at the specified level (258.318 K at 1013 hPa) for a new surface temperature of 257.774 K. BLMULT was only implemented within TPW and is the lowest LPW field in the Gridded NUCAPS. Active development is underway to fully implement BLMULT in the second iteration of the AWIPS capability and the non-AWIPS visualizations. Currently, the surface or 2-meter temperature and relative humidity are found according to Figure 1 after the data are interpolated to standard levels, but BLMULT is not applied. Note that the isothermal layer is removed before vertical interpolation, but BLMULT was not fully implemented due to the complexity of developing the initial AWIPS plugin. The newer version will use BLMULT to adjust the temperature and moisture of the sounding to find 2 m fields and will apply BLMULT prior to performing any vertical interpolation. Active development is underway to test this with non-AWIPS processing and integrate it in updated AWIPS code.

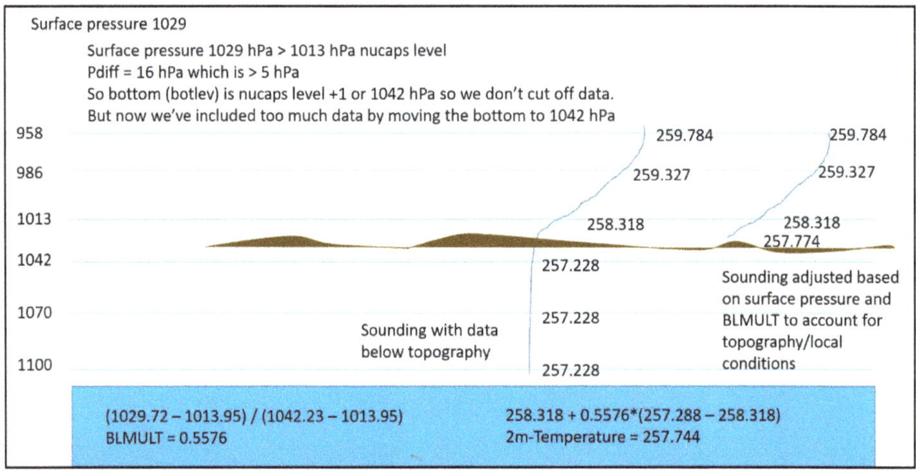

Figure 3. Example of finding the bottom of a sounding, calculation of boundary layer multiplier (BLMULT), and deriving the surface temperature.

2.3. Methods: Gridded NUCAPS-Derived Fields

2.3.1. Lapse Rate

For AWIPS users, the lapse rate is calculated with AWIPS-derived parameters according to the Poisson Equation (Equation (7)). The constant is the result of the division of gravity 9.81 m/s by the gas constant for dry air (287 J/kg·K). T (P) and T_0 (P_0) represent temperature (pressure) at the top and bottom of the layer, respectively. The AWIPS menu includes commonly used lower-level and upper-level lapse rates based on feedback from users to promote ease of access. Less commonly used lapse rates are available through the AWIPS product browser. For non-AWIPS tools which do not automatically compute the lapse rate, the lapse rate is pre-calculated for visualization of the 850–500, 700–500, and 400–200 hPa layers, and additional lapse rate calculations can be flexibly added for processing and display.

$$LR = 0.034167 \times \left[\log \frac{T}{T_0} / \log \frac{P}{P_0} \right]. \tag{7}$$

2.3.2. Haines Index

The Haines Index was first described by [37] and further defined by [38], and is calculated with two terms representing stability and moisture. The stability term is assigned a value, 1–3, based on the lapse rate of the identified layer, which is also calculated with Equation (7). The moisture term is assigned a value of 1–3 based on the dew point depression of the defined level. The two values are added to indicate the potential for large fire growth (e.g., 2–3 = very low, 4 = low, 5 = moderate, 6 = high). Werth and Ochoa [38] give suggested layers/levels to derive the Haines Index to account for topography and reduce the influence of the diurnal variability in the surface temperature and associated surface inversions. The AWIPS-derived parameters calculate the Haines Index given the temperature and relative humidity of the NUCAPS grids. These layers/levels can be adjusted in the AWIPS Haines Index-derived parameter. Current development with the Gridded NUCAPS non-AWIPS visualizations includes the derivation of the Haines Index at the suggested layers/levels based on [38].

2.3.3. Precipitable Water

The derivation of precipitable water (TPW and LPW) was included in the initial AWIPS gridding capability. The TPW and LPW represent the water vapor contained in a vertical column of unit cross-sectional area extending between any two specified levels and is expressed in terms of the height the water would stand if completely condensed into the same unit area, as expressed in Equation (8):

$$TPW = \frac{MW_{H2O}}{N_a} \times \sum_{i=sfc}^{toa} WV_{cd}(i).$$ (8)

The water vapor column density is integrated from the top to bottom level following Equation (8), and then multiplied by the molecular mass of water vapor (MW_{H2O}, 18.0151 g/mol) and divided by Avogadro's number (N_a, 6.02214199 × 10^{23}), yielding a value in cm. Precipitable water is calculated over three additional layers (surface-800 hPa, surface-500 hPa, or surface-300 hPa) in the initial version of the AWIPS implementation, with plans to adjust the layer calculations based on user feedback. In AWIPS, forecasters can view TPW or LPW in cm, m, or inches, depending on user preference and editing user configuration files. BLMULT is applied to adjust the bottom of the TPW and LPW fields. Future AWIPS implementation of LPW will include the derivation of the products with the moisture interpolated to standard levels and LPW calculated over familiar layers similar to other satellite-derived PW products (e.g., surface-850, 850–700, 700–500, and 500–300 hPa). The current Gridded NUCAPS web-visualizations and examples below derived from non-AWIPS code include the new LPW layers.

2.3.4. Ozone-Derived Products

As a result of end user feedback within the JPSS PGRR program, several derived products were included in the Gridded NUCAPS development. Previous work by [19] and [20,21] led to the development of ozone-derived products from hyperspectral infrared sounders to support forecasting rapid cyclogenesis and the development of associated high winds and hurricane extratropical transition. The total column ozone is calculated from the ozone mixing ratio and converted to Dobson Units for gridding and display. The ozone anomaly product was developed to identify regions of climatologically high ozone, indicating the presence of stratospheric air and the potential for tropopause folding [21]. The total column ozone is compared to a latitudinal and monthly climatology database developed by [39] to characterize anomalous ozone values. With the knowledge that stratospheric air can be identified where ozone values are at least 25% greater than climatology [40], the percent of normal between 0% and 200% is calculated and displayed with values 125% or greater in shades of blue. The full product derivation and examples are outlined in [21]. The tropopause level product was created as an innovative method of identifying the tropopause in satellite soundings. Since it can be difficult to ascertain the tropopause height by analyzing vertical temperature and moisture profiles due to the smooth nature of satellite soundings, the use of gridded-plan view ozone products is advantageous. Ozone can be used to identify the height of the tropopause; however, the use of threshold values such as 100 ppb can be misleading due to the seasonal changes in ozone and the tropopause height. Thouret et al. [41] developed a seasonal variation in ozone at the dynamic tropopause, defined as 2 Potential Vorticity Units, using flight observations and model data. The study resulted in the following equation, which is a synthetic definition of the monthly mean climatological ozone value at the tropopause that accounts for the sine seasonal variation with a maximum in May and minimum in November:

$$91 + 28 \times sin(\pi \times (Month - 2)/6. \tag{9}$$

With the NUCAPS soundings, the tropopause level is found by matching the level where the ozone value is greater than or equal to the monthly threshold determined by Equation (9) from Thouret et al. [41]. The tropopause level in hPa is then gridded for display.

3. Results

3.1. Surveilling the Pre-Convective Environment

During the 2019 HWT Spring Experiment, NUCAPS soundings were used in the analysis of convection that developed in central Illinois on 5 June [42,43]. A line of storms developed in southern Iowa, and moved southeast into central Illinois by 1600 UTC. Figure 4a shows the Gridded NUCAPS TPW values around 30 mm over the region, while closer analysis of 700–500 hPa LPW indicates a drier layer in southern Illinois. This same dry signature is also evident in the 700 hPa relative humidity. Although not shown, the near-surface LPW and relative humidity fields indicate a relatively moist near-surface environment. The storms developed along a swath of regionally higher LPW and relative humidity, and increased in intensity during the afternoon hours, before decreasing in overall intensity after moving into the environment with drier air in the mid-levels (e.g., around 2200 UTC, approximately 3.5 h after the NOAA-20 overpass). The Storm Prediction Center storm reports indicate that most of the wind damage associated with the line occurred between 2030 and 2220 UTC [44]. This analysis shows the advantage of plan-view analysis to assess the environment, especially with more reliable fields that are above the boundary layer influence. Interrogating individual NUCAPS profiles can provide valuable temperature and moisture measurements, especially above the boundary layer. However, because soundings are volume measurements and not point observations, near the surface the soundings may underestimate important stability indices or features such as inversions when compared with radiosondes. In the June 5 case, the forecaster found NUCAPS vertical sounding Convective Available Potential Energy (CAPE) values were underestimated in the low to mid-levels when compared to the immediate Lincoln, Illinois sounding, which was valid at 1700 UTC [42]. The gridded fields allow the end user to assess the broad environment quickly and above the boundary layer, with a focus on changes in gradients and patterns.

These activities have led to valuable operations-to-research feedback from end users to tailor products to address the needs of the operational environment. One key area of active research is addressing the representation of the boundary layer in satellite soundings. Forecasters need the accurate representation of surface temperature, moisture, and structures such as inversion layers to diagnose the potential for convective development. The representation of temperature and moisture fields is also necessary, since they are used to derive common stability fields such as CAPE, important for diagnosing convective potential and storm-scale updraft strength. Manual and automated techniques have been applied to improve the boundary layer representation of satellite soundings and have been accepted by forecasters as an improvement in the utility of these data in operations [11,45]. There are ongoing efforts by NUCAPS developers to improve the boundary representation within the retrieval algorithm and as a post-processing step within target applications. An in-depth discussion of these efforts is, however, beyond the scope of this paper.

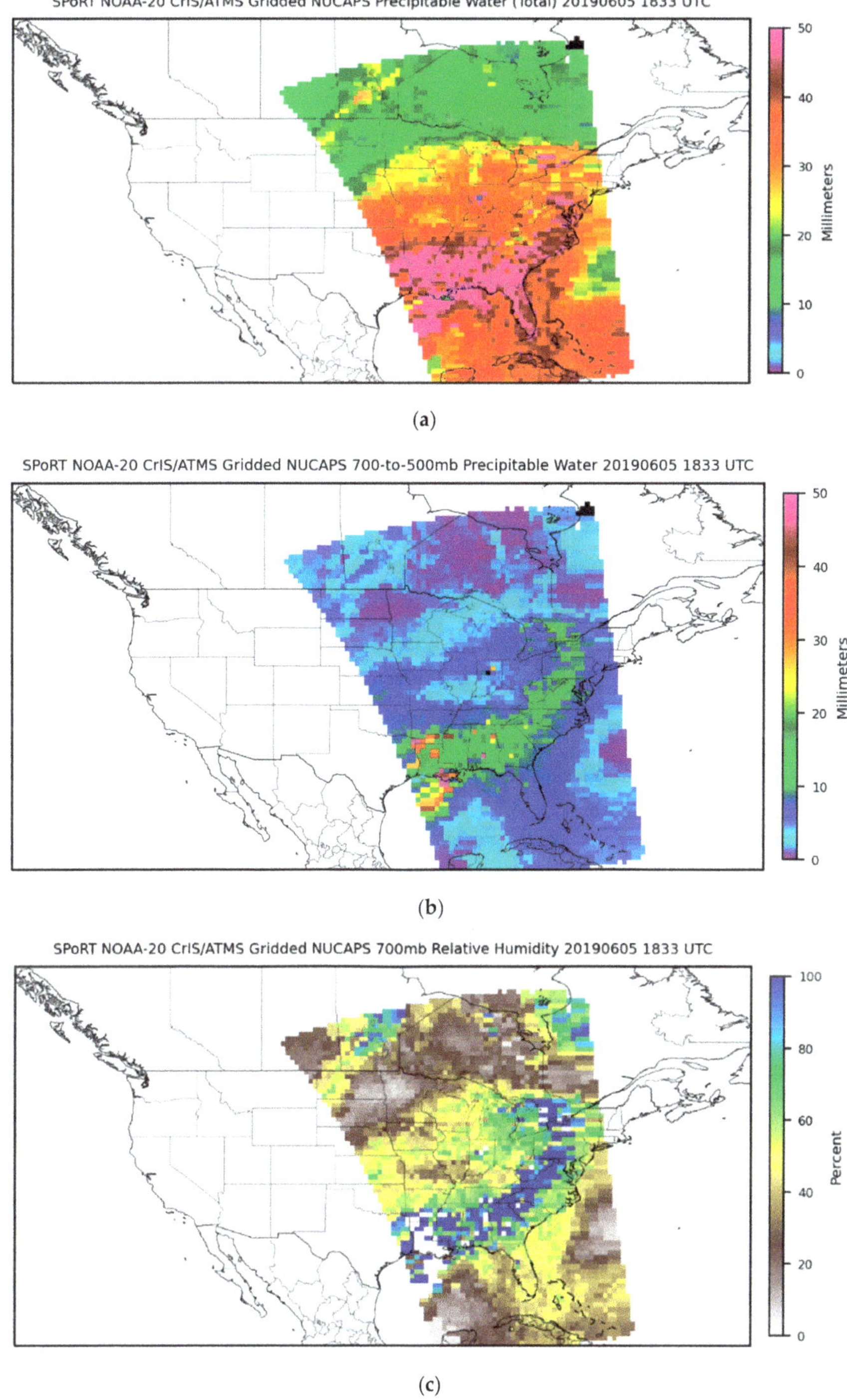

Figure 4. NOAA-20 Gridded NUCAPS on 5 June 2020, 1833 UTC (**a**) total precipitable water (TPW), (**b**) 700–500 hPa layer precipitable water (LPW), and (**c**) 700 hPa relative humidity.

3.2. Fire Weather Analysis

Fire weather is an emerging application to utilize NUCAPS soundings and gridded products to diagnose the thermodynamic characteristics of the environment conducive to the potential for wildfire development and growth, as well as tracking smoke [46,47]. Lindley et al. [48] provide an overview of the common meteorological features associated with wildfires in the southern Great Plains, notably the development of low-level thermal ridges (LLTR). The example presented below highlights the ability of Gridded NUCAPS products to capture the LLTR associated with the 2018 Rhea, Oklahoma fire. In addition, the derived parameters in AWIPS allow for the derivation of the Haines Index.

The Rheafire started around 12 April 2018 and burned approximately 285,196 acres [49]. The region was experiencing an extreme drought and on this particular day a dry line was positioned to the east and an LLTR developed. Lindley et al. [48] suggest the analysis of fields such as mean sea level pressure, 2-meter temperature and relative humidity, 850 hPa temperature, and 500 hPa height to identify the LLTR. Gridded NUCAPS fields from both the S-NPP 1845 and 2025 UTC overpasses can be combined and compared to the 2000 UTC Rapid Update (RAP) model data (Figure 5). The level of 700 hPa was chosen to view data above the influence of topography, as some missing values were apparent at 850 hPa over the Rocky Mountains, impeding broad synoptic analysis. The 2 m temperature and relative humidity fields from NUCAPS indicate warm (25–30 °C temperatures) and dry (10–20% relative humidity) conditions in western Oklahoma (Figure 5a,b). The 10 m RAP winds indicate that these warm, dry conditions are being advected into the region. Analysis of the 700 hPa temperature field indicates the thermal ridge axis over the region (Figure 5c). This feature identified in the Gridded NUCAPS is consistent with the RAP model (Figure 5d), and the RAP 500 hPa height is consistent with the expected pattern of an LLTR.

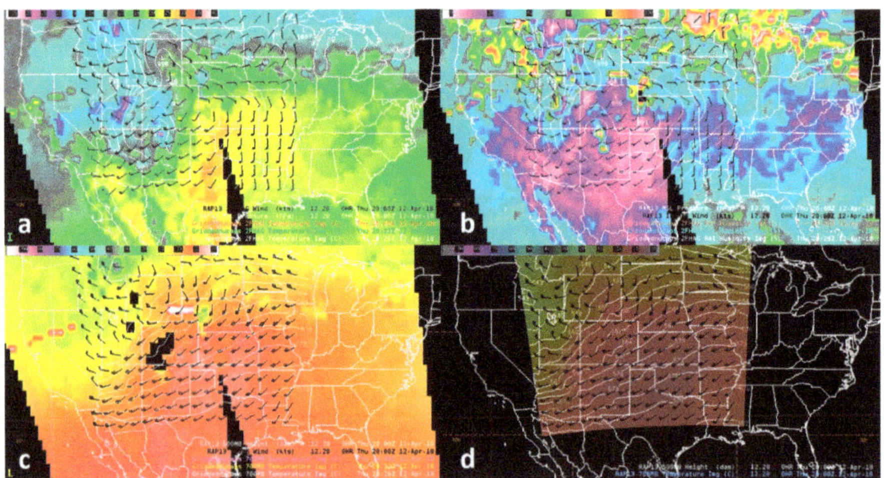

Figure 5. Advanced Weather Interactive Processing System (AWIPS) display of the 12 April 2018 Suomi-National Polar-orbiting Partnership (S-NPP) Gridded NUCAPS 1839 and 2021 UTC overpass and Rapid Update (RAP) model 2000 UTC analysis. (**a**) Gridded NUCAPS 2 m temperature, RAP surface wind, and mean sea level pressure; (**b**) Gridded NUCAPS 2 m relative humidity, RAP surface wind, and mean sea level pressure; (**c**) Gridded NUCAPS 700 hPa temperature and RAP 500 hPa wind and height; (**d**) RAP 700 hPa temperature, 500 hPa wind and height. Note that the AWIPS regional localization prevents the display of RAP on a full conus domain.

On 13 April, these same features continued to persist. The surface thermodynamic fields (Figure 6a,b) reveal the continued persistence of warm, dry conditions, and the well-defined LLTR visually agrees with the RAP analysis (Figure 6c,d). In addition, the Gridded NUCAPS Haines Index did indicate a broad region (orange) of high potential for large fire growth, consistent with the RAP analysis. The Haines Index, calculated with the 850–700 hPa lapse rates and 850 hPa dew point depression, depicted a region of high potential for fire growth over western Oklahoma, but with an axis shifted to the east compared to the RAP model (Figure 7a,b). The Gridded NUCAPS thermodynamic fields and the derived Haines Index demonstrate the application of Gridded NUCAPS to increase the situational awareness of fire weather conditions and the potential for fire growth. The combination of Gridded NUCAPS fields, supplemented by additional model fields such as wind and height, are demonstrated as a viable dataset for the identification of an LLTR. The Gridded NUCAPS fields can provide observations between model runs and are a model-independent observational dataset to confirm model features such as patterns and gradients.

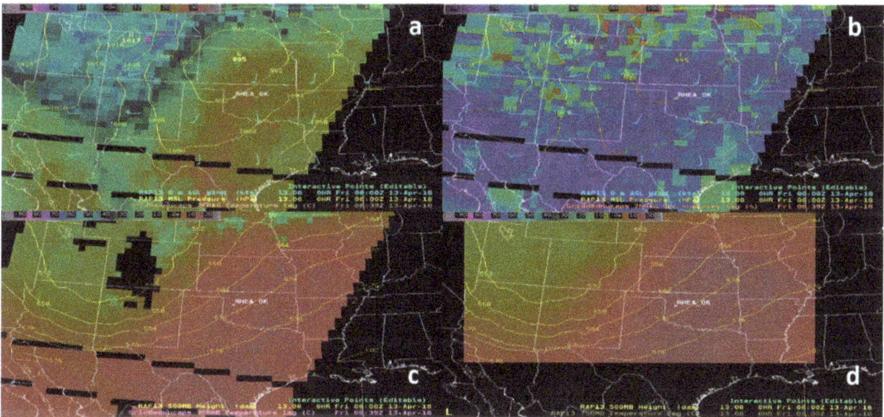

Figure 6. AWIPS display of 13 April 2018 S-NPP Gridded NUCAPS 0839 UTC overpass and RAP 0800 UTC analysis. (**a**) Gridded NUCAPS 2 m temperature, RAP surface wind, and mean sea level pressure; (**b**) Gridded NUCAPS 2 m relative humidity, RAP surface wind, and mean sea level pressure; (**c**) Gridded NUCAPS 750 hPa temperature and RAP 500 hPa height; (**d**) RAP 750 hPa temperature and 500 hPa height.

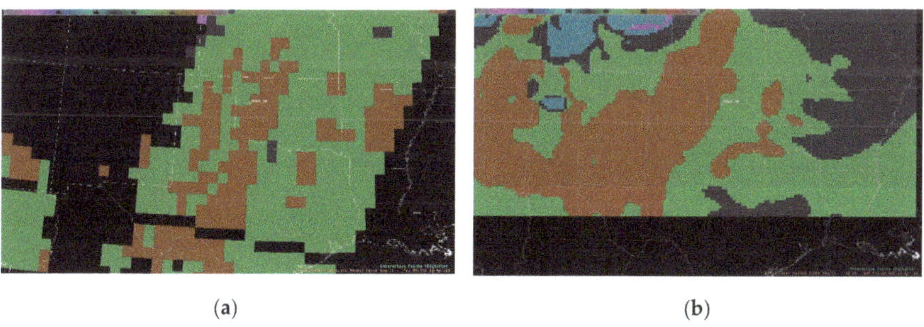

(a)   (b)

Figure 7. AWIPS display of Haines Index calculated from the 850–700 hPa lapse rate and 850 hPa dew point depression. (**a**) S-NPP Gridded NUCAPS Haines Index for 13 April 2019 0839 UTC overpass and (**b**) RAP Haines Index for the 13 April 0800 UTC analysis.

3.3. Monitoring the Saharan Air Layer

The NUCAPS retrievals perform well in clear to partly cloudy conditions; therefore, the Saharan Air Layer (SAL) is an ideal atmospheric phenomenon to observe and monitor. The SAL is an air mass of warm, dry, and often very dusty conditions that originates within the Saharan deserts in northern Africa, then propagates westward for several thousand kilometers, depending on its strength and favorable surrounding environments [50,51]. Using true color imagery, the SAL is identified as a distinct brown (dusty) plume propagating off the northwest coast of Africa, as shown in Figure 8.

Figure 8. NOAA-20 Visible Infrared Imaging Radiometer Suite (VIIRS) True Color imagery on June 22 obtained from NASA Worldview (https://worldview.earthdata.nasa.gov/). The brownish plume that covers the Dominican Republic, Puerto Rico, and the West Indies reveals a strong dust presence associated with the SAL in this region. The bright white rectangular feature toward the top middle portion of the image is sun glint.

From a thermodynamic perspective, [52] used available land-based rawinsonde measurements to provide Skew-T Log P profiles to track the lifespan of a typical SAL outbreak. Near the source region, the SAL outbreak is initially featured with a constant theta (dry adiabatic) profile from the surface to some elevated level, approximately 500 hPa. The accompanying mixing ratio profile starts as very dry at the surface, leading up to 500 hPa to cap the upper extent of the SAL. As the feature propagates westward over the eastern Atlantic basin just offshore of northwest Africa, the surface becomes cut off from the cool and moist marine boundary layer. Finally, the SAL layer greatly becomes diluted by the surrounding cumulus cloud fields and mixing with the boundary layer as it encounters the greater Caribbean and western Atlantic region. Dunion and Marron [53] showed how the mixing ratio at 700 mb is marked by a dry anomaly during a SAL event as compared to the nominal occurrence of the moist tropical environment. The SAL is also accompanied with a low level easterly jet (<10,000 ft). As a result, the slate of NUCAPS sounding products can greatly aid forecasters and analysts in the identification of the SAL, particularly over the data-sparse open water of the Atlantic [54]. The identification of this feature is not only important for impacting hurricane development or suppression [55], but also contributes to adverse health impacts [51,56].

Frequent summertime SAL outbreaks can occur during mid-June through to late August and are of great concern to forecasters and public health agencies throughout the greater Caribbean. Specifically, the population situated within the Caribbean islands, northern South America, the Gulf of Mexico, and the southern United States are particularly impacted by high aerosol content, leading to health hazards associated with poor air quality, as dust concentrations often exceed the United States Environmental Protection Agency standards for PM 2.5 and PM 10. Previous studies [51,56] report that the SAL-related airborne dust impacts Puerto Ricans and its neighboring islands throughout the West Indies, as they suffer from some of the worst global asthma rates, far greater than those of the mainland United States. These results translate into more frequent medical visits and higher mortality rates, especially among the very young and elderly. As the SAL progresses farther west, the feature becomes more diffuse and the satellite identification becomes harder to identify, as the SAL typically encounters cumulus clouds and maritime mixing. The NWS WFO in San Juan, Puerto Rico, monitors and predicts the strength and progression of the SAL in order to issue accurate and timely warnings, and is constantly interested in new environmental resources to improve the accuracy and timeliness of significant SAL event predictions. One of the most sought-after analysis tools is atmospheric soundings, which are greatly lacking in the upstream and data-sparse Atlantic basin. It is here that the NUCAPS Skew-T soundings and gridded formats are currently being investigated.

One of the best opportunities in exploiting the thermodynamic characteristics of the "classic SAL" occurred during the period 17–29 June 2020, where satellite, model, and surface-based measurements highlighted very strong SAL signatures throughout its progression. This episode became a noteworthy global media concern, as human impacts from the Saharan dust were considered an exacerbation of the novel coronavirus pandemic, particularly over the greater Caribbean and southern United States populations. The NOAA-20 Visible Infrared Imaging Radiometer Suite (VIIRS) true color imagery (Figure 8) was used to track the dust plume from its source over northwest Africa through the tropical north Atlantic basin. Note the fairly cloud-free region within the associated dust pattern, which is quite unusual this far from the source region. The strength of the SAL is dramatized as far downwind as off the southeast United States coast on 28 June. Figure 9 provides a mapping of the approximate SAL positions for each day. The "X" within each dot are days that have corresponding plots, as displayed in Figure 10. In Figure 10, the profiles for June 21 and 23 are very similar to a typical SAL event, with the temperature (solid red) lines following constant theta, or dry adiabat from 900 hPa to ~650 hPa. Within the same depth, the mixing ratio profile (dashed red line) follows a slightly drier than constant w profile. The mixing ratio line reaches 600 hPa before reaching another dry layer above 500 hPa.

A number of Gridded NUCAPS products, as sampled in Figure 11, depict strong SAL signatures within each of the products. TPW (Figure 11a) and 700–500 hPa LPW (Figure 11b) exhibit lower precipitable water values in the vicinity of the SAL, as seen in the true color imagery (Figure 8). Although not shown, the near-surface LPW (sfc-850 hPa) indicates moist near-surface conditions, consistent with SAL characteristics. Additionally, warm conditions are evident in the 850 hPa temperature (Figure 11c) Gridded NUCAPS field. Fields such as relative humidity and lapse rate can additionally be analyzed to identify the SAL to further assess dry, stable conditions. Even the Gridded NUCAPS ozone anomaly indicates elevated ozone values in the SAL region, consistent with previous literature [57,58], where elevated ozone mixing ratio values were observed above the SAL. However, additional analysis is needed to determine the efficacy of utilizing the ozone anomaly for SAL identification. As demonstrated here and in other studies [54,59], the NUCAPS vertical soundings and Gridded NUCAPS present new opportunities to analyze the physical process and characteristics of the SAL as it traverses the data-sparse Atlantic basin.

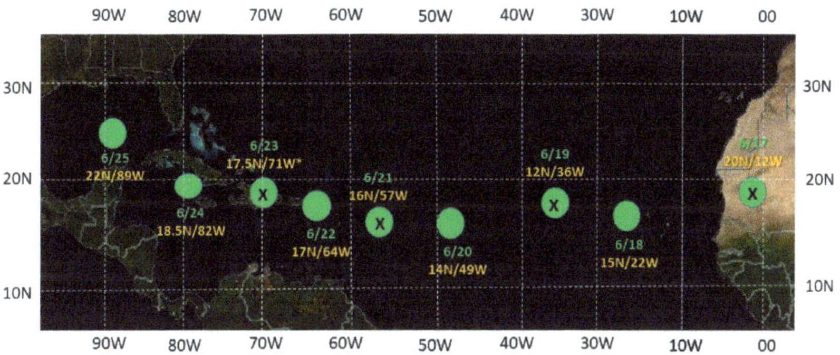

Figure 9. Map of approximate SAL positions (dots) for each afternoon during 17–25 June 2020. Circles with inner "X" annotations are related to the corresponding Skew-T Log P profiles in Figure 10 below.

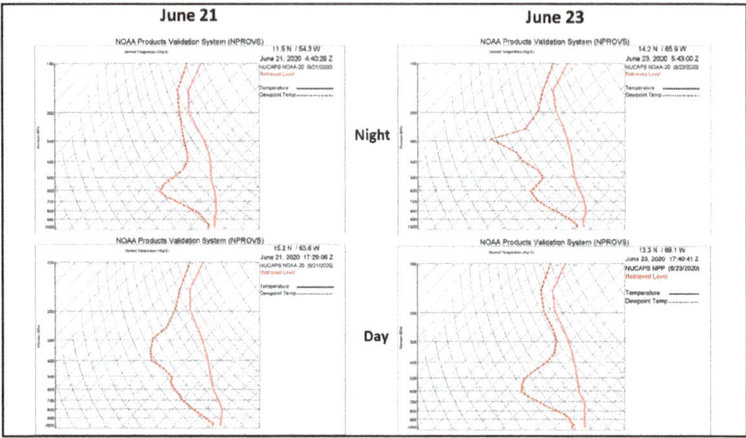

Figure 10. Composites of early morning (night) and afternoon (day) Skew-T plots of temperature (solid lines) and dew point temperature (dashed lines) of the S-NPP and NOAA-20 NUCAPS over the locations mapped in Figure 9 for June 21 and 23.

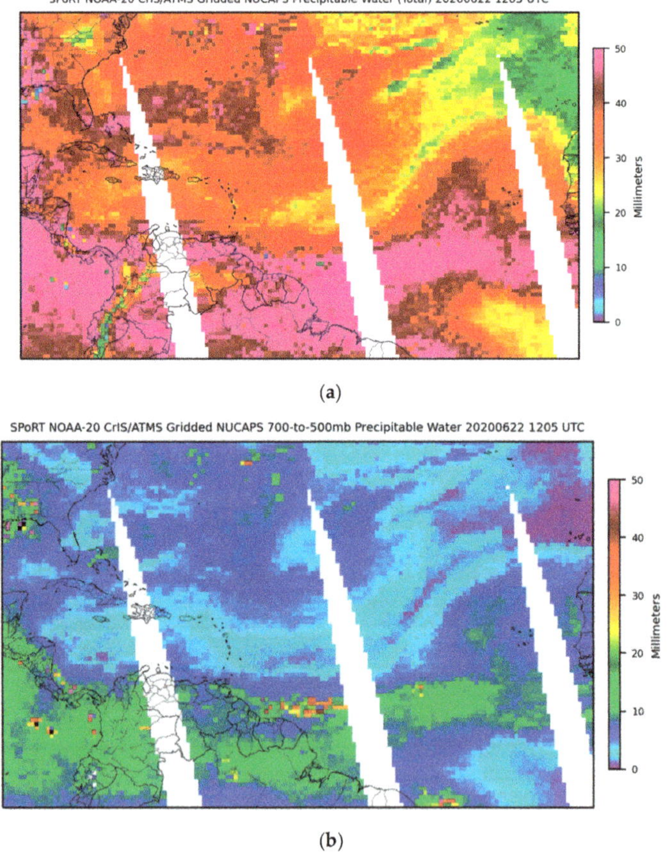

(**a**)

(**b**)

Figure 11. *Cont.*

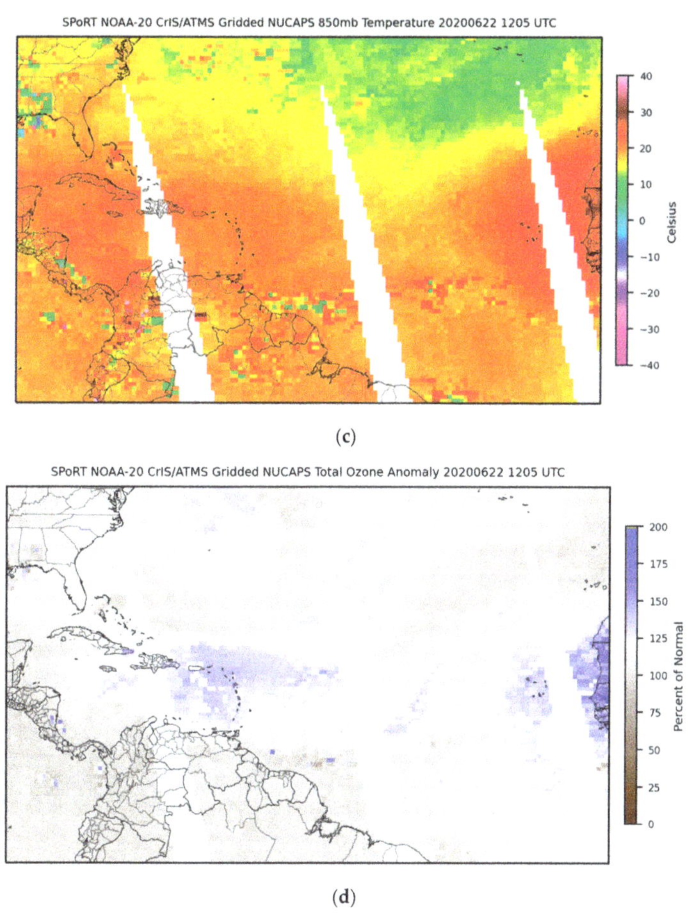

(**c**)

(**d**)

Figure 11. The 22 June 2020, 1205 to 1900 UTC NOAA-20 Gridded NUCAPS (**a**) TPW, (**b**) 700–500 hPa layer precipitable water, (**c**) 850 hPa temperature, and (**d**) ozone anomaly.

3.4. Identifying Stratospheric Air Influence and Tropopause Folding

The Gridded NUCAPS ozone and ozone-derived products can be used to identify the influence of stratospheric air on weather systems and processes such as cyclogenesis, hurricane tropical to extratropical transition, and stratospherically-driven near surface high wind events, predicated on previous work by [19–21]. The ability to identify stratospheric air influence and the potential for tropopause folding can increase situational awareness of the development of hazards (damaging winds, high waves, and heavy rain) associated with these types of events. In addition, the identification of the tropopause can be an important indicator for the potential for turbulence in the vicinity of the jet stream due to the large gradients in temperature and wind [60]. Given the smooth nature of the NUCAPS vertical soundings, the identification of the tropopause features (e.g., isothermal layer and/or inversion) is not always straightforward. Since ozone is a precursor for stratospheric intrusions given the high ozone content of air above the tropopause [61,62], NUCAPS ozone and ozone-derived products can be utilized to identify stratospheric intrusions and the potential for tropopause folding. The example below highlights an instance where the evaluation of Gridded NUCAPS and radiosondes were used to diagnose the tropopause height.

A low-pressure system was traversing the Upper Midwest and Ohio Valley from 31 October to 1 November 2019 and deepening and maturing with time. Figure 12a shows an area of high ozone content associated with the passing cyclone. Since the total column ozone varies climatologically with season and latitude, high ozone values alone are a difficult metric for the identification of anomalous stratospheric air [21], associated with the descent of warm, dry ozone-rich air and its accumulation in the atmospheric column. Figure 12b indeed indicates that the region of high ozone values is associated with anomalous values of the total column ozone for the latitude and season. The darker blue values starting at 125% and greater represent the accumulation of stratospheric air and the potential for tropopause folding. The ozone-derived tropopause level (Figure 12c) indicates that the tropopause was as low as 550–650 hPa over western Illnois and 450–550 hPa over a broad region of the upper Midwest. Although the NOAA-20 overpass was around 1900 UTC on the 31 October, the analysis of the 0000 UTC 1 November sounding at Lincoln, Illnois, confirms a lower tropopause with a double tropopause signature observed at 500 and 300 hPa (Figure 13a). The difference between the Gridded NUCAPS and radiosonde could be explained by the comparison of differing observation types (e.g., points versus an area spanning 50km within the sounding footprint). The radiosonde at Green Bay, Wisconsin, indicates a higher tropopause at about 475 hPa, consistent with the Gridded NUCAPS product (Figure 13b). The comparison of the Gridded NUCAPS ozone-derived products to radiosondes here and in previous literature [19,21] demonstrates the value of ozone-derived fields for assessing the presence of stratospheric air and the potential for tropopause folding. The identification of these features in a plan-view perspective are important in applications such as forecasting rapid cyclogenesis and the development of high winds over data-sparse ocean basins, anticipating hurricane tropical to extratropical transition, and assessing the potential for turbulence near jet streams [63]. With Gridded NUCAPS in AWIPS, these fields are available for testing and demonstrating these applications.

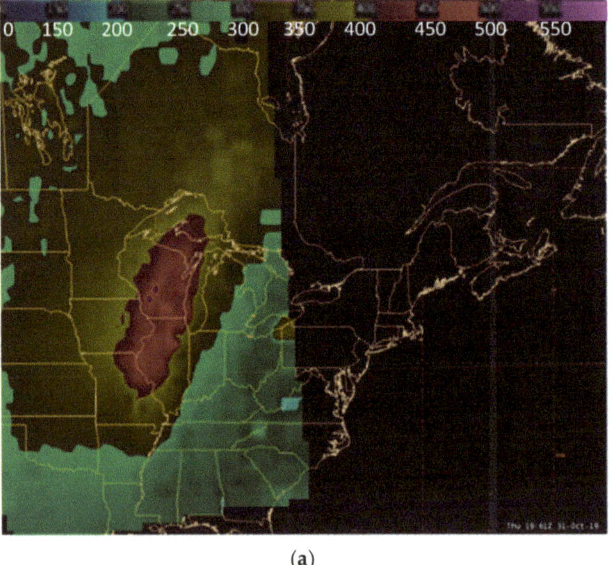

(a)

Figure 12. *Cont.*

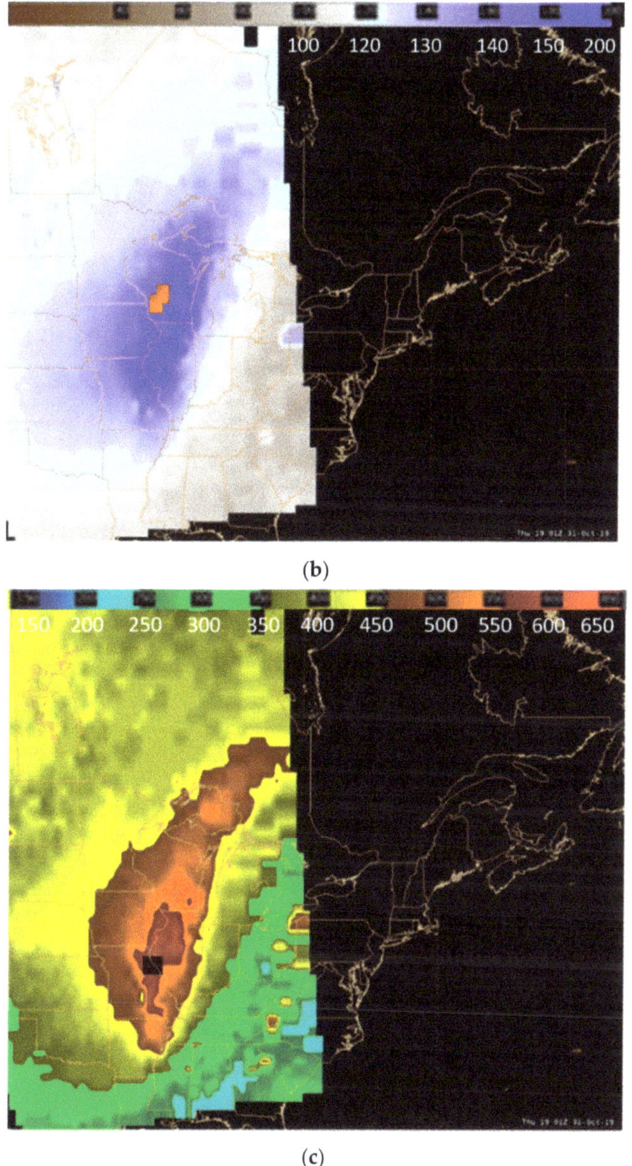

(b)

(c)

Figure 12. AWIPS display of NOAA-20 Gridded NUCAPS on the 31 October 2019, 1901 UTC: (**a**) total column ozone, (**b**) ozone anomaly, and (**c**) tropopause level.

An example of the extratropical transition of Hurricane Arthur in 2014 highlights additional analysis that can increase the situational awareness of changes in the hurricane environment as it relates to anticipating changes in storm intensity. During 4 July, Arthur interacted with an upstream mid-latitude trough and accelerated northeastward. The warm, dry stratospheric air associated with the upper-level trough is colored orange in the air mass composite imagery (Figure 14a; [19,21,65–67]) derived from the Moderate Resolution Imaging Spectroradiometer onboard NASA's Aqua satellite. Figure 14b shows that the stratospheric air is drawn further into the storm over the next 23 h. According to the National

Hurricane Center [68], Arthur began to lose strength as the storm encountered strong upper-level winds and colder sea-surface temperatures. Arthur was classified as a tropical storm by 0600 UTC on 5 July and deemed extratropical by 1200 UTC. Figure 14b is 5.5 h after the extratropical classification.

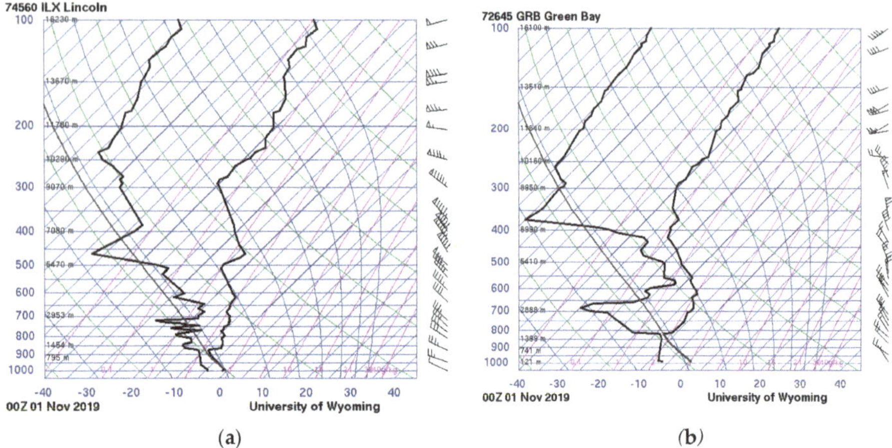

Figure 13. The 1 November 2019, 0000 UTC sounding at (**a**) Lincoln, Illinois, and (**b**) Green Bay, Wisconsin. Images retrieved from the University of Wyoming [64].

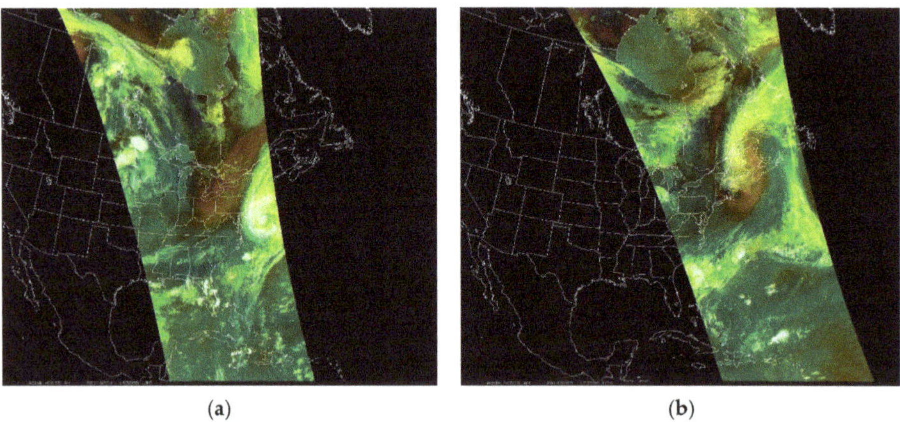

Figure 14. Aqua MODIS Air Mass Composite Imagery on (**a**) 4 July 2014 1835-1845 UTC and (**b**) 5 July 2014 1735-1750 UTC.

Building on the work of Berndt [69], which analyzes the S-NPP overpasses leading up to and following the extratropical transition of Arthur (2014), Gridded NUCAPS can provide additional insights into the hurricane environment and synoptic interactions. The upper-level trough can be identified in the Gridded NUCAPS 500-hPa temperature field, and dry 500 hPa conditions are present in the near-storm environment (Figure 15a,b). The interaction with the 500 hPa trough becomes more pronounced by 0605 UTC, and dry air is closer to the storm center (Figure 15c,d), increasing the situational awareness of the pending extratropical transition. This interaction is much more pronounced by 1735 UTC on 5 July (Figure 15e,f). The ozone anomaly and tropopause height fields can be analyzed to determine the potential for stratospheric intrusion and tropopause folding. The ozone anomaly indicates that a region of stratospheric air is present (Figure 16a; blue colors), but still west of the storm center. In addition, low tropopause heights of 400–500 hPa are associated with this region (Figure 16b). Correspondence with model fields such as potential vorticity can confirm these features. By 0605 UTC on the 5 July, the region of stratospheric air and lower tropopause was much closer to the storm center (Figure 16c,d) and was further drawn into the storm by the afternoon (Figure 16e,f). Events such as extratropical transition and rapid cyclogenesis can create damaging winds, waves, and storm surges that can impact the populous region along the eastern United States or marine activities in the Atlantic and Pacific basins. Gridded NUCAPS, as another observational dataset, can support the thermodynamic and synoptic analysis of these events and complement model analyses to increase the situational awareness of changes in storm intensity that create hazardous conditions.

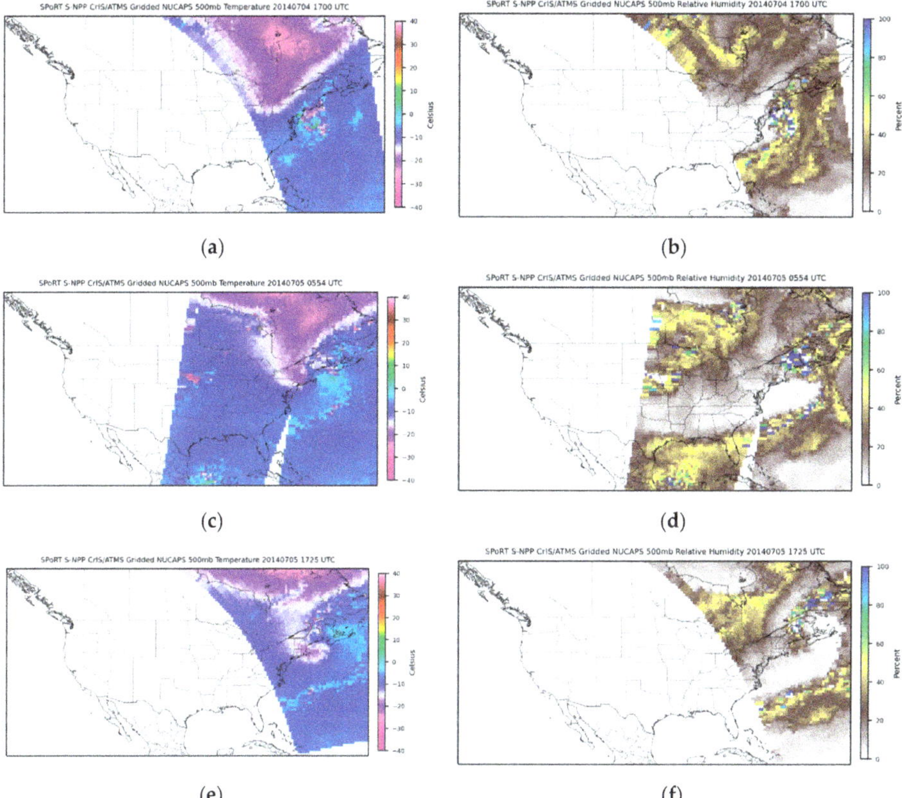

Figure 15. S-NPP Gridded NUCAPS 500 hPa temperature (left) and relative humidity (right): (**a,b**) overpass time on the 4 July 1745-1755 UTC; (**c,d**) combined overpasses on the 5 July with 0605-615 UTC on the right and 0745-0755 UTC on the left; (**e,f**) valid on the 5 July 2014 1725-1735 UTC.

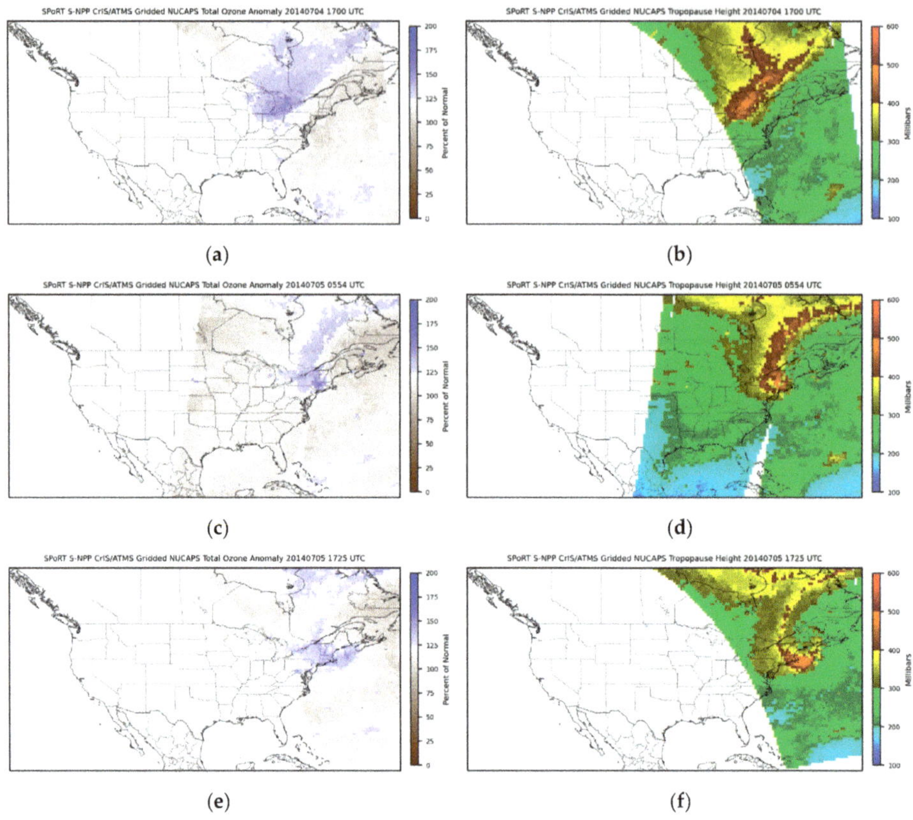

Figure 16. S-NPP Gridded NUCAPS ozone anomaly (left) and tropopause level (right): (**a,b**) overpass time on the 4 July 1745-1755 UTC; (**c,d**) combined overpasses on the 5 July with 0605-615 UTC on the right and 0745-0755 UTC on the left; (**e,f**) valid on the 5 July 2014 1725-1735 UTC.

4. Discussion

New methods and concepts and a standardized approach have been presented to create level 2 gridded and derived products from hyperspectral infrared sounding observations with a focus on NUCAPS observations and short-term weather forecasting. Traditional display tools such as skew-T diagrams, while important, do not fully exploit the strength of satellite soundings (personal communication, C. Barnet), and active user engagement with the weather community led to operations-to-research feedback, ultimately adapting NUCAPS to the operational environment [11]. The development of operationally relevant Gridded NUCAPS fields fills a gap, whereby NUCAPS level 2 gridded products to support short-term weather forecasting have been limited and now allow for the analysis of types of events suitable for thermodynamic analysis [47,63]. This method and capability advance the application and benefit of remote sensing observations, enabling novel analysis and the use of observations beyond their intended use. Few studies have presented methods to create level 2gridded hyperspectral infrared products to support short-term weather forecasting, and the current structure of environmental data records as arrays of vertical soundings require additional data manipulation and processing. Although it is trivial for scientists to process and derive plan-view fields from hyperspectral infrared environmental data records through data processing and manipulation, this data structure has limited the use and application of hyperspectral soundings to the scientific community and advanced users. This work represents new, optimized processing to more effectively visualize the information content of NUCAPS observations, making data more accessible to broader communities and allowing for information compression and the quick analysis of sounding observations [9]. The development of this level 2 gridding method and subsequent integration into baseline AWIPS for NWS-wide distribution was a direct result of operations-to-research feedback and the need to efficiently analyze many soundings in a short period of time, given the constraints, demands, and pace of the operational environment [11–13,17].

This work builds upon the early development of gridding dual-regression algorithm hyperspectral infrared soundings, where data were processed through the polar2grid software [5,29]. These early, experimental methods were adapted to NUCAPS observations in collaboration with the developers, as explained in [13], and further adapted for integration in AWIPS, as explained here. Experimental methods had to be adapted to conform with the constrains of the AWIPS system and available software without requiring burdensome computing expense or resources. The optimization of processing here to create level 2 gridded and derived products with the characteristics of the NUCAPS observations (e.g., footprint size, level vs. layer quantities, retaining data integrity) and the needs of end users in mind (e.g., compatibility with AWIPS, standard levels, fields of interest) represents a new method and technique for processing NUCAPS level 2 products and furthers the accessibility, value, and benefit of these observations to support a wide variety of science and applications. Although level 3 gridded products are routinely produced and available as standard NUCAPS products, there has been a gap in the development of level 2 products or standardized gridding approaches to support short-term weather forecasting. In addition, the derivation of more specialized fields beyond basic temperature, moisture, and trace gases have traditionally not been produced due to the lack of a standard approach to easily process level 2 products. As a feasibility study, [6] demonstrates the information content available to the operational weather community through the derivation of NUCAPS horizontal derived fields of stability indices for convective weather forecasting and emphasizes the advantages and limitations of NUCAPS for this application. The work described here presents the benefit of additional derived fields such as lapse rates, LPW, the Haines Index, and ozone products uniquely developed to optimize the benefit of ozone observations to identify and diagnose the dynamic processes that drive weather. The ozone anomaly and tropopause-level products are developed based on atmospheric dynamics principles relative to how the concentration of ozone varies over time and space as well as the relationship to dynamic variables such as potential vorticity. The processing and derivation of the TPW/LPW fields were designed to facilitate comparison with existing satellite-derived TPW/LPW products.

Few studies have defined or described a methodology for level 2 gridded hyperspectral infrared-derived products for short-term weather forecasting. Gridded NUCAPS products were the result of consciously listening and tailoring NUCAPS towards users' needs and represent a way for the forecaster to quickly assess the environment and highlight baroclinicity and other important features within our soundings to enable the acceptance and, more importantly, value of the NASA and NOAA satellite investments (personal communication, C. Barnet). As described here, this method developed through operations-to-research feedback represents a standard, reproducible approach to effectively visualize NUCAPS observations as level 2 gridded products for more effective analysis and interpretation. As this new approach is now available to all NWS forecasters in the operational AWIPS system and is available online through SPoRT (https://weather.msfc.nasa.gov/cgi-bin/sportPublishData.pl?dataset=griddednucaps), Gridded NUCAPS reaches a broader audience of applied science users for the assessment of novel applications.

5. Conclusions

Interaction with end users and product assessments within the context of the NASA SPoRT research-to-operations/operations-to-research paradigm [1] and collaboration within the NOAA JPSS PGRR Program Sounding Initiative have demonstrated the value of operations-to-research collaborations, specifically to provide insight into the limitations and advantages [11,13] of products to tailor them for the operational environment. A new method and concept for the processing and representation of NUCAPS level 2 gridded and products is presented here, representing the development of approaches to better synthesize remote sensing observations that ultimately increase the availability and usability of NUCAPS observations to benefit scientific analysis and applications. The optimization of basic gridding and interpolation methodologies as appropriately applied to NUCAPS data retains observational characteristics and enables state-of-the-art product development to further support application in weather analysis and forecasting, The derived products presented herein represent the novel development of fields not traditionally derived from hyperspectral infrared sounder observations and new concepts/methods to support applications related to short-term weather forecasting and analysis. The early development and demonstration of Gridded NUCAPS for the cold air aloft aviation hazard and analysis of the pre-convective environment led to the development of a baseline National Weather Service (NWS) capability to create gridded displays of satellite sounding retrievals in the Advanced Weather Interactive Processing System (AWIPS). Gridded NUCAPS was released in AWIPS in 2019, enhancing the capabilities of NUCAPS temperature and moisture soundings that have been available to NWS forecasters as Skew-T's since 2014. The techniques described here were developed to optimally interpolate data to standard levels and grid observations on a 0.5° latitude/longitude grid with minimal interpolation. Each sounding is adjusted to account for changes in the local topography and surface pressure, removing data below the ground surface. Then, they are vertically interpolated to 41 standard meteorological levels from 1100 to 100 hPa every 25 hPa. Temperature is interpolated separately from water vapor and trace gases, which are converted from layer to level quantities and linearized by interpolating the standard logarithm of the column density. Each array of aggregated soundings is added to a 0.5° latitude/longitude grid over a global domain using nearest neighbor and minimal interpolation, masking regions outside of the swath prior to gridding. Horizontal fields are created for temperature and relative humidity on 41 standard levels and at the surface (e.g., 2 m);and derived single-layer products including: quality flags, total precipitable water (TPW) and layer precipitable water (LPW), total ozone, ozone anomaly, and tropopause level. The capabilities additionally include the derivation of lapse rates and the Haines Index. The development of operationally relevant Gridded NUCAPS fields allows for the analysis of types of events suitable to thermodynamic analysis [47,63] and fills a gap whereby NUCAPS level 2 gridded products for supporting short-term weather forecasting have been limited.

The examples presented here demonstrate the analysis possible with the new Gridded NUCAPS capability. Fields such as TPW, LPW, relative humidity, and lapse rates can be used to anticipate the development of convection, where the analysis of gradients and observations between model runs can increase situational awareness, which has already been demonstrated through assessments at the Hazardous Weather Testbed (HWT). The analysis of the 5 June 2019 case demonstrates the value of assessing the broad environment quickly through the identification of the moisture gradient along which the storm developed and produced strong winds. As an emerging application, the assessment of the fire weather environment with NUCAPS soundings is demonstrated here through the identification of the LLTR and analysis of the Haines Index. The near-surface and mid-level temperature and moisture fields were compared to model data to identify the synoptic pattern and LLTR that persisted on the 12–13 April and created weather conditions conducive to the development of the Rhea, Oklahoma fire in 2018. In addition, the derived Haines Index identified a region of high fire potential in the area. The combination of NUCAPS observations with additional model fields such as wind and height demonstrate the value of NUCAPS in supporting fire weather analysis as a model-independent observational dataset to identify thermodynamic features. The visualizations in AWIPS and through a website allow for NWS forecasters and Incident Meteorologists to use NUCAPS products during fire events such as the Rhea fire. Demonstrating the breadth of emerging applications, NUCAPS soundings and Gridded NUCAPS is shown as another observational dataset to identify the Saharan Air Layer (SAL). The June 2020 SAL event is analyzed with NUCAPS vertical profiles, capturing the dry layer on June 21 and 23 in the low to mid-levels. The Gridded NUCAPS TPW and LPW fields were used in the identification of the spatial and vertical extent of the dry, dusty air layer, with the dry pronounced in the 700–500 hPa layer. In addition, elevated temperatures were observed in the NUCAPS 850 hPa temperature field in the SAL region, consistent with typical SAL conditions. Although warranting further analysis and investigation of the efficacy of the approach, an ozone anomaly product with values greater than 125% was observed in the SAL region; thus hinting at elevated ozone mixing ratios associated with the feature. Lastly, the ozone-derived products designed specifically for assessing changes in cyclone or hurricane intensity provide unique information for the identification of such events and anticipating hazards associated with stratospheric air and tropopause folding. The demonstration of the NUCAPS total column ozone, ozone anomaly, and tropopause level for identifying a double tropopause signature from 500 to 300 hPa in the upper Midwest from 31 October to 1 November 2019 captures the ability of Gridded NUCAPS to identify stratospheric intrusions and tropopause folding events. The additional analysis of the extratropical transition of Hurricane Arthur in 2014 was presented to demonstrate the capabilities of the Gridded NUCAPS temperature, moisture, and ozone fields. The Gridded NUCAPS 500 hPa temperature was used to track the development of the upper-level trough and interaction with the storm from 4 to 5 July. The interaction of dry air with the storm, one indicator of many for extratropical transition, was pronounced in the 500 hPa relative humidity fields with dry air infiltrating the storm center by 1735 UTC 5 July, shortly after the extratropical classification. The ozone anomaly and tropopause-level fields observed the region of stratospheric air and lower tropopause heights (400–500 hPa) associated with the upper-level trough, positioned west of the storm on 4 July, moving eastward, and interacting with the storm center by the afternoon of 5 July. Although these fields and applications were previously demonstrated related to the NOAA NWS Ocean Prediction Center analysis of deepening cyclones [19,21] and preliminarily introduced to the NOAA NWS National Hurricane Center [69], these fields are now more widely available to all NWS forecasters to apply to a broader set of applications [63]. The identification of the tropopause and jet stream interactions is important for anticipating changes in storm and hurricane intensity as well as turbulence.

As Gridded NUCAPS is under continued development to add additional derived products and improve the representation of soundings, such as accounting for surface and topography, there are opportunities to discover new applications and how the data can be used for scientific process studies. Additional fields such as trace gases can be processed for display in non-AWIPS visualizations to support additional end users related to tracking smoke plumes important to NWS Incident Meteorologists or researchers conducting field campaigns [46]. Although NUCAPS performs best in clear to partly cloudy conditions, the gridded fields derived from microwave-only soundings have the potential for utility for applications under non-precipitating, cloudy conditions where the microwave retrieval was still successful, such as aviation icing or evaluating the expected precipitation type [70]. There are opportunities to uncover new applications, such as the analysis of the hurricane environment [71,72] and understanding the processes related to the tropical cyclone diurnal cycle [73]. In addition, NUCAPS, especially with multi-satellite assessments and the use of microwave-only soundings, has potential as a proxy to demonstrate the capabilities of the upcoming NASA Time-Resolved Observations of Precipitation structure and storm Intensity with a Constellation of Smallsats (TROPICS; [74]) Mission as a dataset to prepare users for the analysis possible with this new mission. Lastly, the use of multiple satellite platforms or trajectory modeling [17,75] can increase the temporal and spatial coverage of observations, providing insight into the utility of a geostationary hyperspectral infrared sounder in the future.

Author Contributions: Conceptualization, E.B., N.S. and R.E.; software, J.B., N.S., E.B., F.L., J.S. and R.A.; investigation, E.B., K.W., R.A., A.K. and E.D.; resources, J.B., N.S., E.B., R.E., K.W., A.K., E.D., F.L., R.A., J.S.; writing, E.B., R.E., N.S., K.W., A.K., E.D., J.S.; visualization, J.B., E.B., N.S., A.K., F.L., J.S., R.E. and R.A.; supervision, E.B.; project administration, E.B.; funding acquisition, E.B., J.B., N.S., R.E.B. All authors have read and agreed to the published version of the manuscript.

Funding: This research was funded by Mitchell Goldberg through the Joint Polar Satellite System Proving Ground and Risk Reduction Program under the following projects: "The Cold Air Aloft Aviation Hazard: Detection Using Observations from JPSS Satellites and Applications to the Visualization of Gridded Soundings in AWIPS II", "Expanded Application and Demonstration of Gridded NUCAPS in AWIPS", and "NUCAPS Visualization". In addition, this work was partially supported by Tsengdar Lee of the Earth Science Division at NASA Headquarters as part of the NASA Short-term Prediction Research and Transition Center at Marshall Space Flight Center.

Acknowledgments: The authors would like to thank the Joint Polar Satellite System Proving Ground Sounding Initiative for continued support and collaboration. The authors appreciate Christopher Barnet's continual insights and discussions about the complexities of the NUCAPS algorithm. The authors also thank the National Weather Service end users who have participated in assessments and the NUCAPS Users Working Group to give operations-to-research feedback that has improved the products described herein.

Conflicts of Interest: The authors declare no conflict of interest.

References

1. Jedlovec, G. Transitioning Research Satellite Data to the Operational Weather Community: The SPoRT Paradigm [Organization Profiles]. *IEEE Geosci. Remote Sens. Mag.* **2013**, *1*, 62–66. [CrossRef]
2. Nalli, N.R.; Gambacorta, A.; Liu, Q.; Barnet, C.D.; Tan, C.; Iturbide-Sanchez, F.; Reale, T.; Sun, B.; Wilson, M.; Borg, L.; et al. Validation of Atmospheric Profile Retrievals From the SNPP NOAA-Unique Combined Atmospheric Processing System. Part 1: Temperature and Moisture. *IEEE Trans. Geosci. Remote Sens.* **2018**, *56*, 180–190. [CrossRef]
3. Nalli, N.R.; Gambacorta, A.; Liu, Q.; Tan, C.; Iturbide-Sanchez, F.; Barnet, C.D.; Joseph, E.; Morris, V.R.; Oyola, M.; Smith, J.W. Validation of Atmospheric Profile Retrievals from the SNPP NOAA-Unique Combined Atmospheric Processing System. Part 2: Ozone. *IEEE Trans. Geosci. Remote Sens.* **2018**, *56*, 598–607. [CrossRef]
4. Gambacorta, A.; Barnet, C.D. Methodology and Information Content of the NOAA NESDIS Operational Channel Selection for the Cross-Track Infrared Sounder (CrIS). *IEEE Trans. Geosci. Remote Sens.* **2013**, *51*, 3207–3216. [CrossRef]
5. Weisz, E.; Smith, N.; Smith, W.L. The Use of Hyperspectral Sounding Information To Monitor Atmospheric Tendencies Leading to Severe Local Storms: HYPERSPECTRAL SOUNDERS TO MONITOR STORMS. *Earth Space Sci.* **2015**, *2*, 369–377. [CrossRef]
6. Iturbide-Sanchez, F.; da Silva, S.R.S.; Liu, Q.; Pryor, K.L.; Pettey, M.E.; Nalli, N.R. Toward the Operational Weather Forecasting Application of Atmospheric Stability Products Derived From NUCAPS CrIS/ATMS Soundings. *IEEE Trans. Geosci. Remote Sens.* **2018**, *56*, 4522–4545. [CrossRef]
7. Susskind, J.; Barnet, C.D.; Blaisdell, J.M. Retrieval of Atmospheric and Surface Parameters from AIRS/AMSU/HSB data in the Presence of Clouds. *IEEE TGRS* **2003**, *41*, 390–409. [CrossRef]
8. Wheeler, A.; Smith, N.; Gambacorta, A.; Barnet, C.D. Evaluation of NUCAPS Products in AWIPS-II: Results from the 2017 HWT. In Proceedings of the 98th American Meteorological Society Annual Meeting, Austin, TX, USA, 7–11 January 2018.
9. Smith, N.; White, K.D.; Berndt, E.B.; Zavodsky, B.T.; Wheeler, A.; Bowlan, M.A.; Barnet, C.D. NUCAPS in AWIPS–Rethinking Information Compression and Distribution for Fast Decision Making. In Proceedings of the 98th American Meteorological Society Annual Meeting, Austin, TX, USA, 7–11 January 2018.
10. Ackerman, S.A.; Platnick, S.; Bhartia, P.K.; Duncan, B.; L'Ecuyer, T.; Heidinger, A.; Skofronick-Jackson, G.; Loeb, N.; Schmit, T.; Smith, N. Satellites See the World's Atmosphere. *Meteorol. Monogr.* **2018**, *59*, 4.1–4.53. [CrossRef]
11. Esmaili, R.B.; Smith, N.; Berndt, E.B.; Dostalek, J.F.; Kahn, B.H.; White, K.; Barnet, C.D.; Sjoberg, W.; Goldberg, M. Adapting Satellite Soundings for Operational Forecasting within the Hazardous Weather Testbed. *Remote Sens.* **2020**, *12*, 886. [CrossRef]
12. Smith, N.; Berndt, E.B.; Barnet, C.D.; Goldberg, M.D. Why operational meteorologists need more satellite soundings. In Proceedings of the 99th American Meteorological Society Annual Meeting, Phoenix, AZ, USA, 6–10 January 2019.
13. Weaver, G.; Smith, N.; Berndt, E.B.; White, K.D.; Dostalek, J.F.; Zavodsky, B.T. Addressing the Cold Air Aloft Aviation Challenge with Satellite Sounding Observations. *J. Oper. Meteorol.* **2019**, 138–152. [CrossRef]

14. Zavodsky, B.T.; Smith, N.; Dostalek, J.F.; Stevens, E.; Nelson, K.; Weisz, E.; Berndt, E.B.; Line, W.; Barnet, C.D.; Gambacorta, A.; et al. Development and Evaluation of a Gridded CrIS/ATMS Visualization for Operational Forecasting. In Proceedings of the 2016 AGU Fall Meeting, San Francisco, CA, USA, 12–16 December 2016. IN31A–1734.

15. Berndt, E.B.; Smith, N.; White, K.D.; Zavodsky, B.T. Development and Application of Gridded NUCAPS for Operational Forecasting Challenges. In Proceedings of the JPSS Science Seminar Series, Lantham, MD, USA, 23 October 2017.

16. Berndt, E.B. The Evolution of Gridded NUCAPS: Transition of Research to Operations. In Proceedings of the 99th American Meteorological Society Annual Meeting, Phoenix, AZ, USA, 6–10 January 2019.

17. Berndt, E.B.; Smith, N.; Burks, J.; White, K.D.; Allen, R.E. The Evolution of Gridded NUCAPS: Transition of Research to Operations. In Proceedings of the 2019 Joint Satellite Conference, Boston, MA, USA, 28 September–4 October 2019.

18. Ralph, F.M.; Intrieri, J.; Andra, D.; Atlas, R.; Boukabara, S.; Bright, D.; Davidson, P.; Entwistle, B.; Gaynor, J.; Goodman, S.; et al. The Emergence of Weather-Related Test Beds Linking Research and Forecasting Operations. *Bull. Am. Meteorol. Soc.* **2013**, *94*, 1187–1211. [CrossRef]

19. Zavodsky, B.; Molthan, A.; Folmer, M. Multispectral Imagery for Detecting Stratospheric Air Intrusions Associated with Mid-Latitude Cyclones. *J. Oper. Meteorol.* **2013**, *1*, 71–83. [CrossRef]

20. Berndt, E.; Folmer, M. Utility of CrIS/ATMS Profiles to Diagnose Extratropical Transition. *Results Phys.* **2018**, *8*, 184–185. [CrossRef]

21. Berndt, E.B.; Zavodsky, B.T.; Folmer, M.J. Development and Application of Atmospheric Infrared Sounder Ozone Retrieval Products for Operational Meteorology. *IEEE Trans. Geosci. Remote Sens.* **2016**, *54*, 958–967. [CrossRef]

22. Rodgers, C.D. *Inverse Methods for Atmospheric Sounding: Theory and Practice*; Series on Atmospheric, Oceanic and Planetary Physics—Vol. 2; World Scientific Publishing Co.: Singapore, 2000; pp. 1–243. ISBN 978-981-02-2740-1.

23. Irion, F.W.; Kahn, B.H.; Schreier, M.M.; Fetzer, E.J.; Fishbein, E.; Fu, D.; Kalmus, P.; Wilson, R.C.; Wong, S.; Yue, Q. Single-Footprint Retrievals of Temperature, Water Vapor and Cloud Properties from AIRS. *Atmos. Meas. Tech.* **2018**, *11*, 971–995. [CrossRef]

24. Smith, N.; Barnet, C.D. CLIMCAPS Observing Capability for Temperature, Moisture and Trace Gases from AIRS/AMSU and CrIS/ATMS. *Atmos. Meas. Tech. Discuss.* **2020**. [CrossRef]

25. Rosenkranz, P.W. Retrieval of Temperature and Moisture Profiles from AMSU-A and AMSU-B Measurements. *IEEE Trans. Geosci. Remote Sens.* **2001**, *39*, 2429–2435. [CrossRef]

26. Fu, D.; Bowman, K.W.; Worden, H.M.; Natraj, V.; Worden, J.R.; Yu, S.; Veefkind, P.; Aben, I.; Landgraf, J.; Strow, L.; et al. High-Resolution Tropospheric Carbon Monoxide Profiles Retrieved from CrIS and TROPOMI. *Atmos. Meas. Tech.* **2016**, *9*, 2567–2579. [CrossRef]

27. Bowman, K.W.; Rodgers, C.D.; Kulawik, S.S.; Worden, J.; Sarkissian, E.; Osterman, G.; Steck, T.; Lou, M.; Eldering, A.; Shephard, M.; et al. Tropospheric Emission Spectrometer: Retrieval Method And Error Analysis. *IEEE Trans. Geosci. Remote Sens.* **2006**, *44*, 1297–1307. [CrossRef]

28. Goldberg, D.G.; Qu, Y.; McMillim, L.M.; Wolf, W.; Zhou, L.; Divakarla, G. AIRS Near-Real-Time Products and Algorithms in Support of Operational Numerical Weather Prediction. *IEEE TGRS* **2003**, *41*, 379–389. [CrossRef]

29. Smith, W.L.; Weisz, E.; Kireev, S.V.; Zhou, D.K.; Li, Z.; Borbas, E.E. Dual-Regression Retrieval Algorithm for Real-Time Processing of Satellite Ultraspectral Radiances. *JAMC* **2012**, *51*, 1455–1476. [CrossRef]

30. Weisz, E.; Huang, H.L.; Li, J.; Borbas, E.E.; Baggett, K.; Thapliyal, P.; Guan, L. International MODIS and AIRS Processing Package: AIRS Products And Applications. *J. Appl. Remote Sens.* **2007**, *1*. [CrossRef]

31. Weisz, E.; Smith, W.L.; Smith, N. Advances in Simultaneous Atmospheric Profile and Cloud Parameter Regression Based Retrieval from High-Spectral Resolution Radiance Measurements. *J. Geophys. Res. Atmos.* **2013**, *118*, 6433–6443. [CrossRef]

32. Milstein, A.B.; Blackwell, W.J. Neural Network Temperature and Moisture Retrieval Algorithm Validation for AIRS/AMSU and CrIS/ATMS: NEURAL NETWORK T AND Q VALIDATION. *J. Geophys. Res. Atmos.* **2016**, *121*, 1414–1430. [CrossRef]

33. Blackwell, W.J.; Milstein, A.B. A Neural Network Retrieval Technique for High-Resolution Profiling of Cloudy Atmospheres. *IEEE J. Sel. Top. Appl. Earth Obs. Remote Sens.* **2014**, *7*, 1260–1270. [CrossRef]

34. Chahine, M.T. Remote Sensing of Cloud Parameters. *J. Atmos. Sci.* **1982**, *39*, 159–170. [CrossRef]

35. Rosenkranz, P.W.; Barnet, C.D. Microwave Radiative Transfer Model Validation. *J. Geophys. Res.* **2006**, *111*. [CrossRef]

36. Rosenkranz, P.W. Rapid Radiative Transfer Model for AMSU/HSB Channels. *IEEE Trans. Geosci. Remote Sens.* **2003**, *41*, 362–368. [CrossRef]

37. Haines, D.A. A Lower Atmospheric Severity Index for Wildland Fire. *Natl. Weather Dig.* **1988**, *13*, 23–27.

38. Werth, P.; Ochoa, R. The Evaluation of Idaho Wildfire Growth Using the Haines Index. *Wea. Forecast.* **1993**, *8*, 223–234. [CrossRef]

39. Ziemke, J.R.; Chandra, S.; Labow, G.J.; Bhartia, P.K.; Froidevaux, L.; Witte, J.C. A Global Climatology of Tropospheric and Stratospheric Ozone Derived from Aura OMI and MLS Measurements. *Atmos. Chem. Phys.* **2011**, *11*, 9237–9251. [CrossRef]

40. Van Haver, P.; De Muer, D.; Beekmann, M.; Mancier, C. Climatology of Tropopause Folds At Midlatitudes. *Geophys. Res. Lett.* **1996**, *23*, 1033–1036. [CrossRef]

41. Thouret, V.; Cammas, J.-P.; Sauvage, B.; Athier, G.; Zbinden, R.; Nédélec, P.; Simon, P.; Karcher, F. Tropopause Referenced Ozone Climatology and Inter-Annual Variability (1994–2003) from the MOZAIC programme. *Atmos. Chem. Phys.* **2006**, *6*, 1033–1051. [CrossRef]

42. HWT Blog. ILX Upper-Air Sounding Comparison to NUCAPS and Modified NUCAPS. The Satellite Proving Ground at the Hazardous Weather Testbed 2019. Available online: http://goesrhwt.blogspot.com/2019/06/ilx-upper-air-sounding-comparison-to.html (accessed on 19 July 2020).

43. HWT Blog. Springfield Illinois Storms. Spring Experiments and Beyond 2019. Available online: https://inside.nssl.noaa.gov/ewp/2019/06/05/springfield-illinois-storms/ (accessed on 17 July 2020).

44. Storm Prediction Center Storm Prediction Center Storm Reports. Available online: https://www.spc.noaa.gov/climo/reports/190605_rpts.html (accessed on 19 July 2020).

45. Dostalek, J.F.; Haynes, J.; Lindsey, D. NUCAPS Boundary Layer Modifications in Pre-convective Environments: Development of a Timely and More Accurate Product to Assist Forecasters in Making Severe Weather Forecasts. In *2019 JPSS Science Digest*; NOAA NASA Joint Polar Satellite System Program: Greenbelt, MD, USA, 2019.

46. Smith, N.; Esmaili, R.B.; Barnet, C.D.; Frost, G.J.; McKeen, S.A.; Trainer, M.K.; Francoeur, C. Monitoring atmospheric composition and long-range smoke transport with NUCAPS Satellite Soundings in Field Campaigns and Operations. In Proceedings of the 100th American Meteorological Society Annual Meeting, Boston, MA, USA, 12–16 January 2020.

47. Fuell, K. Introduction to NUCAPS Products for Fire Weather Potential. Available online: https://nasasporttraining.wordpress.com/2020/03/10/introduction-to-nucaps-products-for-fire-weather-potential/ (accessed on 27 July 2020).

48. Lindley, T.T.; Bowers, B.R.; Murdoch, G.P.; Smith, B.R.; Gitro, C.M. Fire-effective Low-level Thermal Ridges on the Southern Great Plains. *J. Oper. Meteorol.* **2017**, *5*, 146–160. [CrossRef]

49. National Interagency Coordination Center Wildland Fire Summary and Statistics Annual Report 2018. Available online: https://www.predictiveservices.nifc.gov/intelligence/2018_statssumm/intro_summary18.pdf (accessed on 1 October 2020).

50. Prospero, J.M.; Carlson, T.N. Vertical and Areal Distribution of Saharan dust Over the Western Equatorial North Atlantic Ocean. *J. Geophys. Res.* **1972**, *77*, 5255–5265. [CrossRef]

51. Kuciauskas, A.P.; Xian, P.; Hyer, E.J.; Oyola, M.I.; Campbell, J.R. Supporting Weather Forecasters in Predicting and Monitoring Saharan Air Layer Dust Events as They Impact the Greater Caribbean. *Bull. Am. Meteorol. Soc.* **2018**, *99*, 259–268. [CrossRef]

52. Carlson, T.N. The Saharan Elevated Mixed Layer and its Aerosol Optical Depth. *TOASCJ* **2016**, *10*, 26–38. [CrossRef]

53. Dunion, J.P.; Marron, C.S. A Reexamination of the Jordan Mean Tropical Sounding Based on Awareness of the Saharan Air Layer: Results from 2002. *J. Clim.* **2008**, *21*, 5242–5253. [CrossRef]

54. Kuciauskas, A.; Esmaili, R.B.; Reale, T.; Nalli, N.R. Using NUCAPS to Observe the Thermodynamic Structure of Strong Saharan Air Layer Outbreaks about its Source within the Deserts of Northeast Africa. In Proceedings of the 16th Annual Symposium on New Generation Operational Environmental Satellite Systems Session 11A How JPSS and GOES-R Coupled Resources Improve Forecasting, Boston, MA, USA, 12–16 January 2020; p. 4.

55. Dunion, J.P.; Velden, C.S. The Impact of the Saharan Air Layer on Atlantic Tropical Cyclone Activity. *Bull. Am. Meteorol. Soc.* **2004**, *85*, 353–366. [CrossRef]

56. Lara, M. Heterogeneity of Childhood Asthma Among Hispanic Children: Puerto Rican Children Bear a Disproportionate Burden. *Pediatrics* **2006**, *117*, 43–53. [CrossRef]

57. Jenkins, G.S.; Robjhon, M.L.; Smith, J.W.; Clark, J.; Mendes, L. The influence of the SAL and Lightning on Tropospheric Ozone Variability Over the Northern Tropical Atlantic: Results from Cape Verde during 2010. *Geophys. Res. Lett.* **2012**, *39*, 2012GL053532. [CrossRef]

58. Jenkins, G.S.; Robjhon, M.L.; Demoz, B.; Stockwell, W.R.; Ndiaye, S.A.; Drame, M.S.; Gueye, M.; Smith, J.W.; Luna-Cruz, Y.; Clark, J.; et al. Multi-Site Tropospheric Ozone Measurements Across the North Tropical Atlantic during the summer of 2010. *Atmos. Environ.* **2013**, *70*, 131–148. [CrossRef]

59. Grasso, L.; Bikos, D.; Torres, J.; Dostalek, J.F.; Wu, T.-C.; Forsythe, J.; Cronk, H.; Seaman, C.; Miller, S.; Berndt, E.B.; et al. Satellite Imagery and Products of the 16–17 February 2020 Saharan Air Layer Dust Event over the Eastern Atlantic: Impacts of Water Vapor on Dust Detection and Morphology. *Atmos. Meas. Tech.* **2020**, 1–33, in review. [CrossRef]

60. Reiter, E.R.; Nania, A. Jet-Stream Structure and Clear-Air Turbulence (CAT). *J. Appl. Meteorol.* **1964**, *3*, 247–260. [CrossRef]

61. Shapiro, M.A. Turbulent Mixing within Tropopause Folds as a Mechanism for the Exchange of Chemical Constituents between the Stratosphere and Troposphere. *J. Atmos. Sci.* **1980**, *37*, 994–1004. [CrossRef]

62. Danielsen, E.F. Stratospheric-Tropospheric Exchange Based on Radioactivity, Ozone and Potential Vorticity. *J. Atmos. Sci.* **1968**, *25*, 502–518. [CrossRef]

63. Fuell, K.; LeRoy, A. Gridded NUCAPS Daily Applications of Satellite Soundings. Available online: https://nasasporttraining.wordpress.com/2020/08/25/gridded-nucaps-daily-applications-of-satellite-soundings/ (accessed on 27 August 2020).

64. University of Wyoming College of Engineering Department of Atmospheric Science Upper Air Sounding Archive. Available online: http://weather.uwyo.edu/upperair/sounding.html (accessed on 19 July 2020).

65. EUMETSAT. User Services Division. Best-Practices for RGB Compositing Multi-Spectral Imagery. Available online: http://oiswww.eumetsat.int/~{}idds/html/doc/best_practices.pdf (accessed on 19 July 2020).

66. EUMETSAT. Airmass RGB. Available online: http://oiswww.eumetsat.int/~{}idds/html/doc/airmass_interpretation.pdf (accessed on 19 July 2020).

67. Lensky, I.M.; Rosenfeld, D. Clouds-Aerosols-Precipitation Satellite Analysis Tool (CAPSAT). *Atmos. Chem. Phys.* **2008**, *8*, 6739–6753. [CrossRef]

68. Berg, R. National Hurricane Center Tropical Cyclone Report Hurricane Arthur (AL012014). Available online: https://www.nhc.noaa.gov/data/tcr/AL012014_Arthur.pdf (accessed on 19 July 2020).

69. Berndt, E.B. JPSS Satellite Products for Extratropical Transition. Available online: https://nasasporttraining.wordpress.com/2016/07/21/jpss-satellite-products-for-extratropical-transition/ (accessed on 20 August 2020).

70. Lindstrom, S. CIMSS Satellite Blog. Can You Use NUCAPS Soundings to Determine the Rain/Snow Line? Available online: https://cimss.ssec.wisc.edu/satellite-blog/archives/35383 (accessed on 15 June 2020).

71. Duran, P.; Duran, E. The Wide World of SPoRT. SPoRT Provides Forecasts of Hurricane Isaias' Landfall to NOAA's Hurricane Research Division. Available online: https://nasasport.wordpress.com/2020/08/13/sport-provides-forecasts-of-hurricane-isaias-landfall-to-noaas-hurricane-research-division/ (accessed on 14 August 2020).

72. Lindstrom, S. CIMSS Satellite Blog. Dry Air in the southwest Atlantic. Available online: https://cimss.ssec.wisc.edu/satellite-blog/archives/37767 (accessed on 30 July 2020).

73. Duran, E. The Wide World of SPoRT. Exploring Daily Changes in Tropical Cyclone Temperature and Moisture using NUCAPS Satellite Soundings. Available online: https://nasasport.wordpress.com/2020/08/20/exploring-daily-changes-in-tropical-cyclone-temperature-and-moisture-using-nucaps-satellite-soundings/ (accessed on 21 August 2020).

74. Blackwell, W.J.; Braun, S.; Bennartz, R.; Velden, C.; DeMaria, M.; Atlas, R.; Dunion, J.; Marks, F.; Rogers, R.; Annane, B.; et al. An overview of the TROPICS NASA Earth Venture Mission. *Q. J. R. Meteorol. Soc.* **2018**, *144*, 16–26. [CrossRef]

75. Kalmus, P.; Kahn, B.H.; Freeman, S.W.; van den Heever, S.C. Trajectory-Enhanced AIRS Observations of Environmental Factors Driving Severe Convective Storms. *Mon. Wea. Rev.* **2019**, *147*, 1633–1653. [CrossRef]

Article

Comparison of SLSTR Thermal Emissive Bands Clear-Sky Measurements with Those of Geostationary Imagers

Bingkun Luo *[ID] and Peter J. Minnett[ID]

Rosenstiel School of Marine and Atmospheric Science, University of Miami, 4600 Rickenbacker Causeway, Miami, FL 33149, USA; pminnett@rsmas.miami.edu
* Correspondence: LBK@rsmas.miami.edu

Received: 25 August 2020; Accepted: 7 October 2020; Published: 9 October 2020

Abstract: The Sentinel-3 series satellites belong to the European Earth Observation satellite missions for supporting oceanography, land, and atmospheric studies. The Sea and Land Surface Temperature Radiometer (SLSTR) onboard the Sentinel-3 satellites was designed to provide a significant improvement in remote sensing of skin sea surface temperature (SST_{skin}). The successful application of SLSTR-derived SST_{skin} fields depends on their accuracies. Based on sensor-dependent radiative transfer model simulations, geostationary Geostationary Operational Environmental Satellite (GOES-16) Advanced Baseline Imagers (ABI) and Meteosat Second Generation (MSG-4) Spinning Enhanced Visible and Infrared Imager (SEVIRI) brightness temperatures (BT) have been transformed to SLSTR equivalents to permit comparisons at the pixel level in three ocean regions. The results show the averaged BT differences are on the order of 0.1 K and the existence of small biases between them are likely due to the uncertainties in cloud masking, satellite view angle, solar azimuth angle, and reflected solar light. This study demonstrates the feasibility of combining SST_{skin} retrievals from SLSTR with those of ABI and SEVIRI.

Keywords: SLSTR; evaluation; thermal bands; ABI; SEVIRI

1. Introduction

Skin sea surface temperature (SST_{skin}) is one of the critical variables in the climate system, indicating air–sea interaction patterns near the upper ocean skin layer [1]. The infrared radiometers on earth observation satellites, in both geostationary and polar orbits, have provided retrievals of sea surface temperature (SST) for a half-century [2]. Our choice of satellite radiometers for this analysis was guided by the desire to include one on a polar-orbiting satellite of recent design but with a long planned deployment sequence, a new radiometer type on geostationary satellites again with a long planned deployment duration, and an older radiometer design in geostationary orbit of a type that has been producing data for many years. Thus, the study has relevance not only for the present, but also for the past and future.

The new generation of visible and infrared imaging radiometers, the Sea and Land Surface Temperature Radiometer (SLSTR) onboard Copernicus Sentinel-3A and Sentinel-3B satellites, provide global operational measurements that can be used to derive SST_{skin}, land surface temperature, fire radiative power, aerosol optical depth, etc. [3–5]. The SLSTRs are the fourth and fifth along-track scanning radiometers and are based on the prior along-track scanning radiometers (ATSR; [6]) and advanced ATSR (AATSR; [7]), which have provided valuable measurements to study the Earth's climate system and improve weather forecasting and ocean studies [3,4].

SLSTR was designed to achieve the scientific objective of a mean temporal accuracy of 0.1 K for SST_{skin} products [4]. However, this potential will not be realized without the accurate measurements of top-of-atmosphere radiances. Absolute calibration should be applied to the radiometer, SLSTR radiometric pre-launch calibration is determined at the Rutherford Appleton Laboratory in the United Kingdom (UK) [8,9]. The SLSTR onboard radiometric calibration of the infrared channels is based on two blackbodies with different temperatures (265K and 302K) [8].

With the significant improvements on prior sensors on satellites in geostationary orbits, the new generation of sensors, such as the Advanced Baseline Imager (ABI; [10]) onboard the united states (US) Geostationary Operational Environmental Satellite (GOES) series along with the relatively old sensor Spinning Enhanced Visible and InfraRed Imager (SEVIRI; [11]) onboard the fourth satellite in Meteosat Second Generation (MSG-4), can sample the low- and mid-latitude regions of the Earth's surface and atmosphere and provide valuable data for comparison SLSTR brightness temperatures (BT) in this study.

Sensor-to-sensor comparison can be used to provide assessment on many newly launched sensors. The previous solar reflective band comparisons between the Advanced Himawari Imager (AHI; [12]) and the Visible Infrared Imaging Radiometer Suite (VIIRS; [13]) by Yu and Wu [14] confirmed the linear relationships between them using collocated pairs. The collocated deep convective cloud data have a small difference in the near-infrared bands. Liang, et al. [15] compared measurements and simulations of the AHI, VIIRS and MODerate-resolution Imaging Spectroradiometers (MODIS; [16]) for clear-sky radiances above the sea surface and found the biases in the sensor radiances minus model simulated radiances are relatively stable. Li, et al. [17] have reported a comparison of measurements of MODIS and VIIRS thermal emissive bands using Atmospheric Infrared Sounder (AIRS) hyperspectral radiances convolved with the relative spectral response functions of the MODIS and VIIRS bands, they found the BTs agree relatively well with each other, the differences being within 0.2K. Many other investigators also use this approach to conduct comparisons between various sensors [17–19]. We use conversion functions derived by radiative transfer model simulations to convert the BTs retrieved by geostationary satellite radiometers into SLSTR equivalent versions to perform the analysis reported here. This method has been used by Yu and Wu [14], NASA Langley spectral band difference adjustment [20] and Wu, et al. [21] and others, and found to be useful.

SLSTR, ABI, and SEVIRI provide capabilities for deriving SST_{skin} from the clear sky "atmospheric windows" of wavelengths 3.5–4.1 μm and 8.5–12 μm spectral intervals (which are called thermal emissive bands here). Among the SLSTR, nine spectral channels in the 0.554–12.022 μm spectral range, S7 ($\lambda = 3.74$ μm), S8 ($\lambda = 10.95$ μm), and S9 ($\lambda = 12.00$ μm), can be used for deriving SST_{skin} [22]. For ABI and SEVIRI, the additional bands near $\lambda = 8.5$–8.7 μm are also useful for SST_{skin} retrieval [23–25], as well as in the cloud mask used to eliminate measurements containing radiance emitted or modified by clouds [26]. SST_{skin} derived from measurements in these thermal bands have provided long time-series for various studies [2], the stability of measurements in these bands must be continuously evaluated, especially when they are used to assess the rapidly environmental changes.

This study focused on the preliminary inter-comparison of the new generation of SLSTR radiometers with geostationary radiometers ABI and SEVIRI, in which the performance of the thermal emissive bands were compared. We organize this paper as follows: an overview of the different satellite data is introduced in Section 2. The inter-comparisons between SLSTR and ABI, as well as SLSTR and SEVIR, in three regions, are discussed in Section 3. The reasons for the uncertainties are also introduced in Section 3. Section 4 gives the conclusion of this study.

2. Methods and Materials

2.1. Overview of the Satellite Data

The datasets used in this study include those from radiometers on Sentinel-3A, MSG-4, and GOES-16 satellites are all freely accessible from data servers. Here, we briefly describe the characteristics of the satellite radiometers and their thermal emissive bands that are used to derive SST_{skin}. Relative spectral response functions of SLSTR and the corresponding ABI and SEVIRI channels are given in Figure 1 and Table 1. The gray line in Figure 1 is the atmospheric transmission spectrum for vertical propagation through a standard atmosphere, the spectral response functions of these thermal emissive channels are similar.

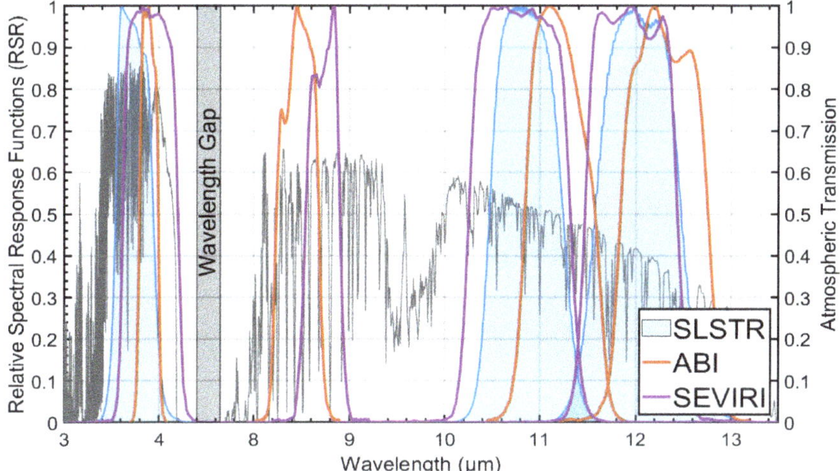

Figure 1. Relative spectral response function of the Sea and Land Surface Temperature Radiometer (SLSTR), with those of Advanced Baseline Imagers (ABI) and Spinning Enhanced Visible and Infrared Imager (SEVIRI) thermal bands around wavelengths of 3.7µm, 8.9µm, 11µm and 12µm. Data are from National Oceanic and Atmospheric Administration Center for Satellite Applications and Research (STAR) National Calibration Center for Visible Infrared Imaging Radiometer Suite (VIIRS)/ABI, from the European Space Agency (ESA)Sentinels Hub for SLSTR and SEVIRI. The gray line is the atmospheric transmission spectrum for vertical propagation through a standard atmosphere.

Table 1. Spectral bands of the SLSTR and geostationary satellite radiometers ABI and SEVIRI. All these bands are usually referred to as thermal emissive bands. Only those with a sea surface temperature (SST) capability are shown.

Band	Band	Center Wavelength (µm)	Band	Center Wavelength (µm)	Band	Center Wavelength (µm)
		GOES-ABI		MSG-4 SEVIRI		Sentinel-3A SLSTR
IR038	7	3.90	4	3.90	S7	3.74
IR087	11	8.50	7	8.70	-	-
IR112	14	11.20	9	10.80	S8	10.95
IR123	15	12.30	10	12.00	S9	12.00

The ability to retrieve the SST_{skin} by making atmospheric corrections is based on different atmospheric transmissions at different infrared wavelengths. The measurements are usually taken in spectral regions with wavelengths from ~3.5 µm to ~4.1 µm and ~10 µm to ~13 µm, where the atmosphere is quite transparent, with variations in clear-sky transmission caused primarily by water vapor, which in itself is highly variable. The widely used SST_{skin} retrieval algorithm, the non-linear SST (NLSST; [27]), is based on the atmospheric transmission window near the IR112 and IR123 bands (Table 1), with other dependences on satellite zenith angle, first-guess SST, coefficient set for latitude bands and month of year [2,28]. The IR038 band near the 3.7–3.9 µm interval can be used to retrieve nighttime SST_{skin} and correct dust aerosol effect [29–31]. Both ABI and SEVIRI are spectrally matched to three SLSTR bands, S7, S8, and S9, respectively. Additionally, ABI and SEVIRI have an IR086 band near 8.5–8.7 µm for deriving SST_{skin}. However, the SLSTR does not include a similar IR086 band in their SST_{skin} retrievals. For this reason, we only consider comparisons of the bands near the SLSTR S7 (λ = 3.74 µm), S8 (λ = 10.95 µm), and S9 (λ = 12.00 µm) spectral ranges in this study.

Figure 2 shows the one-day track of Sentinel-3A as well as the coverage of the GOES-16 and MSG geostationary meteorological satellites that will be used in this study. Table 2 gives the temporal and spatial resolutions of the three satellite retrievals. Details of each radiometer are given in Sections 2.2–2.4.

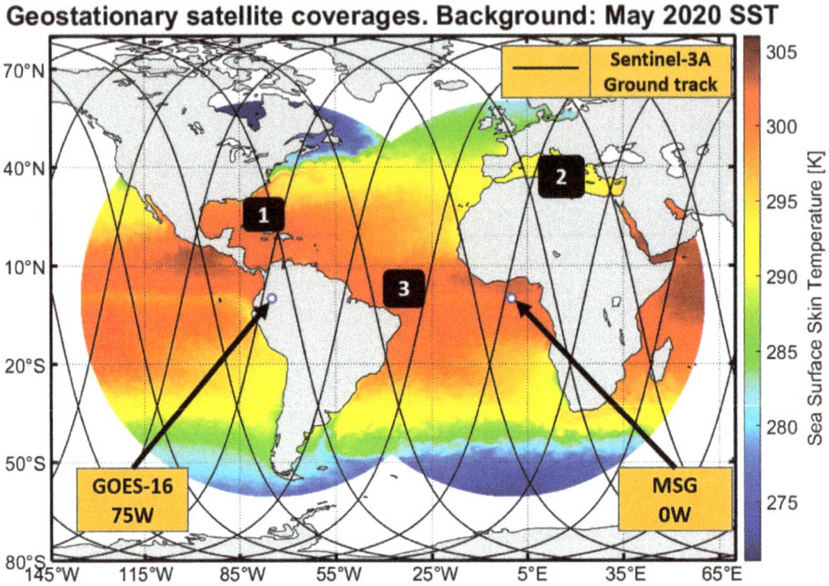

Figure 2. The Sentinel-3A one-day ground tracks along with the coverage areas of the two geostationary meteorological satellites currently in operation and are used in this study. May 2020 monthly mean SST is the background. Three black rectangles indicate the research areas in this study, the numbers correspond to the three parts in Section 4.

Table 2. Characters of each satellite product.

Satellite	Available from	Temporal Resolution	Spatial Resolution
Sentinel-3A SLSTR	EUMETSAT Copernicus Online Data Access (CODA)	every 3 min for Level-1B data	1 km
GOES-ABI	NOAA Amazon Web Services (AWS) Data Centre	every 10 min	1 km
MSG-4 SEVIRI	EUMETSAT Data Centre	every 15 min	3 km

2.2. SLSTR Data

The European Copernicus Sentinel-3A was launched in February 2016 into a polar orbit with descending equator crossing time at 10:00 AM. SLSTR is one of the key instruments for the European Copernicus Sentinel observational system. Unlike the MODIS and VIIRS, which are broad-swath linear-scanners with an atmospheric correction based on the differential atmospheric effects at different wavelengths, the SLSTR onboard Sentinel-3A and Sentinel-3B satellites includes dual view scan systems taking measurements through different atmospheric paths, providing direct measurements of the atmospheric effect, but at the cost of narrower swaths of 740 km. The SLSTR also has a wider nadir view with 1400 km swath. The SLSTRs can provide accurate SST_{skin} derived by radiative transfer model simulated top-of-atmosphere BTs [22,32]. Since the 3.74 μm band can be contaminated during daytime by solar radiation, the SLSTR SST_{skin} is selected from a selection of four algorithms depending on single-view, dual-view, daytime, and nighttime. An initial assessment of the Sentinel-3A SLSTR SST_{skin} accuracy determined by comparisons with measurements of the ship-borne Marine-Atmospheric Emitted Radiance Interferometer (M-AERI; [33]) indicates a median discrepancy of −0.098 K with a robust standard deviation of 0.296 K [3].

2.3. ABI Data

The National Oceanic and Atmospheric Administration (NOAA) geostationary satellite GOES-16, located above 75.2°W, began operation on December 16th, 2017 [10]. ABI has 16 spectral channels, including six visible and near-infrared channels and ten in the infrared. The ABI uses an internal blackbody target and deep space for calibrating the thermal bands. The ABI has improved performance with regard to radiometric calibration accuracy and image navigation/registration compared to prior instruments, it provides full-disk imagery every 10 minutes and the nearest in time Level 2 Cloud and Moisture Imagery Full Disk (CMIPF) data are used to compare with the corresponding SLSTR scenes. The ABI CMIPF files were downloaded from the NOAA data project on Amazon Web Services (AWS) at no cost. The ABI Advanced Clear Sky Processor for Ocean (ACSPO) cloud mask [34] was also used in this study to identify and remove the cloudy pixels.

2.4. SEVIRI Data

Located in geostationary orbit at 0° longitude, SEVIRI onboard MSG-4 can provide full-disk images every 15 minutes. SEVIRI has twelve spectral channels, of which eight are in the infrared. The spatial resolution of the infrared channels is 3 km. The SEVIRI level 1.5 image data were acquired from the European Organisation for the Exploitation of Meteorological Satellites (EUMETSAT) Earth Observation portal. The SEVIRI level 1.5 data are geolocated and have had radiometric calibration applied. As the SEVIRI level 1.5 data include calibrated top-of-atmospheric radiances instead of BTs in each channel, each radiance measurement has been converted into BT according to Planck's equation [35].

2.5. MERRA-2 Data

Sea surface and vertical atmospheric data are needed to drive radiative transfer simulations of top-of-atmosphere BTs to convert those of ABI and SEVIRI into equivalent SLSTR BTs. As the reanalysis ocean surface and atmospheric fields are internally consistent [36], this study uses atmospheric state vectors from the NASA Modern-Era Retrospective Analysis for Research and Applications, version 2 (MERRA-2; [37]). The reanalysis datasets contain geolocated, geophysical variables, including SST_{skin} and air temperature and humidity at 72 standard pressure levels [38,39], these were used to characterize the atmospheric conditions under which the satellite measurements were made for the radiative transfer model simulations of the spectral radiance for each satellite radiometer measurement, and also to derive the formulas to convert the BTs.

2.6. RTTOV Simulation

The radiative transfer model used here is the computationally efficient Radiative Transfer for Tiros Operational Vertical (RTTOV [40]) with sea surface and atmospheric state taken from MERRA-2 reanalysis.

3. Methods

This study used three research areas to perform the comparative analysis of the SLSTR BTs and those measured by geostationary satellite radiometers (Table 3).

Table 3. Details of the SLSTR L1-B data used in this study.

Areas	for	Date	UTC Time
Eastern tropical North Atlantic Ocean	SLSTR with ABI	1 January 2020	Day: 15:21:14 PM Night: 02:55:20 AM
Mediterranean Sea	SLSTR with SEVIRI	23 December 2019	Day: 09:04:56 AM Night: 20:21:51 PM
Cross-covered region	SLSTR with SEVIRI and ABI	27 November 2019	Day: 12:09:44 AM Night: 00:36:20 AM

The first step was to match the SLSTR with ABI and SEVIRI data based on latitude and longitude. Due to the fact that the three instruments have different spatial resolutions, the matched data provide measurements at nearly the same location. We selected the nearest point with the spatial distance between the SLSTR and matched data less than 1 km, which is less than their spatial resolution (Table 2). The time differences between them are usually <5 minutes to mitigate the effects of temporal temperature changes. The satellite viewing geometry is different for each sensor, so to reduce the effect of atmospheric absorption and scattering on radiance measurements, the SLSTR satellite zenith angle is limited to within 45 degrees, then the SLSTR oblique view data will be excluded. Then, the SLSTR Bayesian cloud mask was applied to remove the cloud-contaminated pixels. Additionally, the corresponding ACSPO and SEVIRI cloud masks were used to ensure the clear-sky scenes for the ABI and SEVIRI measurements.

The next step in the analyses was to harmonize the BT measurements taken by each satellite radiometer to account for the relative spectral response functions (Figure 1). The successful harmonization of the BTs obtained from all satellite radiometers is important to this study. Wu, et al. [21] and Yu and Wu [14] assumed the BTs of AHI and Advanced Very High-Resolution Radiometer (AVHRR) at specific channels could be linearly expressed by other similar spectral channels such as VIIRS and MODIS. Wu, et al. [21] and Sohn, et al. [41] used a simple conversion function when comparing MODIS BTs with those of Multifunctional Transport Satellites (MTSAT) or AVHRR. However, they selected the pixels with almost the same viewing geometries, the differences of satellite viewing angle are lower than 50 degrees to reduce the uncertainties caused by different viewing geometries. We updated the conversion functions with the secant of satellite zenith angle terms with respect to the BT changes. Li, et al. [17] and many other investigators used the spectral band difference adjustments based on the NASA Langley Scanning Imaging Absorption Chartography (SCIAMACHY) tool [20]. However, the data flow from SCIAMACHY ended with the failure of Environmental Satellite (Envisat) in April 2012, well before the launch of the Sentinel-3S and the GOES 16 ABI, therefore we derived conversion functions based on radiative transfer simulations to convert the ABI and SEVIRI BTs into SLSTR equivalents:

$$BT_{\text{SLSTR equivalent}} = a \times BT_{ABI\ or\ SEVIRI} + b \times BT_{SLSTR} \times (\sec(\theta_{SLSTR}) - 1) + c$$
$$\times BT_{ABI\ or\ SEVIRI} \times (\sec(\theta_{ABI\ or\ SEVIRI}) - 1) + d$$

The coefficients a, b, c, and d were determined by regressions of the SLSTR BT and ABI/SEVIRI BT of each channel and each geographic area. BT is the BT, θ is the satellite zenith angle. In this study, all of the analyses are based on BTs.

The form of this equation was derived by simulating the spectra of the radiation leaving the top of the atmosphere using RTTOV radiative transfer modeling with the atmospheric state taken from MERRA-2 to derive the simulated satellite radiometer measurements. The SST, 2m air temperature and surface wind data were taken from the MERRA-2 inst1_2d_asm dataset, the three-dimensional air temperature and relative humidity were taken from the MERRA-2 inst3_3d_asm_Nv dataset. The harmonization process is completed across the entire swath of the SLSTR. Satellite zenith angles were set between 0° to 45° to derive the sensitivity to viewing geometry. We did not include the aerosol or cloud effects in the simulations.

4. Results and Discussion

4.1. Eastern Tropical North Atlantic Ocean Region

The variabilities of the oceanographic and atmospheric conditions along the Gulf Stream have drawn a lot of attention for many years. The Florida Current causes complex SST variations as well as a strong atmospheric response in this region [42]. Inter-comparison of Sentinel-3A SLSTR and GOES-16 ABI in this region supports the regional studies of the Gulf Stream and Florida Current.

Selecting a granule with less cloud cover than many others, Figure 3 shows the SLSTR false color infrared image of this area on 1 January 2020, 15:21:14 Coordinated Universal Time (UTC) and the corresponding satellite zenith angles and solar zenith angles. Figures 4 and 5 show the comparison of the pixel-by-pixel matched near-coincident measurements between SLSTR and ABI. All of the ABI values have been converted to SLSTR equivalent BTs. Clearly, there is generally good agreement between all three bands from these scenes. The overall SLSTR BTs are higher in the S8 and S9 comparisons. For SLSTR S7 compared to ABI band 7, there is a negative bias near the Bahamas islands, the S7 band can be contaminated by sun light and there are residual clouds near this region. Figure 5 (third row) shows the histograms of the BT differences in three bands. Their distribution patterns are similar but with many peaks for SLSTR S9 with ABI band 15. Some larger discrepancies, shown in the SLSTR S9 with ABI 15 scatter plot and difference distribution, are caused by large SLSTR satellite zenith angles and cloud edges.

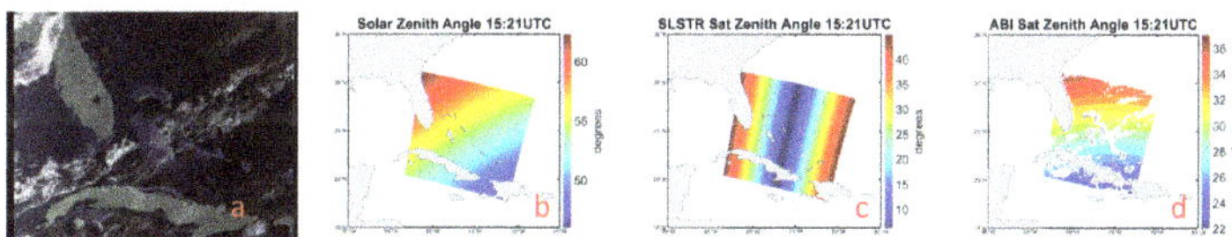

Figure 3. (**a**): SLSTR daytime false color infrared image of the eastern tropical North Atlantic Ocean coast region on 1 January 2020, 15:21:14 UTC. (**b**): Solar zenith angles data at the same time. (**c**): Corresponding SLSTR satellite zenith angles. (**d**): Corresponding ABI satellite zenith angles, only the points with available ABI matched up pairs are shown.

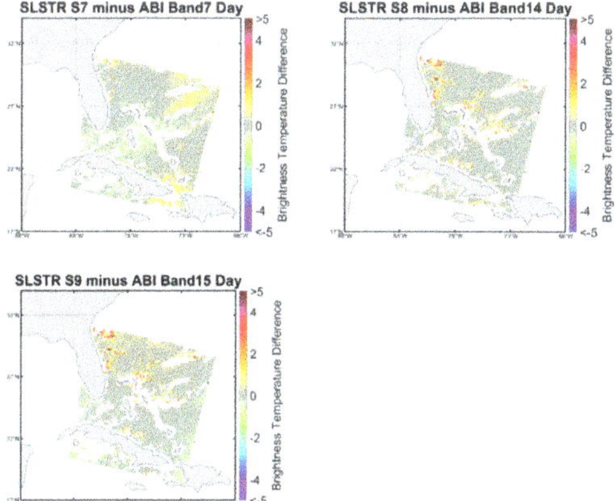

Figure 4. Distributions of the daytime brightness temperature (BT) differences between SLSTR and ABI in the eastern tropical North Atlantic Ocean of SLSTR S7 (**top-left**), S8(**top-right**) and S9 (**bottom-left**). The colors indicate the SLSTR minus ABI equivalent BTs.

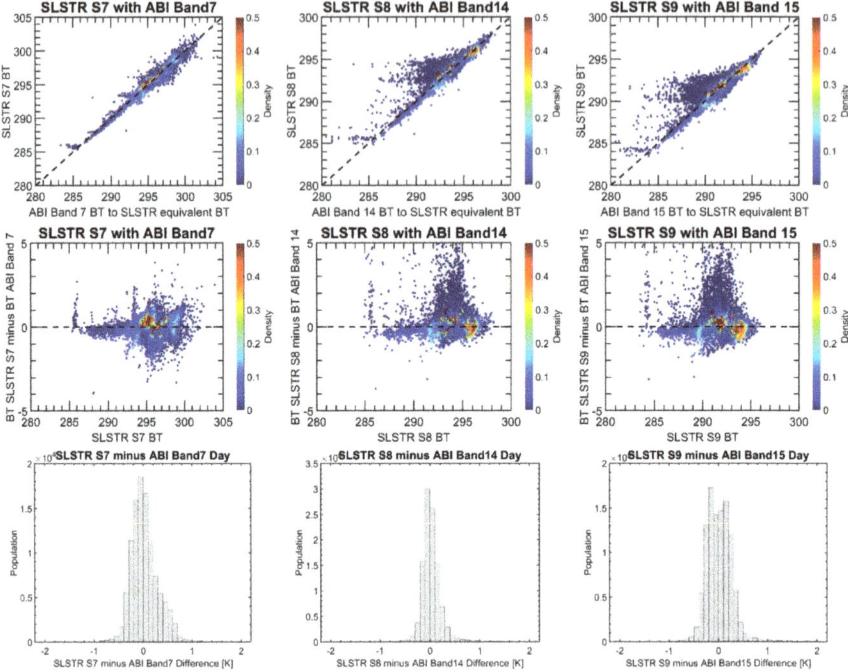

Figure 5. First row: scatter plots of the ABI equivalent BTs as a function of the SLSTR BTs of each channel pair. The colors show the density of the data according to the scale on the right. Second row: scatter plots of the SLSTR minus ABI BTs as a function of the SLSTR BT. Third row: histograms of the SLSTR minus ABI BTs. All of the ABI BTs indicate the transferred SLSTR-equivalent BTs. The BTs are divided into 0.5 K intervals. The density shows the number of matched points within 0.2 K times 0.2 K BT cells divided by the maximum number.

The nighttime false color infrared image is shown in Figure 6. There is a dense cloud cover. Figures 7 and 8 display the nighttime SLSTR BT versus equivalent BTs of ABI. The results of overall comparisons of the nighttime BTs are in better agreement with equivalent BTs than those of daytime.

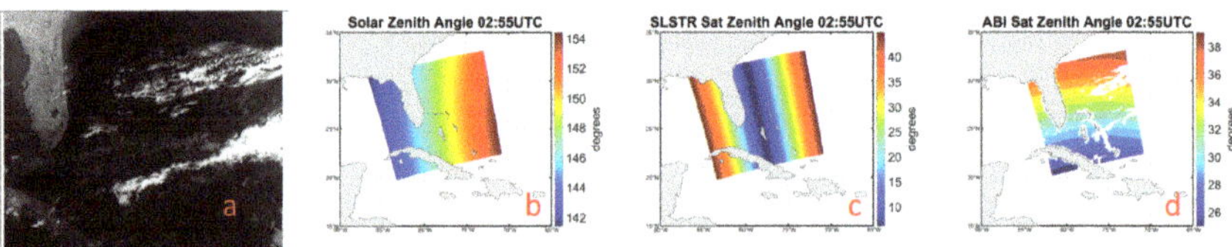

Figure 6. (**a**): SLSTR nighttime false color infrared image of the eastern tropical North Atlantic Ocean coast region on 1 January 2020, 02:55:20 UTC. (**b**): Solar zenith angles at the same time. (**c**): Corresponding SLSTR satellite zenith angles. (**d**): Corresponding ABI satellite zenith angles, only the points with available ABI matched up pairs are shown.

From the geographical distribution of BT differences corresponding to the matching and selection criteria (Figure 7), there is an overall positive bias when comparing the matched SLSTR and equivalent ABI BTs. The dashed lines in the panels in the first row of Figure 8 represent the one-to-one relationship, showing that for SLSTR bands S8 and S9, the BTs < 290 K deviate from the one-to-one lines. These results are consistent with other SLSTR BT comparisons, such as by Shrestha, et al. [43] who also found such discrepancies at the lower SLSTR BTs when compared with those of MODIS. Here, this result may come from the residual contamination at cloud edges and by thin ice clouds, since their BTs are normally lower than those of the sea surface. Although no matchup pairs used to derive the ABI to SLSTR transfer functions are selected with the satellite zenith angle > 45° in this study, the difference distributions based on the selected granule show discrepancies with large viewing angles in S8 and S9 spectral channels. Figure 8 (third row) illustrates the histograms of the BT difference of SLSTR minus ABI during nighttime, which indicates the close similarity of the skewed distributions. Table 4 summarizes the statistics of the SLSTR BTs minus ABI equivalent BTs in this region, the averaged BT differences are on the order of −0.035 K to 0.079 K with the S7 band comparisons having the minimum average difference.

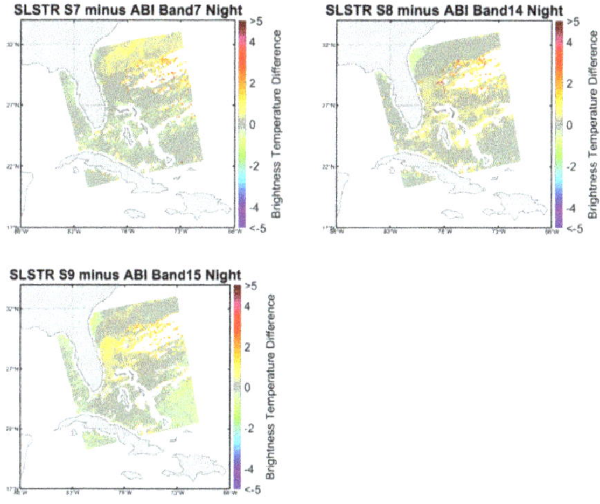

Figure 7. As shown in Figure 4, but for nighttime.

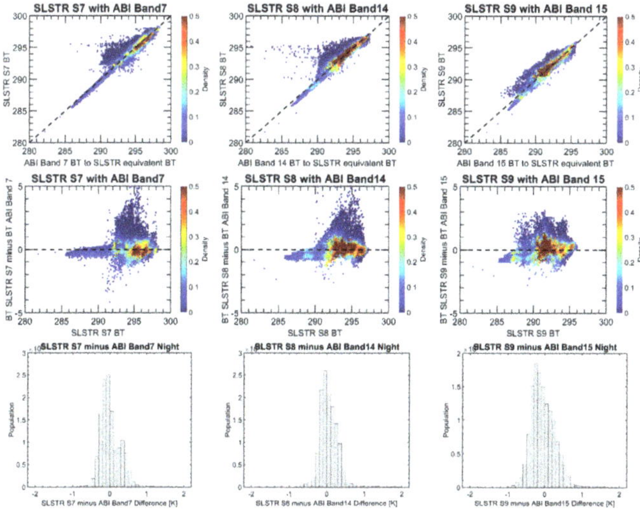

Figure 8. As shown in Figure 5, but for nighttime.

Table 4. Statistics of SLSTR BTs minus ABI equivalent BTs in eastern tropical North Atlantic Ocean. STD: standard deviation. RSD: robust standard deviation.

Eastern Tropical North Atlantic Ocean	Day/Night	Band (SLSTR)	Mean (K)	Median (K)	STD (K)	RSD (K)
		S7	0.028	−0.005	0.296	0.248
	Day	S8	0.054	0.008	0.326	0.145
SLSTR vs ABI		S9	0.042	0.006	0.401	0.260
		S7	0.039	−0.033	0.360	0.281
	Night	S8	0.079	0.028	0.383	0.230
		S9	−0.035	−0.088	0.360	0.330

4.2. Mediterranean Sea Region

As the largest semi-enclosed sea in the world, the Mediterranean Sea has highly specific oceanic characteristics. The SST diurnal cycles in the Mediterranean Sea are more frequent than global regions [44], which can cause marked SST changes. Several studies have estimated the heat budget and their relations to the SST diurnal cycle [45,46]. Satellite measurements can provide high-quality synoptic datasets to study the Mediterranean Sea heat budget, accurate knowledge of their performance is crucial for such research. Figure 9a gives the daytime false color infrared image and satellite geometry data on 23 December 2019, 09:04:56 UTC—the reason for choosing this time is that there is less cloud cover compared to other days.

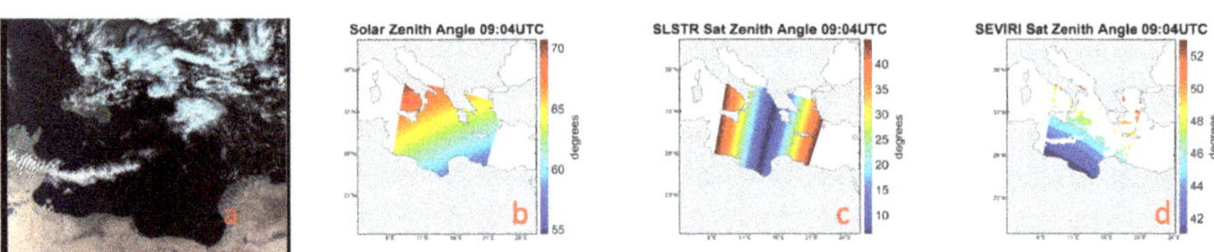

Figure 9. (**a**): SLSTR daytime false color infrared image of the Mediterranean Sea region on 23 December 2019, 09:04:56 UTC. (**b**): Solar zenith angles data at the same time. (**c**): Corresponding SLSTR satellite zenith angles. (**d**): Corresponding SEVIRI satellite zenith angles, only the points with available SEVIRI matched up pairs are shown.

Results of the comparisons in Figures 10 and 11 indicate that for most of the matched points (with high density at the scatter plots), SLSTR BTs agree well with SEVIRI data for S8 and S9 bands, while SLSTR S7 generally has larger differences with SEVIRI band 4 during the daytime due to solar effects. The fact that SLSTR S8 and S9 bands are biased warm may suggest there is residual cloud contamination in the SEVIRI in the Mediterranean Sea region. The SLSTR and SEVIRI cloud masks should be consistent with each other; however, there are large differences near cloud edges, and the difference in viewing angles to the cloud edge causes parallax, which may contribute to these differences.

The most apparent outliers within these channels belong to S7 with large scattering angles of solar radiation and large satellite zenith angles.

Visual inspection of the SLSTR nighttime false color infrared image (Figure 12a) confirms that the cloud edges are the leading cause of the significant warm bias between them. Although the matchup criteria have removed most of the pairs with cloud cover, some of the SEVIRI scenes still have low BTs probably due to the cloud emission instead of from the sea surface.

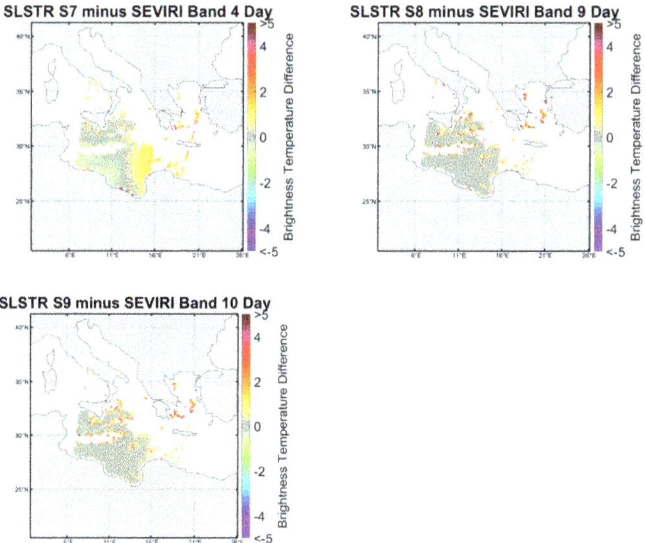

Figure 10. Distributions of the daytime BT differences between SLSTR and SEVIRI in the Mediterranean Sea region of SLSTR S7 (**top-left**), S8 (**top-right**) and S9 (**bottom-left**). The colors indicate the SLSTR minus SEVIRI equivalent BTs.

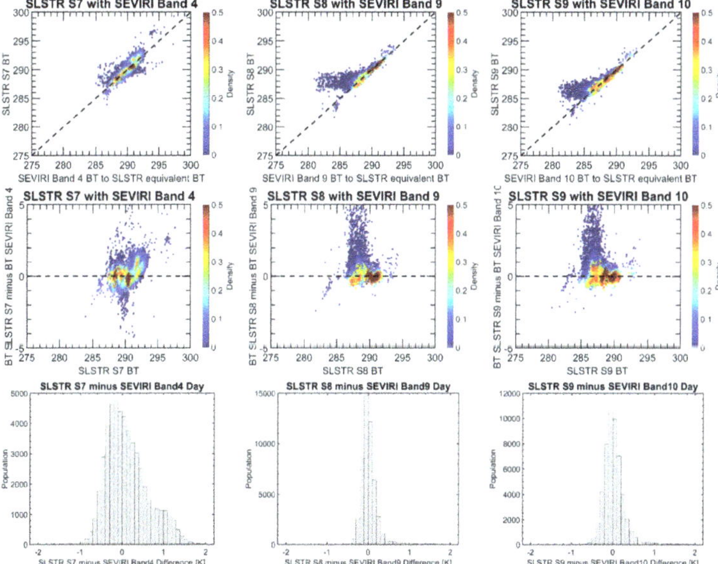

Figure 11. First row: scatter plot of the SLSTR BTs with SEVIRI equivalent BTs of each channel pair. The colors show the density of the data according to the right scale. Second row: scatter plot of the SLSTR BT with SLSTR minus SEVIRI BT of each channel pair. Third row: histograms of the BT differences of SLSTR minus SEVIRI for each channel pair. All of the SEVIRI BTs indicate the transferred SLSTR-equivalent BTs. The BTs are divided into 0.5 K intervals.

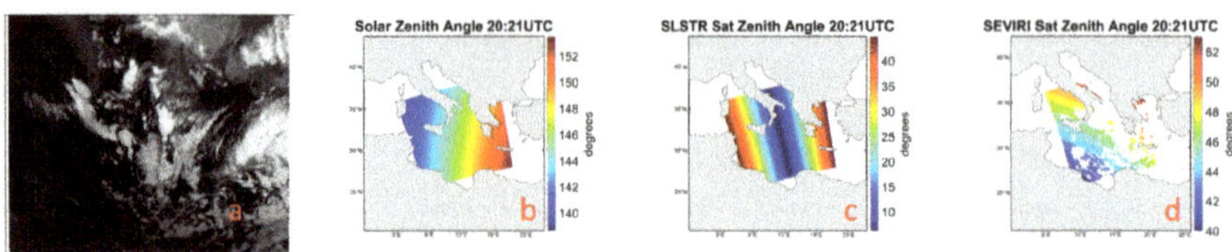

Figure 12. (**a**): SLSTR nighttime false color infrared image of the Mediterranean Sea region on 23 December 2019, 20:21:51 UTC. (**b**): Solar zenith angles at the same time. (**c**): Corresponding SLSTR satellite zenith angles. (**d**): Corresponding SEVIRI satellite zenith angles, only the points with available SEVIRI matched up pairs are shown.

Shown in Figures 13 and 14 are results of the SLSTR and SEVIRI comparison during nighttime. All of the results in these three channels display significant discrepancies at 280–285 K. The most likely distributions of these points with relatively large discrepancies are near coastal regions through visual inspection of Figure 13. Figure 14 (third row) shows the histograms of the BT differences. Each channel of them exhibits very similar difference distributions. Table 5 summarizes the statistics of the SLSTR BTs minus SEVIRI equivalent BTs in the Mediterranean Sea region, the averaged BT differences are over 0.1 K, which is larger than for other regions examined, which may be due to increased number of comparisons indicating cloud-edge effects.

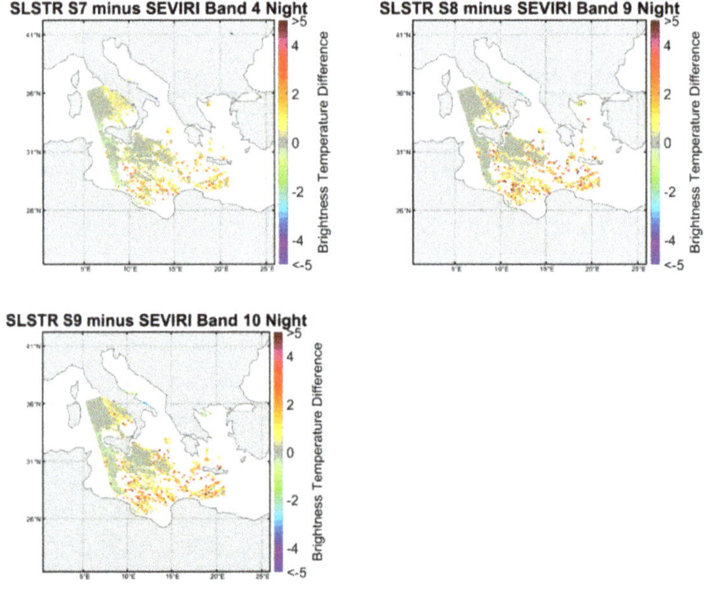

Figure 13. As shown in Figure 10, but for nighttime.

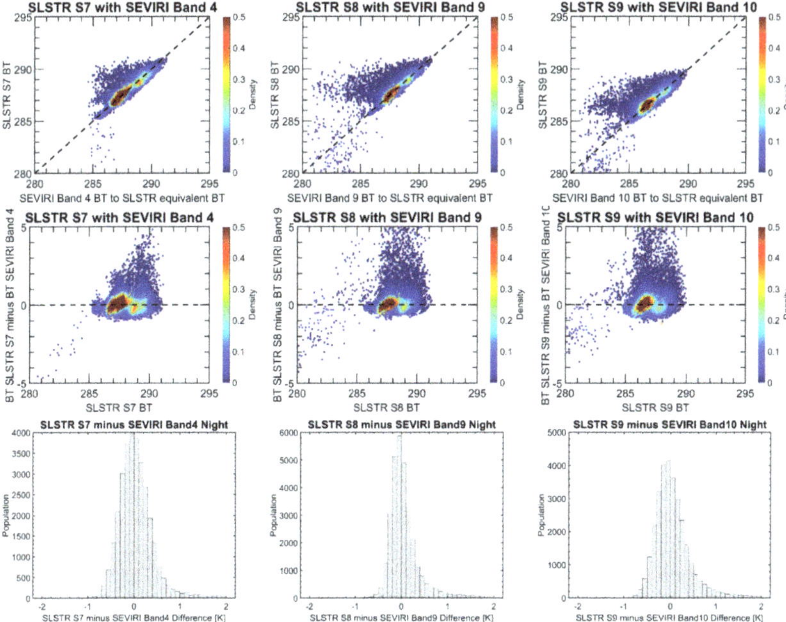

Figure 14. As shown in Figure 11, but for nighttime.

Table 5. Statistics of SLSTR BTs minus SEVIRI equivalent BTs in the Mediterranean Sea area.

Mediterranean Sea	Day/Night	Band (SLSTR)	Mean (K)	Median (K)	STD (K)	RSD (K)
SLSTR vs SEVIRI	Day	S7	0.133	0.045	0.544	0.493
		S8	0.067	−0.005	0.454	0.143
		S9	0.073	0.008	0.440	0.198
	Night	S7	0.077	0.012	0.480	0.320
		S8	0.143	−0.003	0.674	0.240
		S9	0.124	−0.003	0.644	0.328

4.3. Cross-Covered Region

GOES-16 is located above 75.2°W and MSG-4 is located above 0°W; thus, the areas near 37.5°W are under the coverage of three satellites when Sentinel-3A underlies the geostationary satellites. After checking SLSTR true color images, we found this area always includes large amounts of cloud. For this case, two granules of Sentinel-3A SLSTR data from 27 November 2019 are selected to perform the inter-comparison of the three radiometers because of the relatively small cloud coverage compared to other days. The inter-comparison of the thermal emissive bands over this region can further show their performance under the same conditions.

The false color infrared image and satellite geometry data are given in Figure 15. It is clear that the daytime SLSTR image has solar contamination as the area in the right of Figure 15a shows a sun-glitter pattern. There is also a thin cloud cover over this region on 27 November 2019, 12:09 UTC. As the SLSTR S7 near 3.74 μm usually suffers from sunlight contamination during daytime and there is a clear sun-glitter patch in the scene, the SLSTR S8 and S9 BTs are preliminarily evaluated with the corresponding ABI and SEVIRI BTs.

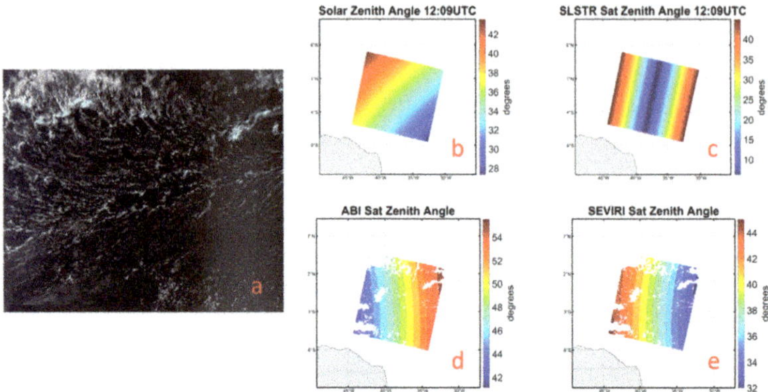

Figure 15. (**a**): SLSTR daytime false color infrared image of the cross-covered region on 27 November 2019, 12:09 UTC. (**b**): Solar zenith angles data at the same time. Corresponding SLSTR (**c**), ABI (**d**), SEVIRI (**e**) satellite zenith angles data.

The cross-comparisons of these three radiometers in this region are limited to SLSTR satellite zenith angles less than 20 degrees as this reduces the range of zenith angle differences to the geostationary satellites. Figure 16 shows the geographical distribution of the daytime BT differences. In comparison to the two infrared channels of SEVIRI, the ABI channels 14 and 15 (shown in the first row of Figure 17) show much larger discrepancies at lower SLSTR BTs, indicating a significant underestimate of BT. Most of the matchup pairs with positive discrepancies at the first row are near the cloud edge, and larger positive discrepancies occur at lower SLSTR BTs, as shown in the second row of Figure 17. All of these discrepancies can also be addressed in the third row, which shows the histograms of the daytime BT differences in three bands.

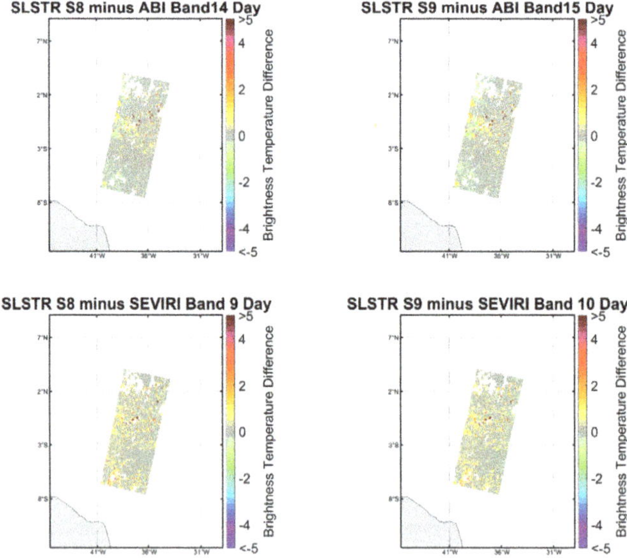

Figure 16. First row: distributions of the daytime BT difference between SLSTR and ABI in the cross-covered region on 27 November 2019, 12:09 UTC. The color indicates the SLSTR minus SEVIRI equivalent BT. Second row: corresponding BT difference distributions between SLSTR and SEVIRI.

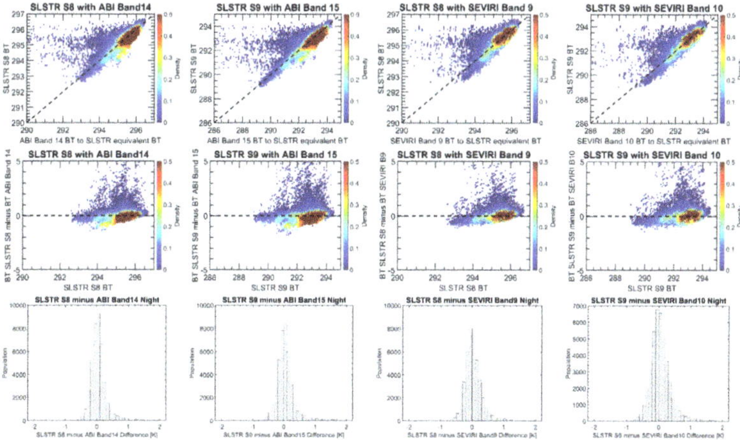

Figure 17. First row: scatter plots of the ABI/SEVIRI equivalent BTs as a function of the SLSTR BTs of each channel pair. The colors show the density of the data according to the scale on the right. Second row: scatter plots of the SLSTR minus ABI/SEVIRI BTs as a function of the SLSTR BT. Third row: histograms of the SLSTR minus ABI/SEVIRI BTs. All of the ABI/SEVIRI BTs indicate the transferred SLSTR-equivalent BTs. The BTs are divided into 0.5 K intervals.

The nighttime false color infrared image and satellite view geometries are shown in Figure 18. Figures 19 and 20 show the nighttime comparisons between SLSTR, ABI and SEVRI. Figure 19 shows the geographic distributions of the BT differences in two bands. As for the eastern tropical North Atlantic Ocean region, the nighttime comparisons show better agreement compared to daytime. The significant positive discrepancies can also be found at the image near the cloud edge. Strong linear relationships between SLSTR and ABI/SEVIRI can be found at most of the matched-up points, as indicated by the first row of Figure 20. However, the overall fitting slopes of high density-points do not agree well with the one-to-one black line. The discrepancies may suggest that the conversion functions have larger uncertainties over this region. Possible reasons are greater water vapor concentrations and large ABI/SEVIRI satellite zenith angles.

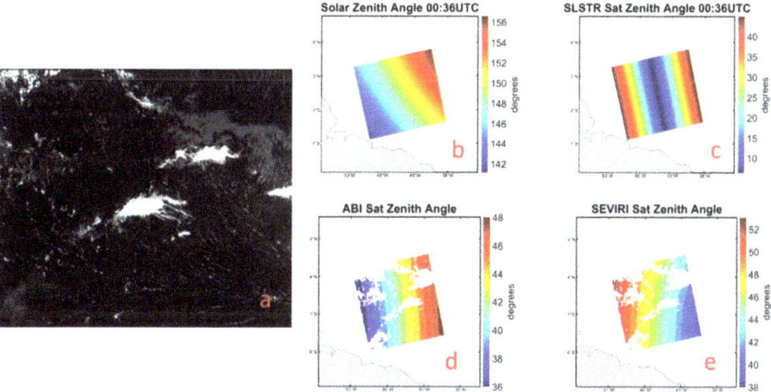

Figure 18. (**a**): SLSTR nighttime false color infrared image of the cross-covered region on 27 November 2019, 00:36 UTC. (**b**): Solar zenith angles data at the same time. Corresponding SLSTR (**c**), ABI (**d**), SEVIRI (**e**) satellite zenith angles data.

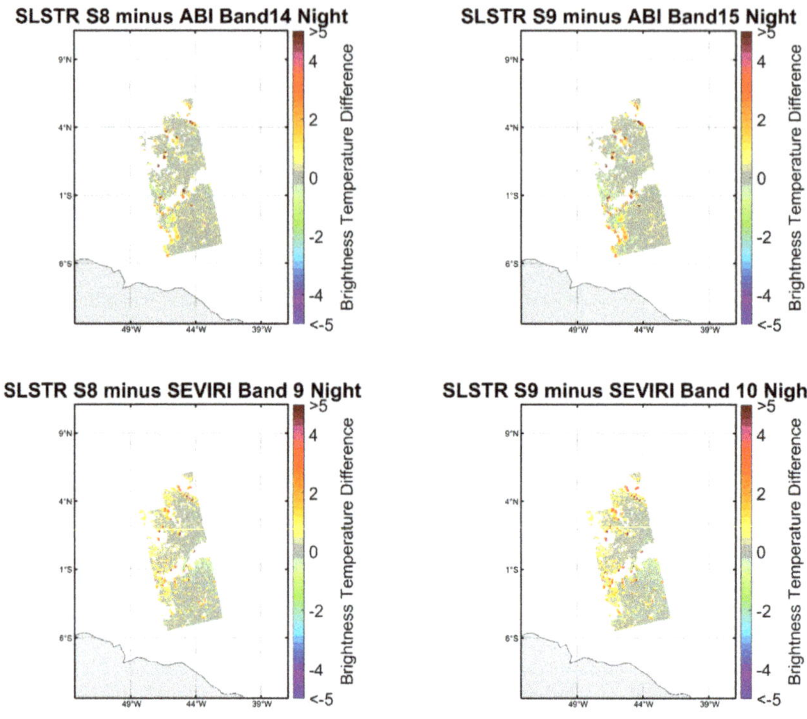

Figure 19. As shown in Figure 16, but for nighttime.

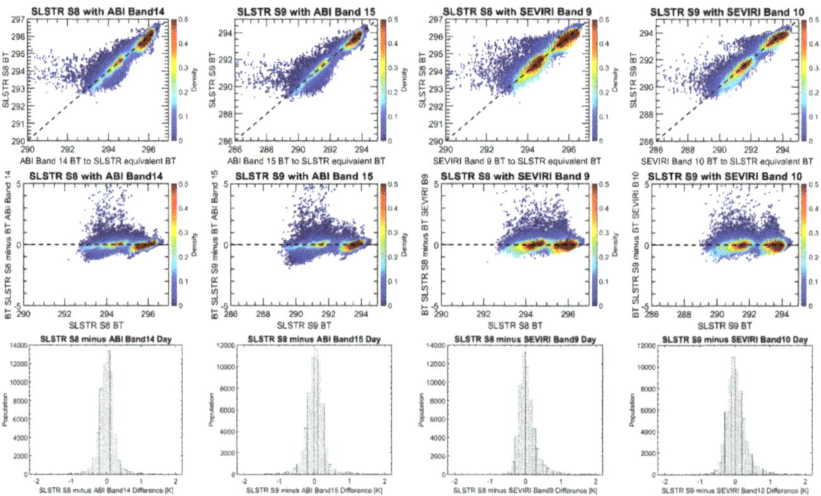

Figure 20. As shown in Figure 17, but for nighttime.

Table 6 shows the statistics of the SLSTR BTs minus ABI/SEVIRI equivalent BTs. The average BT differences are on the order of 0.1 K. Daytime comparisons are better than at nighttime in terms of the average difference and standard deviations. The standard deviations of SLSTR vs ABI are higher than SLSTR vs SEVIRI. The SLSTR S9 band comparisons have larger differences than the S8 band comparisons.

Table 6. Statistics of SLSTR BTs minus ABI/SEVIRI equivalent BTs.

Cross-Covered Region	Day/Night	Band (SLSTR)	Mean (K)	Median (K)	STD (K)	RSD (K)
SLSTR vs ABI	Day	S8	0.035	0.013	0.452	0.184
		S9	0.056	0.030	0.516	0.211
	Night	S8	0.128	0.036	0.891	0.186
		S9	0.143	0.025	1.084	0.207
SLSTR vs SEVIRI	Day	S8	0.087	0.018	0.450	0.202
		S9	0.072	0.010	0.467	0.241
	Night	S8	0.084	0.014	0.465	0.224
		S9	0.105	0.024	0.549	0.265

5. Conclusions

With the significant improvements in design, SLSTRs onboard the Sentinel-3A series of satellites provide observational data in nine visible to infrared bands. Good absolute calibration is required for the accurate derivation of SST_{skin} from radiance measurements, which is achieved by using two onboard blackbodies. Even so, external comparisons of the SLSTR BTs with those of other satellite radiometers are extremely important to ensure the stability and continuity of the long-term satellite climate-related data products, which require the combination of measurements from multiple satellite radiometers, including different designs.

Among the SLSTR nine spectral channels in the 0.554–12.022 µm wavelength spectral range, bands S7 (3.74 µm), S8 (10.95 µm), and S9 (12.00 µm) are used for deriving the SST_{skin}. Here, we compared the BTs of these three SLSTR thermal emission bands with those from geostationary satellite radiometers.

Pixel-by-pixel collocated BTs from SLSTR, ABI, and SEVIRI were used together with their cloud masks to select clear-sky measurements. Empirical regression formulas derived from simulated top-of-atmosphere radiance spectra using the relative spectral response functions of each band were used to convert ABI and SEVIRI BTs to SLSTR-equivalent values, taking into account the satellite zenith angle. The results indicate that SLSTR thermal emissive bands S7, S8 and S9 are comparably well-calibrated as the corresponding ABI and SEVIRI bands, except for S7 bands, which suffer from sunlight contamination during daytime. The measurements from the different satellite radiometers can be combined within the accuracy limits shown in Tables 4–6. Given the occurrence of outliers in the distributions of the BT differences, the robust standard deviation is a better measure of the correspondence of the measurements of the different radiometers. The main differences are due to the residual cloud edges and coast effects, probably land-mask effects, while the other disagreements may be due to different viewing angles and solar contamination of measurements in the mid-infrared atmospheric transmission window. However, it is apparent that the cloud-screening algorithms for all sensors are not identifying all cases of cloud contamination.

It should be noted that the coefficients in the equation to derive SLSTR-equivalent BTs for the geostationary satellite data are dependent on each scene, as a result of the limited ranges of SST and atmospheric conditions in each. Conversion equations applicable to larger areas with greater variability and different times require additional terms, possibly including additional variables, such as the water vapor amount.

Author Contributions: Conceptualization, methodology, software, validation, data curation, writing—original draft preparation, funding acquisition: B.L.; supervision, writing—review and editing, visualization, project administration, funding acquisition: P.J.M. All authors have read and agreed to the published version of the manuscript.

Funding: This research was funded by Rosenstiel School of Marine and Atmospheric Science (RSMAS) Mary Roche endowed Fellowship, EUMETSAT Copernicus Scholarships, Future Investigators in NASA Earth and

Space Science and Technology (FINESST) Program (grant # 80NSSC19K1326), NASA US Participating Investigator Program (grant # NNX17AL69G).

Acknowledgments: This work has benefited from discussions with colleagues at RSMAS.

Conflicts of Interest: The authors declare no conflict of interest.

References

1. Bentamy, A.; Piollé, J.F.; Grouazel, A.; Danielson, R.; Gulev, S.; Paul, F.; Azelmat, H.; Mathieu, P.P.; von Schuckmann, K.; Sathyendranath, S.; et al. Review and assessment of latent and sensible heat flux accuracy over the global oceans. *Remote Sens. Environ.* **2017**, *201*, 196–218. [CrossRef]
2. Minnett, P.J.; Alvera-Azcárate, A.; Chin, T.M.; Corlett, G.K.; Gentemann, C.L.; Karagali, I.; Li, X.; Marsouin, A.; Marullo, S.; Maturi, E.; et al. Half a century of satellite remote sensing of sea-surface temperature. *Remote Sens. Environ.* **2019**, *233*, 111366. [CrossRef]
3. Luo, B.; Minnett, P.J.; Szczodrak, M.; Kilpatrick, K.; Izaguirre, M. Validation of Sentinel-3A SLSTR derived Sea-Surface Skin Temperatures with those of the shipborne M-AERI. *Remote Sens. Environ.* **2020**, *244*, 111826. [CrossRef]
4. Donlon, C.; Berruti, B.; Buongiorno, A.; Ferreira, M.H.; Féménias, P.; Frerick, J.; Goryl, P.; Klein, U.; Laur, H.; Mavrocordatos, C.; et al. The Global Monitoring for Environment and Security (GMES) Sentinel-3 mission. *Remote Sens. Environ.* **2012**, *120*, 37–57. [CrossRef]
5. Coppo, P.; Ricciarelli, B.; Brandani, F.; Delderfield, J.; Ferlet, M.; Mutlow, C.; Munro, G.; Nightingale, T.; Smith, D.; Bianchi, S.; et al. SLSTR: A high accuracy dual scan temperature radiometer for sea and land surface monitoring from space. *J. Mod. Opt.* **2010**, *57*, 1815–1830. [CrossRef]
6. Mutlow, C.T.; Llewellyn-Jones, D.T.; Závody, A.M.; Barton, I.J. Sea-surface temperature measurements by the Along-Track Scanning Radiometer (ATSR) on ESA's ERS-1 Satellite—Early results. *J. Geophys. Res.* **1994**, *99*, 22575–22588. [CrossRef]
7. Llewellyn-Jones, D.; Remedios, J. The Advanced Along Track Scanning Radiometer (AATSR) and its predecessors ATSR-1 and ATSR-2: An introduction to the special issue. *Remote Sens. Environ.* **2012**, *116*, 1–3. [CrossRef]
8. Smith, D.L.; Nightingale, T.J.; Mortimer, H.; Middleton, K.; Edeson, R.; Cox, C.V.; Mutlow, C.T.; Maddison, B.J.; Coppo, P. Calibration approach and plan for the sea and land surface temperature radiometer. *J. Appl. Remote Sens.* **2014**, *8*, 084980. [CrossRef]
9. Smith, D.; Barillot, M.; Bianchi, S.; Brandani, F.; Coppo, P.; Etxaluze, M.; Frerick, J.; Kirschstein, S.; Lee, A.; Maddison, B. Sentinel-3A/B SLSTR pre-launch calibration of the thermal infrared channels. *Remote Sens.* **2020**, *12*, 2510. [CrossRef]
10. Schmit, T.J.; Gunshor, M.M.; Menzel, W.P.; Gurka, J.J.; Li, J.; Bachmeier, A.S. Introducing the Next-Generation Advanced Baseline Imager on GOES-R. *Bull. Am. Meteorol. Soc.* **2005**, *86*, 1079–1096. [CrossRef]
11. Aminou, D.M.A. MSG's SEVIRI instrument. *ESA Bull.* **2002**, *11*, 15–17.
12. Bessho, K.; Date, K.; Hayashi, M.; Ikeda, A.; Imai, T.; Inoue, H.; Kumagai, Y.; Miyakawa, T.; Murata, H.; Ohno, T. An introduction to Himawari-8/9—Japan's new-generation geostationary meteorological satellites. *J. Meteorol. Soc. Jpn. Ser. II* **2016**, *94*, 151–183. [CrossRef]
13. Minnett, P.J.; Evans, R.H.; Podestá, G.P.; Kilpatrick, K.A. Sea-surface temperature from Suomi-NPP VIIRS: Algorithm development and uncertainty estimation. In Proceedings of the SPIE 9111, Ocean Sensing and Monitoring VI, 91110C, Baltimore, MD, USA, 23 May 2014; p. 91110C.
14. Yu, F.; Wu, X. Radiometric inter-calibration between Himawari-8 AHI and S-NPP VIIRS for the solar reflective bands. *Remote Sens.* **2016**, *8*, 165. [CrossRef]
15. Liang, X.; Ignatov, A.; Kramar, M.; Yu, F. Preliminary Inter-Comparison between AHI, VIIRS and MODIS Clear-Sky Ocean Radiances for Accurate SST Retrievals. *Remote Sens.* **2016**, *8*, 203. [CrossRef]
16. Salomonson, V.V.; Barnes, W.L.; Maymon, P.W.; Montgomery, H.E.; Ostrow, H. MODIS: Advanced facility instrument for studies of the earth as a system. *IEEE Trans. Geosci. Remote Sens.* **1989**, *27*, 145–153. [CrossRef]
17. Li, Y.; Wu, A.; Xiong, X. Inter-Comparison of S-NPP VIIRS and Aqua MODIS Thermal Emissive Bands Using Hyperspectral Infrared Sounder Measurements as a Transfer Reference. *Remote Sens.* **2016**, *8*, 72. [CrossRef]
18. Wang, L.; Cao, C.; Ciren, P. Assessing NOAA-16 HIRS radiance accuracy using simultaneous nadir overpass observations from AIRS. *J. Atmos. Ocean. Technol.* **2007**, *24*, 1546–1561. [CrossRef]

19. Wang, L.; Tremblay, D.; Zhang, B.; Han, Y. Fast and accurate collocation of the visible infrared imaging radiometer suite measurements with cross-track infrared sounder. *Remote Sens.* **2016**, *8*, 76. [CrossRef]

20. Scarino, B.R.; Doelling, D.R.; Minnis, P.; Gopalan, A.; Chee, T.; Bhatt, R.; Lukashin, C.; Haney, C. A web-based tool for calculating spectral band difference adjustment factors derived from SCIAMACHY hyperspectral data. *IEEE Trans. Geosci. Remote Sens.* **2016**, *54*, 2529–2542. [CrossRef]

21. Wu, A.; Cao, C.; Xiong, X. Intercomparison of the 11-and 12-um bands of Terra and Aqua MODIS using NOAA-17 AVHRR. In *Earth Observing Systems VIII, Proceedings of the Optical Science and Technology, SPIE's 48th Annual Meeting, San Diego, CA, USA, 3–8 August 2003*; International Society for Optics and Photonics: Bellingham, WA, USA, 2003; Volume 5151, pp. 384–394.

22. Merchant, C. SENTINEL-3 Sea Surface Temperature (SLSTR) Algorithm Theoretical Basis Document. Available online: https://www.eumetsat.int/website/wcm/idc/idcplg?IdcService=GET_FILE&dDocName=PDF_S3_L2_ATBD_SLSTR_SST&RevisionSelectionMethod=LatestReleased&Rendition=Web (accessed on 1 August 2020).

23. Ignatov, A. GOES-R Advanced Baseline Imager (ABI) Algorithm Theoretical Basis Document for Sea Surface Temperature. 2010. Available online: https://www.star.nesdis.noaa.gov/goesr/documents/ATBDs/Baseline/ATBD_GOES-R_SST-v2.0_Aug2010.pdf:NOAA/NESDIS/STAR (accessed on 1 August 2020).

24. Merchant, C.J.; Le Borgne, P.; Roquet, H.; Marsouin, A. Sea surface temperature from a geostationary satellite by optimal estimation. *Remote Sens. Environ.* **2009**, *113*, 445–457. [CrossRef]

25. Romaguera, M.; Sobrino, J.A.; Olesen, F.S. Estimation of sea surface temperature from SEVIRI data: Algorithm testing and comparison with AVHRR products. *Int. J. Remote Sens.* **2006**, *27*, 5081–5086. [CrossRef]

26. Heidinger, A. ABI cloud mask NOAA NESDIS STAR Algorithm Theoretical Basis Doc 2011. Available online: https://www.star.nesdis.noaa.gov/goesr/docs/ATBD/Cloud_Mask.pdf (accessed on 1 August 2020).

27. Walton, C.; Pichel, W.; Sapper, J.; May, D. The development and operational application of nonlinear algorithms for the measurement of sea surface temperatures with the NOAA polar-orbiting environmental satellites. *J. Geophys. Res. Ocean.* **1998**, *103*, 27999–28012. [CrossRef]

28. Kilpatrick, K.A.; Podestá, G.; Walsh, S.; Williams, E.; Halliwell, V.; Szczodrak, M.; Brown, O.B.; Minnett, P.J.; Evans, R. A decade of sea surface temperature from MODIS. *Remote Sens. Environ.* **2015**, *165*, 27–41. [CrossRef]

29. Good, E.J.; Kong, X.; Embury, O.; Merchant, C.J.; Remedios, J.J. An infrared desert dust index for the Along-Track Scanning Radiometers. *Remote Sens. Environ.* **2012**, *116*, 159–176. [CrossRef]

30. Merchant, C.J.; Embury, O.; Le Borgne, P.; Bellec, B. Saharan dust in nighttime thermal imagery: Detection and reduction of related biases in retrieved sea surface temperature. *Remote Sens. Environ.* **2006**, *104*, 15–30. [CrossRef]

31. Luo, B.; Minnett, P.J.; Gentemann, C.; Szczodrak, G. Improving satellite retrieved night-time infrared sea surface temperatures in aerosol contaminated regions. *Remote Sens. Environ.* **2019**, *223*, 8–20. [CrossRef]

32. Merchant, C.J.; Harris, A.R.; Murray, J.; Zavody, A.M. Toward the elimination of bias in satellite retrievals of skin sea surface temperature. 1: Theory. modelling and inter-algorithm comparison. *J. Geophys. Res.* **1999**, *104*, 23565–23578. [CrossRef]

33. Minnett, P.J.; Knuteson, R.O.; Best, F.A.; Osborne, B.J.; Hanafin, J.A.; Brown, O.B. The Marine-Atmospheric Emitted Radiance Interferometer (M-AERI), a high-accuracy, sea-going infrared spectroradiometer. *J. Atmos. Ocean. Technol.* **2001**, *18*, 994–1013. [CrossRef]

34. Petrenko, B.; Ignatov, A.; Kihai, Y.; Heidinger, A. Clear-Sky Mask for the Advanced Clear-Sky Processor for Oceans. *J. Atmos. Ocean. Technol.* **2010**, *27*, 1609–1623. [CrossRef]

35. Tjemkes, S.A. On the Conversion from Radiances to Equivalent Brightness Temperatures. 2005. Available online: http://www.eumetsat.int/groups/ops/documents/document/pdf_msg_seviri_rad2bright.pdf (accessed on 1 August 2020).

36. Luo, B.; Minnett, P. Evaluation of the ERA5 Sea Surface Skin Temperature with Remotely-Sensed Shipborne Marine-Atmospheric Emitted Radiance Interferometer Data. *Remote Sens.* **2020**, *12*, 1873. [CrossRef]

37. Gelaro, R.; McCarty, W.; Suárez, M.J.; Todling, R.; Molod, A.; Takacs, L.; Randles, C.A.; Darmenov, A.; Bosilovich, M.G.; Reichle, R.; et al. The Modern-Era Retrospective Analysis for Research and Applications, Version 2 (MERRA-2). *J. Clim.* **2017**, *30*, 5419–5454. [CrossRef] [PubMed]

38. Luo, B.; Minnett, P.J.; Szczodrak, M.; Nalli, N.R.; Morris, V.R. Accuracy assessment of MERRA-2 and ERA-Interim sea-surface temperature, air temperature and humidity profiles over the Atlantic Ocean using AEROSE measurements. *J. Clim.* **2020**, *33*, 6889–6909. [CrossRef]

39. Bosilovich, M.G.; Santha, A.; Lawrence, C.; Richard, C.; Clara, D.; Ronald, G.; Robin, K.; Qing, L.; Andrea, M.; Peter, N.; et al. *MERRA-2: Initial Evaluation of the Climate*; NASA Goddard Space Flight Center: Greenbelt, MA, USA, 2015; p. 145.

40. Saunders, R.; Hocking, J.; Turner, E.; Rayer, P.; Rundle, D.; Brunel, P.; Vidot, J.; Roquet, P.; Matricardi, M.; Geer, A. An update on the RTTOV fast radiative transfer model (currently at version 12). *Geosci. Model Dev.* **2018**, *11*. [CrossRef]

41. Sohn, B.J.; Park, H.S.; Han, H.J.; Ahn, M.H. Evaluating the calibration of MTSAT-1R infrared channels using collocated Terra MODIS measurements. *Int. J. Remote Sens.* **2008**, *29*, 3033–3042. [CrossRef]

42. Minobe, S.; Miyashita, M.; Kuwano-Yoshida, A.; Tokinaga, H.; Xie, S.-P. Atmospheric response to the Gulf Stream: Seasonal variations. *J. Clim.* **2010**, *23*, 3699–3719. [CrossRef]

43. Shrestha, A.; Angal, A.; Xiong, X. Evaluation of MODIS and Sentinel-3 SLSTR thermal emissive bands calibration consistency using Dome C. In *Algorithms and Technologies for Multispectral, Hyperspectral, and Ultraspectral Imagery XXIV*; International Society for Optics and Photonics: Orlando, FL, USA, 2018; Volume 10644, p. 106441U.

44. Marullo, S.; Santoleri, R.; Ciani, D.; Le Borgne, P.; Péré, S.; Pinardi, N.; Tonani, M.; Nardone, G. Combining model and geostationary satellite data to reconstruct hourly SST field over the Mediterranean Sea. *Remote Sens. Environ.* **2014**, *146*, 11–23. [CrossRef]

45. Marullo, S.; Artale, V.; Santoleri, R. The SST Multidecadal Variability in the Atlantic–Mediterranean Region and Its Relation to AMO. *J. Clim.* **2011**, *24*, 4385–4401. [CrossRef]

46. Marullo, S.; Minnett, P.J.; Santoleri, R.; Tonani, M. The diurnal cycle of sea-surface temperature and estimation of the heat budget of the Mediterranean Sea. *J. Geophys. Res. Ocean.* **2016**, *121*, 8351–8367. [CrossRef]

 *remote sensing*

Article

Evaluating the Magnitude of VIIRS Out-of-Band Response for Varying Earth Spectra

Benjamin Scarino [1,*], David R. Doelling [2], Rajendra Bhatt [1], Arun Gopalan [1] and Conor Haney [1]

[1]   Science Systems and Applications, Inc., Hampton 23666, VA, USA; rajendra.bhatt@nasa.gov (R.B.);
      arun.gopalan-1@nasa.gov (A.G.); conor.o.haney@nasa.gov (C.H.)
[2]   NASA Langley Research Center, Hampton 23666, VA, USA; david.r.doelling@nasa.gov
*   Correspondence: benjamin.r.scarino@nasa.gov

Received: 28 August 2020; Accepted: 5 October 2020; Published: 8 October 2020

Abstract: Prior evaluations of Visible Infrared Imaging Radiometer Suite (VIIRS) out-of-band (OOB) contribution to total signal revealed specification exceedance for multiple key solar reflective and infrared bands that are of interest to the passive remote-sensing community. These assessments are based on laboratory measurements, and although highly useful, do not necessarily translate to OOB contribution with consideration of true Earth-reflected or Earth-emitted spectra, especially given the significant spectral variation of Earth targets. That is, although the OOB contribution of VIIRS is well known, it is not a uniform quantity applicable across all scene types. As such, this article quantifies OOB contribution for multiple relative spectral response characterization versions across the S-NPP, NOAA-20, and JPSS-2 VIIRS sensors as a function of varied SCIAMACHY- and IASI-measured hyperspectral Earth-reflected and Earth-emitted scenes. For instance, this paper reveals measured radiance variations of nearly 2% for the S-NPP VIIRS M5 (~0.67 μm) band, and up to 5.7% for certain VIIRS M9 (~1.38 μm) and M13 (~4.06 μm) bands that are owed solely to the truncation of OOB response for a set of spectrally distinct Earth scenes. If unmitigated, e.g., by only considering the published extended bandpass, such variations may directly translate to scene-dependent scaling discrepancies or subtle errors in vegetative index determinations. Therefore, knowledge of OOB effects is especially important for inter-calibration or environmental retrieval efforts that rely on specific or multiple categories of Earth scene spectra, and also to researchers whose products rely on the impacted channels. Additionally, instrument teams may find this evaluation method useful for pre-launch characterization of OOB contribution with specific Earth targets in mind rather than relying on general models.

Keywords: VIIRS; S-NPP; NOAA-20; JPSS-2; spectral response; out-of-band; in-band; hyperspectral

1. Introduction

Instrument relative spectral response (RSR) characterization is an important element of pre-launch performance specification. Well-characterized spectral performance is critical to the reliable on-orbit operation of Earth-monitoring instruments, whether for routine measurements or for climate studies, and also lends confidence to radiometric calibration efforts and the products reliant on them [1,2]. The Clouds and the Earth's Radiant Energy System (CERES) project, for instance, relies on RSR-dependent calibration adjustments and atmospheric transmissivity calculations to produce accurate cloud products for consistent flux measurements [3–7]. As such, complete pre-launch evaluation of sensor geometric performance, including RSR co-registration and spatial response characterization, is a necessary requirement established to meet the goals of the remote sensing community [8–13]. Moeller et al. and Schwarting et al. conducted extensive laboratory RSR characterization efforts for the Visible Infrared Imaging Radiometer Suite (VIIRS) instrument series using Spectral Measurement Assembly (SpMA) and spherical integrating sphere (SIS) analyses. These laboratory instruments allowed for characterization of the full optical path and any optical or electronic cross talks for nearly all VIIRS bands [1,2,14–19].

Pre-launch, the VIIRS RSRs are specified by their band center, bandpass, extended bandpass, and out-of-band (OOB) response, which are determined from the complete integrated signal. Figure 1 is a schematic recreated from several such figures of Moeller et al. and Schwarting et al. (e.g., "Figure 1" in all listed Moeller et al. references) that illustrates the spectral performance specification metrics for VIIRS, in which the band center is the central wavelength between the 50% response-level bandpass bounds, and the extended bandpass is bound at 1% response levels with associated lower and upper wavelength (λ) thresholds, beyond which are the OOB regions [1,2,16–19]. Moeller et al. and Schwarting et al. conducted these characterization efforts for both the Suomi National Polar-Orbiting Partnership (S-NPP) VIIRS Government Team (GT, consisting of NASA, Aerospace Corp., MIT/Lincoln Lab, and Univ. Wisconsin) and industry (Northrop Grumman, NG) RSR products. The analysis was also performed for versions 1 and 2 (V1 and V2) of the first Joint Polar Satellite System (JPSS-1/NOAA-20) VIIRS RSRs, and V1 and V2 of the future JPSS-2 VIIRS RSRs. With laboratory measurements, they assessed spectral performance metrics with respect to their specified values, results of which are given in tables along with listed bandpass and extended bandpass limits [1,2,16–19].

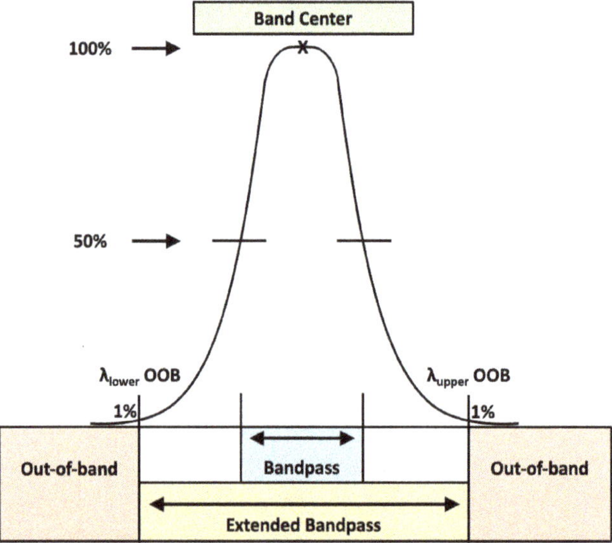

Figure 1. Schematic of "VIIRS spectral performance specification metrics," which is a recreation of "Figure 1" from the works of Moeller et al. and Schwarting et al. [1,2,16–19]. The lower and upper wavelength 1% response limits that separate the extended bandpass and out-of-band (OOB) regions are designated as λ_{lower}OOB and λ_{upper}OOB, respectively.

Although the laboratory results are valuable, they cannot account for the highly varied spectral signatures measured by Earth-observing imagers because OOB contribution to the total scene radiance depends on the spectral shape of the at-sensor radiance. That is, it is difficult to tie the pre-launch spectral performance metrics to OOB behavior for specific Earth-viewed scenes. The OOB radiance signal is dependent on the Earth-reflected spectra and the extended bandpass specifications/measurements unique to each channel. The goal of this study is to quantify the OOB contribution to the total VIIRS signal as a function of instrument version, channel, extended bandpass definition, and Earth scene type. This knowledge is important for scene-dependent inter-calibration efforts, for environmental retrievals, and in regard to error consideration for cloud/aerosol property computations.

This article examines the VIIRS OOB contribution for the S-NPP VIIRS GT and NG RSR products, as well as the V2 releases of the NOAA-20 (V2.1 in the case of band M9) and JPSS-2 VIIRS RSRs. The reason both S-NPP VIIRS RSR products are studied is because despite the post-launch endorsement of the NG RSR release by the Government Team, the GT RSR release, which diverges from the NG RSR "primarily due to analysis differences that only affect the RSR at low response levels," may still be of "investigative interest" to the remote sensing community as an "alternative high quality RSR" [1,2,20]. For the NOAA-20 and JPSS-2 VIIRS, there is no distinction between GT and industry RSR releases because only the former carried out a pre-launch characterization effort [17,19,21]. Users within the inter-calibration community and product teams that rely on VIIRS should find these results useful, even if only for assurance that OOB contribution is within acceptable tolerance for their specific application, which should often be the case especially for the newer VIIRS. Regardless, this work informs users on the impact of limiting spectral integration to published extended bandpass limits versus the full-band RSR for applications that rely on such techniques.

2. Data and Methodology

The Earth-view hyperspectral data used in this study were acquired from the Envisat Scanning Imaging Absorption Spectrometer for Atmospheric Chartography (SCIAMACHY) instrument for visible bands, and from the MetOp Infrared Atmospheric Sounding Interferometer (IASI) instrument for infrared (IR) bands [22–25]. Operating in a 10:00 AM local time sun-synchronous orbit, the SCIAMACHY instrument has a fine spectral resolution across eight channels covering 0.24–2.38 µm; however poor spectral quality in higher channels limits the usable wavelength upper range to ~1.75 µm. It has four 30×240 km^2 nadir fields of view (FOVs) divided along a 960 km swath with footprint-center viewing zenith angles (VZAs) ranging from ~7.5° to ~27.1°. The instrument performed daily solar irradiance measurements via a solar diffuser, remained stable over its lifetime of 1 March 2002–8 April 2012, and maintained an absolute on-orbit calibration accuracy of 2–6% [22,26–28]. The IASI instrument was the first operational interferometer in space measuring 3.6–15.5 µm across 8461 spectral bands with a spectral resolution of 0.5 cm^{-1}, has a 12 km FOV, and, operating on MetOp-A, has a local equator crossing time of 09:30 AM [24,25,29]. The instrument has been relied upon by the Global Space-Based Inter-Calibration System (GSICS) international organization as an absolute calibration reference given the high confidence in IR hyperspectral sensor calibration and the capability of creating pseudo imager radiance signatures by convolving the hyperspectral data with imager RSRs. The imager-RSR-convolved IASI radiance values are used to radiometrically scale the imager to the IASI standard [30–35].

Many studies have employed RSR-integration techniques, involving either simulated or measured hyperspectral radiance information, for the purpose of spectral band adjustment factor (SBAF) computation. An SBAF is used to account for spectral differences between common instrument RSRs, which is an important step of the imager inter-calibration process [36–43]. The background and methodology of the specific SBAF computation pertinent to this work, which is dependent on measured Earth radiance spectra that are relevant to common inter-calibration techniques, were described in detail by Scarino et al. In short, pseudo radiance signatures for a reference and target satellite imager are computed by convolving many hyperspectral radiance footprints with the imager RSRs. A simple ratio of means or regression of the set of pseudo radiance pairs then constitutes the target/reference SBAF for the selected Earth scene. Therefore, applying the SBAF to the true reference radiance data will yield predicted target radiance data that are spectrally consistent with the true target radiance [43]. An online tool (found through https://satcorps.larc.nasa.gov or directly at https://satcorps.larc.nasa.gov/SBAF) was developed to allow users to easily produce Earth-scene-specific SBAFs with the least uncertainty for their carefully chosen inter-calibration conditions [43]. The tool has been recommended by GSICS and each month serves over 4000 requests from the international community.

This SBAF computation methodology was modified to allow for a simple assessment of the OOB contribution to the total signal [43]. Instead of using distinct imagers for the pseudo radiance calculations, the same VIIRS RSR is used as both the reference and the target. The reference pseudo radiance values are integrated from either SCIAMACHY or IASI using the full-band VIIRS RSR (i.e., the extended bandpass plus OOB radiance L_{total}), whereas the target pseudo radiance values are integrated only within the range of the extended bandpass (i.e., the in-band radiance L_{in}). The OOB contribution is examined in terms of both the specified and measured lower and upper 1% extended bandpass limits for S-NPP VIIRS GT, S-NPP VIIRS NG, NOAA-20 VIIRS V2, and JPSS-2 VIIRS V2, which are provided by Moeller et al. and are also listed in Table 1 [1,2,17,19]. Table 2 provides the version descriptions of the VIIRS RSR products used in this study. Note that in the case of S-NPP VIIRS RSR products, operational calibration of the VIIRS radiances does not employ either of the versions listed in Table 2, but rather relies on Modulated RSR Release 1.0 [44]. The impact of this discrepancy is discussed at length in Section 4. The OOB contribution γ can be measured by the ratio of L_{in} to L_{total}, and then expressed as a percentage as follows:

$$\gamma = \left| \frac{L_{in}}{L_{total}} - 1 \right| \times 100 \tag{1}$$

Values of γ close to 0 suggest minimal OOB contribution for the evaluated scene type. Note that Moeller et al. define a maximum integrated out-of-band (MIOOB) response, described as the ratio of integrated out-of-band response to integrated in-band response, which is formulaically different than Equation (1) but leads to similar conclusions [17].

Table 1. Lower and upper 1% extended bandpass limits (μm) as provide by Moeller et al. [1,2,17,19].

Band	VIIRS Specified		S-NPP VIIRS GT Measured		S-NPP VIIRS NG Measured		NOAA-20 VIIRS V2 Measured		JPSS-2 VIIRS V2 Measured	
	Lower	Upper	Lower	Upper	Lower	Upper	Lower	Upper	Lower	Upper
I1	0.5650	0.7150	0.5832	0.6866	0.5830	0.6868	0.5944	0.6915	0.5941	0.6878
I2	0.8020	0.9280	0.8287	0.8979	0.8285	0.8978	0.8427	0.8923	0.8359	0.8981
I3	1.5090	1.7090	1.5431	1.6641	1.5413	1.6628	1.5443	1.6677	1.5486	1.6880
I4	3.3400	4.1400	3.4730	4.0090	3.4725	4.0093	3.4741	4.0152	3.4900	4.0405
I5	9.9000	12.9000	10.1910	13.0813	10.1702	13.0355	10.1708	13.0906	10.4751	12.7011
M1	0.3760	0.4440	0.3949	0.4268	0.3948	0.4267	0.3956	0.4251	0.3976	0.4235
M2	0.4170	0.4730	0.4314	0.4585	0.4313	0.4585	0.4292	0.4577	0.4345	0.4565
M3	0.4550	0.5210	0.4725	0.5065	0.4725	0.5026	0.4729	0.5044	0.4761	0.5013
M4	0.5230	0.5890	0.5298	0.5728	0.5298	0.5727	0.5402	0.5737	0.5418	0.5687
M5	0.6380	0.7060	0.6484	0.6938	0.6484	0.6937	0.6497	0.6851	0.6513	0.6937
M6	0.7210	0.7710	0.7302	0.7606	0.7302	0.7605	0.7342	0.7582	0.7364	0.7585
M7	0.8010	0.9290	0.8293	0.8980	0.8293	0.8979	0.8428	0.8925	0.8362	0.8983
M8	1.2050	1.2750	1.2135	1.2652	1.2105	1.2652	1.2140	1.2649	1.2257	1.2564
M9	1.3510	1.4050	1.3621	1.3900	1.3613	1.3899	1.3620	1.3900	1.3691	1.3977
M10	1.5090	1.7090	1.5426	1.6648	1.5420	1.6645	1.5457	1.6676	1.5487	1.6877
M12	3.4100	3.9900	3.5162	3.8900	3.5153	3.8905	3.5191	3.8938	3.5290	3.8749
M13	3.7900	4.3100	3.9005	4.2137	3.9004	4.2408	3.9091	4.2247	3.8665	4.1710
M14	8.0500	9.0500	8.3335	8.8759	8.3322	8.8755	8.3363	8.8793	8.2331	8.9251
M15	9.7000	11.7400	9.9187	11.6499	9.9162	11.6502	9.9169	11.6387	10.0329	11.3481
M16A	11.0600	13.0500	11.0951	12.6700	11.0684	12.6681	11.1041	12.6925	11.2984	12.6509
M16B	11.0600	13.0500	11.0983	12.6787	11.0727	12.6766	11.1015	12.6985	11.2986	12.6576
DNBM	0.4700	0.9600	-	-	-	-	0.4878	0.9069	0.4909	0.9003
DNBL	0.4700	0.9600	-	-	-	-	0.4910	0.9001	0.4907	0.9012

Table 2. Version descriptions of the VIIRS relative spectral response (RSR) products used in this study.

VIIRS RSR	Version Description
S-NPP GT	Government Team "Best" Spacecraft Level RSR for F1 VisNIR M Bands (7 Apr 2011) and I Bands (27 Jun 2011)
S-NPP NG	Northrop Grumman October 2011 RSR Release
NOAA-20	Data Analysis Working Group Release of J1 VIIRS RSR Version 2 (Version 2.1 for Band M9)
JPSS-2	Data Analysis Working Group Release of J2 VIIRS RSR Version 2

It should be acknowledged that examination of OOB contribution with respect to the specified 1% extended bandpass limits is inherently contradictory, given that OOB is only defined in terms of the true measured limits. That is, because specified limits are provided by the manufacturer before the instrument is built, true determination of the 1% response levels is inseparable from physical measurements. Therefore, it should be recognized that, in this paper, any OOB contribution that is said to be examined based on specified 1% extended bandpass limits in fact necessarily relies on measured response values that are inside of the specified lower and upper bounds. In other words, regardless of the actual response value associated with the specified limits, measured values within those limits are treated as part of the in-band region. Thus, the "1%" designation of the specified extended bandpass is in name only, and actual response levels for specified OOB contribution cases are less than 1%, which contradicts the OOB definition. As a result, examinations offered in this manuscript that are in terms of specified limits should be considered theoretical. The value of such examinations is in understanding the sensitivity of γ to a narrower set of response limits that signify a theoretical, alternative definition of OOB, provided it is understood that the results of specified limit examinations are inherently biased and serve only as a reference for relative interpretation. That is, this view fosters a means for analysis that allows one to visualize OOB signals with respect to different sets of limits, which is a way of illustrating how energy contributions within or outside of a defined extended bandpass change the integrated radiances L_{in} and L_{total}—a technique that is used in Section 4. To this end, testing with the limit values of the published specified 1% extended bandpass rather than some other arbitrary set of limits is a matter of convenience, and also reflects the presentation structure of the works of Moeller et al. and Schwarting et al., which offered the inspiration for this effort [1,2,16–19].

No VIIRS data, aside from RSR information, are used in this study. All L_{in} and L_{total} data are based on integrated hyperspectral radiance measurements. For solar reflective bands, the average OOB contribution $\overline{\gamma}$ is determined from RSR-integrated SCIAMACHY Level-1b Version-7.03 radiances from August 2002 through December 2010, where $\overline{L_{in}}$ and $\overline{L_{total}}$ are the mean values of hundreds to thousands (depending on the scene) of pseudo radiance pairs computed for each scene-relevant footprint:

$$\overline{\gamma} = \left| \frac{\overline{L_{in}}}{\overline{L_{total}}} - 1 \right| \times 100 \tag{2}$$

The solar reflective bands include I-bands I1–I3, M-bands M1–M10, and Day/Night Band (DNB) mid (MGS) and high (HGS) gain stages (where applicable). Band M11, with a central wavelength near 2.25 μm, is not evaluated due to poor SCIAMACHY spectral calibration quality [28]. Note that the S-NPP VIIRS DNB contribution is not investigated in this study owing to the absence of associated measured 1% extended bandpass limits [1,2]. A flowchart of the complete $\overline{\gamma}$ determination methodology is given in Figure 2. Alternatively, Figure 3 is a notated scatter plot of pseudo radiance L_i/L_{total} pairs that illustrates $\overline{\gamma}$ determination for an all-sky tropical ocean (ATO) scene. $\overline{L_{in}}$ and $\overline{L_{total}}$ are the y-axis and x-axis averaged datapoints, respectively, the ratio of which is equal to the slope of a linear regression that is forced through the origin. The linear nature of the scatter datapoints is expected for comparable Earth-scene spectra given the rather subtle difference between in-band and full-band RSR structure, which is found to be true for all scenes investigated in this study. That is, the small difference in signal contribution between in-band and full-band RSR integration, which this study aims to quantify, can be expressed as a constant value for the given scene type, with small uncertainty.

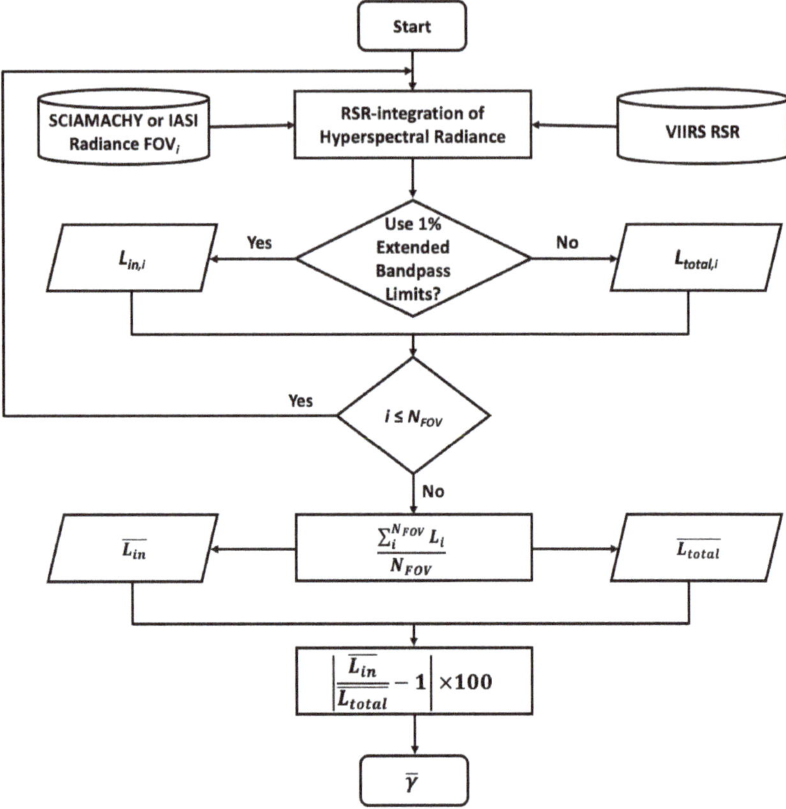

Figure 2. Flow chart of methodology for $\overline{\gamma}$ determination using i of N SCIAMACHY or IASI hyperspectral radiance fields of view (FOVs).

Eleven Earth-reflected scene types were evaluated, which are based on the distinct spectra offerings of the SBAF tool [43]. The scenes include deep convective clouds (DCC), ATO, clear-sky tropical ocean (CTO), the Libya-4 Pseudo Invariant Calibration Site (Lib-4 PICS) the Uyuni Salt Flats, and six land classifications defined by the International Geosphere–Biosphere Programme (IGBP) [45]. The Forest classification consists of IGBP IDs 1–5, Shrubland consist of IDs 6 and 7, Woodland is ID 8, Grassland consists of IDs 9 and 10, Wetland is ID 11, and Cropland consists of IDs 12 and 14. These classification groupings are based on approximate spectral similarity as relevant to OOB contribution assessment, which was determined empirically. Any remaining, unused IGBP IDs are either redundant with already considered scenes, or there was an insufficient number of SCIAMACHY measurements for that land type given the large FOV size. Accurate representation of the identified IGBP type by the SCIAMACHY footprint is ensured by requiring that the center and all four corners of the 30×240 km^2 SCIAMACHY FOV be of the same ID. Note that Bhatt et al. showed that the large size of the SCIAMACHY footprint does accurately represent the spectra of the Lib-4 PICS by comparing the influence of an SBAF determined using the Lib-4 PICS boundaries to that from a Libyan Desert PICS one-third the original Lib-4 size. They found the spectral radiance difference between the two Libyan Desert domains to be less than 0.6%, indicating minimal impact from the spatial disparity [43,45].

For IR bands, $\bar{\gamma}$ was determined from footprint-mean $\overline{L_{in}}$ and $\overline{L_{total}}$ using IASI Level-1c radiances acquired from the NOAA Comprehensive Large Array-Data Stewardship System (CLASS) archive, based on the combined time periods of January, April, July, and October 2008. The evaluated bands include I-bands I4–I5 and M-bands M12–M16, where M16 is separated into time-delay-integrated bands M16A and M16B [11,46–48]. Rather than by scene type, IR $\bar{\gamma}$ evaluation is separated by thermal infrared brightness temperature (IR BT) based on the integration of IASI hyperspectral radiance over the Aqua MODIS 11-μm band (band 31), that is then converted to temperature using the Planck function and the Aqua MODIS band 31 central wavelength (11.02 μm). The average OOB contribution is assessed for measurements with MODIS-integrated IASI IR BT that is less than 205 K and greater than 295 K, as well as the full dynamic range (FDR). The remote sensing calibration community should find the described method effective for scene-specific evaluation of OOB contribution to total signal, which is possible to perform pre-launch.

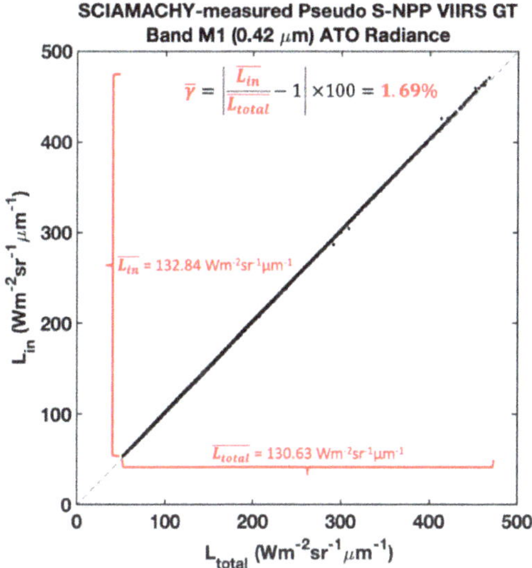

Figure 3. Notated scatter plot illustrating $\bar{\gamma}$ determination for an all-sky tropical ocean (ATO) scene. The ratio $\overline{L_{in}}/\overline{L_{total}}$ can also be thought of as the slope of a pseudo radiance pairs linear regression that is forced through 0 Wm^{-2}sr^{-1}μm^{-1}.

3. Results

Tables 3–10 summarize the SCIAMACHY-based Earth-reflected radiance scene-specific $\bar{\gamma}$ results for solar reflective bands within the reliable SCIAMACHY hyperspectral range of 0.24–1.75 μm, dependent on the selected RSR characterization versions of each VIIRS instrument. The tables are separated by instrument and version, and also by whether OOB contribution is evaluated using specified or measured lower and upper 1% extended bandpass limits [1,2,17,18]. For example, Table 3 presents results for S-NPP VIIRS GT RSR characterization for specified 1% extended bandpass limits, and Table 4 shows the same for measured 1% extended bandpass limits. Results for S-NPP VIIRS NG, NOAA-20 V2, and JPSS-2 V2 RSR characterization follow in the same manner. Entries in each table with bold text signify $\bar{\gamma}$ of at least 0.5%. This somewhat arbitrary 0.5% significance level was chosen loosely based on unofficial practices of the CERES Imager and Geostationary Calibration Group (IGCG) to achieve inter-calibration consistency that is better than 0.5%. It is a threshold that works well for this study in exemplifying the relative performance of the different VIIRS instruments with regard to $\bar{\gamma}$.

Table 3. S-NPP VIIRS GT RSR Earth-reflected scene-dependent radiance average OOB contribution ($\bar{\gamma}$) for specified 1% extended bandpass limits, shown in %.

Band	DCC	ATO	CTO	Lib-4	Uyuni	Forest	Shrubland	Woodland	Grassland	Wetland	Cropland
I1	0.00	0.00	0.00	0.00	0.00	0.00	0.00	0.00	0.00	0.00	0.00
I2	0.13	0.10	0.23	0.01	0.05	0.07	0.01	0.04	0.02	0.04	0.04
I3	0.01	0.01	0.02	0.02	0.01	0.02	0.01	0.02	0.01	0.01	0.02
M1	**0.83**	**1.67**	**2.46**	**1.28**	0.39	**1.05**	0.08	**1.21**	**0.62**	**1.54**	**0.87**
M2	**0.52**	**0.59**	**0.91**	**0.51**	0.14	**0.52**	0.05	**0.50**	0.22	**0.66**	0.37
M3	**0.57**	**0.63**	**0.89**	0.19	0.30	0.17	0.08	0.32	0.16	0.41	0.20
M4	0.26	0.18	0.06	0.25	0.22	0.43	0.25	0.19	0.10	0.42	0.13
M5	0.01	0.06	0.47	0.37	0.13	**1.87**	0.26	**0.71**	0.02	**1.13**	**0.55**
M6	0.07	0.06	0.25	0.14	0.02	0.31	0.17	0.19	0.14	0.20	0.21
M7	0.15	0.12	0.26	0.00	0.06	0.09	0.01	0.05	0.02	0.05	0.05
M8	0.02	0.05	0.05	0.05	0.04	0.06	0.04	0.06	0.05	0.05	0.05
M9	0.00	0.02	0.20	**1.23**	0.08	0.38	**0.94**	**0.76**	**0.71**	**0.57**	**0.79**
M10	0.01	0.01	0.02	0.02	0.01	0.02	0.01	0.02	0.02	0.01	0.02

Table 4. S-NPP VIIRS GT RSR Earth-reflected scene-dependent $\bar{\gamma}$ for measured 1% extended bandpass limits, shown in %.

Band	DCC	ATO	CTO	Lib-4	Uyuni	Forest	Shrubland	Woodland	Grassland	Wetland	Cropland
I1	0.00	0.00	0.00	0.00	0.00	0.00	0.00	0.00	0.00	0.00	0.00
I2	0.10	0.05	0.16	0.08	0.03	0.16	0.07	0.11	0.08	0.11	0.11
I3	0.00	0.04	0.03	0.03	0.02	0.05	0.02	0.04	0.03	0.03	0.04
M1	**0.83**	**1.69**	**2.50**	**1.27**	0.40	**1.08**	0.10	**1.24**	**0.64**	**1.58**	**0.89**
M2	0.35	**0.61**	**0.94**	0.49	0.16	**0.55**	0.03	**0.52**	0.24	**0.67**	0.39
M3	0.48	**0.64**	**0.90**	0.20	0.30	0.18	0.08	0.32	0.16	0.41	0.21
M4	0.27	0.19	0.04	0.24	0.22	0.38	0.27	0.16	0.10	0.37	0.10
M5	0.01	0.04	0.45	0.40	0.15	**1.94**	0.28	**0.74**	0.02	**1.19**	**0.56**
M6	0.02	0.06	0.09	0.31	0.12	**0.51**	0.33	0.37	0.30	0.38	0.40
M7	0.12	0.06	0.18	0.08	0.04	0.18	0.07	0.13	0.08	0.13	0.12
M8	0.03	0.08	0.10	0.10	0.06	0.12	0.10	0.11	0.10	0.11	0.11
M9	0.00	0.02	0.22	**1.27**	0.09	0.41	**1.01**	**0.79**	0.79	**0.59**	**0.85**
M10	0.01	0.04	0.04	0.03	0.02	0.05	0.03	0.04	0.03	0.02	0.04

Table 5. S-NPP VIIRS NG RSR Earth-reflected scene-dependent $\bar{\gamma}$ for specified 1% extended bandpass limits, shown in %.

Band	DCC	ATO	CTO	Lib-4	Uyuni	Forest	Shrubland	Woodland	Grassland	Wetland	Cropland
I1	0.02	0.06	0.18	0.08	0.01	0.21	0.06	0.10	0.00	0.16	0.07
I2	0.08	0.08	0.19	0.02	0.03	0.04	0.01	0.02	0.01	0.02	0.02
I3	0.01	0.01	0.01	0.02	0.01	0.02	0.01	0.02	0.01	0.02	0.01
M1	**0.90**	**1.52**	**2.16**	**0.97**	0.46	**0.77**	0.09	**1.00**	**0.54**	**1.27**	**0.70**
M2	**0.57**	0.37	**0.53**	0.23	0.13	0.20	0.01	0.24	0.12	0.33	0.17
M3	**0.59**	**0.60**	**0.83**	0.15	0.29	0.12	0.06	0.28	0.15	0.35	0.18
M4	0.26	0.16	0.10	0.28	0.23	0.47	0.23	0.21	0.10	0.46	0.15
M5	0.01	0.07	**0.51**	0.38	0.14	**1.90**	0.27	**0.73**	0.02	**1.16**	**0.55**
M6	0.09	0.11	0.37	0.15	0.01	0.31	0.18	0.19	0.14	0.19	0.22
M7	0.11	0.10	0.25	0.02	0.04	0.06	0.02	0.03	0.02	0.03	0.04
M8	0.03	0.04	0.05	0.04	0.03	0.05	0.04	0.05	0.04	0.04	0.05
M9	0.00	0.03	0.21	**1.37**	0.11	0.43	**1.11**	**0.87**	**0.86**	**0.70**	**0.92**
M10	0.01	0.01	0.02	0.02	0.01	0.02	0.01	0.02	0.01	0.01	0.02

Table 6. S-NPP VIIRS NG RSR Earth-reflected scene-dependent $\overline{\gamma}$ for measured 1% extended bandpass limits, shown in %.

Band	DCC	ATO	CTO	Lib-4	Uyuni	Forest	Shrubland	Woodland	Grassland	Wetland	Cropland
I1	0.02	0.06	0.19	0.09	0.02	0.22	0.07	0.10	0.00	0.17	0.07
I2	0.06	0.02	0.12	0.09	0.00	0.12	0.07	0.09	0.07	0.10	0.09
I3	0.00	0.04	0.03	0.03	0.02	0.05	0.02	0.04	0.02	0.03	0.03
M1	**0.94**	**1.57**	**2.21**	**0.93**	**0.50**	**0.82**	0.14	**1.05**	**0.58**	**1.32**	**0.75**
M2	0.26	0.38	**0.55**	0.21	0.15	0.22	0.01	0.26	0.14	0.34	0.19
M3	0.44	**0.60**	**0.83**	0.15	0.29	0.12	0.06	0.28	0.15	0.35	0.17
M4	0.27	0.18	0.08	0.26	0.23	0.42	0.25	0.19	0.10	0.42	0.12
M5	0.01	0.04	0.48	0.42	0.15	**1.97**	0.29	**0.75**	0.01	**1.22**	**0.57**
M6	0.04	0.02	0.22	0.33	0.12	**0.52**	0.34	0.37	0.30	0.37	0.40
M7	0.08	0.04	0.17	0.10	0.02	0.15	0.08	0.11	0.08	0.11	0.11
M8	0.04	0.08	0.09	0.10	0.07	0.11	0.10	0.10	0.09	0.10	0.10
M9	0.04	0.08	0.31	**1.53**	0.16	**0.54**	**1.25**	**1.00**	**1.00**	**0.82**	**1.07**
M10	0.00	0.04	0.04	0.03	0.02	0.05	0.02	0.04	0.03	0.02	0.04

Table 7. NOAA-20 VIIRS V2 RSR Earth-reflected scene-dependent $\overline{\gamma}$ for specified 1% extended bandpass limits, shown in %.

Band	DCC	ATO	CTO	Lib-4	Uyuni	Forest	Shrubland	Woodland	Grassland	Wetland	Cropland
I1	0.00	0.00	0.00	0.00	0.00	0.00	0.00	0.00	0.00	0.00	0.00
I2	0.07	0.01	0.01	0.00	0.01	0.01	0.00	0.00	0.00	0.00	0.00
I3	0.01	0.01	0.02	0.02	0.01	0.02	0.01	0.02	0.01	0.02	0.02
M1	0.06	0.10	0.14	0.08	0.02	0.04	0.00	0.05	0.03	0.08	0.03
M2	0.11	0.17	0.21	0.04	0.09	0.03	0.01	0.07	0.05	0.09	0.05
M3	0.18	0.07	0.08	0.00	0.06	0.03	0.01	0.01	0.01	0.01	0.00
M4	0.06	0.05	0.04	0.04	0.05	0.07	0.01	0.02	0.01	0.04	0.01
M5	0.06	0.06	0.06	0.06	0.06	0.23	0.04	0.06	0.01	0.12	0.05
M6	0.00	0.00	0.04	0.02	0.01	0.03	0.03	0.02	0.02	0.02	0.02
M7	0.11	0.03	0.05	0.01	0.02	0.02	0.01	0.01	0.00	0.00	0.01
M8	0.03	0.04	0.04	0.04	0.03	0.05	0.04	0.04	0.04	0.04	0.04
M9	0.00	0.02	0.20	**1.26**	0.10	0.40	**1.01**	**0.80**	**0.78**	**0.63**	**0.84**
M10	0.01	0.01	0.02	0.02	0.01	0.02	0.01	0.02	0.01	0.01	0.01
DMBMGS	0.00	0.00	0.00	0.00	0.00	0.00	0.00	0.00	0.00	0.00	0.00
DMBLGS	0.00	0.00	0.00	0.00	0.00	0.00	0.00	0.00	0.00	0.00	0.00

Table 8. NOAA-20 VIIRS V2 RSR Earth-reflected scene-dependent $\overline{\gamma}$ for measured 1% extended bandpass limits, shown in %.

Band	DCC	ATO	CTO	Lib-4	Uyuni	Forest	Shrubland	Woodland	Grassland	Wetland	Cropland
I1	0.00	0.00	0.00	0.01	0.00	0.01	0.01	0.01	0.00	0.02	0.01
I2	0.01	0.00	0.01	0.00	0.01	0.01	0.00	0.01	0.00	0.01	0.01
I3	0.01	0.03	0.03	0.02	0.02	0.05	0.02	0.04	0.02	0.02	0.03
M1	0.10	0.13	0.17	0.05	0.05	0.07	0.03	0.09	0.06	0.11	0.07
M2	0.12	0.18	0.22	0.02	0.11	0.03	0.02	0.08	0.06	0.10	0.06
M3	0.07	0.07	0.07	0.00	0.05	0.05	0.02	0.01	0.00	0.02	0.01
M4	0.06	0.05	0.04	0.05	0.05	0.06	0.01	0.01	0.01	0.04	0.01
M5	0.07	0.08	0.08	0.08	0.07	0.22	0.06	0.05	0.03	0.11	0.03
M6	0.04	0.06	0.03	0.10	0.06	0.12	0.10	0.10	0.09	0.09	0.10
M7	0.03	0.02	0.04	0.00	0.02	0.03	0.00	0.01	0.00	0.01	0.01
M8	0.03	0.07	0.08	0.09	0.06	0.10	0.08	0.09	0.08	0.10	0.09
M9	0.01	0.02	0.22	**1.32**	0.08	0.44	**1.08**	**0.85**	**0.85**	**0.69**	**0.91**
M10	0.01	0.03	0.03	0.02	0.02	0.04	0.02	0.04	0.02	0.02	0.03
DMBMGS	0.00	0.01	0.02	0.01	0.00	0.00	0.01	0.00	0.00	0.00	0.00
DMBLGS	0.00	0.00	0.01	0.01	0.00	0.00	0.00	0.00	0.00	0.01	0.00

Table 9. JPSS-2 VIIRS V2 RSR Earth-reflected scene-dependent $\bar{\gamma}$ for specified 1% extended bandpass limits, shown in %.

Band	DCC	ATO	CTO	Lib-4	Uyuni	Forest	Shrubland	Woodland	Grassland	Wetland	Cropland
I1	0.00	0.00	0.00	0.00	0.00	0.00	0.00	0.00	0.00	0.00	0.00
I2	0.00	0.00	0.00	0.00	0.00	0.00	0.00	0.00	0.00	0.00	0.00
I3	0.01	0.01	0.02	0.02	0.01	0.02	0.02	0.02	0.02	0.02	0.02
M1	0.03	0.03	0.04	0.01	0.01	0.02	0.01	0.02	0.01	0.02	0.02
M2	0.02	0.03	0.02	0.01	0.02	0.00	0.01	0.01	0.01	0.00	0.01
M3	0.05	0.02	0.02	0.01	0.01	0.01	0.00	0.01	0.01	0.02	0.01
M4	0.00	0.00	0.01	0.01	0.01	0.01	0.00	0.00	0.00	0.01	0.00
M5	0.06	0.07	0.09	0.06	0.06	0.17	0.04	0.04	0.02	0.07	0.03
M6	0.03	0.03	0.04	0.03	0.03	0.04	0.03	0.03	0.03	0.03	0.03
M7	0.04	0.01	0.02	0.00	0.01	0.01	0.00	0.01	0.00	0.01	0.01
M8	0.01	0.02	0.03	0.03	0.02	0.02	0.02	0.02	0.02	0.04	0.02
M9	0.01	0.09	**1.14**	**5.66**	0.21	**2.17**	**3.29**	**3.76**	**2.42**	**2.85**	**3.31**
M10	0.01	0.01	0.02	0.01	0.01	0.02	0.01	0.01	0.01	0.02	0.01
DMBMGS	0.00	0.00	0.00	0.00	0.00	0.00	0.00	0.00	0.00	0.00	0.00
DMBLGS	0.00	0.00	0.00	0.00	0.00	0.00	0.00	0.00	0.00	0.00	0.00

Table 10. JPSS-2 VIIRS V2 RSR Earth-reflected scene-dependent $\bar{\gamma}$ for measured 1% extended bandpass limits, shown in %.

Band	DCC	ATO	CTO	Lib-4	Uyuni	Forest	Shrubland	Woodland	Grassland	Wetland	Cropland
I1	0.00	0.00	0.00	0.00	0.00	0.00	0.00	0.00	0.00	0.00	0.00
I2	0.00	0.03	0.04	0.04	0.01	0.05	0.03	0.04	0.03	0.04	0.04
I3	0.01	0.03	0.03	0.02	0.02	0.05	0.02	0.04	0.03	0.03	0.03
M1	0.23	0.04	0.05	0.00	0.02	0.04	0.02	0.04	0.03	0.04	0.03
M2	0.02	0.03	0.02	0.02	0.03	0.01	0.01	0.01	0.01	0.01	0.01
M3	0.01	0.02	0.02	0.01	0.01	0.01	0.00	0.01	0.00	0.02	0.01
M4	0.01	0.00	0.01	0.01	0.01	0.00	0.00	0.00	0.00	0.01	0.00
M5	0.07	0.09	0.11	0.08	0.07	0.20	0.06	0.04	0.03	0.09	0.03
M6	0.05	0.06	0.08	0.07	0.06	0.09	0.07	0.08	0.07	0.08	0.08
M7	0.01	0.02	0.03	0.03	0.01	0.06	0.03	0.05	0.03	0.03	0.04
M8	0.01	0.03	0.04	0.04	0.03	0.04	0.04	0.04	0.04	0.04	0.04
M9	0.01	0.10	**1.15**	**5.70**	0.23	**2.21**	**3.39**	**3.80**	**2.50**	**2.88**	**3.38**
M10	0.01	0.03	0.02	0.02	0.02	0.05	0.02	0.04	0.02	0.03	0.03
DMBMGS	0.00	0.00	0.01	0.01	0.00	0.00	0.01	0.00	0.00	0.00	0.00
DMBLGS	0.00	0.00	0.01	0.01	0.00	0.00	0.00	0.00	0.00	0.01	0.00

The cause of the scene-dependent variation in $\bar{\gamma}$ can be interpreted from visualization of the selected RSR, with its associated lower and upper 1% extended bandpass boundaries, overlaid with hyperspectral radiance spectra of the various Earth scenes. Such visualizations have been prepared as Figures 4–6. For example, Figure 4 highlights the significant OOB signal of the S-NPP VIIRS M1 (~0.42 μm) band, in this case based on the NG RSR characterization effort, and how that might respond to the Earth-reflected spectra of common inter-calibration targets like DCC, ATO, CTO, and the Lib-4 PICS. Figure 5 reveals scene-dependent $\bar{\gamma}$ behavior for the S-NPP VIIRS NG M5 (~0.67 μm) band, which is an important spectral channel for inter-calibration and parameterization efforts [6,7,49]. As seen in Tables 7–10, only the M9 (~1.38 μm) bands offer $\bar{\gamma}$ values in exceedance of 0.5% for both NOAA-20 and JPSS-2, the nature of which can be evaluated in Figure 6 for select scenes.

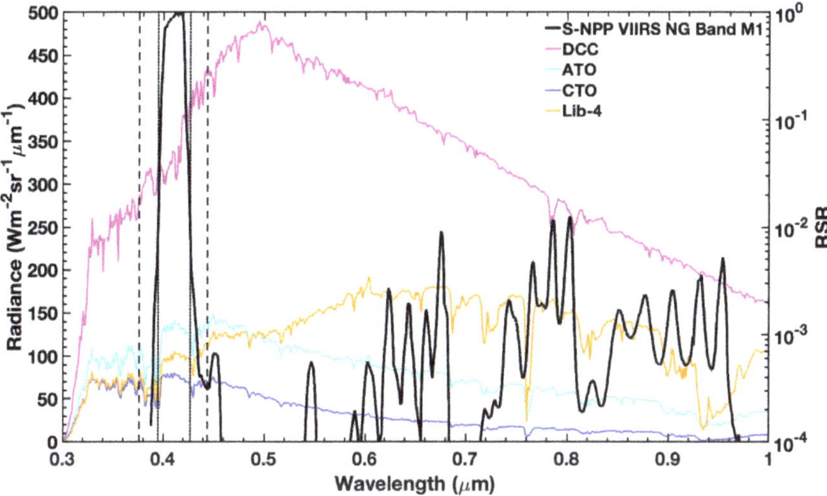

Figure 4. S-NPP VIIRS NG band M1 RSR overlaid with SCIAMACHY Earth-reflected radiance. Response is displayed on a logarithmic scale in order to better highlight the OOB magnitudes. Vertical dashed lines indicate the specified lower and upper 1% response limits, and the vertical dotted lines indicate the measured lower and upper 1% response limits.

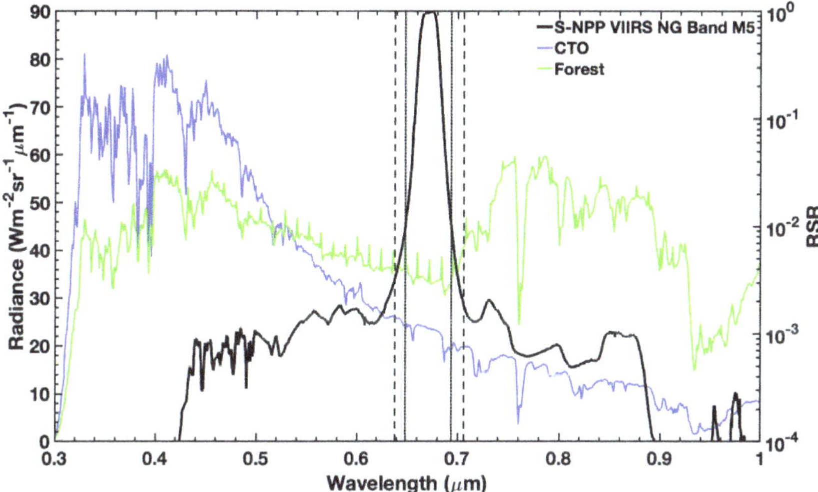

Figure 5. Same as Figure 4, except with an S-NPP VIIRS NG band M5 example.

Table 11 shows the IASI-based Earth-emitted and Earth-reflected radiance scene-specific $\overline{\gamma}$ values for VIIRS bands with central wavelengths greater than 3.6 μm, dependent on the selected RSR characterization versions of each VIIRS instrument. Note that the Earth-reflected contribution is only relevant for the mid-wave IR (~3–4 μm) bands I4 (~3.74 μm), M12 (~3.69 μm), and M13 (~4.06 μm) during daytime. The "<205 K" and ">295 K" columns signify $\overline{\gamma}$ results for IASI footprints in which the Aqua MODIS 11-μm band-integrated (i.e., band 31) IASI IR BT is greater than or less than the indicated value, such as to separate evaluation based on Earth-emitted temperature. The column label "FDR" signifies that the full dynamic range of IASI measurements was considered, i.e., without any BT-based truncation. The associated Figure 7a highlights the OOB signal of the S-NPP VIIRS GT band M13 RSR, and how the impact of in-band vs. OOB magnitude can change depending on scene temperature. Figure 7a allows for close analysis of the in-band measurement range, whereas Figure 7b uses alternative axes scaling limits in order to grant full view of the band M13 OOB measurement range.

Table 11. VIIRS RSR Earth-emitted (and Earth-reflected for daytime I4, M12, and M13 bands) temperature-dependent radiance average OOB contribution ($\overline{\gamma}$) for specified and measured 1% extended bandpass limits, shown in %, analyzed separately where Aqua MODIS 11-μm band-integrated IASI infrared brightness temperature (IR BT) is less than 205 K, is greater than 295 K, and with no BT limit specified (i.e., full dynamic range or FDR).

Instrument	Band	Specified			Measured		
		<205 K	FDR	>295 K	<205 K	FDR	>295 K
S-NPP VIIRS GT	I4	0.00	0.00	0.00	0.07	0.00	0.00
	I5	0.14	0.28	0.29	0.20	0.23	0.23
	M12	0.02	0.01	0.01	0.15	0.02	0.03
	M13	0.09	0.04	0.04	**2.39**	**0.59**	**0.62**
	M14	0.01	0.02	0.02	0.02	0.01	0.01
	M15	0.08	0.09	0.09	0.05	0.11	0.11
	M16A	0.11	0.03	0.03	0.11	0.04	0.04
	M16B	0.10	0.03	0.02	0.09	0.03	0.03
S-NPP VIIRS NG	I4	0.00	0.00	0.00	0.06	0.00	0.00
	I5	0.03	0.03	0.03	0.00	0.00	0.00
	M12	0.02	0.01	0.01	0.15	0.02	0.03
	M13	0.08	0.03	0.04	**0.76**	0.17	0.19
	M14	0.00	0.00	0.00	0.01	0.01	0.01
	M15	0.07	0.07	0.07	0.05	0.10	0.10
	M16A	0.08	0.00	0.01	0.08	0.01	0.00
	M16B	0.08	0.00	0.01	0.07	0.01	0.00
NOAA-20 VIIRS V2	I4	0.00	0.00	0.00	0.06	0.00	0.00
	I5	0.04	0.05	0.05	0.00	0.00	0.00
	M12	0.03	0.01	0.01	0.15	0.03	0.04
	M13	0.06	0.02	0.03	**0.91**	0.20	0.21
	M14	0.00	0.00	0.00	0.01	0.01	0.01
	M15	0.07	0.07	0.07	0.04	0.09	0.09
	M16A	0.08	0.00	0.01	0.08	0.01	0.00
	M16B	0.08	0.00	0.01	0.07	0.01	0.00
JPSS-2 VIIRS V2	I4	0.00	0.00	0.00	0.07	0.00	0.00
	I5	0.00	0.00	0.00	0.00	0.00	0.00
	M12	0.01	0.01	0.01	0.13	0.03	0.03
	M13	0.02	0.01	0.01	0.22	0.14	0.15
	M14	0.01	0.00	0.00	0.01	0.01	0.01
	M15	0.00	0.00	0.00	0.02	0.02	0.02
	M16A	0.01	0.00	0.00	0.01	0.00	0.00
	M16B	0.01	0.00	0.00	0.01	0.00	0.00

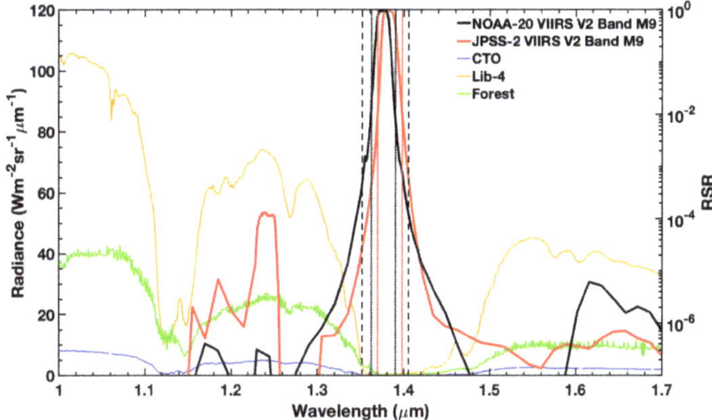

Figure 6. Same as Figure 4, except with NOAA-20 VIIRS V2 and JPSS-2 VIIRS V2 band M9 examples and the right y-axis limit extended to 10^{-7}. Here, the red vertical dotted lines indicate the measured lower and upper 1% response limits for JPSS-2 VIIRS. Both instruments have the same specified lower and upper 1% response limits (black vertical dashed lines).

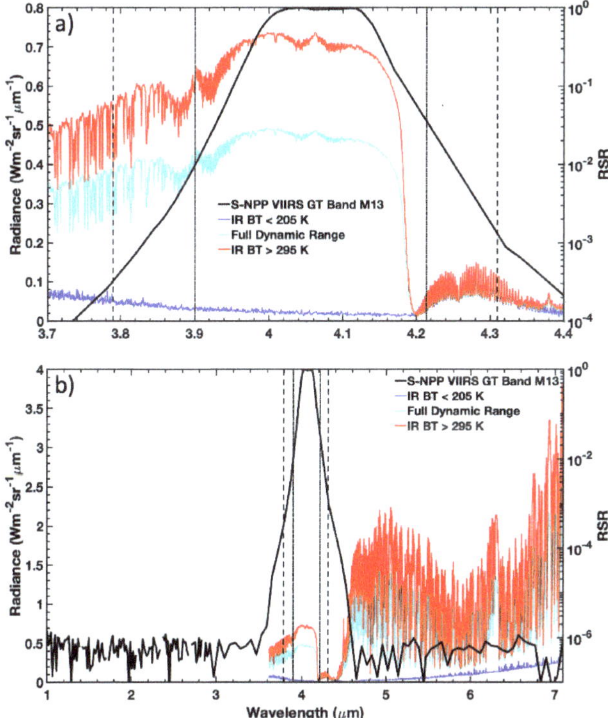

Figure 7. (**a**) Same as Figure 4, except with S-NPP VIIRS GT Band M13 RSR overlaid with IASI Earth-reflected (day only) and Earth-emitted (day and night) radiance for hyperspectral footprint measurements where Aqua MODIS 11-µm band-integrated IASI IR BT is less than 205 K, is greater than 295 K, and with no BT limit specified (i.e., full dynamic range). (**b**) Rescaled version of (a) with broader x-axis, left y-axis, and right y-axis limits in order to reveal the full OOB measurement range of S-NPP VIIRS GT band M13. Note that 3.6 µm is the lower wavelength limit of IASI.

Table 12 summarizes $\bar{\gamma}$ for VIIRS mid-wave IR bands I4, M12, and M13 based on measured 1% extended bandpass limits, separated by day and night IASI measurements. As in Table 11, Aqua MODIS 11-μm band-integrated IASI IR BT limits are also considered. With an overall smaller magnitude compared to Figure 7a, Figure 8a reveals the nighttime-only, temperature-dependent Earth-emitted radiance spectra with JPSS-2 VIIRS V2 band M13 RSR and extended bandpass limit information overlaid. Even though the JPSS-2 VIIRS V2 band M13 OOB measurement range is significantly narrower than that of S-NPP VIIRS GT, Figure 8b offers the same rescaled view as that of Figure 7b for the sake of comparison.

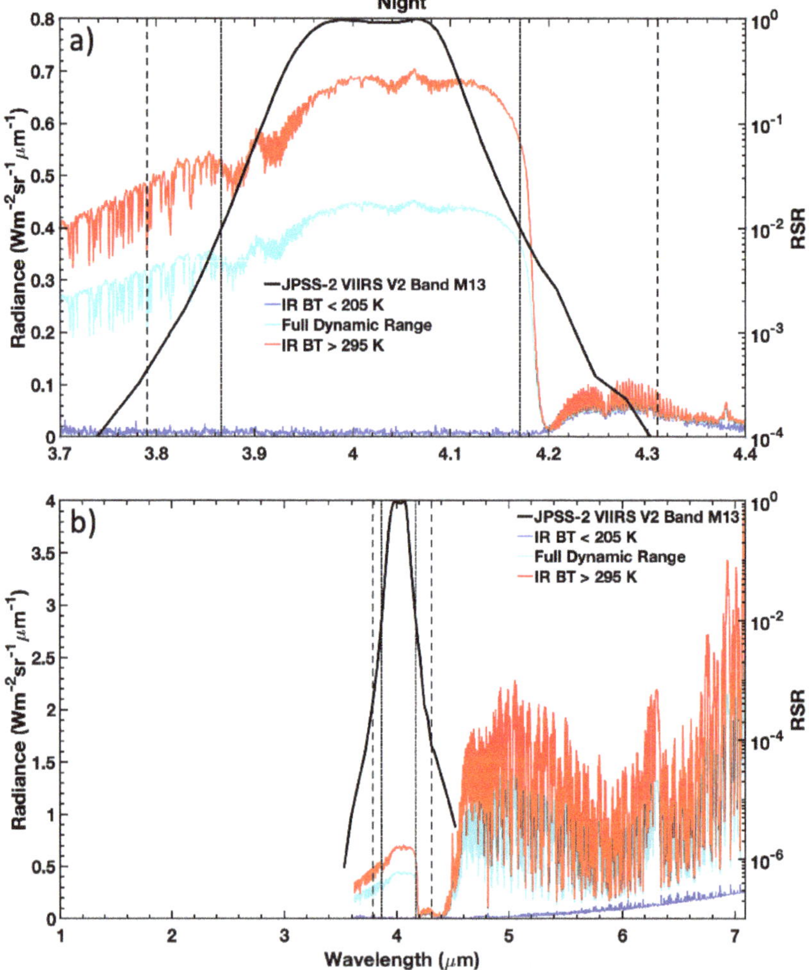

Figure 8. (**a**) Same as Figure 7a, except with JPSS-2 VIIRS V2 band M13 RSR overlaid with nighttime-only IASI Earth-emitted radiance. (**b**) Rescaled version of (a) for comparative study with Figure 7b.

Table 12. Same as Table 11, except separated by day and night, and based only on measured 1% extended bandpass limits for bands I4, M12, and M13.

Instrument	Band	Day			Night		
		<205 K	**FDR**	**>295 K**	**<205 K**	**FDR**	**>295 K**
S-NPP VIIRS GT	I4	0.07	0.01	0.00	0.02	0.01	0.01
	M12	0.16	0.00	0.02	0.10	0.05	0.05
	M13	**1.52**	**0.57**	**0.61**	**5.71**	**0.61**	**0.64**
S-NPP VIIRS NG	I4	0.06	0.00	0.00	0.05	0.05	0.05
	M12	0.15	0.01	0.02	0.09	0.05	0.05
	M13	**0.56**	0.16	0.18	**1.53**	0.19	0.20
NOAA-20 VIIRS V2	I4	0.07	0.01	0.00	0.05	0.01	0.01
	M12	0.16	0.01	0.03	0.09	0.06	0.05
	M13	**0.64**	0.19	0.21	**1.96**	0.22	0.22
JPSS-2 VIIRS V2	I4	0.07	0.01	0.00	0.04	0.01	0.01
	M12	0.14	0.01	0.02	0.07	0.05	0.05
	M13	0.16	0.14	0.15	0.49	0.15	0.15

4. Discussion

The overall magnitude of OOB Earth-scene-dependent radiance contribution is much more significant for the S-NPP VIIRS instrument compared to the later NOAA-20 and JPSS-2 sensors. This finding is true for solar reflective bands and IR bands, as evidenced by the greater frequency of bold table entries, which signify $\bar{\gamma}$ of 0.5% or greater, for S-NPP compared to the other platforms. For example, the S-NPP VIIRS GT and NG characterization versions have six solar reflective bands with $\bar{\gamma}$ of at least 0.5% for at least one scene type (although up to ten scene types for a single band in case of NG band M1), based on measured lower and upper 1% extended bandpass limits (Table 4, Table 6). For NOAA-20 and JPSS-2 VIIRS V2 characterization, by comparison, only solar reflective band M9 exhibits $\bar{\gamma}$ of at least 0.5%, in this case across six (NOAA-20) or eight (JPSS-2) scenes based on measured extended bandpass limits (Table 8, Table 10). Finally, although overall $\bar{\gamma}$ magnitudes for NOAA-20 and JPSS-2 VIIRS are small compared to those of S-NPP VIIRS (also including where $\bar{\gamma}$ does not exceed 0.5%), the largest OOB contribution is found for JPSS-2 VIIRS V2 band M9 over the Lib-4 PICS, with a magnitude of 5.70% (5.66%) based on measured (specified) extended bandpass limits (Tables 9 and 10). The smallest $\bar{\gamma}$ for this band and scene is that of S-NPP VIIRS GT, with a magnitude of 1.27% (1.23%) for measured (specified) extended bandpass limits, which is still a rather significant OOB influence (Tables 3 and 4). The reason $\bar{\gamma}$ values are notably large for this band is related to the low spectral signal at these wavelengths (~1.38 μm).

The cause of the larger S-NPP VIIRS band M1 $\bar{\gamma}$ values across most scene types can be interpreted from Figure 4. In this figure, significant OOB response is observed up to a wavelength of nearly 1.0 μm, with the maximum OOB response peak exceeding 0.01 twice near 0.8 μm. By comparison, maximum OOB response for NOAA-20 and JPSS-2 VIIRS V2 band M1 is ~0.0006, with fewer OOB response peaks beyond 0.5 μm (not shown). Of the four selected scenes of Figure 4, the greatest OOB influence occurs for CTO, the darkest scene having a maximum $\bar{\gamma}$ of ~2.2%, whereas the least influence occurs for the for DCC and Lib-4, the brightest scenes having $\bar{\gamma}$ values of ~0.9–1.0% (Tables 5 and 6). It is intuitive that darker scenes are most influenced by OOB radiance contribution given that even minimal additional OOB energy measured by the sensor is significant compared to the already low signal of the in-band measurement. That is, a low signal is susceptible to noise. This concept does not discount the importance of OOB consideration for bright scenes, however, as evident by the ~1.0% $\bar{\gamma}$ values. Furthermore, the fact that S-NPP VIIRS band M1 $\bar{\gamma}$ values for CTO and DCC each differ from that for ATO (Tables 3–6) suggests that ATO OOB contribution is not strictly represented by a constant factor (as otherwise suggested in Figure 3), but rather is sensitive to the radiance magnitude and the changing spectral composition of the scene. Nevertheless, ATO retrieval applications, such as those used for CERES calibration, are designed to accommodate average conditions with an acceptable level of uncertainty, and therefore the ATO average OOB contribution as given is appropriate, especially because the ATO $\bar{\gamma}$ value is, as expected, found to be roughly the average of that for CTO and DCC [49].

Even though $\bar{\gamma}$ results based on the specified 1% extended bandpass considerations should not be treated as absolute, but rather are theoretical as previously discussed in Section 2, their examination provides a point of reference that helps illustrate how energy contributions relative to varying OOB definitions affect integrated radiance. Thereby it is interesting to note that if considering specified 1% extended bandpass limits, Lib-4 $\bar{\gamma}$ exceeds that of DCC by 0.07%, with Lib-4 at 0.97% and DCC at 0.90% (Table 5). For measured extended bandpass limits, however, the $\bar{\gamma}$ are within 0.01%, with Lib-4 at 0.93% and DCC at 0.94% (Table 6). In other words, between the use of measured limits to specified limits, an increase in OOB radiance contribution of 0.04% is observed for DCC. This increase is owed to the greater values of integrated radiance allowed by the further extension of the specified limit in the increasing-wavelength direction from the central wavelength of ~0.42 μm (Figure 4). That is, the greater integrated radiance allowed by the specified extended bandpass limits results in better agreement with the integrated contribution from OOB response at high wavelengths, and thus $\bar{\gamma}$ based on the specified extended bandpass limits is less than that based on the measured limits. Similarly, a decrease in OOB radiance contribution of 0.04% is observed for Lib-4, going from specified to measured limits. In this case, the further extension of the specified limit allows for more integration of low radiance values at ultraviolet (UV) wavelengths of less than 0.4 μm compared to that for the measured limits. This additional integration of low UV-reflected radiance in part offsets the integrated radiance gained at the higher OOB wavelengths, which results in a larger $\bar{\gamma}$ for specified 1% extended bandpass limits compared to that of measured limits.

For Earth-monitoring efforts like CERES, satellite records must be combined seamlessly in order to avoid discontinuities in retrievals that arise from either radiometric scaling errors or varying algorithm assumptions. Therefore, it is important to particularly examine the average OOB contribution for the I1 (~0.64 μm) and M5 bands given the historic and continued proliferation of similar channels on Earth-observing imagers and their importance to inter-calibration and cloud parameterization efforts [6,7,49]. Scene-dependent $\bar{\gamma}$ significance should, it appears, generally not be a concern for the NOAA-20 and JPSS-2 VIIRS instruments in these bands (Tables 7–10). For S-NPP VIIRS, OOB contributions are also similarly minimal for band I1, with maximum magnitudes of ~0.2% for CTO and Forest scene types. Although these values could be significant in any application that relies exclusively on such views, the effect is mitigated when radiance contribution from these scenes is combined with other scenes likely found within the instrument FOV, e.g., CERES inter-calibration relies on ATO rather than CTO. Note that in some cases, CERES inter-calibration and other applications do rely on a single scene, such as DCC [49–52]. Fortunately, OOB contributions in band I1 and M5 are minimal in such a case. Nevertheless, it is advisable that CERES applications of VIIRS measurements consider the full-band RSR when inter-calibrating or for other applications in which spectral integration is necessary, e.g., atmospheric transmissivity determinations [3].

Although the I1 band appears to be relatively unimpacted by scene-dependent OOB contribution, the S-NPP VIIRS M5 band does exhibit rather large $\bar{\gamma}$ for ocean and vegetative scene types. In Figure 5, the S-NPP VIIRS NG M5 band is examined particularly with CTO (~0.5%) and Forest (~2%) given that CTO is a subset of ATO, which is relied upon in CERES inter-calibration, and Forest has the largest $\bar{\gamma}$ value of all M5 band scene assessments. Even though the OOB response for M5 does not reach the magnitude of that for M1, the response coupled with relatively high CTO radiance values near ~0.42–0.50 μm results in 0.5% OOB contribution, even considering counter-acting OOB contribution from measurements at wavelengths greater than the upper 1% extended bandpass limit. In the case of Forest spectra, the M5 in-band contributions are situated within in a relative minimum, i.e., a spectral "valley," compared to immediate lower and higher wavelength ranges. As such, OOB integrated radiance contribution is compounded by strong signals on either side of the central wavelength with comparable response on the order of about 0.001. How this average OOB contribution might influence environmental retrievals can be examined through normalized difference vegetation index (NDVI) determinations, which are calculated from visible (VIS: ~0.65 μm) and near infrared (NIR: ~0.86 μm) radiance measurement ratios [NDVI = (NIR − VIS)/(NIR + VIS)] [53–56]. When considering measured 1% extended bandpass limits (Table 6), the S-NPP VIIRS NG bands M5 and M7 (~0.86 μm) 1.97% and 0.15%, respectively, OOB contributions would amount to less than a 1% change in NDVI for a rough median value of measured Forest radiance (i.e., ~24 $Wm^{-2}sr^{-1}\mu m^{-1}$ for VIS and ~55 $Wm^{-2}sr^{-1}\mu m^{-1}$ for NIR). The influence is small, but if relying on S-NPP VIIRS for NDVI determination, bands I1 and I2 (~0.86 μm) should be favored over the comparable M-bands when considering OOB-contributed error, because any reduction in contrast between the VIS and NIR bands owing to the OOB signal is undesirable. Although these OOB effects may only rarely require consideration within specialized, environmental retrieval subsets of the remote sensing community, there is, nonetheless, value to be found in improved understanding of scene-specific OOB contribution to the total signal, especially with regard to pre-launch evaluation.

It is interesting to note that despite the prominence of significant OOB contribution across many S-NPP VIIRS bands and scene types, band-specific behaviors, in terms of overall $\bar{\gamma}$ for different targets, do not necessarily reflect the performance metric findings of the Moeller et al. measured MIOOB analyses. Although $\bar{\gamma}$ and MIOOB are not equivalent metrics, there is value, nevertheless, in acknowledging how failed specification as determined by measured MIOOB assessment, i.e., designated as out-of-specification in Moeller et al.'s Table 6, relates to the scene-dependent OOB contribution results presented in this manuscript [2]. As an example, of the VIS/NIR bands (i.e., M1–M7, I1, and I2) the overall highest MIOOB values are found for band M4 (~0.55 μm) at 3.80% and 3.65% for the S-NPP VIIRS GT and NG RSR products, respectively. In terms of $\bar{\gamma}$, however, OOB effects in the M4 band remain consistently low (in a relative sense) across all scene types for both S-NPP VIIRS GT and NG. Conversely, whereas band M2 (~0.44 μm) measured MIOOB metrics meet specification for both S-NPP VIIRS RSR products, $\bar{\gamma}$ for this band exceeds the chosen 0.5% significance level for five scenes in the case of the GT RSR product, and for one scene in the case of the NG RSR product (Table 4, Table 6). Again, MIOOB and $\bar{\gamma}$ magnitudes are not directly comparable, but from a relative perspective, these findings support the idea that, with regard to performance specification, there is value in the scene-dependent evaluation of OOB contribution, knowledge of which complements the understanding of MIOOB metrics.

Before continuing with a closer examination of the NOAA-20 and JPSS-2 VIIRS and IR band results, it should be acknowledged that strong, spectrally dependent degradation in the S-NPP VIIRS mirror reflectance has modulated the VIIRS RSRs in the solar reflective bands, which prompted the generation and release of degradation modulated S-NPP VIIRS RSRs by the NASA VIIRS Characterization Support Team (VCST). The modulated RSRs have been used for operational sensor data record production since 5 April 2013 [15,44,57,58]. In order to assess the impact of a modulated RSR with regard to OOB contribution, $\bar{\gamma}$ was evaluated for the eleven Earth-reflected scene types using band M1, which of all bands is predicted to have the largest radiance error after the four VIIRS mirrors are completely degraded [14]. The evaluation is based on the measured 1% extended bandpass limits of the S-NPP-VIIRS NG spectral performance characterization effort, because the VCST used the NG product as the baseline RSR in coming up with the modulated RSR [2,44]. On average, use of the M1 modulated RSR reduced scene-dependent $\bar{\gamma}$ by 25% of the original Table 6 values (ranging from 14% for Lib-4 to 50% for Shrubland). These percentages translate to an average reduction in $\bar{\gamma}$ magnitude of 0.22% (ranging from 0.07% for Shrubland to 0.50% for CTO). Shrubland has both the largest percentage reduction and the smallest magnitude reduction owing to its initially small $\bar{\gamma}$ value of 0.14 (Table 6). Overall, these are significant reductions, which demonstrate the benefit of the modulated RSR in terms of $\bar{\gamma}$. Nevertheless, even with this reduction in $\bar{\gamma}$ across all evaluated scenes, only two of the ten categories that initially had a $\bar{\gamma}$ value above the 0.5% significance level fell below that threshold when using the modulated RSR (i.e., Uyuni, which dropped from 0.50% to 0.35%, and Grassland, which dropped from 0.58% to 0.44%). Therefore, given that the Moeller et al. published extended bandpass limits predate formulation of the modulated RSRs, and because the M1 band is the most influenced by the mirror degradation, further examination of the remaining bands is left for future efforts. The M1 band results suggest that scenes with the most influence from OOB contribution still remain significantly affected even with this update to the operational RSR, and lesser impact from the modulated RSR implementation is expected for the other bands [15].

Although NOAA-20 and JPSS-2 VIIRS V2 average OOB contribution is largely better than that of S-NPP VIIRS, Figure 6 offers close examination of the band M9 exception. For either VIIRS, the in-band integrated radiance for Lib-4 spectra is on the order of 0.6 $Wm^{-2}sr^{-1}\mu m^{-1}$, and less than that for Forest and CTO scenes. Given such a small in-band signal, it is intuitive that even minimal OOB energy contribution could have a significant impact on the radiance measurement, especially if that OOB contribution occurs at wavelengths with substantially more energy. This is exactly the case for JPSS-2 VIIRS V2, for which a relatively small (~0.0001) OOB peak at ~1.24 µm can amount to over 5% in OOB radiance signal for the Lib-4 PICS. Furthermore, because CTO, Forest, Shrubland, Woodland, Grassland, Wetland, and Cropland share a similar spectral signature with regard to a low signal near ~1.38 µm and an increased signal near ~1.24 µm, their respective $\bar{\gamma}$ values are also high, although not as high as that for Lib-4. This substantial signal-to-noise sensitivity for ~1.38 µm bands favors high cloud detection, which is why CERES inter-calibration efforts cannot use PICS methods for such imager channels, and instead employ DCC-based calibration techniques to characterize the signal at a much higher magnitude near ~50 $Wm^{-2}sr^{-1}\mu m^{-1}$ [51]. Finally, although NOAA-20 VIIRS does not have the same OOB peak near ~1.24 µm as JPSS-2 VIIRS has, it does, nevertheless, have a strong OOB signal just outside the lower specified 1% extended bandpass limit, which coincides with the rapid increase in the Lib-4 (as well as other scenes) radiance spectral signature. Additionally, compared to JPSS-2, NOAA-20 VIIRS has a stronger OOB signal beyond 1.6 µm, although at a rather weak response of less than 10^{-5}. Albeit not as severe as the JPSS-2 VIIRS case, the small leak beyond the extended bandpass combined with the heightened OOB signal in the 1.6–1.75 µm range (shown only up to the range of 1.7 µm in Figure 6) is enough to cause significantly more than 1% in OOB radiance contribution to the imager measurement.

For VIIRS IR bands, with each instrument except for that on JPSS-2, a significant OOB influence of at least 0.5% is found only for band M13, especially for cold scenes, i.e., ~11-µm IR BT IASI footprint values smaller than 205 K (Table 11). The cause once again is due to a low in-band radiance signal (especially for cold measurements) being dominated by OOB leaks at high relative energy, which can be interpreted from Figure 7a following previously described analysis methods. In short, a significant portion of the OOB contribution is sourced from the energy in the wavelength range between the M13 specified and measured 1% extended bandpass limits, as evidenced by the at least 0.55% difference in $\bar{\gamma}$ for all three BT ranges, with all $\bar{\gamma}$ for the specified column being 0.09% or less (Table 11). The energy contribution from outside of this in-band-focused view of S-NPP VIIRS GT band M13 is insignificant. Nevertheless, a broader view of the complete OOB signal is given in Figure 7b, with x-axis limits based on the published OOB measurement range of 1000–7096 nm [1]. The RSR axis (right y-axis) lower limit has been extended from 10^{-4} to 10^{-7} in order to fit the band M13 response. The amount of $\bar{\gamma}$ sourced from this level of response, even for the relatively broad wavelength range covered and considering the higher values of integrated radiance compared to those of the in-band region, is small relative to the impact of OOB contribution from the higher response regions nearer to the extended bandpass limits, as discussed above. Specifically, $\bar{\gamma}$ based on integration starting at 4.4 µm (a relative radiance minimum) and ending at the upper OOB measurement range amounts to less than 0.006% for all three BT ranges.

The fact that M13 is a mid-wave IR band suggests there could be a diurnal dependency owed to a daytime solar contribution. This idea is explored in Table 12 for all mid-wave IR bands, in which significantly greater $\bar{\gamma}$ values are found for nighttime measurements due to the overall decrease in energy at night, again with most significance for the coldest measurements. Daytime $\bar{\gamma}$ values are expectedly smaller in magnitude than those for the combined day and night analysis of Table 11, and although no additional bands, regardless of temperature, exceed the chosen 0.5% $\bar{\gamma}$ significance level at night, neither do previously significant bands drop below the 0.5% $\bar{\gamma}$ level during the day. Figure 8a not only reveals a significantly smaller in-band radiance for IR BT < 205 K, which is the cause of the 5.71% $\bar{\gamma}$ level for S-NPP VIIRS GT at night, but also offers evidence as to why these effects are not observed for JPSS-2 VIIRS V2. For S-NPP VIIRS GT, the measured upper 1% extended bandpass limit is at 4.2137 μm, which is near where the radiance signal begins to increase and the RSR is relatively strong. For JPSS-2 VIIRS V2, the measured upper 1% extended bandpass limit is at 4.171 μm, which is short of the signal increase that begins near 4.2 μm. Furthermore, the JPSS-2 VIIRS V2 RSR in this (>4.2 μm) OOB region is steeper and of lesser magnitude compared to those for S-NPP VIIRS GT. Figure 8b is given with the same axes limits as those of Figure 7b in order to highlight the OOB signal differences between the S-NPP VIIRS GT and JPSS-2 VIIRS V2 band M13, the latter case being significantly more constrained. For the remaining IR bands where radiance magnitudes are higher overall, OOB contribution to the total signal is minimal.

5. Conclusions

On-orbit operational and climate-monitoring measurements by Earth-overserving instruments rely on well-characterized spectral performance—a critical aspect of imager pre-launch testing. Understanding spectral performance is necessary for recognizing the proper implementation and accuracy of radiometric calibration efforts and the products and research endeavors that rely on them, e.g., CERES. Critical laboratory experiments allow for characterization of the full optical path and any optical or electronic cross talks for nearly all VIIRS bands, but they cannot account for the specific spectral signatures measured by Earth-observing imagers. Therefore, it is difficult to tie pre-launch spectral performance metrics to OOB behavior for many varied Earth-viewed scenes. This study quantifies the OOB contribution to the total VIIRS signal and how it changes based on instrument version, channel, extended bandpass limits, and scene type, knowledge of which is important for scene-dependent inter-calibration efforts and environmental and cloud product retrievals. The results inform users of the target-dependent impact of published extended bandpass limits for methods that allow for selective RSR integration. It is appropriate for the remote sensing community to rely on the VIIRS channel-measured 1% extended bandpass, which will be sufficient in most applications. Inter-calibration or retrieval efforts that are dependent on certain scene types for which OOB is significant, however, may require consideration of the full-band VIIRS RSR for improved accuracy or at least for understanding of the potential sources of bias.

The OOB contribution to total signal was assessed using modified methodologies for SCIAMACHY- and IASI-based SBAF computation, employing ratio analysis of VIIRS RSR in-band and full-band integrated radiance FOVs. This method can be used to evaluate scene-specific OOB contribution in pre-launch spectral characterization efforts, which may be of interest to product teams that rely on specific scene conditions. This paper not only quantifies the scene dependence and influence of specified vs. measured 1% extended bandpass limits, but also provides visualization of the OOB contribution for selected targets and VIIRS bands. It was shown that S-NPP VIIRS is, overall, subject to a greater magnitude of OOB Earth-scene-dependent radiance contribution compared to that for the later NOAA-20 and JPSS-2 VIIRS instruments. That said, the OOB contribution for JPSS-2 VIIRS V2 band M9 has a magnitude of 5.7% for the Lib-4 PICS, which is the largest $\bar{\gamma}$ value found for solar reflective bands. The fact that dark scenes, as is the case for Lib-4 in the M9 band, are most influenced by OOB radiance contribution is intuitive, given that a small amount of outside energy is significant compared to the low signal of the in-band measurement, i.e., a low signal dominated by noise. This signal-to-noise consideration was similarly the main cause of OOB-contributed error for the M13 band, particularly for cold scenes at night, reaching a $\overline{\gamma}$ magnitude of 5.7% in the case of S-NPP VIIRS GT.

It is important that satellite records be combined seamlessly for long-term Earth-monitoring efforts like CERES. The aim is to minimize retrieval discontinuities, and thus particular examination of OOB contribution for the VIIRS I1 and M5 bands is valuable. Of the VIIRS instruments, only that on S-NPP elicits a need for meaningful consideration of OOB contribution in these channels, and only in the case of the M5 band for ocean and vegetative scene types. Such cases of OOB influence may impact certain environmental retrieval applications, e.g., NDVI determinations, but these negative effects can be largely avoided by using either another VIIRS instrument, if possible, or by utilizing the comparable I-band or M-band alternatives to the OOB-influenced channels, assuming the potential drawbacks are otherwise acceptable. For similar reasoning, the results of this study support NOAA-20 VIIRS over S-NPP VIIRS as a CERES inter-calibration reference when considering OOB-contributed uncertainty. Furthermore, it is advisable for CERES VIIRS and similar retrieval groups to consider the full-band RSR when spectral integration is required, or otherwise be aware of potential bias, thereby being able to account for potentially impactful OOB signal contribution depending on the scene and application. Although the spectral performance of VIIRS is adequate in the majority of applications, there is, nevertheless, value in understanding the scene-dependent OOB response, even if only for quality assurance purposes.

Author Contributions: B.S., D.R.D., R.B., A.G., and C.H. conceptualized the project and carried out methodology design. B.S. and R.B. conducted formal analysis and investigation efforts. B.S. and A.G. handled resources and software development. D.R.D. was the primary supervisor. B.S. performed all validation and visualization efforts. B.S. prepared the manuscript drafts with reviewing and editing contributions from all co-authors. All authors have read and agreed to the published version of the manuscript.

Funding: This research received no external funding.

Acknowledgments: This research is funded by the NASA CERES project. The authors would like to thank Amit Angal for his valuable insight regarding the many VIIRS spectral characterization efforts.

Conflicts of Interest: The authors declare no conflict of interest.

References

1. Moeller, C.; McIntire, J.; Schwarting, T.; Moyer, D. VIIRS F1 "best" relative spectral response characterization by the government team. In *Earth Observing Systems XVI*; International Society for Optics and Photonics: Bellingham, WA, USA, 2011; Volume 8153.

2. Moeller, C.; McIntire, J.; Schwarting, T.; Moyer, D.; Costa, J. Suomi NPP VIIRS spectral characterization: Understanding multiple RSR releases. In *Earth Observing Systems XVII*; International Society for Optics and Photonics: Bellingham, WA, USA, 2012; Volume 8510.

3. Kratz, D.P. The correlated k-distribution technique as applied to the AVHRR channels. *J. Quant. Spectrosc. Radiat. Transf.* **1995**, *53*, 501–507. [CrossRef]

4. Scarino, B.; Minnis, P.; Palikonda, R.; Reichle, R.H.; Morstad, D.; Yost, C.; Shan, B.; Liu, Q. Retrieving clear-sky surface skin temperature for numerical weather prediction applications from geostationary satellite data. *Remote Sens.* **2013**, *5*, 342–366. [CrossRef]

5. Scarino, B.R.; Minnis, P.; Chee, T.; Bedka, K.M.; Yost, C.R.; Palikonda, R. Global clear-sky surface skin temperature from multiple satellites using a single-channel algorithm with angular anisotropy corrections. *Atmos. Meas. Tech.* **2017**, *10*, 351–371. [CrossRef]

6. Trepte, Q.Z.; Minnis, P.; Sun-Mack, S.; Yost, C.R.; Chen, Y.; Jin, Z.; Hong, G.; Chang, F.-L.; Smith, W.L.; Bedka, K.M.; et al. Global cloud detection for CERES edition 4 using Terra and Aqua MODIS data. *IEEE Trans. Geosci. Remote Sens.* **2019**, *57*, 9410–9449. [CrossRef]

7. Minnis, P.; Sun-Mack, S.; Chen, Y.; Chang, F.-L.; Yost, C.R.; Smith, W.L., Jr.; Heck, P.W.; Arduini, R.F.; Bedka, S.T.; Yi, Y.; et al. CERES MODIS cloud product retrievals for Edition 4, Part I: Algorithm changes. *IEEE Trans. Geosci. Remote Sens.* **2020**. [CrossRef]

8.   Scalione, T.; DeLuccia, F.; Cymerman, J.; Johnson, E.; McCarthy, J.K.; Olejnicza, D. VIIRS initial performance verification—subassembly, early integration and ambient phase I testing of EDU. In Proceedings of the 2005 IEEE International Geoscience and Remote Sensing Symposium, IGARSS'05, Seoul, Korea, 25–29 July 2005.

9.   Lin, G.; Wolfe, R.E.; Nishihama, M. NPP VIIRS geometric performance status. In *Earth Observing Systems XVI*; International Society for Optics and Photonics: Bellingham, WA, USA, 2011; Volume 8153.

10.   Lei, N.; Wang, Z.; Fulbright, J.; Lee, S.; McIntire, J.; Chiang, K.; Xiong, X. Initial on-orbit radiometric calibration of the Suomi NPP VIIRS reflective solar bands. In *Earth Observing Systems XVII*; International Society for Optics and Photonics: Bellingham, WA, USA, 2012; Volume 8510.

11.   Lin, G.; Tilton, J.C.; Wolfe, R.E.; Tewari, K.P.; Nishihama, M. SNPP VIIRS spectral bands co-registration and spatial response. In *Earth Observing Systems XVIII*; International Society for Optics and Photonics: Bellingham, WA, USA, 2013; Volume 8866.

12.   Oudrari, H.; McIntire, J.; Xiong, X.; Butler, J.; Lee, S.; Lei, N.; Schwarting, T.; Sun, J. Pre-launch radiometric characterization and calibration of the S-NPP VIIRS sensor. *IEEE Trans. Geosci. Remote Sens.* **2015**, *53*, 2195–2210. [CrossRef]

13.   Lei, N.; Wang, Z.; Xiong, X. On-orbit radiometric calibration of Suomi NPP VIIRS reflective solar bands through observations of a sunlit solar diffuser panel. *IEEE Trans. Geosci. Remote Sens.* **2015**, *53*, 5983–5990. [CrossRef]

14.   Schueler, C.; Clement, J.E.; Ardanuy, P.; Welsch, C.; DeLuccia, F.; Swenson, H. NPOESS VIIRS sensor design overview. In *Earth Observing Systems VI*; International Society for Optics and Photonics: Bellingham, WA, USA, 2001; Volume 4483.

15.   Cao, C.; De Luccia, F.J.; Xiong, X.; Wolfe, R.; Weng, F. Early on-orbit performance of the visible infrared imaging radiometer suite onboard the Suomi National Polar-Orbiting Partnership (S-NPP) satellite. *IEEE Trans. Geosci. Remote Sens.* **2014**, *52*, 1142–1156. [CrossRef]

16.   Moeller, C.; Schwarting, T.; McIntire, J.; Moyer, D. JPSS-1 VIIRS prelaunch spectral characterization and performance. In *Earth Observing Systems XX*; International Society for Optics and Photonics: Bellingham, WA, USA, 2015; Volume 9607.

17.   Moeller, C.; Schwarting, T.; McIntire, J.; Moyer, D.; Zeng, J. JPSS-1 VIIRS version 2 at-launch relative spectral response characterization and performance. In *Earth Observing Systems XXI*; International Society for Optics and Photonics: Bellingham, WA, USA, 2016; Volume 9972.

18.   Schwarting, T.; Moeller, C.; Moyer, D.; McIntire, J.; Oudrari, H. JPSS-2 VIIRS version 1 spectral characterization and performance assessment. In *Earth Observing Systems XXIV*; International Society for Optics and Photonics: Bellingham, WA, USA, 2019; Volume 11127.

19.   Moeller, C.; Schwarting, T.; McCorkel, J.; Moyer, D.; McIntire, J. JPSS-2 VIIRS version 2 at-launch relative spectral response characterization. In *Earth Observing Systems XXIV*; International Society for Optics and Photonics: Bellingham, WA, USA, 2019; Volume 11127.

20.   Moeller, C.; Schwarting, T.; McIntire, J.; Oudrari, H.; Moyer, D. NPP VIIRS Flight 1 Relative Spectral Response (RSR) Overview. Available online: https://ncc.nesdis.noaa.gov/documents/documentation/f1-rsr-overview-gt-final-jan2012.pdf (accessed on 21 September 2020).

21.   Angal, A.; SSAI/NASA/GSFC/VCST, Lanham, MD, USA. Personal communication, 2020.

22.   Bovensmann, H.; Burrows, J.P.; Buchwitz, M.; Frerick, J.; Noël, S.; Rozanov, V.V.; Chance, K.V.; Goede, A.P.H. SCIAMACHY: Mission objectives and measurement modes. *J. Atmos. Sci.* **1999**, *56*, 127–150. [CrossRef]

23.   Blumstein, D.; Chalon, G.; Carlier, T.; Buil, C.; Hebert, P.; Maciaszek, T.; Ponce, G.; Phulpin, T.; Tournier, B.; Simeoni, D.; et al. IASI instrument: Technical overview and measured performances. In *Infrared Spaceborne Remote Sensing XII*; International Society for Optics and Photonics: Bellingham, WA, USA, 2004; Volume 5543.

24.   Blumstein, D.; Tournier, B.; Cayla, F.R.; Phulpin, T.; Fjortoft, R.; Buil, C.; Ponce, G. In-flight performance of the infrared atmospheric sounding interferometer (IASI) on Metop-A. In *Atmospheric and Environmental Remote Sensing Data Processing and Utilization III*; International Society for Optics and Photonics: Bellingham, WA, USA, 2007; Volume 6684.

25.   Klaes, K.D.; Cohen, M.; Buhler, Y.; Schlüssel, P.; Munro, R.; Luntama, J.-P.; von Engeln, A.; Clérigh, E.Ó.; Bonekamp, H.; Ackermann, J.; et al. An introduction to the EUMETSAT polar system. *Bull. Am. Meteorol. Soc.* **2007**, *88*, 1085–1096. [CrossRef]

26. Skupin, J.; Noël, S.; Wuttke, M.W.; Gottwald, M.; Bovensmann, H.; Weber, M.; Burrows, J.P. SCIAMACHY solar irradiance observation in the spectral range from 240 to 2380 nm. *Adv. Space Res.* **2005**, *35*, 370–375. [CrossRef]

27. Lichtenberg, G.; Kleipool, Q.; Krijger, J.M.; van Soest, G.; van Hees, R.; Tilstra, L.G.; Acarreta, J.R.; Aben, I.; Ahlers, B.; Bovensmann, H.; et al. SCIAMACHY Level 1 data: Calibration concept and in-flight calibration. *Atmos. Chem. Phys.* **2006**, *6*, 5347–5367. [CrossRef]

28. SPPA Engineer. Disclaimer for SCIAMACHY Level 1b data version SCIAMACHY/7.04. In *ENVISAT-1 Products Specifications: SCIAMACHY Products Specifications PO-RS-MDA-GS-2009*; European Space Agency: Paris, France, 2010; Volume 15.

29. Wang, L.; Wu, X.; Li, Y.; Goldberg, M.; Sohn, S.-H.; Cao, C. Comparison of AIRS and IASI radiances using GOES imagers as transfer radiometers toward climate data records. *J. Appl. Meteorol. Climatol.* **2010**, *49*, 478–492. [CrossRef]

30. Tobin, D.C.; Revercomb, H.E.; Moeller, C.C.; Pagana, T.S. Use of atmospheric infrared sounder high-spectral resolution spectra to assess the calibration of moderate resolution imaging spectroradiometer on EOS aqua. *J. Geophys. Res.* **2006**, *111*. [CrossRef]

31. Gunshor, M.M.; Schmit, T.J.; Menzel, W.P.; Tobin, D.C. Intercalibration of broadband geostationary imagers using AIRS. *J. Atmos. Ocean. Technol.* **2009**, *26*, 746–758. [CrossRef]

32. Goldberg, M.; Ohring, G.; Butler, J.; Cao, C.; Datla, R.; Doelling, D.; Gärtner, V.; Hewison, T.; Iacovazzi, B.; Kim, D.; et al. The global space-based inter-calibration system. *Bull. Am. Meteorol. Soc.* **2011**, *92*, 467–475. [CrossRef]

33. Hewison, T.J.; Wu, X.; Yu, F.; Tahara, Y.; Hu, X.; Kim, D.; Koenig, M. GSICS inter-calibration of infrared channels of geostationary imagers using Metop/IASI. *IEEE Trans. Geosci. Remote Sens.* **2013**, *51*, 1160–1170. [CrossRef]

34. Hewison, T.J. An evaluation of the uncertainty of the GSICS SEVIRI-IASI intercalibration products. *IEEE Trans. Geosci. Remote Sens.* **2013**, *51*, 1171–1181. [CrossRef]

35. Hewison, T.J.; Doelling, D.R.; Lukashin, C.; Tobin, D.O.; John, V.; Joro, S.; Bojkov, B. Extending the Global Space-Based Inter-Calibration System (GSICS) to Tie Satellite Radiances to an Absolute Scale. *Remote Sens.* **2020**, *12*, 1782. [CrossRef]

36. Slater, P.N.; Biggar, S.F.; Holm, R.G.; Jackson, R.D.; Mao, Y.; Moran, M.S.; Palmer, J.M.; Yuan, B. Reflectance- and radiance-based methods for the in-flight absolute calibration of multispectral sensors. *Remote Sens. Environ.* **1987**, *22*, 11–37. [CrossRef]

37. Slater, P.N.; Biggar, S.F.; Thome, K.J.; Gellman, D.I.; Spyak, P.R. Vicarious radiometric calibrations of EOS sensors. *J. Atmos. Ocean. Technol.* **1996**, *13*, 349–359. [CrossRef]

38. Le Marshal, J.F.; Simpson, J.J.; Jin, Z. Satellite calibration using a collocated nadir observation technique: Theoretical basis and application to the GMS-5 pathfinder benchmark period. *IEEE Trans. Geosci. Remote Sens.* **1999**, *37*, 499–507. [CrossRef]

39. Teillet, P.M.; Fedosejevs, G.; Gauthier, R.P.; O'Neill, N.T.; Thome, K.J.; Biggar, S.F.; Ripley, H.; Meygret, A. A generalized approach to the vicarious calibration of multiple Earth observation sensors using hyperspectral data. *Remote Sens. Environ.* **2001**, *77*, 304–327. [CrossRef]

40. Chander, G.; Hewison, T.J.; Fox, N.; Wu, X.; Xiong, X.; Blackwell, W.J. Overview of intercalibration of satellite instruments. *IEEE Trans. Geosci. Remote Sens.* **2013**, *51*, 1–25. [CrossRef]

41. Chander, G.N. Applications of spectral band adjustment factors (SBAF) for cross-calibration. *IEEE Trans. Geosci. Remote Sens.* **2013**, *51*, 1267–1281. [CrossRef]

42. Henry, P.; Chander, G.; Fougnie, B.; Thomas, C.; Xiong, X. Assessment of spectral band impact on intercalibration over desert sites using simulation based on EO-1 Hyperion data. *IEEE Trans. Geosci. Remote Sens.* **2013**, *51*, 1297–1308. [CrossRef]

43. Scarino, B.R.; Doelling, D.R.; Minnis, P.; Gopalan, A.; Chee, T.; Bhatt, R.; Lukashin, C.; Haney, C.O. A web-based tool for calculating spectral band difference adjustment factors derived from SCIAMACHY hyperspectral data. IEEE Trans. *Geosci. Remote Sens.* **2016**, *54*, 2529–2542. [CrossRef]

44. Moeller, C.; Lei, N. SNPP VIIRS Modulated RSR Release 1.0. 2013. Available online: https://ncc.nesdis.noaa.gov/documents/documentation/README_Modulated_RSR_Release_1.0 (accessed on 21 September 2020).

45. Bhatt, R.; Doelling, D.R.; Morstad, D.; Scarino, B.R.; Gopalan, A. Desert-based absolute calibration of successive geostationary visible sensors using a daily exoatmospheric radiance model. *IEEE Trans. Geosci. Remote Sens.* **2013**, *52*, 3670–3682. [CrossRef]
46. Wolfe, R.; Lin, G.; Nishihama, M.; Tewari, K.; Montano, E. NPP VIIRS early on-orbit geometric performance. In *Earth Observing Systems XVII*; International Society for Optics and Photonics: Bellingham, WA, USA, 2012; Volume 8510.
47. Wolfe, R.; Lin, G.; Nishihama, M.; Tewari, K.P.; Tilton, J.C.; Isaacman, A.R. Suomi NPP VIIRS prelaunch and on-orbit geometric calibration and characterization. *J. Geophys. Res.* **2013**, *118*, 11508–11521. [CrossRef]
48. Cao, C.; Xiong, X.; Wolfe, R.; DeLuccia, F.; Liu, Q.; Blonski, S.; Lin, G.; Nishihama, M.; Pogorzala, D.; Oudrari, H.; et al. Visible infrared imaging radiometer suite (VIIRS) sensor data record (SDR) user's guide. In *NOAA Technical Report NESDIS 142*; National Oceanic and Atmospheric Administration: Washington, DC, USA, 2017.
49. Doelling, D.R.; Haney, C.; Bhatt, R.; Scarino, B.; Gopalan, A. Geostationary visible imager calibration for the CERES SYN1deg Edition 4 product. *Remote Sens.* **2018**, *10*, 288. [CrossRef]
50. Bhatt, R.; Doelling, D.R.; Wu, A.; Xiong, X.; Scarino, B.R.; Haney, C.O.; Gopalan, A. Initial stability assessment of S-NPP VIIRS reflective solar band calibration using invariant desert and deep convective cloud Targets. *Remote Sens.* **2014**, *6*, 2809–2826. [CrossRef]
51. Bhatt, R.; Doelling, D.R.; Scarino, B.; Haney, C.; Gopalan, A. Development of seasonal BRDF models to extend the use of deep convective clouds as invariant targets for satellite SWIR-band calibration. *Remote Sens.* **2017**, *9*, 1061. [CrossRef]
52. Bhatt, R.; Doelling, D.R.; Angal, A.; Xiong, X.; Scarino, B.; Gopalan, A.; Haney, C.; Wu, A. Characterizing response versus scan-angle for MODIS reflective solar bands using deep convective clouds. *J. Appl. Remote Sens.* **2017**, *11*. [CrossRef]
53. STAR—Global Vegetation Health Products: Technique Background. Available online: https://www.star.nesdis.noaa.gov/smcd/emb/vci/VH/vh_TechniqueBackground.php (accessed on 13 August 2020).
54. STAR—Global Vegetation Health Products: VIIRS Bands. Available online: https://www.star.nesdis.noaa.gov/smcd/emb/vci/VH/npp_bandsForVH.php (accessed on 13 August 2020).
55. Measuring Vegetation (NDVI and EVI): Normalized Difference Vegetation Index (NDVI). Available online: https://earthobservatory.nasa.gov/features/MeasuringVegetation/measuring_vegetation_2.php (accessed on 13 August 2020).
56. Didan, K.; Munoz, A.B.; Comptom, T.J.; Pinzon, J.E. *Suomi National Polar-Orbiting Partnership Visible Infrared Imaging Radiometer Suite Vegetation Index Product Suite User Guide and Abridged Algorithm Theoretical Basis Document Version 2.0*; The University of Arizona Vegetation Index and Phenology Lab: Tucson, Arizona, USA, 2017.
57. Lei, N.; Wang, Z.; Guenther, B.; Xiong, X.; Gleason, J. Modeling the detector radiometric response gains of the Suomi NPP VIIRS reflective solar bands. In *Sensors, Systems, and Next-Generation Satellites XVI*; International Society for Optics and Photonics: Bellingham, WA, USA, 2012; Volume 8533.
58. Xiong, X.; Butler, J.; Chiang, K.; Efremova, B.; Fulbright, J.; Lei, N.; McIntire, J.; Oudari, H.; Sun, J.; Wang, Z.; et al. VIIRS on-orbit calibration methodology and performance. *J. Geophys. Res. Atmos.* **2014**, *119*, 5065–5078. [CrossRef]

 *remote sensing*

Article

Validation of Carbon Trace Gas Profile Retrievals from the NOAA-Unique Combined Atmospheric Processing System for the Cross-Track Infrared Sounder

Nicholas R. Nalli [1,*] , Changyi Tan [1], Juying Warner [2], Murty Divakarla [1],
Antonia Gambacorta [3], Michael Wilson [1], Tong Zhu [1], Tianyuan Wang [1], Zigang Wei [1],
Ken Pryor [4], Satya Kalluri [4], Lihang Zhou [5], Colm Sweeney [6], Bianca C. Baier [6,7],
Kathryn McKain [6,7], Debra Wunch [8], Nicholas M. Deutscher [9], Frank Hase [10], Laura T. Iraci [11],
Rigel Kivi [12], Isamu Morino [13], Justus Notholt [14], Hirofumi Ohyama [13], David F. Pollard [15],
Yao Té [16], Voltaire A. Velazco [9], Thorsten Warneke [14], Ralf Sussmann [17] and Markus Rettinger [17]

[1] IMSG, Inc. at NOAA/NESDIS Center for Satellite Applications and Research (STAR), College Park, MD 20740, USA; Changyi.Tan@noaa.gov (C.T.); Murty.Divakarla@noaa.gov (M.D.); Michael.Wilson@noaa.gov (M.W.); Tong.Zhu@noaa.gov (T.Z.); Tianyuan.Wang@noaa.gov (T.W.); Zigang.Wei@noaa.gov (Z.W.)

[2] UMD Department of Atmospheric and Oceanic Science, College Park, MD 20740, USA; juying@atmos.umd.edu

[3] NASA Goddard Space Flight Center, Climate and Radiation Laboratory, Greenbelt, MD 20771, USA; antonia.gambacorta@nasa.gov

[4] NOAA/NESDIS/STAR, College Park, MD 20740, USA; Ken.Pryor@noaa.gov (K.P.); Satya.Kalluri@noaa.gov (S.K.)

[5] NOAA JPSS Program Office, Lanham, MD 20706, USA; Lihang.Zhou@noaa.gov

[6] NOAA Global Monitoring Laboratory (GML), Boulder, CO 80305-3328, USA; Colm.Sweeney@noaa.gov (C.S.); Bianca.Baier@noaa.gov (B.C.B.); kathryn.mckain@noaa.gov (K.M.)

[7] Cooperative Institute for Research in Environmental Sciences, University of Colorado Boulder, Boulder, CO 80305, USA

[8] Department of Physics, University of Toronto, Toronto, ON M5S 1A7, Canada; dwunch@atmosp.physics.utoronto.ca

[9] Centre for Atmospheric Chemistry, School of Earth, Atmospheric and Life Sciences, University of Wollongong, Wollongong, NSW 2522, Australia; ndeutsch@uow.edu.au (N.M.D.); voltaire@uow.edu.au (V.A.V.)

[10] Institute of Meteorology and Climate Research (IMK-ASF), Karlsruhe Institute of Technology, 76137 Karlsruhe, Germany; frank.hase@kit.edu

[11] NASA Ames Research Center, Atmospheric Science Branch, Mail Stop 245-5, Moffett Field, CA 94035, USA; laura.t.iraci@nasa.gov

[12] Finnish Meteorological Institute, Space and Earth Observation Centre, Tähteläntie 62, 99600 Sodankylä, Finland; Rigel.Kivi@fmi.fi

[13] Satellite Remote Sensing Section and Satellite Observation Center, National Institute for Environmental Studies (NIES), Onogawa 16-2, Tsukuba, Ibaraki 305-8506, Japan; morino@nies.go.jp (I.M.); oyama.hirofumi@nies.go.jp (H.O.)

[14] Institute of Environmental Physics, University of Bremen, 28195 Bremen, Germany; jnotholt@iup.physik.uni-bremen.de (J.N.); warneke@iup.physik.uni-bremen.de (T.W.)

[15] National Institute of Water and Atmospheric Research Ltd., Hamilton 3216, New Zealand; Dave.Pollard@niwa.co.nz

[16] Laboratoire d'Etudes du Rayonnement et de la Matière en Astrophysique et Atmosphères (LERMA-IPSL), Sorbonne Université, CNRS, Observatoire de Paris, PSL Université, 75005 Paris, France; yao-veng.te@upmc.fr

[17] Karlsruhe Institute of Technology, IMK-IFU, 82467 Garmisch-Partenkirchen, Germany; ralf.sussmann@kit.edu (R.S.); markus.rettinger@kit.edu (M.R.)

* Correspondence: Nick.Nalli@noaa.gov

Received: 24 August 2020; Accepted: 29 September 2020; Published: 6 October 2020

Abstract: This paper provides an overview of the validation of National Oceanic and Atmospheric Administration (NOAA) operational retrievals of atmospheric carbon trace gas profiles, specifically carbon monoxide (CO), methane (CH_4) and carbon dioxide (CO_2), from the NOAA-Unique Combined Atmospheric Processing System (NUCAPS), a NOAA enterprise algorithm that retrieves atmospheric profile environmental data records (EDRs) under global non-precipitating (clear to partly cloudy) conditions. Vertical information about atmospheric trace gases is obtained from the Cross-track Infrared Sounder (CrIS), an infrared Fourier transform spectrometer that measures high resolution Earth radiance spectra from NOAA operational low earth orbit (LEO) satellites, including the Suomi National Polar-orbiting Partnership (SNPP) and follow-on Joint Polar Satellite System (JPSS) series beginning with NOAA-20. The NUCAPS CO, CH_4, and CO_2 profile EDRs are rigorously validated in this paper using well-established independent truth datasets, namely total column data from ground-based Total Carbon Column Observing Network (TCCON) sites, and in situ vertical profile data obtained from aircraft and balloon platforms via the NASA Atmospheric Tomography (ATom) mission and NOAA AirCore sampler, respectively. Statistical analyses using these datasets demonstrate that the NUCAPS carbon gas profile EDRs generally meet JPSS Level 1 global performance requirements, with the absolute accuracy and precision of CO 5% and 15%, respectively, in layers where CrIS has vertical sensitivity; CH_4 and CO_2 product accuracies are both found to be within $\pm1\%$, with precisions of $\approx1.5\%$ and $\lesssim0.5\%$, respectively, throughout the tropospheric column.

Keywords: satellite cal/val; error analysis; greenhouse gases; carbon monoxide; methane; carbon dioxide; trace gas; remote sensing; retrieval algorithms; satellite applications

---

1. Introduction

The U.S. National Oceanic and Atmospheric Administration (NOAA) Joint Polar Satellite System (JPSS) is a NOAA-operational low earth orbit (LEO) satellite series that features the hyperspectral infrared (IR) Cross-track Infrared Sounder (CrIS) [1] and Advanced Technology Microwave Sounder (ATMS) [2] systems. Four satellites are planned to fly in the same orbit over the next two decades beginning with the NOAA-20 satellite (which was referred to as JPSS-1 or J-1 prior to launch in late 2017), and was preceded by the Suomi National Polar-orbiting Partnership (SNPP) satellite launched in late 2011. The CrIS instrument is an advanced IR Fourier transform spectrometer (FTS) that obtains sensor data records (SDRs) consisting of well-calibrated IR Earth emission spectra over three bands (longwave 650–1095 cm^{-1}, midwave 1210–1750 cm^{-1}, and shortwave 2155–2550 cm^{-1}), with 2211 channels in full spectral-resolution (FSR) mode (maximum optical path difference of 0.8 cm for all three bands and spectral resolution $\Delta v = 0.625$ cm^{-1}, with 713, 865 and 633 channels in the longwave, midwave and shortwave bands, respectively) [3]. The CrIS spectra allow for retrieval of atmospheric vertical profile environmental data records (EDRs) with the best possible vertical resolution (≈2–7 km for temperature and water vapor throughout the troposphere) comparable to predecessor sounding systems, namely the European Organisation for the Exploitation of Meteorological Satellites (EUMETSAT) Metop-series Infrared Atmospheric Sounding Interferometer (IASI) [4,5] and the National Aeronautics and Space Adminstration (NASA) EOS-Aqua Atmospheric Infrared Sounder (AIRS) [6,7]. The NOAA-operational EDR retrieval algorithm for operational hyperspectral thermal IR sounders (viz., CrIS and IASI) is the NOAA-Unique Combined Atmospheric Processing System (NUCAPS) [8,9]. The NUCAPS algorithm is based upon the heritage methodology developed for the EOS-Aqua AIRS and is a modular implementation of the multi-step NASA AIRS Science Team retrieval algorithm Version 5 [10,11].

NUCAPS SNPP previously ran on CrIS spectra with reduced resolution in the midwave and shortwave bands (1.25 cm^{-1} and 2.5 cm^{-1}, respectively) due to truncated interferograms in those bands during operational processing [3]; these reduced-resolution spectra have been referred to as "nominal" or "normal" resolution as this was the originally planned operational resolution of the CrIS SDR. However, offline production of CrIS FSR began in December 2014 [3,12], with operational Interface Data Processing Segment production starting in March 2017. The move to FSR was motivated in part by a demonstration study showing the impact of the CrIS spectral resolution on the retrieval of the carbon monoxide EDR [13]. Given that the CrIS FSR mode has been operational since then (i.e., for the remainder of the SNPP lifetime as well as the follow-on JPSS satellite series, beginning with NOAA-20), the NUCAPS system was upgraded to run in FSR mode using the Stand-Alone Radiative Transfer Algorithm (SARTA) [14] delivered by the University of Maryland Baltimore County (UMBC). For more details on the NUCAPS algorithm theoretical basis and user applications, the reader is referred to other papers [9,10] and/or the Algorithm Theoretical Basis Document (ATBD) available online [15].

The Earth emission spectra (i.e., SDRs) measured by CrIS, IASI and AIRS contain information about the atmospheric temperature (T) and moisture (q) profiles, along with trace gases including O_3, CO, CH_4, CO_2, SO_2, HNO_3 and N_2O. The NUCAPS physical retrieval module [15] retrieves these individual parameters in a sequential fashion, using channels rigorously determined to be sensitive to each parameter [16], beginning with cloud-cleared radiance spectra (i.e., clear-column IR spectra which are derived with the help of the collocated ATMS data) [10], followed by T, q, ozone (O_3) and the remaining trace gases, with the results output on the radiative transfer model (RTM) (or radiative transfer algorithm, RTA) 100 layer grid (T output is on layer boundaries or "levels"). The NUCAPS algorithm solves for the trace gases in an effort to optimize the retrieved thermodynamic (T and q) profile EDRs [9], but the long-term investments in the CrIS and IASI sounders onboard future operational NOAA and EUMETSAT LEO satellite missions (as indicated above) has motivated the exploitation of these space assets for the routine production of the carbon trace gas EDRs, namely carbon monoxide (CO) [17,18], methane (CH_4) [19], and carbon dioxide (CO_2) [20].

The validation of the NUCAPS T, q and O_3 profile EDRs with respect to high-quality reference datasets has been previously reported in Nalli et al. [21,22], where it was demonstrated that the SNPP EDRs meet JPSS Level 1 requirements; additional independent assessments of the SNPP T and q EDRs versus other reference datasets have been reported elsewhere [23,24]. Similar performances have been established for the EDR products from the NOAA-20 satellite, launched since the original SNPP validation effort. Since that time, the NUCAPS algorithm development team has focused on improvements to the operational carbon trace gas EDR products mentioned above. The improvements include updated a priori profiles (based on current zonal climatologies) and RTA tuning (empirically removing residual biases between the model and observations), along with optimized quality assurance (QA) flags (based upon the algorithm chi-square, χ^2, degrees-of-freedom, and other quality measures). Thus, in this paper we focus our attention on validating the operational NUCAPS (offline v2.8) CO, CH_4 and CO_2 trace gas EDRs; additional details on the NUCAPS carbon trace gas retrievals can be found in a forthcoming paper (Warner et al., manuscript in prep for Atmos. Chem. Phys.).

2. Methodology

Carbon trace gas EDR validation was a new requirement within the JPSS calibration/validation (cal/val) program [25] beginning with the transition to the full spectral-resolution (FSR) CrIS NUCAPS. The JPSS Level 1 requirements for carbon trace gas profile EDRs are given in Table 1, which are defined for the global ensemble of total column, cloud-cleared cases. These requirements serve as the program metrics by which the system is considered to have reached Validated Maturity and meets mission requirements.

Table 1. JPSS Level 1 Requirements * for (CrIS) Carbon Trace Gas Column EDRs.

Statistic	Threshold	Objective
Carbon Monoxide EDR		
CO Precision [†]	15%	3%
CO Accuracy [‡]	±5%	±5%
Methane EDR		
CH_4 Precision	1% (20 ppbv)	N/A
CH_4 Accuracy	±4% (80 ppbv)	N/A
Carbon Dioxide EDR		
CO_2 Precision	0.5% (2 ppmv)	1.05 to 1.4 ppmv
CO_2 Accuracy	±1% (4 ppmv)	N/A

* Source: Joint Polar Satellite System (JPSS) Program Level 1 Requirements Supplement—Final, Version 2.10, 25 June 2014, JPSS-REQ-1002, NOAA/NESDIS, pp. 39–41, 98; "Level 1" is a programmatic term that refers to the "highest level" program requirement; [†] Measurement precision is defined as the standard deviation (one sigma) of the sample measurement errors; [‡] Measurement accuracy is defined as the magnitude of the mean measurement error.

Satellite sounder validation methodology has been well-established for T, q and O_3 profile EDRs within previous validation work, with the various coarse-layer statistical uncertainty characterizations conducted relative to baseline reference datasets (i.e., "truth") roughly classified within a "hierarchy" [26]. Profile statistics for layer gas concentrations are defined in terms of fractional errors, including systemic (i.e., bias or "accuracy"), random (i.e., 1σ variability or "precision"), and total combined error (i.e., root mean square error, RMSE). For carbon trace gases we have adopted a similar hierarchical approach based upon available reference datasets, consisting of (1) numerical model global comparisons, (2) satellite EDR intercomparisons, (3) surface-based observing network assessments, and (4) intensive field campaign in situ data assessments. Those at the base of the hierarchy may be readily employed during the early cal/val stages (or anytime thereafter) of a satellite's lifetime, whereas those near the top are employed during later stages. These are briefly overviewed below.

Numerical model output (analysis and/or forecast interpolated to NUCAPS footprints) enables the rapid comparison with large, global datasets obtained during "Focus Days" (i.e., days selected for the acquisition of global SDRs that are used for retrieving EDRs using the latest versions of offline code) and as such are extremely useful for early evaluation of the algorithm and identifying gross problem areas [27]. Numerical models used for such comparisons include the European Center for Medium-Range Weather Forecast (ECMWF), the NOAA CarbonTracker [28], and the Copernicus Atmosphere Monitoring Service (CAMS) [29]. Such analyses are useful in identifying regional or spectral biases. However, dynamical models (e.g., ECMWF) do not constitute independent correlative data given that they assimilate radiances, and generally do not model chemistry and/or surface fluxes.

Trace gas EDRs obtained from other satellite sensors or algorithms provide quasi-independent observations for global intercomparisons. Like numerical model comparisons, this approach also allows for the acquisition of large, global data samples that can facilitate early consistency checks. In addition, such data (depending on the sensor/algorithm) may be more reliable than model analyses, especially in the case of previously validated EDR products, thus providing additional global confidence. AIRS is extremely useful for this purpose given it is a mature, high-resolution IR sounder that runs an end-to-end algorithm similar to NUCAPS, with the Aqua satellite flying in the same 01:30, 13:30 local equator crossing time orbit. Other satellite sounder EDR datasets include Tropospheric Monitoring Instrument (TROPOMI) onboard the Copernicus Sentinel-5, the NASA Orbiting Carbon Observatory (OCO-2), Greenhouse gases Observing Satellite (GOSAT), and the Aura Microwave Limb Sounder (MLS). However, a limitation of these data for validation is that they may possess similar retrieval error characteristics (in the case of AIRS [27]) or different vertical sensitivity, and thus ultimately would require proper treatment of each sensor's averaging kernels [30] (cf. Appendix A).

Ground-based, remotely sensed observations obtained periodically from surface-based observing networks provide independent truth datasets with a global distribution reasonably representing global latitude zones roughly analogous to radiosonde observations (RAOB) for temperature and moisture. The most notable example of such a dataset is the Total Carbon Column Observing Network (TCCON) [31], a ground-based network of uplooking solar-spectrum FTS instruments that obtain total column measurements of traces gases (discussed more in Section 3.1). A newer source of in situ data are vertical profiles obtained from the balloon-borne AirCore sampling system [32,33]. NUCAPS EDR collocations with these ground-based networks thus provide independent datasets for statistical assessments [26]. However, limitations in these datasets include the time latencies needed for acquiring reasonable collocation sample sizes, uncertainties in unit conversions, and different sensitivities to atmospheric layers.

At the top of the validation data hierarchy are intensive aircraft campaigns that provide episodic, but generally comprehensive sets of in situ and remotely sensed vertical profile data from multiple ascents and descents of dedicated aircraft flying over a specified region. Aircraft campaigns thus allow for detailed performance specification over regions of interest. Examples of trace gas campaigns suitable for SNPP validation include the Atmospheric Tomography (ATom) mission [34] (discussed in Section 3.3) and, previously, the HIAPER Pole-to-Pole Observations (HIPPO) [35] campaigns. The specific datasets used for NUCAPS trace gas validation are detailed in Section 3 below.

3. Data

Following the hierarchical approach described in Section 2, multiple complementary correlative truth datasets are relied upon to provide independent measurements for validation. We have leveraged three datasets for this purpose, namely uplooking spectrometer total-column data from TCCON, balloon-borne profiles from AirCore, and finally aircraft in situ vertical profile data from ATom. Existing satellite EDR datasets from other platforms (viz., Aqua AIRS and TROPOMI) have also been utilized for global intercomparisons to demonstrate the NUCAPS products look qualitatively reasonable and geographically consistent (Warner et al., manuscript in prep for Atmos. Chem. Phys.). Specifics of the layer and unit conversions required for conducting quantitative statistical assessments of NUCAPS (Section 4) are explicitly described for completeness and reproducibility.

3.1. TCCON

The Total Carbon Column Observing Network (TCCON) [31] is a ground-based network of Bruker 125HR uplooking solar-spectrum FTS that obtain spectral measurements in the near-IR region that encompasses the CO, CH_4, CO_2, N_2O, and O_2 absorption bands [31], thus comprising an independent data source for validating NUCAPS. The interferograms are collected with a 45 cm optical path difference (45 cm was chosen deliberately to optimize retrievals of CO_2 in the 6000 cm^{-1} band) yielding a spectral resolution of $\approx$0.02 cm^{-1}. The total column retrievals of these trace gases is achieved via a retrieval algorithm (called GFIT) that includes both the forward and inverse model calculations. The inverse algorithm employs least-squares fitting by scaling an a priori [31]. The a priori profiles used in the GFIT system, x_0, were obtained from the NOAA National Centers for Environmental Prediction, National Center for Atmospheric Research (NCEP/NCAR) analysis. TCCON station data can facilitate intercomparisons (acting as a "transfer-standard") between retrievals from multiple satellites.

The total column trace gases retrieved by TCCON are in dry mole fractions (DMF), whereas the NUCAPS algorithm retrieves trace gas layer abundances (in molecules/cm^2) on the 100 RTA model layers. Thus, a conversion scheme must be implemented. Furthermore, because TCCON and NUCAPS have fundamental differences in vertical sensitivity, it is desirable that the TCCON column averaging kernels (AKs) be utilized in the integration of the NUCAPS observation. For explicitness, the conversion scheme and application of TCCON column AKs in the integration of NUCAPS retrieved trace gas profile EDRs are detailed in Appendix A.

In practice the column AKs are dependent only on the solar zenith angle, $\theta_\odot$, of the measurements [31]. Thus a single set of column AKs from the Lamont Site, $\mathbf{a}(\theta_\odot)$, are provided for gridded values of $\theta_\odot = 10°, 15°, \ldots, 85°$ (shown in Figure 1), which can then be interpolated to the solar zenith angle of the measurement. From Figure 1 a fundamental limitation in the utility of TCCON data for IR sounder (e.g., NUCAPS) validation becomes evident, namely the TCCON tendency for higher sensitivity in the upper layers of the atmosphere, except at larger $\theta_\odot$ for CH_4 and CO_2. The TCCON vertical sensitivity must be taken into account when comparing against the sounder retrieved EDRs, which typically have peak sensitivity in the mid-troposphere (discussed more in Section 4).

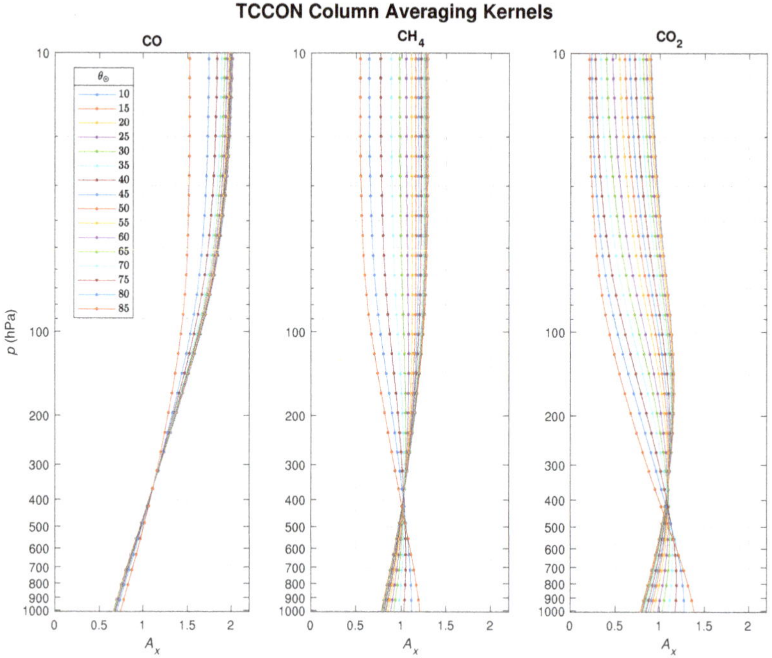

Figure 1. TCCON tropospheric column AKs [31] as a function of solar zenith angle, $\theta_\odot$, for (**left**) carbon monoxide, (**middle**) methane, and (**right**) carbon dioxide.

3.2. AirCore

The NOAA Global Monitoring Laboratory (GML) AirCore sampling system [32,33,36] is an innovative in situ sampling approach that employs long, coated stainless-steel tubes to collect a sample of the ambient atmospheric air column (i.e., a "core" analogous to an ice-core). The tubes are open at one end, filled with a "fill gas" (with known levels of CO_2, CH_4, and CO), and configured in a tight coil so that they can be deployed upon a suitable platform, notably helium or hydrogen-filled 3000 g balloons. The AirCore is evacuated upon balloon-borne ascent and then fills with ambient air upon parachuted descent. However, unlike a radiosonde, the sampling package is tracked during its return from ≈30 km altitude to the surface (e.g., via a parachute in the case of a balloon) and subsequently sealed and recovered, where it can then be brought back to the lab for analysis using a laboratory-grade trace gas analyzer (e.g., Picarro, Inc.). AirCore thus allows in situ measurement of mole fraction samples for various trace gases (e.g., CO, CH_4 and CO_2) without requiring an aircraft or onboard data transmission system (e.g., as with an ozonesonde). A distinct advantages of AirCore is the capability for multiple deployments with a distributed geographic coverage over land. In this capacity AirCore has promise to be a surface-based network like TCCON, albeit with calibrated, high-resolution profiles measured in samples that survey $\gtrsim 98\%$ column, somewhat analogous to an ozonesonde network.

Because the length scale of diffusion is <0.5 m over the time it takes to analyze an AirCore sample (≈ 4 h), >100 discrete samples can be measured in a 100 m AirCore tube. The resultant profile resolution surpasses the 100-layer forward model grid employed in the retrieval. Thus we follow the approach documented in Nalli et al. [26] (Appendix B op. cit.), performing molecular-integrations of column densities for each trace gas constituent, allowing us to redivide the atmospheric path to the 101 layer boundaries, which then allows the computation of the effective layer values in a physically rigorous manner. The conversions to NUCAPS RTA layer abundances therefore requires concurrent measurements of temperature and water vapor, which are obtained from a radiosonde package flown on the AirCore payload.

3.3. ATom

The Atmospheric Tomography (ATom) mission [34] deployed an extensive gas and aerosol measurement payload on the NASA DC-8 aircraft for global-scale sampling of the atmosphere, profiling continuously from 0.2–12 km altitude. Flights occurred during all four seasons, originating from the Armstrong Flight Research Center in Palmdale, California, USA, flying north to the western Arctic, south to the South Pacific, east to the Atlantic, north to Greenland, before returning to California across central North America or the North American Arctic. Figure 2 shows the flight paths for the 2016–2018 sampling periods (ATom-1, -2, and -4). ATom-1 and -2 data were first used for SNPP NUCAPS development and validation prior to our J-1 validation effort; we subsequently obtained ATom-4 data for NOAA-20 validation (note that ATom-3 was still pre NOAA-20), and simply combined it with our existing ATom-1 and -2 collocation data for SNPP going forward.

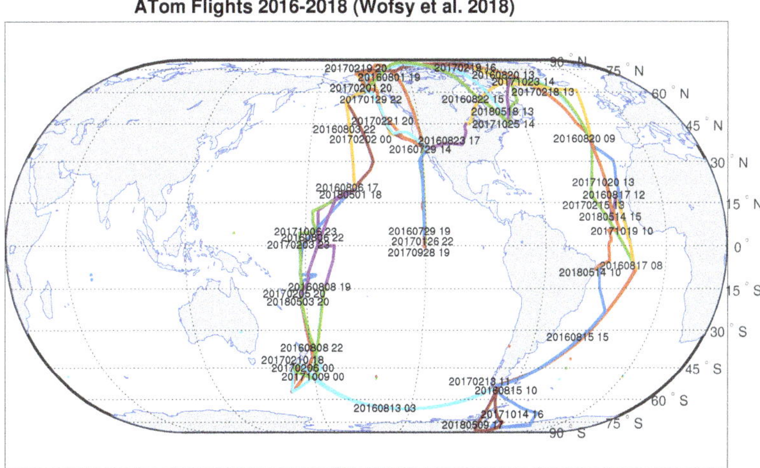

Figure 2. Atmospheric Tomography (ATom) flights for the period of 2016–2018 (ATom-1, -2, and -4) [34]; individual flights are distinguished from one another using different colors. Map projection is equal-area.

During ATom flights, the aircraft repeatedly ascended to 10–12 km, leveled-off, then descended to different heights at different rates. The raw aircraft data are recorded as a function of time, with reported altitudes featuring small-to-medium scale fluctuations throughout any given flight. Correlative truth profiles must thus be extracted only from smooth, continuous ascent/descents, disregarding small-scale altitude fluctuations and periods when the aircraft leveled off. Through trial and error, we devised an approach for extracting these profiles from the flight data based upon three criteria, namely the ascent/descent rate, $\Delta z / \Delta t$ (to find actual ascents/descents), the time difference between successive ascending/descending profiles (given that level-off periods separate the ascents and descents), and the thickness of the interquartile range covered by the profile (to ensure reasonably complete tropospheric profiles).

In this work we use the NOAA Picarro G2401-m in situ measurements of CO, CH_4 and CO_2 from ATom. More information on the ATom NOAA Picarro data can be found at https://espo.nasa.gov/atom/instrument/NOAA_Picarro. Because the 10 second average ATom Picarro data are given in mixing ratios in ppm (or dry air mole fractions) at a vertical resolution comparable to the RTM/RTA layering, we simply interpolate these data to the RTA levels (layer boundaries), then convert to layer abundances (molecules/cm^2) for the statistical assessment of the NUCAPS EDRs. This is in contrast to the molecule-conservation approach required for high-resolution data (e.g., AirCore) as briefly alluded to in the previous section.

Relevant to the JPSS requirements, the ATom statistics on total columns in Section 4 are computed as follows. NUCAPS performs retrievals of CO and CH_4 concentrations (as well as H_2O and O_3) in layer abundance space (molecules/cm^2). Therefore column assessments for CO and CH_4 are performed for total column quantities by integrating the NUCAPS retrieved layer abundances; CO_2, on the other hand, is retrieved in mixing ratios (ppm), and thus we need only take the mean for the total column (CO_2 is treated differently given that CO_2 channels are used first in the physical retrieval steps for the T profile retrievals). The column abundance for atmospheric species X (viz., CO and CH_4) is defined as the vertical integral of the number density N_x from the top measurement z_t to the measurement level height z

$$\Sigma_z(X) \equiv \int_{z_t}^{z} N_x(z')\, dz' . \tag{1}$$

For the NUCAPS retrieval on the RTA layers, the total column may be computed from the finite difference formula [26]

$$\Sigma_{z_s}(X) \approx \mathcal{F}_{BL}\, \overline{N}_{x,L_b}\, \delta z_{L_b} + \sum_{L}^{L_b-1} \overline{N}_{x,L}\, \delta z_L , \tag{2}$$

where z_s is the surface altitude and the quantities $\overline{N}_{x,L}\, \delta z_L$ are the NUCAPS retrieved layer abundance for gas species x and RTA layer L (of thickness δz_L), L_b is the bottom partial layer, and $\mathcal{F}_{BL}$ is the bottom-layer multiplier factor defined as

$$\mathcal{F}_{BL} \equiv \frac{p_s - P_{l_b-1}}{P_{l_b} - P_{l_b-1}} , \tag{3}$$

where p_s is the surface level (boundary) pressure, P_{l_b} and P_{l_b-1} are the bottom-layer boundary pressures (i.e., the pressures of the bottom two levels, l_b and $l_b - 1$).

3.4. NUCAPS Retrievals

The NUCAPS retrieval sensitivity to state profile parameters (e.g., trace gas concentration) can be inferred from the retrieval AKs. The AK matrix is theoretically defined as $\mathbf{A} \equiv \partial \hat{\mathbf{x}}/\partial \mathbf{x}$ [37–40], where $\mathbf{A}$ is a square matrix dimensioned $m \times m$, m being the number of layers for the retrieved (i.e., estimated) and "true" (correlative) profiles, $\hat{\mathbf{x}}$ and $\mathbf{x}$, respectively. Note that the retrieval $\hat{\mathbf{x}}$ is related to $\mathbf{x}$ via the measurement equation $\hat{\mathbf{x}} = I[F(\mathbf{x}, \mathbf{b}), \mathbf{b}, \mathbf{c}]$, where F is the forward model with parameters $\mathbf{b}$ (e.g., spectroscopy), and I is the inverse model (i.e., retrieval), with parameters $\mathbf{c}$ not included in F (i.e., unrelated to the measurement) [26,39]. In the case of the NUCAPS algorithm, trapezoidal basis functions are used in the physical retrievals of each parameter (e.g., CO, CH_4, CO_2), and thus the corresponding $\mathbf{A}$ matrices must be transformed to "effective AKs" on the RTA layers, $\mathbf{A_e}$ (dimensioned $n \times n$, where $n \equiv 100 > m$ is the number of RTA layers), the details of which can be found in earlier papers [26,40].

Figure 3 shows zonal-mean NUCAPS effective AKs taken from a global Focus Day (23 January 2020) for the tropics, northern and southern hemisphere (NH and SH) midlatitude, and polar zones.

The Focus Day includes on the order of 220,000 NUCAPS retrievals over the entire globe, which, generally speaking, is considered representative of the range of global atmospheric conditions. The plots show the RTA column (or area, i.e., the row-sum along the first dimension) effective AKs for the CO, CH_4, and CO_2 channels [16] (subplots a–c, respectively). It can be seen that the peak sensitivities comprise broad layers. For CO, this roughly spans 600 to 300 hPa and that the peak remains fairly constant from the poles to the tropics. However, sensitivity is markedly less in the polar zones, this related to the lower tropopause, with sensitivity lowest in the SH plausibly due to lower ambient concentrations associated with substantially reduced source regions. There may also be some seasonal variability not accounted for in the Focus Day sample (which was during boreal winter). For CH_4 and CO_2, the peak sensitivities are somewhat lower in magnitude and higher in altitude than CO ($\approx$400–200 hPa and 300–200 hPa, respectively), with the height and sensitivity likewise decreasing with latitude zone.

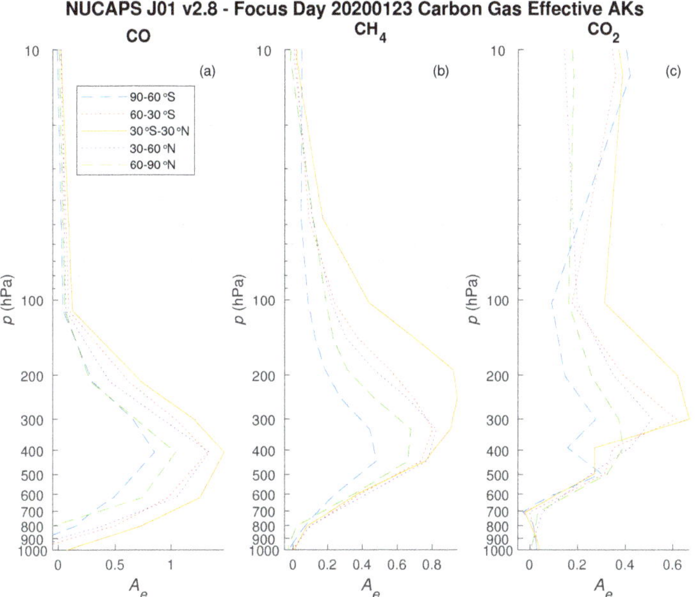

Figure 3. Zonal-mean NUCAPS RTA column (or row-sum) effective averaging kernels A_e for full spectral-resolution NOAA-20 CrIS carbon trace gas retrievals from a global Focus Day, 23 January 2020: (**a**) carbon monoxide, (**b**) methane, and (**c**) carbon dioxide. The solid lines are tropics (30°S to 30°N), dotted lines are midlatitudes (30–60°S and °N) and dashed lines are polar (60–90°S and °N).

The ability of the CrIS sensor to provide information about the trace gas profiles is also demonstrated by considering the NUCAPS algorithm degrees-of-freedom (DoF), defined as the sum of the **A** matrix diagonals, representing the total vertical information content [9]. Figure 4 shows the NUCAPS DoF for CO, CH_4 and CO_2 for the same Focus Day as in Figure 3. DoF for CO are mostly $\geq$1 (i.e., contain $\geq$1 independent pieces of information from the CrIS spectra) for most of the globe equatorward of the polar zones, with the exception of some high altitude locations, whereas areas with DoF $\geq$1 for CH_4 and CO_2 are primarily limited to the tropics (where the tropopause is at a higher altitude). Generally we expect greater retrieval skill in the regions with higher DoF.

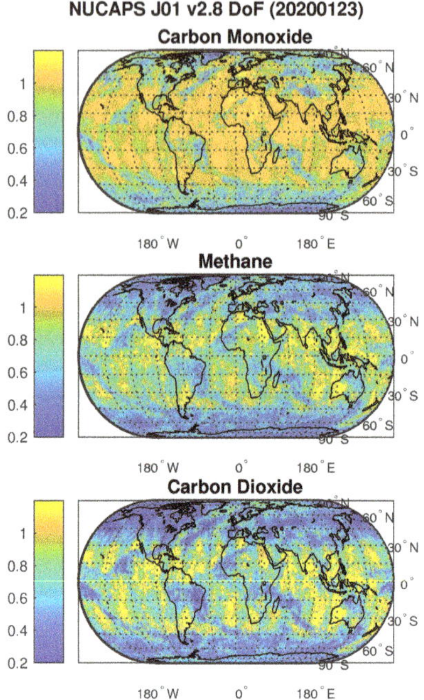

Figure 4. NUCAPS algorithm degrees-of-freedom (DoF) for NOAA-20 carbon trace gas retrievals from a global Focus Day, 23 January 2020: (**top**) carbon monoxide, (**middle**) methane, and (**bottom**) carbon dioxide. Map projections are equal-area.

4. Results and Discussion

In the following sections, the NUCAPS carbon trace gas retrievals are statistically validated versus the collocated baseline datasets described in Sections 3.1–3.3. In these analyses, we apply essentially the same collocation methodology as that used for our earlier ozone profile validation [22], whereby we impose a space-time collocation criterion in an effort to strike a balance between collocation mismatch uncertainty and sample size.

4.1. Statistical Analysis versus TCCON Baseline

For NUCAPS carbon gas validation using TCCON site observations, we ran offline SNPP and NOAA-20 NUCAPS retrievals for 6 global focus days spanning the annual cycle (1 April, 15 June, 20 August, 15 October and 15 December 2018, and 15 February 2019), then collocated the NUCAPS fields-of-regard (FORs, which consist of 3×3 CrIS fields-of-view used for cloud-clearing [9,10]) within $\Delta r \leq 125$ km radius and Δt within ± 2 h of the TCCON measurements. The global focus day runs also allowed for numerical model and satellite EDR comparisons (as discussed in Section 2), but these will be highlighted in a future paper (Warner et al., manuscript in prep for Atmos. Chem. Phys.). Figure 5 shows the locations of TCCON stations with available data that collocated with the SNPP data (NUCAPS QA-accepted cases) during these focus days.

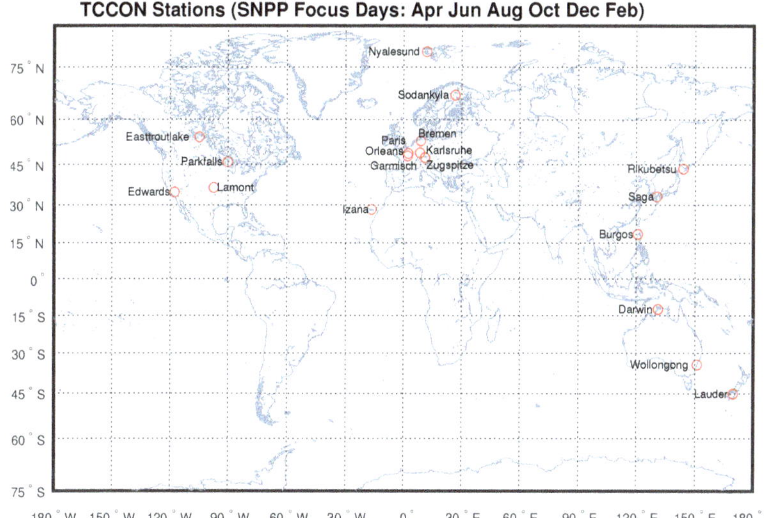

Figure 5. TCCON stations [31] collocated with SNPP NUCAPS (QA -accepted cases, $\Delta r \leq 125$ km and ± 2 h) for 6 global focus days (1 April 2018, 15 June 2018, 20 August 2018, 15 October 2018, 15 December 2018, 15 February 2019); stations shown are (south to north): Lauder (NZ) [41,42], Wollongong (AU) [43], Darwin (AU) [44], Burgos, Ilocos Norte (PH) [45], Izana (ES) [46], Saga (JP) [47], Edwards (US) [48], Lamont (US) [49], Rikubetsu (JP) [50], Park Falls (US) [51], Zugspitze (DE) [52], Garmisch (DE) [53], Orléans (FR) [54], Paris (FR) [55], Karlsruhe (DE) [56], Bialystok (PL) [57], East Trout Lake, SK (CA) [58], Sodankylä (FI) [59,60], and Ny Ålesund, Spitsbergen (NO) [61].

As mentioned in Section 3.1, statistical comparisons of NUCAPS with TCCON requires unit conversions, as well as integration of the NUCAPS 100 layer profiles. In this case, the NUCAPS profiles (in layer abundances, molecules/cm^2) are first converted to dry mole fractions and then integrated into a total column value with or without the TCCON AKs applied, as detailed in Appendix A. The results for SNPP NUCAPS retrievals versus individual TCCON stations, ordered from south to north, are summarized in Figure 6; these plots show reasonable consistency of the SNPP NUCAPS retrievals versus individual TCCON stations (similar results were obtained for NOAA-20, but not shown here due to space constraints). The positive bias evident in the CO results may in part be due to the different vertical sensitivities between the NUCAPS and TCCON measurements, as evidenced by column AK peak altitudes shown in Figures 1 and 3. The TCCON vertical sensitivities for CO are weighted toward the upper troposphere and above, whereas sensitivities for CH_4 and CO_2 transition to the troposphere for larger solar zenith angles, with a crossover point roughly in the mid-troposphere (≈ 450 hPa). NUCAPS retrievals, on the other hand, being derived from passive thermal IR spectra, tend on having peak sensitivity weighted toward the mid-troposphere. In addition to the different instrument sensitivities, however, there is also a known problem in the TCCON X_{CO} scaling, wherein the TCCON data were scaled down by $\approx 6.7\%$ to match older aircraft data. There is now less confidence placed in this value given that it has changed as more recent in situ profiles have been added for comparison [62], and thus it is believed that this scaling factor also contributes to the observed discrepancy between the NUCAPS and TCCON CO.

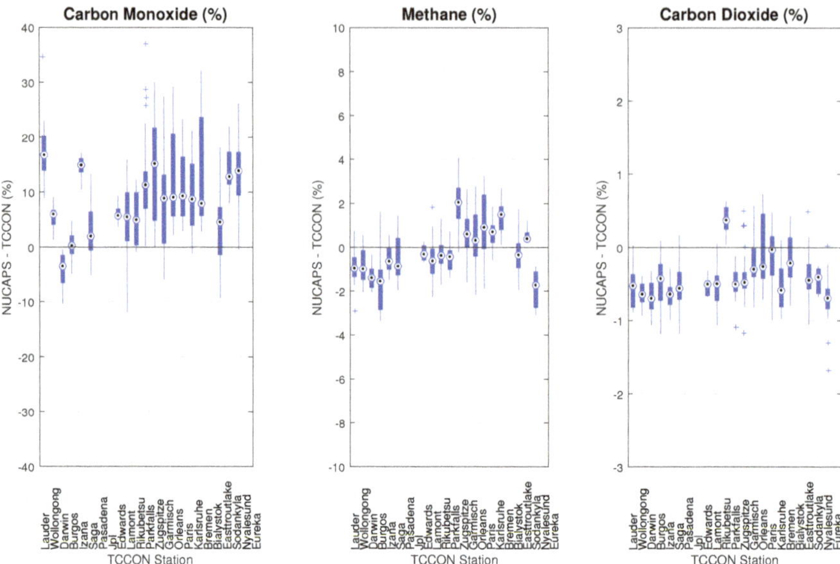

Figure 6. Box-whisker robust error statistics (%) of total column SNPP NUCAPS trace gas retrievals (QA-accepted cases including trace gas QA) versus means at individual collocated ($\Delta r \leq 125$ km and ± 2 h) TCCON stations, ordered from south to north: (**left**) CO, (**center**) CH_4, and (**right**) CO_2. Circles and blue boxes depict medians and interquartile range, respectively; blue "whiskers" depict remaining data spread excluding outliers, and + signs designate outliers. TCCON column AKs were applied in the NUCAPS column integrations (cf. Appendix A). Note that available data from Caltech, Pasadena (US) [63], Jet Propulsion Laboratory (US) [64], Bremen (DE) [65], and Eureka (CA) [66] ultimately did not meet the collocation criteria.

The results for the complete QA-ed samples ($N = 472, 422, 540$, yields = 74%, 67%, 85% for CO, CH_4 and CO_2, respectively) are summarized as scatterplots and histograms in Figures 7 and 8, respectively. The scatterplots show reasonable correlation between the total column retrievals and TCCON measurements ($r = 0.89, 0.63, 0.86$ for CO, CH_4 and CO_2, respectively), and the histograms show roughly Gaussian distributions in the errors. Featured in the histograms are results with (blue) and without (red) the TCCON AKs applied to the integrations, which for CH_4 and CO_2 basically show very little difference when the AKs are applied. However, for CO a somewhat larger bias is seen when TCCON AKs are applied. At first this may seem counterintuitive, but this is likely because, as already mentioned, the TCCON AKs for CO (unlike CH_4 and CO_2) all peak above the UT/LS, whereas the NUCAPS AKs for CO peak in the mid-troposphere. Thus, greater weight is given to the upper-troposphere/lower-stratosphere (UT/LS) when TCCON AKs are applied to NUCAPS, and given that NUCAPS has no skill above 100 hPa, we therefore would expect less agreement in the total column results.

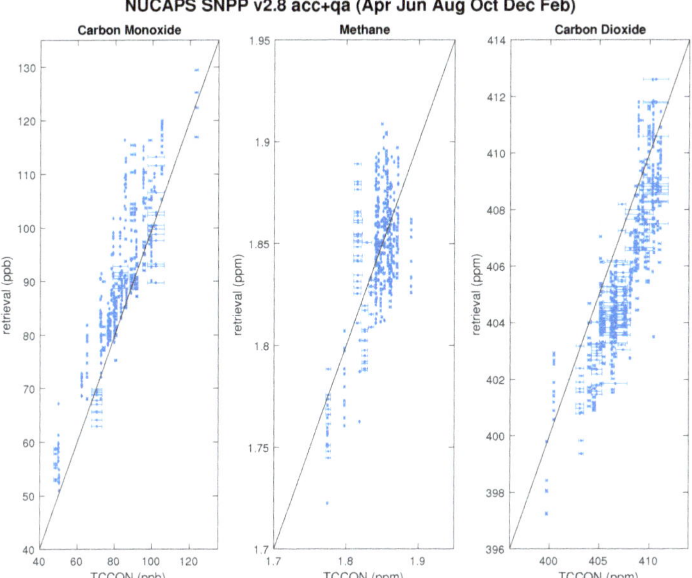

Figure 7. Scatterplots of total column SNPP NUCAPS trace gas retrievals versus means at individual collocated TCCON stations ($\Delta r \leq 125$ km radius and Δt within ± 2 h): (left) CO, (center) CH_4, and (right) CO_2; horizontal errorbars denote the 3σ uncertainties in the mean collocated TCCON measurements. TCCON column AKs were applied in the NUCAPS column integrations (cf. Appendix A); "acc+qa" indicates QA-accepted retrievals including trace gas QA.

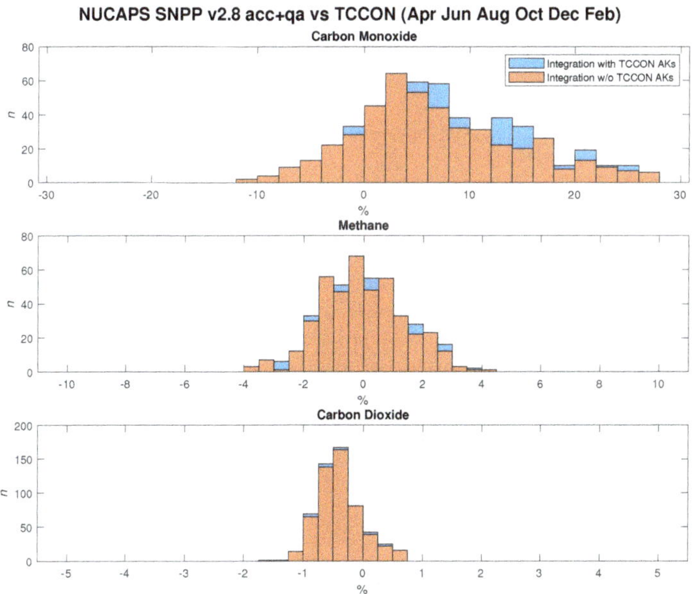

Figure 8. Histograms of total column differences (%) between SNPP NUCAPS trace gas retrievals and individual TCCON station means: (**top**) CO, (**center**) CH_4, and (**bottom**) CO_2. The blue and red histograms show results with and without TCCON column AKs applied in the NUCAPS integrations, respectively; "acc+qa" indicates QA-accepted retrievals including trace gas QA.

4.2. Statistical Analysis versus AirCore Baseline

Like TCCON, AirCore data can provide spot-checks and an additional evaluation method for comparing results from multiple satellites (viz. SNPP and NOAA-20). NOAA/GML provided us with 42 complete AirCore profiles launched over the period of 22 March 2018 to 30 January 2020. The AirCore balloon launches were timed for LEO satellite overpasses, specifically the Orbiting Carbon Observatory-2 (OCO-2) within the NASA A-Train constellation (01:30 and 13:30 local equator crossing time orbit), which fortuitously collocate with the SNPP and NOAA-20 overpasses in the same afternoon orbit. NUCAPS FORs are included within $\Delta r \leq 100$ km radius and Δt within ± 2 h of the AirCore launches; Figure 9 shows the locations of collocated NOAA-20 NUCAPS FOR along with the AirCore launch sites. One may see that the samples are primarily located over North America, with a handful located in Europe.

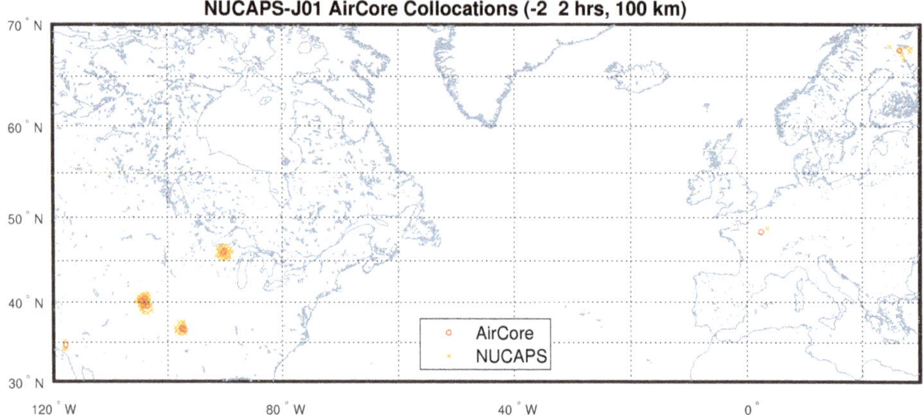

Figure 9. NOAA-20 NUCAPS collocations with AirCore launches ($\Delta r \leq 100$ km radius and Δt within ± 2 h).

The original "high density" profiles were reduced to the NUCAPS 100 RTA layer abundances as described in Section 3.2. To perform the unit conversions it was necessary to utilize the AirCore payload InterMet-1 radiosonde temperature and relative humidity (RH) measurements, but negative RH values were sometimes reported by the sonde in the UT/LS. To get around this problem, we simply adjusted these values to a small positive number (1%), as the stated accuracy of the InterMet-1 RH sensor is ± 5%. To justify this, we performed a simple sensitivity test to determine the error in layer abundance for a +1% RH perturbation (performing conversions with 1% added to the entire RH profile and then subtracting the unperturbed values). The results are presented as a function of ambient water vapor mixing ratio (ppmv) and pressure altitude in Figure 10, where it can be seen that the error from a 1% RH adjustment is negligible.

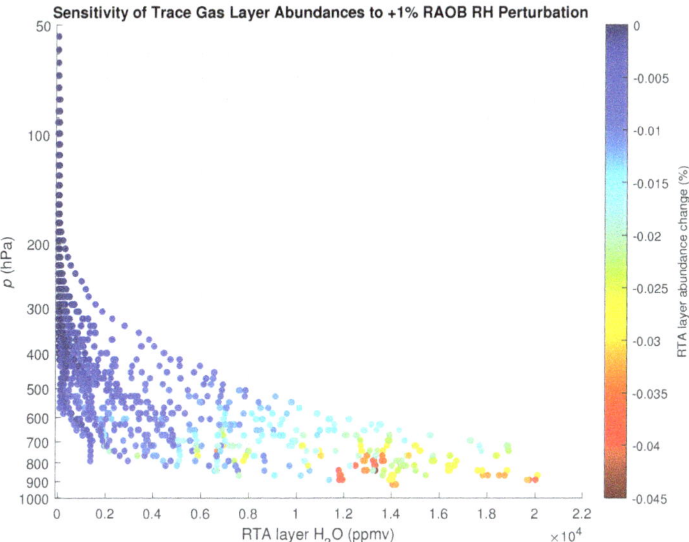

Figure 10. Sensitivity of computed trace gas (carbon monoxide) NUCAPS RTA layer abundances (% error). The *x*-axis is the RTA layer water vapor volume mixing ratio (ppmv) and the *y*-axis is pressure altitude (hPa).

Because of the limited DoF and vertical sensitivity of the instrument, we also use AKs in our evaluation of the NUCAPS carbon trace gas profile retrievals. Thus the analysis will include results based upon "smoothed" correlative truth data, $\mathbf{x_s}$, which is obtained by applying the NUCAPS effective-AKs $\mathbf{A_e}$ to the original high-resolution truth profile $\mathbf{x}$ [26,39,40]

$$\ln(\mathbf{x_s}) = \ln(\mathbf{x_0}) + \mathbf{A_e}\left[\ln(\mathbf{x}) - \ln(\mathbf{x_0})\right], \tag{4}$$

where $\mathbf{x_0}$ is the a priori profile. Using $\mathbf{x_s}$ in place of $\mathbf{x}$ in the statistical analyses effectively removes the null-space error associated with the limited vertical resolution inherent in the radiances used by the retrieval algorithm. However, caution must be exercised when using this approach. When the algorithm possess little-to-no sensitivity to a profile parameter $\mathbf{x}$, the AK matrix becomes a null matrix, $\mathbf{A_e} \to \mathbf{0}_{n,n}$, and the second term on the right in Eqution (4) goes to zero. In this case, both the smoothed truth profile and the retrieval reduce to the a priori, $\mathbf{x_0}$. Although the result would indicate that the retrieval system is self-consistent and working properly, it would also give the misleading appearance of a perfect retrieval of the true atmospheric state, which is definitely not the case.

Figures 11 and 12 show the resulting statistical comparisons of the collocated NOAA-20 NUCAPS retrievals versus the AirCore profiles, without and with AKs applied, respectively. The results for CO, CH$_4$ and CO$_2$ are shown in the left, middle and right plots, respectively. From Figure 9, we recall that the AirCore profiles are all located over Northern Hemisphere (NH) land-based sites (viz., North America and Europe). We subsequently found that several of these profiles exhibited very large gradients that are well outside the theoretical vertical resolution limitations of the CrIS sensor, with vertical gradients in the AirCore high-resolution profiles not well-captured by the NUCAPS climatological a priori. The profile statistics for AirCore are shown on the coarse-layers defined by the NUCAPS algorithm trapezoidal basis functions, similar to the statistical analyses of NUCAPS $T(p)$/H$_2$O/O$_3$ profiles [21,22].

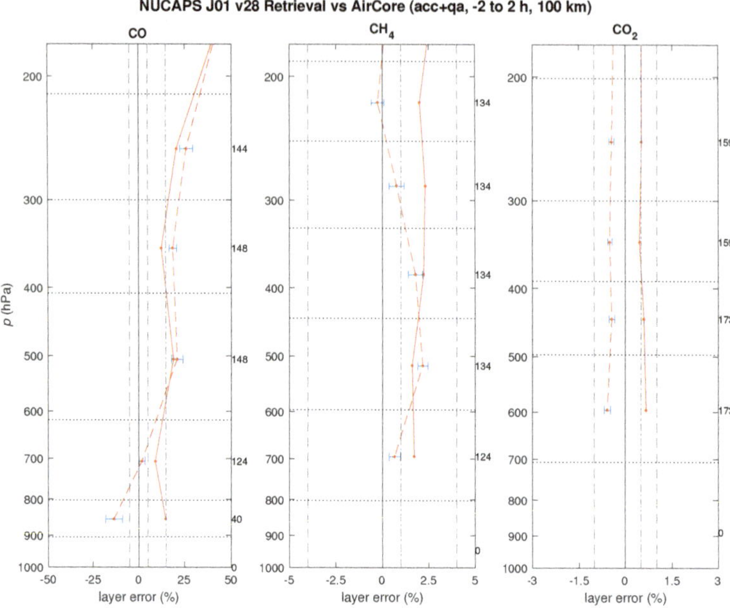

Figure 11. NOAA-20 NUCAPS coarse-layer accuracy (bias $\pm 2\sigma$ uncertainty in the sample mean; dotted red line and blue hatches) and precision (1σ variability; solid dark red line) statistics versus AirCore profiles: (**left**) carbon monoxide, (**center**) methane, and (**right**) carbon dioxide. Layer sample sizes are indicated on the right margins; "acc+qa" indicates QA-accepted retrievals including trace gas QA.

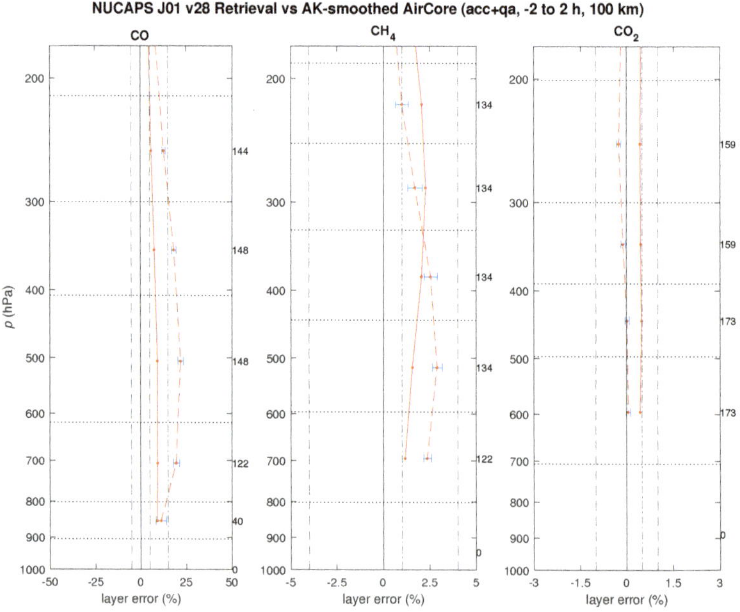

Figure 12. As Figure 11 except with NUCAPS AKs applied to the high resolution AirCore data as indicated by Eqution (4).

For this small NH continental sample, the NUCAPS retrievals are found to exhibit somewhat a positive bias in CO and CH_4, both without (Figure 11) and with AKs (Figure 12) applied. The latter indicates biases not arising from null-space errors ($\lesssim$25% and $\lesssim$2.5%, respectively) and thus there is systematic error in the layers where NUCAPS has sensitivity. This results from AirCore profiles with higher observed concentrations (not shown here) in the lower troposphere ($\gtrsim$700 hPa), decreasing rapidly to the mid-troposphere ($\approx$500 hPa), then increasing again to the upper troposphere ($\approx$200 hPa). NUCAPS, on the other hand, has sensitivity in the mid-troposphere (Figure 3), and the a priori concentrations generally decrease with height. In contrast, the CO_2 retrievals exhibit a very small negative bias ($\approx$0.5%), with most of that apparently null-space error as seen in the results with AKs applied (Figure 12 right), indicating that the retrieval is accurate in the layers of sensitivity (Figure 3). Precision magnitudes (random errors) for all three gases are somewhat comparable to their accuracies (systematic errors), with some of those errors originating from their null-spaces, especially carbon monoxide, and to a lesser extent, carbon dioxide. Given the limited size and geographic representation of the sample, these results should not be considered definitive or globally representative, but they do offer insight into the challenges inherent in retrieving regional profiles over land. But more importantly, these first-use results demonstrate the potential utility of the AirCore sampling system for operational trace gas validation.

4.3. Statistical Analysis versus ATom Baseline

The in situ global data from the ATom intensive campaigns are considered to be at the top of our validation "hierarchy" (cf. Section 2). Thus, while we relied more on the TCCON analyses for the developmental phases of the trace gas algorithms (per the hierarchal approach), we give higher weight to the ATom data for a final quantitative evaluation of the NUCAPS carbon gas EDR product performance relative to the metrics defined by the JPSS Level 1 requirements summarized in Table 1. Although JPSS requirements are applicable to the total system error (including null-space error), it is nevertheless imperative to include AKs in the validation of the carbon trace gases as in Section 4.2, with the caveats discussed above in that section. Similar to the analyses for TCCON and AirCore, NUCAPS FORs are collocated within $\Delta r \leq 100$ km radius and Δt within ± 1.5 h of the ATom measurements. Figure 13 shows the dates and locations of SNPP and NOAA-20 NUCAPS FOR collocated with the midpoint of extracted profiles from the ATom-1, -2, and -4 campaigns. These maps show the excellent global zonal representation of the validation sample, albeit primarily over oceans. Although the NUCAPS retrievals may generally be "easier" (i.e., more accurate) over ocean surfaces (i.e., where the surface emission/reflectance properties are relatively uniform and well characterized relative to the retrieval uncertainties) [67], this is not always the case [68], and operational satellite data have been demonstrated to make their greatest impact over the data-sparse oceans [69]. Thus the ATom data are of singular value for our validation.

Based on the NUCAPS-ATom collocation samples, the global profile error statistics for the NUCAPS retrievals (IR accepted cases, clear to partly cloudy, with trace gas QA applied (Warner et al., manuscript in prep for Atmos. Chem. Phys.)) are computed versus ATom NOAA Picarro baseline; as before (cf. Section 4.2), the results are summarized within Figures 14–17, with CO, CH_4 and CO_2 statistics shown in the left, center, and rightmost plots. Because the ATom profiles generally exhibit smaller vertical gradients and are closer to the NUCAPS a priori profiles (as opposed to AirCore), we display these results on the original 100 RTA layers (as opposed to trapezoidal coarse-layers). For reference, the JPSS Level 1 global specification requirements (Table 1) for accuracy (bias) and precision (variability) are included in the plots with dashed gray lines. Figures 14 and 15 show results for NOAA-20 and SNPP, respectively, and these are followed by Figures 16 and 17, which show the results with the NUCAPS AKs applied to the ATom profiles.

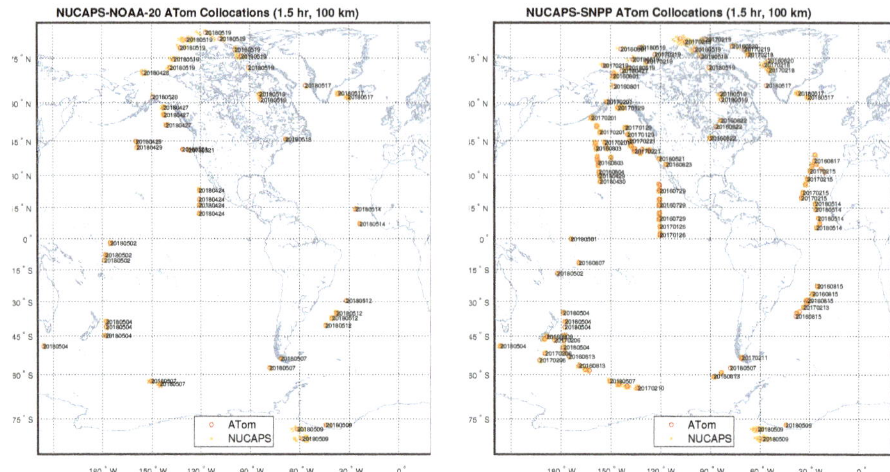

Figure 13. ATom collocation samples for NUCAPS carbon trace gas profile validation: NUCAPS FOR (gold ×) are shown collocated with extracted ATom profiles (red circles) within space-time collocation windows of $\Delta r \leq 100$ km radius and Δt within ± 1.5 h for (**left**) ATom-4 with NOAA-20, and (**right**) ATom-1, -2, -4 with SNPP.

In the leftmost plots of Figures 14 and 15 we find that the CO accuracy (biases) for the broad layer between 400–600 hPa (which corresponds to the region where the algorithm has maximum sensitivity) are reasonably close to, or within, JPSS requirements; CH_4 and CO_2 biases, on the other hand, are well within requirements throughout the troposphere, with CH_4 bias statistically close to zero (at the 2σ level) below 400 hPa. Precision (variabilities) for CO and CH_4 fall somewhat outside the requirements, whereas the CO_2 precision meets requirements throughout the entire tropospheric column.

When the AKs are applied to the truth data via Eqution (4) (Figures 16 and 17), the retrievals are seen to be within JPSS requirements throughout the tropospheric column, with the exception of CH_4. While these are not the actual total-system accuracy and precision relative to the correlative measurement, they indicate that the algorithm is performing properly within its theoretical limits, which includes the vertical resolution afforded by the radiances, cloud-clearing, RTA tuning, a priori, algorithm damping factor (an optimization parameter that limits noise propagation into the solution [40]) and QA flags. In particular we can see that errors falling outside of requirements in the CO and CO_2 retrievals are the result of the null-space error; thus these errors are indicative of a fundamental limitation in the vertical resolving power of the CrIS sensor. The methane precision, on the other hand, poses an enduring problem, given that the results still fall outside requirements throughout the troposphere (with the exception of the lower troposphere, where NUCAPS has little skill; Figure 3b), even with AKs applied. We also found this to be the case with tighter space-time collocation criteria (not shown here), which would reduce potential mismatch errors, but also decreases sample size and thus redundancy. Thus, these comparisons to a large swath of recent in situ measurements (ATom, AirCore and TCCON) suggest that the CH_4 precision threshold (viz., 1%), may in fact be unrealistically stringent, especially when one considers the far more relaxed requirement for accuracy (viz., 4%).

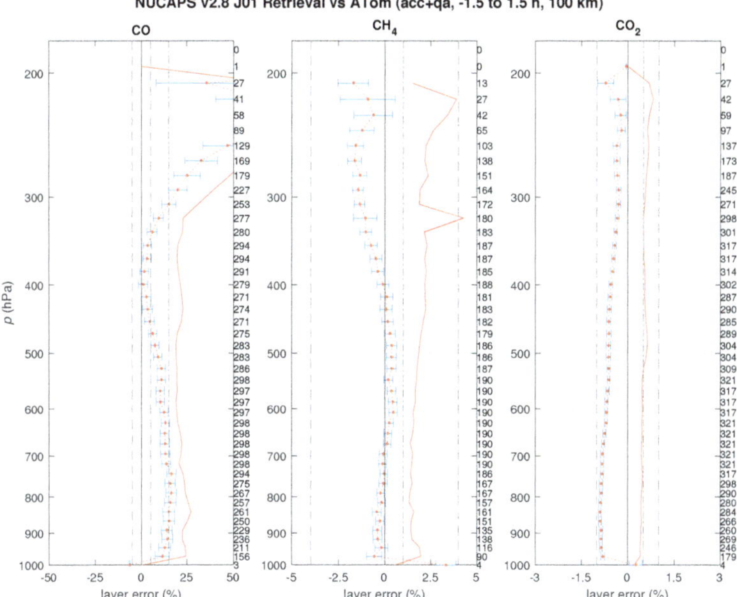

Figure 14. NOAA-20 NUCAPS 100-RTA layer accuracy (bias $\pm 2\sigma$ uncertainty in the sample mean; dotted red line and blue hatches) and precision (1σ variability; solid dark red line) statistics versus ATom-4 in situ aircraft data (NOAA Picarro measurements): (**left**) carbon monoxide, (**center**) methane, and (**right**) carbon dioxide. Layer sample sizes are indicated on the right margins. The vertical dashed and dot-dashed gray lines indicate the JPSS Level 1 requirements for accuracy (bias) and precision (variability), respectively; "acc+qa" indicates QA-accepted retrievals including trace gas QA.

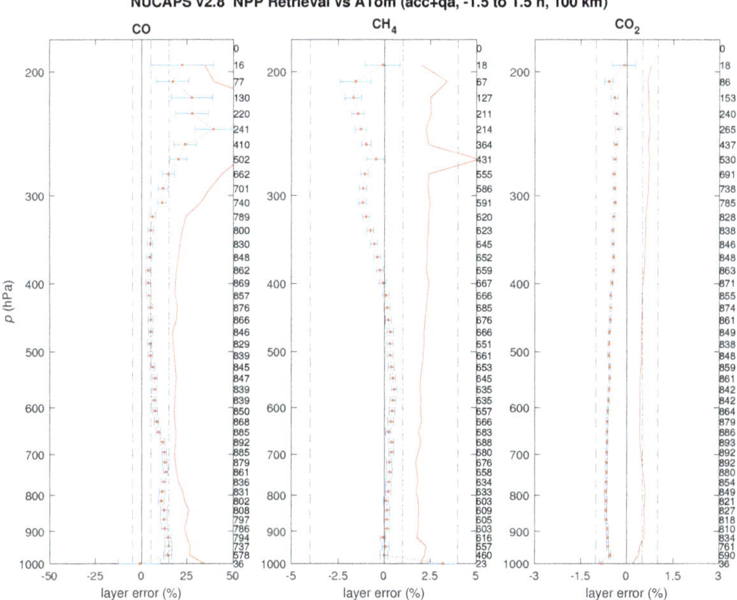

Figure 15. As Figure 14 except for the SNPP satellite versus ATom-1, -2 and -4.

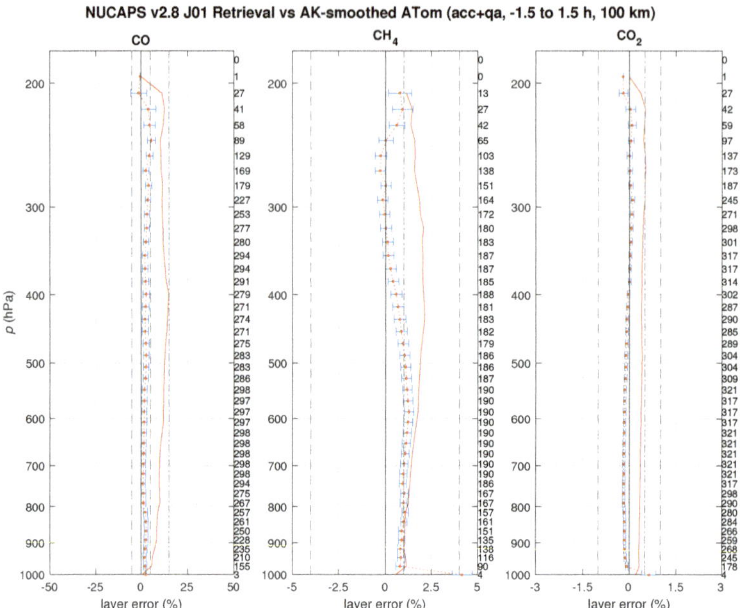

Figure 16. As Figure 14 except with NUCAPS AKs applied to the ATom data as indicated by Eqution (4).

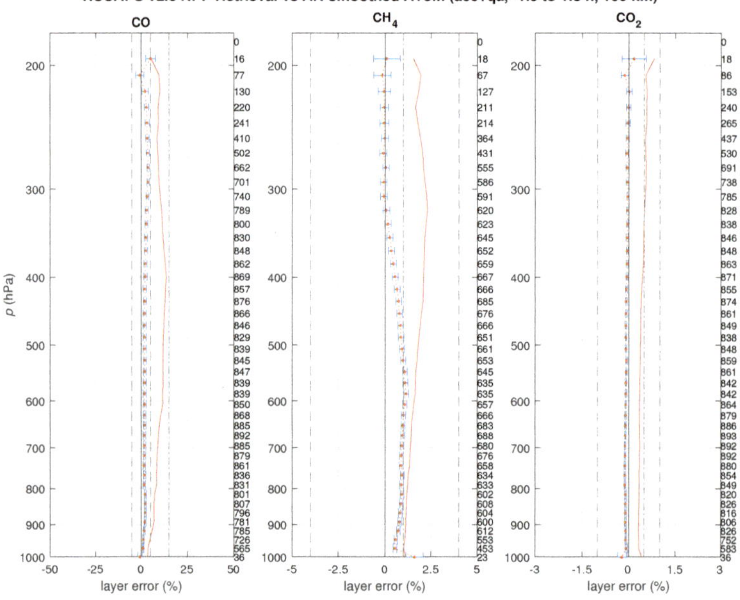

Figure 17. As Figure 14 except for the SNPP satellite.

The total column results relevant to the JPSS requirements (cf. Section 3.3) for both the NOAA-20 and SNPP data samples are summarized in Table 2 in terms of fractional accuracy (bias), precision (1σ variability), total combined uncertainty (RMSE), correlation coefficient (r) and associated p-values, sample sizes (N), and yield. For completeness, we include results with and without NUCAPS AKs applied to the ATom truth profiles, indicated by "AK" and "raw"columns, respectively. As commented above, with the exception of the CH_4 precision, results are generally within requirements, especially when NUCAPS AKs are applied. Likewise, the NUCAPS CO and CO_2 products both exhibit good total column correlation with ATom measurements (>0.75), with CH_4 on the order of 0.5.

Table 2. Validated NUCAPS-CrIS Trace Gas EDR Total Column Measurement Uncertainty (ATom Baseline).

Trace Gas	Bias (%)		σ (%)		RMSE (%)		r		p		N	Yield
	Raw	AK	Raw	AK	Raw	AK	Raw	AK	Raw	AK		
					NOAA-20							
CO	+10.5	+2.0	18.6	9.6	21.4	9.8	0.92	0.92	0	0	298	59%
CH_4	−0.2	+0.8	1.4	1.3	1.4	1.6	0.61	0.61	0	0	190	38%
CO_2	−0.7	−0.1	0.4	0.3	0.8	0.3	0.81	0.84	0	0	321	63%
					Suomi NPP							
CO	+7.8	+1.9	15.6	8.3	17.5	8.5	0.91	0.89	0	0	901	64%
CH_4	+0.0	+0.7	1.6	1.3	1.6	1.5	0.38	0.38	0	0	696	49%
CO_2	−0.6	−0.1	0.4	0.3	0.7	0.3	0.78	0.79	0	0	969	69%

5. Conclusions and Future Work

This work has presented the formal validation of NOAA-20 and SNPP NUCAPS IR atmospheric carbon trace gas profile EDRs (CO, CH_4 and CO_2), in continuation of the validation of the T, q and O_3 profile EDRs described in earlier papers [21–24]. Because of the NUCAPS cloud-clearing methodology, the NUCAPS atmospheric profile EDRs, including trace gases, are retrieved under global, non-precipitating conditions, allowing the benefit and advantage of twice-per-day (per satellite) global yields on the order of 40–70%.

The NUCAPS IR sounder validation strategy employs a "hierarchical" approach drawing upon multiple independent baseline truth datasets [26], including TCCON ground-based spectrometers, AirCore profiles, and ATom aircraft-based in situ profiles. Based upon these globally representative data, we have conducted ongoing statistical analyses (per the JPSS Cal/Val Program) that have provided guidance for the recent NUCAPS trace gas algorithm improvements validated in this work (Warner et al., manuscript in prep for Atmos. Chem. Phys.). The NUCAPS optimal estimation (OE) physical retrievals generally improve upon the climatological a priori (not shown here due to space limitations) where CrIS has sensitivity (Figure 3). We have subsequently shown here that the carbon trace gas EDRs (CO, CH_4, and CO_2) from the latest version of NUCAPS are performing reasonably within expectations. It is noted that the truth data used in these analyses span all global climate zones (tropical, midlatitude and polar), as well as land and ocean locations (Figures 5, 9 and 13). Based upon our analysis comparing to global in situ vertical profiles from the ATom campaigns, it has been shown that the NUCAPS CrIS-FSR carbon trace gas profile EDRs generally meet JPSS Level 1 global performance requirements (Tables 1 and 2), with the exception of the stringent 1% CH_4 precision specification, which may be extremely difficult to achieve in practice.

Future work on the NUCAPS trace gas products include optimization of the damping parameters, implementation of QA for the CO_2 retrievals, improvements to the SARTA forward model surface emissivity first-guess (land, ocean and snow/ice), as well as exploring additional trace gas products (e.g., NH_3, SO_2, Isoprene, PAN) and collaborations with in situ data providers (e.g., NOAA/GML). The NUCAPS AKs are planned to be included in a future version as standard output in the operational NetCDF files (currently the AKs are output only to offline binary files), and the NUCAPS algorithm will also operationally be supported for data from the EUMETSAT Metop-B, -C and Metop-SG hyperspectral IASI systems.

Author Contributions: Conceptualization, N.R.N., A.G., J.W. and L.Z.; methodology, N.R.N., J.W., C.T., A.G.; software, C.T., M.W., N.R.N., A.G., T.Z.; validation, N.R.N., C.T., J.W., T.Z., Z.W., T.W. (Tianyuan Wang); formal analysis, N.R.N.; investigation, N.R.N.; resources, L.Z., S.K., M.D., M.W., C.T., C.S., B.C.B., K.M., D.W., N.D., F.H., L.T.I., R.K., I.M., J.N., H.O., D.F.P., Y.T., V.V., T.W. (Thorsten Warneke), R.S., M.R.; data curation, N.R.N., C.T., T.W. (Tianyuan Wang), Z.W., A.G., C.S., B.C.B., K.M., D.W., N.M.D., F.H., L.T.I., R.K., I.M., J.N., H.O., D.F.P., Y.T., V.A.V., T.W. (Thorsten Warneke), R.S., M.R.; writing—original draft preparation, N.R.N.; writing—review and editing, N.R.N., A.G., T.Z., B.C.B., K.M., D.W., N.M.D., F.H., L.T.I., R.K., I.M., J.N., H.O., D.F.P., Y.T., V.A.V., T.W. (Thorsten Warneke); visualization, N.R.N.; supervision, K.P., S.K., L.Z., M.D.; project administration, S.K., K.P., L.Z.; funding acquisition, L.Z., S.K. All authors have read and agreed to the published version of the manuscript.

Funding: This research was supported by the NOAA Joint Polar Satellite System (JPSS-STAR) Office and Cal/Val Program (M. D. Goldberg), along with the NOAA/NESDIS/STAR Satellite Meteorology and Climatology Division. The Atmospheric Tomography Mission (ATom) is a NASA Earth Venture Suborbital 2 project funded by NASA's Earth Science Division; NOAA Picarro measurements on ATom were supported by NASA grant #NNX16AL92A to the University of Colorado. The Paris TCCON site has received funding from Sorbonne Université, the French research center CNRS, the French space agency CNES, and Région Île-de-France. The TCCON stations at Rikubetsu, and Burgos are supported in part by the GOSAT series project. Local support for Burgos is provided by the Energy Development Corporation (EDC, Philippines). NMD is funded by ARC Future Fellowship FT180100327. Darwin and Wollongong TCCON stations are supported by ARC grants DP160100598, LE0668470, DP140101552, DP110103118 and DP0879468. The TCCON stations Garmisch and Zugspitze have been supported by the European Space Agency (ESA) under grant 4000120088/17/I-EF and by the German Bundesministerium für Wirtschaft und Energie (BMWi) via the DLR under grant 50EE1711D.

Acknowledgments: The TCCON data were obtained from the TCCON Data Archive hosted by CaltechDATA at https://tccondata.org. Atmospheric Tomography (ATom) Mission data were obtained at https://doi.org/10.3334/ORNLDAAC/1613. We are grateful to M. Kopacz (NOAA/UCAR), and other contributors to the NUCAPS development/validation effort, especially C. D Barnet, N. Smith and R. Esmaili (STC), X. Xiong (NASA/LaRC), E. Maddy (Riverside Technology, Inc.), and W. W. Wolf and A. K. Sharma (NOAA/NESDIS/STAR). Finally, we express our appreciation to the 4 anonymous reviewers who provided constructive feedback that we used to improve the quality of these papers. The views, opinions, and findings contained in this paper are those of the authors and should not be construed as an official NOAA or U.S. Government position, policy, or decision.

Conflicts of Interest: The authors declare no conflict of interest.

Abbreviations

The following abbreviations are used in this manuscript:

AIRS	Atmospheric Infrared Sounder
AK(s)	averaging kernel(s)
ATMS	Advanced Technology Microwave Sounder
ATom	Atmospheric Tomography mission
CAMS	Copernicus Atmosphere Monitoring Service
CrIS	Cross-track Infrared Sounder
DoF	degrees-of-freedom
ECMWF	European Center for Medium Range Weather Forecast
EDR(s)	environmental data record(s)
EUMETSAT	European Organisation for the Exploitation of Meteorological Satellites
FOR(s)	field(s)-of-regard (NUCAPS)
FSR	full spectral-resolution (CrIS)
FTS	Fourier transform spectrometer
IASI	Infrared Atmospheric Sounding Interferometer
JPSS	Joint Polar Satellite System
J-1 or J01	JPSS-1 satellite (i.e., NOAA-20 pre-launch, still used as a designator in operational files)
LEO	low earth orbit
NOAA	National Oceanic and Atmospheric Adminstration
NUCAPS	NOAA-Unique Combined Atmospheric Processing System
OE	optimal estimation
QA	quality assurance
RH	relative humidity

RMSE	root mean square error
RTA	radiative transfer algorithm (alternatively, rapid transmittance algorithm)
RTM	radiative transfer model
SARTA	Stand-Alone Radiative Transfer Algorithm
SDR(s)	sensor data record(s)
SNPP	Suomi National Polar-orbiting Partnership (satellite)
TCCON	Total Carbon Column Observing Network
UT/LS	upper-troposphere/lower-stratosphere

Appendix A. NUCAPS to TCCON Conversions

Appendix A.1. Column Integration Formulas

The TCCON measurement for a given atmospheric profile constitutes an integrated column measurement of dry mole fraction (DMF), X_d. The integrated mole fraction is related to the constituent profile (as a function of pressure, p) as [70]

$$\Sigma_p(X) \equiv \int_0^{p_s} \frac{X(p)}{M(p)\, g(p,\varphi)}\, dp \, , \tag{A1}$$

where p_s is the surface pressure, g is the gravitational acceleration, φ is the latitude, X is the constituent mole fraction, and M is the molecular mass of air. $X(p)$ is related to the DMF, X_d, as

$$X(p) = X_d(p)\left[1 - Q(p)\right] , \tag{A2}$$

where Q is the mole fraction of water vapor. $M(p)$ may be broken into moist (q) and dry (d) components as

$$M(p) = m_q\, Q(p) + m_d\left[1 - Q(p)\right] . \tag{A3}$$

Substituting Equtions (A2) and (A3) into Eqution (A1) yields [70]

$$\Sigma_p(X) \equiv \int_0^{p_s} \frac{X_d(p)}{m_d\, g(p,\varphi)\left[1 + \epsilon\, Q_d(p)\right]}\, dp \, , \tag{A4}$$

where $\epsilon \equiv m_q/m_d$ and from Eqution (A2) the water vapor DMF is given by

$$Q_d(p) = \frac{Q(p)}{1 - Q(p)} \, . \tag{A5}$$

Eqution (A4) forms the basis integrating the NUCAPS retrievals, which we will return to below in Appendix A.3. It may also be seen that dry mole fractions are required for comparing a given trace gas measurement against TCCON. The conversion of NUCAPS retrievals to DMFs are discussed in the next section.

Appendix A.2. NUCAPS Layer Conversions

In the current application using NUCAPS, the X-constituent dry mole fraction X_d is derived from the NUCAPS retrieved volume mixing ratios, X_v (which in turn are computed from the retrieved layer abundances), as follows. The retrieved mole fraction may be calculated from the gas partial pressure as

$$\widehat{X}(P) = \frac{\hat{p}_x(P)}{P} \, , \tag{A6}$$

where carrots (e.g., $\widehat{X}$) denote measurement estimates (here being the NUCAPS retrievals), P is the RTA atmospheric effective-layer pressure and $\hat{p}_x$ is the retrieved partial pressure of the gas computed from

$$\hat{p}_x(P) = 10^6 \cdot \widehat{X}_v(P) \left[P - \hat{p}_q(P) \right] \tag{A7}$$

and the water vapor partial pressure is computed from

$$\hat{p}_q(P) = P \, \frac{\widehat{Q}_v(P) \cdot 10^{-6}}{1 + \widehat{Q}_v(P) \cdot 10^{-6}} \tag{A8}$$

where $\widehat{Q}_v$ is the volume mixing ratio in in parts per million (ppmv). The retrieval dry mole fraction is then computed from Eqution (A2) as

$$\widehat{X}_d(P) = \frac{\widehat{X}(P)}{1 - \widehat{Q}(P)}, \tag{A9}$$

where from Eqution (A6)

$$\widehat{Q}(P) = \frac{\hat{p}_q(P)}{P}. \tag{A10}$$

Appendix A.3. Application of TCCON Column AKs

Rodgers and Connor [30] formulated the theoretical basis for performing rigorous intercomparisons of remotely sensed atmospheric soundings obtained by instruments with differing measurement characteristics. For total column estimates from two observing systems, $\widehat{C}_1$ and $\widehat{C}_2$, the expected difference is given by [30]

$$\widehat{C}_1 - \widehat{C}_2 = (\mathbf{a_1} - \mathbf{a_2})^T (\mathbf{x} - \mathbf{x_c}) + (\varepsilon_1 - \varepsilon_2), \tag{A11}$$

where $\mathbf{a_1}$, $\mathbf{a_2}$ are the column averaging kernels, and ε_1, ε_2 are the column measurement errors, for each sensor, respectively, $\mathbf{x}$ is the "true" atmospheric profile state (implicitly in dry mole fraction, omitting the subscript d in vector notation for convenience), and $\mathbf{x_c}$ is the central tendency of the ensemble (assumed to be Gaussian); we take the subscripts "1" and "2" to denote NUCAPS and TCCON, respectively. The corresponding variance, σ^2, is given by [30]

$$\sigma^2 (\widehat{C}_1 - \widehat{C}_2) = (\mathbf{a_1} - \mathbf{a_2})^T \mathbf{S_c} (\mathbf{a_1} - \mathbf{a_2}) + (\sigma_1^2 + \sigma_2^2), \tag{A12}$$

where $\mathbf{S_c}$ is the background covariance matrix. Given a known "true" profile state, $\mathbf{x}$, along with $\mathbf{S_c}$, Equtions (A11) and (A12) can be used to verify rigorously whether a collocated NUCAPS and TCCON column observation are consistent within their theoretical measurement limitations.

However, in the current application we are given only a profile estimate (NUCAPS retrieval, $\hat{\mathbf{x}}_1$), and a column estimate (TCCON observation, $\widehat{C}_2$) for the purpose of evaluating NUCAPS using TCCON as a reference, while the "true" profile state remains unknown. Given the significant differences between each system's AKs (cf. Figures 1 and 3), and that the NUCAPS $\hat{\mathbf{x}}_1$ is an OE retrieval, we estimate the TCCON observation of that state by integrating $\hat{\mathbf{x}}_1$ using the TCCON AKs [30]

$$\widehat{C}_{12} = C_0 + \mathbf{a_2}^T (\hat{\mathbf{x}}_1 - \mathbf{x_0}), \tag{A13}$$

where C_0 and $\mathbf{x_0}$ denote the TCCON column and profile a priori, respectively. This equation roughly follows from Equation (A11) by assuming $\mathbf{a_1} \equiv \mathbf{i}$ (the unit vector), $\mathbf{x} \equiv \hat{\mathbf{x}}_1$ (the NUCAPS retrieved profile is used in lieu of the unknown truth), $\varepsilon_2 \approx 0$ (the TCCON measurement is accurate), and $\mathbf{x_c} \equiv \mathbf{x_0}$ (i.e., the ensemble central tendency is captured by the TCCON a priori). $\widehat{C}_{12}$ can then be used in place of $\widehat{C}_1$ for comparisons against the TCCON observations, $\widehat{C}_2$, in empirically estimating ε_1.

The righthand side of Eqution (A13) integrates the NUCAPS profile in a manner approximating what TCCON would have observed (under the same ambient environmental conditions) by applying TCCON AKs within the integration. The two terms are computed as follows:

$$
C_0 = \frac{\Sigma_p(X_0)}{\int_0^{p_s} \left\{ m_d\, g(p, \varphi)\left[1 + \epsilon\, \widehat{Q}_d(p)\right] \right\}^{-1} dp}, \tag{A14}
$$

where Σ_p is shorthand for the vertical sum, and

$$
\mathbf{a}^T\left(\hat{\mathbf{x}} - \mathbf{x_0}\right) = \frac{\Sigma_p\left[A \cdot (\widehat{X} - X_0)\right]}{\int_0^{p_s} \left\{ m_d\, g(p, \varphi)\left[1 + \epsilon\, \widehat{Q}_d(p)\right] \right\}^{-1} dp}, \tag{A15}
$$

where the denominators are the integrated columns for dry air and

$$
\Sigma_p(X_0) \equiv \int_0^{p_s} \frac{X_{d_0}(p)}{m_d\, g(p,\varphi)\left[1 + \epsilon\widehat{Q}_d(p)\right]}\, dp, \tag{A16}
$$

$$
\Sigma_p\left[A \cdot (\widehat{X} - X_0)\right] \equiv \int_0^{p_s} A(p)\, \frac{\widehat{X}_d(p) - X_{d_0}(p)}{m_d\, g(p,\varphi)\left[1 + \epsilon\widehat{Q}_d(p)\right]}\, dp. \tag{A17}
$$

Rigorous application of Equtions (A11) and (A12) toward NUCAPS and TCCON intercomparisons using an independent set of collocated truth profiles **x** (e.g., high-resolution AirCore profiles) will be the subject of future collaborative work.

References

1. Han, Y.; Revercomb, H.; Cromp, M.; Gu, D.; Johnson, D.; Mooney, D.; Scott, D.; Strow, L.; Bingham, G.; Borg, L.; et al. Suomi NPP CrIS measurements, sensor data record algorithm, calibration and validation activities, and record data quality. *J. Geophys. Res. Atmos.* **2013**, *118*, 12734–12748, doi:10.1002/2013JD020344.
2. Weng, F.; Zou, X.; Wang, X.; Yang, S.; Goldberg, M.D. Introduction to Suomi national polar-orbiting partnership advanced technology microwave sounder for numerical weather prediction and tropical cyclone applications. *J. Geophys. Res.* **2012**, *117*, D19112, doi:10.1029/2012JD018144.
3. Han, Y.; Chen, Y. Calibration algorithm for Cross-Track Infrared Sounder full spectral resolution measurements. *IEEE Trans. Geosci. Remote Sens.* **2018**, *56*, 1008–1016.
4. Cayla, F.R. IASI infrared interferometer for operations and research. In *High Spectral Resolution Infrared Remote Sensing for Earth's Weather and Climate Studies*; NATO ASI Series; Chedin, S., Ed.; Springer: Berlin/Heidelberg, Germany, 1993; Volume 19, pp. 9–19.
5. Hilton, F.; Armante, R.; August, T.; Barnet, C.; Bouchard, A.; Camy-Peyret, C.; Capelle, V.; Clarisse, L.; Clerbaux, C.; Coheur, P.F.; et al. Hyperspectral Earth observation from IASI: Five years of accomplishments. *Bull. Am. Meteorol. Soc.* **2012**, *93*, 347–370, doi:10.1175/BAMS-D-11-00027.1.
6. Aumann, H.H.; Chahine, M.T.; Gautier, C.; Goldberg, M.D.; Kalnay, E.; McMillin, L.M.; Revercomb, H.; Rosenkranz, P.W.; Smith, W.L.; Staelin, D.H.; et al. AIRS/AMSU/HSB on the Aqua Mission: Design, science objectives, data products, and processing systems. *IEEE Trans. Geosci. Remote Sens.* **2003**, *41*, 253–264.
7. Chahine, M.T.; Pagano, T.S.; Aumann, H.H.; Atlas, R.; Barnet, C.; Blaisdell, J.; Chen, L.; Divakarla, M.; Fetzer, E.J.; Goldberg, M.; et al. AIRS: Improving weather forecasting and providing new data on greenhouse gases. *Bull. Am. Meteorol. Soc.* **2006**, *87*, 911–926.
8. Gambacorta, A.; Barnet, C.; Wolf, W.; Goldberg, M.; King, T.; Nalli, N.; Maddy, E.; Xiong, X.; Divakarla, M. The NOAA Unique CrIS/ATMS Processing System (NUCAPS): First light retrieval results. In Proceedings of the ITSC-XVIII International TOVS Working Group (ITWG), Toulouse, France, 21–27 March 2012.
9. Smith, N.; Barnet, C.D. Uncertainty Characterization and Propagation in the Community Long-Term Infrared Microwave Combined Atmospheric Product System (CLIMCAPS). *Remote Sens.* **2019**, *11*, 1227, doi:10.3390/rs11101227.

10. Susskind, J.; Barnet, C.D.; Blaisdell, J.M. Retrieval of atmospheric and surface paramaters from AIRS/AMSU/HSB data in the presence of clouds. *IEEE Trans. Geosci. Remote Sens.* **2003**, *41*, 390–409.

11. Susskind, J.; Blaisdell, J.; Iredell, L.; Keita, F. Improved temperature sounding and quality control methodology using AIRS/AMSU data: The AIRS Science Team version 5 retrieval algorithm. *IEEE Trans. Geosci. Remote Sens.* **2011**, *49*, 883–907, doi:10.1109/TGRS.2010.2070508.

12. Han, Y.; Chen, Y.; Xiong, X.; Jin, X. *S-NPP CrIS Full Spectral Resolution SDR Processing and Data Quality Assessment*; Annual Meeting; American Meteorological Society: Phoenix, AZ, USA, 2015. Available online: https://ams.confex.com/ams/95Annual/webprogram/Paper261524.html (accessed on 28 September 2020).

13. Gambacorta, A.; Barnet, C.; Wolf, W.; King, T.; Maddy, E.; Strow, L.; Xiong, X.; Nalli, N.; Goldberg, M. An experiment using high spectral resolution CrIS measurements for atmospheric trace gases: Carbon monoxide retrieval impact study. *IEEE Geosci. Remote Sens. Lett.* **2014**, *11*, 1639–1643, doi:10.1109/LGRS.2014.2303641.

14. Strow, L.L.; Hannon, S.E.; Souza-Machado, S.D.; Motteler, H.E.; Tobin, D. An overview of the AIRS Radiative Transfer Model. *IEEE Trans. Geosci. Remote Sens.* **2003**, *41*, 303–313.

15. Gambacorta, A.; Nalli, N.R.; Barnet, C.D.; Tan, C.; Iturbide-Sanchez, F.; Zhang, K. *The NOAA Unique Combined Atmospheric Processing System (NUCAPS): Algorithm Theoretical Basis Document (ATBD)*; ATBD v2.0; NOAA/NESDIS/STAR Joint Polar Satellite System: College Park, MD, USA, 2017. Available online: https://www.star.nesdis.noaa.gov/jpss/documents/ATBD/ATBD_NUCAPS_v2.0.pdf (accessed on 28 September 2020).

16. Gambacorta, A.; Barnet, C. Methodology and information content of the NOAA NESDIS operational channel selection for the Cross-Track Infrared Sounder (CrIS). *IEEE Trans. Geosci. Remote Sens.* **2013**, *51*, 3207–3216, doi:10.1109/TGRS.2012.2220369.

17. Warner, J.X.; Wei, Z.; Strow, L.L.; Barnet, C.D.; Sparling, L.C.; Diskin, G.; Sachse, G. Improved agreement of AIRS tropospheric carbon monoxide products with other EOS sensors using optimal estimation retrievals. *Atmos. Chem. Phys.* **2010**, *10*, 9521–9533, doi:10.5194/acp-10-9521-2010.

18. Warner, J.X.; Carminati, F.; Wei, Z.; Lahoz, W.; Attié, J.L. Tropospheric carbon monoxide variability from AIRS under clear and cloudy conditions. *Atmos. Chem. Phys.* **2013**, *13*, 12469–12479, doi:10.5194/acp-13-12469-2013.

19. Xiong, X.; Barnet, C.; Maddy, E.S.; Gambacorta, A.; King, T.S.; Wofsy, S.C. Mid-upper tropospheric methane retrieval from IASI and its validation. *Atmos. Meas. Tech.* **2013**, *6*, 2255–2265, doi:10.5194/amt-6-2255-2013.

20. Maddy, E.S.; Barnet, C.D.; Goldberg, M.; Sweeney, C.; Liu, X. CO_2 retrievals from the Atmospheric Infrared Sounder: Methodology and validation. *J. Geophys. Res.* **2008**, *113*, D11301, doi:10.1029/2007JD009402.

21. Nalli, N.R.; Gambacorta, A.; Liu, Q.; Barnet, C.D.; Tan, C.; Iturbide-Sanchez, F.; Reale, T.; Sun, B.; Wilson, M.; Borg, L.; et al. Validation of atmospheric profile retrievals from the SNPP NOAA-Unique Combined Atmospheric Processing System. Part 1: Temperature and moisture. *IEEE Trans. Geosci. Remote Sens.* **2018**, *56*, 180–190, doi:10.1109/TGRS.2017.2744558.

22. Nalli, N.R.; Gambacorta, A.; Liu, Q.; Tan, C.; Iturbide-Sanchez, F.; Barnet, C.D.; Joseph, E.; Morris, V.R.; Oyola, M.; Smith, J.W. Validation of atmospheric profile retrievals from the SNPP NOAA-Unique Combined Atmospheric Processing System. Part 2: Ozone. *IEEE Trans. Geosci. Remote Sens.* **2018**, *56*, 598–607, doi:10.1109/TGRS.2017.2762600.

23. Sun, B.; Reale, A.; Tilley, F.; Pettey, M.; Nalli, N.R.; Barnet, C.D. Assessment of NUCAPS S-NPP CrIS/ATMS Sounding Products Using Reference and Conventional Radiosonde Observations. *IEEE J. Sel. Top. Appl. Earth Obs.* **2017**, *10*, 2499–2509, doi:10.1109/JSTARS.2017.2670504.

24. Feltz, M.L.; Borg, L.; Knuteson, R.O.; Tobin, D.; Revercomb, H.; Gambacorta, A. Assessment of NOAA NUCAPS upper air temperature profiles using COSMIC GPS radio occultation and ARM radiosondes. *J. Geophys. Res. Atmos.* **2017**, *122*, 9130–9153, doi:10.1002/2017JD026504.

25. Zhou, L.; Divakarla, M.; Liu, X. An Overview of the Joint Polar Satellite System (JPSS) Science Data Product Calibration and Validation. *Remote Sens.* **2016**, *8*, doi:10.3390/rs8020139.

26. Nalli, N.R.; Barnet, C.D.; Reale, A.; Tobin, D.; Gambacorta, A.; Maddy, E.S.; Joseph, E.; Sun, B.; Borg, L.; Mollner, A.; et al. Validation of satellite sounder environmental data records: Application to the Cross-track Infrared Microwave Sounder Suite. *J. Geophys. Res. Atmos.* **2013**, *118*, 13628–13643, doi:10.1002/2013JD020436.

27. Fetzer, E.; McMillin, L.M.; Tobin, D.; Aumann, H.H.; Gunson, M.R.; McMillan, W.W.; Hagan, D.E.; Hofstadter, M.D.; Yoe, J.; Whiteman, D.N.; et al. AIRS/AMSU/HSB validation. *IEEE Trans. Geosci. Remote Sens.* **2003**, *41*, 418–431.

28. Jacobson, A.R.; Schuldt, K.N.; Miller, J.B.; Oda, T.; Tans, P.; Arlyn, A.; Mund, J.; Ott, L.; Collatz, G.J.; Aalto, T.; et al. CarbonTracker CT2019. 2020. doi:10.25925/39M3-6069. Available online: https://www.esrl.noaa.gov/gmd/ccgg/carbontracker/CT2019/ (accessed on 28 September 2020).

29. Inness, A.; Ades, M.; Agustí-Panareda, A.; Barré, J.; Benedictow, A.; Blechschmidt, A.M.; Dominguez, J.J.; Engelen, R.; Eskes, H.; Flemming, J.; et al. The CAMS reanalysis of atmospheric composition. *Atmos. Chem. Phys.* **2019**, *19*, 3515–3556, doi:10.5194/acp-19-3515-2019.

30. Rodgers, C.D.; Connor, B.J. Intercomparison of remote sounding instruments. *J. Geophys. Res.* **2003**, *108*, 4116, doi:10.1029/2002JD002299.

31. Wunch, D.; Toon, G.C.; Blavier, J.F.L.; Washenfelder, R.A.; Notholt, J.; Connor, B.J.; Griffith, D.W.T.; Sherlock, V.; Wennberg, P.O. The Total Carbon Column Observing Network. *Philos. Trans. R. Soc. A* **2011**, *369*, 2087–2112, doi:10.1098/rsta.2010.0240.

32. Karion, A.; Sweeney, C.; Tans, P.; Newberger, T. AirCore: An Innovative Atmospheric Sampling System. *J. Atmos. Ocean. Technol.* **2010**, *27*, 1839–1853, doi:10.1175/2010JTECHA1448.1.

33. Membrive, O.; Crevoisier, C.; Sweeney, C.; Danis, F.; Hertzog, A.; Engel, A.; Bönisch, H.; Picon, L. AirCore-HR: A high-resolution column sampling to enhance the vertical description of CH_4 and CO_2. *Atmos. Meas. Tech.* **2017**, *10*, 2163–2181.

34. Wofsy, S.; Afshar, S.; Allen, H.; Apel, E.; Asher, E.; Barletta, B.; Bent, J.; Bian, H.; Biggs, B.; Blake, D.; et al. ATom: Merged Atmospheric Chemistry, Trace Gases, and Aerosols. ORNL Distributed Active Archive Center. 2018. doi:10.3334/ornldaac/1581. Available online: https://daac.ornl.gov/cgi-bin/dsviewer.pl?ds_id=1581 (accessed on 28 September 2020).

35. Wofsy, S.C. HIAPER Pole-to-Pole Observations (HIPPO): Fine-grained, global-scale measurements of climatically important atmospheric gases and aerosols. *Philos. Trans. R. Soc. A* **2011**, *369*, 2073–2086, doi:10.1098/rsta.2010.0313.

36. Tans, P.P. System and Method for Providing Vertical Profile Measurements of Atmospheric Gases. U.S. Patent 7,597,014, 6 October 2009.

37. Backus, G.; Gilbert, F. Uniqueness in the Inversion of Inaccurate Gross Earth Data. *Philos. Trans. R. Soc. Lond. A Math. Phys. Sci.* **1970**, *266*, 123–192.

38. Conrath, B.J. Vertical resolution of temperature profiles obtained from remote radiation measurements. *J. Atmos. Sci.* **1972**, *29*, 1262–1271.

39. Rodgers, C.D. Characterization and error analysis of profiles retrieved from remote sounding measurements. *J. Geophys. Res.* **1990**, *95*, 5587–5595.

40. Maddy, E.S.; Barnet, C.D. Vertical resolution estimates in Version 5 of AIRS operational retrievals. *IEEE Trans. Geosci. Remote Sens.* **2008**, *46*, 2375–2384.

41. Pollard, D.F.; Robinson, J.; Shiona, H. TCCON Data from Lauder (NZ), Release GGG2014.R0. TCCON Data Archive, Hosted by CaltechDATA. 2019. doi:10.14291/tccon.ggg2014.lauder03.R0. Available online: https://data.caltech.edu/records/1220 (accessed on 28 September 2020).

42. Sherlock, V.; Connor, B.J.; Robinson, J.; Shiona, H.; Smale, D.; Pollard, D. TCCON data from Lauder (NZ), 125HR, Release GGG2014R0. TCCON Data Archive, Hosted By CaltechDATA. 2017. doi:10.14291/tccon.ggg2014.lauder02.R0/1149298. Available online: https://data.caltech.edu/records/281 (accessed on 28 September 2020).

43. Griffith, D.W.; Velazco, V.A.; Deutscher, N.M.; Murphy, C.; Jones, N.; Wilson, S.; Macatangay, R.; Kettlewell, G.; Buchholz, R.R.; Riggenbach, M. TCCON data from Wollongong (AU), Release GGG2014R0. TCCON Data Archive, Hosted by CaltechDATA. 2014. doi:10.14291/tccon.ggg2014.wollongong01.R0/1149291. Available online: https://ro.uow.edu.au/data/47/ (accessed on 28 September 2020).

44. Griffith, D.W.; Deutscher, N.M.; Velazco, V.A.; Wennberg, P.O.; Yavin, Y.; Aleks, G.K.; Washenfelder, R.A.; Toon, G.C.; Blavier, J.F.; Murphy, C.; et al. TCCON Data from Darwin (AU), Release GGG2014R0. TCCON Data Archive, Hosted by CaltechDATA. 2017. doi:10.14291/tccon.ggg2014.darwin01.R0/1149290. Available online: https://data.caltech.edu/records/269 (accessed on 28 September 2020).

45. Morino, I.; Velazco, V.A.; Akihiro, H.; Osamu, U.; Griffith, D.W.T. TCCON data from Burgos, Ilocos Norte (PH), Release GGG2014.R0. TCCON Data Archive, Hosted by CaltechDATA. 2018. doi:10.14291/tccon.ggg2014.burgos01.R0. Available online: https://data.caltech.edu/records/1090 (accessed on 28 September 2020).

46. Blumenstock, T.; Hase, F.; Schneider, M.; Garcia, O.E.; Sepulveda, E. TCCON data from Izana (ES), Release GGG2014R0. TCCON data archive, hosted by CaltechDATA. 2017. doi:10.14291/tccon.ggg2014.izana01.R0/1149295. Available online: https://data.caltech.edu/records/275 (accessed on 28 September 2020).

47. Kawakami, S.; Ohyama, H.; Arai, K.; Okumura, H.; Taura, C.; Fukamachi, T.; Sakashita, M. TCCON Data from Saga (JP), Release GGG2014R0. TCCON Data Archive, Hosted by CaltechDATA. 2017. doi:10.14291/tccon.ggg2014.saga01.R0/1149283. Available online: https://data.caltech.edu/records/288 (accessed on 28 September 2020).

48. Iraci, L.T.; Podolske, J.; Hillyard, P.W.; Roehl, C.; Wennberg, P.O.; Blavier, J.F.; Allen, N.; Wunch, D.; Osterman, G.B.; Albertson, R. TCCON Data from Edwards (US), Release GGG2014R1. TCCON Data Archive, Hosted by CaltechDATA. 2017. doi:10.14291/tccon.ggg2014.edwards01.R1/1255068. Available online: https://data.caltech.edu/records/270 (accessed on 28 September 2020).

49. Wennberg, P.O.; Wunch, D.; Roehl, C.; Blavier, J.F.; Toon, G.C.; Allen, N.; Dowell, P.; Teske, K.; Martin, C.; Martin, J. TCCON data from Lamont (US), Release GGG2014R1. TCCON Data Archive, Hosted by CaltechDATA. 2017. doi:10.14291/tccon.ggg2014.lamont01.R1/1255070. Available online: https://data.caltech.edu/records/279 (accessed on 28 September 2020).

50. Morino, I.; Yokozeki, N.; Matzuzaki, T.; Horikawa, M. TCCON data from Rikubetsu (JP), Release GGG2014R2. TCCON Data Archive, Hosted by CaltechDATA. 2017. doi:10.14291/tccon.ggg2014.rikubetsu01.R2. Available online: https://data.caltech.edu/records/287 (accessed on 28 September 2020).

51. Wennberg, P.O.; Roehl, C.; Wunch, D.; Toon, G.C.; Blavier, J.F.; Washenfelder, R.A.; Keppel-Aleks, G.; Allen, N.; Ayers, J. TCCON data from Park Falls (US), Release GGG2014R0. TCCON Data Archive, Hosted by CaltechDATA. 2017. doi:10.14291/tccon.ggg2014.parkfalls01.R0/1149161. Available online: https://data.caltech.edu/records/204 (accessed on 28 September 2020).

52. Sussmann, R.; Rettinger, M. TCCON data from Zugspitze (DE), Release GGG2014R1. TCCON Data Archive, Hosted by CaltechDATA. 2018. doi:10.14291/tccon.ggg2014.zugspitze01.R1. Available online: https://data.caltech.edu/records/923 (accessed on 28 September 2020).

53. Sussmann, R.; Rettinger, M. TCCON Data from Garmisch (DE), Release GGG2014R0. TCCON Data Archive, Hosted by CaltechDATA. 2014. doi:10.14291/tccon.ggg2014.garmisch01.R0/1149299. Available online: https://data.caltech.edu/records/273 (accessed on 28 September 2020).

54. Warneke, T.; Messerschmidt, J.; Notholt, J.; Weinzierl, C.; Deutscher, N.M.; Petri, C.; Grupe, P.; Vuillemin, C.; Truong, F.; Schmidt, M.; et al. TCCON Data from Orléans (FR), Release GGG2014R0. TCCON Data Archive, Hosted by CaltechDATA. 2017. doi:10.14291/tccon.ggg2014.orleans01.R0/1149276. Available online: https://data.caltech.edu/records/283 (accessed on 28 September 2020).

55. Té, Y.; Jeseck, P.; Janssen, C. TCCON Data from Paris (FR), Release GGG2014R0. TCCON Data Archive, Hosted by CaltechDATA. 2017. doi:10.14291/tccon.ggg2014.paris01.R0/1149279. Available online: https://data.caltech.edu/records/284 (accessed on 28 September 2020).

56. Hase, F.; Blumenstock, T.; Dohe, S.; Gross, J.; Kiel, M. TCCON Data from Karlsruhe (DE), Release GGG2014R1. TCCON Data Archive, Hosted by CaltechDATA. 2017. doi:10.14291/tccon.ggg2014.karlsruhe01.R1/1182416. Available online: https://data.caltech.edu/records/278 (accessed on 28 September 2020).

57. Deutscher, N.M.; Notholt, J.; Messerschmidt, J.; Weinzierl, C.; Warneke, T.; Petri, C.; Grupe, P.; Katrynski, K. TCCON Data from Bialystok (PL), Release GGG2014R1. TCCON Data Archive, Hosted by CaltechDATA. 2017. doi:10.14291/tccon.ggg2014.bialystok01.R1/1183984. Available online: https://data.caltech.edu/records/267 (accessed on 28 September 2020).

58. Wunch, D.; Mendonca, J.; Colebatch, O.; Allen, N.; Blavier, J.F.L.; Roche, S.; Hedelius, J.K.; Neufeld, G.; Springett, S.; Worthy, D.E.J.; et al. TCCON Data from East Trout Lake (CA), Release GGG2014R1. TCCON Data Archive, Hosted by CaltechDATA. 2017. doi:10.14291/tccon.ggg2014.easttroutlake01.R1. Available online: https://data.caltech.edu/records/362 (accessed on 28 September 2020).

59. Kivi, R.; Heikkinen, P.; Kyrö, E. TCCON Data from Sodankyla (FI), Release GGG2014R0. TCCON Data Archive, Hosted by CaltechDATA. 2017. doi:10.14291/tccon.ggg2014.sodankyla01.R0/1149280. Available online: https://data.caltech.edu/records/289 (accessed on 28 September 2020).

60. Kivi, R.; Heikkinen, P. Fourier transform spectrometer measurements of column CO2 at Sodankylä, Finland. *Geosci. Instrum. Method. Data Syst.* **2016**, *5*, 271–279, doi:10.5194/gi-5-271-2016.

61. Notholt, J.; Warneke, T.; Petri, C.; Deutscher, N.M.; Weinzierl, C.; Palm, M.; Buschmann, M. TCCON data from Ny Ålesund, Spitsbergen (NO), Release GGG2014.R0. TCCON Data Archive, Hosted by CaltechDATA. 2017. doi:10.14291/tccon.ggg2014.nyalesund01.R0/1149278. Available online: https://data.caltech.edu/records/301 (accessed on 28 September 2020).

62. Hedelius, J.K.; He, T.L.; Jones, D.B.A.; Baier, B.C.; Buchholz, R.R.; De Mazière, M.; Deutscher, N.M.; Dubey, M.K.; Feist, D.G.; Griffith, D.W.T.; et al. Evaluation of MOPITT Version 7 joint TIR–NIR X_{CO} retrievals with TCCON. *Atmos. Meas. Tech.* **2019**, *12*, 5547–5572, doi:10.5194/amt-12-5547-2019.

63. Wennberg, P.O.; Wunch, D.; Roehl, C.; Blavier, J.F.; Toon, G.C.; Allen, N. TCCON Data from Caltech (US), Release GGG2014R1. TCCON Data Archive, Hosted by CaltechDATA. 2017. doi:10.14291/tccon.ggg2014.pasadena01.R1/1182415. Available online: https://data.caltech.edu/records/285 (accessed on 28 September 2020).

64. Wennberg, P.O.; Roehl, C.; Blavier, J.F.; Wunch, D.; Landeros, J.; Allen, N. TCCON Data from Jet Propulsion Laboratory (US), 2011, Release GGG2014R1. TCCON Data Archive, Hosted by CaltechDATA. 2017. doi:10.14291/tccon.ggg2014.jpl02.R1/1330096. Available online: https://data.caltech.edu/records/277 (accessed on 28 September 2020).

65. Notholt, J.; Petri, C.; Warneke, T.; Deutscher, N.M.; Buschmann, M.; Weinzierl, C.; Macatangay, R.; Grupe, P. TCCON Data from Bremen (DE), Release GGG2014R0. TCCON Data Archive, Hosted by CaltechDATA. 2017. doi:10.14291/tccon.ggg2014.bremen01.R0/1149275. Available online: https://data.caltech.edu/records/268 (accessed on 28 September 2020).

66. Strong, K.; Mendonca, J.; Weaver, D.; Fogal, P.; Drummond, J.; Batchelor, R.; Lindenmaier, R. TCCON Data from Eureka (CA), Release GGG2014R1. TCCON Data Archive, Hosted by CaltechDATA. 2017. doi:10.14291/tccon.ggg2014.eureka01.R1/1325515. Available online: https://data.caltech.edu/records/271 (accessed on 28 September 2020).

67. Nalli, N.R.; Joseph, E.; Morris, V.R.; Barnet, C.D.; Wolf, W.W.; Wolfe, D.; Minnett, P.J.; Szczodrak, M.; Izaguirre, M.A.; Lumpkin, R.; et al. Multi-year observations of the tropical Atlantic atmosphere: Multidisciplinary applications of the NOAA Aerosols and Ocean Science Expeditions (AEROSE). *Bull. Am. Meteorol. Soc.* **2011**, *92*, 765–789, doi:10.1175/2011BAMS2997.1.

68. Nalli, N.R.; Barnet, C.D.; Reale, T.; Liu, Q.; Morris, V.R.; Spackman, J.R.; Joseph, E.; Tan, C.; Sun, B.; Tilley, F.; et al. Satellite sounder observations of contrasting tropospheric moisture transport regimes: Saharan air layers, Hadley cells, and atmospheric rivers. *J. Hydrometeorol.* **2016**, *17*, 2997–3006, doi:10.1175/JHM-D-16-0163.1.

69. Le Marshall, J.; Jung, J.; Goldberg, M.; Barnet, C.; Wolf, W.; Derber, J.; Treadon, R.; Lord, S. Using cloudy AIRS fields of view in numerical weather prediction. *Aust. Met. Mag.* **2008**, *57*, 249–254.

70. Wunch, D.; Toon, G.C.; Wennberg, P.O.; Wofsy, S.C.; Stephens, B.B.; Fischer, M.L.; Uchino, O.; Abshire, J.B.; Bernath, P.; Biraud, S.C.; et al. Calibration of the Total Carbon Column Observing Network using aircraft profile data. *Atmos. Meas. Tech.* **2010**, *3*, 1351–1362.

remote sensing

Letter

Using the BFAST Algorithm and Multitemporal AIRS Data to Investigate Variation of Atmospheric Methane Concentration over Zoige Wetland of China

Yuanyuan Yang [1] and Yong Wang [2,*]

[1] School of Resources and Environment, University of Electronic Science and Technology of China (UESTC), 2006 Xiyuan Avenue, West Hi-tech Zone, Chengdu 611731, China; 201511180105@std.uestc.edu.cn

[2] Department of Geography, Planning, and Environment, East Carolina University, Greenville, NC 27858, USA

* Correspondence: wangy@ecu.edu

Received: 11 September 2020; Accepted: 28 September 2020; Published: 30 September 2020

Abstract: The monitoring of wetland methane (CH_4) emission is essential in the context of global CH_4 emission and climate change. The remotely sensed multitemporal Atmospheric Infrared Sounder (AIRS) CH_4 data and the Breaks for Additive Season and Trend (BFAST) algorithm were used to detect atmospheric CH_4 dynamics in the Zoige wetland, China between 2002 and 2018. The overall atmospheric CH_4 concentration increased steadily with a rate of 5.7 ± 0.3 ppb/year. After decomposing the time-series of CH_4 data using the BFAST algorithm, we found no anomalies in the seasonal and error components. The trend component increased with time, and a total of seven breaks were detected within four cells. Six were well-explained by the air temperature anomalies primarily, but one break was not. The effect of parameter h on decomposition outcomes was studied because it could influence the number of breaks in the trend component. As h increased, the number of breaks decreased. The interplays of the observations of interest, break numbers, and statistical significance should determine the h value.

Keywords: atmospheric infrared sounder (AIRS); breaks for additive season and Trend (BFAST) algorithm; methane (CH_4); multitemporal data; Zoige wetland; China

---

1. Introduction

Among all natural and anthropogenic sources, wetlands are the single largest methane (CH_4) source and contribute 20%~40% of the total global CH_4 emission [1]. Wetland CH_4 emissions result from interactions between several biological, chemical, and physical processes that primarily include CH_4 production, transportation, and oxidation. Methanogenic bacteria carry out the production by decomposing a limited number of relatively simple substrates under strictly anaerobic conditions. Thus, the production rate is limited by the availability of substrate and regulated by climatic and edaphic factors such as temperature, water table position, and pH [2–6]. CH_4 can be transported to the atmospheric through various pathways: molecular diffusion, ebullition, and via vascular plant stems [7]. The produced CH_4 is mostly oxidized by methanotrophs present at the oxic-anoxic boundary in the soil before emitting into the atmosphere [8]. Thus, the difference between the production and oxidation rates determines the rate of CH_4 emission into the atmosphere.

Paleo records and recent studies suggest vital positive feedback of wetlands to global warming through CH_4 emissions [9,10]. Therefore, the long-term variation and abrupt changes in wetland CH_4 emissions are essential elements to understand the present conditions of the global CH_4 emissions and climate changes [11,12]. An abrupt change or break usually denotes a rupture in the established range of observations. In this study, a breakpoint occurs when the wetland CH_4 emission is beyond a given threshold value, as observed in the remote-sensing time-series and delineated by an algorithm, triggering a discontinuous transition where a new starting point and rate are initiated.

Within the context of climate change, continuous monitoring of wetland CH_4 emissions is essential. With available multitemporal remote-sensing observations and datasets, various long-term change detection methods have been proposed. Temporal decomposition techniques have shown the ability to account for seasonal, gradual, and abrupt changes or breaks in terrestrial ecosystems. An early LandTrendr (Landsat-based detection of Trends in Disturbance and Recovery) algorithm divides long-term trends into piecewise-linear segments to characterize long-term changes in forest properties [13,14]. The algorithm captures changes at an annual scale but not at an intra-annual one. The Detecting Breakpoints and Estimating Segments in Trend (DBEST) can detect both abrupt and non-abrupt changes [15]. All the above methods are generally used to detect changes in the trend components, while seasonality is ignored.

The Seasonal-Trend decomposition based on a locally weighted regression smoother (STL) can identify both the phenological cycle and gradual change [16]. The STL cannot detect abrupt changes, as it assumed that the trend component varies smoothly [17]. Based on the STL algorithm, Verbesselt et al. [18] developed the Breaks for Additive Season and Trend (BFAST) algorithm that detects seasonal, gradual, and abrupt changes in a time-series simultaneously. The algorithm has been used and validated in many studies. For instance, Verbesselt et al. [19] detected drought-related vegetation disturbances. Saatchi et al. [20] examined the impact of the water deficit on the Amazon forest. Watts and Laffan [21] assessed the effectiveness of the algorithm in semi-arid regions, where the vegetation response is typically aseasonal. Hamunyela et al. [22] studied deforestation from the same data in dry and humid tropical forest areas.

CH_4 studies have been conducted in the Zoige wetland, mainly using in situ measurements [23,24]. However, the measurements are sparse and cannot be representative on a large scale. Systematic observation of the vertical variation of CH_4 is scarce. Therefore, space-borne measures become crucial as they provide broad spatial and multitemporal coverage, helping to understand better variations (e.g., abrupt changes or breaks) of the wetlands CH_4 emission and its impact on global climate change. Although the BFAST method has received much attention, no study has been conducted to use the technique coupled with the multitemporal remote-sensing data to understand the variations of the atmospheric CH_4 concentrations over wetlands. Thus, our aims are (i) to capture the CH_4 dynamic in the Zoige wetland using the BFAST algorithm coupled with remote-sensing observations of a time-series and (ii) to investigate the role of air temperature in altering a CH_4 time-series. Like any study using an algorithm, the algorithm parameterization is anticipated. The parameterization of h, a key parameter in the BFAST algorithm, is evaluated as the third objective. Thus, the impact of h on the outcome is studied.

2. Study Area, Datasets, and Methodology

2.1. Study Area

The Zoige Plateau (100°34′–103°45′ E, 31°40′–34°48′ N) is at the eastern edge of the Qinghai-Tibetan Plateau, China. Elevations of the plateau range from about 2400 to 5000 m above the mean sea level. The mean is ~3500 m (Figure 1). The wetland in the Zoige Plateau, approximately 4600 km^2, consists mainly of peatland that is about 40% of the peat stock in China. The peatland is regarded as one of the largest alpine peatlands in the world [25]. The area is within the high-altitude temperate humid climate region. The annual precipitation ranges from 400 to 800 mm [26]. The temperature varies considerably, with a yearly mean near 0 °C. The long cold-dry winters but short warm-humid summers generally make the accumulation rate of organic matter in soil higher than the decomposition rate. Methanogens use organic matter to generate CH_4.

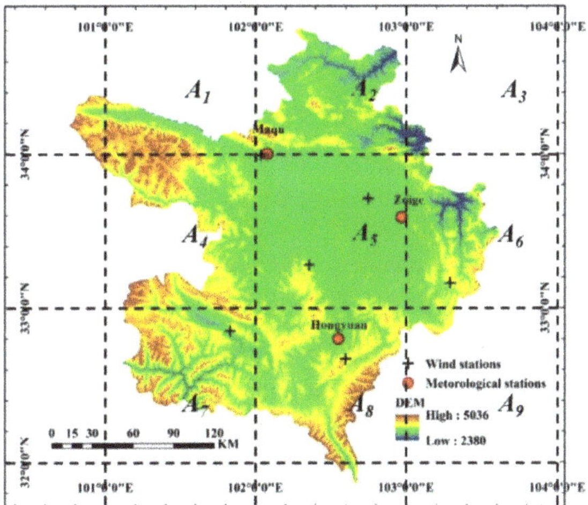

Figure 1. Digital elevation model (DEM) of the Zoige wetland, China. Three meteorological stations (red dots) and six wind measurement sites (black crosses) are identified.

2.2. Datasets

2.2.1. Meteorological and GLDAS Datasets

Three meteorological stations are located at Maqu, Zoige, and Hongyuan (Figure 1). The air pressure and temperature, wind direction and speed, humidity, and precipitation are measured. The data between September of 2002 and March of 2017 are available and downloadable at the China Meteorological Data Service Center (http://data.cma.cn/site/index.html).

The wind speed and direction are also measured at 90 m above the ground surface at six sites (Figure 1). After analyzing all the wind data from September of 2002 to March of 2018, we found that the wind direction changes annually with an inter-annual cyclic variation. The wind speed varies annually but may not have a clear high or low period intra-annually. The average wind speed between 2002 and 2018 was ~4 m/s. There was not a noticeable trend of increase or decrease. With the spatial resolution of the rasterized CH_4 data of 1° (longitude) × 1° (latitude) or ~100 km by ~100 km in the study area, the CH_4 diffusion and transport caused by winds were not considered.

The Global Land Data Assimilation System (GLDAS) (https://ldas.gsfc.nasa.gov/gldas/) is a global land-data assimilation system established in recent years, aimed at using satellite- and ground-based observation data products, advanced land surface models, and data assimilation technology to generate optimal surface conditions and flux data. The GLDAS data are downloadable at the NASA Goddard Earth Science Data and Information Services Center (http://disc.sci.gsfc.nasa.gov/datasets). The soil moisture (0–10 cm, 10–40 cm, 40–100 cm, and 100–200 cm) and soil temperature (0–10 cm, 10–40 cm, 40–100 cm, and 100–200 cm) data of GLDAS-Noah Version 2 between September of 2002 and March of 2018 were downloaded. They are monthly datasets with a spatial resolution of 1° (longitude) × 1° (latitude).

2.2.2. CH_4 and Landcover Datasets

The Atmospheric Infrared Sounder (AIRS) instrument on-board the NASA Earth Observing System Aqua satellite was launched into space in May of 2002. The AIRS is hyperspectral, having 2378 detectors in the infrared spectra from 3.7 to 15.4 μm [27]. The spatial resolution of AIRS is 13.5 km at nadir. Within a 24-h period, AIRS usually observes the globe twice. AIRS methane retrievals are broadly sensitive, ranging between 850 hPa (hectopascal) and the lower stratosphere, with peak sensitivity around 300–400 hPa. The AIRS Standard Version 6 Level 3 monthly data (AIRS3STM) of the atmospheric CH_4 concentration [28] were chosen. The data were divided into twenty-four layers corresponding to different atmospheric pressures or heights above the mean sea level. Here, the CH_4 concentration (parts per billion, ppb) at 600 hPa atmospheric pressure was extracted. The equivalent elevation is ~3600 m, which is about 100 m higher than the mean elevation of the study area. The data between September of 2002 and March of 2018 was downloaded from the NASA Goddard Data and Information Services Center at https://disc.gsfc.nasa.gov/datasets. Since the data at a single pixel was analyzed, a 3×3 AIRS sub-image covered the Zoige wetland spatially (Figure 1). The cells are named as A_1–A_9 from left to right and then from top to bottom.

We used the recently created China land cover products provided by the Resource and Environment Science and Data Center. Multiyear products in 2000, 2005, 2010, 2015, and 2018 are available and downloadable at http://www.resdc.cn/Default.aspx. The original land cover type is designed as a hierarchical classification scheme that allows one to adjust the thematic detail that describes each land cover class. Here, we first grouped the "level 2" classes into four categories: cropland, grassland, forest, and water body. Peatland in "level 2" remained as a category. The rest land cover types were classified as other. Thus, we had six land cover types. At such an aggregation level, the six land cover types did not change much from 2000 to 2018. The land cover data in 2010, which was near the middle of the studied time-series, was chosen. It should be noted that the peatland with high grassland coverage (>20%) might be classified into grassland in the downloaded datasets. Thus, some grassland, when the ground is wet, can be considered as wetland, per se.

Within the overlapped area of the study area and each cell, percentages of the six cover types were calculated and are shown in Table 1. Grassland and forest are the primary land cover in A_1, whose ground area is only about 25% within the study area. A_2 is the mixture of peatland, grassland, and forest, with the grassland cover type being dominant. About 65% area of A_2 is inside of the study area. A_3 is the grassland area having the lowest elevation in the study area (i.e., Figure 1). The majority of A_3 is outside of the study area. In A_4, the grassland, forest, and peatland are the major land cover types. More than one-half of the site is within the study area. A_5, entirely within the study area, is covered by the peatland, grassland, and forest, with a small part of the waterbody. A_6 is the mixture of peatland and grassland, with the grassland being dominant. About 70% of the area of A_6 is inside the study area. A_7 is covered by the grassland and forest, with a percentage of the peatland. More than one-half of the site is within the study area. The land cover types in A_8 are like those in A_7. More than 50% area of A_8 is located inside of the study area. A small northwestern corner of A_9 is inside the study area. Since A_1, A_3, or A_9 is mostly outside of the study area, the atmospheric CH_4 concentrations over each cell were not studied. Thus, we focused on A_2, A_4, A_5, A_6, A_7, and A_8.

Table 1. Percent of each land cover type of the overlapped area within the study area and each cell.

	A_2	A_4	A_5	A_6	A_7	A_8
Cropland	0.2	1.3	0.4	0.5	0.4	0.5
Grassland	77.3	78.4	74.1	70.4	48.8	77.4
Forest	11.0	12.4	4.7	13.0	45.9	18.6
Peatland	9.1	1.5	17.3	15.7	0.2	3.3
Water body	1.4	0.2	0.9	0.0	0.0	0.0
others	1.0	6.2	2.6	0.4	4.7	0.2

2.3. The BFAST Algorithm

The algorithm decomposes the multitemporal AIRS CH_4 data, $Y(t)$, into three components, as

$$Y(t) = S(t) + T(t) + E(t) \tag{1}$$

where $S(t)$ is the seasonal component, $T(t)$ is the trend component, and $E(t)$ the error one. All are functions of time t. If there is no single abrupt change point or breakpoint, $S(t)$ is continuous over the entire period. If one breakpoint occurs, $S(t)$ becomes two piecewise functions. If multiple breakpoints exist, one piecewise function is developed between two adjacent breakpoints. With the anticipated periodic characteristics of $S(t)$, a harmonic function is used. Assume p breakpoints occur at times $\tau_1^{\#}, \ldots, and \tau_p^{\#}$, with $\tau_0^{\#}$ being the start of the time-series and $\tau_{p+1}^{\#}$ the end of the series. Then, between $\tau_{j-1}^{\#}$ and $\tau_j^{\#}$ ($j = 1, 2, \ldots, p$), one can express $S(t)$ as

$$S(t) = \sum_{k=1}^{K} \left[\gamma_{j,k} \sin(\frac{2\pi kt}{f}) + \theta_{j,k} \cos(\frac{2\pi kt}{f}) \right] \tag{2}$$

where k is the kth number of the harmonic term. K is the highest-order harmonic term used in the algorithm. f is the frequency. Since the period for $S(t)$ is annual, f is one cycle per year. $\gamma_{j,k} = a_{j,k} \cos(\delta_{j,k})$ and $\theta_{j,k} = a_{j,k} \sin(\delta_{j,k})$. $a_{j,k}$ is amplitude and $\delta_{j,k}$ phase, and both are segment-specific parameters. In this study, we are interested in the $S(t)$ on an annual basis. The highest order of harmonic terms used in (2) cannot be greater than three [17,29,30], such that we can focus on changes using an entire season as the smallest timespan and eliminate unnecessary high-frequency variations in the AIRS data. Thus, K is set to 3.

$T(t)$ is continuous over the entire period if a single breakpoint does not occur. If breaks happen, $T(t)$ is expressed as piecewise functions as well. Assume m breakpoints happen at times $\tau_1^*, \ldots, and \tau_m^*$, with τ_0^* being the start of the time-series and τ_{m+1}^* the end of the series. A piecewise linear function within $\tau_{i-1}^* < t \leq \tau_i^* (i = 1, 2, \ldots, m)$ is

$$T(t) = \alpha_i + \beta_i t \tag{3}$$

where α_i is the ith intercept and β_i the ith slope. Breakpoints that occur in $T(t)$ or $S(t)$ can generally differ in time or magnitude. Finally, the error term of $E(t)$ is obtained in the decomposition.

2.4. The Effect of the h Parameter on the Decomposition

The BFAST algorithm uses the ordinary least squares residuals-based moving sum (OLS-MOSUM) to evaluate whether one or more breakpoints happen in the trend component or seasonal component [31]. The sum of a fixed number of residuals in a moving data window, whose size was determined by the bandwidth parameter, $h \in (0, 1)$, moving over the whole sample period, was analyzed. If the evaluation indicated a significant change (with the significance level of $p < 0.05$), the break was estimated [32]. As implemented ([33]), the Bayesian information criterion determined the number of breakpoints. The date and confidence interval (CI) of each breakpoint was estimated at 95%. Additionally, h determined the minimal segment size between two potential breakpoints in the time-series and was the ratio of the number of observations within a segment divided by the total length of a time-series. The two-end points of the segment or the entire time-series were excluded before the division. Although $h \in (0, 1)$, the maximum h was ≤ 0.5 if one breakpoint was to occur [21]. Per recommendations in [21] and [34], the minimum h was at least $\geq 5\%$ of the observations within the time-series. Therefore, we varied h between 0.05 and 0.5 to understand its impact on the outcome and determine an h value to link the breaks with abnormal natural events (e.g., temperature).

3. Results

3.1. Increase of Atmospheric CH₄ Concentration Derived From AIRS Data

The time-series of the atmospheric CH_4 concentrations over A_2, A_4, A_5, A_6, A_7, and A_8 were studied individually. Figure 2 demonstrates an annual cyclic pattern and a persistent increase in the CH_4 concentrations at the Zoige wetland during 2002–2018. In the figure, each dot is one observation or one month. A linear fit line was added and is shown as a red, dashed line. The parameters of the linear fit lines over the six cells are listed in Table 2. The slopes of the lines are between 0.015 and 0.017 ppb/day. With each fit line, the atmospheric CH_4 concentrations in September of 2002 and March of 2018 were calculated. The concentration values ranged from 1811.357 to 1854.134 ppb in September of 2002 and from 1901.917 to 1944.989 ppb in March of 2018. The increases, after nearly 16 years, were 96.220, 90.560, 90.560, 84.900, 84.900, and 84.900 ppb in A_2, A_4, A_5, A_6, A_7, and A_8, respectively. The average annual rate was 5.7 ± 0.3 ppb/year. The globally averaged annual rate was 5.1 ± 0.6 ppb/year using the marine surface data between 2002 and 2017 (E. Dlugokencky, National Oceanic and Atmospheric Administration (NOAA)/Earth System Research Laboratory (ESRL), https://www.esrl.noaa.gov/gmd/ccgg/trends_ch4/). Regarding the global data, an increase in the atmospheric CH_4 concentration at the Zoige wetland is likely true quantitatively.

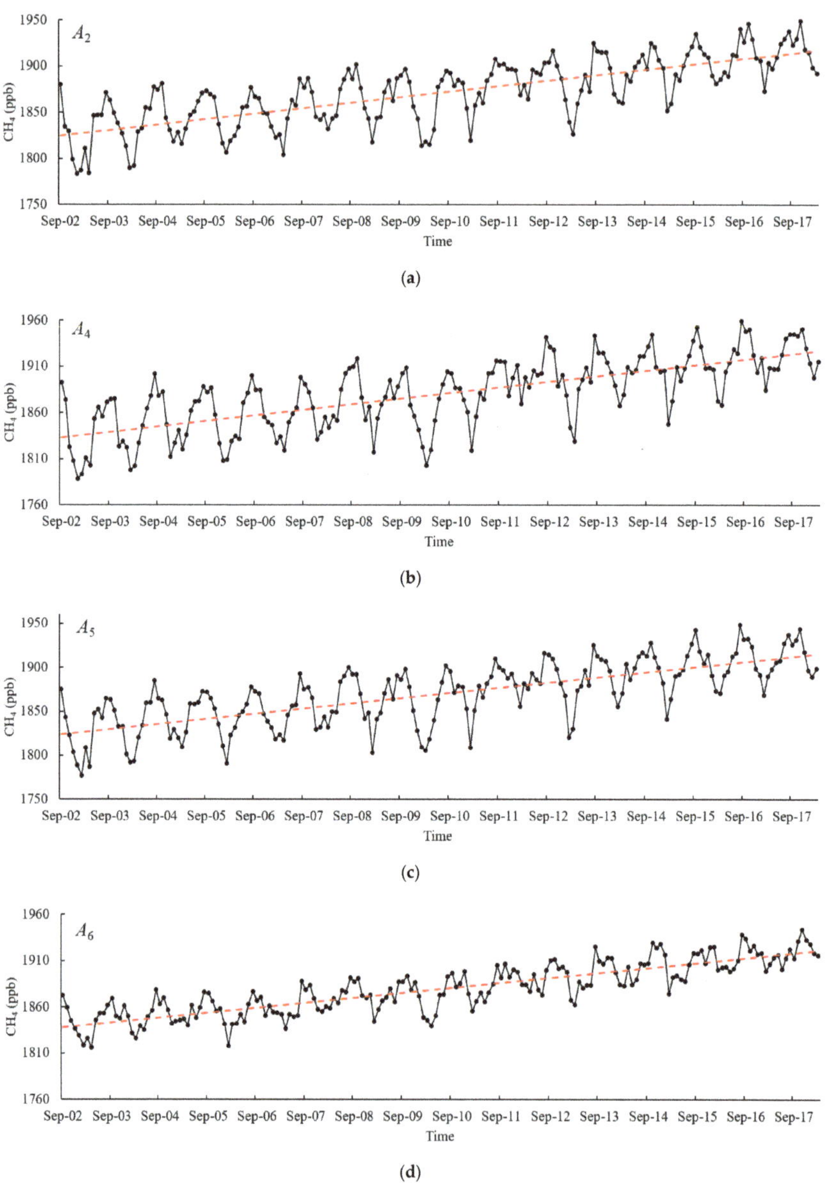

Figure 2. *Cont.*

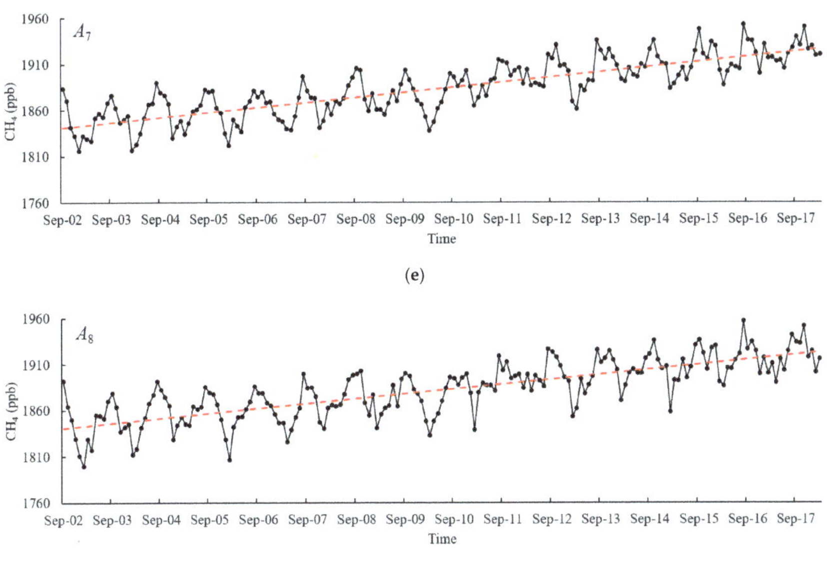

Figure 2. Atmospheric Infrared Sounder (AIRS) atmospheric CH_4 concentration data between 2002 and 2018, shown as a black curve. Each dot is an observation or one month in the monthly AIRS data. The red, dashed line is a linear fit line. (**a**) A_2, (**b**) A_4, (**c**) A_5, (**d**) A_6, (**e**) A_7, and (**f**) A_8.

Table 2. Intercept and slope values of linear fit lines for A_2, A_4, A_5, A_6, A_7, and A_8.

	A_2	A_4	A_5	A_6	A_7	A_8
Slope (ppb/day)	0.017	0.016	0.016	0.015	0.015	0.015
Intercept (ppb)	1211.269	1211.357	1219.759	1289.315	1273.904	1291.634

3.2. Decomposition of the CH_4 Time-Series

The decomposition for each of the six cells was conducted individually. No breaks in $S(t)$ for each cell were found. Values of the crest, trough, height, and mean are tabulated in Table 3. The seasonal component of each cell has a mean value near zero, which suggests normality. As an example, Figure 3 shows an $S(t)$ of A_5 between 2002 and 2018. It is annually cyclic, which is anticipated. The parameters of $S(t)$ for A_5 are given in Table 4. In short, no further analysis of $S(t)$ in each cell was carried out.

Table 3. Descriptive summary of the seasonal and error components of A_2, A_4, A_5, A_6, A_7, and A_8. The unit is ppb. $S(t)$: seasonal component and $E(t)$: trend component.

		$S(t)$			$E(t)$	
	Crest	**Trough**	**Height**	**Mean**	**Mean**	**St. Dev.**
A_2	26.857	−31.275	58.132	0.002	0.000	10.933
A_4	36.272	−32.849	69.121	−0.014	0.000	12.784
A_5	30.139	−37.296	67.435	−0.003	0.000	10.703
A_6	17.652	−16.801	34.453	−0.010	0.000	9.344
A_7	25.744	−23.094	48.838	−0.005	0.000	10.294
A_8	25.745	−23.095	48.840	−0.008	0.000	11.971

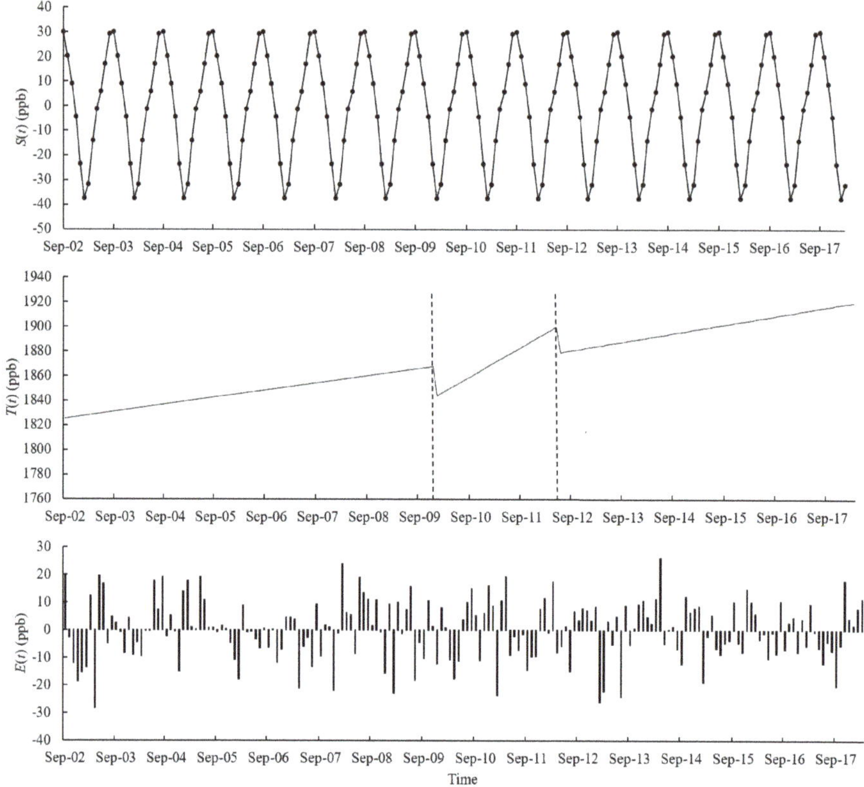

Figure 3. Decomposition of the atmospheric CH_4 concentration data over A_5. $S(t)$, $T(t)$, and $E(t)$ are the seasonal, trend, and error components, respectively. Two breakpoints were detected in the trend component. The vertical dotted lines in $T(t)$ indicate the beginnings of breakpoints (December of 2009 and May of 2012).

Table 4. Parameters of $S(t)$ shown in Figure 3a and θ_i ($i = 1, 2, 3$) are the coefficients of the sine and cosine terms, respectively. The unit for γ_i and θ_i is ppb. f is the frequency (once per year).

γ_1	θ_1	γ_2	θ_2	γ_3	θ_3	f
0.066	0.002	−0.070	0.042	17.540	−24.890	1

The mean and standard deviation of the error component of the six cells were also analyzed, respectively. The annual mean value for each of the six cells is 0.000 ppb. The standard deviations range from 9.344 to 12.784 ppb and are ~0.57% of the average atmospheric CH_4 concentration within the study period. Thus, $E(t)$ is considered normal. The $E(t)$ of A_5 between 2002 and 2018 are shown in Figure 3 as well. No particular patterns exist.

In the trend components, two breakpoints were detected at A_2, A_4, and A_5. A_7 had one. No breaks were found at A_6 and A_8. The time and magnitude of the changes are shown in Table 5. Timewise, two breakpoints occurred in December of 2009, one in January of 2010, one in October of 2010, and three in May of 2012. Additionally, a negative magnitude value indicates a drop, whereas a positive value, an increase. Thus, there are six decreases and one increase. Figure 3 shows the trend component of A_5 between 2002 and 2018. There are two drops and three segments or piecewise functions.

Table 5. Atmospheric CH_4 changes in the trend components over A_2, A_4, A_5, and A_7 at each breakpoint. At a breakpoint, an increase is positive, but a decrease negative.

YYYY/MM–YYYY/MM	A_2 (ppb)	A_4 (ppb)	A_5 (ppb)	A_7 (ppb)
2009/12–2010/01		−27.821	−23.861	
2010/01–2010/02	−21.427			
2010/10–2010/11				12.851
2012/05–2012/06	−25.544	−22.925	−20.843	

Each piecewise linear function within each segment was derived for A_2, A_4, A_5, A_6, A_7, and A_8, respectively. The intercept and slope of each function for each cell are tabulated in Table 6. Of A_2, A_4, A_5, and A_7, there is at least one break in the trend components. The intercept values vary and are linked to the breaks. The intercept value of A_6, 1285.509, is similar to that in the fit line (1289.315, Table 2). The similarity repeats for A_8 (Table 6 confer, c.f., Table 2). All slope values are positive, showing an increasing trend. The values range from 0.011 to 0.076 ppb/day. Moreover, the slope in the final segment for A_2, A_4, A_5, or A_7 is always steeper than the counterpart in the first segment. The timing of the acceleration is mostly in agreement with previous studies. The growth rate is plateaued in the mid-2000s, and then, the rate accelerates onwards [35,36].

Table 6. Intercept and slope values of each segment of $T(t)$ of A_2, A_4, A_5, A_6, A_7, and A_8.

	YYYY/MM–YYYY/MM	Intercept (ppb)	Slope (ppb/Day)
	2002/09–2010/01	1217.768	0.016
A_2	2010/02–2012/05	−832.480	0.067
	2012/06–2018/03	975.031	0.022
	2002/09–2009/12	1246.428	0.016
A_4	2010/01–2012/05	−1216.253	0.076
	2012/06–2018/03	1122.718	0.019
	2002/09–2009/12	1221.172	0.016
A_5	2010/01–2012/05	−789.977	0.066
	2012/06–2018/03	1089.205	0.019
A_7	2002/09–2010/10	1441.086	0.011
	2010/11–2018/03	1350.378	0.013
A_6	2002/09–2018/03	1285.509	0.015
A_8	2002/09–2018/03	1286.080	0.015

3.3. Parameterization of h and Its Impact on the Decomposition

As one knows that the number of breakpoints decreases when h increases, the negative and monotonic relationship suggests two aspects. First, an h value cannot be too large. An excessively large one can unnecessarily smoothen the trend component. Second, an h value or values exist after considering the interplays of the observations of interest, break numbers, and statistical significance. In this study, we were interested in CH_4 variations in the trend components using monthly remote-sensing time-series data. Factors such as an abnormal temperature event or events very likely affecting the observed atmospheric CH_4 concentrations were of interest.

As discussed previously, h was between 0.05 and 0.5. At $h = 0.05$, seven breakpoints over A_2, A_4, A_5, and A_7 occurred from September of 2002 to March of 2018 (Table 7). The total number of monthly observations within September of 2002 and March of 2018 was 185. Thus, at $h = 0.05$, the corresponding number of observations within two breakpoints was 9.3. Statistically, the number is too small. One needs to increase the h, boosting the number of observations while keeping the same number of breakpoints, if possible. Then, an exploratory approach is taken at an increment step of $h = 0.01$. As h changes from 0.05 to 0.13, the number of observations, n_o, and the seven breakpoints remain. Once $h \geq 0.14$, the number of breakpoints decreases. All the breakpoints disappear when $h \geq 0.17$ (Table 7).

Table 7. Of an h value, number of observations (n_o), and number of breakpoints detected in the trend components of A_2, A_4, A_5, and A_7 of the time-series. h is the ratio of the number of observations within a segment divided by the total length of a time-series, excluding the beginning and end observations.

h	n_o	A_2	A_4	A_5	A_7
0.05	9.3	2	2	2	1
0.13	24.1	2	2	2	1
0.14	25.9	0	1	2	0
0.16	29.6	0	1	2	0
0.17	31.5	0	0	0	0

Using the times that breaks happened in Table 5, we calculated the number of observations between two breakpoints in A_2, A_4, and A_5. The numbers are 26, 27, and 27, respectively. For the three cells, h at 0.13 is the maximum value if all six breaks are desired. Furthermore, $h \geq 0.17$ should not be considered if one breakpoint is wanted (Table 7). Therefore, h does influence decomposition. To maintain the maximum number of breakpoints in the trend components at A_2, A_4, and A_5, we should set $h \leq 0.13$. Unfortunately, the h value of 0.17 and a corresponding number of observations of 31.5 could not be used to explain the disappearance of the breakpoint at A_7. The number of monthly observations between September of 2002 (the starting month of the time-series) and October of 2010 was 96. The number of observations between November of 2010 and March of 2018 (the end of the time-series) was 87.

4. Discussion

4.1. Interpretation of Breaks With Air Temperature Variations

With the occurrence of a breakpoint in the trend component, one is interested to know what the possible causes are. In Table 5, there are six breakpoints linked to decreasing values but one increasing value. As illustrated and stated previously, the topography and land cover types differ among A_2, A_4, A_5, and A_7. Logically, one reason is whether a more or less uniform physical feature or event exists and predominantly causes the change.

The soil temperature and water table level are two main factors that influence CH_4 emissions from the wetlands into the atmosphere [2,3]. The soil moisture at the surface is usually positively related to the water table position [37]. Thus, the abnormal changes in soil temperature and moisture content may be responsible for the abrupt changes in the trend components. With the available GLDAS monthly soil moisture and soil temperature data, three data points (before the breakpoint, breakpoint, and after the breakpoint) at each breakpoint seem normal. The data cannot be used to interpret the delineated breaks. The aggregation of both types of GLDAS data at a monthly scale might overly smooth the intra-month variations.

We have the daily air temperature data at Maqu (located in A_2), Zoige (A_5), and Hongyuan (A_8) meteorological stations and articulate the following to use variations of the air temperature to explain the breaks. First, to establish the relationship between the air temperature data and soil temperature data, we aggregated the available daily air temperature data into monthly data between 2002 and 2017. Then, the correlation analyses between the monthly air temperature data at Maqu and soil temperature at A_2, between the monthly air temperature data at Zoige and soil temperature at A_5, and between the monthly air temperature data at Hongyuan and soil temperature at A_8 were, respectively, conducted. The correlation coefficients are summarized in Table 8. In the table, the soil temperature data at 0–100-cm depths are the average of the available temperature data at 0–10 cm, 10–40 cm, and 40–100 cm. The correlation coefficients are ≥ 0.921. For the top layer (0–10 cm), the coefficients are 0.979 or higher.

Thus, the air temperature can be a surrogate for the soil temperature. If the daily soil temperature data are not available, one can use the daily air temperature alternatively.

Table 8. The correlations between the monthly air temperature at three stations and the monthly soil temperature and moisture of the cell where the station is located. Data at two soil depths are analyzed.

		Maqu and A_2		Zoige and A_5		Hongyuan and A_8	
		Corr. Coef.	Sig.	Corr. Coef.	Sig.	Corr. Coef.	Sig.
Soil	0–10 cm	0.979	0.000	0.983	0.000	0.989	0.000
temperature	0–100 cm	0.921	0.000	0.934	0.000	0.955	0.000
Soil	0–10 cm	0.837	0.000	0.812	0.000	0.713	0.000
moisture	0–100 cm	0.787	0.000	0.720	0.000	0.261	0.001

Similarly, we analyzed the correlation of the aggregated air temperature data and soil moisture monthly data. The correlation coefficients are tabulated in Table 8. The coefficients between the air temperature and soil moisture contents at the 0–10-cm soil depth were at least 0.713 or higher, although the coefficients decreased as the depth increased (Table 8). Again, with the missing daily soil moisture data, the daily air temperature is an alternative.

A cold front moved across the area in the middle of December 2009, causing a significant temperature drop. The mean air temperature between the 16th and 31st of December was 5.5 °C lower than that from the 1st to 15th of December (Table 9). The *t*-test using the 1–15 temperature data ($n_1 = 45$, the sample size) versus 16–31 temperature data ($n_2 = 48$) at three meteorological stations resulted in a *p*-value = 0.000. Thus, the temperature drop was significant.

Table 9. Descriptive statistics of daily temperatures (°C) in December of 2009 after combining the daily datasets at the Maqu, Zoige, and Hongyuan meteorological stations.

	Mean	St. Dev.
1–15 Dec ($n_1 = 45$)	−4.2	2.3
16–31 Dec ($n_2 = 48$)	−9.7	1.4
Temperature drop	5.5	

To verify whether the temperature drop of 16–31 December 2009 was abnormal between 2002 and 2018, the daily temperature data of 16–31 December 2009, the averaged 16–31 December daily temperature data between 2002 and 2018, and the averaged 16–31 December daily temperature data without the 16–31 December 2009 temperature were plotted at the meteorological stations. As shown in Figure 4, the temperature data of 16–31 December 2009 differs from the other two averaged daily temperature datasets in December. The matched-pairs *t*-test of the former versus either of the latter two results in a *p*-value = 0.000. Thus, the 16–31 December 2009 temperature drop was abnormal.

Most of the microorganisms of methanogens are thermophilic. The abnormally cold weather in the second part of December 2009 could slow down their CH_4 production generally. The CH_4 emissions into the atmosphere decreased, and so did the atmospheric CH_4 concentrations. Although the drop in soil temperature is typically lagged compared with the air temperature decrease [38], the below-0 °C air temperature in December of 2009, and mainly, the colder air temperature in the late month (Table 9) further froze the soil column downward. Then, the chance that CH_4 escaped from the frozen soil column into the atmosphere decreased. With the cold temperature (mean = −6.3 °C and standard deviation = 2.2 °C) in January of 2010, the frozen soil column remained, or even deepened into the column, reducing the CH_4 escape further. Consequently, the trend components of A_4 and A_5 dropped from December of 2009 to January of 2010. The drop of atmospheric CH_4 concentrations from January to February of 2010 over A_2 could be attributed to the temperature drop in December of 2009, coupled with the elevation difference. The average elevation of A_2 is about 500 m lower than that of A_4

or A_5. The air temperature at A_2 is typically about 3 °C warmer than that at A_4 or A_5. The warmer temperature postponed the above-discussed processes, including the CH_4 production reductions in the soil column and the CH_4 emissions decrease from the soil into the atmosphere. Therefore, the drop was delayed for one month.

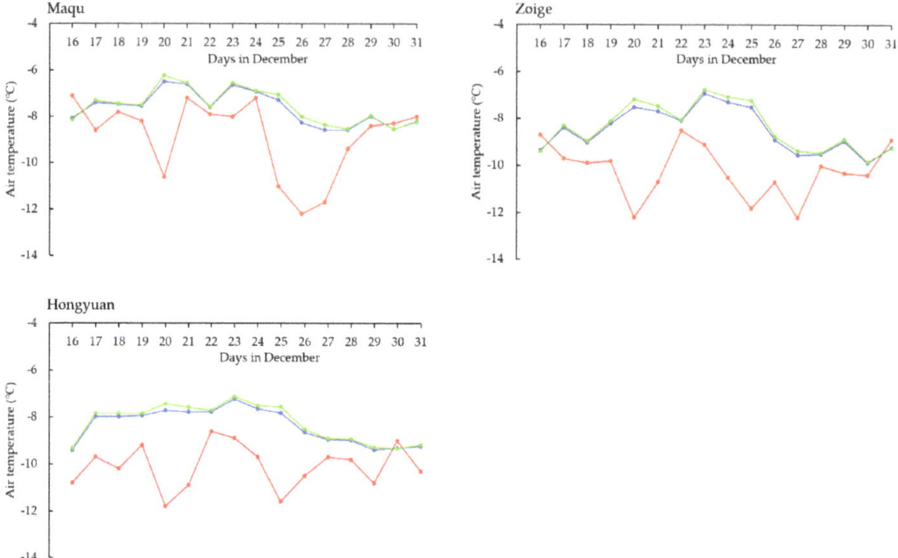

Figure 4. The daily temperature data of 16–31 December 2009 (red line), average 16–31 December temperature data between 2002 and 2018 (green line), and average 16–31 December temperature data without the 16–31 December 2009 temperature data (blue line) at the three meteorological stations at Maqu, Zoige, and Hongyuan.

Similarly, the abnormal temperature variation was used to interpret the drop between May and June of 2012 at A_2, A_4, and A_5. Descriptive statistics of the air temperatures in April of 2011, 2012, and 2013 and in May of 2012 are given in Table 10. The mean values in April of 2012 were lower than those in April of 2011 or April of 2013. Then, due to the lower air temperature, CH_4 in the soil column of the subsurface that was frozen in early spring might not have easily escaped into the air in April of 2012, as compared to CH_4 in April of 2011 or 2013 did. Additionally, the mean value of the daily air temperature in May of 2012 was 7.4 °C. The temperature in May 2012 was almost 5 °C higher than that in April 2012. The colder temperature in April of 2012 dampened the CH_4 emissions from the soil into the atmosphere, but the subsequent warmer temperatures in May sped up the escape processes of CH_4 [39]. An elevated atmospheric CH_4 concentration in May of 2012 occurred. Therefore, relative to the spike of the trend components in May of 2012, the atmospheric CH_4 concentration in June of 2012 dropped.

Table 10. Descriptive statistics of the daily temperature (°C) in April of 2011, 2012, and 2013 and in May of 2012. Daily air temperature data at the three stations were combined.

	Mean	**St. Dev.**
Apr. 2011 ($n = 90$)	3.2	2.9
Apr. 2012 ($n = 90$)	2.6	2.0
May 2012 ($n = 93$)	7.4	2.5
Apr. 2013 ($n = 90$)	2.9	2.6

An increase in the atmospheric CH_4 concentration over A_7 was detected (Table 5). No anomaly was identified in September, October, and November of 2010. The change in atmospheric CH_4 concentrations for A_2, A_4, A_5, A_6, or A_8 between September of 2002 and March of 2018 did not show any apparent anomaly. Therefore, the cause of the increase is not clear.

4.2. Atmospheric CH_4 Concentration Differences of Western Cells Versus Eastern Cells

One or two breakpoints occurred in trend components of A_2, A_4, A_5, and A_7 but no breakpoints in A_6 and A_8. Concerning Figure 1, A_6 and A_8 were east of A_2, A_4, A_5, and A_7. Then, component-by-component, average values of A_6 and A_8 and average values of A_2, A_4, and A_7 were calculated. The matched-pairs t-test of the average values of A_6 and A_8 versus the average ones of A_2, A_4, and A_7 was conducted. As shown in Table 11, the trend components differed significantly, but the seasonal and error components did not. The matched-pairs t-test for the average AIRS data was significantly different as well. Further studies will be pursued to understand not only the difference but also the possible causes.

Table 11. The p-values of the matched-pairs t-test for the western and eastern cells. AIRS: Atmospheric Infrared Sounder.

		Average Value of A_6 and A_8			
		$T(t)$	$S(t)$	$E(t)$	AIRS
	$T(t)$	0.000			
Average values	$S(t)$		0.856		
of A_2, A_4, and A_7	$E(t)$			1.000	
	AIRS				0.000

4.3. Possible Impact on the Decomposition if Random Noisy Observations Exist in the Time-Series

The time-series of the observed monthly atmospheric CH_4 concentration may consist of random noise that can affect the decomposition results. One feasible way to evaluate the impact of the noisy observations is to remove some observations randomly first and then replace them through the imputation [40] of the time-series. The imputation is not only widely used statistically in handling missing data but also is needed before running the BFAST algorithm. Without imputing a missing observation, the algorithm would move to the next observation in the time-series to fill in the missing one. The moving and filling continue until reaching the end of the time-series. Then, mixed matches of CH_4 observations and months occur and can eventually invalidate the time-series analysis. Therefore, two types of removal and imputation were considered, with A_5 as an example.

At $h = 0.13$, two breakpoints were identified (Table 5). The number of observations was 24.1 (Table 7). As 24 was the nearest integer for 24.1, we chronologically split the 24 observations with the 1st–12th observations and the breakpoint and the 13th–24th observations after the breakpoint. Using the Statistical Package for the Social Sciences (SPSS) software (https://www.ibm.com/analytics/spss-statistics-software), one observation in the 1st–12th observations and one in the 13th–24th observations were randomly selected and removed. The selection and then removal were performed twice, one for each breakpoint. The observation months and corresponding CH_4 values are given in Table 12. Then, we used the multiple imputation algorithm of the SPSS software to impute the four missing values. In the imputation, m was set to 5, and the Markov Chain Monte Carlo (MCMC) method was chosen. The imputed values are tabulated in Table 12. The original value and imputed value differed in each of the four cases. After replacing the original value with the related imputed one, the BFAST algorithm was applied to the new time-series, A_{5_r}, again. No breakpoint was found in the seasonal component, whereas two breakpoints were delineated (Table 13). Then, the piecewise functions were derived within each segment. The intercepts and slopes are shown in Table 14. The drop at the 2009/12–2010/01 breakpoint was comparable, and so was the drop at the 2012/05–2012/06 breakpoint (Table 13 c.f. Table 5). The piecewise linear equations were similar as well (Table 14 c.f. Table 6). Therefore, although the removal and imputation of four observations altered the time-series values of A_5, the changes might not significantly affect the outcomes from the BFAST algorithm decomposition.

Table 12. Four randomly removed observations, and the observed and imputed atmospheric CH_4 concentration values (ppb) of A_5.

Breakpoint	Removed obs.	Original Value	Imputed Value
2009/12–2010/01	(1) Sept. 2009	1886.510	1862.952
	(2) Aug. 2010	1902.343	1844.674
2012/05–2012/06	(3) Oct. 2011	1896.755	1869.282
	(4) Dec. 2012	1881.970	1894.124

Table 13. Atmospheric CH_4 concentration changes at each breakpoint in the trend component of A_{5_r}.

YYYY/MM–YYYY/MM	A_{5_r} (ppb)
2009/12–2010/01	−27.335
2012/05–2012/06	−18.510

Table 14. Intercept and slope of each segment of $T(t)$ in A_{5_r}.

YYYY/MM–YYYY/MM	Intercept (ppb)	Slope (ppb)
2002/09–2009/12	1241.015	0.016
2010/01–2012/05	−926.188	0.069
2012/06–2018/03	1110.085	0.019

To further explore the removal and imputation influences on the decomposition outcomes, we randomly removed and imputed observations at 5%, 6%, ... , of the entire time-series of A_5. The increment was 1%. At $h = 0.13$, two breakpoints were still obtained until 7% or the removal and imputation of 13 observations. The drop values at the breakpoints and piecewise linear functions in the three segments varied. At 8% or above, the decomposition outcomes fluctuated with the disappearance of one or both breaks. Thus, caution should be exercised if numerous erroneous observations exist. In this study, without a significant impact on the outcomes, the ceiling number of erroneous observations for A_5 was 14.96 (15 as an integer). It should be noted that Watts et al. [41] reported the BFAST algorithm was sensitive to the time-series datasets of vegetation indices collected by the Moderate Resolution Imaging Spectroradiometer (MODIS) Aqua and Terra remote sensors, although the time-series datasets themselves were highly correlated. Thus, one should consider the impact of erroneous data points on outcomes, and one possible way to conduct the sensitivity study to reveal and quantify the effect was suggested.

5. Conclusions

The multitemporal remotely sensed methane (CH_4) data from the Atmospheric Infrared Sounder (AIRS) instrument, on-board the NASA Earth Observing System Aqua satellite, were studied to understand the variations of atmospheric CH_4 over the Zoige wetland, China. The time spanned from September of 2002 to March of 2018. The Breaks for Additive Season and Trend (BFAST) algorithm was used to decompose the remotely sensed CH_4 data into the seasonal, trend, and error components. The meteorological and GLDAS datasets were auxiliary. They were used to interpret multitemporal CH_4 data and decomposition outcomes.

The overall pattern of the atmospheric CH_4 concentrations was derived from the AIRS data first. The concentrations increased steadily during 2002 and 2018. The average annual rate was on par with the globally average yearly rate. Thus, the annual rate at the Zoige wetland is likely valid.

Cell-by-cell, the seasonal, trend, and error components were next delineated by the BFAST algorithm. We analyzed the seasonal component, considering the crest, trough, height, and mean values. The components showed annual cycles with a mean value of 0. No anomaly was found. The error components varied from 2002 to 2018, but no particular intra- or inter-annual patterns were found. The mean and standard deviations were calculated. With the mean value of 0 and one standard deviation of ~0.5% or less of the average AIRS data, no anomaly was detected in the error components. The trend components increased gradually, with no breakpoints delineated in A_6 and A_8 but seven breakpoints collectively in A_2, A_4, A_5, and A_7. The timing and magnitude of each breakpoint were analyzed. After establishing significant correlations between the air temperature and soil temperature, and between the air temperature and soil moisture, we concluded that the temperature anomalies were primarily responsible for six breakpoints decomposing. However, the temperature anomaly could not explain the occurrence of one breakpoint.

The parameterization of the h parameter in the BFAST algorithm can be the most critical because it considerably influences the detection of breaks and the number of breakpoints. The minimum h is ≥ 0.05 statistically, but it cannot be >0.5 if one wants to identify one breakpoint. As h increases, the number of breakpoints decreases. Thus, a large h value can adversely smoothen the decomposed component. One may miss the critical and explainable breakpoints in the time-series. An optimized h value may be found after studying the interplays of the observation of interest, break numbers, and statistical significance. In this study, $h = 0.13$ was found.

Finally, erroneous observations can exist in time-series data, impacting the outcomes using the BFAST algorithm. The removal of one observation or observations and then the imputation of the removed observation or observations, can be one possible approach to conduct a sensitivity study. Thus, one can delineate and quantify the potential impact of the erroneous data point or points.

Author Contributions: Y.Y. and Y.W. conceived the study. Y.Y. studied the BFAST algorithm and analyzed the multitemporal AIRS data in consultation with Y.W. Both wrote the paper. All authors have read and agreed to the published version of the manuscript.

Funding: This research was funded by the National Natural Science Foundation of China under grants #41771401 and #41471361 to the University of Electronic Science and Technology of China.

Conflicts of Interest: The authors declare no conflict of interest.

References

1. Ciais, P.; Sabine, C.; Bala, G.; Bopp, L.; Brovkin, V.; Canadell, J.; Chhabra, A.; DeFries, R.; Galloway, J.; Heimann, M.; et al. *Carbon and Other Biogeochemical Cycles*; Cambridge University Press: Cambridge, UK; New York, NY, USA, 2013; pp. 465–570. Available online: https://www.ipcc.ch/site/assets/uploads/2018/02/WG1AR5_all_final.pdf (accessed on 29 September 2020).
2. Conrad, R. Control of methane production in terrestrial ecosystems. In *Exchange of Trace Gases between Terrestrial Ecosystems and the Atmosphere*; Andreae, M.O., Schimel, D.S., Eds.; Wiley: Chichester, UK; New York, NY, USA, 1989; pp. 39–58.
3. Valentine, D.W.; Holland, E.A.; Schimel, D.S. Ecosystem and physiological controls over methane production in northern wetlands. *J. Geophys. Res. Space Phys.* **1994**, *99*, 1563. [CrossRef]
4. Christensen, T.R.; Ekberg, A.; Ström, L.; Mastepanov, M.; Panikov, N.; Öquist, M.; Svensson, B.H.; Nykänen, H.; Martikainen, P.J.; Oskarsson, H. Factors controlling large scale variations in methane emissions from wetlands. *Geophys. Res. Lett.* **2003**, *30*, 1414. [CrossRef]
5. Christensen, T.R.; Prentice, I.C.; Kaplan, J.; Haxeltine, A.; Sitch, S. Methane flux from northern wetlands and tundra. *Tellus B Chem. Phys. Meteorol.* **1996**, *48*, 652–661. [CrossRef]
6. Moore, T.; Roulet, N.T.; Waddington, J. Uncertainty in predicting the effect of climatic change on the carbon cycling of Canadian peatlands. *Clim. Chang.* **1998**, *40*, 229–245. [CrossRef]
7. Walter, B.P.; Heimann, M. A process-based, climate-sensitive model to derive methane emissions from natural wetlands: Application to five wetland sites, sensitivity to model parameters, and climate. *Glob. Biogeochem. Cycles* **2000**, *14*, 745–765. [CrossRef]
8. King, G.M. Regulation by light of methane emissions from a wetland. *Nature* **1990**, *345*, 513–515. [CrossRef]
9. Nisbet, E.G.; Chappellaz, J. Shifting gear, quickly. *Science* **2009**, *324*, 477–478. [CrossRef]

10. Petrenko, V.V.; Smith, A.M.; Brook, E.J.; Lowe, D.; Riedel, K.; Brailsford, G.; Hua, Q.; Schaefer, H.; Reeh, N.; Weiss, R.F.; et al. 14CH4 measurements in greenland ice: Investigating last glacial termination CH4 sources. *Science* **2009**, *324*, 506–508. [CrossRef]

11. Forster, P.; Ramaswamy, V.; Artaxo, P.; Berntsen, T.; Betts, R.; Fahey, D.W.; Haywood, J.; Lena, J.; Lowe, D.C.; Myhre, G.; et al. Changes in atmospheric constituents and in radiative forcing. In *Climate Change 2007: The Physical Science Basis. Contribution of Working Group I to the Fourth Assessment Report of the Intergovernmental Panel on Climate Change*; Solomon, S., Qin, D., Manning, M., Chen, Z., Marquis, M., Averyt, K.B., Tignor, M.T., Miller, H.L., Eds.; Cambridge University Press: Cambridge, UK; New York, NY, USA, 2007; pp. 129–234. Available online: http://www.ipcc.ch/pdf/assessment-report/ar4/wg1/ar4-wg1-chapter2.pdf (accessed on 29 September 2020).

12. Shindell, D.; Faluvegi, G.; Koch, R.M.; Schmidt, G.; Unger, N.; Bauer, S.E. Improved attribution of climate forcing to emissions. *Science* **2009**, *326*, 716–718. [CrossRef]

13. Cohen, W.B.; Yang, Z.; Kennedy, R. Detecting trends in forest disturbance and recovery using yearly Landsat time series: 2. TimeSync—Tools for calibration and validation. *Remote Sens. Environ.* **2010**, *114*, 2911–2924. [CrossRef]

14. Kennedy, R.E.; Yang, Z.; Cohen, W.B. Detecting trends in forest disturbance and recovery using yearly Landsat time series: 1. LandTrendr—Temporal segmentation algorithms. *Remote Sens. Environ.* **2010**, *114*, 2897–2910. [CrossRef]

15. Jamali, S.; Jönsson, P.; Eklundh, L.; Ardö, J.; Seaquist, J. Detecting changes in vegetation trends using time series segmentation. *Remote Sens. Environ.* **2015**, *156*, 182–195. [CrossRef]

16. Jacquin, A.; Sheeren, D.; Lacombe, J.-P. Vegetation cover degradation assessment in Madagascar savanna based on trend analysis of MODIS NDVI time series. *Int. J. Appl. Earth Obs. Geoinf.* **2010**, *12*, S3–S10. [CrossRef]

17. Verbesselt, J.; Hyndman, R.J.; Zeileis, A.; Culvenor, D. Phenological change detection while accounting for abrupt and gradual trends in satellite image time series. *Remote Sens. Environ.* **2010**, *114*, 2970–2980. [CrossRef]

18. Verbesselt, J.; Hyndman, R.J.; Newnham, G.; Culvenor, D. Detecting trend and seasonal changes in satellite image time series. *Remote Sens. Environ.* **2010**, *114*, 106–115. [CrossRef]

19. Verbesselt, J.; Zeileis, A.; Herold, M. Near real-time disturbance detection using satellite image time series. *Remote Sens. Environ.* **2012**, *123*, 98–108. [CrossRef]

20. Saatchi, S.; Asefi-Najafabady, S.; Malhi, Y.; Aragão, L.E.O.C.; Anderson, L.O.; Myneni, R.B.; Nemani, R. Persistent effects of a severe drought on Amazonian forest canopy. *Proc. Natl. Acad. Sci. USA* **2012**, *110*, 565–570. [CrossRef]

21. Watts, L.M.; Laffan, S.W. Effectiveness of the BFAST algorithm for detecting vegetation response patterns in a semi-arid region. *Remote Sens. Environ.* **2014**, *154*, 234–245. [CrossRef]

22. Hamunyela, E.; Verbesselt, J.; Herold, M. Using spatial context to improve early detection of deforestation from Landsat time series. *Remote Sens. Environ.* **2016**, *172*, 126–138. [CrossRef]

23. Chen, H.; Yao, S.; Wu, N.; Wang, Y.; Luo, P.; Tian, J.; Gao, Y.; Sun, G. Determinants influencing seasonal variations of methane emissions from alpine wetlands in Zoige Plateau and their implications. *J. Geophys. Res. Space Phys.* **2008**, *113*. [CrossRef]

24. Chen, H.; Wu, N.; Gao, Y.; Wang, Y.; Luo, P.; Tian, J. Spatial variations on methane emissions from Zoige alpine wetlands of Southwest China. *Sci. Total. Environ.* **2009**, *407*, 1097–1104. [CrossRef] [PubMed]

25. Wang, M.; Yang, G.; Gao, Y.; Chen, H.; Wu, N.; Peng, C.; Zhu, Q.; Zhu, D.; Wu, J.; He, Y.; et al. Higher recent peat C accumulation than that during the Holocene on the Zoige Plateau. *Quat. Sci. Rev.* **2015**, *114*, 116–125. [CrossRef]

26. Li, J.; Wang, W.; Hu, G.; Wei, Z. Changes in ecosystem service values in Zoige Plateau, China. *Agric. Ecosyst. Environ.* **2010**, *139*, 766–770. [CrossRef]

27. Xiong, X.; Barnet, C.; Maddy, E.; Sweeney, C.; Liu, X.; Zhou, L.; Goldberg, M. Characterization and validation of methane products from the Atmospheric Infrared Sounder (AIRS). *J. Geophys. Res. Space Phys.* **2008**, *113*, 00 01. [CrossRef]

28. Tian, B.; Manning, E.; Fetzer, E.; Olsen, E.; Wong, S.; Susskind, J.; Iredell, L. *AIRS/AMSU/HSB Version 6 Level 3 Product User Guide*; Jet Propulsion Laboratory: Pasadena, CA, USA, 2013. Available online: http://disc.sci.gsfc.nasa.gov/AIRS/documentation/v6_docs/v6releasedocs1/V6_L3_User_Guide.pdf (accessed on 29 September 2020).

29. Geerken, R.A. An algorithm to classify and monitor seasonal variations in vegetation phenologies and their inter-annual change. *ISPRS J. Photogramm. Remote Sens.* **2009**, *64*, 422–431. [CrossRef]

30. Julien, Y.; Sobrino, J.A. Comparison of cloud-reconstruction methods for time series of composite NDVI data. *Remote Sens. Environ.* **2010**, *114*, 618–625. [CrossRef]

31. Zeileis, A. A unified approach to structural change tests based on ML scores, FStatistics, and OLS residuals. *Econ. Rev.* **2005**, *24*, 445–466. [CrossRef]

32. Bai, J.; Perron, P. Computation and analysis of multiple structural change models. *J. Appl. Econ.* **2003**, *18*, 1–22. [CrossRef]

33. Zeileis, A.; Leisch, F.; Hornik, K.; Kleiber, C. Strucchange: An R package for testing for structural change in linear regression models. *J. Stat. Softw.* **2002**, *7*, 1–38. [CrossRef]

34. Chu, C.-S.J.; Kaun, C.-M.; Hornik, K. MOSUM tests for parameter constancy. *Biometrika* **1995**, *82*, 603–617. [CrossRef]

35. McKain, K.; Down, A.; Raciti, S.M.; Budney, J.; Hutyra, L.R.; Floerchinger, C.; Herndon, S.C.; Nehrkorn, T.; Zahniser, M.S.; Jackson, R.B.; et al. Methane emissions from natural gas infrastructure and use in the urban region of Boston, Massachusetts. *Proc. Natl. Acad. Sci. USA* **2015**, *112*, 1941–1946. [CrossRef] [PubMed]

36. Zou, M.; Xiong, X.; Wu, Z.; Li, S.; Zhang, Y.; Li, S. Increase of atmospheric methane observed from space-borne and ground-based measurements. *Remote Sens.* **2019**, *11*, 964. [CrossRef]

37. Price, J. Soil moisture, water tension, and water table relationships in a managed cutover bog. *J. Hydrol.* **1997**, *202*, 21–32. [CrossRef]

38. Gabriel, C.-E.; Kellman, L. Investigating the role of moisture as an environmental constraint in the decomposition of shallow and deep mineral soil organic matter of a temperate coniferous soil. *Soil Boil. Biochem.* **2014**, *68*, 373–384. [CrossRef]

39. Feng, X.; Deventer, M.J.; Lonchar, R.; Ng, G.H.C.; Sebestyen, S.D.; Roman, D.T.; Griffis, T.J.; Millet, D.B.; Kolka, R.K. Climate sensitivity of peatland methane emissions mediated by seasonal hydrologic dynamics. *Geophys. Res. Lett.* **2020**, *47*, e2020GL088875. [CrossRef]

40. Graham, J.W. Missing Data: Analysis and Design. Springer: New York, NY, USA; Heidelberg, Germany; Dordrecht, The Netherlands; London, UK, 2012; 323p.

41. Watts, L.M.; Laffan, S.W. Sensitivity of the BFAST Algorithm to MODIS Satellite and Vegetation Index. In Proceedings of the 20th International Congress on Modelling and Simulation, Adelaide, Australia, 1–6 December 2013; pp. 1638–1644. Available online: https://mssanz.org.au/modsim2013/H2/watts.pdf (accessed on 29 September 2020).

remote sensing

Technical Note

Development of a Machine Learning-Based Radiometric Bias Correction for NOAA's Microwave Integrated Retrieval System (MiRS)

Yan Zhou * and Christopher Grassotti

Cooperative Institute for Satellite and Earth System Studies, Earth System Science Interdisciplinary Center, University of Maryland, College Park, MD 20742, USA; christopher.grassotti@noaa.gov
* Correspondence: yanzhou@umd.edu

Received: 28 August 2020; Accepted: 24 September 2020; Published: 26 September 2020

Abstract: We present the development of a dynamic over-ocean radiometric bias correction for the Microwave Integrated Retrieval System (MiRS) which accounts for spatial, temporal, spectral, and angular dependence of the systematic differences between observed and forward model-simulated radiances. The dynamic bias correction, which utilizes a deep neural network approach, is designed to incorporate dependence on the atmospheric and surface conditions that impact forward model biases. The approach utilizes collocations of observed Suomi National Polar-orbiting Partnership/Advanced Technology Microwave Sounder (SNPP/ATMS) radiances and European Centre for Medium-Range Weather Forecasts (ECMWF) model analyses which are used as input to the Community Radiative Transfer Model (CRTM) forward model to develop training data of radiometric biases. Analysis of the neural network performance indicates that in many channels, the dynamic bias is able to reproduce realistically both the spatial patterns of the original bias and its probability distribution function. Furthermore, retrieval impact experiments on independent data show that, compared with the baseline static bias correction, using the dynamic bias correction can improve temperature and water vapor profile retrievals, particularly in regions with higher Cloud Liquid Water (CLW) amounts. Ocean surface emissivity retrievals are also improved, for example at 23.8 GHz, showing an increase in correlation from 0.59 to 0.67 and a reduction of standard deviation from 0.035 to 0.026.

Keywords: machine learning; neural network; bias correction; MiRS

1. Introduction

1.1. MiRS

The Microwave Integrated Retrieval System (MiRS, https://www.star.nesdis.noaa.gov/mirs) has been the official operational microwave retrieval algorithm of the National Oceanic and Atmospheric Administration (NOAA) since 2007. Compared to visible and infrared radiation, microwaves have a longer wavelength, and thus can penetrate through the atmosphere more effectively. This feature allows microwave observations under almost all weather conditions, including in cloudy and rainy atmospheres. MiRS follows a one-dimensional variational (1DVAR) methodology [1,2]. The inversion is an iterative physical algorithm in which the fundamental physical attributes affecting the microwave observations are retrieved physically, including the profiles of atmospheric temperature, water vapor, non-precipitating cloud, hydrometeors, as well as surface emissivity and skin temperature [3]. The Joint Center for Satellite Data Assimilation (JCSDA) Community Radiative Transfer Model (CRTM) [4,5] is used as the forward and Jacobian operator to simulate the radiances at each iteration prior to fitting the measurements to within the combined instrument and forward model noise level. After the core parameters of the state vector are retrieved in the 1DVAR step, an additional post-processing is performed to retrieve derived parameters based on inputs from the core 1DVAR retrieval. The post processing products include total precipitable water (TPW), snow water equivalent (SWE), snowfall rate (SFR), surface precipitation rate, etc. [6]

MiRS has also been integrated into the Community Satellite Processing Package (CSPP), developed at the University of Wisconsin/Space Science and Engineering Center for users in the NOAA Direct Broadcast/Readout community. MiRS retrieval products are used routinely in operational weather analyses and forecasts, and also serve as inputs to downstream applications that are also used in operations. For example, MiRS water vapor profiles and TPW are used to generate the multi-satellite blended layer precipitable water and blended TPW products [7]. MiRS profiles of temperature and water vapor are also used as inputs to the tropical cyclone (TC) intensity estimation algorithm, the hurricane intensity and structure algorithm (HISA), developed at the Colorado State University/Cooperative Institute for Research in the Atmosphere (CSU/CIRA) [8] which is used operationally at the National Hurricane Center. Finally, MiRS precipitation rates are used as one of several satellite-based precipitation inputs to the NOAA Climate Prediction Center (CPC) Morphing Technique Algorithm (CMORPH) [9,10]. A schematic of the MiRS processing components and data flow is shown in Figure 1.

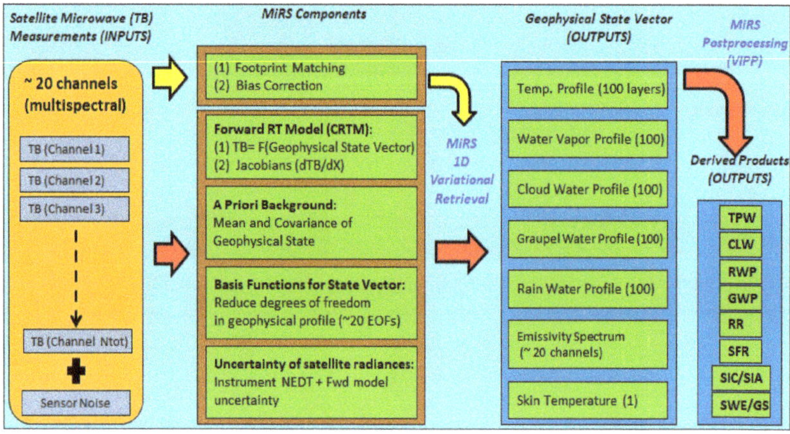

Figure 1. Schematic of MiRS processing components and data flow showing MiRS core retrieval and post-processing components. Core products are retrieved simultaneously as part of the state vector. Post-processing products are derived through vertical integration (water vapor, hydrometeors), catalogs (SIC, SWE), or fast regressions (rain rate). Post-processed hydrometeor retrieval products are Rain Rate, Graupel Water Path, Rain Water Path and Cloud Liquid Water.

1.2. Radiometric Bias Correction

The mathematical basis for the inversion can be formulated as a minimization of a cost function. Two important assumptions are made for the minimization process, the local-linearity of the forward operator and the Gaussian nature of both the geophysical state vector and the simulated radiometric vector around the measured vector. However, the differences between the simulated radiometric vectors and the actual measurements can show significant biases, which can come from several sources. These include deficiencies in the forward model linking the atmospheric state to the radiative measurements (e.g., due to errors in the physics or spectroscopy), measurement errors (e.g., due to inadequacies in the characterization of instruments), and biases in the atmospheric state used as input to the forward model. Here, we assume that atmospheric state biases are small and focus on quantifying and removing biases due to forward model and measurement errors prior to use in the physical retrieval process.

Many efforts have been made to develop bias correction schemes for the numerical weather prediction (NWP) data assimilation (DA) systems and the physical retrieval systems which share the same bias removal scheme because both the DA and the retrieval systems are based on similar variational approaches and cost function minimization processes. For example, Auligné et al. and Zhu et al. [11,12] discussed the adaptive radiance observation bias correction scheme applied in the National Centers for Environmental Prediction's (NCEP) Gridpoint Statistical Interpolation (GSI) data assimilation system, in which the variational air-mass bias component is estimated at the same time as the analysis control variables. A similar variational bias correction scheme for radiance data has been implemented and operational since 12 September 2006 at the European Centre for Medium-Range Weather Forecasts (ECMWF) [13,14].

1.3. Radiometric Bias Correction in MiRS

The current operational bias correction in MiRS is a procedure that applies a histogram adjustment to the radiative measurements to produce bias corrected radiances ready for inversion using a physical forward model (in this case, CRTM). This method, as inferred by its name, adjusts the histogram of the brightness temperature difference between simulated and observed radiances to make it centered around zero, which can reduce systematic errors in the retrievals related to forward and model and measurement biases. The histogram adjustment method specifies bias as a function of channel and scan position for each instrument, which means the bias does not change over time, and is static. However, in practice, the bias associated with a given instrument and frequency band generally varies in space and time, and may be air-mass or surface dependent at the time of the observation. For example, Figure 2 shows the global brightness temperature biases of Suomi National Polar-orbiting Partnership/Advanced Technology Microwave Sounder (SNPP/ATMS) over ocean, for two days (9 June 2019 and 1 October 2019) at two different frequencies (31.4 GHz and 183.31 ± 7 GHz, i.e., ATMS channels 2 and 18, respectively), showing variability with spatial, temporal and spectral dependence. When the same scan-dependent biases are applied regardless of location or air-mass characteristics, the local variations of systematic errors would not be accounted for, which can then propagate into the retrieval variables. Therefore, replacement of the static bias correction scheme with a dynamic one that changes geographically and varies with atmospheric conditions can potentially reduce the errors of the retrieved parameters.

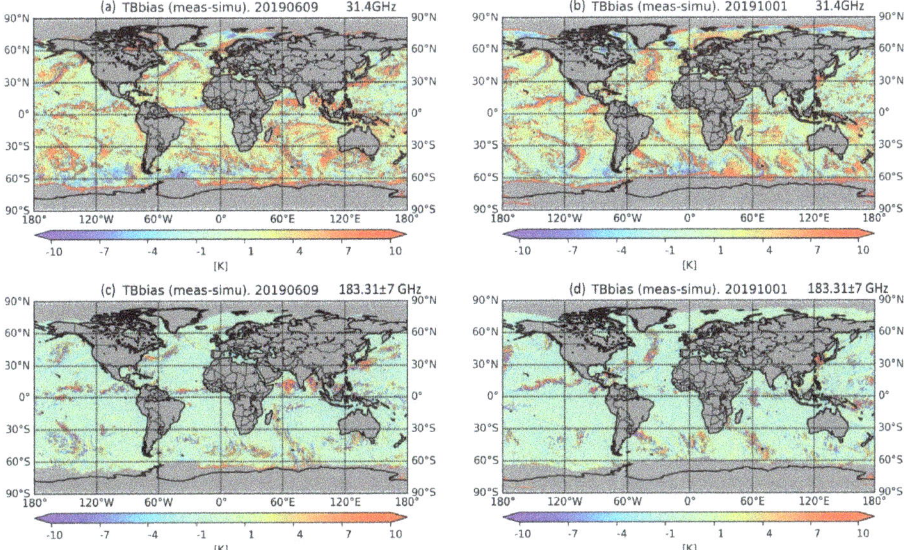

Figure 2. Global brightness temperature measurements minus simulations [K] over ocean, for 9 June 2019 at 31.4 GHz and 183.31 ± 7 GHz frequencies (**a,c**), and similarly for 1 October 2019 (**b,d**).

1.4. Neural Networks

Neural networks (NNs) have been widely used in the retrieval of geophysical parameters based on remote sensing data and other atmosphere science fields in recent years [15–20]. He et al. [21] studied two radiative bias correction methods developed through the correlation analysis between the microwave humidity and temperature sounder (MWHTS) measurements and air-mass. One method is linear regression, and the other is the neural network correction representing a nonlinear method. The authors found that the neural network correction outperformed the linear regression method in obtaining the desired bias; and by incorporating brightness temperatures corrected using a neural network approach in a one-dimensional variational system they could obtain higher retrieval accuracies of atmospheric temperature and humidity profiles. Considering the probable nonlinear nature of the difference between simulated and measured radiances, we choose neural networks as a new approach to implement a radiometric bias correction in MiRS system. The basic idea is to use NNs to learn the bias structure from historical collocations of simulated and measured brightness temperatures, along with the estimated corresponding atmospheric and surface state. The NN model, once trained, can then be used in near real time for bias correction during the retrieval process.

The remainder of this paper the contains the following outline: Section 2 contains a discussion of the datasets and methodology used in the study. The experimental design, including a description of the neural network and the MiRS algorithm, is contained in Section 3. Experimental results are highlighted in Section 4, which includes an evaluation of the neural network performance, an assessment of the impact of the neural network-derived bias corrections on MiRS retrieval performance, and some assessment of the neural network algorithm stability and robustness. Section 5 contains a summary and conclusions.

2. Materials and Methods

A Neural Network is a type of machine learning algorithm that is usually used in supervised learning. In supervised learning, a training dataset is given in which each set of input variables (or predictors) is corresponding to an already known output. The purpose of the neural network is to find the relationship between the predictors and the outputs in the training dataset. When a new dataset is provided (testing dataset), predictions are made by applying the learned relationship on predictors from the testing dataset. A deep neural network (DNN) is used in this research to simulate and predict the brightness temperature difference (between simulated radiance and actual measurements, labeled here as TBbias) for the advanced technology microwave sounder (ATMS) onboard the Suomi National Polar-orbiting Partnership (SNPP) satellite. A description of the SNPP/ATMS will be provided in Section 2.1, followed by discussion of how the training and testing datasets were assembled in Section 2.2, and a description of the validation dataset in Section 2.3.

2.1. Satellite and Sensor

The SNPP/ATMS data quality has been carefully evaluated [22,23], and its impact on the European Centre for Medium-Range Weather Forecasts (ECMWF) system and the United Kingdom's Met Office numerical weather forecast was reported by Bormann et al. [24] and Doherty et al. [25], respectively. SNPP is the first of a series of the next-generation U.S. polar-orbiting satellites, which was launched in October 2011 and continues to be operated by NOAA until present. SNPP is the result of a partnership between NOAA and the National Aeronautics and Space Administration (NASA). It is designed to collect data on long-term climate change and short-term weather conditions to extend and improve upon data records established by the NASA's Earth Observing System. SNPP was designed as a preparatory mission for the Joint Polar Satellite System (JPSS) series of satellites, all of which will also fly an ATMS instrument. The first satellite of the JPSS series (NOAA-20) was launched in November 2017 and is currently operational along with SNPP.

ATMS is a cross-track scanning instrument, with 22 channels at frequencies ranging from 23 to 183 GHz, which allows for profiling the atmospheric temperature/moisture, as well as providing information on clouds, non-precipitating clouds, and surface characteristics under clear-sky and cloudy conditions. In precipitating conditions, several channels can also provide information on liquid and frozen hydrometeors, which is indirectly related to the surface precipitation rate. In clear and cloudy (non-precipitating) conditions, channels at 23, 31 50, 88, and 165 GHz can provide information on total column water vapor, surface conditions and cloudiness. Channels between 50 and 60 GHz are used for atmospheric temperature sounding from the surface to about 1 hPa, while channels around 183 GHz provide information on the water vapor profile from the surface to about 200 hPa. Channels at 88 and 165 GHz provide significant information on the presence of rain and ice hydrometeors. Table 1 provides channel characteristics of all 22 ATMS channels, including central frequency, polarization, beam width, noise equivalent differential temperature (NEΔT), and the peak weight function (WF).

Table 1. Channel Characteristics of ATMS.

Channel	Central Frequency (GHz)	Polarization	Beam Width (deg)	NEΔT (K)	Peak WF (hPa)
1	23.8	V	5.2	0.9	Window
2	31.4	V	5.2	0.9	Window
3	50.3	H	2.2	1.2	Window
4	51.76	H	2.2	0.75	950
5	52.8	H	2.2	0.75	850
6	53.596 ± 0.115	H	2.2	0.75	700
7	54.4	H	2.2	0.75	400
8	54.94	H	2.2	0.75	250
9	55.5	H	2.2	0.75	200
10	57.290344	H	2.2	0.75	100
11	57.290344 ± 0.217	H	2.2	1.2	50
12	57.290344 ± 0.322 ± 0.048	H	2.2	1.2	25
13	57.290344 ± 0.322 ± 0.022	H	2.2	1.5	10
14	57.290344 ± 0.322 ± 0.010	H	2.2	2.4	5
15	57.290344 ± 0.322 ± 0.0045	H	2.2	3.6	2
16	88.2	V	2.2	0.5	Window
17	165.5	H	1.1	0.6	Window
18	183.31 ± 7.0	H	1.1	0.8	800
19	183.31 ± 4.5	H	1.1	0.8	700
20	183.31 ± 3.0	H	1.1	0.8	500
21	183.31 ± 1.8	H	1.1	0.8	400
22	183.31 ± 1.0	H	1.1	0.9	300

2.2. NN Training and Testing Datasets

Brightness temperature bias along with the concurrent SNPP/ATMS measured brightness temperature for 22 channels, satellite viewing angle, latitude, and other geophysical parameters including cloud liquid water (CLW), total precipitable water (TPW), and skin temperature (Tskin) have been used to establish the NN training and testing datasets. All of these parameters are collected over ocean. A data screening was applied in the training dataset which required that only brightness temperature biases less than 30K were included in the training set. The reason for the 30K limit is that such a large difference between observation and simulation indicates likely scattering or precipitation. In these cases, we do not have confidence that the NWP representation of the rain or ice particles is accurate enough to provide a reliable input to the CRTM simulation. (As noted in Section 3.2, the MiRS retrieval approach allows for the determination of highly scattering (precipitating) conditions and in these cases a bias correction is not applied due to large uncertainties in the forward modeling.) The training dataset contains 12 days with one-day from each month of either 2018 or 2019 (Table 2). The CLW, TPW, and Tskin for training were from the ECMWF analyses. Once the NN bias correction model was trained, the impact assessment was done on an independent day, 1 October 2019 for all ocean scenes between 55 S to 55N latitude. The purpose of the latitude limit is to avoid sea ice covered areas where surface emissivity is not well simulated. Since in operational application MiRS does not use any real-time data from NWP model forecasts or analyses, such as those from ECMWF, in the impact assessment testing experiments these three parameters were calculated either by a regression scheme (CLW) or by neural networks constructed for TPW and Tskin, respectively. The inputs for calculating them were brightness temperature measurements and geolocation parameters like satellite viewing angle and latitude. The training and testing data were selected based on the availability of the SNPP/ATMS measurements and our computational resources. Further details of the NN training and testing are contained in Sections 3 and 4.

Table 2. Days Used for NN Training.

14-January-2019	15-July-2018
15-February-2019	1-August-2018
25-March-2019	1-September-2019
1-April-2019	20-October-2018
11-May-2019	1-November-2019
4-June-2019	1-December-2019

2.3. MiRS Retrieval Validation Dataset

To evaluate the impact of the new radiometric bias correction scheme, most of the MiRS retrievals, including atmospheric temperature profiles, water vapor profiles, CLW, TPW, and Tskin were validated with the ECMWF analyses. The ECMWF analyses are originally specified on 90 vertical pressure layers and at a 0.25-degree horizontal resolution. MiRS interpolates the analyses vertically into the 100 CRTM layers (from the surface to 0.01 hPa), and horizontally averages to 1 degree for the collocation with the ATMS measurement locations. Validation for the MiRS surface emissivity over the ocean at all channels was performed against the fast microwave emissivity model (FASTEM) [26] that takes the ECMWF analyses of wind speed, frequency, and observation zenith angle as inputs. In MiRS, ECMWF and other operational NWP data sets are used only for calibration in the radiance processing and in the retrieved product monitoring. They are not used in the 1DVAR inversion process. As the bias correction was developed for over-ocean measurements only, this paper evaluates MiRS retrievals performance of SNPP/ATMS over ocean only. Over other surfaces, the operational static bias correction scheme remained unchanged, therefore producing no impact on the MiRS retrievals over land, snow, and sea-ice scenes.

3. Algorithm and Experiment Design

3.1. MiRS Algorithm

The MiRS is an iterative, physically-based retrieval system based on 1DVAR inversion. The 1DVAR physical principle is to minimize a cost function (Equation (1)). The first item on the right side of Equation 1 represents the departure of state vector X to be retrieved from background X_0, weighted by background error covariance matrix B. The second item represents the departure of simulated radiance Y from the observed radiance Y^m, weighted by instrument and radiative transfer modelling error E. CRTM is the forward and adjoint operator used to generate simulated radiance Y, as well as the Jacobians (derivatives) which is the radiance response to a unit perturbation of the state vector. Minimization of the cost function is an iterative process with convergence reached if chi-squared metric, χ^2, is less than or equal to 1 (Equation (2)). The iterative loop is also ended if the chi-squared metric does not meet the convergence criterion within 7 iterations. In practice, the global convergence rates approach 95% [27].

$$J(X) = [\frac{1}{2}(X - X_0)^T \times B^{-1} \times (X - X_0)] + [\frac{1}{2}(Y^m - Y(X))^T \times E^{-1} \times (Y^m - Y(X))] \tag{1}$$

$$\chi^2 = (Y^m - Y(X))^T \times E^{-1} \times (Y^m - Y(X)) \tag{2}$$

3.2. Experiment Design

This research chose a 4-layer neural network (with two hidden layers) to predict brightness temperature biases (22 channels). There are 200 neurons in each hidden layer, with a Rectified Linear Unit (ReLU) as the activation function, which is the most successful and widely-used activation function thus far in the deep learning community [28]. This configuration of layers, nodes and activation function was selected after extensive testing with smaller and larger numbers of layers and nodes, and with different activation functions such as Sigmoid and Leaky ReLU. The NN design used here produced the best results in terms of reproducing the observed biases. The optimizer used in this NN is RMSprop, with the learning rate of 0.001. Another problem with training neural networks is to choose the number of training epochs. Too many epochs can lead to overfitting, while too few could result in an underfit model. In the present study, early stopping was the method used to terminate training before overfitting occurred. This method split the training dataset and used a subset (in this study, 20%) as a validation dataset to monitor performance of the neural network model during training. An arbitrary large number of training epochs (or maximum number of epochs) was specified, and the training would be stopped if the loss on the validation dataset did not change over a given number of epochs (or patience). The maximum number of epochs and the patience used in this study were 1000 and 100, respectively.

The input layer has 27 variables over ocean, which includes SNPP/ATMS measured brightness temperature (22 channels), satellite viewing angle, latitude, CLW, TPW, and Tskin. Normalization of the input variables to a standard scale would allow the neural network to more quickly learn the optimal weights and biases. All of the 27 input variables were normalized by their respective mean and standard deviation calculated from the training dataset. Data screening was also applied which required that only observations where brightness temperature biases less than 30K were included in the training dataset. The output layer has 22 variables, representing brightness temperature biases for each of the 22 channels. The neural network was applied for all ocean scenes on the testing day. MiRS has the flexibility to choose the bias correction method for each channel. NN predicted brightness temperature biases were applied to SNPP/ATMS channels 1–15 plus 17 over ocean. All other channels used the static bias correction method. This choice of which channels the NN correction was applied to was based on a large number of sensitivity tests where the impact on retrievals was quantified. Finally, the MiRS retrieval approach structures the retrieval and state vector based on whether hydrometeor scattering is determined to be significant. In this study, bias corrections are only applied in conditions of little to no scattering (i.e., clear and cloudy/light rain scenes), as development of bias corrections for scenes with significant scattering and/or precipitation is a more challenging task.

4. Results

4.1. Neural Network Performance

The neural network prediction of the brightness temperature bias of SNPP/ATMS was first validated with the true bias (measurements minus simulations) for one single day, 1 October 2019, as shown in Figure 3. The bars represent the mean true bias (blue) and NN predicted bias (orange) for each of the 22 SNPP/ATMS channels, with left y-axis showing their values. The difference between the mean NN predicted and true biases is presented by the red line with right y-axis showing its value. Only profiles with biases less than 30 K and located between 55 S and 55 N latitude over ocean are used in the averaging. Except for Channel 15, the differences between the NN prediction and true bias are less than 0.38. Channel 15 differences may be larger due to the peak height of its weighting function, which is approximately 2 hPa (~ 45 km altitude) where both the ECMWF model analyses and CRTM simulations may be less reliable.

The performance for Channels 1–12 is generally good, with Channels 1, 6, and 7 having the smallest difference. For example, the averaged difference between NN predicted and true bias is about 0.01 K for Channel 1 (23.8 GHz), and their spatial patterns are very similar (Figure 4a,b). Figure 4 shows the brightness temperature bias maps of Channel 1 (23.8 GHz) for true values (a), NN prediction (b), and the difference between NN and true values (c), over ocean for latitudes between 55 S and 55 N. Quantitatively, the NN prediction matches the true bias special pattern very well. However, the NN estimates miss some cold features (blue color) in the midlatitudes and show warmer patterns over the south Pacific Ocean. It also shows slight scan angle dependency over the tropics despite using scan angle as one of the NN inputs aimed at minimizing this dependency. The histogram of the NN prediction is compared with that of the true bias (red vs. black in Figure 4d). The NN prediction does not capture the extreme cold bias less than –10 K and has fewer points between 0 K to –10 K, while it has more profiles around the peak (about 2 K) and contains a small number of points with a predicted warm bias higher than 30 K.

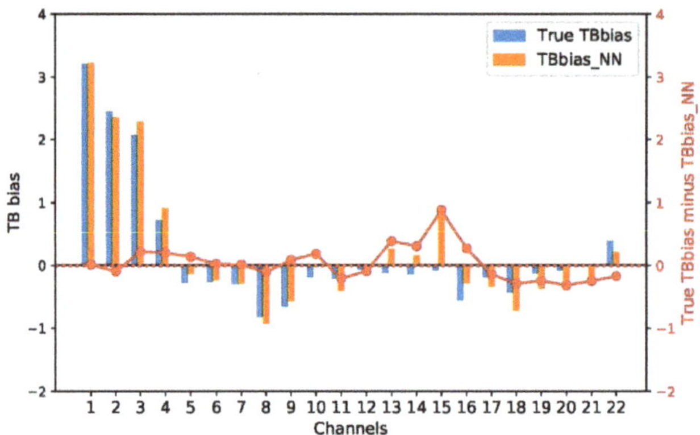

Figure 3. Mean brightness temperature bias [K] of the 22 SNPP/ATMS channels over ocean with latitude between 55S and 55N for the true values (blue bars), NN prediction (orange bars) and their difference (red line). 1 October 2019.

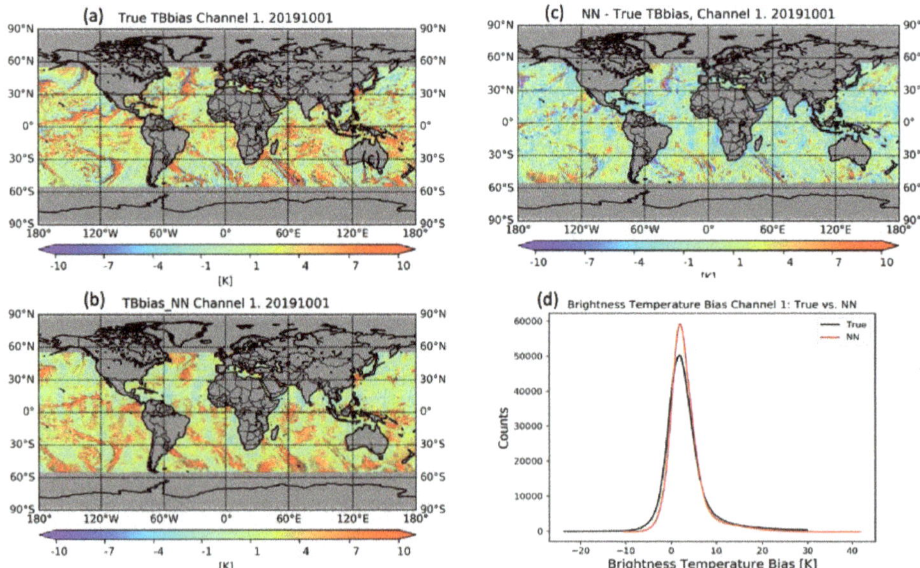

Figure 4. Brightness temperature bias [K] maps of SNPP/ATMS 23.8 GHz on 1 October 2019, over ocean with latitude between 55 S and 55 N, for (**a**) true value, (**b**) NN prediction, the difference between NN and the true value (**c**), and histograms of the true value and the NN prediction (**d**).

4.2. MiRS Retrievals

With MiRS using static bias correction as the baseline experiment (named as Static), an assessment of the impact of the neural network-derived bias corrections on MiRS retrieval performance over ocean is presented (named as NN). The MiRS retrieval performance of atmospheric temperature profiles for Static (black) and NN (red) are shown in Figure 5, with solid lines for bias and dashed lines for standard deviation, verified with ECMWF analyses between 1000–100 hPa. This verification was stratified by CLW amount into three scenarios: clear (CLW ≤ 0.05 mm), cloudy (0.05 ≤ CLW ≤ 0.275 mm), and heavy cloud or light rain (CLW > 0.275 mm), corresponding to Figure 5a–c. The sample sizes for each experiment under each scenario are given in the legend. In the clear and cloudy scenarios, impact of NN bias correction was similar, both with slightly reduced bias below 700 hPa and with slightly increased standard deviation at almost all levels. In the heavy cloud or light rain scenario, NN shows significantly reduced standard deviation under 300 hPa, with about 0.5 K smaller standard deviation at 650 hPa. Since the baseline static bias correction is developed using clear sky measurements only, it is perhaps expected that the largest and most positive impact of the NN bias correction is for scenes with significant cloudiness and/or light precipitation.

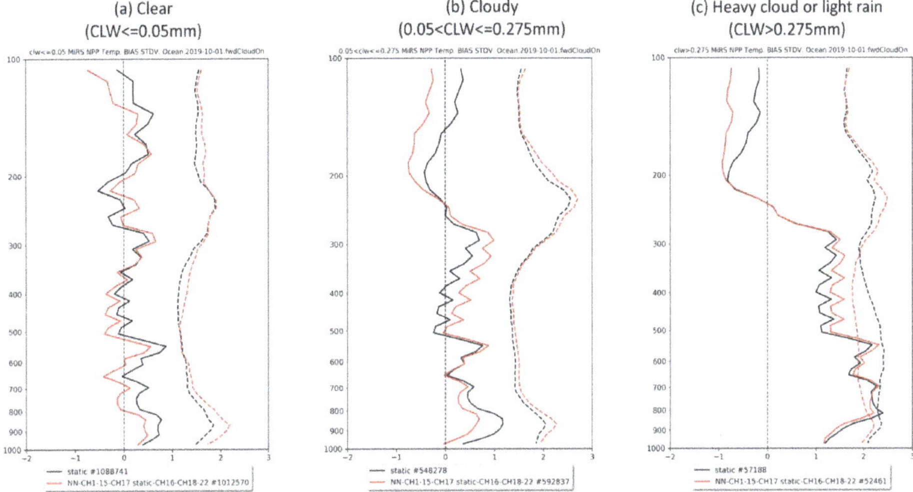

Figure 5. MiRS temperature profiles (K) validated with ECMWF analyses for SNPP/ATMS over ocean for (**a**) clear, (**b**) cloudy, and (**c**) heavy cloud or light rain conditions on 1 October 2019. The black lines are for MiRS using the static bias correction method, and red for the NN bias correction method. Solid lines are for bias, and dashed lines for standard deviation.

Similar profile plots for water vapor are shown in Figure 6, while the x-axis represents the percentage changes of bias and standard deviation with respect to ECMWF analysis at each layer. In the clear scenario, the NN bias percentage increased in magnitude about 5–10% between 700 hPa and 400 hPa compared with static experiment, while the standard deviation percentage decreased about 3% near 500 hPa. In the cloudy scenario, the bias percentage magnitude of the NN experiment is larger than Static at almost all levels, while the standard deviation is about 5–10% less than static between 600 hPa and 400 hPa. And in the heavy cloud or light rain scenario, NN water vapor bias slight decreased between 600 hPa and 350 hPa but increased between 350 hPa and 200 hPa. The most dramatic change is the NN standard deviation, which reduced from 80% to 60% between 600 hPa and 300 hPa. In summary, the MiRS water vapor retrieval using NN bias correction showed significantly reduced standard deviation at the middle levels. Similar, to the temperature profile results the largest positive impacts appear to be for cases with significant cloudiness and/or light rain.

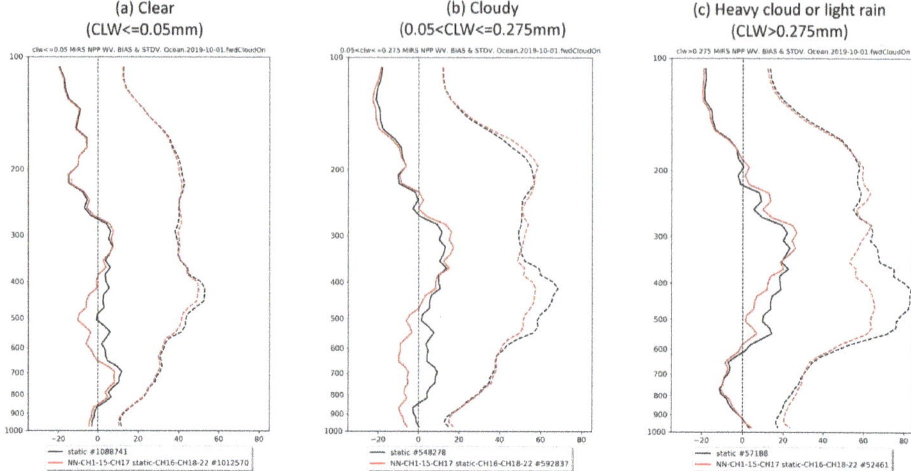

Figure 6. Similar to Figure 5 but for water vapor profiles. Bias and standard deviation are the percentage mixing ratio with respect to the mean ECMWF analysis at each layer.

MiRS retrieval performance for two-dimensional variables TPW, Tskin, and surface emissivity is presented in the following figures and tables. The TPW map of ECMWF analysis collocated with the SNPP/ATMS ascending node over ocean is shown in Figure 7a. The MiRS TPW bias with ECMWF analysis from static and NN is shown in Figure 7b,d, and the histograms of ECMWF (blue), static (black), and NN (dashed red) are given in Figure 7c. Relative to ECMWF, static has a moist bias, while a drier bias is observed on the NN bias map, especially over high mid-latitude ocean in the Southern Hemisphere. In the histogram plot, the NN TPW is drier and closer to ECMWF analysis, especially between 0-5, 15–25, and 55–60 mm. The validation statistics of static and NN TPW are shown in the top left panel of Table 3. TPW from NN experiment shows dramatically smaller bias and a slight increase in standard deviation.

Table 3. Performance metrics of MiRS retrievals including TPW, Tskin, emissivity (EM) at 23.8 GHz and 88.2 GHz, validated using ECMWF analyses (and FASTEM5 for emissivity) for SNPP/ATMS 1 October 2019 over ocean, ascending node. The numbers inside parentheses are sample sizes. The bias change percentages refer to their magnitude changes.

TPW (mm)	Static (868,412)	NN (875,423)	Change (%)	EM 23.8 GHz	Static (868,299)	NN (875,351)	Change (%)
Correlation	0.99	0.99	0%	Correlation	0.5862	0.6740	+15.0%
Bias	1.52	0.60	−60.5%	Bias	0.0071	0.0099	+42.9%
Std. Dev	2.33	2.62	+12.9%	Std. Dev	0.0353	0.0255	−25.7%
Tskin (K)	Static (868,299)	NN (875,351)	Change (%)	EM 88.2 GHz	Static (868,299)	NN (875,351)	Change (%)
Correlation	0.96	0.97	+1.0%	Correlation	0.7135	0.7229	+1.3%
Bias	0.38	−0.05	−86.8%	Bias	0.0022	−0.0004	−80.0%
Std. Dev	3.01	3.02	+0.3%	Std. Dev	0.0311	0.0301	−3.2%

The MiRS SNPP/ATMS TPW local zenith angle dependency over ocean is presented in Figure 8, with the local zenith angle within −70 to 70 degrees along the x-axis and the bias between static (black) or NN (red) and ECMWF on the y-axis. NN has smaller biases than static at all angles, and it almost has no scan angle dependency. In contrast, TPW bias from static is larger at and near nadir and quickly drops from 1.75 mm to 0.75 mm when it reaches the edges of the scan. This significant improvement is likely due to the explicit accounting of scan angle and other geophysical variables in the NN bias correction model.

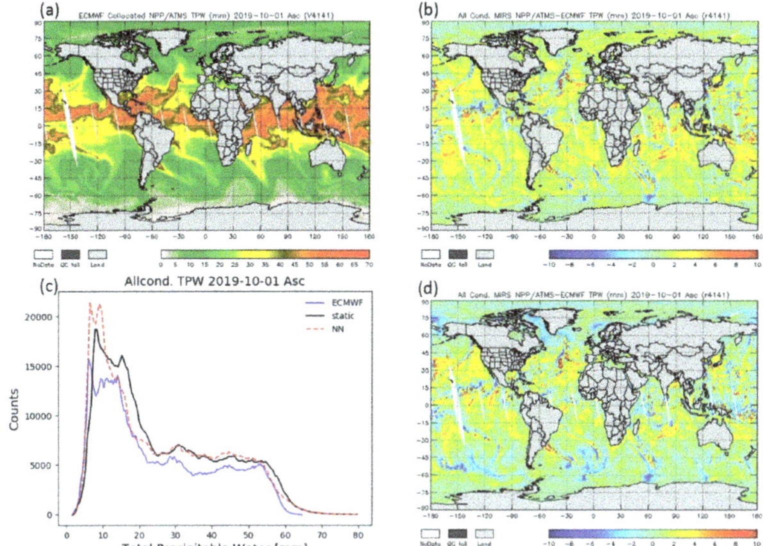

Figure 7. Total precipitable water (mm) of SNPP/ATMS on 1 October 2019, ascending node over ocean for (**a**) ECMWF analysis, (**b**) MiRS retrieval difference with ECMWF using static bias correction, (**d**) MiRS retrieval difference with ECMWF using NN bias correction, and (**c**) histograms of ECMWF and MiRS retrieval experiments shown in (**a**), (**b**), and (**d**).

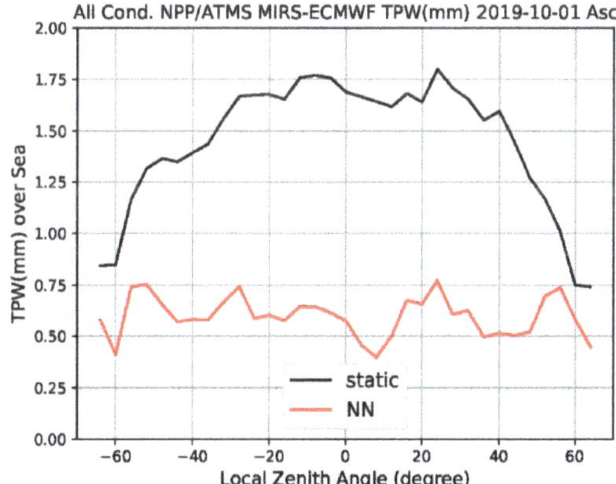

Figure 8. MiRS total precipitable water (mm) local zenith angle dependence of SNPP/ATMS on 1 October 2019 over ocean, ascending node. The black line is the result using a static bias correction and red line is the result using the NN bias correction.

Table 3 shows the performance metrics of MiRS TPW, Tskin, and emissivity at 23.8 GHz and 88.2 GHz for SNPP/ATMS on 1 October 2019 over ocean, ascending node. The metrics include correlation, bias, and standard deviation validated against the ECMWF analysis (and FASTEM5 for emissivity), as well as the percentage change of these metrics from NN to Static. The numbers inside parentheses are sample sizes for each parameter and each experiment. Except for emissivity at

23.8 GHz, other parameters from NN have 0% or 1% increase in correlation comparing with Static, and a 60–80% decrease in bias. The standard deviation of TPW from NN is 12.9% greater, while there is a very small impact on Tskin and emissivity at 88.2 GHz.

Results for performance metrics of descending node are similar to Table 3 except for Tskin, which has strong diurnal cycle. In Figure 9, MiRS Tskin retrievals are verified with ECMWF analyses and shown by density scatterplots. Tskin for static (left) and NN (right) are shown for both ascending (top) and descending (bottom) nodes. In ascending node, the bias from NN decreases from ~0.4 K to ~0.1 K. However, in descending node, the bias increases from ~−0.2 K to −0.6 K.

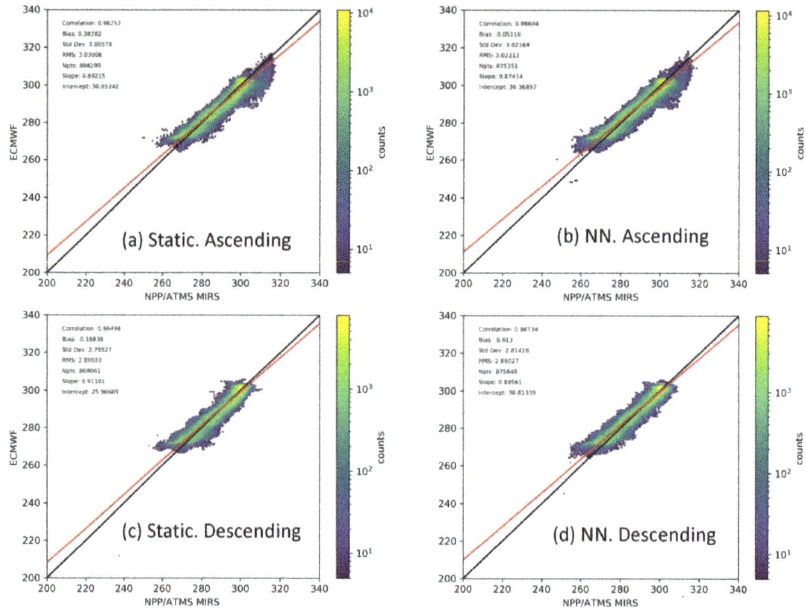

Figure 9. Skin temperature (K) scatterplots of SNPP/ATMS validated with ECMWF on 1 October 2019 over ocean, with ascending (top) and descending (bottom) nodes, using static (left) or NN (right) bias correction. The x-axis is MiRS skin temperature, and the y-axis represents ECMWF. The scatterplots are colored by the density of points.

Overall, the impact of applying the NN bias correction has a positive impact on the retrievals of several key retrieval parameters, depending on the performance statistic, but atmospheric and surface conditions appear to modulate significantly the magnitude and sign (improvement or degradation) of the impacts.

5. Conclusions

We report on preliminary results of applying a machine learning approach to estimation of the radiometric bias correction of passive microwave measurements from the SNPP/ATMS instrument. The bias correction was based on collocations of ATMS data with ECMWF operational analyses in conjunction with the FASTEM5 ocean surface emissivity model. A neural network model was used to estimate the bias corrections, and the model explicitly includes impacts from surface and atmospheric conditions, as well as scan angle and frequency dependence. Furthermore, the NN bias corrections were tested in the MiRS retrieval system to assess the impact of the bias corrections relative to retrievals using the operational static bias corrections. Because of this NN formulation the bias correction model is dynamic, adjusting the bias prediction with each scene or field of view, in contrast to the static bias correction.

The impact study examined retrieval performance of vertical temperature and water vapor profiles, total precipitable water, skin temperature, and surface emissivity. For temperature, largest positive impacts of using the NN bias correction appeared focused in scenes with higher amounts of cloud (>0.275 mm) and/or light precipitation. Lower tropospheric temperature bias was also reduced for scenes with fewer clouds. For water vapor, the largest positive impacts were on the error standard deviation concentrated in the 600–300 hPa layer. The impact on the water vapor bias was mixed, with some layers showing improvement, and other showing higher bias. The impact on the global TPW bias was significant and positive with a large reduction in the bias and only a small increase in the error standard deviation. For emissivity, the impact depended on frequency. At 23.8 GHz, correlation and standard deviation improved significantly, but there was also an increase in the bias. At 88.2 GHz, a significant reduction in the bias was seen, with only small impacts on the correlation and standard deviation noted. Impacts on Tskin depended on orbital node with both increases and decreases in the bias seen.

The experiments conducted clearly demonstrate the sensitivity of the MiRS retrieval system to the type of radiometric bias correction that is applied. By explicitly accounting for surface and atmospheric conditions in the formulation of the NN bias correction model, it appears that the largest positive impacts relative to the static bias formulation are in conditions that deviate significantly from the assumptions and training data of the static bias (i.e., scenes with clouds and light oceanic precipitation). Further investigations are underway to refine the approach, for example, using channel predictors more specific for each individual channel in the bias prediction model, as opposed to using the same channels as inputs, regardless of the channel bias prediction in question. Using additional independent days for retrieval experiments will clarify the seasonal dependence of the bias correction impacts. Finally, the approach is being extended to other satellite measurements, namely, from NOAA-20/ATMS.

Author Contributions: Conceptualization, C.G.; methodology, Y.Z and C.G.; software, Y.Z.; validation, Y.Z.; formal analysis, Y.Z and C.G.; investigation, Y.Z. and C.G.; resources, C.G.; data curation, Y.Z. and C.G.; writing—original draft preparation, Y.Z. and C.G.; writing—review and editing, Y.Z. and C.G.; visualization, Y.Z. and C.G.; supervision, C.G.; project administration, C.G.; funding acquisition, C.G. All authors have read and agreed to the published version of the manuscript.

Funding: This work was supported by the NOAA under NA19NES4320002 at the Cooperative Institute for Satellite and Earth System Studies (CISESS) at the University of Maryland/Earth System Science Interdisciplinary Center (ESSIC).

Acknowledgments: We thank Ryan Honeyager for providing the MiRS reformatting software, Xingming Liang for fruitful discussions on optimizing approaches to machine learning applications, and Kevin Garrett for providing the CLW regression algorithm.

Conflicts of Interest: The authors declare no conflict of interest. The funders had no role in the design of the study; in the collection, analyses, or interpretation of data; in the writing of the manuscript, or in the decision to publish the results.

References

1. Boukabara, S.A.; Garrett, K.; Chen, W.C.; Iturbide-Sanchez, F.; Grassotti, C.; Kongoli, C.; Chen, R.Y.; Liu, Q.H.; Yan, B.H.; Weng, F.Z.; et al. MiRS: An All-Weather 1DVAR Satellite Data Assimilation and Retrieval System. *IEEE Trans. Geosci. Remote Sens.* **2011**, *49*, 3249–3272. [CrossRef]
2. Boukabara, S.A.; Garrett, K.; Grassotti, C.; Iturbide-Sanchez, F.; Chen, W.; Jiang, Z.; Clough, S.A.; Zhan, X.; Liang, P.; Liu, Q.; et al. A Physical Approach for a Simultaneous Retrieval of Sounding, Surface, Hydrometeor and Cryospheric Parameters from SNPP/ATMS. *J. Geophys. Res.* **2013**, *118*, 12600–12619. [CrossRef]
3. Boukabara, S.A.; Garrett, K.; Grassotti, C. Dynamic inversion of global surface microwave emissivity using a 1DVAR approach. *Remote Sens.* **2018**, *10*, 679–696. [CrossRef]
4. Han, Y.; Van Delst, P.; Liu, Q.H.; Weng, F.Z.; Yan, B.; Treadon, R.; Derber, J. *Community Radiative Transfer Model (CRTM),—Version 1*; NOAA Technical Report 122; Dept. of Commerce/NOAA/NESDIS: Washington, DC, USA, 2006; 33p.

5. Ding, S.; Yang, P.; Weng, F.Z.; Liu, Q.H.; Han, Y.; Van Delst, P.; Li, J.; Baum, B. Validation of the community radiative transfer model. *J. Quant. Spectrosc. Radiat. Transf.* **2011**, *112*, 1050–1064. [CrossRef]

6. Liu, S.; Grassotti, C.; Chen, J.; Liu, Q.H. GPM Products from the Microwave-Integrated Retrieval System. *IEEE J. Sel. Top. Appl. Earth Obs. Remote Sens.* **2017**, *10*, 2565–2574. [CrossRef]

7. Forsythe, J.M.; Kidder, S.Q.; Fuell, K.K.; LeRoy, A.; Jedlovec, G.J.; Jones, A.S. A multisensor, blended, layered water vapor product for weather analysis and forecasting. *J. Oper. Meteorol.* **2015**, *3*, 41–58. [CrossRef]

8. Chirokova, G.; DeMaria, M.; DeMaria, R.; Dostalek, J.; Beven, J. Use of JPSS ATMS-MiRS retrievals to improve tropical cyclone intensity forecasting. The 20th Conference on Satellite Meteorology and Oceanography, Phoenix, AZ, Amer. *Meteor. Soc.* **2015**, 157. Available online: https://ams.confex.com/ams/95Annual/webprogram/Paper263652.html (accessed on 25 September 2020).

9. Joyce, R.J.; Xie, P. Kalman filter based CMORPH. *J. Hydrometeorol.* **2011**, *12*, 1547–1563. [CrossRef]

10. Joyce, R.J.; Janowiak, J.E.; Arkin, P.A.; Xie, P. CMORPH: A method that produces global precipitation estimates from passive microwave and infrared data at high spatial and temporal resolution. *J. Hydrometeorol.* **2004**, *5*, 487–503. [CrossRef]

11. Auligné, T.; McNally, A.P.; Dee, D.P. Adaptive bias correction for satellite data in a numerical weather prediction system. *Q. J. R. Meteorol. Soc.* **2007**, *133*, 631–642. [CrossRef]

12. Zhu, Y.; Derber, J.; Collard, A.; Dee, D.P.; Treadon, R.; Gayno, G.; Jung, J.A. Enhanced radiance bias correction in the National Centers for Environmental Prediction's Gridpoint Statistical Interpolation data assimilation system. *Q. J. R. Meteorol. Soc.* **2014**, *140*, 1479–1492. [CrossRef]

13. Dee, D.P. Bias and data assimilation. *Q. J. R. Meteorol. Soc.* **2005**, *131*, 3323–3343. [CrossRef]

14. Dee, D.P.; Uppala, S. Variational bias correction of satellite radiance data in the ERA-Interim reanalysis. *Q. J. R. Meteorol. Soc.* **2009**, *135*, 1830–1841. [CrossRef]

15. Blackwell, W.J.; Chen, F.W. *Neural Networks in Atmospheric Remote Sensing*; Artech House: Norwood, MA, USA, 2009.

16. Gangwar, R.K.; Mathur, A.K.; Gohil, B.S.; Basu, S. Neural network based retrieval of atmospheric temperature profile using AMSU-A observations. *Int. J. Atmos. Sci.* **2014**. [CrossRef]

17. Krasnopolsky, V.M. Neural network emulations for complex multidimensional geophysical mappings: Applications of neural network techniques to atmospheric and oceanic satellite retrievals and numerical modeling. *Rev. Geophys.* **2007**, *45*. [CrossRef]

18. Krasnopolsky, V.M.; Fox-Rabinovitz, M.S.; Belochitski, A.A. Decadal climate simulations using accurate and fast neural network emulation of full, longwave and shortwave, radiation. *Monthly Weather Rev.* **2008**, *136*, 3683–3695. [CrossRef]

19. Lee, Y.; Han, D.; Ahn, M.H.; Im, J.; Lee, S.J. Retrieval of total precipitable water from Himawari-8 AHI data: A comparison of random forest, extreme gradient boosting, and deep neural network. *Remote Sens.* **2019**, *11*, 1741. [CrossRef]

20. Manzato, A. Hail in Northeast Italy: A neural network ensemble forecast using sounding-derived indices. *Weather Forecast* **2013**, *28*, 3–28. [CrossRef]

21. He, Q.; Wang, Z.; He, J. Bias Correction for Retrieval of Atmospheric Parameters from the Microwave Humidity and Temperature Sounder Onboard the Fengyun-3C Satellite. *Atmosphere* **2016**, *7*, 156. [CrossRef]

22. Weng, F.; Zou, X.; Wang, X.; Yang, S.; Goldberg, M.D. Introduction to Suomi national polar-orbiting partnership advanced technology microwave sounder for numerical weather prediction and tropical cyclone applications. *J. Geophys. Res.* **2012**, *117*. [CrossRef]

23. Weng, F.; Zou, X.; Sun, N.; Yang, H.; Tian, M.; Blackwell, W.J.; Wang, X.; Lin, L.; Anderson, K. Calibration of Suomi national polar-orbiting partnership advanced technology microwave sounder. *J. Geophys. Res.* **2013**, *118*, 11–187. [CrossRef]

24. Bormann, N.; Fouilloux, A.; Bell, W. Evaluation and assimilation of ATMS data in the ECMWF system. *J. Geophys. Res.* **2013**, *118*, 12–970. [CrossRef]

25. Doherty, A.; Atkinson, N.; Bell, W.; Smith, A. An assessment of data from the advanced technology microwave sounder at the Met Office. *Adv. Meteorol.* **2015**. [CrossRef]

26. Liu, Q.; Weng, F.; English, S. An improved fast microwave water emissivity model. *IEEE Trans. Geosci. Remote Sens.* **2011**, *49*, 1238–1250. [CrossRef]

27. Liu, S.; Grassotti, C.; Liu, Q.; Lee, Y.K.; Honeyager, R.; Zhou, Y.; Fang, M. The NOAA Microwave Integrated Retrieval System (MiRS): Validation of Precipitation From Multiple Polar-Orbiting Satellites. *IEEE J. Sel. Top. Appl. Earth Obs. Remote Sens.* **2020**, *13*, 3019–3031. [CrossRef]
28. Ramachandran, P.; Zoph, B.; Le, Q.V. Searching for activation functions. *arXiv* **2017**, arXiv:1710.05941.

 *remote sensing*

Article

Bias Correction of the Ratio of Total Column CH_4 to CO_2 Retrieved from GOSAT Spectra

Haruki Oshio [1,*], Yukio Yoshida [1], Tsuneo Matsunaga [1], Nicholas M. Deutscher [2], Manvendra Dubey [3], David W. T. Griffith [2], Frank Hase [4], Laura T. Iraci [5], Rigel Kivi [6], Cheng Liu [7,8], Isamu Morino [1], Justus Notholt [9], Young-Suk Oh [10], Hirofumi Ohyama [1], Christof Petri [9], David F. Pollard [11], Coleen Roehl [12], Kei Shiomi [13], Ralf Sussmann [14], Yao Té [15], Voltaire A. Velazco [2], Thorsten Warneke [9] and Debra Wunch [16]

[1]   Center for Global Environmental Research, National Institute for Environmental Studies, Tsukuba, Ibaraki 305-8506, Japan; yoshida.yukio@nies.go.jp (Y.Y.); matsunag@nies.go.jp (T.M.); morino@nies.go.jp (I.M.); oyama.hirofumi@nies.go.jp (H.O.)
[2]   Centre for Atmospheric Chemistry, School of Earth, Atmospheric and Life Sciences, Faculty of Science, Medicine and Health, University of Wollongong, Wollongong 2522, Australia; ndeutsch@uow.edu.au (N.M.D.); griffith@uow.edu.au (D.W.T.G.); voltaire@uow.edu.au (V.A.V.)
[3]   Los Alamos National Laboratory, Los Alamos, NM 87545, USA; dubey@lanl.gov
[4]   Institute of Meteorology and Climate Research IMK-ASF, Karlsruhe Institute of Technology, 76021 Karlsruhe, Germany; frank.hase@kit.edu
[5]   Atmospheric Science Branch, NASA Ames Research Center, Moffett Field, CA 94035, USA; laura.t.iraci@nasa.gov
[6]   Space and Earth Observation Centre, Finnish Meteorological Institute, 99600 Sodankylä, Finland; rigel.kivi@fmi.fi
[7]   Key Laboratory of Environmental Optics and Technology, Anhui Institute of Optics and Fine Mechanics, Chinese Academy of Sciences, Hefei 230031, China; Chliu81@ustc.edu.cn
[8]   Department of Precision Machinery and Precision Instrumentation, University of Science and Technology of China, Hefei 230026, China
[9]   Institute of Environmental Physics, University of Bremen, 28359 Bremen, Germany; jnotholt@iup.physik.uni-bremen.de (J.N.); christof_p@iup.physik.uni-bremen.de (C.P.); warneke@iup.physik.uni-bremen.de (T.W.)
[10]   Climate Research Division, National Institute of Meteorological Sciences, Seogwipo, Jeju-do 63568, Korea; ysoh306@korea.kr
[11]   National Institute of Water and Atmospheric Research, Lauder, Omakau 9352, New Zealand; dave.pollard@niwa.co.nz
[12]   Division of Geology and Planetary Science, California Institute of Technology, Pasadena, CA 91125, USA; coleen@caltech.edu
[13]   Japan Aerospace Exploration Agency (JAXA), Tsukuba, Ibaraki 305-8505, Japan; shiomi.kei@jaxa.jp
[14]   Institute of Meteorology and Climate Research—Atmospheric Environmental Research (IMK-IFU), Karlsruhe Institute of Technology, 82467 Garmisch-Partenkirchen, Germany; ralf.sussmann@kit.edu
[15]   Laboratoire d'Etudes du Rayonnement et de la Matière en Astrophysique et Atmosphères (LERMA-IPSL), Sorbonne Université, CNRS, Observatoire de Paris, PSL Université, 75005 Paris, France; yao-veng.te@upmc.fr
[16]   Department of Physics, University of Toronto, Toronto, ON M5S 1A7, Canada; dwunch@atmosp.physics.utoronto.ca
*   Correspondence: oshio.haruki@nies.go.jp; Tel.: +81-29-850-2416

Received: 21 August 2020; Accepted: 24 September 2020; Published: 25 September 2020

Abstract: The proxy method, using the ratio of total column CH_4 to CO_2 to reduce the effects of common biases, has been used to retrieve column-averaged dry-air mole fraction of CH_4 from satellite data. The present study characterizes the remaining scattering effects in the CH_4/CO_2 ratio component of the Greenhouse gases Observing SATellite (GOSAT) retrieval and uses them for bias correction. The variation of bias between the GOSAT and Total Carbon Column Observing Network (TCCON) ratio component with GOSAT data-derived variables was investigated. Then, it was revealed that the variability of the bias could be reduced by using four variables for the bias correction—namely, airmass, 2 μm band radiance normalized with its noise level, the ratio between the partial column-averaged dry-air mole fraction of CH_4 for the lower atmosphere and that for the upper atmosphere, and the difference in surface albedo between the CH_4 and CO_2 bands. The ratio of partial column CH_4 reduced the dependence of bias on the cloud fraction and the difference between hemispheres. In addition to the reduction of bias (from 0.43% to 0%), the precision (standard deviation of the difference between GOSAT and TCCON) was reduced from 0.61% to 0.55% by the correction. The bias and its temporal variation were reduced for each site: the mean and standard deviation of the mean bias for individual seasons were within 0.2% for most of the sites.

Keywords: methane; proxy method; GOSAT; TCCON

---

1. Introduction

Atmospheric methane (CH_4) is the most significant anthropogenic greenhouse gas after carbon dioxide (CO_2) and is emitted from both anthropogenic and natural sources. Satellite observation, which can obtain data over wide areas, is effective to elucidate the CH_4 budget over the globe. In the last 15 years, the column-averaged dry-air mole fraction of methane (XCH_4) has been retrieved from the spectra of the backscattered Short-Wavelength InfraRed (SWIR) sunlight measured by sensors onboard satellites [1–6]. Satellite-derived XCH_4 data have been applied to the inverse modeling of CH_4 sources and sinks [7–11]. High precision and small bias in spatiotemporal variation are required for the XCH_4 data to be used in inverse modeling [12,13]. It is possible that even a small regional bias (0.5%) in the XCH_4 data can lead to significant errors in regional source and sink estimation [12]. Optical path length modification due to the light scatterings by aerosols and clouds is a large source of error for the satellite-based SWIR retrievals [14–16]. The degree of optical path length modification depends on the abundance, optical properties, and vertical distributions of aerosols and clouds and the reflectance of ground surfaces.

Two retrieval methods have been used to reduce systematic biases due to atmospheric scatterings: the full-physics method and the proxy method. In the full-physics method, the existence of aerosols is described in the forward model, and the aerosol-related parameters are simultaneously retrieved with the gas abundance [17,18]. In the proxy method, information on the optical path length modification for the CH_4 absorption band is obtained from that for the adjacent CO_2 absorption band [2] (Equation (1)),

$$XCH_4 = \frac{XCH_{4,clr}}{XCO_{2,clr}} \times XCO_{2,mdl},\tag{1}$$

where $XCH_{4,clr}$ and $XCO_{2,clr}$ are XCH_4 and XCO_2 retrieved under a clear-sky assumption (no aerosols and clouds are assumed) and $XCO_{2,mdl}$ is XCO_2 from the numerical model. It is assumed that, in the ratio component ($XCH_{4,clr}/XCO_{2,clr}$), the impacts of aerosol and clouds cancel each other out between $XCH_{4,clr}$ and $XCO_{2,clr}$. It is also assumed that the relative variation of XCO_2 is much smaller than that of XCH_4, and that XCO_2 is well represented by the numerical model. The proxy method is expected to offer a larger amount of useful retrieved data than the full-physics method, since highly cloud- and aerosol-loaded scenes are difficult to handle in the current full-physics algorithms [19–22].

Both errors in the ratio component and the model XCO_2 lead to errors in the resulting XCH_4. Butz et al. [23] showed that the scattering-related errors are not perfectly canceled out in the ratio component depending on atmospheric and ground surface conditions (i.e., cirrus and aerosol load and surface albedo). Schepers et al. [19] discussed the possibility that the temporal variation of bias in proxy XCH_4 corresponds to that of bias in the ratio component. In the case of inverse modeling, Parker et al. [24] suggested that it is beneficial to use the ratio component with each own XCO_2

model that is consistent with the XCH_4 model used in the inversion. The ratio component can also be directly inverted [25–27]. The ratio component derived from the Greenhouse gases Observing SATellite (GOSAT) data has been validated by comparing it with that derived from the Total Carbon Column Observing Network (TCCON) data [19,24,26,28,29]. However, it has not been fully investigated what range of cloud and aerosol load permits the canceling out of scattering-related errors in the ratio component. In general, the criteria for the cloud and aerosol screening have been chosen by the algorithm developer. Scattering-related errors are expected to remain and to cause bias in the resulting XCH_4, especially when relaxing the data-screening criteria, although this increases the data throughput. Bias corrections of the proxy-based XCH_4 have been conducted; however, a simple linear relationship between the bias and the surface albedo [30] and a simple global bias correction [29,31] have been used.

The present study sought to characterize the bias in the GOSAT ratio component and develop a method for correcting the bias while considering the atmospheric scattering effects. GOSAT has been operating for more than 10 years, allowing us to investigate the variation of the bias with time and space and its factors. The ratio component derived from TCCON data was used as the ground truth. In Section 2, the data used and its processing are described. In Section 3, the relationship between the bias and the related variables derived from GOSAT data is investigated. In Section 4, the bias correction is conducted based on the results of Section 3, and the corrected results are evaluated.

2. Materials and Data Processing

2.1. GOSAT Data

GOSAT was launched on 23 January 2009 and is on a sun-synchronous orbit at 666 km altitude with 3-day recurrence and a descending node around 13:00 local time. It is equipped with two instruments: the Thermal And Near-infrared Sensor for carbon Observation–Fourier Transform Spectrometer (TANSO-FTS) and the Cloud and Aerosol Imager (TANSO-CAI). The TANSO-FTS has three bands in the Short-Wavelength InfraRed (SWIR) region (an O_2 A band, a weak CO_2 absorption band, and a strong CO_2 absorption band (Bands 1, 2, and 3) centered at 0.76, 1.6, and 2.0 µm, respectively) and records two orthogonal polarization components (hereafter called P/S components). For the signal processing of the TANSO-FTS, the amplifier gain level can be controlled at different levels, high (H) and medium (M), according to the brightness of the target. Gain M is used over bright surfaces such as the Sahara and central Australia. The instantaneous field of view (IFOV) of the TANSO-FTS is 15.8 mrad, which corresponds to a circular surface footprint of about 10.5 km in diameter at nadir. The TANSO-FTS L1B product (radiance spectral data) version V210.210 was used in the present study. We used spectra acquired from April 2009 to December 2018. The sensitivity degradation of the TANSO-FTS was corrected using a radiometric degradation model that is based on the on-orbit solar calibration data [32]. The P and S polarization components of the observed spectra were synthesized to produce a total intensity spectrum [5]. The TANSO-CAI is a push-broom imager and has four narrow bands in the near-ultraviolet to near-infrared regions centered at 0.38, 0.674, 0.87, and 1.6 µm with spatial resolutions of 0.5, 0.5, 0.5, and 1.5 km, respectively, for nadir pixels. The TANSO-CAI L2 cloud flag product (integrated clear confidence level for each TANSO-CAI pixel) version V02.00 was used in the present study. The integrated clear confidence level expresses the cloudy area with 0, the clear area with 1, and the ambiguous area with a numerical value between 0 and 1 [33].

2.2. Retrieval

The spectral windows of 1.626–1.695 µm and 1.567–1.618 µm within the TANSO-FTS Band 2 were used to retrieve $XCH_{4,clr}$ and $XCO_{2,clr}$ under the clear-sky assumption, respectively, using the same retrieval scheme as in the NIES TANSO-FTS L2 SWIR full-physics retrieval [5,21,34]. The atmospheric column was divided into 15 layers from the surface to 0.1 hPa with a constant pressure difference, and the average gas concentration for each layer was retrieved. As an indicator of optical path length modification, the surface pressure (P_{srf}) under the clear-sky assumption was also retrieved from the

TANSO-FTS Band 1 spectra. The state vector for the retrieval of each band also included the surface albedo and the wavenumber dispersion. The surface albedo was retrieved at several wavenumber grid points within each band (CH_4, CO_2, and O_2 A) [5]. The mean value was calculated for each band and was used in the following analysis. Data satisfying all the following criteria were used for the subsequent analysis: (1) several data-quality flags stored in the TANSO-FTS L1B product and spectrum quality check utilizing the out-of-band spectra [34] are set as OK; (2) the solar zenith angle is <70°; (3) the land fraction of the TANSO-FTS footprint is ≥60%; (4) the mean squared value of the residual spectrum of the CO_2 band is≤ 4; and (5) the degree of freedom for signals (DFS) is ≥1 for both $XCH_{4,clr}$ and $XCO_{2,clr}$.

2.3. Variables for Explaining the Bias in the Ratio Component

Butz et al. [23] evaluated the accuracy of the ratio component and analyzed the error sources (the reasons why the scattering-related errors were not canceled in the ratio component) using simulated satellite measurements. They used cloud-free aerosol-loaded and cirrus-loaded scenes that were assumed to be the targets of the full-physics method. They showed that the primary sources of error were the difference in surface albedo between the CH_4 band and the CO_2 band and the difference in the retrieval sensitivity to scattering effects at each height level between these bands. The impact of these sources is expected to vary according to the amount and vertical distribution of the scattering materials.

In the studies validating the ratio component derived from the actual GOSAT data [19,24,26,28], the cloud screening was conducted using the cloud fraction within the IFOV of the TANSO-FTS provided by the TANSO-CAI onboard GOSAT. The TANSO-CAI is prone to fail to detect optically thin cirrus clouds [35]. In several studies (e.g., [19]), cirrus-loaded scenes were screened out using the information from the TANSO-FTS 2 μm band. More specifically, if the TANSO-FTS signal level at the strong water vapor absorption channels exceeds the noise level, elevated scattering materials (mainly cirrus cloud) are expected [5,36]. However, it has not been fully addressed how well the cloud fraction and the 2 μm band signal are related to the bias in the ratio component (i.e., the systematic difference between GOSAT and TCCON). Therefore, we investigated the variation of the bias with the cloud fraction and the 2 μm band signal. The cloud fraction (f_c) was defined as the ratio of TANSO-CAI pixels with an integrated clear confidence level lower than 0.33 and all TANSO-CAI pixels within the TANSO-FTS IFOV in this study. The TANSO-CAI tends to identify the pixels over snow and ice surfaces as cloudy pixels. Thus, we calculated the Normalized Difference Snow Index (NDSI = $(\rho_{O2} - \rho_{CH4})/(\rho_{O2} + \rho_{CH4})$, where ρ_{O2} and ρ_{CH4} are the retrieved surface albedo for the O_2 A band and the CH_4 band, respectively), and used data having NDSI ≤ 0.4 for investigating the relationship between the bias and f_c. The threshold value of 0.4 was empirically determined (Figure S1). The radiance at the strong water vapor absorptive channels normalized with its noise level in the TANSO-FTS Band 3 ($I_{2\mu m}$) was calculated in a manner similar to [5,34].

In the bias correction of the proxy method, possible error sources (e.g., those indicated by Butz et al. [23]) have hardly been considered. The correction has been conducted based on the relationship between the bias and the retrieved surface albedo [30] and by a simple global bias correction [29,31]. The bias in the ratio component has also been corrected using the surface albedo [26]. Then, we investigated the relationship between the bias in the ratio component and the related variables while considering f_c and $I_{2\mu m}$. For the related variables, the difference in the surface albedo between the CH_4 and CO_2 bands, the surface albedo, the vertical profile of CH_4 and CO_2, the airmass, and the deviation of the clear-sky surface pressure from its prior value were considered. The details of the variables are described below.

The differences in surface albedo and the vertical profile have been revealed as important error sources [23]. For the difference in surface albedo, the retrieved albedo values (ρ_{CH4} and ρ_{CO2}) were used ($\Delta_{alb} = \rho_{CH4} - \rho_{CO2}$). Albedo itself ($\rho_{CH4}$) was also used, since it has been used to explain the bias of the proxy method [26,30] and the full-physics method [37]. For the vertical profile, the partial column-averaged dry-air mole fractions were calculated for Layers 1–7 (upper atmosphere,

$0–0.47P_{srf}$ hPa ($XCH_{4,upper}$ and $XCO_{2,upper}$)) and for Layers 12–15 (lower atmosphere, $0.73P_{srf}–P_{srf}$ hPa ($XCH_{4,lower}$ and $XCO_{2,lower}$)). See Section 2.2 for the definition of layers. The range of the calculation (1–7 layers and 12–15 layers) was decided by considering the averaging kernel (Figure S2). It is well-known that the SWIR retrieval has a slight sensitivity to the detailed profile but has a known sensitivity to the lower atmosphere (e.g., [38]); therefore, the ratios ($R_{CH4} = XCH_{4,lower}/XCH_{4,upper}$ and $R_{CO2} = XCO_{2,lower}/XCO_{2,upper}$) were used to represent the characteristics of the vertical profile. Airmass is expected to be related to the impact of optical path length modification on the retrieval of $XCH_{4,clr}$ and $XCO_{2,clr}$. The approximate airmass was calculated as (1/cos(solar zenith angle) + 1/cos(observing zenith angle)), which was similar to the previous studies that conducted bias correction for the full-physics method by considering airmass [37,39]. For the full-physics method, the deviation of the retrieved surface pressure from its prior value was also used for the bias correction [37,39]. In the case of the clear-sky retrieval, the deviation simply indicates the degree of optical path length modification for the O_2 A band and has been used for the cloud screening [40]. The deviation is expected to contain information about the scattering effect by aerosols that can be hardly accounted for by f_c and $I_{2\mu m}$. Then, the deviation of the clear-sky surface pressure from its prior value was calculated as ($\Delta P_{srf} = P_{srf,retrieve} − P_{srf,prior}$).

2.4. TCCON Data and Matching with GOSAT Data

TCCON XCH_4 and XCO_2 data (GGG2014) from 26 sites [41–67] were used as the ground truth to validate the GOSAT ratio component. The map of the sites is shown in Figure A1, and the overview of the sites is shown in Table A1. The TCCON XCH_4 and XCO_2 data used in the present study ($XCH_{4,TCCON}$ and $XCO_{2,TCCON}$) represent the mean values measured at each TCCON site within ±30 min of the GOSAT observation time. GOSAT data were selected within a ±2° latitude/longitude box centered at each TCCON site and within the difference in altitude between GOSAT (average within footprint) and TCCON site of 400 m. The ratio component of TCCON was then calculated ($XCH_{4,TCCON}/XCO_{2,TCCON}$). The relative difference (Δ_{ratio}) was calculated to evaluate the GOSAT ratio component as

$$\Delta_{ratio} = 100 \times (X_{ratio,G} − X_{ratio,T})/X_{ratio,T}, \tag{2}$$

where $X_{ratio,G}$ and $X_{ratio,T}$ are the ratio components of GOSAT and TCCON, respectively. Most of the matched GOSAT data were acquired with gain H. Therefore, the data acquired with gain H were used in the following analysis. The data acquired with gain M are briefly addressed in the latter part of the analysis.

3. Investigating the Bias in the Ratio Component

3.1. Comparison Between GOSAT and TCCON Under Cloud-Free Conditions

First, in order to confirm the baseline of the bias, $X_{ratio,G}$ was compared with $X_{ratio,T}$ under the conditions in which cloud-free scenes were expected ($f_c = 0$ and $I_{2\mu m} \leq 1$). Figure 1 shows the scatterplot of the ratio component and the latitudinal variation of the bias, precision of single scan, and interseasonal bias for each TCCON site. The bias and precision are defined as the mean and standard deviation of Δ_{ratio}, respectively. This precision value was used with the number of data to calculate the standard error of the mean value. The interseasonal bias is the standard deviation of bias values of the four seasons (DJF, MAM, JJA, SON) regardless of year, which has been used to represent the seasonal variability of bias [68]. The intersite bias is the standard deviation of bias values for individual TCCON sites, which is related to the spatial variability of bias. Only sites having more than nine data points were included in the calculation of intersite bias. The intersite bias of 0.14% might indicate that the spatial variation of bias is sufficiently small; however, the influence of loosening the cloud screening criteria and the temporal variation of bias should be investigated. Previous studies validating the GOSAT ratio component using TCCON data reported that the bias, precision, and intersite bias were about 0.2–0.6%, 0.5–0.7%, and 0.15–0.2%, respectively [19,24,26,28]. Although the retrieval scheme, the version of the TANSO-FTS L1B product, the number of TCCON sites used, the data matching criteria, the cloud screening method, and other details of the data screening differ between the present study and the previous studies, the overall results were comparable.

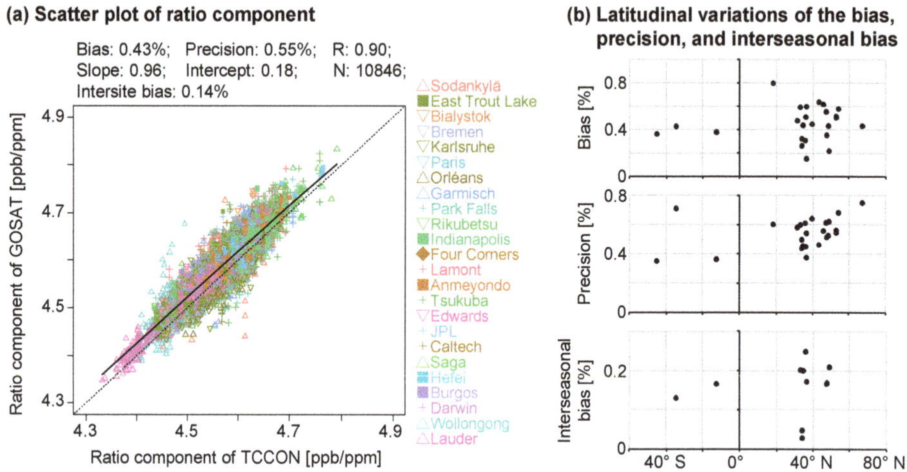

Figure 1. Comparison of the ratio component between Greenhouse gases Observing SATellite (GOSAT) and Total Carbon Column Observing Network (TCCON): (**a**) Scatter plot of individual data; (**b**) result for each site. GOSAT data obtained under conditions where cloud-free scenes were expected (cloud fraction = 0 and normalized 2 µm band radiance ≤1) were used. In (**a**), the correlation coefficient (R), the slope and intercept of the linear regression, and the number of data points (N) are also shown. In (**b**), the latitudinal variations of the bias, precision, and interseasonal bias for TCCON sites having more than nine data points are presented.

Table A2 shows that there was no clear variation of bias with tightening of the matching criteria of GOSAT and TCCON, but the precision was improved (there was a decrease in the standard deviation). Then, we assessed the influence of the matching criteria on our analysis and confirmed that the results shown below were hardly affected by the matching criteria (Appendix A). We also assessed the relationship between Δ_{ratio} and Fractional Variation in Solar Intensity (FVSI). FVSI is stored in TCCON data, and low FVSI values (≤1%) indicate a reasonably clear sky, where larger FVSI values could indicate some cirrus cloud presence. Only TCCON data having small FVSI values (≤5%) were provided to ensure the quality of XCH_4 and XCO_2 data. Figures S3 and S4 show that similar results were obtained between data with FVSI ≤ 1% and that with FVSI > 1%. This suggests that TCCON data can be used to validate the GOSAT ratio component regardless of FVSI values (0–5%).

3.2. Relationship Between Bias and Related Variables

3.2.1. Cloud Fraction and Normalized 2 µm Band Radiance

Figure 2 shows the variation of Δ_{ratio} with f_c and $I_{2\mu m}$. The data with NDSI ≤ 0.4 were used. Δ_{ratio} increases with the increase in $I_{2\mu m}$. Δ_{ratio} decreases with the increase in f_c for the data with small $I_{2\mu m}$. The large Δ_{ratio} is observed for data with both f_c and $I_{2\mu m}$ exceeding certain levels, although the amount of data is small for such cases. To interpret the information from the TANSO-CAI and the TANSO-FTS 2 µm band clearly, we mainly used the data with $f_c = 0$ and that with $I_{2\mu m} \leq 1$ in the following analysis. Most of the data fell within these cases (leftmost column and bottom row of Figure 2d). In the case of $f_c = 0$, $I_{2\mu m}$ is related to the optically thin elevated scattering materials (mainly cirrus cloud) that were not identified by the TANSO-CAI. In the case of $I_{2\,\mu m} \leq 1$, f_c is related to the middle- or low-altitude clouds with a certain level of optical thickness.

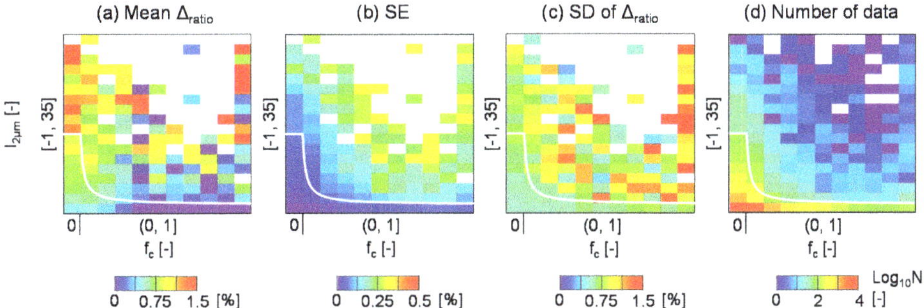

Figure 2. Variation of the relative difference in the ratio component between GOSAT and TCCON (Δ_{ratio}) according to the cloud fraction (f_c) and the normalized 2 µm band radiance ($I_{2\,\mu m}$). Data with $0 < f_c \leq 1$ were divided into 10 bins (horizontal axis), and data with $-1 \leq I_{2\,\mu m} \leq 35$ were divided into 18 bins (vertical axis). The mean Δ_{ratio}, standard error (SE) calculated by the precision of single data ((**a**,**b**)), standard deviation (SD) of Δ_{ratio} (**c**), and number of data points (**d**) for each bin are presented. Data with Normalized Difference Snow Index (NDSI) ≤ 0.4 were used. Only bins having more than two data points are colored except in panel (**d**). The white line indicates the cloud screening criterion (Section 4.3).

Figure 3a shows the variation of Δ_{ratio} with $I_{2\,\mu m}$. Δ_{ratio} increases with $I_{2\,\mu m}$. This can be attributed to the light path enhancement, which is greater for the CH_4 band than for the CO_2 band because the surface albedo of the CH_4 band is generally higher than that of the CO_2 band. The difference in vertical profile between CH_4 and CO_2 also seems to contribute to the results. More specifically, the light path enhancement means that the light repeatedly passes the area where the CH_4 concentration is higher than the column average, since the CH_4 concentration significantly decreases in the upper atmosphere. It is considered that the influence of the difference in the albedo and vertical profile becomes large with the increase in $I_{2\,\mu m}$.

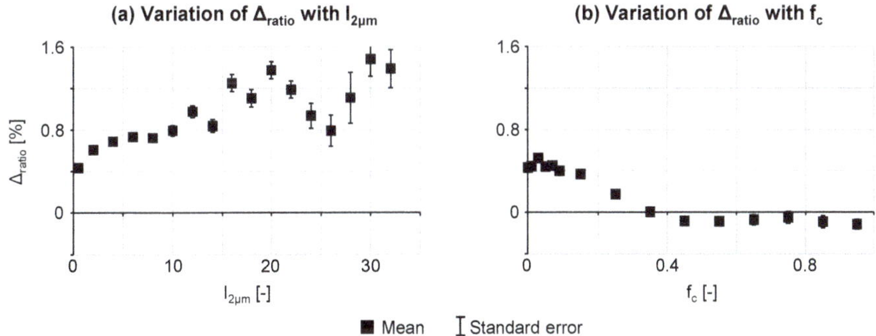

Figure 3. Variation of the relative difference in the ratio component between GOSAT and TCCON (Δ_{ratio}) according to (**a**) the normalized 2 µm band radiance ($I_{2\,\mu m}$) and (**b**) the cloud fraction ($f_{c,}$). Only data with $f_c = 0$ were used for (**a**), and only data with $I_{2\,\mu m} \leq 1$ and NDSI ≤ 0.4 were used for (**b**).

Figure 3b shows the variation of Δ_{ratio} with f_c. The data with NDSI ≤ 0.4 were used. Δ_{ratio} is 0.4% to 0.5% for the data with $f_c \leq 0.2$ and decreases with increasing f_c. This is because the effect of the difference in the retrieved surface albedo between the CH_4 and CO_2 bands becomes small with the increase in f_c, and the influence of the upper atmosphere where the CH_4 concentration is low becomes large because of the increase in the amount of light passing through a short path (scattered at the upper part of the cloud and reaching the sensor). Although the influence of clouds depends on their height and optical thickness, it is considered that the influence of the ground surface becomes small with the increase in cloud cover. Δ_{ratio} is almost stable for the data with $f_c \geq 0.4$. It seems that the above-mentioned effects of the decreasing Δ_{ratio} and light path enhancement by clouds (multiple scattering within clouds) are balanced. Note that we obtained GOSAT data having large f_c value, which fell within the matching criteria of GOSAT and TCCON. This means that GOSAT data was obtained under cloudy conditions even when TCCON data was obtained under clear-sky conditions since cloud conditions varied within a $\pm 2°$ latitude/longitude box.

3.2.2. Surface Albedo, Difference in Surface Albedo, Airmass, and Deviation of Surface Pressure

Figure 4 shows the relationship between Δ_{ratio} and the related variables (ρ_{CH4}, Δ_{alb}, airmass, and ΔP_{srf}). Section 3.2.1 showed that Δ_{ratio} was stable for data with $f_c \leq 0.2$ and increased with $I_{2\mu m}$ continuously; therefore, the results are separately plotted according to the f_c and $I_{2\mu m}$ values. Case 1: $f_c \leq 0.2$ and $I_{2\mu m} \leq 1$; Case 2: $f_c \leq 0.2$ and $I_{2\mu m} > 1$; Case 3: $f_c > 0.2$ and $I_{2\mu m} \leq 1$. The results for Case 1 are discussed in this paragraph. Δ_{ratio} increases with ρ_{CH4}, although the variation is gentler than that for the other variables (Figure 4b–d). One possible reason is that the high surface albedo is prone to bring light path enhancement, by which the influence of the difference in the vertical profile between CH_4 and CO_2 becomes large (even if the surface albedo is similar between the CH_4 and CO_2 bands). For the difference in albedo, Δ_{ratio} clearly increases with the increase in Δ_{alb}. As Butz et al. [23] indicated in their theoretical study and as discussed in the former section, the difference in albedo causes a difference in optical path length modification between the CH_4 and CO_2 bands, significantly affecting the ratio component. Δ_{ratio} decreases with an increase in airmass. The influence of optical path length modification seems to be relatively small for the cases with large airmass (the modified light path is relatively short when the geometric path is long). The characteristics of the retrieval (e.g., errors in spectroscopy) might also affect the airmass dependence. The airmass dependence of retrieved XCH_4 and XCO_2 has been corrected empirically for satellite data [37,39] and TCCON data [69,70]. Recently, Mendonca et al. [71] found that using speed-dependent Voigt line shapes for retrieval of the O_2 total column reduces the airmass dependence of TCCON XCO_2. Δ_{ratio} significantly increases with ΔP_{srf}, although the number of data with large ΔP_{srf} is small. Although ΔP_{srf} is related to the optical path length modification for the O_2 A band (TANSO-FTS Band 1), the large ΔP_{srf} indicates the possibility that the light path enhancement effect, rather than light path shortening, is dominant for Band 2. More specifically, multiple scattering might occur within the area where the concentration of CH_4 is relatively high compared to the column average. In contrast to the cases with high f_c, the fraction of light passing a short path is expected to be low, and the influence of the ground surface is expected to be large, yielding a positive bias in the ratio component.

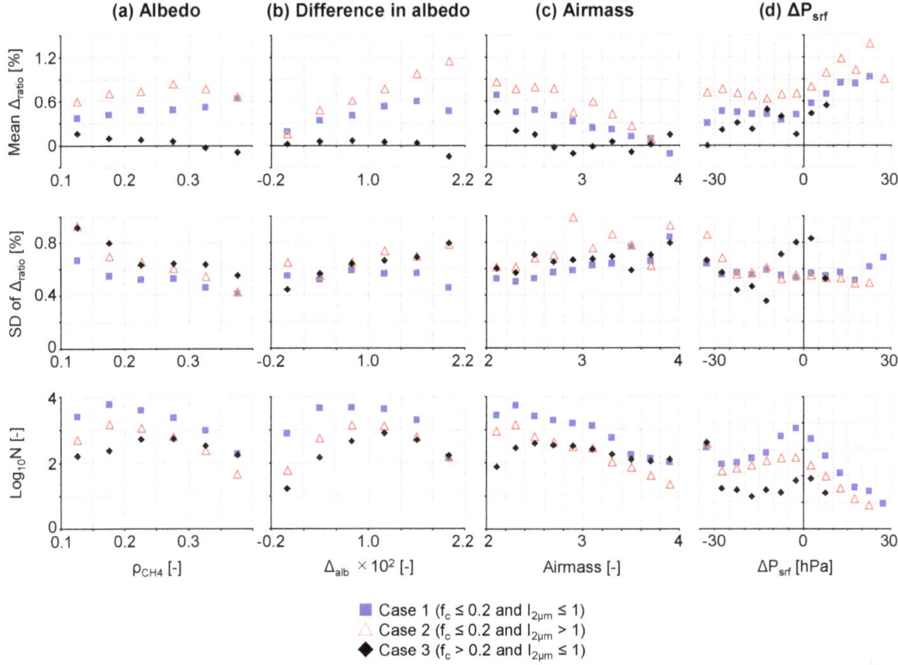

Figure 4. Variation of the relative difference in the ratio component between GOSAT and TCCON (Δ_{ratio}) according to the related variables: (**a**) retrieved surface albedo of the CH_4 band (ρ_{CH4}); (**b**) difference in the retrieved surface albedo (CH_4 band minus CO_2 band, Δ_{alb}); (**c**) airmass; (**d**) deviation of the retrieved clear-sky surface pressure from its prior (ΔP_{srf}). In each panel, the mean and standard deviation (SD) of Δ_{ratio} and the number of data points (N) for each bin are presented. The results are separately plotted according to the cloud fraction (f_c) and the normalized 2 μm band radiance ($I_{2\mu m}$).

For Case 2, Δ_{ratio} is larger than that for Case 1. Although the dependence of Δ_{ratio} on Δ_{alb} is in the same direction between Cases 1 and 2, the increase in Δ_{ratio} with Δ_{alb} becomes steep, meaning that the influence of the difference in surface albedo becomes large when elevated scattering materials exist. In contrast, for Case 3, the variation trend of Δ_{ratio} with ρ_{CH4} and Δ_{alb} differs significantly from that of Cases 1 and 2. This is because clouds affected the retrieved albedo (ρ_{CH4} and Δ_{alb} increased with the increase in f_c).

3.2.3. Vertical Profile of CH_4 and CO_2

Figure 5 shows the variation of Δ_{ratio} with R_{CH4} and R_{CO2} (see Section 2.3 for the definition). The variation of Δ_{ratio} with R_{CH4} is small if the possibility of the existence of elevated scattering materials is low (Case 1). In contrast, for Case 2, Δ_{ratio} increases with the increase in R_{CH4}. This corresponds to the qualitative discussion of the influence of the CH_4 profile on Δ_{ratio} in Section 3.2.1. Although the retrieval cannot reproduce the real-world profile in detail, it is considered that the retrieved R_{CH4} represents the real-world R_{CH4} (R_{CH4}^{act}) well and can be used to explain Δ_{ratio}. In contrast to CH_4, Δ_{ratio} was expected to decrease with an increase in R_{CO2} because CO_2 is the denominator of the ratio component. However, no clear relationship between Δ_{ratio} and R_{CO2} is seen in Figure 5. Two possible reasons are considered: (1) the influence of the CO_2 profile on the ratio component is small, since R_{CO2}^{act} is smaller than R_{CH4}^{act} in general; (2) the sensitivity of the retrieval to the vertical profile for CO_2 is lower than that for CH_4 (DFS for $XCO_{2,clr}$ was 1.0–1.5 and that for $XCH_{4,clr}$ was 1.7–2.3 in the present study). When only the retrieved data having DFS for $XCO_{2,clr} \geq 1.3$ (almost half of all data) were used, the results were not noticeably changed (Figure S5). For Case 3, the range of R_{CH4} and R_{CO2} differs from that for Cases 1 and 2, since R_{CH4} and R_{CO2} were affected by clouds. R_{CH4} was negatively correlated with f_c, which is considered to contribute to the increase in Δ_{ratio} with R_{CH4}. Although a variation of Δ_{ratio} with R_{CO2} was expected for Case 3, since R_{CO2} was also negatively correlated with f_c, the variation was small. The correlation between the variables (as mentioned in Sections 3.2.2 and 3.2.3) was accounted for in the variable selection for the bias correction (Section 4.2).

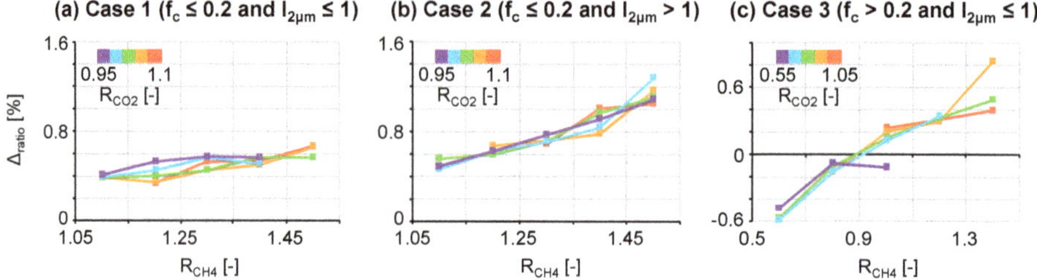

Figure 5. Variation of the relative difference in the ratio component between GOSAT and TCCON (Δ_{ratio}) according to the ratio between the partial column-averaged dry-air mole fractions for the lower atmosphere and that for the upper atmosphere (R_{CH4} and R_{CO2}): (**a**) Case 1 (the cloud fraction (f_c) ≤ 0.2 and the normalized 2 μm band radiance ($I_{2\mu m}$) ≤ 1); (**b**) Case 2 ($f_c \leq 0.2$ and $I_{2\mu m} > 1$); (**c**) Case 3 ($f_c > 0.2$ and $I_{2\mu m} \leq 1$).

4. Bias correction

4.1. Method

Linear regression was used as in many previous studies [20,37,39,72] as

$$\Delta_{ratio}^{pred} = \sum_{i=1}^{i=n} C_i x_i + C_{n+1}, \tag{3}$$

where Δ_{ratio}^{pred} is the predicted Δ_{ratio}, n is the number of variables, C_1–C_{n+1} are the regression coefficients, and x_i represents the explanatory variable. In the calculation of coefficients (least squares method), each data point (matched GOSAT and TCCON data) was weighted according to the amount of total matched GOSAT and TCCON data of the site to which the data point belonged. More specifically, the weight was given as R_{site}/N_j and $1/N_j$ for the data from sites in the northern hemisphere and that from sites in the southern hemisphere, respectively, where N_j is the total number of matched GOSAT data for each site, and R_{site} is the number of sites in the southern hemisphere divided by that in the northern hemisphere. For each GOSAT data point, the correction was calculated as

$$X_{ratio,G}^{cor} = X_{ratio,G} / \left(1 + \Delta_{ratio}^{pred} / 100\right), \tag{4}$$

where $X_{ratio,G}^{cor}$ is the corrected ratio component.

4.2. Selecting Explanatory Variables

The correlation between the variables and the correlation between Δ_{ratio} and the variables were evaluated to select the variables for the bias correction (Figure 6). In Figure 6, the data with NDSI ≤ 0.4 were used to calculate the correlation coefficient with respect to f_c. Figure 6a shows that the variables were correlated with each other. In particular, f_c was highly correlated with the other variables, as mentioned in Sections 3.2.2 and 3.2.3. Therefore, the variation of Δ_{ratio} was expected to be explained by the use of fewer than the total number of variables. We confirmed that the correlation between Δ_{ratio} and the variables did not become small by the correction using one variable. Therefore, corrections by multiple linear regression were tested. Figure 6b shows the correlation coefficient between the corrected Δ_{ratio} and the variables for the corrections using different numbers of variables. First, $I_{2\mu m}$ and airmass were used, since Δ_{ratio} varied significantly with these variables (Figures 3 and 4), and the correlation between them was low (Figure 6a). When three variables including R_{CH4} were used, the correlation coefficient became close to zero, except in the case of Δ_{alb}. This can be attributed to the fact that R_{CH4} included information on optical path length modification by scattering by clouds and aerosols. R_{CH4} is considered to be a useful variable for correcting Δ_{ratio}, although the relationship between Δ_{ratio} and R_{CH4} is somewhat empirical, especially when f_c is high.

(a) Correlation coefficient between individual variables

	$I_{2\mu m}$	f_c	ρ_{CH4}	Δ_{alb}	A_m	ΔP_{srf}	R_{CH4}	R_{CO2}
$I_{2\mu m}$		-0.09	-0.01	0.09	-0.05	-0.27	0.22	-0.13
f_c			0.43	0.33	0.30	-0.57	-0.64	-0.71
ρ_{CH4}				0.41	-0.10	-0.21	-0.29	-0.22
Δ_{alb}					0.16	-0.13	-0.17	-0.30
A_m						-0.22	-0.23	-0.24
ΔP_{srf}							0.48	0.71
R_{CH4}								0.72
R_{CO2}								

(b) Correlation coefficient between Δ_{ratio} and each variable

	$I_{2\mu m}$	f_c	ρ_{CH4}	Δ_{alb}	A_m	ΔP_{srf}	R_{CH4}	R_{CO2}
Uncorrected	0.16	-0.19	-0.07	0.07	-0.24	0.15	0.26	0.18
Corrected ($I_{2\mu m}$, A_m)		-0.10	-0.07	0.12		0.13	0.21	0.13
Corrected ($I_{2\mu m}$, A_m, R_{CH4})		0.00	0.01	0.15		0.03		-0.01
Corrected ($I_{2\mu m}$, A_m, R_{CH4}, Δ_{alb})		-0.01	-0.05			0.02		-0.01

Figure 6. Correlation coefficient (**a**) between variables and (**b**) between the relative difference in the ratio component between GOSAT and TCCON (Δ_{ratio}) and variables. The color in the figure corresponds to the coefficient value (blue: large negative coefficient; white: zero; red: large positive coefficient). The meaningless pixels are filled by gray. Eight variables were used: the normalized 2 μm band radiance ($I_{2\mu m}$), cloud fraction (f_c), retrieved surface albedo of CH_4 band (ρ_{CH4}), difference in the retrieved surface albedo (CH_4 band minus CO_2 band, Δ_{alb}), airmass (A_m), deviation of the retrieved clear-sky surface pressure from its prior (ΔP_{srf}), and ratio between the partial column-averaged dry-air mole fractions for the lower atmosphere and that for the upper atmosphere (R_{CH4} and R_{CO2}). In (**b**), results for bias-uncorrected Δ_{ratio} and bias-corrected Δ_{ratio} (corrections using two, three, and four variables) are shown. Data with $f_c = 0$ and that with $I_{2\mu m} \leq 1$ were used. Only data with NDSI ≤ 0.4 were used to calculate the correlation coefficient for f_c.

In addition to the correlation coefficient, the relationship between Δ_{ratio} and the variables was further evaluated. Figure 7 shows the variation of Δ_{ratio} with the variables. The results were separately plotted for the different hemispheres and seasons in order to confirm that the spatiotemporal variation of bias was reduced. Data with NDSI ≤ 0.4 were used to obtain the variation of Δ_{ratio} with f_c. Even when the correction was conducted using $I_{2\mu m}$ and one other variable, the difference in variation of the Δ_{ratio} with $I_{2\mu m}$ between the northern and southern hemispheres remained. Although the number of TCCON sites and the number of data points for the southern hemisphere were smaller than those for the northern hemisphere (larger standard error for the southern hemisphere), the difference in variation of the Δ_{ratio} with $I_{2\mu m}$ between hemispheres was larger than the standard error. The cause of this difference is that the degree of influence of elevated scattering materials on Δ_{ratio} varies according to the vertical profile of CH_4. When R_{CH4} was used in the correction, the difference in variation of the Δ_{ratio} with $I_{2\mu m}$ between the hemispheres was significantly reduced. For f_c, the corrected Δ_{ratio} varies around zero for both hemispheres and both seasons. For other variables, the difference between the hemispheres becomes small. Then, we decided to use the four variables airmass, $I_{2\mu m}$, R_{CH4}, and Δ_{alb} for the bias correction.

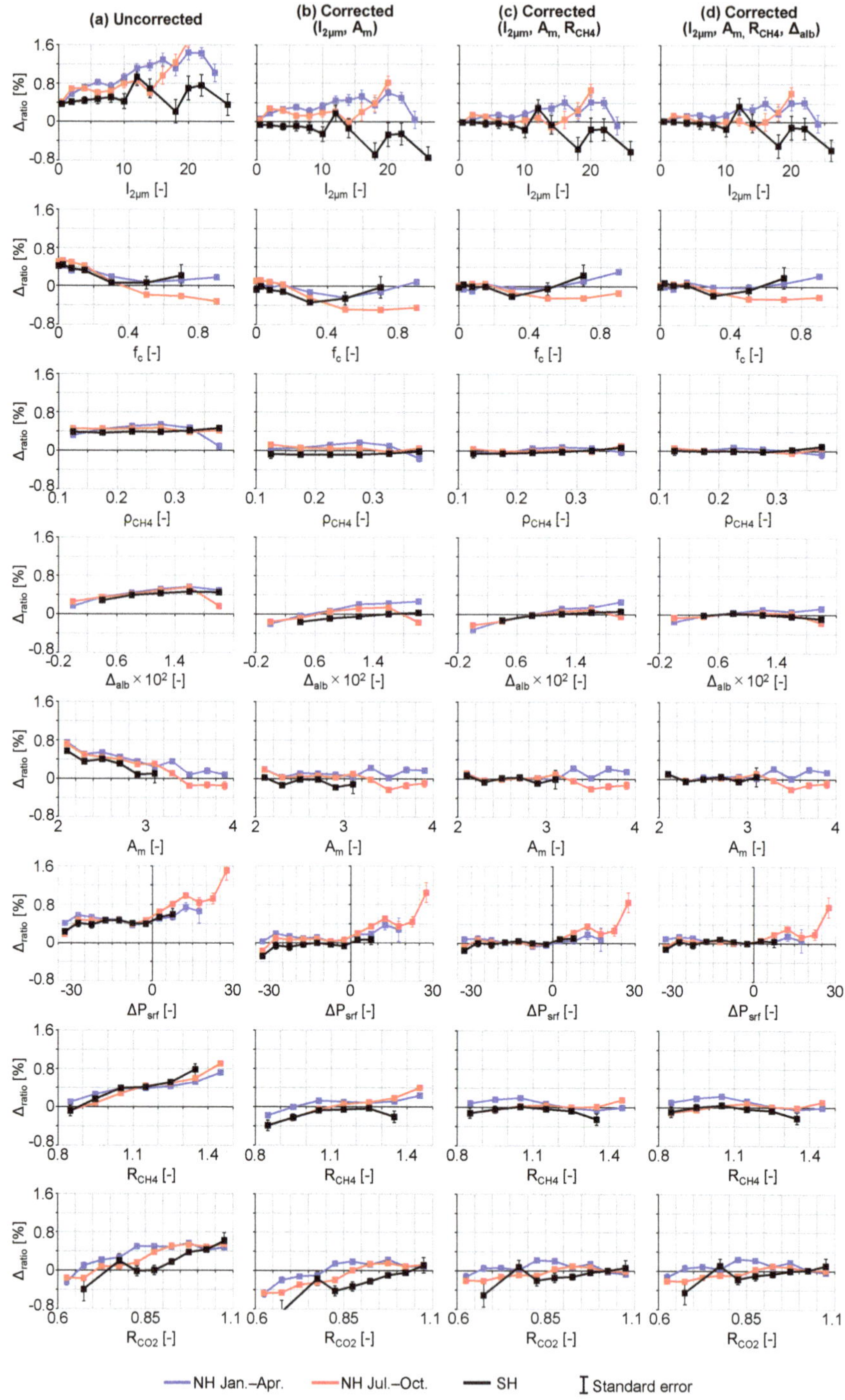

Figure 7. Variation of the relative difference in the ratio component between GOSAT and TCCON (Δ_{ratio}) according to the explanatory variables: normalized 2 μm band radiance ($I_{2\mu m}$); cloud fraction

(f_c); retrieved surface albedo of CH_4 band (ρ_{CH4}); difference in the retrieved surface albedo (CH_4 band minus CO_2 band, Δ_{alb}); airmass (A_m); deviation of the retrieved clear-sky surface pressure from its prior (ΔP_{srf}); and ratio between the partial column-averaged dry-air mole fractions for the lower atmosphere and that for the upper atmosphere (R_{CH4} and R_{CO2}). Results for (**a**) bias-uncorrected Δ_{ratio} and bias-corrected Δ_{ratio} (corrections using (**b**) two, (**c**) three, and (**d**) four variables) are presented. Results for the winter (January to April) northern hemisphere, the summer (July to October) northern hemisphere, and the southern hemisphere are plotted with different colors. Data with $f_c = 0$ and that with $I_{2\mu m} \leq 1$ were used. Only data with NDSI ≤ 0.4 were used to obtain the variation of Δ_{ratio} with f_c.

4.3. Quality Control

As quality control before evaluating the bias-corrected ratio component, cloud screening was investigated. Figure 8 shows the variation of the corrected Δ_{ratio} by the four variables according to f_c and $I_{2\mu m}$. The data with NDSI ≤ 0.4 were used. Although the data with $f_c = 0$ and those with $I_{2\mu m} \leq 1$ were used for the results shown in the previous sections (Figures 3–7), all data were used to generate Figure 8 (similar to Figure 2). Although Δ_{ratio} is close to zero for many cases where f_c is > 0 and $I_{2\mu m}$ is > 1 (i.e., data not used for the regression), the large mean and standard deviation of Δ_{ratio} remains. Therefore, the cloud screening was conducted as follows. We considered a function of f_c as $g(f_c) = (af_c + b)/(cf_c + 1)$, where a, b, and c are coefficients, so that the tolerance range of $I_{2\mu m}$ decreases with the increase in f_c. The coefficients were determined to make the criteria be $I_{2\mu m} \leq 15$ when $f_c = 0$ and $I_{2\mu m} \leq 1$ when $f_c = 1$, and to screen out the data in areas where the mean and/or the standard deviation of Δ_{ratio} were large in Figure 8. The criteria of $I_{2\mu m} \leq 15$ when $f_c = 0$ is based on the result that the difference in Δ_{ratio} between the hemispheres became large with the increase in $I_{2\mu m}$ (Figure 7). Then, we decided to reject the data with $I_{2\mu m} > g(f_c)$, where coefficients a, b, and c were 46, 15, and 60, respectively. The criteria are depicted by the white line in Figure 8.

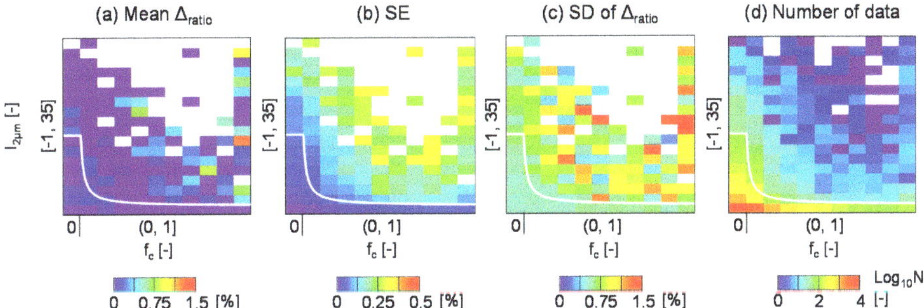

Figure 8. Similar to Figure 2, but for the bias-corrected ratio component: (**a**) mean relative difference in the ratio component between GOSAT and TCCON (Δ_{ratio}); (**b**) standard error (SE); (**c**) standard deviation (SD) of Δ_{ratio}; (**d**) number of data points. The white line indicates the cloud screening criterion.

4.4. Evaluating the Corrected Results

To examine the usefulness and applicability of the bias correction, the matched GOSAT and TCCON data acquired in the even-numbered years were used to obtain the regression coefficients, and then the correction was applied to the GOSAT data acquired in the odd-numbered years. The obtained coefficients for airmass, $I_{2\mu m}$, R_{CH4}, and Δ_{alb} and the intercept were −0.28, 0.019, 1.06, 17.69, and −0.31, respectively (note that the four variables are dimensionless; see Section 2.3 for the definition of the variables). The cloud screening described in Section 4.3 was conducted. Figure 9 shows the comparison of the bias, precision, and interseasonal bias for each TCCON site between the bias-uncorrected and the bias-corrected ratio component. After the bias correction, the interseasonal bias was reduced, and the variation between sites in the northern hemisphere and the difference between hemispheres became small. The precision and intersite bias were also improved. The reduction of precision value was statistically significant ($p < 0.001$ from F test). The reduction of intersite bias value was less significant, but the 75% confidence intervals (CI) showed almost no overlap (0.177–0.234% and 0.143–0.179% for before and after the bias correction, respectively). The CI of intersite bias was estimated by the nonparametric bootstrap method. For each site, a bootstrap sample was taken, and a bootstrap estimate of bias (mean Δ_{ratio} of the sample) was obtained. Then, the intersite bias (standard deviation of the estimated bias values for individual sites) was calculated. This calculation was repeated 2000 times, and the CI was obtained from the 2000 intersite bias values.

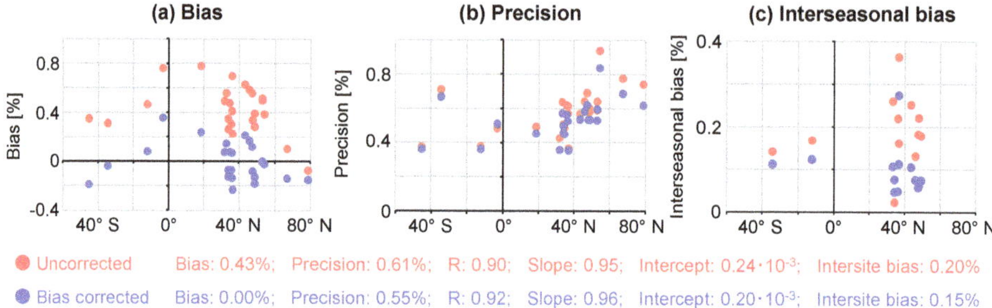

Figure 9. Latitudinal plot of (**a**) the bias, (**b**) precision, and (**c**) interseasonal bias of the uncorrected and the bias-corrected GOSAT ratio component for each TCCON site. Only sites having more than nine data points are shown. The bias, precision, correlation coefficient (R), and the slope and intercept of the linear regression calculated using all data (in a manner similar to Figure 1a) are presented below the graphs.

Figure 10 shows the temporal variation of the uncorrected and corrected Δ_{ratio} for each TCCON site. The results for TCCON sites with long-term observation and a large number of data points are shown. For other sites, it was difficult to discuss the temporal variation of Δ_{ratio}, but no results contradicting the following discussion were obtained. On the whole, seasonality was seen for the uncorrected Δ_{ratio}, but it was significantly reduced by the correction. Δ_{ratio} was small even before the correction for the high-latitude site (Sodankylä [73]). Such a general seasonal and latitudinal pattern of Δ_{ratio} seems to be caused by the airmass dependence of Δ_{ratio} (Figure 4). In addition to this general pattern, the characteristics of each site were observed as described below. For each site, the temporal variation of Δ_{ratio} was determined as follows: the mean Δ_{ratio} was calculated for each season (DJF, MAM, JJA, SON) in each year (four seasons × five years), and then the mean and the standard deviation of the mean Δ_{ratio} values were calculated (Figure 11). For comparison, results for the corrected Δ_{ratio} with two and three variables are also shown.

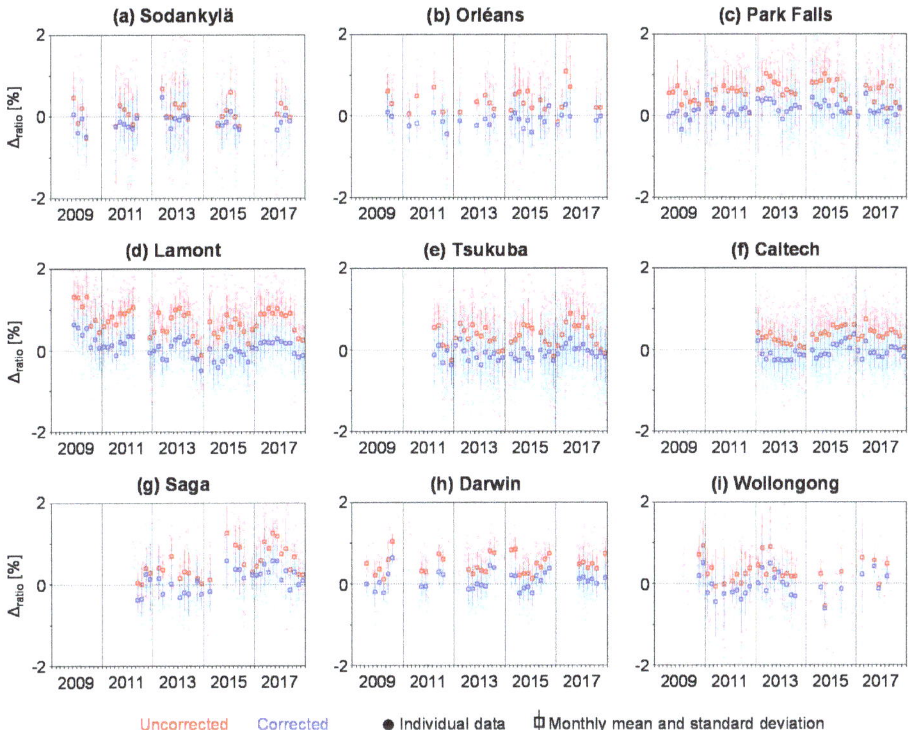

Figure 10. Temporal variation of the relative difference in the ratio component between GOSAT (bias-uncorrected and bias-corrected) and TCCON (Δ_{ratio}) for each TCCON site: (**a**) Sodankylä; (**b**) Orléans; (**c**) Park Falls; (**d**) Lamont; (**e**) Tsukuba; (**f**) Caltech; (**g**) Saga; (**h**) Darwin; (**i**) Wollongong. Individual data and monthly mean are plotted. Results for the data acquired in the odd-numbered years are shown to assess the applicability of the bias correction using the independent data (coefficients for the correction were obtained using the data acquired in the even-numbered years).

(a) Mean Δ_{ratio} [%]

	Uncorrected	Corrected ($I_{2\mu m}$, A_m)	Corrected ($I_{2\mu m}$, A_m, R_{CH4})	Corrected ($I_{2\mu m}$, A_m, R_{CH4}, Δ_{alb})
Sodankylä	0.07	-0.20	-0.15	-0.13
Orléans	0.33	0.02	-0.08	-0.10
Park Falls	0.58	0.25	0.18	0.18
Lamont	0.63	0.21	0.07	-0.01
Tsukuba	0.35	-0.06	-0.15	-0.07
Caltech	0.34	-0.05	-0.16	-0.06
Saga	0.56	0.12	0.07	0.14
Darwin	0.50	-0.01	0.06	0.09
Wollongong	0.27	-0.14	-0.11	-0.09

(b) SD of Δ_{ratio} [%]

	Uncorrected	Corrected ($I_{2\mu m}$, A_m)	Corrected ($I_{2\mu m}$, A_m, R_{CH4})	Corrected ($I_{2\mu m}$, A_m, R_{CH4}, Δ_{alb})
Sodankylä	0.11	0.10	0.07	0.07
Orléans	0.25	0.14	0.11	0.10
Park Falls	0.22	0.16	0.14	0.13
Lamont	0.20	0.16	0.17	0.16
Tsukuba	0.23	0.15	0.11	0.11
Caltech	0.12	0.13	0.13	0.13
Saga	0.32	0.28	0.23	0.23
Darwin	0.19	0.15	0.14	0.16
Wollongong	0.20	0.19	0.19	0.19

Figure 11. (**a**) Mean and (**b**) standard deviation (SD) of the mean values of relative difference in the ratio component between GOSAT and TCCON (Δ_{ratio}) calculated for individual seasons (each season in each year) for each TCCON site. Results for the bias-uncorrected and bias-corrected ratio component (corrections using two, three, and four variables) are presented. The color in the figure corresponds to the Δ_{ratio} value (blue: large negative value; white: zero; red: large positive value) and the SD value (blue: zero; red: large value) for the mean and SD, respectively.

For Sodankylä, the uncorrected Δ_{ratio} was smaller than that for the other sites owing to the large airmass. The low possibility of elevated scattering materials ($I_{2\mu m}$ was small for most of the data) also seems to have contributed to the small Δ_{ratio}. Δ_{ratio} was overcorrected by the correction using two variables. f_c tended to be high for the data over this site, and R_{CH4}^{act} seemed to be large according to the model (prior value). Then, the overcorrection was reduced by using four variables. For Orléans, occasional large $I_{2\mu m}$, in addition to the airmass effect, seemed to cause a large standard deviation; however, the deviation decreased after the correction. For Park Falls, the uncorrected Δ_{ratio} was considered to be affected by many factors. The temporal variation of Δ_{alb} was large, since this site was covered by snow during winter and by vegetation during summer. Large $I_{2\mu m}$ was observed during spring. R_{CO2}^{act} is expected to be small during summer due to photosynthesis by vegetation. R_{CH4}^{act} seemed to be large during summer according to the model (prior value). Correction using R_{CH4} reduced Δ_{ratio}, although a relatively large Δ_{ratio} remained, since the complicated conditions were not perfectly accounted for by the correction. For Lamont, large Δ_{alb} generally brought large Δ_{ratio}. The correction using Δ_{alb} reduced Δ_{ratio}. A large uncorrected Δ_{ratio} was occasionally seen in the early months of the year (Figure 10). We found that the large Δ_{ratio} corresponded to the large $I_{2\mu m}$ in February 2013 and 2017. According to Figure 10, these large errors were properly corrected. The use of $I_{2\mu m}$ in the correction had only a small effect on the averaged results (Figure 11), since $I_{2\mu m}$ was not always large; however, the effect could be confirmed for individual cases. For Tsukuba, a large $I_{2\mu m}$ was observed during spring to summer, which enhanced the seasonality of the uncorrected Δ_{ratio}. Δ_{alb} of this site was small, and thus the overcorrection was reduced by using Δ_{alb} in the correction. For Caltech, Δ_{alb} was small, and the temporal variations of R_{CH4}^{act} and R_{CO2}^{act} were small according to the model. The difference between R_{CH4}^{act} and R_{CO2}^{act} was also small. Then, the uncorrected Δ_{ratio} showed small temporal variation, and the corrected Δ_{ratio} varied around zero with the small temporal variation retained. For Saga, the airmass effect and the large $I_{2\mu m}$ during spring to summer enhanced the seasonality, and Δ_{ratio} was significantly reduced by the correction using two variables. The temporal variation of R_{CH4}^{act} was expected to be large according to the model. Thus, using R_{CH4} in the correction improved both the mean and standard deviation (Figure 11). For the sites in the southern hemisphere (Darwin and Wollongong), although R_{CH4}^{act} and R_{CO2}^{act} were expected to be small, the amplitude of temporal variation of uncorrected Δ_{ratio} was comparable to that for the sites in the northern hemisphere (Figure 10). For Darwin, the uncorrected Δ_{ratio} showed large temporal variation, although the variation of airmass was small. For Wollongong, occasional large uncorrected Δ_{ratio} did not correspond to the large $I_{2\mu m}$. The standard deviation was reduced only slightly by the correction for these sites (Figure 11). Although using R_{CH4} in the correction reduced the difference between the northern and southern hemispheres (Figure 7), unaccounted factors might remain. On the whole, the optimal variables differ between sites; but the correction using four variables brought the best result in terms of the balance between sites (the mean of absolute Δ_{ratio} values for individual sites was close to zero and the variation between sites was small).

We also conducted a bias correction for the data acquired with gain M using the four variables and the above-mentioned coefficients (i.e., the coefficients were obtained using the gain H data). The bias correction functioned properly for the gain M data, although the bias was small even for the uncorrected data (Appendix B). This result supports the applicability of the bias correction.

5. Conclusions

The relationship between the bias in the GOSAT ratio component and the variables derived from GOSAT data was investigated, and the bias correction and its evaluation were performed. The bias between the uncorrected GOSAT ratio component and the TCCON ratio component was 0.43% when the cloud-free condition was expected (normalized 2 μm band radiance ($I_{2\mu m}$) ≤ 1 and cloud fraction (f_c) = 0). The bias increased with the increase in $I_{2\mu m}$ and reached 1.5% when $I_{2\mu m}$ was 30. The variation of bias with f_c was small when f_c was small or large (the bias was 0.4% to 0.5% for the data with $f_c \leq 0.2$ and -0.1% to 0% for the data with $f_c \geq 0.4$). The variation of bias according to $I_{2\mu m}$ and f_c

could be interpreted based on the difference in the detection target between $I_{2\mu m}$ and f_c, the difference in surface albedo between the CH_4 and CO_2 bands (Δ_{alb}), and the vertical profile of CH_4 and CO_2. The relationship between the bias and other related variables was also investigated. We used the retrieved profile and the ratios between the upper and lower atmosphere CH_4 and CO_2 (R_{CH4} and R_{CO2}), in addition to the variables that have been used in the bias correction for the full-physics method (airmass) and in the cloud screening (deviation of the retrieved clear-sky surface pressure from its prior (ΔP_{srf})) and that have been revealed to be a large error source for the proxy method (Δ_{alb}). The bias showed clear variations with the variables except for R_{CO2}.

Then, airmass, $I_{2\mu m}$, R_{CH4}, and Δ_{alb} were selected as explanatory variables for a linear regression of the bias by considering the correlation between the variables and the correlation between the variables and the bias. Using R_{CH4} in the correction reduced the dependence of the bias on f_c and ΔP_{srf}. The difference in bias between the northern and the southern hemispheres was also reduced. These results are attributed to the information on the CH_4 vertical profile and the effect of atmospheric scattering included in R_{CH4}. Although the relationship between bias and R_{CH4} is somewhat empirical, R_{CH4} is an important variable for correcting the bias in the ratio component. Before evaluating the corrected results, cloud screening was applied. The criteria were determined as the threshold value of $I_{2\mu m}$ decreases with the increase in f_c ($I_{2\mu m} \leq 15$ when $f_c = 0$ and $I_{2\mu m} \leq 1$ when $f_c = 1$), by investigating the variation of bias in the corrected ratio component with f_c and $I_{2\mu m}$. Although f_c and $I_{2\mu m}$ have been used for the cloud screening of the GOSAT proxy retrievals, our results give a quantitative basis for the screening.

Comparison between the corrected ratio component and TCCON showed that the precision (standard deviation of the difference between GOSAT and TCCON) was reduced from 0.61% to 0.55%, and the intersite bias was reduced from 0.20% to 0.15%. The temporal variation of bias was further investigated for the sites having a long-term record and a large amount of data. The uncorrected bias showed a seasonality with a large bias in summer. The difference in monthly mean bias between summer and winter exceeded 1% for several sites. In addition to such seasonality, the months with a large mean bias corresponded to the months with a large mean $I_{2\mu m}$. The temporal variation of bias was significantly reduced by the correction. Although the optimal variables differed between sites, the mean and standard deviation of the mean bias values for individual seasons (each season in each year) were within 0.2% for most of the sites, when the four variables were used for the correction.

The bias-corrected and cloud-screened ratio component data in the present study reduce the concern that the residual errors related to the atmospheric scattering and the property of ground surfaces affect the inverse modeling of CH_4 sources and sinks. In future work, the impact of utilizing the bias-corrected data in the inverse modeling will be investigated. GOSAT has been operating for more than 10 years. GOSAT-2, a successor mission to the GOSAT, was launched on 29 October 2018. Providing long-term, consistent, and high-quality XCH_4 data set by the GOSAT series is expected to contribute to the studies on CH_4 budgets over the globe. To construct such a data set, GOSAT and GOSAT-2 data will be continuously compared with each other and with the data from other satellites and TCCON.

Supplementary Materials: The following are available online at http://www.mdpi.com/2072-4292/12/19/3155/s1, Figure S1: Relationship among the Normalized Difference Snow Index (NDSI), relative difference in the ratio component between GOSAT and TCCON (Δ_{ratio}), and month of observation. Figure S2: Averaging kernel of the GOSAT clear-sky retrieval for (a) XCH_4 and (b) XCO_2 (monthly mean values of GOSAT data matched the TCCON data). Figure S3: Similar to Figure 1, but for the data with TCCON Fractional Variation in Solar Intensity (FVSI) $\leq 1\%$ and the data with FVSI $> 1\%$. Figure S4: Relationship between the relative difference in the ratio component between GOSAT and TCCON (Δ_{ratio}) and the TCCON Fractional Variation in Solar Intensity (FVSI). Figure S5: Similar to Figure 5, but using GOSAT data with DFS for $XCO_2 \geq 1.3$.

Author Contributions: Conceptualization, Y.Y.; methodology, H.O. (Haruki Oshio) and Y.Y.; software, H.O. (Haruki Oshio) and Y.Y.; validation, H.O. (Haruki Oshio); formal analysis, H.O. (Haruki Oshio) and Y.Y.; investigation, H.O. (Haruki Oshio); resources, T.M., N.M.D., M.D., D.W.T.G., F.H., L.T.I., R.K., C.L., I.M., J.N., Y.-S.O., H.O. (Hirofumi Ohyama), C.P., D.F.P., C.R., K.S., R.S., Y.T., V.A.V., T.W., and D.W.; data curation, Y.Y. and T.M.; writing—original draft preparation, H.O. (Haruki Oshio); writing—review and editing, all; visualization,

H.O. (Haruki Oshio); supervision, Y.Y.; project administration, T.M.; funding acquisition, T.M. All authors have read and agreed to the published version of the manuscript.

Funding: The Garmisch TCCON station has been supported by the European Space Agency (ESA) under grant 4000120088/17/I-EF and by the German Bundesministerium für Wirtschaft und Energie (BMWi) via the Deutsche Zentrum für Luft- und Raumfahrt (DLR) under grant 50EE1711D. TCCON stations at Tsukuba, Rikubetsu and Burgos are supported in part by the GOSAT series project. Burgos is also partially supported by the Energy Development Corp, Philippines. Darwin and Wollongong TCCON sites and NMD are supported by Australian Research Council funding via FT180100327, DP160101598, DP140101552, DP110103118, DP0879468 and LE0668470 and NASA grants NAG5-12247 and NNG05-GD07G and the University of Wollongong. The Paris TCCON site has received funding from Sorbonne Université, the French research center CNRS, the French space agency CNES, and Région Île-de-France.

Acknowledgments: GOSAT clear-sky retrieval in the present study was conducted at the GOSAT-2 Research Computation Facility.

Conflicts of Interest: The authors declare no conflict of interest.

Appendix A. TCCON Sites and the Influence of the Matching Criteria of GOSAT and TCCON

We used TCCON data from 26 sites [41–67]. Figure A1 shows the map of TCCON sites, and Table A1 shows an overview of the sites. Basically, the sites are located in areas with relatively uniform surface properties and are reasonably far from anthropogenic sources. However, several sites are in areas with nonflat topography or are located near or in urban areas. Such characteristics are summarized in the rightmost column of Table A1.

In the main text, GOSAT data within a ±2° latitude/longitude box centered at each TCCON site were used for the comparison. The influence of the matching criteria on the analysis was assessed. Table A2 shows the mean and standard deviation of Δ_{ratio} for each TCCON site. Results for the different matching criteria are tabulated. The mean value showed almost no clear trend, but the deviation value decreased as the matching criteria were tightened. Figure A2 shows the relationship between Δ_{ratio} and the related variables for the matching criteria of ±2° and that of ±0.5°. Although more deviated values are seen for the matching criteria of ±2°, the regression lines were similar between the criteria. The mean Δ_{ratio} values for the bins also show similar trends between the criteria (Figure A3). The skewness around 0 indicates the small asymmetry for the distribution of Δ_{ratio} values for each bin. Large negative kurtosis values were hardly seen, meaning there was no high frequency for either side (large and small Δ_{ratio}).

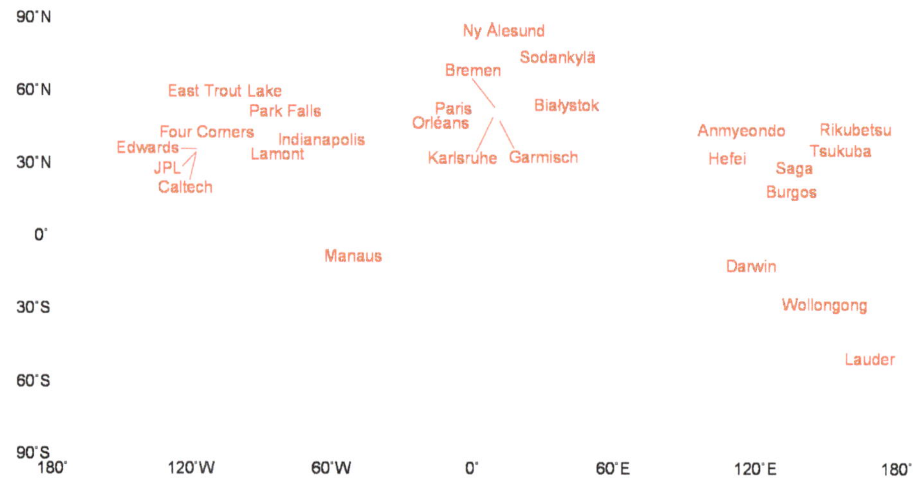

Figure A1. Map of the TCCON sites used.

Table A1. Overview of the TCCON sites used.

Site	Country	Latitude (deg.)	Longitude (deg.)	Altitude(km)	Observation Period of the Data Used	Specific Characteristics
Ny Ålesund	Spitzbergen, Norway	78.92N	11.92E	0.02	6 April 2014–31 December 2018	Useful data are not obtained during winter due to the high solar zenith angle.
Sodankylä	Finland	67.37N	26.63E	0.19	16 May 2009–31 December 2018	Useful data are not obtained during winter due to the high solar zenith angle.
East Trout Lake	Canada	54.35N	104.99W	0.50	7 October 2016–31 December 2018	
Bialystok	Poland	53.23N	23.03E	0.18	23 April 2009–1 October 2018	
Bremen	Germany	53.10N	8.85E	0.03	22 January 2010–31 December 2018	The site is in the middle-sized city (population ~550 000).
Karlsruhe	Germany	49.10N	8.44E	0.12	19 April 2010–31 December 2018	The site is near the middle-sized city (population ~300 000).
Paris	France	48.85N	2.36E	0.06	23 September 2014–31 December 2018	The site is in the large city (population ~2.15 million).
Orléans	France	47.97N	2.11E	0.13	29 August 2009–31 December 2018	
Garmisch	Germany	47.48N	11.06E	0.74	23 April 2009–31 December 2018	
Park Falls	USA	45.95N	90.27W	0.44	23 April 2009–31 December 2018	
Rikubetsu	Japan	43.46N	143.77E	0.38	16 November 2013–31 December 2018	
Indianapolis	USA	39.86N	86.00W	0.27	23 August 2012–1 December 2012	The site is in the suburban area of a middle-sized city (population ~880 000).
Four Corners	USA	36.80N	108.48W	1.64	16 March 2013–4 October 2013	The site is far from the city area but observes plant plumes and methane from mine shafts and fugitive leaks [74,75]. The observed total column peaks in the late morning.
Lamont	USA	36.60N	97.49W	0.32	23 April 2009–31 December 2018	
Anmyeondo	Korea	36.54N	126.33E	0.03	2 February 2015–18 April 2018	The site is located on the west coast of the Korean Peninsula.
Tsukuba	Japan	36.05N	140.12E	0.03	4 August 2011–31 December 2018	The site is in the middle-sized city (population ~240 000).
Edwards	USA	34.96N	117.88W	0.70	20 July 2013–31 December 2018	The site is adjacent to a very bright playa.
JPL	USA	34.20N	118.18W	0.39	19 May 2011–14 May 2018	The site is near the large city (population ~17 million).
Caltech	USA	34.14N	118.13W	0.23	20 September 2012–31 December 2018	The site is near the large city (population ~17 million).

Table A1. *Cont.*

Site	Country	Latitude (deg.)	Longitude (deg.)	Altitude(km)	Observation Period of the Data Used	Specific Characteristics
Saga	Japan	33.24N	130.29E	0.01	28 July 2011– 31 December 2018	The site is in the middle-sized city (population ~230 000).
Hefei	China	31.91N	117.17E	0.03	18 September 2015– 31 December 2016	The site is near the large city (population ~5 million).
Burgos	Philippines	18.53N	120.65E	0.04	3 March 2017– 31 December 2018	The site is at the northernmost point of Luzon Island in the Philippines.
Manaus	Brazil	3.21S	60.60W	0.05	1 October 2014– 24 June 2015	
Darwin	Australia	12.42S	130.89E	0.03	23 April 2009– 31 December 2018	
Lauder	New Zealand	45.04S	169.68E	0.37	23 April 2009– 31 December 2018	The site is in the midst of rolling hills.
Wollongong	Australia	34.41S	150.88E	0.03	23 April 2009– 31 December 2018	The site is between the ocean and a sharp escarpment.

Table A2. Mean (μ) and standard deviation (SD) of the relative difference in the ratio component between GOSAT and TCCON and the number of matched GOSAT and TCCON data points (N) for each TCCON site. Results for four different matching criteria are shown: GOSAT data within $\pm 2°$, $\pm 1°$, $\pm 0.5°$, and $\pm 0.1°$ latitude/longitude boxes centered at each TCCON site. Only sites having more than 29 data points are shown. The GOSAT data obtained under conditions where cloud-free scenes were expected (cloud fraction = 0 and normalized 2 μm band radiance ≤ 1) were used. Intersite bias is calculated using the same sites as in the case of matching criteria of $\pm 0.1°$.

	$\pm 2°$			$\pm 1°$			$\pm 0.5°$			$\pm 0.1°$		
Site	μ (%)	SD (%)	N	μ (%)	SD (%)	N	μ (%)	SD (%)	N	μ (%)	SD (%)	N
Sodankylä	0.43	0.74	328	0.26	0.85	88						
East Trout Lake	0.57	0.68	40									
Bialystok	0.50	0.56	263	0.46	0.51	87	0.53	0.45	50			
Bremen	0.51	0.54	81									
Karlsruhe	0.22	0.62	267	0.19	0.57	102						
Paris	0.43	0.52	102	0.50	0.50	59						
Orléans	0.35	0.51	518	0.35	0.51	224	0.43	0.52	75			
Garmisch	0.55	0.61	286	0.53	0.56	184	0.54	0.52	56			
Park Falls	0.61	0.55	1017	0.61	0.55	728	0.59	0.55	575	0.60	0.55	536
Rikubetsu	0.63	0.46	135	0.73	0.44	56						
Indianapolis	0.44	0.64	67									
Lamont	0.59	0.54	1949	0.59	0.51	855	0.61	0.45	461	0.62	0.45	305
Anmeyondo	0.51	0.45	50	0.52	0.37	33						
Tsukuba	0.31	0.61	1407	0.31	0.59	1186	0.28	0.60	675	0.41	0.49	179
Edwards	0.43	0.45	126	0.60	0.45	59						
JPL	0.26	0.49	609	0.27	0.43	490	0.24	0.39	438	0.25	0.38	103
Caltech	0.32	0.43	2067	0.32	0.43	1924	0.32	0.43	1412	0.31	0.43	853
Saga	0.59	0.59	438	0.58	0.55	288	0.65	0.53	222	0.72	0.52	66
Darwin	0.37	0.36	555	0.38	0.36	500	0.56	0.36	46	0.48	0.34	30
Wollongong	0.42	0.71	349	0.31	0.58	195	0.34	0.61	100			
Lauder	0.36	0.35	124	0.35	0.34	121	0.37	0.31	92	0.36	0.30	85
All sites	0.43	0.55	10846	0.41	0.52	7263	0.40	0.51	4357	0.45	0.49	2249
Intersite bias	0.14			0.13			0.16			0.15		

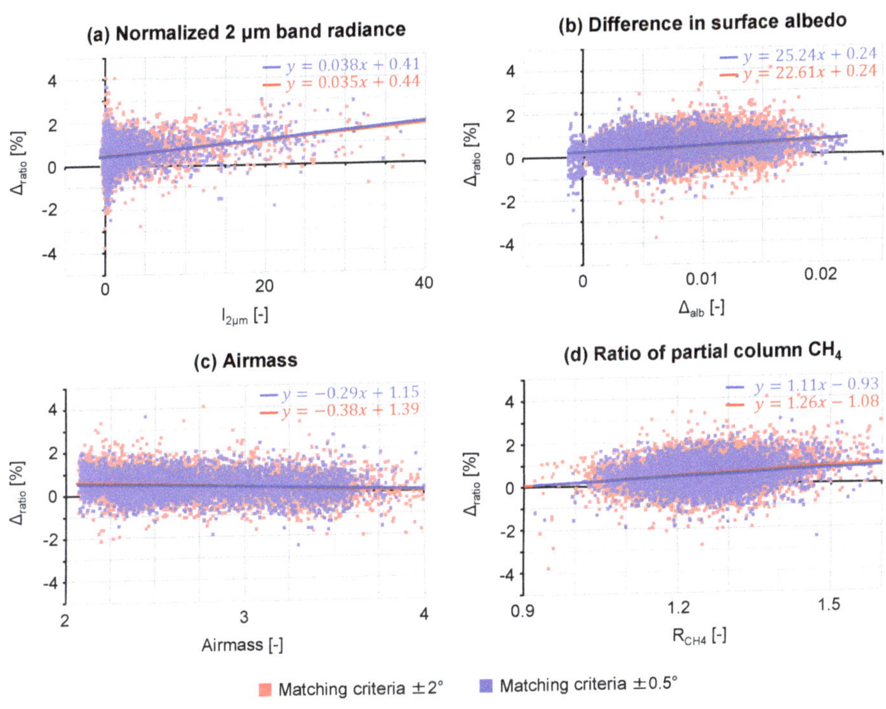

Figure A2. Relationship between the relative difference in the ratio component between GOSAT and TCCON (Δ_{ratio}) and the related variable: (**a**) normalized 2μm band radiance ($I_{2\mu m}$); (**b**) difference in retrieved surface albedo (CH_4 band minus CO_2 band, Δ_{alb}); (**c**) airmass; (**d**) ratio between the partial column-averaged dry-air mole fractions for the lower atmosphere and that for the upper atmosphere (R_{CH4}). Results for the different matching criteria are plotted (GOSAT data within ±2° and ±0.5° latitude/longitude boxes centered at each TCCON site). Linear regressions are depicted by solid lines. For $I_{2\mu m}$ and R_{CH4}, the data with cloud fraction = 0 were used. For Δ_{alb} and airmass, the data with cloud fraction = 0 and $I_{2\mu m} \leq 1$ were used.

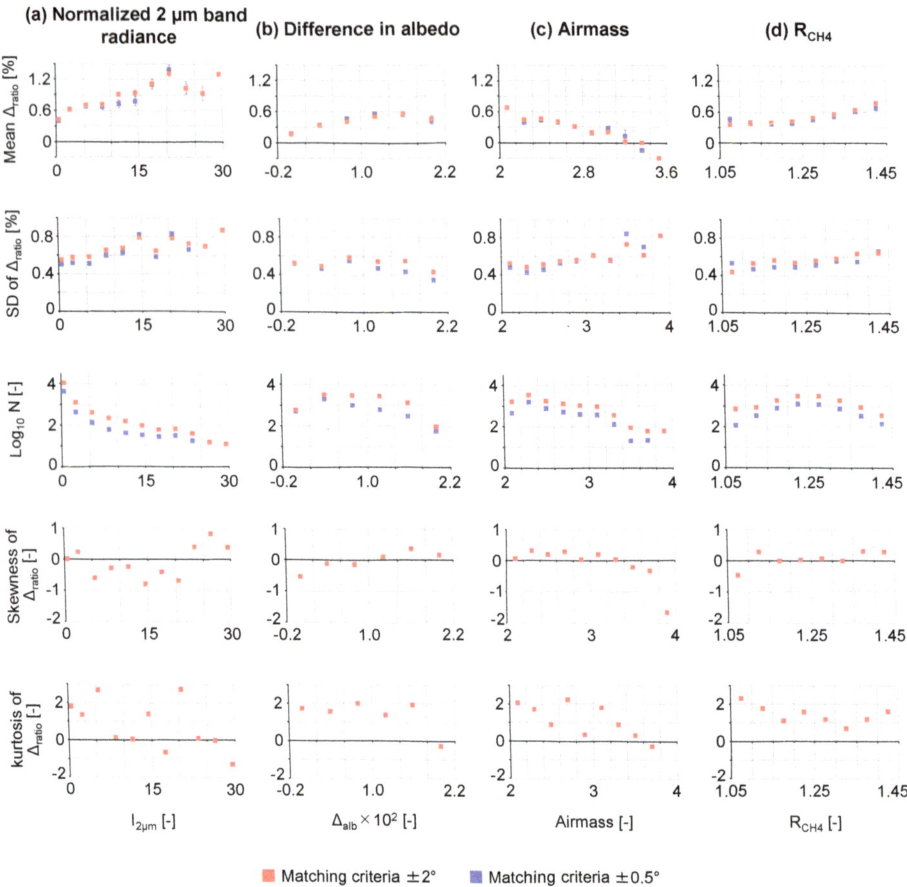

Figure A3. Relationship between the relative difference in the ratio component between GOSAT and TCCON (Δ_{ratio}) and the related variables. The variables (**a–d**) are similar to those in Figure A2, but in each panel, the mean and standard deviation (SD) of Δ_{ratio}, number of data points (N), skewness, and kurtosis for each bin are shown. Results for the different matching criteria are plotted (GOSAT data within ±2° and ±0.5° latitude/longitude boxes centered at each TCCON site). For $I_{2\mu m}$ and R_{CH4}, the data with cloud fraction = 0 were used. For Δ_{alb} and airmass, the data with cloud fraction = 0 and $I_{2\mu m} \leq 1$ were used.

Appendix B. Bias Correction for the Data Acquired by Gain M

The different gain settings of the TANSO-FTS (H and M) only affect the spectral radiance of the Band 1 [76]. Therefore, the ratio component should show similar characteristics between the data acquired with gain H and that acquired with gain M. Then, the bias correction equation established using the gain H data (four variables with the coefficients shown in Section 4.4) was applied to the gain M data. Figure A4 shows the scatterplot of the ratio component derived from GOSAT gain M data and that derived from TCCON. The results for each site are shown in Table A3. The bias-uncorrected and bias-corrected results are presented in the left and right sides in Figure A4 and Table A3, respectively. Figure A5 shows the dependence of Δ_{ratio} on the related variables for the uncorrected and the corrected data. The matching criteria were similar to those in the main text. Only GOSAT data over the areas east of 118°W were used to reject the data over Bakersfield where there are oil fields. According to

.e bias was smaller than that for the gain H data for the uncorrected data, and the bias gntly overcorrected. Table A3 shows that the overcorrection was seen for the data of JPL and .utech. This is reasonable for the following reasons. Almost all the GOSAT data acquired with gain M in this region were located around the Edwards site. For the time period during which TCCON observation was not being performed at the Edwards site, the GOSAT data were matched with the TCCON data at JPL or Caltech. The Edwards site is located in the north of San Gabriel Mountains, while the other two sites are located in the south of the mountains. Therefore, it was possible that the air at the other two sites was occasionally different from the air at the Edwards site. Then, the results for the Edwards site (bias of −0.07% after the correction) indicate that the GOSAT data can be used for the proxy retrieval without exercising caution with respect to the gain settings and support the applicability of the bias correction.

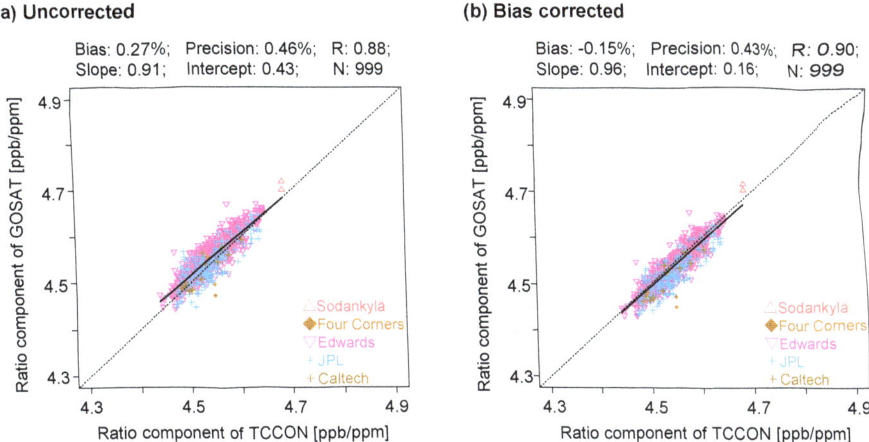

Figure A4. Similar to Figure 1a but for GOSAT data acquired with gain M: (**a**) bias-uncorrected; (**b**) bias-corrected.

Table A3. The relative difference in the ratio component between GOSAT (gain M) and TCCON (Δ_{ratio}) for each site: mean (μ), standard deviation (SD), and interseasonal bias (S_{SD}) of Δ_{ratio} and the number of matched GOSAT and TCCON data points (N) are shown.

Site	Uncorrected				Bias-Corrected			
	μ (%)	SD (%)	S_{SD} (%)	N	μ (%)	SD (%)	S_{SD} (%)	N
Edwards	0.35	0.41	0.11	766	−0.07	0.39	0.04	766
JPL	0.01	0.50	0.11	214	−0.39	0.46	0.09	214
Caltech	0.04	0.35		15	−0.40	0.34		15

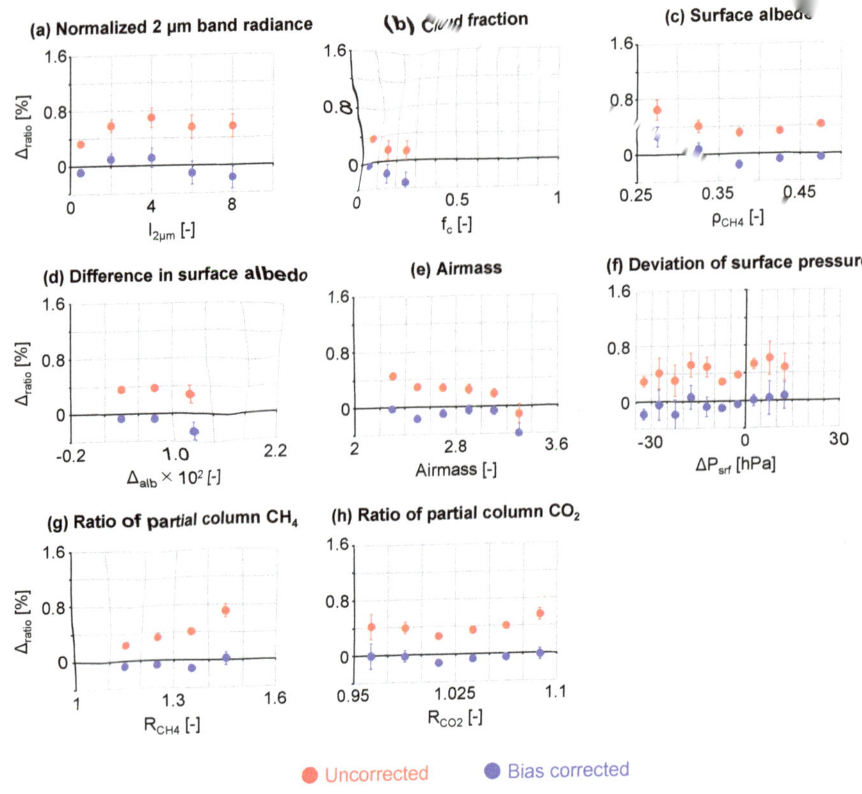

Figure A5. Relationship between the relative difference in the ratio component between GOSAT and TCCON (Δ_{ratio}) and the related variables for the GOSAT data acquired with gain M: (**a**) normalized 2 μm band radiance ($I_{2\mu m}$); (**b**) cloud fraction (f_c); (**c**) retrieved surface albedo of CH_4 band (ρ_{CH4}); (**d**) difference in the retrieved surface albedo (CH_4 band minus CO_2 band, Δ_{alb}); (**e**) airmass (A_m); (**f**) deviation of the retrieved clear-sky surface pressure from its prior (ΔP_{srf}); (**g**,**h**) ratio between the partial column-averaged dry-air mole fractions for the lower atmosphere and that for the upper atmosphere (R_{CH4} and R_{CO2}). Mean Δ_{ratio} for each bin is plotted for the bias-uncorrected and bias-corrected data.

References

1. Buchwitz, M.; de Beek, R.; Burrows, J.P.; Bovensmann, H.; Warneke, T.; Notholt, J.; Meirink, J.F.; Goede, A.P.H.; Bergamaschi, P.; Körner, S.; et al. Atmospheric methane and carbon dioxide from SCIAMACHY satellite data: Initial comparison with chemistry and transport models. *Atmos. Chem. Phys.* **2005**, *5*, 941–962. [CrossRef]

2. Frankenberg, C.; Meirink, J.F.; van Weele, M.; Platt, U.; Wagner, T. Assessing methane emissions from global space-borne observations. *Science* **2005**, *308*, 1010–1014. [CrossRef] [PubMed]

3. Parker, R.; Boesch, H.; Cogan, A.; Fraser, A.; Feng, L.; Palmer, P.I.; Messerschmidt, J.; Deutscher, N.; Griffith, D.W.T.; Notholt, J.; et al. Methane observations from the Greenhouse Gases Observing SATellite: Comparison to ground-based TCCON data and model calculations. *Geophys. Res. Lett.* **2011**, *38*, L15807. [CrossRef]

4. Butz, A.; Guerlet, S.; Hasekamp, O.; Schepers, D.; Galli, A.; Aben, I.; Frankenberg, C.; Hartmann, J.-M.; Tran, H.; Kuze, A.; et al. Toward accurate CO_2 and CH_4 observations from GOSAT. *Geophys. Res. Lett.* **2011**, *38*, L14812. [CrossRef]

shida, Y.; Ota, Y.; Eguchi, N.; Kikuchi, N.; Nobuta, K.; Tran, H.; Morino, I.; Yokota, T. Retrieval algorithm for CO_2 and CH_4 column abundances from short-wavelength infrared spectral observations by the Greenhouse gases observing satellite. *Atmos. Meas. Tech.* **2011**, *4*, 717–734. [CrossRef]

6. Hu, H.; Landgraf, J.; Detmers, R.; Borsdorff, T.; de Brugh, J.A.; Aben, I.; Butz, A.; Hasekamp, O. Toward global mapping of methane with TROPOMI: First results and intersatellite comparison to GOSAT. *Geophys. Res. Lett.* **2018**, *45*, 3682–3689. [CrossRef]

7. Bergamaschi, P.; Frankenberg, C.; Meirink, J.F.; Krol, M.; Dentener, F.; Wagner, T.; Platt, U.; Kaplan, J.O.; Körner, S.; Heimann, M.; et al. Satellite chartography of atmospheric CH_4 from SCIAMACHY on board ENVISAT: 2. Evaluation based on inverse model simulations. *J. Geophys. Res.-Atmos.* **2007**, *112*, D02304. [CrossRef]

8. Monteil, G.; Houweling, S.; Butz, A.; Guerlet, S.; Schepers, D.; Hasekamp, O.; Frankenberg, C.; Scheepmaker, R.; Aben, I.; Röckmann, T. Comparison of CH_4 inversions based on 15 months of GOSAT and SCIAMACHY observations. *J. Geophys. Res.-Atmos.* **2013**, *118*, 11807–11823. [CrossRef]

9. Cressot, C.; Chevallier, F.; Bousquet, P.; Crevoisier, C.; Dlugokencky, E.J.; Fortems-Cheiney, A.; Frankenberg, C.; Parker, R.; Pison, I.; Scheepmaker, R.A.; et al. On the consistency between global and regional methane emissions inferred from SCIAMACHY, TANSO-FTS, IASI and surface measurements. *Atmos. Chem. Phys.* **2014**, *14*, 577–592. [CrossRef]

10. Alexe, M.; Bergamaschi, P.; Segers, A.; Detmers, R.; Butz, A.; Hasekamp, O.; Guerlet, S.; Parker, R.; Boesch, H.; Frankenberg, C.; et al. Inverse modelling of CH_4 emissions for 2010–2011 using different satellite retrieval products from GOSAT and SCIAMACHY. *Atmos. Chem. Phys.* **2015**, *15*, 113–133. [CrossRef]

11. Turner, A.J.; Jacob, D.J.; Wecht, K.J.; Maasakkers, J.D.; Lundgren, E.; Andrews, A.E.; Biraud, S.C.; Boesch, H.; Bowman, K.W.; Deutscher, N.M.; et al. Estimating global and North American methane emissions with high spatial resolution using GOSAT satellite data. *Atmos. Chem. Phys.* **2015**, *15*, 7049–7069. [CrossRef]

12. Meirink, J.F.; Eskes, H.J.; Goede, A.P.H. Sensitivity analysis of methane emissions derived from SCIAMACHY observations through inverse modelling. *Atmos. Chem. Phys.* **2006**, *6*, 1275–1292. [CrossRef]

13. GHG-CCI project team. User Requirements Document (URD) for the Essential Climate Variable (ECV) Greenhouse Gases (GHG), Version 2.1 (URDv2.1). 2016. Available online: http://cci.esa.int/sites/default/files/URDv2.1_GHG-CCI_Final.pdf (accessed on 15 May 2020).

14. O'Brien, D.M.; Rayner, P.J. Global observations of the carbon budget 2. CO_2 column from differential absorption of reflected sunlight in the 1.61 μm band of CO_2. *J. Geophys. Res.-Atmos.* **2002**, *107*, 4354.

15. Kuang, Z.; Margolis, J.; Toon, G.; Crisp, D.; Yung, Y. Spaceborne measurements of atmospheric CO_2 by high-resolution NIR spectrometry of reflected sunlight: An introductory study. *Geophys. Res. Lett.* **2002**, *29*, 1716. [CrossRef]

16. Dufour, E.; Bréon, F.-M. Spaceborne estimate of atmospheric CO_2 column by use of the differential absorption method: Error analysis. *Appl. Optics* **2003**, *42*, 3595–3609. [CrossRef]

17. Connor, B.J.; Boesch, H.; Toon, G.; Sen, B.; Miller, C.; Crisp, D. Orbiting Carbon Observatory: Inverse method and prospective error analysis. *J. Geophys. Res.-Atmos.* **2008**, *113*, D05305. [CrossRef]

18. Butz, A.; Hasekamp, O.P.; Frankenberg, C.; Aben, I. Retrievals of atmospheric CO_2 from simulated space-borne measurements of backscattered near-infrared sunlight: Accounting for aerosol effects. *Appl. Opt.* **2009**, *48*, 3322–3336. [CrossRef]

19. Schepers, D.; Guerlet, S.; Butz, A.; Landgraf, J.; Frankenberg, C.; Hasekamp, O.; Blavier, J.-F.; Deutscher, N.M.; Griffith, D.W.T.; Hase, F.; et al. Methane retrievals from Greenhouse Gases Observing Satellite (GOSAT) shortwave infrared measurements: Performance comparison of proxy and physics retrieval algorithms. *J. Geophys. Res.-Atmos.* **2012**, *117*, D10307. [CrossRef]

20. Cogan, A.J.; Boesch, H.; Parker, R.J.; Feng, L.; Palmer, P.I.; Blavier, J.-F.L.; Deutscher, N.M.; Macatangay, R.; Notholt, J.; Roehl, C.; et al. Atmospheric carbon dioxide retrieved from the Greenhouse gases Observing SATellite (GOSAT): Comparison with ground-based TCCON observations and GEOS-Chem model calculations. *J. Geophys. Res.-Atmos.* **2012**, *117*, D21301. [CrossRef]

21. Yoshida, Y.; Kikuchi, N.; Morino, I.; Uchino, O.; Oshchepkov, S.; Bril, A.; Saeki, T.; Schutgens, N.; Toon, G.C.; Wunch, D.; et al. Improvement of the retrieval algorithm for GOSAT SWIR XCO2 and XCH4 and their validation using TCCON data. *Atmos. Meas. Tech.* **2013**, *6*, 1533–1547. [CrossRef]

22. O'Dell, C.W.; Eldering, A.; Wennberg, P.O.; Crisp, D.; Gunson, M.R.; Fisher, B.; Frankenberg, C.; Lindqvist, H.; Mandrake, L.; et al. Improved retrievals of carbon dioxide from Orbiting Carbon Observatory with the version 8 ACOS algorithm. *Atmos. Meas. Tech.* **2018**, *11*, 6539–6576. [CrossRef]

23. Butz, A.; Hasekamp, O.P.; Frankenberg, C.; Vidot, J.; Aben, I. CH_4 retrievals from space-based solar backscatter measurements: Performance evaluation against simulated aerosol and cirrus loaded scenes. *J. Geophys. Res.-Atmos.* **2010**, *115*, D24302. [CrossRef]

24. Parker, R.J.; Boesch, H.; Byckling, K.; Webb, A.J.; Palmer, P.I.; Feng, L.; Bergamaschi, P.; Chevallier, F.; Notholt, J.; Deutscher, N.; et al. Assessing 5 years of GOSAT Proxy XCH_4 data and associated uncertainties. *Atmos. Meas. Tech.* **2015**, *8*, 4785–4801. [CrossRef]

25. Fraser, A.; Palmer, P.I.; Feng, L.; Bösch, H.; Parker, R.; Dlugokencky, E.J.; Krummel, P.B.; Langenfelds, R.L. Estimating regional fluxes of CO_2 and CH_4 using space-borne observations of XCH_4: XCO_2. *Atmos. Chem. Phys.* **2014**, *14*, 12883–12895. [CrossRef]

26. Pandey, S.; Houweling, S.; Krol, M.; Aben, I.; Chevallier, F.; Dlugokencky, E.J.; Gatti, L.V.; Gloor, E.; Miller, J.B.; Detmers, R.; et al. Inverse modeling of GOSAT-retrieved ratios of total column CH_4 and CO_2 for 2009 and 2010. *Atmos. Chem. Phys.* **2016**, *16*, 5043–5062. [CrossRef]

27. Feng, L.; Palmer, P.I.; Bösch, H.; Parker, R.J.; Webb, A.J.; Correia, C.S.C.; Deutscher, N.M.; Domingues, L.G.; Feist, D.G.; Gatti, L.V.; et al. Consistent regional fluxes of CH_4 and CO_2 inferred from GOSAT proxy XCH_4: XCO_2 retrievals, 2010–2014. *Atmos. Chem. Phys.* **2017**, *17*, 4781–4797. [CrossRef]

28. GHG-CCI project team. Algorithm inter-comparison and error characterization & analysis report (AIECAR) of the GHG-CCI project of ESA's Climate Change Initiative, Version 1 (AIECARv1). 2012. Available online: http://www.iup.uni-bremen.de/~{}buch/ghgcci_public/AIECARv1_GHG-CCI_Final.pdf (accessed on 15 May 2020).

29. Parker, R.J.; Webb, A.; Boesch, H.; Somkuti, P.; Barrio Guillo, R.; Di Noia, A.; Kalaitzi, N.; Anand, J.; Bergamaschi, P.; Chevallier, F.; et al. A Decade of GOSAT Proxy Satellite CH_4 Observations. *Earth Syst. Sci. Data Discuss.* **2020**, in review.

30. Detmers, R.; Hasekamp, O. Product User Guide (PUG) for the RemoTeC XCH_4 PROXY GOSAT Data Product v2.3.8. 2016. Available online: ftp://ftp.sron.nl/pub/pub/RemoTeC/TEMIS/CRDP4/DOCS/PUGv4p1_CH4_GOS_SRPR.pdf (accessed on 8 March 2020).

31. GHG-CCI project team. Product User Guide (PUGv4) for the University of Leicester Proxy XCH_4 GOSAT Data Product (CH4_GOS_OCPR). 2016. Available online: https://catalogue.ceda.ac.uk/uuid/4774bc5719754c44add5c6f209fc25ae (accessed on 8 March 2020).

32. Yoshida, Y.; Kikuchi, N.; Yokota, T. On-orbit radiometric calibration of SWIR bands of TANSO-FTS onboard GOSAT. *Atmos. Meas. Tech.* **2012**, *5*, 2515–2523. [CrossRef]

33. Ishida, H.; Nakajima, T.Y. Development of an unbiased cloud detection algorithm for a spaceborne multispectral imager. *J. Geophys. Res.-Atmos.* **2009**, *114*, D07206. [CrossRef]

34. Yoshida, Y.; Eguchi, N.; Ota, Y.; Kikuchi, N.; Nobuta, K.; Aoki, T.; Yokota, T. Algorithm theoretical basis document (ATBD) for CO_2, CH_4 and H_2O column amounts retrieval from GOSAT TANSO-FTS SWIR. 2017. Available online: http://data2.gosat.nies.go.jp/doc/documents/ATBD_FTSSWIRL2_V2.0_en.pdf (accessed on 8 March 2020).

35. Nakajima, T.; Nakajima, T.Y.; Higurashi, A.; Sano, I.; Takamura, T.; Ishida, H.; Schutgens, N. A study of aerosol and cloud information retrievals from CAI imager on board GOSAT satellite. *J. Remote Sens. Soc. Japan.* **2008**, *28*, 178–189, (in Japanese with English abstract and figure captions).

36. Guerlet, S.; Butz, A.; Schepers, D.; Basu, S.; Hasekamp, O.P.; Kuze, A.; Yokota, T.; Blavier, J.-F.; Deutscher, N.M.; Griffith, D.W.T. Impact of aerosol and thin cirrus on retrieving and validating XCO_2 from GOSAT shortwave infrared measurements. *J. Geophys. Res.-Atmos.* **2013**, *118*, 4887–4905. [CrossRef]

37. Inoue, M.; Morino, I.; Uchino, O.; Nakatsuru, T.; Yoshida, Y.; Yokota, T.; Wunch, D.; Wennberg, P.O.; Roehl, C.M.; Griffith, D.W.T.; et al. Bias corrections of GOSAT SWIR XCO_2 and XCH_4 with TCCON data and their evaluation using aircraft measurement data. *Atmos. Meas. Tech.* **2016**, *9*, 3491–3512. [CrossRef]

38. Christi, M.J.; Stephens, G.L. Retrieving profiles of atmospheric CO_2 in clear sky and in the presence of thin cloud using spectroscopy from the near and thermal infrared: A preliminary case study. *J. Geophys. Res.-Atmos.* **2004**, *109*, D04316. [CrossRef]

39. Wunch, D.; Wennberg, P.O.; Toon, G.C.; Connor, B.J.; Fisher, B.; Osterman, G.B.; Frankenberg, C.; Mandrake, L.; O'Dell, C.; Ahonen, P.; et al. A method for evaluating bias in global measurements of CO_2 total columns from space. *Atmos. Chem. Phys.* **2011**, *11*, 12317–12337. [CrossRef]

40. Taylor, T.E.; O'Dell, C.W.; O'Brien, D.M.; Kikuchi, N.; Yokota, T.; Nakajima, T.Y.; Ishida, H.; Crisp, D.; Nakajima, T. Comparison of cloud-screening methods applied to GOSAT near-infrared spectra. *IEEE Trans. Geosci. Remote Sens.* **2012**, *50*, 295–309. [CrossRef]

41. Notholt, J.; Schrems, O.; Warneke, T.; Deutscher, N.; Weinzierl, C.; Palm, M.; Buschmann, M.; AWI-PEV Station Engineers. TCCON data from Ny Alesund, Spitzbergen, Norway, Release GGG2014R1, TCCON Data Archive. Available online: https://data.caltech.edu/records/1289 (accessed on 8 March 2020).

42. Kivi, R.; Heikkinen, P.; Kyro, E. TCCON data from Sodankyla, Finland, Release GGG2014R0, TCCON Data Archive. Available online: https://data.caltech.edu/records/289 (accessed on 8 March 2020).

43. Wunch, D.; Mendonca, J.; Colebatch, O.; Allen, N.; Blavier, J.-F.L.; Springett, S.; Worthy, D.; Kessler, R.; Strong, K. TCCON data from East Trout Lake, Canada, Release GGG2014R1, TCCON Data Archive. Available online: https://data.caltech.edu/records/362 (accessed on 8 March 2020).

44. Deutscher, N.; Notholt, J.; Messerschmidt, J.; Weinzierl, C.; Warneke, T.; Petri, C.; Grupe, P. TCCON data from Bialystok, Poland, Release GGG2014R2, TCCON Data Archive. Available online: https://data.caltech.edu/records/267 (accessed on 8 March 2020).

45. Notholt, J.; Petri, C.; Warneke, T.; Deutscher, N.; Buschmann, M.; Weinzierl, C.; Macatangay, R.; Grupe, P. TCCON data from Bremen, Germany, Release GGG2014R1, TCCON Data Archive. Available online: https://data.caltech.edu/records/268 (accessed on 8 March 2020).

46. Hase, F.; Blumenstock, T.; Dohe, S.; Groß, J.; Kiel, M. TCCON data from Karlsruhe, Germany, Release GGG2014R1, TCCON Data Archive. Available online: https://data.caltech.edu/records/278 (accessed on 8 March 2020).

47. Té, Y.; Jeseck, P.; Janssen, C. TCCON data from Paris (FR), Release GGG2014.R0, TCCON Data Archive. Available online: https://data.caltech.edu/records/284 (accessed on 8 March 2020).

48. Warneke, T.; Messerschmidt, J.; Notholt, J.; Weinzierl, C.; Deutscher, N.; Petri, C.; Grupe, P.; Vuillemin, C.; Truong, F.; Schmidt, M.; et al. TCCON data from Orleans, France, Release GGG2014R1, TCCON Data Archive. Available online: https://data.caltech.edu/records/283 (accessed on 8 March 2020).

49. Sussmann, R.; Rettinger, M. TCCON data from Garmisch, Germany, Release GGG2014R2, TCCON Data Archive. Available online: https://data.caltech.edu/records/956 (accessed on 8 March 2020).

50. Wennberg, P.O.; Roehl, C.; Wunch, D.; Toon, G.C.; Blavier, J.-F.; Washenfelder, R.; Keppel-Aleks, G.; Allen, N.; Ayers, J. TCCON data from Park Falls, Wisconsin, USA, Release GGG2014R1, TCCON Data Archive. Available online: https://data.caltech.edu/records/295 (accessed on 8 March 2020).

51. Morino, I.; Yokozeki, N.; Matzuzaki, T.; Horikawa, M. TCCON Data from Rikubetsu, Hokkaido, Japan, Release GGG2014R2, TCCON Data Archive. Available online: https://data.caltech.edu/records/957 (accessed on 8 March 2020).

52. Iraci, L.; Podolske, J.; Hillyard, P.; Roehl, C.; Wennberg, P.O.; Blavier, J.-F.; Landeros, J.; Allen, N.; Wunch, D.; Zavaleta, J.; et al. TCCON data from Indianapolis, Indiana, USA, Release GGG2014R1, TCCON Data Archive. Available online: https://data.caltech.edu/records/274 (accessed on 8 March 2020).

53. Dubey, M.; Lindenmaier, R.; Henderson, B.; Green, D.; Allen, N.; Roehl, C.; Blavier, J.-F.; Butterfield, Z.; Love, S.; Hamelmann, J.; et al. TCCON data from Four Corners, NM, USA, Release GGG2014R0, TCCON data archive, TCCON Data Archive. Available online: https://data.caltech.edu/records/272 (accessed on 8 March 2020).

54. Wennberg, P.O.; Wunch, D.; Roehl, C.; Blavier, J.-F.; Toon, G.C.; Allen, N.; Dowell, P.; Teske, K.; Martin, C.; Martin, J. TCCON data from Lamont, Oklahoma, USA, Release GGG2014R1, TCCON Data Archive. Available online: https://data.caltech.edu/records/279 (accessed on 8 March 2020).

55. Goo, T.-Y.; Oh, Y.-S.; Velazco, V.A. TCCON data from Anmeyondo, South Korea, Release GGG2014R0, TCCON Data Archive. Available online: https://data.caltech.edu/records/266 (accessed on 8 March 2020).

56. Morino, I.; Matsuzaki, T.; Horikawa, M. TCCON data from Tsukuba, Ibaraki, Japan, 125HR, Release GGG2014R2, TCCON Data Archive. Available online: https://data.caltech.edu/records/958 (accessed on 8 March 2020).

57. Iraci, L.; Podolske, J.; Hillyard, P.; Roehl, C.; Wennberg, P.O.; Blavier, J.-F.; Landeros, J.; Allen, N.; Wunch, D.; Zavaleta, J.; et al. TCCON data from Armstrong Flight Research Center, Edwards, CA, USA, Release GGG2014R1, TCCON Data Archive. Available online: https://data.caltech.edu/records/270 (accessed on 8 March 2020).

58. Wennberg, P.O.; Roehl, C.; Blavier, J.-F.; Wunch, D.; Landeros, J.; Allen, N. TCCON data from Jet Propulsion Laboratory, Pasadena, California, USA, Release GGG2014R1, TCCON Data Archive. Available online: https://data.caltech.edu/records/277 (accessed on 8 March 2020).

59. Wennberg, P.O.; Wunch, D.; Roehl, C.; Blavier, J.-F.; Toon, G.C.; Allen, N. TCCON data from California Institute of Technology, Pasadena, California, USA, Release GGG2014R1, TCCON Data Archive. Available online: https://data.caltech.edu/records/285 (accessed on 8 March 2020).

60. Kawakami, S.; Ohyama, H.; Arai, K.; Okumura, H.; Taura, C.; Fukamachi, T.; Sakashita, M. TCCON data from Saga, Japan, Release GGG2014R0, TCCON Data Archive. Available online: https://data.caltech.edu/records/288 (accessed on 8 March 2020).

61. Liu, C.; Wang, W.; Sun, Y. TCCON data from Hefei, China, Release GGG2014R0, TCCON Data Archive. Available online: https://data.caltech.edu/records/1092 (accessed on 8 March 2020).

62. Morino, I.; Velazco, V.A.; Hori, A.; Uchino, O.; Griffith, D.W.T. TCCON data from Burgos, Philippines, Release GGG2014R0, TCCON Data Archive. Available online: https://data.caltech.edu/records/1090 (accessed on 8 March 2020).

63. Dubey, M.; Henderson, B.; Green, D.; Butterfield, Z.; Keppel-Aleks, G.; Allen, N.; Blavier, J.-F.; Roehl, C.; Wunch, D.; Lindenmaier, R. TCCON data from Manaus, Brazil, Release GGG2014R0, TCCON Data Archive. Available online: https://data.caltech.edu/records/282 (accessed on 8 March 2020).

64. Griffith, D.W.T.; Deutscher, N.; Velazco, V.A.; Wennberg, P.O.; Yavin, Y.; Keppel Aleks, G.; Washenfelder, R.; Toon, G.C.; Blavier, J.-F.; Paton, W.C.; et al. TCCON data from Darwin, Australia, Release GGG2014R0, TCCON Data Archive. Available online: https://data.caltech.edu/records/269 (accessed on 8 March 2020).

65. Griffith, D.W.T.; Velazco, V.A.; Deutscher, N.; Paton, W.C.; Jones, N.; Wilson, S.; Macatangay, R.; Kettlewell, G.; Buchholz, R.R.; Riggenbach, M. TCCON data from Wollongong, Australia, Release GGG2014R0, TCCON Data Archive. 2017. Available online: https://data.caltech.edu/records/291 (accessed on 8 March 2020).

66. Sherlock, V.; Connor, B.; Robinson, J.; Shiona, H.; Smale, D.; Pollard, D. TCCON data from Lauder, New Zealand, 120HR, Release GGG2014R0, TCCON Data Archive. Available online: https://data.caltech.edu/records/280 (accessed on 8 March 2020).

67. Sherlock, V.; Connor, B.; Robinson, J.; Shiona, H.; Smale, D.; Pollard, D. TCCON data from Lauder, New Zealand, 125HR, Release GGG2014R0, TCCON Data Archive. Available online: https://data.caltech.edu/records/281 (accessed on 8 March 2020).

68. Dils, B.; Buchwitz, M.; Reuter, M.; Schneising, O.; Boesch, H.; Parker, R.; Guerlet, S.; Aben, I.; Blumenstock, T.; Burrows, J.P.; et al. The Greenhouse Gas Climate Change Initiative (GHG-CCI): Comparative validation of GHG-CCI SCIAMACHY/ENVISAT and TANSO-FTS/GOSAT CO_2 and CH_4 retrieval algorithm products with measurements from the TCCON. *Atmos. Meas. Tech.* **2014**, *7*, 1723–1744. [CrossRef]

69. Wunch, D.; Toon, G.C.; Blavier, J.-F.L.; Washenfelder, R.A.; Notholt, J.; Connor, B.J.; Griffith, D.W.T.; Sherlock, V.; Wennberg, P.O. The Total Carbon Column Observing Network. *Phil. Trans. R. Soc. A* **2011**, *369*, 2087–2112. [CrossRef]

70. Deutscher, N.M.; Griffith, D.W.T.; Bryant, G.W.; Wennberg, P.O.; Toon, G.C.; Washenfelder, R.A.; Keppel-Aleks, G.; Wunch, D.; Yavin, Y.; Allen, N.T.; et al. Total column CO_2 measurements at Darwin, Australia—Site description and calibration against in situ aircraft profiles. *Atmos. Meas. Tech.* **2010**, *3*, 947–958. [CrossRef]

71. Mendonca, J.; Strong, K.; Wunch, D.; Toon, G.C.; Long, D.A.; Hodges, J.T.; Sironneau, V.T.; Franklin, J.E. Using a speed-dependent Voigt line shape to retrieve O_2 from Total Carbon Column Observing Network solar spectra to improve measurements of XCO_2. *Atmos. Meas. Tech.* **2019**, *12*, 35–50. [CrossRef]

72. Schneising, O.; Heymann, J.; Buchwitz, M.; Reuter, M.; Bovensmann, H.; Burrows, J.P. Anthropogenic carbon dioxide source areas observed from space: Assessment of regional enhancements and trends. *Atmos. Chem. Phys.* **2013**, *13*, 2445–2454. [CrossRef]

73. Kivi, R.; Heikkinen, P. Fourier transform spectrometer measurements of column CO_2 at Sodankylä, Finland. *Geosci. Instrum. Method. Data Syst.* **2016**, *5*, 271–279. [CrossRef]

74. Lindenmaier, R.; Dubey, M.K.; Henderson, B.G.; Butterfield, Z.T.; Herman, J.R.; Rahn, T.; Lee, S.-H. Multiscale observations of CO_2, $^{13}CO_2$, and pollutants at Four Corners for emission verification and attribution. *Proc. Natl. Acad. Sci. USA* **2014**, *111*, 8386–8391. [CrossRef] [PubMed]
75. Kort, E.A.; Frankenberg, C.; Costigan, K.R.; Lindenmaier, R.; Dubey, M.K.; Wunch, D. Four corners: The largest US methane anomaly viewed from space. *Geophys. Res. Lett.* **2014**, *41*, 6898–6903. [CrossRef]
76. Kuze, A.; Suto, H.; Shiomi, K.; Urabe, T.; Nakajima, M.; Yoshida, J.; Kawashima, T.; Yamamoto, Y.; Kataoka, F.; Buijs, H. Level 1 algorithms for TANSO on GOSAT: Processing and on-orbit calibrations. *Atmos. Meas. Tech.* **2012**, *5*, 2447–2467. [CrossRef]

 *remote sensing*

Article

Calibration and Validation of Antenna and Brightness Temperatures from Metop-C Advanced Microwave Sounding Unit-A (AMSU-A)

Banghua Yan [1,*], Junye Chen [2], Cheng-Zhi Zou [1], Khalil Ahmad [2], Haifeng Qian [3], Kevin Garrett [1], Tong Zhu [2], Dejiang Han [4] and Joseph Green [5]

[1] Satellite Meteorology and Climatology Division, Center for Satellite Applications and Research, National Oceanic and Atmospheric Administration (NOAA), 5830 University Research Court, College Park, MD 20740, USA; cheng-zhi.zou@noaa.gov (C.-Z.Z.); kevin.garrett@noaa.gov (K.G.)

[2] Global Science Technologies, Greenbelt, MD 20770, USA; Junye.Chen@noaa.gov (J.C.); khalil.ahmad@noaa.gov (K.A.); tong.zhu@noaa.gov (T.Z.)

[3] Earth System Science Interdisciplinary Center, University of Maryland, 5825 University Research Ct Suite 4001, College Park, MD 20740, USA; haifeng.qian@noaa.gov

[4] Office of Satellite and Product Operations (OSPO), NOAA, 4231 Suitland Road Suitland, MD 20746, USA; dejiang.han@noaa.gov

[5] Office of Projects, Planning, and Analysis (OPPA), NOAA, 1335 East West Highway, Silver Spring, MD 20910, USA; Phil.Green@noaa.gov

* Correspondence: banghua.Yan@noaa.gov; Tel.: +1-301-683-3602

Received: 9 June 2020; Accepted: 22 July 2020; Published: 14 September 2020

Abstract: This study carries out the calibration and validation of Antenna Temperature Data Record (TDR) and Brightness Temperature Sensor Data Record (SDR) data from the last National Oceanic and Atmospheric Administration (NOAA) Advanced Microwave Sounding Unit-A (AMSU-A) flown on the Meteorological Operational satellite programme (MetOp)-C satellite. The calibration comprises the selection of optimal space view positions for the instrument and the determination of coefficients in calibration equations from the Raw Data Record (RDR) to TDR and SDR. The validation covers the analyses of the instrument noise equivalent differential temperature (NEDT) performance and the TDR and SDR data quality from the launch until 15 November 2019. In particular, the Metop-C data quality is assessed by comparing to radiative transfer model simulations and observations from Metop-A/B AMSU-A, respectively. The results demonstrate that the on-orbit instrument NEDTs have been stable since launch and continue to meet the specifications at most channels except for channel 3, whose NEDT exceeds the specification after April 2019. The quality of the Metop-C AMSU-A data for all channels except channel 3 have been reliable since launch. The quality at channel 3 is degraded due to the noise exceeding the specification. Compared to its TDR data, the Metop-C AMSU-A SDR data exhibit a reduced and more symmetric scan angle-dependent bias against radiative transfer model simulations, demonstrating the great performance of the TDR to SDR conversion coefficients. Additionally, the Metop-C AMSU-A data quality agrees well with Metop-A/B AMSU-A data, with an averaged difference in the order of 0.3 K, which is confirmed based on Simultaneous Nadir Overpass (SNO) inter-sensor comparisons between Metop-A/B/C AMSU-A instruments via either NOAA-18 or NOAA-19 AMSU-A as a transfer.

Keywords: Metop-C advanced microwave sounding unit-A; radiometry; calibration and validation; inter-sensor calibration among Metop-A to -C; simultaneous nadir overpass (SNO)

1. Introduction

The European Meteorological Operational satellite program C (Metop-C) satellite, which was launched into low Earth orbit on 6 November 2018, carries the last NOAA Advanced Microwave Sounding Unit-A (AMSU-A). On 15 November 2018, nine days after the launch of the Metop-C satellite, the first day AMSU-A science data were received. The AMSU-A provides temperature soundings from the Earth's near surface to an altitude of about 42 km through measurements of Raw Data Record (RDR) with 15 channels from 23.8 to 89 GHz. Table 1 lists the AMSU-A main channel characteristics, which include the channel frequency, bandwidth, and radiometric temperature sensitivity or Noise Equivalent Differential Temperature (NEDT) for each of the 15 channels. Intensive calibration activities for the AMSU-A Raw Data Record (RDR) to derive Earth antenna Temperature Data Record (TDR) data have been conducted in the NOAA Center for Satellite Applications and Research (STAR) [1–4]. Since April 2019, the TDR data have been distributed to the user community through both the NOAA Office of Satellite and Product Operations (OSPO) and Production Distribution Access (PDA) for near-real time applications and the NOAA Comprehensive Large Array-data Stewardship System (CLASS) for long-term data analysis and applications. Additionally, the conversion coefficients from TDR to Sensor Data Record (SDR) (brightness temperatures) data were derived in [5]. Today, Metop-C AMSU-A TDR and SDR data are successfully applied to a series of Environmental Data Record (EDR) retrieval systems and are also assimilated into the NOAA National Weather Service Global Forecast System (personal communication with Andrew Collard), the U.S. Navy Global Environmental Model (NAVGEM) (personal communication with Ruston Ben) and the European Centre for Medium-Range Weather Forecasts (ECMWF) global forecast system (personal communication with Niels Bormann).

Table 1. Advanced Microwave Sounding Unit-A (AMSU-A) instrument specifications [6].

Channel Index	Center Frequency (MHz)	Central Frequency Stability (MHz)	Bandwidth (MHz)	Polarization	Measured 3-db Beamwidth [1,2] (°)	Temperature Sensitivity ($NE\Delta T$) (K)
1	23,800	±10	270	V	3.48	0.3
2	31,400	±10	180	V	3.52	0.3
3	50,300	±10	180	V	3.64	0.4
4	52,800	±5	400	V	3.40	0.25
5	53,596 ± 115 [3]	±5	170	H	3.60	0.25
6	54,400	±5	400	H	3.44	0.25
7	54,940	±5	400	V	3.44	0.25
8	55,500	±10	330	H	3.44	0.25
9	f_0 = 57,290.344	±0.5	330	H	3.32	0.25
10	f_0 ± 217 [3]	±0.5	78	H	3.325	0.4
11	f_0 ± 322.2 ± 48 [4]	±1.2	36	H	3.32	0.4
12	f_0 ± 322.2 ± 22 [4]	±1.2	16	H	3.32	0.6
13	f_0 ± 322.2 ± 10 [4]	±0.5	8	H	3.32	0.8
14	f_0 ± 322.2 ± 4.5 [4]	±0.5	3	H	3.32	1.2
15	89,000	±130	1500	V	3.56	0.5

[1] Specifications of 3 db bandwidth are within 3.3° ± 10%; [2] 3-db bandwidth data correspond to beam position 15; [3] the channel has double bands; [4] the channel has four bands.

The data, either in TDR or SDR, from AMSU-A instruments onboard various legacy satellites from NOAA-15 to NOAA-19, and from Metop-A to Metop-B, play an important role in EDR retrieval systems [7–11], climate analysis [12,13], and Numerical Weather Prediction (NWP) models [14–17]. Metop-C AMSU-A data continue be used in those important fields. The sufficient calibration and validation of Metop-C AMSU-A data becomes very necessary to ensure the accuracy of the data. Another important parameter that could affect the performance of AMSU-A observations is the instrument noise [18,19], i.e., Noise Equivalent Differential Temperature ($NE\Delta T$ or NEDT), which represents the smallest temperature difference that an instrument can distinguish when looking at Earth scenes. This parameter also helps weight satellite data by channel in the observation error covariance matrix used by satellite EDR product retrieval systems [7], as well as by NWP data assimilation systems [15,17]. In climate studies, instrument noise affects the detection of long-term climate trends of Earth scene temperature data [20–22]. In addition, the AMSU-A instrument possesses four separate space view (SV) positions, resulting in the selection of an optimal cold space view position among them, prior to measurements for operational use. Therefore, this work describes, in detail, the calibration and validation process for Metop-C AMSU-A from the RDR to SDR via TDR, including, but not limited to, the following analyses: optimal cold space position selection, cold space calibration correction, calibration coefficients, postlaunch instrument NEDT performance, TDR and SDR data quality validation. The lunar intrusion correction algorithm, which is another important portion of the calibration, is studied separately. The derivation process of the conversion coefficients from TDR to SDR is presented in detail in [5], although the conversion equation is briefly described in this study.

This paper is organized as follows. The next section provides a brief description of AMSU-A instruments along with an optimal cold space position selection based on Metop-C AMSU-A System In-Orbit Verification (SIOV) data. In Section 3, we establish the Metop-C AMSU-A calibration methodology, comprising the radiometric calibration equation from RDR to TDR, the conversion equation from TDR to SDR, and the determination of required coefficients and parameters in the equations. In Section 4, we analyze the instrument NEDT trend by using the current Integrated Calibration/Validation System (ICVS) developed at STAR [7] and compare it with those of legacy AMSU-A instruments. Meanwhile, a new NEDT estimation method is implemented for comparison. Regarding the data quality assessment, we conduct this analysis in Section 5. This is formed first by using the Joint Center for Satellite Data Assimilation (JCSDA) Community Radiative Transfer Model (CRTM) [23] to investigate the AMSU-A antenna and brightness temperatures bias features. Inter-sensor comparisons are further given of AMSU-A antenna temperatures between Metop-C and each of Metop-A and Metop-B AMSU-A instruments. These are performed by using each of the NOAA-18 and NOAA-19 AMSU-A instruments as a transfer based on the existing Simultaneous Nadir Overpass (SNO) method [24] and some proper quality schemes applicable for microwave satellite measurements at surface-sensitive channels [25,26]. The final section summarizes the overall Metop-C AMSU-A calibration and validation results.

2. AMSU-A Instrument Description and Optimal Cold Space Position Selection

The Metop-C AMSU-A instrument, which was built by Northrop Grumman, is composed of two modules, A1 and A2, with three antenna systems, A1-1, A1-2 and A2. The A1-1 system contains channels 6–7 and 9–15; A1-2 contains channels 3–5 and 8; and the A2 system contains channels 1 and 2. During each scan cycle, which lasts 8 s, the instrument samples 30 Earth scene cells (beam positions) within a satellite scan angle of 48.333° from nadir on each side of the sub-satellite path, each of which is separated by 3.33° in a stepped-scan fashion [27]. These scan patterns and geometric resolutions translate to a 48-km diameter cell at nadir and a 2343-km swath width from an 870-km nominal orbital altitude. In addition, the instrument measures the radiation from two calibration targets in every scan cycle, i.e., the cosmic background radiation or cold space that is viewed immediately after the Earth has been scanned, and the internal blackbody calibration target or warm load that is viewed immediately after the cold space. As a result, every scan cycle contains three consecutive views: 30 Earth scenes, cold space and blackbody warm calibration measurements (see Figure 1).

As illustrated in Figure 1, the AMSU-A instrument possesses four separate space view (SV) positions, i.e., 83.3° (SV1), 81.67° (SV2), 80.0° (SV3), and 76.67° (SV4). In practice, however, an optimal cold space view position among them needs to be determined prior to measurements for operational use. This optimal SV position is assumed to produce cold counts with minimum contamination radiating from the spacecraft and Earth's limb, thus mostly providing a minimum averaged cold count per SV period. The optimal SV position for Metop-C AMSU-A is selected during the instrument SIOV early on-orbit verification (OV) period. This period covers 08:15am `Coordinated Universal Time (UTC)` on 19 November, 2018 to 14:21 UTC on 30 November, 2018, and is made up of observations of approximately 30 consecutive orbits (2 days) for each of the four positions (SV1 to SV4), with the exception of position 4, which has three consecutive days of measurements. In addition, two scanning modes were set up for the SV1 position, so the second was defined as 'SV1n' to distinguish it from the first. In total, five sets of data were collected (refer to Table 2). In addition, a few types of signals unrelated to the change in SV position need to be removed in the measured cold count data sets during the above OV period, e.g., lunar contamination events, count outliers, variations due to instrument noise, diurnal and orbital variations due to instrument temperature change, and trends due to instrument warm-up. The data sets were pre-processed to catch features of cold counts, primarily due to the change in SV position.

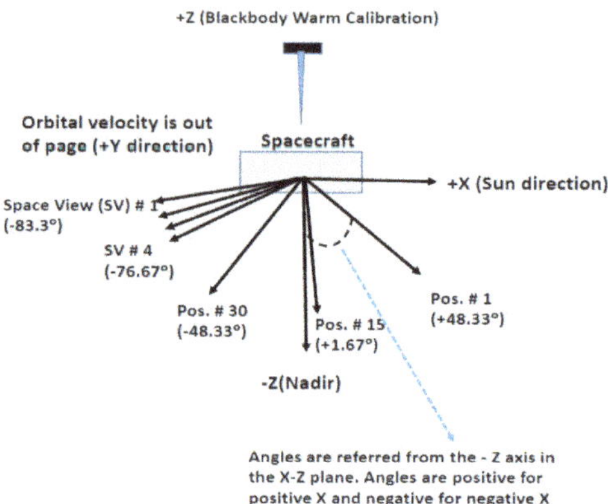

Figure 1. AMSU-A geometry sketch of a coordinator placed at the center of the instrument antenna [28].

Table 2. Daily averaged space view counts corresponding to each space view (SV) position and optimal selection for each channel of Metop-C AMSU-A instrument.

Channel	SV1 *	SV2 *	SV4 *	SV 3 *	SV1n *	Optimal SV
1	11,862.49	11,862.90	11,865.61	11,863.34	11,862.85	SV1
2	11,351.19	11,351.23	11,353.70	11,350.86	11,351.80	SV3
3	11,794.02	11,791.19	11,794.24	11,789.59	11,785.46	SV1n
4	12,694.09	12,694.11	12,696.86	12,695.06	12,692.76	SV1n
5	13,123.09	13,123.91	13,125.99	13,125.03	13,125.80	SV1
6	12,335.72	12,335.36	12,336.37	12,337.13	12,337.19	SV2
7	12,813.11	12,807.50	12,809.56	12,810.05	12,813.40	SV2
8	12,196.05	12,196.08	12,196.09	12,200.60	12,199.81	SV1
9	12,284.01	12,281.09	12,278.17	12,278.76	12,280.33	SV4
10	12,184.44	12,183.84	12,183.69	12,186.46	12,189.38	SV4
11	13,059.05	13,061.89	13,060.14	13,062.13	13,066.32	SV1
12	12,820.44	12,821.88	12,819.51	12,819.92	12,823.72	SV4
13	13,287.03	13,288.95	13,286.26	13,285.88	13,288.63	SV3
14	12,760.81	12,764.91	12,764.76	12,766.90	12,771.17	SV1
15	13,843.62	13,843.38	13,842.72	13,848.32	13,850.08	SV4

* On-orbit verification dates corresponding to five SV data sets: 19 November 2018 to 21 November for SV1; 21 November 2018 to 23 November for SV2; 23 November 2018 to 26 November for SV4; 26 November 2018 to 28 Nov for SV3; 28 November 2018 to 30 November for SV1n (second cycle for SV1).

For demonstration, Figure 2 displays a time series of the data sets at channel 8, from original cold count measurements to the 'cold counts' after the corrections of the abovementioned signals, step by step, i.e., (a) original cold count measurements, (b) cold counts after lunar contamination removal (i.e., Lu-Rm), (c) 'cold counts' after removing count outliers from (b) (i.e., Lu-Ot-Rm), (d) 'cold counts' after filtering high frequent noise components from (c) (i.e., Lu-Ot-Hf-Rm), (e) 'cold counts' after mitigating diurnal and orbital variations due to instrument temperature change from (d) (i.e., Lu-Ot-Hf-Cy-Rm), and (f) 'cold counts' after removing trend due to instrument warm-up from (e) (i.e., Lu-Ot-Hf-Cy-Trd-Rm). Note that the impact of the lunar intrusion on the overall cold counts is small during this period. The maximum magnitude of the lunar contamination is about

10 counts, which occurred on 27 November 2018. A similar procedure is applied to other channels. Therefore, for a given channel, the resulting data after the corrections are averaged for each SV position to produce the mean count per SV, as given in Table 2.

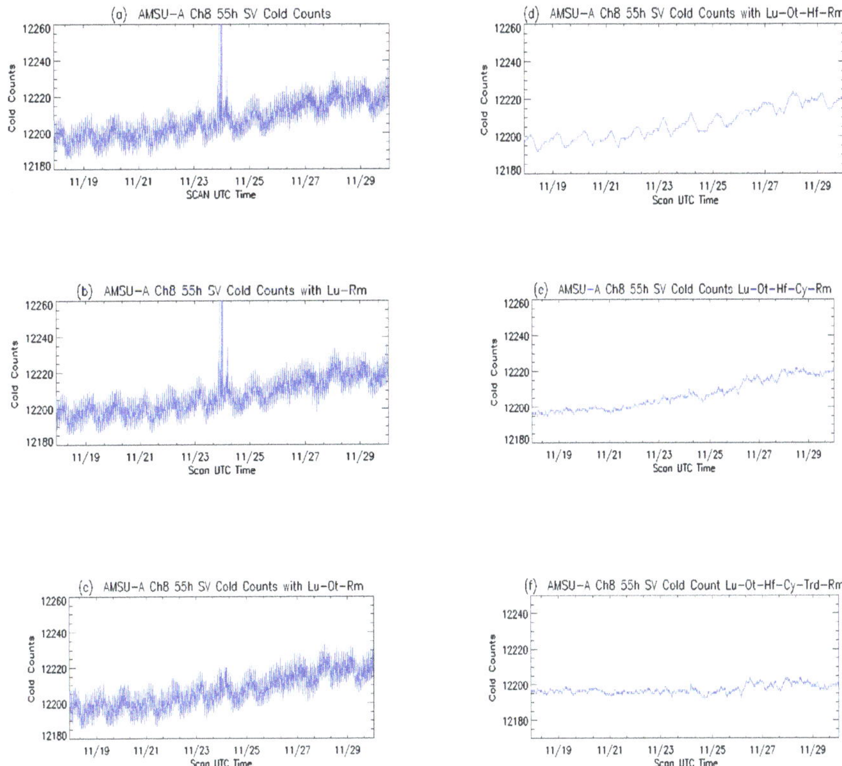

Figure 2. Cold count time series during November 19 0815 UTC and November 30 1421 UTC, 2018 for Metop-C AMSU-A channel 8 after a series of corrections of signals, as mentioned in the main text. (**a**) Original cold count measurements. (**b**) Cold counts after lunar contamination removal (i.e., Lu-Rm). (**c**) 'Cold counts' after removing count outliers from (**b**) (i.e., Lu-Ot-Rm). (**d**) 'Cold counts' after filtering high frequent noise components from (**c**) (i.e., Lu-Ot-Hf-Rm). (**e**) 'Cold counts' after mitigating diurnal and orbital variations due to instrument temperature change (i.e., Lu-Ot-Hf-Cy-Rm). (**f**) 'Cold counts' after removing trend due to instrument warm-up from (**e**) (i.e., Lu-Ot-Hf-Cy -Trd-Rm).

As shown in Table 2, for AMSU A1 channels, the frequency of the occurrence with the lowest averaged count is six times for SV1(n), twice for SV2, once for SV3 and four times for SV4. For AMSU-A2 channels, the averaged counts are very similar for all SV positions, although the counts for SV3 are slightly lower. As a result, the optimal cold position is SV1 (which is nearest to the satellite platform) for AMSU-A1 and SV3 for AMSU-A2. Since 30 November 1421 UTC, AMSU-A1 and A2 cold space positions have been switched to positions 1 and 3, respectively, to start regular measurements, which were conducted by the Metop-C flight team of the National Aeronautics and Space Administration (NASA) Goddard Space Flight Center (GSFC). Note that the choice for AMSU-A1 is the same as that of legacy AMSU-A instruments onboard NOAA-16, 18, Metop-A and –B, but the choice for AMSU-A2 is different from that of legacy AMSU-A2 instruments. The data used in the following analyses correspond to the selected optimal SV positions unless otherwise noted.

3. AMSU-A Calibration Methodology Description

The calibration methodology consists of a radiometric equation from RDR to TDR and a conversion equation from TDR to SDR. Basic equations are the same as previous studies for legacy AMSU-A instruments flown onboard NOAA-15–19, Metop-A and –B [29–32], except for the different calibration coefficients, nonlinearity and cold target calibration corrections, so the equations are only briefly described below. The equations are relevant to the channel frequency (v) and beam position (satellite zenith angle β), but those indices are typically omitted in this study for clarity unless otherwise noted.

3.1. Calibration Equations

Two calibration measurements, i.e., cold space and warm load, are used to determine antenna temperatures via a radiometric calibration equation, as illustrated in the calibration scheme in Figure 3. In particular, the radiometric calibration equation converts the measured digitized radiometric scene counts C_S (i.e., scene counts) to radiance R_S for the Earth scene target using the following equation [33,34].

$$R_S = R_W + (R_W - R_C)\frac{\left(C_S - \overline{C}_W\right)}{\left(\overline{C}_w - \overline{C}_c\right)} + Q = R_{SL} + Q \tag{1}$$

with

$$R_{SL} = R_W + (R_W - R_C)\frac{\left(C_S - \overline{C}_W\right)}{\left(\overline{C}_w - \overline{C}_c\right)} = R_W + \frac{\left(C_S - \overline{C}_W\right)}{G} \tag{2}$$

$$Q = \mu(R_W - R_C)^2\frac{\left(C_S - \overline{C}_W\right)\left(C_S - \overline{C}_C\right)}{\left(\overline{C}_W - \overline{C}_C\right)^2} \tag{3}$$

$$G = \frac{\left(\overline{C}_W - \overline{C}_C\right)}{(R_W - R_C)} \tag{4}$$

where R_S represents the radiometric scene radiance of individual channels, accounting for the nonlinear contribution due to an imperfect square law detector (see the line $\overline{CSW}$ in Figure 3); R_{SL} denotes a linear two-point calibration equation with the assumption of a perfect detector (see the dash line $\overline{CS_L W}$ in Figure 3); C_S is the radiometric count from the Earth scene target; G is the channel calibration gain; $\overline{C}_W$ and $\overline{C}_C$ denote the averaged blackbody and space counts, respectively, over several calibration cycles (Appendix A). In addition, R_W and R_C denote the radiance corresponding to T_W and T_c, respectively. T_W denotes the platinum resistance thermometer (PRT) temperature of the warm load converted from measured radiometric counts, and its calculation and calibration are given in [1] and is also referred to Appendix B. The conversion coefficients from counts to PRT temperature are included in TDR data. T_c is the cosmic temperature after certain correction, and Q is the nonlinearity of the instrument's square law detector, which is a function of nonlinearity parameter μ (see discussions in Section 3.2). The variables, i.e., R_C, R_W, and R_S, in the above equations denote the radiance, represented in mW/(m^2·sr·cm). In reality, by following the processing procedure for legacy AMSU-A measurements flown on the NOAA-15, -16, -17, and -18, -19, Metop-A and -B satellite platforms, the final output from TDR data for Metop-C are presented as the temperature instead of the radiance. Therefore, in the operational processing of Metop-C AMSU-A TDR data, a conversion between temperature and radiance is needed and is achieved by using the inverse of the Planck function in [35].

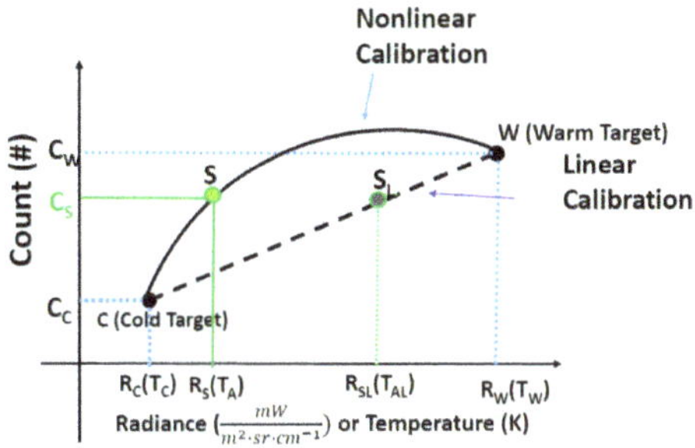

Figure 3. AMSU-A instrument linear and nonlinear calibration scheme.

The AMSU-A instrument is composed of two units (A1 and A2), and has three antenna systems, A1-1, A1-2 and A2, where the A1-1 system contains channels 6–7 and 9–15, the A1-2 contains channels 3–5 and 8 and the A2 system contains channels 1 and 2. Each of these systems consists of an offset parabolic reflector housed in a cylindrical shroud [27]. The temperature corresponding to R_S is actually the antenna temperature (T_A), which is provided in the TDR data. T_A usually contains antenna sidelobe contributions [5,28], antenna emissions and other radiation perturbations [5,34]. Hence, the brightness temperature of Earth scene T_B needs be obtained from the antenna temperature after removing antenna sidelobe contributions, antenna emissions and other radiation perturbations, which are usually collectively defined as the antenna pattern correction for AMSU-A [28].

To understand the conversion from antenna temperature to brightness temperature, Equation (1) is expressed in temperature, i.e.,

$$T_A = T_W + (T_W - T_C)\frac{\left(C_S - \overline{C}_W\right)}{\left(\overline{C}_w - \overline{C}_c\right)} + Q_T, \tag{5}$$

where Q_T is the nonlinearity of the instrument square law detector in temperature, converted from Q in Equation (1).

The antenna pattern corrections and the recovery of brightness temperatures from measured antenna temperatures obtained from legacy AMSU-A radiometers were studied previously [28]. By taking advantage of existing algorithms [28,34], we have established a similar conversion from Earth scene antenna temperature in TDR to Earth scene brightness temperature in SDR for Metop-C AMSU-A [5]. According to [5], brightness temperatures are computed from antenna temperature using the following expression.

$$T_B(\beta) = \alpha_0(\beta)T_A(\beta) - \alpha_1(\beta) \tag{6}$$

with

$$\alpha_0(\beta) = 1.0 + \frac{f_C(\beta)}{f_E(\beta)} + \frac{\sigma f_{SAT}(\beta)}{f_E(\beta)} \tag{7}$$

$$\alpha_1(\beta) = \frac{f_C(\beta)T_C + \sigma f_{SAT}(\beta)T_{SAT}}{f_E(\beta)} \tag{8}$$

In these equations, $f_E(\beta)$, $f_C(\beta)$ and $f_{SAT}(\beta)$ represent the antenna pattern efficiencies over regions of Earth (main and sidelobes), cold space (sidelobes) and satellite spacecraft (sidelobes), respectively. The efficiency values are provided upon request from the authors. The satellite zenith angle β (see Figure 1) is included to highlight that the three efficiencies are not constant with the beam position. The σ is a scale factor to take into account the approximation of the near-field effect of the satellite platform in the antenna pattern correction, varying from 0.01 at channel 1 to 0.11 at channel 15, depending on the channel [28,36].

3.2. Determinations of Cold Space Calibration Correction and Nonlinearity

Three important variables, i.e., T_c, Q and μ, which are used in Equation (1), are determined in prelaunch, as introduced below.

T_C represents the cold space brightness temperature, and is the cosmic temperature (T_{Cosmic}) after removing the correction of the antenna side lobe interference on cold space temperature via the Earth limb and spacecraft, as well the nonlinearity of the instrument square law detector. It is estimated by adding two correction terms to T_{Cosmic} [6,37]:

$$T_C = T_{Cosmic} + \Delta T_C^{RJ} + \Delta T_C^{ER} \tag{9}$$

where T_{Cosmic} is 2.72 K with an uncertainty of ±0.02 K. ΔT_C^{RJ}, representing a correction using Planck's Radiation Law for the error introduced by the Rayleigh–Jeans (RJ) approximation and is given in the second column of Table 3 according to the analysis in [6,37].

Table 3. Bias correction for the cosmic cold background. In the table, ΔT_C^{ER} is computed using (10) corresponding to the selected optimal SV position.

Channel Index	ΔT_C^{RJ} (K)	ΔT_C^{ER} (K)
1	0.040	1.162
2	0.069	1.107
3	0.176	1.994
4	0.194	2.269
5	0.200	2.089
6	0.206	1.253
7	0.210	1.615
8	0.214	1.903
9–14	0.228	1.138
15	0.537	0.754

ΔT_C^{ER} in (9) signifies the contribution from the antenna side lobe interference with the Earth limb and spacecraft. By following the methodology in [5], ΔT_C^{ER} is computed using the following equation

$$\Delta T_C^{ER} = \left(1 - \epsilon_{Ref}\right)\left\{ \frac{1}{N_\sigma}\left(f_E^{SV_i}\overline{T}_{ELB} + f_C^{SV_i}T_{Cosmic} + f_{SAT}^{SV_i}\sigma\overline{T}_{SAT} \right) - T_{Cosmic} \right\} + \epsilon_{Ref}T_{SAT} , \tag{10}$$

Approximately,

$$\Delta T_C^{ER} = \frac{\left(1 - \epsilon_{Ref}\right)}{N_\sigma}\left(f_E^{SV_i}\overline{T}_E + f_{SAT}^{SV_i}\sigma\overline{T}_{SAT} \right) + \epsilon_{Ref}\overline{T}_{SAT} \tag{11}$$

where ϵ_{Ref} denotes the emissivity of the AMSU-A reflector; $\overline{T}_{ELB}$ (=210 K) denotes an averaged Earth limb brightness temperature; $\overline{T}_{SAT}$ denotes an averaged instrument temperature. The quantity $N_\sigma = f_E(\beta) + f_C(\beta) + f_{SAT}(\beta)\sigma$ normalizes the contribution of energy by each radiation component. $f_E^{SV_i}$, $f_C^{SV_i}$, and $f_{SAT}^{SV_i}$ are the antenna efficiencies over the region of cold space, where the subscript 'i' in SV_i corresponds to a specific SV position with a defined beam position where i = 1~4. The values for 15 channels are provided in Table 4, which are computed using Metop-C AMSU-A pattern function data. The estimated ΔT_C^{ER} at all channels using (10) is shown in the third column of Table 3, where $\overline{T}_{SAT} = 300\tilde{K}$; $\overline{T}_E = 210$ K; $\epsilon_{Ref} \approx 0.0002, 0.0006, 0.0004, 0.0004, 0.0005, 0.0003, 0.0004, 0.0006,$ 0.0003 and 0.0005 for the ten channels in the table according to the Northrop Grumman Electronic Systems (NGES) Calibration Log Book [6,37].

Table 4. Metop-C AMSU-A Antenna efficiencies (%) at four cold space view (SV) positions for 15 channels over regions of cold space, Earth, and satellite spacecraft. The selected SV positions are highlighted in the table.

Ch.	FC (%)				FE (%)				FSAT (%)			
	SV1	SV2	SV3	SV4	SV1	SV2	SV3	SV4	SV1	SV2	SV3	SV4
1	99.06	99.00	99.08	99.09	0.49	0.54	0.5	0.53	0.52	0.49	0.48	0.44
2	99.23	99.17	99.25	99.26	0.36	0.43	0.39	0.41	0.42	0.39	0.37	0.34
3	98.45	98.31	98.45	98.45	0.85	1.00	0.88	0.90	0.70	0.68	0.67	0.64
4	98.19	98.04	98.21	98.22	0.97	1.14	1.00	1.02	0.84	0.81	0.80	0.77
5	98.52	98.40	98.54	98.55	0.87	1.01	0.90	0.92	0.63	0.60	0.58	0.55
6	98.99	98.93	99.00	99.00	0.53	0.61	0.55	0.57	0.49	0.47	0.46	0.44
7	98.67	98.66	98.70	98.72	0.68	0.71	0.71	0.73	0.66	0.63	0.61	0.57
8	98.76	98.60	98.76	98.76	0.79	0.95	0.82	0.83	0.48	0.46	0.44	0.42
9–14	99.07	99.02	99.09	99.09	0.48	0.54	0.50	0.52	0.46	0.45	0.43	0.41
15	99.52	99.49	99.52	99.52	0.27	0.27	0.27	0.27	0.27	0.26	0.25	0.25

Note that some uncertainties remain with the estimation of ΔT_C^{ER}. Particularly, the calculation of ΔT_C^{ER} relies on an averaged Earth scene brightness temperature ($\overline{T}_{ELB}$) per channel. However, the Earth scene temperature can vary by location over the Earth. For example, the brightness temperature at channel 1 varies primarily between 180 K and 310 K, which could cause an error of up to one quarter of the original estimation. Theoretically, the ΔT_C^{ER} should be computed using the actual Earth scene temperature per location, thus producing a changeable correction along with the location. However, in the current operational processing system for all AMSU-A TDR observations, a fixed correction is used to reduce the contamination from the antenna sidelobe interference with the Earth limb and spacecraft.

Regarding the quantity Q, this represents the nonlinear contribution to R_S due to an imperfect square law detector being a function of parameter μ. To quantify the magnitude of the instrument linearity performance, the Q was estimated using prelaunch Metop-C AMSU-A Thermal Vacuum Chamber (TVAC) data sets. The TVAC data were taken at three instrument temperatures (see Table 5) and the scene target was cycled at each instrument temperature through six temperatures 84, 130, 180, 230, 280, and 330 K, respectively [6,37]. According to the analysis in [1], the maximum (absolute) Q values for Metop-C AMSU-A instruments are about 0.6 K. Consequently, Metop-C AMSU-A instrument nonlinearities at all channels exceed the specification since the specification requires Q = 0.5 K for channels 1, 2, and 15, and $Q = 0.375$ K for other channels. This indicates the significance of applying the nonlinearity correction in the instrument calibration process. Table 5 shows the values of parameter μ at three instrument temperatures (low, nominal, and high). After launch, the μ values at the actual on-orbit instrument temperatures are interpolated from these three values. For channels 9–14 (AMSU-A1-1), two sets of the μ parameters are provided; one set is for the primary Phase Locked-Loop Oscillators (PLLO) #1 and the other one for the redundant PLLO #2 phase.

Therefore, the Metop-C AMSU-A Earth scene antenna and brightness temperatures can be determined using Equation (1) and Equation (6), along with the prelaunch-determined coefficients and corrections that are provided in Tables 4 and 5. The following two sections focus on the assessment of the instrument noise performance and the derived TDR and SDR data quality, correspondingly, since launch to 15 November 2019.

Table 5. Nonlinearity parameters μ in dimension of $(m^2\text{-sr-cm}^{-1})/mW$ for 15 Metop-C AMSU-A channels, which were derived in [1].

Ch. #	1st Instrument Temperature		2nd Instrument Temperature		3rd Instrument Temperature	
1	5.802		5.600		5.769	
2	2.236		2.192		2.145	
3	0.096		0.100		−0.076	
4	0.881		1.005		0.969	
5	0.597		0.724		0.597	
6	3.309		2.849		2.146	
7	3.180		2.698		2.011	
8	0.574		0.670		0.569	
	PLLO#1	PLLO#2	PLLO#1	PLLO#2	PLLO#1	PLLO#2
9	3.011	2.988	2.598	2.594	2.02	2.248
10	3.391	3.298	2.915	2.927	2.27	2.517
11	3.031	3.047	2.748	2.801	2.225	2.461
12	3.115	3.184	2.915	2.942	2.426	2.659
13	3.106	3.107	2.817	2.944	2.43	2.66
14	3.075	3.157	3.007	3.035	2.4	2.773
15	1.216		0.990		0.710	

Notes: For AMSU-A1, the three instrument temperatures are −2, 18, and 38 °C; for AMSU-A2, the three instrument temperatures are −7, 11.5, and 30 °C.

4. Instrument Noise Performance Assessment

Currently, the on-orbit NEDT performance of AMSU-A and other microwave instruments is characterized typically using gain-based statistical methods. In gain-based methods, the NEDT is defined as the quotient of the fluctuation (standard deviation or overlapping Allan deviation) of warm counts and the calibration gain during one orbit of observations [18,38]. The gain denotes the averaged sensitivity of calibration counts per Kelvin [32] (also refer to Appendix C). In particular, the overlapping Allan deviation [18,34,39,40] is employed in the NOAA Integrated Calibration/Validation System (ICVS), which is briefly described in Appendix B. Recently, the gain-based methods have been revealed to over-estimate the instrument noise because of an overrated temperature sensitivity to warm counts due to the use of the gain [19]. Nevertheless, the ICVS gain method (namely the ICVS method) has been widely applied to all AMSU-A and Microwave Humidity Sounder (MHS) instruments onboard NOAA-15, -16, -17, and -18, -19, Metop-A, and -B AMSU-A. To comply with legacy AMSU-A instrument noise analysis, in this study, the ICVS method continues to be applied to Metop-C AMSU-A for the one-year noise performance assessment, albeit the new method is used for comparison. Several important conclusions are discovered from our results, as described below.

Firstly, the Metop-C AMSU-A instrument has a stable noise performance for all channels except for channel 3. For demonstration, Figure 4 displays the AMSU-A specification, prelaunch and on-orbit NEDT at 15 channels on the first day (5 November 2018), the 90th day (15 February 2019), and one year (15 November 2019) after the launch. The AMSU-A channel noises from 1 to 2 and 4 to 15 are within the specification and are also lower than or comparable to the prelaunch values. However, the channel 3 NEDT is unstable and gradually exceeds the specification. To better understand this feature, Figure 5a displays the time series of the channel 3 NEDT from 15 November 2018 to 15 November 2019. The NEDT was mostly within the specification (0.4 K) prior to 7 April 2019, but it rises to the order of 1 K, which exceeds the specification from this point onwards. This feature is attributed to noisy calibration target counts. Figure 5b,c display the time series of daily mean and standard deviation for the same time period for warm load counts and cold counts, respectively. The warm count standard deviation apparently rises with time after March 2019, which directly causes a high overlapping Allan deviation. Meanwhile, the cold counts increase more rapidly than the warm counts, thus producing a degraded gain with time. For example, as of 15 November 2019, the daily mean warm count, cold count and gain at channel 3 have been changed by approximately 22.7%, 40.3% and −38.2% (decrease), respectively, compared with the first day of the data (i.e., 15 November 2018). Therefore, the increased overlapping Allan deviation, but decreased gain, produces a high NEDT, as shown in Figure 5a.

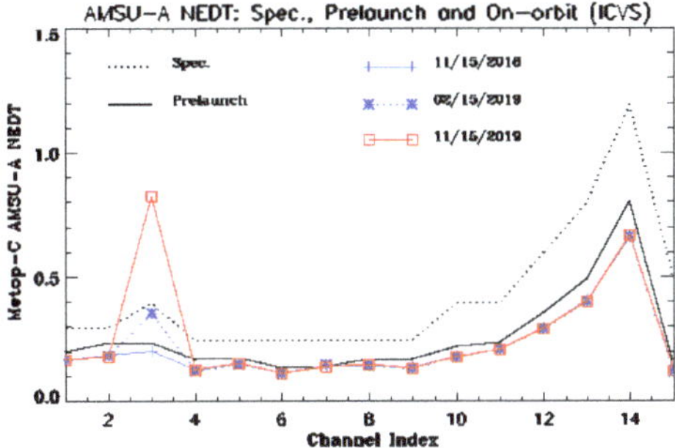

Figure 4. Metop-C AMSU-A specification, prelaunch and on-orbit noise equivalent differential temperature (NEDT) at 15 channels on the first day (15 November 2018), 90th day (15 February 2019), and one year (15 November 2019) after the launch.

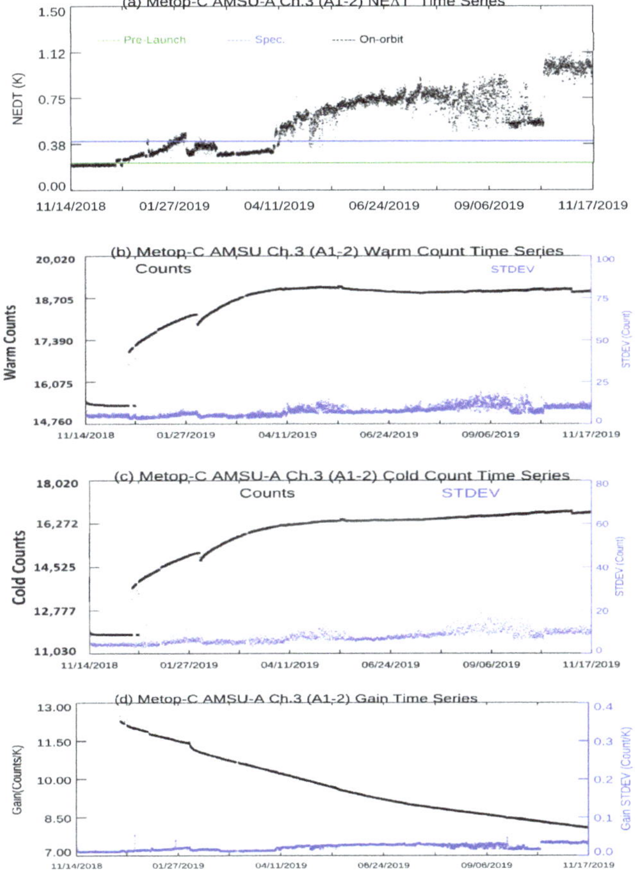

Figure 5. Time series of on-orbit NEDT, warm counts, cold counts and calibration gain for Metop-C AMSU-A channel 3 from 15 November 2018 to 15 November 2019, where the standard deviation of daily variations per parameter is included as the second Y-axis in the figures. (**a**) On-orbit NEDT. (**b**) Warm counts. (**c**) Cold counts. (**d**) Calibration gain.

Secondly, the Metop-C instrument exhibits slightly smaller noise values than two legacy AMSU-A instruments onboard Metop-A and -B satellites with some exceptions at channel 3. For demonstration, Figure 6a shows the results on 15 November 2019 among Metop-A to -C AMSU-A instruments, where Metop-A channels 7 and 8 are not available. It is also noted that the AMSU-A channel 3 for Metop-A/B/C has a higher NEDT value than the specification, which indicates that certain systematic performance issues remain with this channel.

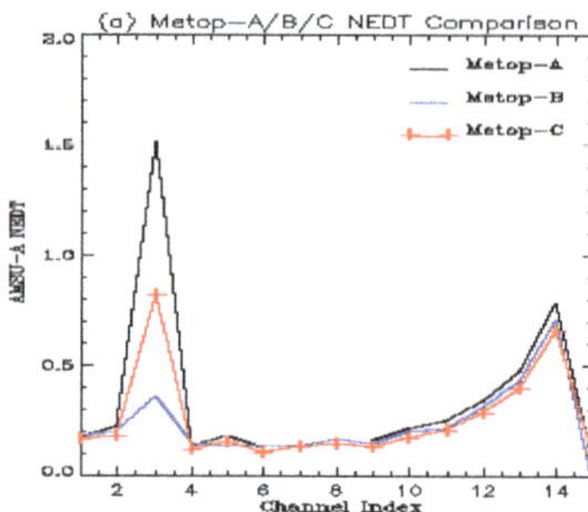
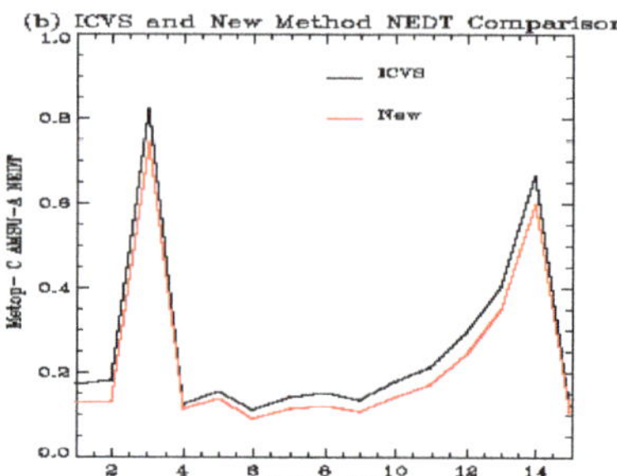

Figure 6. (**a**) NEDTs comparison on 15 November 2019 among Metop-A to -C AMSU-A instruments, where the NEDTs are computed using the Integrated Calibration/Validation System (ICVS) method and Metop-A AMSU-A channels 7 and 8 are not available. (**b**) NEDT value comparison estimated using the ICVS method and the new method.

Thirdly, the NEDT values estimated using the ICVS method are slightly higher than those using the new method (see Figure 6b) because of an overrated temperature sensitivity to warm counts [19]. Among all channels, the ICVS method produces relatively large errors in noise estimation in the first three channels compared with the new method. For example, the ICVS method causes an absolute error of 0.07 K in channel 3. The upper temperature sounding channels 10–14 are important for applications in NWP models especially. The ICVS method overestimated an error of around 0.05 K in those channels. In other words, the new method improves the accuracy of the noise estimate by 0.05 K. More discussions on the new method are conducted in [19].

Overall, channels 1–2 and 4–15 have demonstrated a stable noise performance within the specification since the launch. However, channel 3 displays an unstable noise feature and its NEDT constantly failed to meet the specification due to highly fluctuating warm counts and degraded channel gain over time.

5. AMSU-A TDR and SDR Quality Assessment

The quality assessment of Earth scene antenna (TDR) and brightness temperature (SDR) data are conducted by, respectively, using CRTM simulations and the inter-sensor comparison with legacy AMSU-A instruments flown on Metop-A and -B.

5.1. Comparisons with CRTM Simulations

This study focuses on a long-term stability assessment of the Metop-C AMSU-A TDR and SDR data quality by monitoring a one-year time series of AMSU-A observation (O) minus RTM simulation (B) differences from 15 November 2018 to 15 November 2019. Our observations represent either antenna temperatures (T_A) or SDR (T_B). The model simulations are computed using version 2.3 of the JCSDA CRTM [23,41,42], where we used the Fast Microwave Water Emissivity Model version 6 (FASTEM6) as the the oceanic microwave emissivity model [43,44]. The CRTM instrument characteristics for Metop-C AMSU-A are based on the specifications shown in Table 1 above. It is noted that the measured central frequency stability at channel 6 is from −4 to +10 MHz (not listed in Table 1) [27], slightly exceeding the upper limit of the specification (+5 MHz). The 10-MHz shift corresponds to the instrument temperature at 263 K, which is lower than the on-orbit Metop-C AMSU-A1-1 instrument temperature (typically above 282 K). In addition, our sensitivity test also shows that the shift of 10 MHz causes an error in the order of 0.05 K when simulating brightness temperatures (the figure is omitted). Thus, the shift beyond the specification is neglected in the following simulations. As ancillary data of atmospheric and surface properties for the CRTM model, this study uses ECMWF analysis data for surface conditions and atmospheric profiles [45,46]. For consistency, the simulations were only carried out over oceans under clear skies for both window and sounding channels. A legacy algorithm for cloud liquid water content (LWC) estimates over oceans [47] is employed to exclude cloud-contaminated data, where LWC smaller than 0.1 mm is considered a clear sky condition.

For demonstration, Figure 7a–d display four types of results about Metop-C AMSU-A antenna temperature (T_A) and brightness temperature (T_B) biases against CRTM simulations for the data spanning from 15 November 2018 to 15 November 2019. The graph in Figure 7a is the yearly mean T_A (black color) and T_B (pink color) biases vs. the channel. Generally, the T_A mean biases at sounding channels 4–14 are within −1 K, where the CRTM simulations are relatively accurate since they are less affected by errors in surface emissivity. However, the biases at three window channels (1, 2 and 15) and dirty sounding channel 3 are higher than 1.5 K. This inconsistency in the upper sounding channels is mostly due to RTM simulation errors because the simulation accuracy is very sensitive to errors in surface emissivity. For example, an emissivity error of 0.01 could cause an error in the order of 2 K at the abovementioned window and dirty sounding channels.

Compared with the T_A biases, the T_B biases are typically smaller for all AMSU-A channels except for the above window and dirty sounding channels due to inaccurate CRTM simulations. The reduced bias feature demonstrates the good performance of the conversion coefficients from TDR to SDR data. On the other hand, the standard deviations of all daily T_A and T_B mean biases during the same period are also included in Figure 7a, distributed from 0.05 to 0.3 K depending on the channel, with the largest standard deviation at channel 3. The relatively small standard deviation implies the decent stability of the data quality with time, while the largest standard deviation occurs at channel 3 due to its highly variable NEDT value with time. Regarding the standard deviation of the biases for all available pixels per day, they are large and are within the range from 0.2 K (upper sounding channels) to 2 K (window channels) (the figure is omitted).

The graph in Figure 7b illustrates the scan angle dependency of the yearly mean T_A and T_B biases at window channel 3 and sounding channels 5 and 10. It is well known that satellite microwave radiance (either T_A or T_B) can show a strong angle-dependent feature towards the two ends of the scanning swath, partly due to changes in the optical path length through the Earth's atmosphere between the Earth and the satellite [48]. A certain angle dependency still remains within both T_A and T_B biases at all channels. As shown in (b), the T_A biases from the nadir to the (left or right) end scanning positions show differences of more than 0.8 K for the sounding channels and more than 2 K for the window channel. The T_B biases typically exhibit a reduced and more uniform scan dependent bias compared to T_A. For example, at channel 5, the T_A biases change from −0.1 K at the nadir to −1.3 K at the right ending position (scan index 30), but the T_B biases change from 0.2 K to −0.6 K. A similar angle dependency feature exists at other channels (the figure is omitted).

To give a full picture of the magnitude and angle dependency of the biases with time, Figure 7c,d display the time series of daily mean T_A biases vs. time (X-axis) and scan position (Y-axis) for channels 3 and 5, respectively, covering the period from 15 November 2018 to 15 November 2019. Both Figure 7e,f are the same as Figure 7c,d except for the daily mean T_B biases. Channel 5 has a stable bias pattern for both T_A and T_B along with angles and time, although the channel 3 bias is slightly variable with day, which is partially caused by the NEDT feature in Figure 5a. Again, the T_B biases typically exhibit a reduced and more uniform scan-dependent bias compared to T_A, albeit the RTM simulation uncertainties remain at window channels. Similar conclusions are made at other channels (the figures are omitted). Currently, the derived antenna pattern correction (APC) coefficients have been delivered to a series of important users, including, but not limited to, the NOAA Microwave

Integrated Retrieval System (MiRS) [7], the NOAA Unique Combined Atmospheric Processing System (NUCAPS), the NOAA Environmental Modeling Center (EMC), the European Organisation for the Exploitation of Meteorological Satellites (EUMETSAT), the US Naval Research Laboratory (NRL), ECMWF, and the ATOVS (Advanced Television and infrared operational satellite Operational Vertical Sounder) and AVHRR (Advanced Very High Resolution Radiometer) Pre-Processing Package (AAPP).

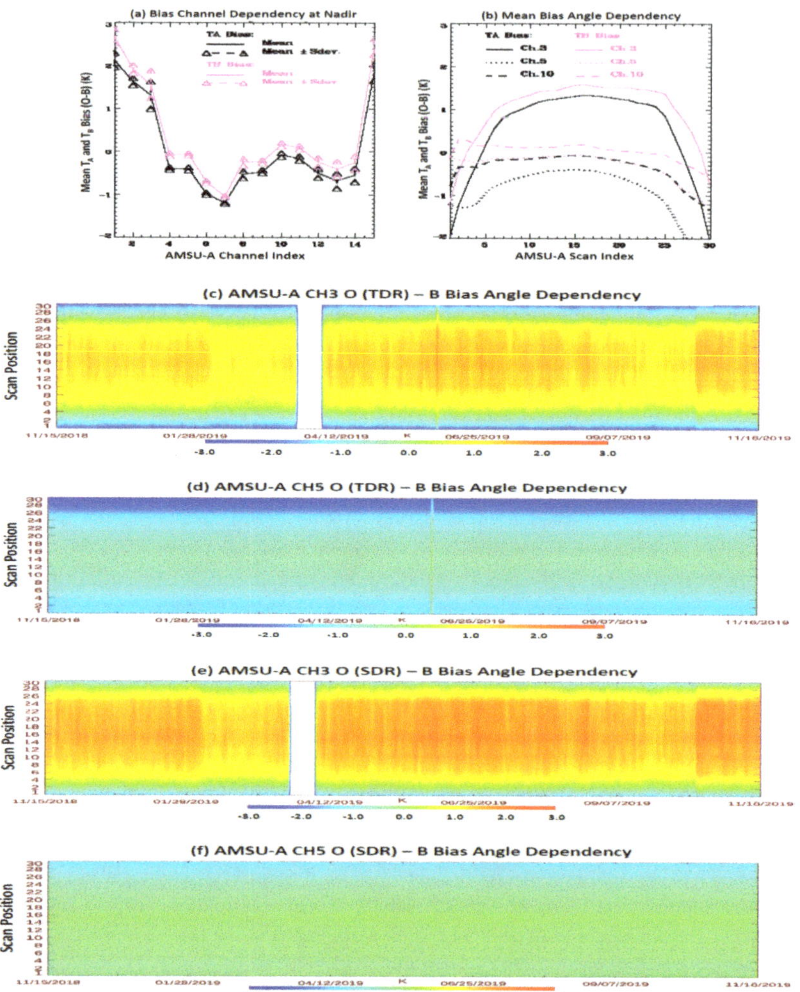

Figure 7. Metop-C AMSU-A antenna (T_A) and brightness (T_B) temperature biases against Community Radiative Transfer Model (CRTM) simulations from 15 November 2018 to 15 November 2019. (**a**) Yearly mean T_A (black) and T_B (pink) biases and standard deviation of all daily mean biases at the nadir direction vs. AMSU-A channel. (**b**) Yearly mean T_A (black) and T_B (pink) biases at channels 3, 5, and 10 vs. scan position. (**c**) Time series of channel 3 daily mean T_A bias vs. scan position from 15 November 2018 to 15 November 2019. (**d**) Same as (**c**) except for channel 5. (**e**) Same as (**c**) except for T_B bias. (**d**) Same as (**e**) except for channel 5.

The long-term stability of Metop-C AMSU-A TDR and SDR data quality has been validated by comparing the data to CRTM simulations from 15 November 2018 to 15 November 2019, showing a stable angular dependency feature against model simulations. Next, we investigated whether

Metop-C AMSU-A data quality is comparable to the data quality of legacy AMSU-A instruments flown on Metop-A and -B.

5.2. Metop-A, -B and -C Inter-Sensor Comparisons Using SNO Method

More than a decade ago, a technique was developed for accurately predicting the Simultaneous Nadir Overpasses (SNOs) of two Earth-orbiting satellites [24], which is referred as the SNO method. At each SNO, radiometers from both satellites view the same place at the same time at nadir, providing an ideal scenario for the intercalibration of radiometers aboard the two satellites. This technique was further improved to achieve the collocation of two passive-microwave satellite instrument SNO datasets with quality-controlled bilinear interpolation for window and surface-sensitive channels [25]. In this study, the inter-sensor comparisons among Metop-A, -B, and -C AMSU-A observations are performed based on double differences (DD) of SNO pairs between Metop-A, -B, and -C and each of NOAA-18 and -19 AMSU-A, where NOAA-18 or NOAA-19 AMSU-A is used as a transfer, as described below.

$$DD_{M_3-M_x}(N_{18}) = (M_3 - N_{18})_{SNO} - (M_x - N_{18})_{SNO}, \quad with\ x = 1,2 \tag{12}$$

and

$$DD_{M_3-M_x}(N_{19}) = (M_3 - N_{19})_{SNO} - (M_x - N_{19})_{SNO}, \quad with\ x = 1,2, \tag{13}$$

where M_1, M_2, and M_3 denote the Metop-B, -A, and -C individually for simplifying the length of the equations; and N_{18} and N_{19} are for NOAA-18 and -19, respectively.

All collocated AMSU-A SNO data sets are produced from the TDR data from Metop-A to -C and NOAA-18 and -19 from 30 November 2018 to 15 November 2019, all of which existed primarily in polar regions near 80° N and 80° S. To obtain more observations, each SNO pair is generated using 80-s temporal and 30-km spatial windows between two sensor observations. As discovered in previous studies, the large antenna temperature bias estimation uncertainties might remain within the SNO data sets for window and lower sounding channels, particularly over highly variable Earth scenes or cloudy conditions. Hence, an additional quality control (QC) is applied to check the inhomogeneity within field-of-view (FOV) for SNO pairs, as done in [26]. All pairs within an SNO event are removed from the collocated data sets if their standard deviation is greater than 2 K. Note that channels 7 and 8 for Metop-A AMSU-A and channel 15 for Metop-B AMSU-A are not operational during the selected data sets.

Figure 8a displays the averaged inter-sensor differences at 13 AMSU-A channels between Metop-C and Metop-A using either NOAA-18 (named N18 in the graphs for clarity) or NOAA-19 (named in the graphs as N19 for clarity) AMSU-A as a transfer. The results demonstrate that antenna temperatures from Metop-C AMSU-A are very comparable with those from Metop-A AMSU-A at the available channels. The differences (absolute values) at all channels, except for channel 3, are typically smaller than 0.3 K, by using either NOAA-18 or NOAA-19 AMSU-A as a transfer. Meanwhile, the differences are very comparable with two SNO references of NOAA-18 and NOAA-19 AMSU-A, except for channel 3. This is partially due to the large NEDT of Metop-A AMSU-A channel 3, which has a much high NEDT value (about 1.5 K), exceeding the specification and showing an unstable measurement performance.

Figure 8b shows the averaged inter-sensor differences at the 14 channels between Metop-C and Metop-B using either NOAA-18 or NOAA-19 AMSU-A as a transfer. Similar to the conclusion for Metop-C and -A, antenna temperatures from Metop-C AMSU-A are very comparable with those from Metop-B AMSU-A at all channels except for channel 15, which failed. The absolute differences at all available channels are typically smaller than 0.3 K and the differences are very comparable from two SNO references of NOAA-18 and NOAA-19 AMSU-A, except for channels 3 and 8. This deviation between two transfers is related to the noisy channel 8 of NOAA-19 with its high NEDT (0.9–1.2 K).

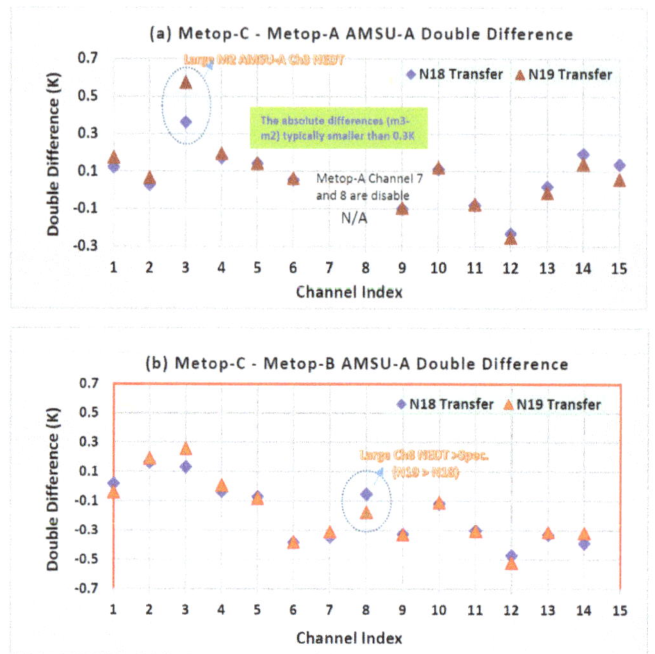

Figure 8. (**a**) Averaged inter-sensor differences at 13 AMSU-A channels between Metop-C and Metop-A using either NOAA-18 (named as N18 in the graphs for clarity) or NOAA-19 (named as N19 in the graphs for clarity) AMSU-A as a transfer. (**b**) Averaged inter-sensor differences at 14 AMSU-A channels between Metop-C and Metop-B using either NOAA-18 or NOAA-19 AMSU-A as a transfer.

Overall, the observed TDR and SDR data at all AMSU-A channels have shown a relatively stable quality since the launch. The higher NEDT at channel 3 has not had a critical impact on TDR and SDR data. The derived APC coefficients from TDR to SDR data have demonstrated a good performance in both deriving brightness temperatures and improving the asymmetrical bias features at most of the channels against the CRTM simulations. Moreover, the Metop-C AMSU-A data quality is comparable to Metop-A/B AMSU-A data, showing a decent quality and consistency with Metop-A/B AMSU-A.

6. Summary and Conclusions

This study presents an end-to-end Metop-C AMSU-A calibration and validation analysis. The calibration consists of the radiometric equation from Earth scene radiometric counts to antenna temperature and the conversion equation from antenna temperature to brightness temperature by removing side lobe contaminations resulting from cold space and satellite spacecraft. In the radiometric equation, the cold space temperature calibration correction due to antenna side lobe contaminations are derived using prelaunch antenna pattern functions, albeit the nonlinearity parameter is derived using the prelaunch TVAC data. Moreover, the optimal cold space view (SV) positions for Metop-C AMSU-A are determined based on initial OV data from 18 November 2018 to 30 November 2018, where SV1 (i.e., the satellite zenith angle of 83.3°) is determined for AMSU-A1 and SV3 (i.e., 80.0°) for AMSU-A2.

Next, the instrument noise performance is characterized using the NEDT, which is calculated primarily by using the current ICVS method [18] to enable a consistent analysis with that of legacy AMSU-A instruments, albeit the EUMETSAT and United Kingdom Met Office gain-based methods and a new method [19] are implemented for comparison. Channels 1–2 and 4–15 have demonstrated a stable noise performance within the specifications since the launch and up to 15 November 2019. However, channel 3 displays an unstable noise feature and is frequently higher than the specification (recently in the order of 1.0 K) due to highly fluctuating warm counts and degraded channel gain with time. Regarding the accuracy of the NEDT estimation using the ICVS method, the ICVS method is found to overestimate the NEDT against the new method by approximately 1–10%, depending on the channel.

Finally, the quality of Metop-C AMSU-A TDR and SDR data is comprehensively assessed by using the CRTM simulations and inter-sensor comparison with legacy AMSU-A onboard Metop-A and -B. Against the CRTM simulations, Metop-C AMSU-A TDR and SDR data at all AMSU-A channels have shown a relatively stable quality since the launch. The higher NEDT at channel 3 has not caused a vital impact on TDR and SDR data quality. The derived APC coefficients have demonstrated a good performance in both deriving brightness temperatures and improving the asymmetrical bias features at most of the channels against the CRTM simulations. On the other hand, the inter-sensor comparisons between Metop instruments, via either NOAA-18 or NOAA-19 AMSU-A as a transfer, have demonstrated that Metop-C AMSU-A data quality is comparable to Metop-A/B AMSU-A data, showing that Metop-C AMSU-A fits into the family of Metop series AMSU-A instruments.

However, residual biases remain in the calibration process. Particularly, brightness temperature biases at some channels are not close to zero and show certain residual symmetric angle dependence, where the biases towards the two ends of the scanning swath are slightly different. This feature is a common issue for all AMSU-A instruments. A few radiation perturbation components could contribute to the residual biases, which are neglected in the TDR to SDR conversion algorithm, e.g., antenna emissions and heterogeneity effects due to the difference in the Earth's radiation at different viewing angles [5]. In addition, possible instrument polarization misalignment might be an additional cause of the asymmetric feature [49]. In addition, the current calibration equation (see Equation (1) or (2)) is established to derive the Earth scene radiance or antenna temperature by using the warm load temperature as the starting point in the interpolation. This approach becomes questionable if the warm load PRT temperature is unstable with time. For example, a couple of Kelvin variations have remained in Metop-C AMSU-A since the launch. This instability with time could result in some errors in the derived Earth scene antenna and brightness temperatures. Alternatively, the calibration equation should be revised to use the cold space temperature as the starting point in the interpolation. This is a common issue for all AMSU-A instruments. Therefore, it is worth conducting a separate study to understand these common issues in more depth and to further improve the AMSU-A TDR and SDR data quality.

Author Contributions: Conceptualization, B.Y., J.C., and C.-Z.Z.; methodology, B.Y., J.C., C.-Z.Z., and K.A.; software, B.Y., J.C., K.A., H.Q., and T.Z.; validation, B.Y., J.C., C.-Z.Z., and K.A.; investigation, B.Y., J.C., C.-Z.Z., K.A. and K.G.; data curation, D.H. and J.G.; writing—original draft preparation, B.Y.; writing—review and editing, B.Y., C.-Z.Z., K.G., and J.G.; visualization, B.Y., J.C.; supervision, B.Y.; project administration, B.Y.; funding acquisition, B.Y. and J.G. All authors have read and agreed to the published version of the manuscript.

Funding: This research was funded by the NOAA Office of Projects, Planning and Analysis (OPPA).

Acknowledgments: The authors would like to thank Walter Asplund and Michael Honaker of the NASA Goddard Space Flight Center for providing AMSU-A System In-Orbit Verification (SIOV) information for the optimal cold space position analysis. The useful comments of Changyong Cao are also gratefully acknowledged. Thanks also go to Ninghai Sun and Zhaohui Zhang for their contributions to the previous and current AMSU-A CRTM simulations. We would also like to thank the four anonymous reviewers for providing many valuable suggestions. Finally, yet importantly, thanks go to Northrop Grumman Electronic Systems for providing a number of technical reports, which are available upon request from the authors.

Conflicts of Interest: The authors declare no conflict of interest.

Appendix A. Radiometric Calibration Counts (Blackbody and Cold Counts)

There are two samples of cold and warm count measurements per scan for AMSU-A1 and -A2 [27]. For each scan, the blackbody counts C_W and the space counts C_C are the averages of two samples of the internal black body and the space view, respectively.

$$\overline{C}_X(i) = \frac{C_{X_1}(i) + C_{X_2}(i)}{2}, \tag{A1}$$

where $C_X(i)$ (where $X = W$ or C) for the ith scan line. If any two samples differ by more than a preset limit of blackbody count variation ΔC_X (the initial limit is set to 3σ, where the standard deviation, σ, is calculated from the prelaunch calibration data C_X for each channel), the data in the scan should not be used. To further reduce the noise in the calibrations, C_X (where $X = W$ or C) for each scan line is convoluted over several neighboring scan lines according to the weighting function [29]

$$\overline{C}_X = \frac{\sum_{i=-n}^{n} W_i C_X(t_i)}{\sum_{i=-n}^{n} W_i} \tag{A2}$$

where t_i (when $i \ldots 0$) represents the time of the scan lines just before or after the current scan line and t_0 is the time of the current scan line. One can write $t_i = t_0 + i\Delta t$, where $\Delta t = 8$ s for AMSU-A. The $2n + 1$ values are equally distributed about the scan line to be calibrated. Following the NOAA-KLM operational preprocessor software, the value of $n = 3$ is chosen for all AMSU-A antenna systems. A set of triangular weights of 1, 2, 3, 4, 3, 2, and 1 are chosen for the weight factor W_i that appears in Equation (A2) for the seven scans with $i = -3, -2, -1, 0, 1, 2,$ and 3.

Appendix B. Blackbody Target Temperatures

Radiances for both AMSU-A1 and -A2 Earth views are derived from the radiometric counts and the calibration coefficients inferred from the internal blackbody and space view data. The physical temperatures of the internal blackbody targets are measured by platinum resistance thermometers (PRTs). As shown in [29], the PRTs were calibrated against 'standard' ones traceable to the National Institute of Standards and Technology (NIST) to measure the temperatures of the internal blackbody targets and have an accuracy of 0.1 K. The outputs of the telemetry are PRT counts, which must be converted to PRT temperatures. The normal approach for deriving the PRT temperatures from counts is a two-step process, in which the resistance of each PRT (in ohms) is computed by a count-to-resistance look-up table provided by its manufacturer. Then, the individual PRT temperature (in degrees) is obtained from an analytic PRT equation. Here, this has been compressed to a single step in a polynomial form, with negligible errors, using an existing method [29], i.e.,

$$T_{Wk} = \sum_{j=0}^{3} f_{kj} C_k^j \tag{A3}$$

where T_{Wk} and C_k^j represent the temperature and count of each PRT. The coefficients f_{kj} are provided for each PRT.

The mean blackbody temperature used in the calibration in Equation (1) (in the main body of the manuscript) T_W is a weighted average of all samples of the PRT temperatures per scan:

$$T_W = \frac{\sum_{k=1}^{m} W_k T_{Wk}}{\sum_{k=1}^{m} W_k} + \Delta T_W \tag{A4}$$

where m represents the number of PRTs for each antenna system and the scan index 'i' is omitted in the equation for clarification. For AMSU-A1, which includes channels 3–15, there are five measurement samples of warm load PRT temperatures per scan. For AMSU-A2, there are seven samples of warm load PRT temperatures per scan [27]. W_k is the weight assigned to each PRT and ΔT_w is the warm load correction factor for each channel, derived from the TVAC calibration data for three instrument temperatures (low, nominal, and high). Values for ΔT_W are provided for each instrument. For AMSU-A1-1, ΔT_W values for Phase Locked-Loop Oscillators (PLLO) #1 and PLLO #2 are provided separately. The W_k value, which equals 1(0) if the PRT is determined to be good (bad) before launch, will be provided for each flight model. If any of the PRT temperatures T_{Wk} differ by more than 0.2 K from their value in the previous scan line, then T_{Wk} should be omitted from the average in Equation (A4).

Appendix C. On-Orbit AMSU-A NEDT Methods

In the following descriptions, for clarity, the calculation method for AMSU-A instrument on-orbit NEDT in the ICVS is expressed as $NE\Delta T^{ICVS}$, whereas the new method in [19] is called $NE\Delta T^{New}$. A brief introduction without the channel index is given below, but detailed descriptions can be found in [18,19], correspondingly.

According to [18],

$$NE\Delta T^{ICVS} = \sqrt{\frac{1}{4(N-2)} \sum_{i=1}^{N-1} \frac{1}{\overline{G(i)}^2}\left[\left(C_{W_1}(i+1) - C_{W_1}(i)\right)^2 + \left(C_{W_2}(i+1) - C_{W_2}(i)\right)^2\right]} \qquad (A5)$$

with

$$\overline{G(i)} = \left|\frac{\left(\overline{C_W(i)} - \overline{C_C(i)}\right)}{\left(\overline{T_w(i)} - \overline{T_C(i)}\right)}\right|, \qquad (A6)$$

where N is the number of scans per orbit; 'i' is the scan index per orbit; $\overline{C_C(i)}$ and $\overline{C_W(i)}$ are the averages of two samples of cold and warm counts per scan, respectively, as defined in (A1); $\overline{T_W(i)}$ is the average of five samples (for AMSU-A1 channels) or seven samples (for AMSU-A2) of warm load PRT temperatures per scan; and $\overline{G(i)}$ is the averaged calibration gain per scan.

According to [19], the new NEDT method is described as follows.

$$NE\Delta T^{New} = \sqrt{\left(NE\Delta T_{C_W}\right)^2 + \left(NE\Delta T_{C_C}\right)^2 + \delta_{Cov(C_W, C_C)}} \qquad (A7)$$

where

$$\left(NE\Delta T_{C_W}\right)^2 = \frac{1}{4(N-2)} \sum_{i=1}^{N-1} \overline{\left(\frac{\partial T_A(i)}{\partial C_W(i)}\right)}^2 \left[\left(C_{W_1}(i+1) - C_{W_1}(i)\right)^2 + \left(C_{W_2}(i+1) - C_{W_2}(i)\right)^2\right], \qquad (A8)$$

$$\left(NE\Delta T_{C_C}\right)^2 = \frac{1}{4(N-2)} \sum_{i=1}^{N-1} \overline{\left(\frac{\partial T_A(i)}{\partial C_C(i)}\right)}^2 \left[\left(C_{C_1}(i+1) - C_{C_1}(i)\right)^2 + \left(C_{C_2}(i+1) - C_{C_2}(i)\right)^2\right], \qquad (A9)$$

$$\delta_{Cov(C_W,C_C)} = \frac{1}{4(N-2)} \sum_{i=1}^{N-1} \overline{\frac{\partial T_A(i)}{\partial C_W(i)} \cdot \frac{\partial T_A(i)}{\partial C_C(i)}} \times \left\{\sum_{k=1}^{2}\left(C_{W_k}(i+1) - C_{W_k}(i)\right)\left(C_{C_k}(i+1) - C_{C_k}(i)\right)\right\} \qquad (A10)$$

$$\frac{\overline{\partial T_A(i)}}{\partial C_W(i)} = \frac{\left(\overline{T_w(i)} - \overline{T_C(i)}\right)\left(\overline{C_C(i)} - \overline{C_S(i)}\right)}{\left(\overline{C_W(i)} - \overline{C_C(i)}\right)^2}, \qquad (A11)$$

$$\frac{\overline{\partial T_A(i)}}{\partial C_C(i)} = \frac{\left(\overline{T_w(i)} - \overline{T_C(i)}\right)\left(\overline{C_S(i)} - \overline{C_W(i)}\right)}{\left(\overline{C_W(i)} - \overline{C_C(i)}\right)^2}, \qquad (A12)$$

where $\overline{\frac{\partial T_A}{\partial C_W(i)}}$ and $\overline{\frac{\partial T_A}{\partial C_C(i)}}$ denote the scan-averaged derivatives. More detail about the above equations can be found in [19].

References

1. Zou, C.; Qian, H. *Prelaunch Calibration of the Advanced Microwave Sounding Unit-A Radiometer for MetOp-C*; NOAA/STAR Technical Report; NOAA STAR: Silver Spring, MD, USA, 2018.
2. Yan, B.; Chen, J.; Ahmad, K.; Zou, C.; Qian, H. *Metop-C Advanced Microwave Sounding Unit-A (AMSU-A) TDR/SDR Beta Maturity Status*; NOAA Office of Projects, Planning, and Analysis (OPPA): College Park, MD, USA, February 2019.
3. Yan, B.; Chen, J.; Zou, C.; Ahmad, K.; Qian, H. *Metop-C Advanced Microwave Sounding Unit-A (AMSU-A) TDR/SDR Provisional Maturity Status*; NOAA Office of Projects, Planning, and Analysis (OPPA): College Park, MD, USA, April 2019.
4. Yan, B.; Chen, J.; Zou, C.; Ahmad, K.; Qian, H. *NOAA-20 Advanced Microwave Sounding Unit-A (AMSU-A) TDR/SDR Validated Maturity Status*; NOAA Office of Projects, Planning, and Analysis (OPPA): College Park, MD, USA, August 2019.
5. Yan, B.; Ahmad, K. Derivation and validation of sensor brightness temperatures for advanced microwave sounding unit-A instruments. *IEEE Trans. Geosci. Remote Sens.* **2020**. [CrossRef]
6. Northrop Grumman Electronic Systems (NGES). *Calibration Log Book for AMSU-A1 SN 105*; Report No. 16729; Northrop Grumman: Azusa, CA, USA, 2010.
7. Boukabara, S.; Garrett, K.; Chen, W.; Iturbide-Sanchez, F.; Grassotti, C.; Kongoli, C.; Chen, R.; Liu, Q.; Yan, B.; Weng, F.; et al. MiRS: An all-weather 1dvar satellite data assimilation and retrieval system. *IEEE Trans. Geosci. Remote Sens.* **2011**, *49*, 3249–3272. [CrossRef]
8. Weng, F.; Yan, B. A microwave snow emissivity model. In Proceedings of the International TOVS Study Conference-XIII, Sainte-Adèle, QC, Canada, 29 October–4 November 2003. Available online: http://library.ssec.wisc.edu/research_Resources/publications/pdfs/ITSC13/weng01_ITSC13_2003.pdf (accessed on 3 August 2020).
9. Weng, F.; Zhu, T.; Yan, B. Satellite data assimilation in numerical weather prediction models. Part II: Uses of rain-affected radiances from microwave observations for hurricane vortex analysis. *J. Atmos. Sci.* **2007**, *64*, 3910–3925. [CrossRef]
10. Yan, B.; Weng, F.; Meng, H. Retrieval of snow surface microwave emissivity from the advanced microwave sounding unit. *J. Geophys. Res.* **2008**, *113*, D19206. [CrossRef]
11. Meng, H.; Dong, J.; Ferraro, R.; Yan, B.; Zhao, L.; Kongoli, C.; Wang, N.-Y.; Zavodsky, B. A 1DVAR-based snowfall rate retrieval algorithm for passive microwave radiometers. *J. Geophys. Res. Atmos.* **2017**, *122*, 6520–6540. [CrossRef]
12. Zhu, T.; Zhang, D.; Weng, F. Impact of the advanced microwave sounding unit measurements on hurricane prediction. *Mon. Wea. Rev.* **2002**, *130*, 2416–2432. [CrossRef]
13. Zou, X.L. Climate trend detection and its sensitivity to measurement precision. *Adv. Meteorol. Sci. Technol.* **2012**, *2*, 41–43.
14. Yan, B.; Weng, F. Effects of microwave desert surface emissivity on AMSU-A data assimilation. *IEEE Trans. Geosci. Remote. Sens.* **2011**, *49*, 1263–1276. [CrossRef]
15. Kelly, G.; Thépaut, J.-N. Evaluation of the impact of the space component of the Global Observation System through observing system experiments. *ECMWF Newsl.* **2017**, *113*, 16–28. Available online: http://www.ecmwf.int/publications/newsletters/pdf/113.pdf (accessed on 3 August 2020).
16. Zou, X.L.; Qin, Z.; Weng, F. Impacts from assimilation of one data stream of AMSU-A and MHS radiances on quantitative precipitation forecasts, Q.J.R. *Meteorol. Soc.* **2017**, *143*, 731–743. [CrossRef]
17. Zhu, K.; Xue, M.; Pan, Y.; Hu, M.; Benjamin, S.G.; Weygandt, S.S.; Lin, H. The impact of satellite radiance data assimilation within a frequently updated regional forecast system using a GSI-based ensemble kalman filter. *Adv. Atmos. Sci.* **2019**, *36*, 1308–1326. [CrossRef]
18. Tian, M.; Zou, X.; Weng, F. Use of allan deviation for characterizing satellite microwave sounder noise equivalent differential temperature (NEDT). *IEEE Geosci. Remote Sens. Lett.* **2015**, *12*, 2477–2480. [CrossRef]

19. Yan, B.; Kireev, S. A new methodology about characterizing on-orbit advanced microwave sounding unit—A (AMSU-A) earth-scene temperature noise equivalent differential temperature. *IEEE Trans. Geosci. Remote Sens.* **2020**. in revision.

20. Ohring, G.; Wielicki, B.; Spencer, R.; Emery, B.; Datla, R. Satellite instrument calibration for measuring global climate change: Report of a workshop. *Bull. Amer. Meteorol. Soc.* **2005**, *86*, 1303–1313. [CrossRef]

21. Wielicki, B.; Young, D.F.; Mlynczak, M.G.; Mlynczak, M.G. Achieving climate change absolute accuracy in orbit. *Bull. Am. Meteorol. Soc.* **2013**, *94*, 1519–1539. [CrossRef]

22. Zou, C.Z.; Wang, W. Intersatellite calibration of AMSU-A observations for weather and climate applications. *J. Geophys. Res.* **2011**, *116*, D23113. [CrossRef]

23. Han, Y. JCSDA community radiative transfer model (CRTM)—Version 1. *NOAA Tech. Rep.* **2006**, *122*, 1–33.

24. Cao, C.; Weinreb, M.; Xu, H. Predicting simultaneous nadir overpasses among polar-orbiting meteorological satellites for the intersatellite calibration of radiometers. *J. Atmos. Ocean. Technol.* **2004**, *21*, 537–542. [CrossRef]

25. Iacovazzi, R.; Cao, C.; Cao, C. Reducing uncertainties of sno-estimated intersatellite amsu-a brightness temperature biases for surface-sensitive channels. *J. Atmos. Oceanic Technol.* **2007**, *25*, 1048–1054. [CrossRef]

26. Yan, B.; Weng, F. Intercalibration between special sensor microwave imager/sounder and special sensor microwave imager. *IEEE Trans. Geosci. Remote Sens.* **2008**, *46*, 984–995.

27. Northrop Grumman Electronic Systems (NGES). *AMSU-A System Operation and Maintenance Manual for METSAT/METOP*; Technique Report NAS 5-32314 CDRL 307; NGES: Linthicum Heights, MD, USA, 2010; pp. 105–109.

28. Mo, T. AMSU-A antenna pattern corrections. *IEEE Trans. Geosci. Remote Sens.* **1999**, *37*, 103–112.

29. Mo, T. Prelaunch calibration of the advanced microwave sounding unit-A for NOAA-K. *IEEE Trans. Microw. Theory Technol.* **1996**, *44*, 1460–1469. [CrossRef]

30. Mo, T. *Calibration of the Advanced Microwave Sounding Unit-A Radiometer for MetOp-A*; NOAA Technical Report NESDIS 121; NOAA NESDIS: Silver Spring, MD, USA, 2006.

31. Mo, T.; Kigawa, S. A study of lunar contamination and onorbit performance of the NOAA-18 advanced microwave sounding unit-A. *J. Geophys. Res.* **2007**, *112*, D20124. [CrossRef]

32. Mo, T. Postlaunch calibration of the MetOp-A advanced microwave sounding unit-A. *IEEE Trans. Geosci. Remote Sens.* **2008**, *46*, 3581–3600. [CrossRef]

33. Mo, T.; Weinreb, M.; Grody, N.; Wark, D. *AMSU-A Engineering Model Calibration*; NOAA Technical Report NESDIS 68; NOAA NESDIS: Silver Spring, MD, USA, 1993.

34. Weng, F.; Zou, X.; Sun, N.; Yang, H.; Tian, M.; Blackwell, W.J.; Wang, X.; Lin, L.; Anderson, K. Calibration of Suomi national polar-orbiting partnership advanced technology microwave sounder. *J. Geophys. Res.* **2013**, *118*, 1–14. [CrossRef]

35. Weng, F.; Zou, X. Errors from Rayleigh–Jeans approximation in satellite microwave radiometer calibration systems. *Appl. Opt.* **2013**, *52*, 505–508. [CrossRef]

36. Northrop Grumman Electronic Systems (NGES). *Calibration Log Book for the Advanced Microwave Sounding Unit-A (AMSU-A)*; Report#10481A; Northrop Grumman Electronic Systems: Aerojet, Azusa, CA, USA, 1995.

37. Northrop Grumman Electronic Systems (NGES). *Calibration Log Book for AMSU-A2 SN 107*; Report No. 15267; Northrop G Grumman: Azusa, CA, USA, 2008.

38. Hans, I.; Burgdorf1, M.; John, V.O.; Mittaz, J.; Buehler, S.A. Noise performance of microwave humidity sounders over their life time. *Atmos. Meas. Tech. Discuss.* **2017**, *10*, 4927–4945. [CrossRef]

39. Allan, D.W. Statistics of atomic frequency standards. *Proc. IEEE* **1966**, *54*, 221–230. [CrossRef]

40. Allan, D.W. Should classical variance be used as a basic measure in standards metrology? *IEEE Trans. Instrum. Meas.* **1987**, *36*, 646–654. [CrossRef]

41. Ding, S.; Yang, P.; Weng, F.; Liu, Q.; Han, Y.; van Delst, P.; Li, J.; Baum, B. Validation of the community radiative transfer model. *J. Quant. Spectrosc. Radiat. Transf.* **2011**, *112*, 1050–1064. [CrossRef]

42. Johnson, B.; Zhu, T.; Chen, M.; Ma, Y.; Auligné, T. The Community Radiative Transfer Model (CRTM). In Proceedings of the 21st International TOVS Study Conference, Darmstadt, Germany, 29 November 2017. Available online: https://cimss.ssec.wisc.edu/itwg/itsc/itsc21/program/29november/0900_1.01_JCSDA_CRTM_ITSC-21_BJohnson_final.pdf (accessed on 3 August 2020).

43. Liu, Q.; Weng, F.; English, S.J. An improved fast microwave water emissivity model. *IEEE Trans. Geosci. Remote Sens.* **2011**, *49*, 1238–1250. [CrossRef]

44. Bormann, N.; Geer, A.; English, S.J. *Evaluation of the Microwave Ocean Surface Emissivity Model FASTEM-5 in the IFS*; Technical Report Technical Memorandum 667; ECMWF: Reading, UK, 2012.
45. Ingleby, B. An Assessment of Different Radiosonde Types 2015/2016. ECMWF Technical Memoranda, ECMWF Shinfield Park Reading RG2 9AX United Kingdom: 2017. Available online: https://www.ecmwf.int/en/publications (accessed on 3 August 2020).
46. Carminati, F.; Migliorini1, S.; Ingleby, B.; Bell, W.; Lawrence, H.; Newman, S.; Hocking, J.; Smith, A. Using reference radiosondes to characterise NWP model uncertainty for improved satellite calibration and validation. *Atmos. Meas. Tech.* **2019**, *12*, 83–106. [CrossRef]
47. Weng, F.; Zhao, L.; Ferraro, R.; Poe, G.; Li, X.; Grody, N. Advanced microwave sounding unit cloud and precipitation algorithms. *Radio Sci.* **2003**, *38*, 8086–8096. [CrossRef]
48. Goldberg, M.D.; Crosby, D.S.; Zhou, L.H. The limb adjustment of AMSU-A observations. *J. Appl. Meteorol.* **2001**, *40*, 70–83. [CrossRef]
49. Yang, W.; Meng, H.; Ferraro, R. Cross-scan asymmetry of AMSU-A window channels: Characterization, correction, and verification. *IEEE Trans. Geosci. Remote Sens.* **2013**, *51*, 1514–1530. [CrossRef]

 *remote sensing*

Article

Inter-Calibration of AMSU-A Window Channels

Wenze Yang [1],*, Huan Meng [2] **, Ralph R. Ferraro [2]** **and Yong Chen [3]**

1   I. M. Systems Group, 5830 University Research Court, College Park, MD 20740, USA
2   Center for Satellite Applications and Research, National Environmental Satellite Data and Information
    Service, National Oceanic and Atmospheric Administration, 5825 University Research Court, Suite 4001,
    College Park, MD 20740-3823, USA; Huan.Meng@noaa.gov (H.M.); Ralph.R.Ferraro@noaa.gov (R.R.F.)
3   Global Science and Technology, Inc., 5830 University Research Court, College Park, MD 20740, USA;
    Yong.Chen@noaa.gov
*   Correspondence: Yang.Wenze@imsg.com

Received: 4 August 2020; Accepted: 9 September 2020; Published: 14 September 2020

Abstract: More than one decade of observations from the Advanced Microwave Sounding Unit-A
(AMSU-A) onboard the polar-orbiting satellites NOAA-15 to NOAA-19 and European Meteorological
Operational satellite program-A (MetOp-A) provided global information on atmospheric temperature
profiles, water vapor, cloud, precipitation, etc. These observations were primarily intended for
weather related prediction and applications, however, in order to meet the requirements for climate
application, further reprocessing must be conducted to first eliminate any potential satellites biases.
After the geolocation and cross-scan bias corrections were applied to the dataset, follow-on research
focused on the comparison amongst AMSU-A window channels (e.g., 23.8, 31.4, 50.3 and 89.0 GHz)
from the six different satellites to remove any inter-satellite inconsistency. Inter-satellite differences
can arise from many error sources, such as bias drift, sun-heating-induced instrument variability in
brightness temperatures, radiance dependent biases due to inaccurate calibration nonlinearity, etc.
The Integrated microwave inter-calibration approach (IMICA) approach was adopted in this study
for inter-satellite calibration of AMSU-A window channels after the appropriate standard deviation
(STD) thresholds were identified to restrict Simultaneous Nadir Overpass (SNO) data for window
channels. This was a critical step towards the development of a set of fundamental and thematic
climate data records (CDRs) for hydrological and climatological applications. NOAA-15 served as
the main reference satellite for this study. For ensuing studies that expand to beyond 2015, however,
it is recommended that a different satellite be adopted as the reference due to concerns over potential
degradation of NOAA-15 AMSU-A.

Keywords: inter satellite calibration; microwave radiometry; passive microwave remote sensing;
AMSU-A

---

1. Introduction

Satellite measurements and derived meteorological products from the Polar Operational
Environmental Satellite (POES) system have demonstrated their capability in Numerical Weather
Prediction (NWP). The long time series of these measurements make them candidates for use in
climate monitoring and assessment. In 2004, a panel of experts convened and developed a strategy
for the generation of Climate Data Records (CDR) from satellite observations [1]. In that document,
CDR was defined as "a time series of measurements of sufficient length, consistency, and continuity
to determine climate variability and change", and segmented into Fundamental CDR (FCDR), at the
radiance/reflectance level, and Thematic CDR (TCDR), at the geophysical product level. Thereafter,
National Oceanic and Atmospheric Administration (NOAA)'s National Climatic Environmental

Information (NCEI) launched the NOAA's CDR program, which included several types of CDRs, including atmospheric and sea surface temperatures, snow and ice conditions, precipitation and clouds.

In order to achieve CDR quality from a satellite time series like AMSU-A, which has spanned seven different satellites since 1998, inter-satellite calibration is perhaps the most critical step, and is commonly accomplished through direct comparisons of collocated observations from pairs of satellite instruments (i.e., NOAA-15 with NOAA-16, etc.). Through these comparisons, systematic calibration transfer functions can be generated to correct the radiometric biases of the monitored sensors in References [2–4]. Numerous inter-satellite calibration approaches have emerged recently, and can be classified into two broad categories according to their sensor scanning types: simultaneous nadir overpass (SNO) for cross-track scanning sensor pairs like AMSU-A [5] and simultaneous conical overpass (SCO) for conical scanning sensor pairs like the Special Sensor Microwave Imager (SSM/I) [6]. SCO uses relatively stable on-Earth targets as references for sensor comparisons, e.g., vicarious cold target [7] and warm target [8], because of the relatively stable earth incidence angle of the conically scanning instruments. To introduce cross-platform calibration (XCAL), a unified calibration was developed to adapt comparison results from a group of approaches [9]—most of these approaches have been applied to the SSM/I to develop its FCDR [10].

The space-borne passive microwave measurements, especially from Advanced Microwave Sounding Unit-A (AMSU-A), first launched on May 13, 1998 onboard NOAA-15 (N15), and later NOAA-16 (N16) through NOAA-19 (N19), and the European Meteorological Operational satellite program-A (MetOp-A), have been identified as the largest contributor to the improvement of the global 24-h forecast skills by about 25% [11]. Among the fifteen AMSU-A channels, Channels 1–3 and 15, with the nominal central frequencies at 23.8, 31.4, 50.3 and 89.0 GHz, are collectively referred to as window channels, as their weighting functions peak at or near the surface, and they are more transparent in the atmosphere than other channels. Radiances and Brightness Temperatures (TB's) from these channels are operationally used by two product systems at NOAA's National Environmental Satellite, Data and Information Service (NESDIS): the Microwave Surface and Precipitation Products System (MSPPS) [12], and the Microwave Integrated Retrieval System (MIRS) [13], to retrieve a suite of hydrological products, including cloud liquid water (CLW), total precipitable water (TPW), sea ice, snow cover, as well as relevant variables, including land surface temperature and surface emissivity of 23.8, 31.4 and 50.3 GHz. Radiance observations of these channels are also routinely assimilated into NWP and reanalysis systems such as Climate forecasting System Reanalysis (CSFR) [14] and the Modern-Era Retrospective Analysis for Research and Applications (MERRA) [15]. It should be noted that, even though the AMSU-A temperature sounding data is assimilated in ERA-Interim [16], which is produced by European Center for Medium-Range Weather Forecasts (ECMWF), AMSU-A window channels are not included [16].

As a follow on study from our previous geolocation [17] and cross-scan bias corrections [18], the main goal of this work is to inter-calibrate the window channels of AMSU-A instruments onboard six satellites to a common calibration standard for use in developing CDRs. After an exhaustive trial of several methods, such as vicarious cold reference, and various SNO approaches in the earlier stage of the study, we selected the IMICA algorithm for the intersatellite calibration of AMSU-A window channels. This approach, developed by Zou et al. [19,20], extensively uses SNO to solve calibration coefficients and remove/minimize inter-satellite biases, captures the major causes of the inter-satellite inconsistency, and accounts for scene homogeneity, etc. Note that until now almost all published inter-calibration work related to AMSU-A have focused on the sounding channels in the oxygen absorption region, which are Channels 4–14, and their corresponding physical variables. For instance, the work in Reference [19] inter-calibrated FCDR for these sounding channels, and produced TCDR of atmospheric temperature at different atmospheric levels from the FCDR, while the work in References [21,22] produced temperature TCDR using operationally calibrated brightness temperatures. In comparison, physical variables from window channels are more closely related to the surface (e.g., see ice and wind speed over the sea), precipitation, and integrated atmospheric water since these channels are sensitive to water vapor and liquid water emission. Inconsistency in the data set may cause various problems in the follow on applications. For instance, according to Robertson et al. [23], it is primarily through the window channels that the AMSU-A instrument can affect the water vapor increments in MERRA.

This paper is organized as follows. Section 2 presents the materials and methods, including the simultaneous nadir overpass (SNO) method and how it is used in our application, the Integrated Microwave Inter-Calibration Approach (IMICA) used to correct the warm target contamination and non-linearity problem, and two other error sources: bias drift and frequency shift. Section 3 demonstrates the effectiveness of the inter-satellite calibration approaches through time series of both FCDR and TCDR. Further discussions and concluding remarks are presented in Section 4.

2. Materials and Methods

2.1. Sno Overview

Originally proposed by Cao et al. [24], the simultaneous nadir overpass (SNO) approach is widely used to perform inter-satellite calibration for both microwave sensors [5,19] and infrared sensors [25,26] onboard polar orbiting satellites. "Nadir" refers to the observations with a zero degree of sensor scan angle or earth incidence angle, so as to eliminate uncertainties associated with difference of atmospheric paths and viewing geometries. Since AMSU-A does not have nadir observations, the two observations closest to nadir were combined as the nadir scene. "Simultaneous" requires temporal and spatial restrictions to describe the event of two satellites meeting. These requirements typically vary depending on the sensor configuration and orbital overpass times and may limit the number of SNO's. There are some general rules for selecting the spatial and temporal thresholds for SNO. In considering spatial threshold, it is optimal to adopt a distance between the nadir scenes of the two satellites close to the sensor spatial resolution [20]. For AMSU-A, the spatial threshold should be about 50–75 km since its nadir field of view is 48 km. The major consideration for temporal restriction is the tradeoff between observational change and the number of matches. A temporal threshold of 50 s was used in this study to provide an adequate number of matchups and still maintains any meteorological changes at a minimum. Further discussion about this threshold is provided later in this paper. Additionally, for the AMSU-A window channels, special consideration must be taken to ensure that the pair of nadir observations is homogenous, as they are much more sensitive to the surface compared to the sounding channels [20]. Sensitivity tests were performed, following the investigation of SNO performance of all the satellite pairs, which confirms that our selections of SNO thresholds are valid.

2.1.1. Temporal Features and Number of SNO Pairs

Due to the different orbital configuration of each NOAA and MetOp satellite, such as the time of local ascending node, altitude, inclination, orbital drift rate, and central frequencies of window channels (Tables 1 and 2, and more details in Reference [27]), patterns of SNO events vary between different pairs of satellites. For example, SNO events occur between N15 and N16, roughly every 8 days, while SNOs between N18 and N19 occur in 8 consecutive days when their orbits are close, but then, SNO's do not occur for over 300 days. A comprehensive list of the SNO overlap and interval patterns is given in Table 3.

Note the numbers in Table 3 are average numbers for the period of record. There are slight variations over the period of satellite operation due to changes in satellite drifts. For example, in early 2001, the SNO interval time for the N15 vs. N16 pair was 8.07 days, this interval time increased to 8.20 days when it had a global SNO in August 2008, and stepped down to 8.17 days in late 2010.

Table 1. Satellite orbital parameters.

	Launch Date	Decommission Date	Altitude (km)	Period (min)	Inclination (deg)	Precession Rate (min/mon)
NOAA-15	05/13/1998		807	101.10	98.5	1.05
NOAA-16	09/21/2000	06/09/2014	849	102.00	99.0	3.00
NOAA-17	06/24/2002	04/10/2013	810	101.20	98.7	−4.62
NOAA-18	05/20/2005		854	102.12	98.7	3.52
MetOP-A	10/19/2006		817	101.36	98.7	
NOAA-19	02/06/2009		870	102.14	98.7	0.77

Table 2. Satellite central frequencies.

	Central Frequency (GHz)			
	Channel 1	Channel 2	Channel 3	Channel 15
NOAA-15	23.800013593	31.399992238	50.299988043	89.000016571
NOAA-16	23.800013593	31.399992238	50.299988043	89.000016571
NOAA-17	23.799204154	31.399662466	50.299178603	89.000076529
NOAA-18	23.799204154	31.399662466	50.299178603	88.999986591
MetOP-A	23.799204154	31.399662466	50.299178603	89.000076529
NOAA-19	23.799204154	31.399662466	50.299178603	89.000076529

Table 3. Overlap and interval days (in parentheses) for each satellite pairs.

	NOAA-16	NOAA-17	NOAA-18	MetOP-A	NOAA-19
NOAA-15	1 (8.16)	4.5 (104)	1 (7.31)	1 (31.7)	1 (7.14)
NOAA-16		1 (8.44)	3 (82.0)	1 (11.2)	2 (66.0)
NOAA-17			1 (7.66)	2 (40.0)	1 (7.52)
NOAA-18				1 (9.81)	8 (326.0)
MetOP-A					1 (9.62)

2.1.2. Spatial Features

Due to the orbital properties of sun synchronous, polar orbiting satellites, most SNO events between satellite pairs occur near polar regions over the 70°N to 80°N and 70°S to 80°S bands [20,24]. In some instances, an SNO could occur at lower latitudes, which provides additional information on inter-satellite differences in warmer atmospheric conditions and allows for a wider dynamic range to perform the inter-satellite calibration. [28]. Figure 1 shows the global SNO for three satellite pairs: N15 vs. N16 in August, 2008, N17 vs. MetOp A in April and May, 2009, and N18 vs. N19 in September, 2009. Using the time and space criteria of 50 s and 50 km, respectively (N18 vs. N19 pair needs a larger distance of 75 km to allow for enough number of observations due to their very close equatorial crossing times), SNOs occurred at all latitudes between 81°S to 81°N. As can be seen, the differences at 23.8 GHz brightness temperatures are mostly within 3 K, and the largest TB differences come from coast regions. Global SNO greatly enhances the spatial coverage and number of observations for further analysis.

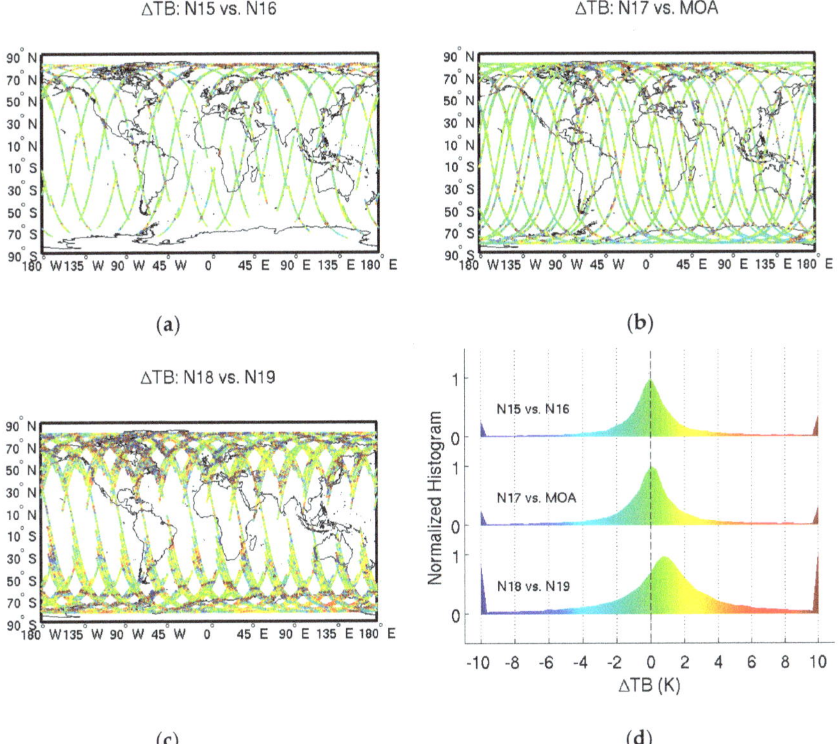

(a)   (b)

(c)   (d)

Figure 1. Geographical distribution of 23.8 GHz brightness temperature difference where global SNOs occur between (**a**) N15 and N16, (**b**) N17 and MOA, (**c**) N18 and N19. (**d**) shows normalized histogram of the brightness temperature difference of the three pairs. The SNO is defined as observation from two satellites with time difference within 50 s, and distance within 50 km (NOAA-18 vs. NOAA-19 pair needs larger distance as 75 km). The color bar of ΔTB and associated normalized histogram of these three cases are illustrated in the lower right panel. In the labels of the figures, N15 is short for NOAA-15, and MOA is short for MetOP-A.

2.1.3. Brightness Temperature Time Series of SNO Pairs

After investigating the overlap pattern and spatial distribution of the SNO, the next step is to evaluate the TB difference time series, which provides the opportunity to gain a better understanding of the inter-satellite differences. Figure 2 shows the SNO time series between AMSU-A onboard N15 with other satellites, i.e., N16 through the N19 and MetOp-A. N17 AMSU-A1 modules (containing 50.3 and 89 GHz) stopped working in November 2003. However, the AMSU-A2 module continued to operate, and thus, the time series of 23.8 and 31.4 GHz channels could extend through 2010. Nevertheless, the N17 time series generate the most outliners, especially for 23.8 and 31.4 GHz channels. In most cases, the TB difference is within 3 K, which is consistent with the observations in Section 2.1.2 and Figure 1.

The brightness temperature difference (ΔTB) of SNO pairs may come from various sources. For instance, Table 1 also lists the central frequencies of these AMSU-A window channels onboard difference satellites, and one can find that they fall into two groups: the first group includes NOAA-15 and NOAA-16, and the second group includes the other satellites. The channels within a group are generally the same, but there is a slight difference between different groups, e.g., 0.8 MHz difference for 23.8 and 50.3 GHz channel, 0.3 MHz difference for 31.4 GHz channel, and 60 KHz difference for the 89.0 GHz channel. In addition, although we have applied a geolocation correction [17], there still might be residual variations in view angles. These sensor factors inevitably lead to systematic ΔTB, although they might be small and hard to quantify as separate error components.

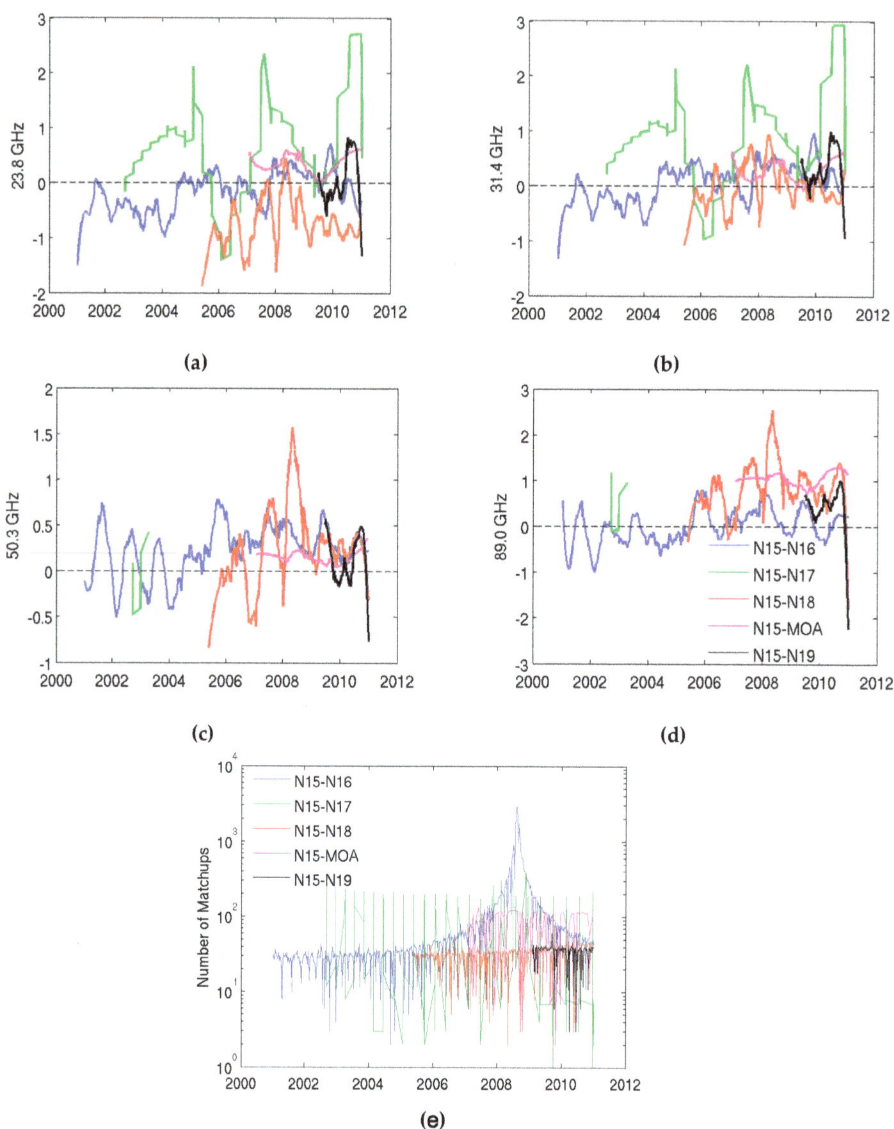

Figure 2. Smoothed brightness temperature differences of SNO pairs between AMSU-A onboard N15 with N16 through N19 and MetOP-A, from their operational time till the end of 2010 for (**a**) 23.8, (**b**) 31.4, (**c**) 50.3 and (**d**) 89 GHz. The typical uncertainty on the difference is about the same order of the difference. (**e**) Shows number of SNO matchups between N15 and other satellites, in the logarithm scale.

2.1.4. Sensitivity Test of Brightness Temperature Difference

Figures 1 and 2 display the spatial and temporal distributions of brightness temperature differences (ΔTB) between the nadir scenes of the SNO satellite pairs. Since the AMSU-A window channels, i.e., 23.8, 31.4, 50.3, and 89 GHz, are sensitive to surface conditions, we need to examine the impact of certain factors on the ΔTB between satellite pairs. These factors include the distance between the nadir scenes of the satellite pairs, time difference between the satellite pairs passing over the same location, and brightness temperature contrasts (BTC) between the two central observations, i.e., beam positions 15 and 16 of AMSU-A, which together form the nadir scene. Note that BTC is computed for each individual satellite within the SNO satellite pairs. A large value of BTC indicates that the observations from the two beam positions may come from different surface types, e.g., one beam position is mainly observed over ocean while the other is mainly over land. This is the primary difference between window channels and sounding channels, since the latter are not sensitive to various surface types. Table 4 shows the average correlation coefficients between SNO ΔTB and various factors from more than 50,000 SNO occurrences. Higher absolute values in Table 4 indicate more important factors.

Table 4. Correlation coefficients between ΔTB and related factors.

Correlation Coefficients	23.8 GHz	31.4 GHz	50.3 GHz	89 GHz
(NOBS)	(53,531)	(53,531)	(53,534)	(53,506)
Distance	0.19	0.18	0.15	0.18
S1 * BTC	0.53	0.53	0.47	0.43
S2 * BTC	0.55	0.55	0.5	0.44
Time Difference	−0.01	−0.01	−0.01	−0.01

* S1 and S2 indicate the two satellites in a SNO pair, e.g., in the pair NOAA-15 vs. NOAA-16, S1 refers to NOAA-15 and S2 refers to NOAA-16.

The sensitivity test reveals that the BTCs of both satellites produce the largest correlation coefficients with ΔTB so BTC serves as the most important factor to explain the variation of ΔTB. A small BTC usually suggests that the scene is more homogeneous and leads to small ΔTB and vice versa. For this reason, the BTC information is used to screen out the potential inhomogeneous SNO pairs. In contrast, the distance between satellite pairs can only explain a fraction of the ΔTB variation since a longer distance may potentially increase the possibility of having different surface types between observed scenes. Finally, the time difference between the satellites essentially plays no role in explaining any differences, as both the atmospheric and surface conditions are relatively stable within the time frame of 50 s. For this reason, the time difference threshold is more relaxed in some studies, e.g., it was 300 s in the study by John et al. [28]. In our study, the time difference relates more to the number of matchup observations. As the number of polar SNOs was large enough, and the number of global SNO was mainly determined by the meeting collocation pattern of satellite pairs (Section 2.1.2), we kept the time difference at 50 s.

Optimizing the uncertainties of inter-satellite SNO ΔTB requires a proper threshold for BTC that is small enough to ensure scene homogeneity, yet large enough to allow for a sufficient number of observations (NOBS). Since these channels have different sensitivities to the surface, the BTC threshold can vary with channels. A series of sensitivity tests were performed based on NEΔT using the SNO matchups between N15 and N16 (Figure 3). Based on the analysis, the threshold was set at 10xNEΔT.

2.2. Warm Target Contamination and Correction

The inter-satellite differences of ΔTB revealed some inter-calibration issues, such as bias drift, sun-heating-induced instrument variability in TB, scene temperature dependent biases due to inaccurate calibration nonlinearity, and frequency shifts [19]. Among them, the warm target induced bias and the inaccurate calibration nonlinearity impacted the measurements in the entire life span of the instrument—these were the primary biases to investigate and correct.

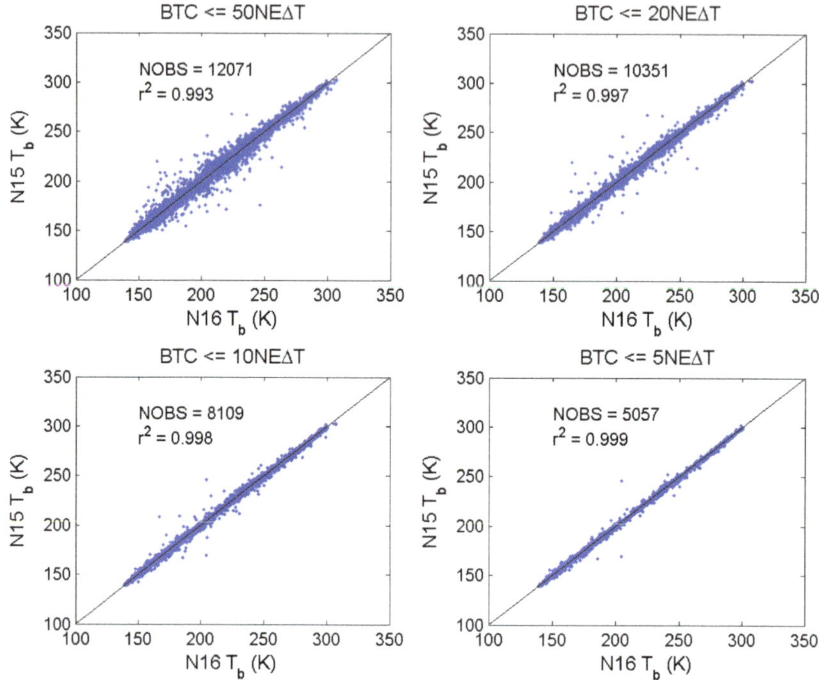

Figure 3. Scatter plot between NOAA-15 and NOAA-16 SNO brightness temperature for 23.8 GHz channel under various ranges of NEΔT, to test the brightness temperature contrast (BTC) threshold.

2.2.1. Identification of Warm Target Contamination

The AMSU-A instrument onboard each satellite measured counts of three objects in each full scan: cold target (C_c), warm target (C_w), and earth scene (C_e). The radiances of cold target (R_c) and warm target (R_w) were also measured to calculate earth scene radiance (R) through the following series of equations [29]:

$$R = R_L + \mu Z \tag{1}$$

$$R_L = R_c + S(C_e - C_c) \tag{2}$$

$$S = \frac{R_w - R_c}{C_w - C_c} \tag{3}$$

$$Z = S^2(C_e - C_c)(C_e - C_w) \tag{4}$$

Equation (1) decomposes the earth scene radiance into a linear part (R_L) and a non-linear part (μZ). The two parts were further described in Equations (2) and (4). The non-linear coefficient μ was calculated from the reference instrument temperature in a pre-defined range in the Level 1b data before inter-satellite calibration.

Cold and warm targets were designed to provide reference radiances for on-orbit calibration to obtain accurate earth scene radiance, so it was ideal for these radiances and TBs to be stable. The cold target temperature was relatively stable, given that the cosmic background should not vary very much in the life span of the sensors. An exception is lunar contamination, which was well detected and corrected [30]. The contamination of the cold space view from antenna side/back-lobes/spillover may also introduce bias, and was corrected beforehand [31]. The warm target, however, may be heated by the sun and is susceptible to seasonal and inter-annual variability, as shown in Figure 4. The variability of warm target temperature was explained well using the solar beta angle in Reference [19]. Figure 4 shows that 23.8 and 31.4 GHz channels have similar variabilities as they both reside on the AMSU-A2 module, while the variability of 50.3 and 89 GHz are similar because their sub modules (AMSU-A1-2 for 50.3 GHz channel, and AMSU-A1-1 for 89 GHz channel) are close to one another on the spacecraft. The NOAA-16 warm target temperature was rather stable before 2008, mainly due to the fact that the satellite solar shield worked well during that period and reflected a large portion of solar heat.

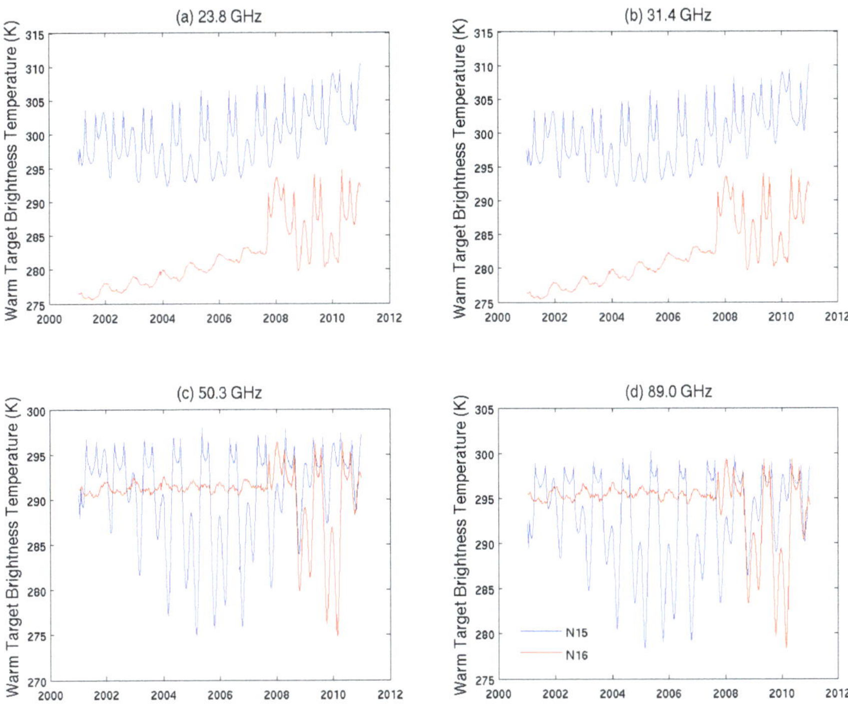

Figure 4. Warm target temperature variation between AMSU-A onboard N15 and N16, from their operational time through 2010 for (**a**) 23.8, (**b**) 31.4, (**c**) 50.3, and (**d**) 89 GHz. Note: the only observations during their SNO period are shown to decrease the sampling.

2.2.2. Correction Utilizing Integrated Microwave Inter-Calibration Approach

The Integrated Microwave Inter-Calibration Approach (IMICA), developed by Zou et al. [19,20,32,33], has demonstrated the calibration coefficients obtained from regressions minimize three biases: constant offsets between satellite pairs, scene temperature dependent biases between satellite pairs, and the sun-heating induced seasonal variability in brightness temperature. When the coefficients are time-varying, the coefficients also remove bias drifts over time between satellite pairs. It is effective for AMSU-A temperature sounding channels. The IMICA approach was adopted in this study for the inter-satellite calibration of AMSU-A window channels in order to remove the biases in an empirical way by adjusting the transfer function that converts counts to radiance, after certain modifications are made, e.g., the application of appropriate standard deviation (STD) thresholds to filter the SNO time series.

In the IMICA approach, Equation (1) is modified to

$$R = R_L - \delta R + \mu Z \tag{5}$$

where the radiance offset δR allows for adjustments of the inter-satellite radiance difference, and the non-linear coefficient μ is redefined and calculated, as described below following [19]. It is noted that the non-linear coefficient in the text refers to this inter-calibrated, redefined μ. Firstly, the non-linear term, Z, of the two satellites can be linearly related as:

$$Z_j = \beta Z_k + \alpha + \zeta \tag{6}$$

where β and α are slope and offset, respectively, ζ is a white noise, and j and k indicate the two satellites, respectively. Secondly, the variability of Equation (5) can be obtained through a least square calculation as:

$$\begin{cases} \sum\limits_{i=1}^{N} \Delta R_{L,i} = a_0 + a_1 \sum\limits_{i=1}^{N} Z_{k,i} \\ \sum\limits_{i=1}^{N} Z_{k,i} \Delta R_{L,i} = a_0 \sum\limits_{i=1}^{N} Z_{k,i} + a_1 \sum\limits_{i=1}^{N} Z_{k,i}^2 \end{cases} \tag{7}$$

where i represents different observations, ΔR_L refers to the difference of R_L between the two SNO matchup satellites, and the newly introduced coefficients a_0 and a_1 can be linked back to other variables such as

$$\begin{cases} a_0 = \Delta\delta R + \alpha\mu_j \\ a_1 = -\mu_k + \beta\mu_j \end{cases} \tag{8}$$

By obtaining the coefficients α, β, a_0 and a_1, one can calculate $\Delta\delta R$ (the difference of δR between the two SNO satellites), μ_j and μ_k Then, δR_j can be determined by setting the reference δR_k to zero.

In order to achieve the optimum results for all six satellites of interest, a sequential adjusting process was applied and is described below:

1. Generate intermediate SNO data set from the launch of the newer satellite of the satellite pair to 2010, which includes multiple intermediate variables, including time, latitude, longitude, brightness temperature, counts, radiance, etc., for each SNO event;
2. Filter this data set by the BTC threshold described in Section 2.1;
3. Calculate SNO coefficients (α, β, a_0 and a_1);
4. Set $\delta R_{N15} = 0$, and μ_{N15}, calculate δR_j, μ_j, j = 1 to 5;
5. Generate level-1c radiances for all six satellites using recalibration coefficients;
6. Compute tropical ocean mean time series of ΔTb for available overlaps between pairs;
7. Change the value of μ_{N15} and repeat steps 3, 4, and 5;
8. Stop when the summation of the standard deviation of ΔTb is a minimum.

The iterative adjustment of μ_{N15} is illustrated in Figure 5. By setting μ_{N15} from −25 to 25 at an interval of 2.5 (sr m^2 cm^{-1})(mW)$^{-1}$, δR_j and μ_j can be calculated, as described in step 4 above. Then, a set of level 1c radiances over the tropical ocean is generated, and compared with that of NOAA-15. Each satellite pair would reach minimum an STD of ΔTb, but the minimum average STD of ΔTb is selected.

The optimal μ and δR are given in Table 5 after further corrections and modifications, as detailed in the next section.

Table 5. Calibration coefficients for AMSU-A Channels 1–3 and 15 [a].

	Ch #	NOAA-15	NOAA-16	NOAA 17	NOAA-18	MetOP A	NOAA-19
μ	1	−3.00870	−7.25050	−7.22996	−0.88067	−0.98053	0.10012
	2	−1.05123	−3.35409	−2.84701	1.51212	−1.28394	−2.30045
	3	−2.37781	−2.31567	−2.20964	−2.09040	−2.62705	−1.28555
	15	0	−0.16528	−0.25743	0.36618	0.21446	0.25637
δR_0	1	0	-3.874×10^{-7}	-5.459×10^{-7}	1.675×10^{-6}	-4.635×10^{-7}	-3.931×10^{-7}
	2	0	-6.009×10^{-7}	-6.199×10^{-7}	-2.792×10^{-7}	-5.270×10^{-7}	-4.772×10^{-7}
	3	0	-1.496×10^{-6}	-1.750×10^{-6}	1.051×10^{-5}	-5.953×10^{-6}	-4.744×10^{-6}
	15	0	0	-7.220×10^{-7}	-2.927×10^{-6}	-6.715×10^{-6}	-2.017×10^{-6}

Table 5. *Cont.*

	Ch #	NOAA-15	NOAA-16	NOAA 17	NOAA-18	MetOP A	NOAA-19
κ	1	0	0	0	0	0	0
	2	0	0	0	0	0	0
	3	0	1.448×10^{-6}	0	0	0	0
	15	0	0	0	0	0	0

[a] Units for μ, δR_0 and κ are $(\text{sr m}^2 \text{ cm}^{-1})(\text{mW})^{-1}$, $(\text{mW})(\text{sr m}^2 \text{ cm}^{-1})^{-1}$, and $(\text{mW})(\text{sr m}^2 \text{ cm}^{-1})^{-1}\text{yr}^{-1}$.

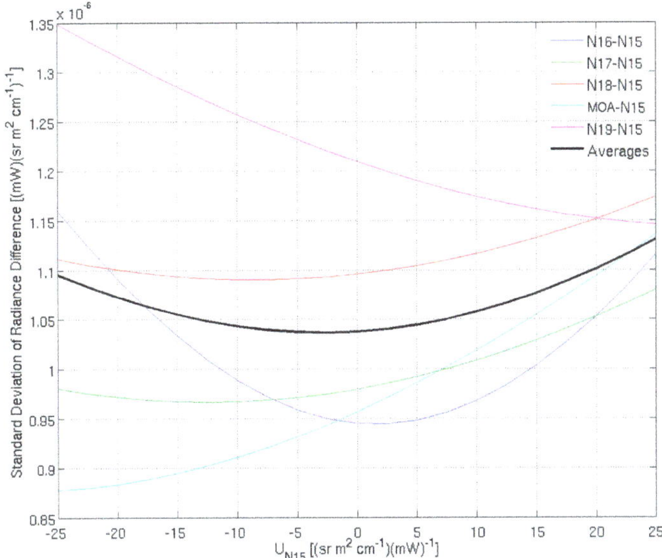

Figure 5. Iterative search for μ_{N15} of 23.8 GHz channel. The average line is the major criterion to select the optimal μ_{N15}.

2.3. Other Satellite Specific Corrections

After the warm target contamination bias is removed through the iterative process, two channels, 50.3 GHz on NOAA-16 and 89 GHz on NOAA-15, still exhibit some issues that need further attention before a final FCDR data set can be generated, as detailed in the following sections.

2.3.1. Slope Correction on 50.3 GHz, NOAA-16

Compared to the tropical ocean mean TB of the 50.3 GHz from NOAA-15, NOAA-16 exhibits a clear upward trend of 0.06 K per year (Figure 6). This type of trend does not appear in any other satellite pairs. Apparently, this trend can only be explained by an unidentified sensor bias drift.

NOAA-16 sounding channels also show similar bias drift, as reported in References [19,22]. In Reference [22], the entire record of NOAA-16 is excluded in the time series due to the bias drift, while in Reference [19] this bias drift is corrected using the following equation:

$$\delta R = \delta R_0 + \kappa(t - t_0) \tag{9}$$

where δR_0 is a constant offset, κ is the rate of changes in the offset, t is time and t_0 is a reference time. This study utilized Equation (9) to remove the bias drift, and set t_0 to the NOAA-16 launch time. The corrected result is shown as the blue curve in Figure 6.

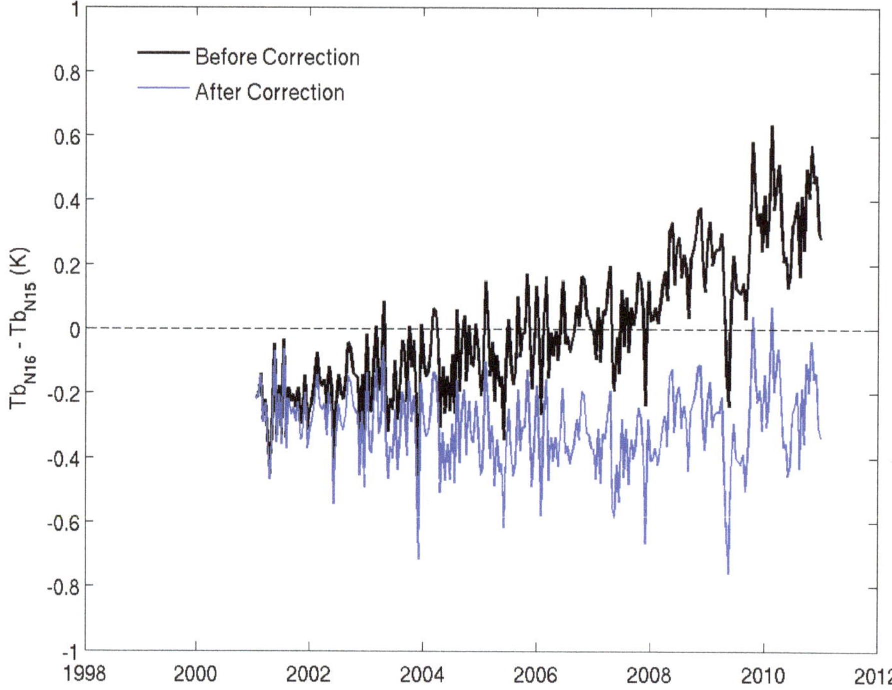

Figure 6. The tropical ocean mean brightness temperature time series of 50.3 GHz channel onboard NOAA-16 versus NOAA-15 displays an upper trend (black curve), a new time series (blue curve) is formed through removing the trend line.

2.3.2. Possible Frequency Shift in 89 GHz, NOAA-15

The sequential adjusting process described in Section 2.2.2 failed when applied to the 89 GHz channel on NOAA-15, as no optimal μ and δR combination could be found to form a reasonable time series. Figure 7a shows the inter-satellite difference on the antenna temperature T_a and the linear component T_l from the SNO observations. Even though the difference of T_a is generally around zero, the T_l difference is around 3 K, and displays a well-defined seasonal variation. It is expected that their nonlinear difference would display a similar variation pattern. Figure 7b further shows the nonlinear coefficient, μ, derived from Equation (1). μ of NOAA-16 89 GHz channel is nearly a constant at around 0.1, while that of NOAA-15 89 GHz is around 3. The latter also displays a distinctive variation, which is apparently compensating for the TB difference in Figure 7a. Though not shown, μ of 89 GHz of other satellites are approximately constant, which matches the IMICA assumption of the fixed optimal μ.

Figure 7a,b show that a constant optimal μ for the 89 GHz of NOAA-15 will only result in a varied ΔT_a time series similar to the red curve in Figure 7a. It means that the error in this channel cannot be corrected using IMICA. In addition, Zou and Wang stated that "Radiance biases caused by the frequency errors may mix with the nonlinear calibration errors" [19]. These observations led to our postulation that the bias arrives from an erroneous central frequency or frequency shift.

To verify this assumption and determine the magnitude of the frequency shift, we examined the change of the standard deviation of ΔTb′ by adjusting the frequency shift df, where ΔTb′ is the TB difference and is defined as a function of the difference between the SNO pair and simulated TBs:

$$
\begin{aligned}
\Delta T_b' &= T_b(N15) - \Delta T_b^s(N15, df) - T_b(N16) \\
&= [T_b(N15) - T_b^s(N15, f_m + df)] + [T_b^s(N15, f_m) - T_b(N16)]
\end{aligned}
\tag{10}
$$

where f_m is the nominal central frequency of 89 GHz channel of NOAA-15. Following both References [19,34], the assumption of this scheme is that the minimum standard deviation of $\Delta Tb'$ corresponds to the value of the frequency shift. To achieve this, the global SNO pairs of NOAA-15 and NOAA-16, which occurred in August 2008, as shown in Figure 1a, were adopted as the basic comparison data set. The data is further restricted to ocean observations between 30°S and 30°N to make the simulation more reliable, as this channel is sensitive to surface conditions. Following References [18,19], CRTM [35] is selected as the radiative transfer model, and the ERA-Interim [16] as the model input. Different experimental frequencies, centered at 89 GHz on NOAA-15, were examined in the CRTM simulations with the frequency shift, df. The shift is relative to the prelaunch measurement and range from −1.5 GHz to 1.5 GHz with an interval of 25 MHz.

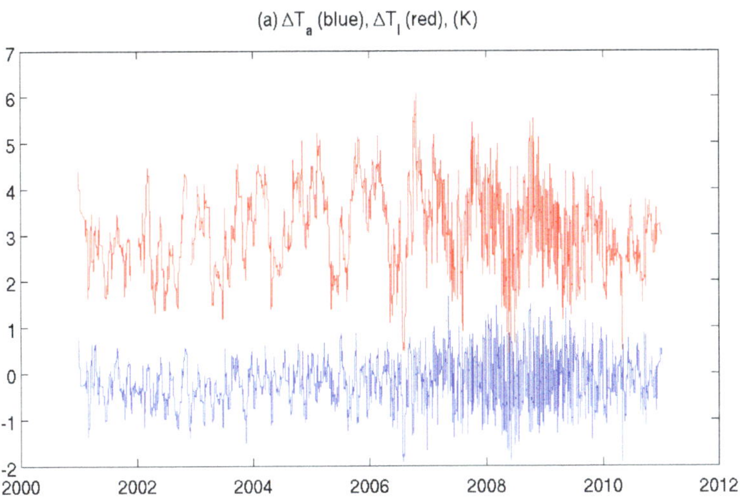

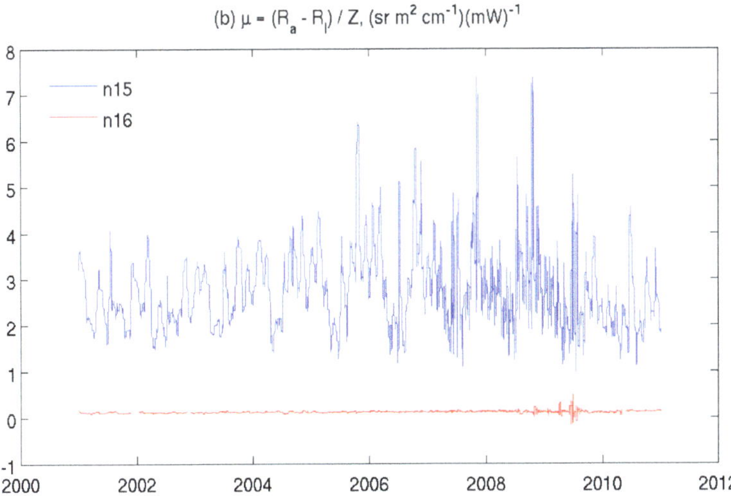

Figure 7. Characterization of the nonlinearity of 89 GHz channel of NOAA-15 from SNO pairs of NOAA-15 and 16. (**a**) T_a and T_l difference between NOAA-15 and 16. (**b**) μ of NOAA-15 and -16, calculated from Equation (1).

The comparison between the SNO observation and simulation is shown in Figure 8. The minimum standard deviation of ΔTb' (STD) was achieved between $-350{\sim}-325$ MHz, though the change with frequency shift was small. An additional test was carried out using the ocean emissivity, calculated using the radiative transfer model developed by the Remote Sensing System (RSS) [36]. However, the result concerning the frequency shift remains the same.

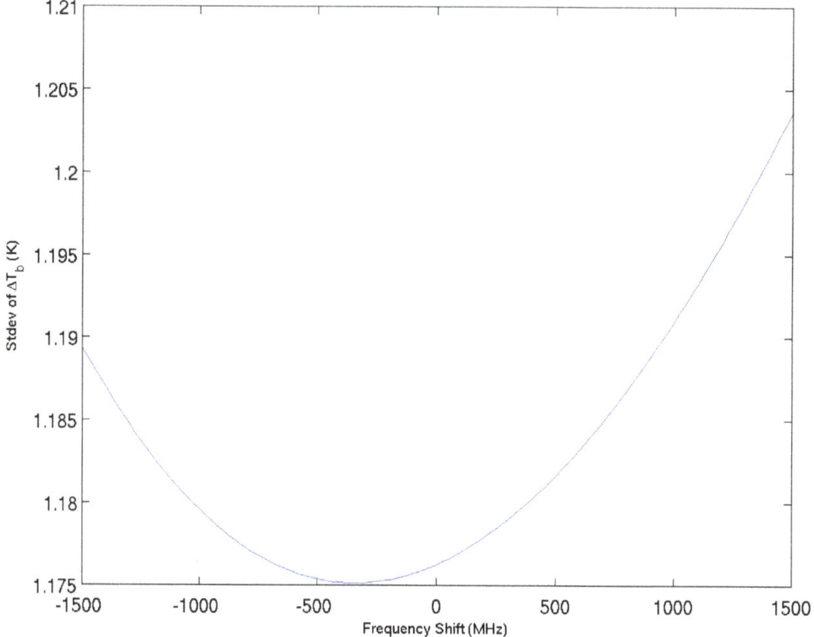

Figure 8. Amplitude of the standard deviation of ΔTb' (Equation (10)) using SNO pairs over the tropical ocean versus the 89 GHz channel of NOAA-15 frequency shift, df.

Figure 8 reveals that STD is rather insensitive to frequency shifts with a 0.03K change corresponding to about a 1.8 GHz shift. The result shows that frequency shift cannot explain the large STD, and thus our hypothesis is not valid. The root cause of the variable coefficient μ of the NOAA-15 89 GHz channel (Figure 7b) remains unclear, which may leave room for future research. For this reason, a frequency correction is not applied to this channel in the AMSU-A FCDR.

Since no optimal μ and δR combination could be found for the 89 GHz channel of NOAA-15 AMSU-A, NOAA-16 was used as the reference satellite for this channel in the iterative approach described in Section 2.2.2.

3. Results

Table 5 lists a series of coefficients, μ, δR_0, and κ, which are obtained by using the approaches described in Sections 2.2 and 2.3. Since NOAA-15 is the reference satellite for Channels 1~3, their corresponding δR_0's are set to zero, so is δR_0 for Channel 15 of NOAA-16. The coefficients μ and δR_0 for Channel 15 of NOAA-15 were zero because the corrections shown by Equations (5) and (9) were not applied to this channel. Most values of μ were within the range of $[-4, 4]$, except Channel 1 of NOAA-16 and NOAA-17, which was around -7 and indicates a strong non-linearity. In general, the offsets δR_0 of Channels 1 and 2 were on the order of 10^{-7} (mW)(sr m^2 cm^{-1})$^{-1}$, corresponding to ΔTb on the order of 0.02 and 0.013 K, respectively, and δR_0 of Channels 3 and 15 were on the order of 10^{-6} (mW)(sr m^2 cm^{-1})$^{-1}$, corresponding to ΔTb on the order of 0.045 and 0.019 K, respectively. The exceptions were Channels 1 and 3 of NOAA-18, as both were one order of magnitude higher than the other satellites. As indicated in Section 2.3.1, only Channel 3 of NOAA-16 suffered from bias drift with a non-zero κ.

3.1. Fundamental CDR

The Fundamental CDR (FCDR) was obtained by applying geolocation correction [17], asymmetry correction [18], and inter-satellite calibration, as described in Sections 2.2 and 2.3. The impacts of the former two corrections have been documented in the corresponding papers [17,18]. This study investigates the impact of the inter-satellite calibration.

To compare the time series performance before and after inter-satellite calibration, we extracted the TB time series on the level 1b data before inter-satellite calibration and newly corrected FCDR from satellite launch time to 2015, over tropical oceans between 30°S and 30°N. Further, to avoid the residual impact of the asymmetry correction, only nadir pixels (e.g., AMSU-A beam positions 15 and 16) were selected. As seen in Figures 9a–d and 10a–d, the level 1b data before inter-satellite calibration suffered no obvious inter-satellite offsets, and the FCDR Tb time series shows similar features. However, upon closer examination carried out by subtracting the TB time series from those of the referenced satellite, (shown in Figures 9e–h and 10e–h), we noticed that the ΔTb of FCDR was more constant over the entire time series, which indicated that the impact of warm target contamination had been minimized. To further confirm this, Table 6 shows a comparison of the STD of ΔTb calculated from Figures 9e–h and 10e–h. As an example, the STD of Channel 1 (23.8 GHz) between NOAA-16 and NNOAA-15 was 0.374 K before correction, which decreased to 0.217 K after inter-satellite calibration. Generally, the improvement to all the channels was approximately 50%.

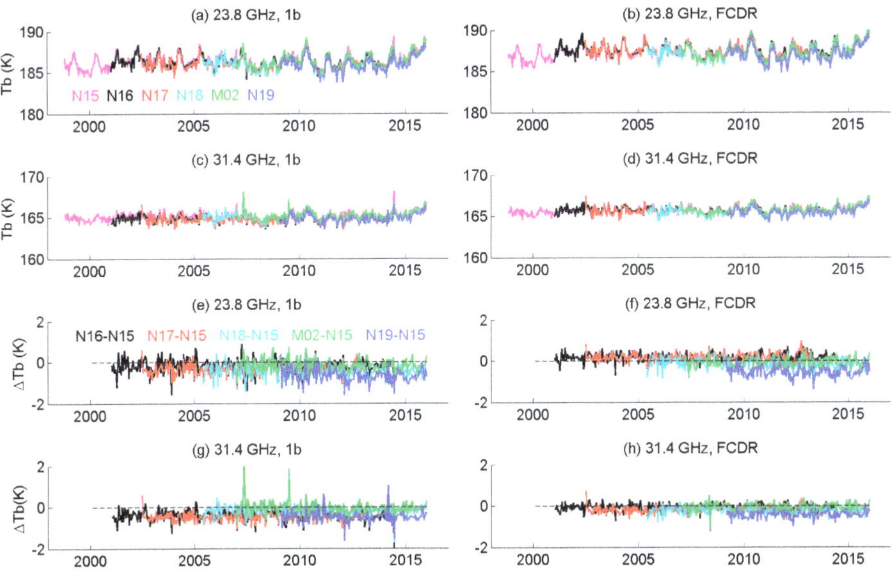

Figure 9. Tropical ocean mean Tb (**a–d**) and ΔTb (**e–h**) for 23.8 and 30.4 GHz channels. Left panels display the values of 1b before inter-satellite calibration, while the right panels are FCDRs after inter-satellite calibration.

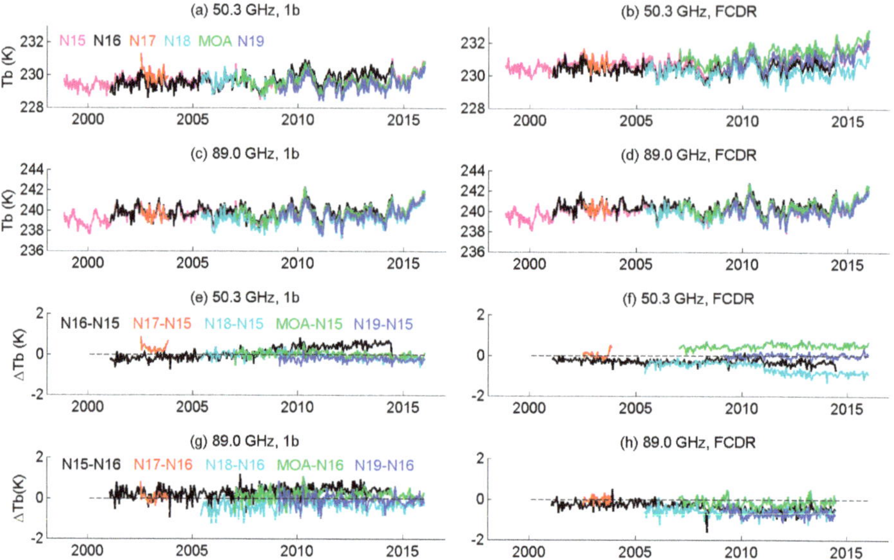

Figure 10. Tropical ocean mean Tb (**a–d**) and ΔTb (**e–h**) for 50.3 and 89.0 GHz channels. Left panels display the values of 1b before inter-satellite calibration, while the right panels are FCDRs after inter-satellite calibration.

Table 6. Standard deviation of Tb difference of specified channels and satellite pairs, before and after inter-satellite calibration [a].

	Before				After			
Channel	**1**	**2**	**3**	**15**	**1**	**2**	**3**	**15**
N16-N15	0.374	0.263	0.267	0.315	0.217	0.193	0.126	0.227
N17-N15	0.285	0.217	0.191	0.225	0.191	0.191	0.171	0.132
N18-N15	0.386	0.259	0.168	0.337	0.239	0.197	0.13	0.242
M02-N15	0.37	0.384	0.167	0.328	0.215	0.207	0.108	0.227
N19-N15	0.424	0.276	0.174	0.374	0.263	0.187	0.115	0.208

[a] Unit is K. "Before" corresponds to the uncorrected L1b data and "After" refers to the FCDR.

It is worth noting that the spurious upward trend of the 50.3 GHz Tb difference between NOAA-16 and NOAA-15 (black line in Figure 10e) has been corrected (Figure 10f). There is a 0.6 K offset between Metop-A and NOAA-15, and a −0.6 K offset between NOAA-18 and NOAA-15, as indicated in Figure 10f, which can be attributed to the following reasons: 1. In order to obtain more accurate calibration coefficients solutions (α, β, a_0 and a_1) with the IMICA method, it is better to include more SNO pairs. 2. In the sequential adjusting process to obtain μ_{N15}, it is better to extend the training data. In the current study, both SNO pairs and the global mean difference data are only limited to 2010 to obtain the coefficients, yet the coefficients are applied to the time series up to 2015, thus the FCDR results after 2010 are less ideal, future reprocessing may improve the performance. 3. The choice of the reference satellite. In the current study, NOAA-15 was selected as the reference satellite, mainly due to its data length. This channel from NOAA-16 contains a serious bias drift, as mentioned in Section 2.3.1; the NOAA-17 AMSU-A1 module operated for only a limited time from June 2002 to the end of October 2003 due to scan motor failure; NOAA-18 drifted for nearly 8 h during its operations more than 15 years ago. Thus, future reference satellites, which perform better in terms of drifting, should include only Metop-A and NOAA-19 as candidates. However, as the diurnal drift for NOAA-18 was large, the large differences between NOAA-18 and NOAA-15 could be due to NOAA-18 diurnal drift, and it could be unrealistic to expect to have small biases between NOAA-18 and NOAA-15. However, the biases between MetOp-A and NOAA-15 might be improved when we have an SNO of up to 2015.

3.2. Thematic CDR

The suite of AMSU-A Thematic CDR (TCDR) from window channels was obtained by applying the corresponding FCDR's to the MSPPS product suite [12]. If the FCDR was generated properly at all four AMSU-A window channels, its improvements were expected to extend into the derived products of the TCDR, including CLW, TPW, sea ice, and snow cover, etc. The derivations or regressions of all the TCDR's were based on multiple channels of FCDR's, and thus any erroneous correction or calibration of the FCDR's would be inherited and magnified. Thus, both CDR's need to be developed in a synergistic manner before the FCDR is finalized and this concept was adopted during various aspects of the FCDR development, including the geolocation and scan bias corrections. More detailed aspects of the performance of TCDR (i.e., comparisons with in-situ data, spatial analysis of seasonal to annual climate variations, etc.) will be investigated in ongoing research. Here, we only show the product consistency resulting from the inter-satellite calibration.

Figure 11 shows the time series of the MSPPS and TCDR's of NOAA-15 for two ocean products, TPW and CLW, over tropical ocean between 30°S and 30°N. Only their nadir retrievals are shown to further isolate the impact of the inter-satellite corrections. Figure 12 shows similar comparison of two land products, land surface temperature (Ts), and land surface emissivity of 50.3 GHz (Emis$_{50}$), over tropical land.

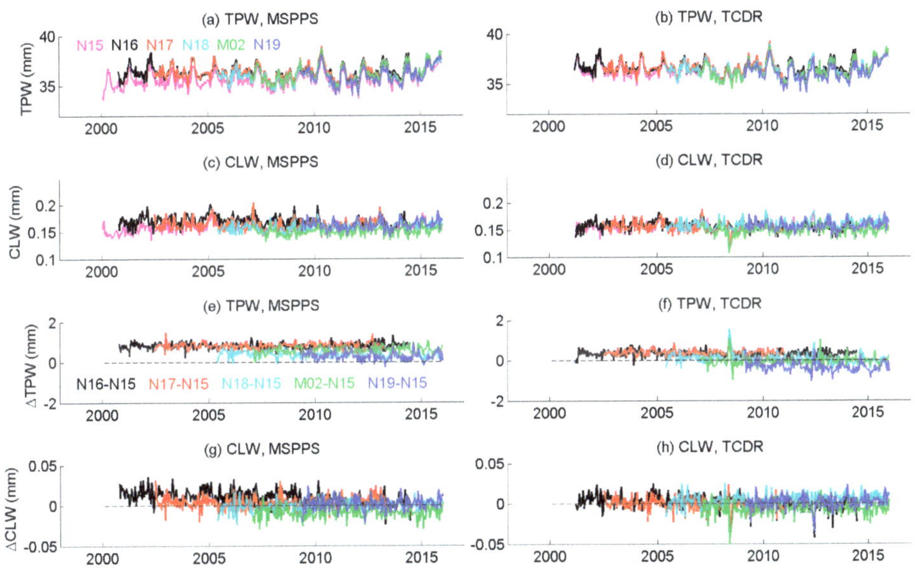

Figure 11. Retrieved tropical ocean products comparison, note different variables are used in different panels. Left panels display the products produced by MSPPS from 1b before inter-satellite calibration, while right panels are TCDR after inter-satellite calibration. Time series of total precipitable water (TPW) is shown in (**a,b**), the difference versus N15 is shown in (**e,f**); time series of cloud liquid water (CLW) is shown in (**c,d**), the difference versus N15 is shown in (**g,h**).

From Figure 11a, it is noticeable that the TPW before inter-satellite calibration displays a wet bias relative to NOAA-15, which is more clearly shown in Figure 11e, where positive values are seen. The wet bias of TPW was corrected by the TCDR, as shown in Figure 11b,f, with the overall difference being closer to zero. Similarly, the TCDR CLW series displays a better agreement among the six satellites than that of the uncorrected CLW products.

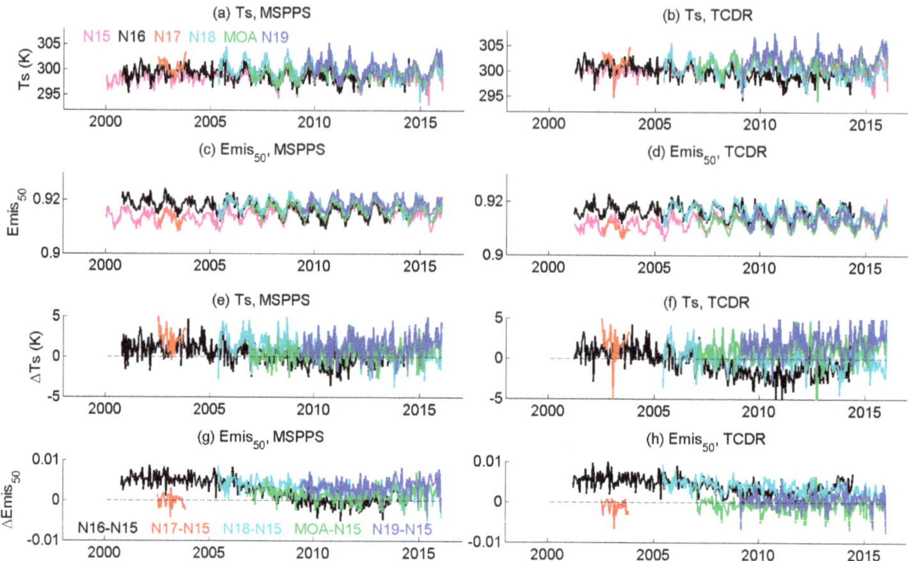

Figure 12. Retrieved tropical land products comparison, note different variables are used in different panels. Left panels display the products produced by MSPPS from 1b before inter-satellite calibration, while right panels are TCDR after inter-satellite calibration. Time series of land surface temperature (Ts) is shown in (**a,b**), the difference versus N15 is shown in (**e,f**); time series of land surface emissivity of 50.3 GHz channel (Emis$_{50}$) is shown in (**c,d**), the difference versus N15 is shown in (**g,h**).

As opposed to the ocean products, the land products display distinctive diurnal effects in both versions before and after inter-satellite calibration. Specifically, Figure 12g,h displays nearly constant offsets in 50.3 GHz emissivity for different satellite pairs. The offsets are within the range of 0.01, and agree very well with other studies [37]. It should be noted that during the entire inter-satellite calibration process, land observations were only used in the calculation of SNO coefficients (α, β, a_0 and a_1), mostly in high latitudes. No land observation was used in the iterative adjusting of μ_{N15}. Additionally, no diurnal correction was applied; collectively, evaluation of the land products in regions where no data was used in the adjustments further demonstrates the robustness of the FCDR's.

4. Discussion and Conclusions

With over 15 years of observations from the AMSU-A sensor and through a series of six satellites—NOAA-15, NOAA-16, NOAA-17, NOAA-18, NOAA-19 and Metop-A—global monitoring of important parameters such as total precipitable water, cloud liquid water and land surface emissivity has been achieved and well documented in the open literature. The excellent temporal sampling by the NOAA POES constellation has provided excellent compliments to the passive microwave imagers (i.e., SSM/I, AMSR-E and TMI), thus yielding a long time series dating back to around 1998. While the scientific community took note of the huge potential of passive microwave "all weather" capability and generally stable radiometric measurements to generate climate data sets in support of both national and international programs, the data sets contain deficiencies that prevent them from being classified as a true CDR, which is provided with "sufficient length, consistency, and continuity" [1]. This is mainly due to the lack of routinely updated satellite calibration information and inter-satellite calibrations that are not factored into the data processing, and which are primarily driven by the main purpose of these measurements (e.g., NWP model assimilation and derived products to support weather forecasts). Additionally, resource limitations and a lack of a true climate "requirement" within the satellite programs at NOAA have prohibited any sort of routine reprocessing as part of its core satellite operations. This gap has been bridged for AMSU-A window channels of 23.8, 31.4, 50.3 and 89.0 GHz, with the introduction of the inter-satellite calibration, after the application of geolocation [17] and cross-scan bias corrections [18]. As a result, the FCDR and TCDR time series from 1998 to 2015 have been generated and utilized to test the effectiveness of the calibration.

The Integrated Microwave Inter-Calibration Approach (IMICA) approach, which was developed for the inter-satellite calibration of AMSU-A sounding channels, was adopted in this study for the AMSU-A window channels. To minimize the impact of surface heterogeneity on the window channel observations, the appropriate standard deviation thresholds have been identified through sensitivity tests of various factors to restrict Simultaneous Nadir Overpass (SNO) data.

One critical element of the IMICA approach is how to select the proper reference satellite for each channel. It is desirable that the channel on the reference satellite has a long enough time span, and is free of, or at least suffers from few, error sources such as bias drift, frequency shift, etc. In considering the length of the record, NOAA-15 could be the best candidate as a reference satellite. Yet, when error sources are taken into account, Channel 15 (89 GHz) on NOAA-15 is not ideal due to a significant varying non-linear coefficient μ, as mentioned in Section 2.3.2. The selection of the reference satellite for Channel 3 (50.3 GHz channel) is not ideal either, as discussed in Section 3.1. Further, since NOAA-15 was launched in 1998 and is the oldest satellite carrying AMSU-A, there are some concerns over its orbital and radiometric stabilities in recent years. It is recommended that ensuing studies utilize a newer satellite, e.g., Metop-A, as the reference satellite for the years beyond 2015, while still adopting the same inter-calibration techniques. In the future, the selection of the reference channel/satellite may be improved by introducing other techniques, such as the two-sample Allan deviation to an international standard (SI) traceable noise characterization [38].

After applying the inter-satellite calibration, the consistency of AMSU-A window channels brightness temperatures have improved by 50%. In addition, the spurious trend in the NOAA-16 50.3 GHz channel has been removed. Though further assessment work, such as the detection of climate change, is needed before claiming the dataset "climate quality", these improvements will ensure the quality of the TCDR that are generated from the AMSU-A data, such as the hydrological bundle CDR based on AMSU [39]. With the removal of most inter-satellite biases, the inter-calibrated AMSU-A radiances (FCDR's) are expected to improve the related NWP reanalysis and other applications with regard to consistency and accuracy. Similar calibrations can be applied to measurements from more recent sensors, such as Metop-B and S-NPP ATMS. Another set of TCDR can also be generated by utilizing the operational MiRS system. In summary, this study is a critical step towards the development of a set of fundamental and thematic CDRs for hydrological and meteorological applications, and the data sets, in turn, provide a beneficial contribution to a wider CDR community [40].

Author Contributions: All authors contributed significantly to this manuscript. Specific contributions include Data Collection, W.Y. and H.M.; Data Analysis, W.Y. and Y.C.; Methodology, W.Y., H.M. and R.R.F.; Project Management, H.M. and R.R.F.; Manuscript Preparation, W.Y., H.M. and R.R.F. All authors have read and agreed to the published version of the manuscript.

Funding: This research was funded by NOAA/NCEI CDR program through grant NA09NES4400006 (Cooperative Institute for Climate and Satellites -CICS) at the University of Maryland, Earth System Science Interdisciplinary Center (ESSIC).

Acknowledgments: The authors would like to thank Cheng-zhi Zou for fruitful discussions and guidance in this study. Three anonymous reviewers are thanked for their constructive comments and suggestions. The views, opinions, and findings contained in this report are those of the authors and should not be construed as an official National Oceanic and Atmospheric Administration or U.S. Government position, policy, or decision.

Conflicts of Interest: The authors declare no conflict of interest.

References

1. National Research Council. *Climate Data Records from Environmental Satellites*; National Research Council of the National Academies: Washington, DC, USA, 2004; p. 150.
2. Goldberg, M.; Ohring, G.; Butler, J.; Cao, C.; Datla, R.; Doelling, D.; Gärtner, V.; Hewison, T.; Iacovazzi, B.; Kim, D.; et al. The global space-based inter-calibration system. *Bull. Am. Meteorol. Soc.* **2011**, *92*, 467–475. [CrossRef]
3. Hewison, T.J.; Wu, X.; Yu, F.; Tahara, Y.; Koenig, M. GSICS Inter-Calibration of Infrared Channels of Geostationary Imagers using Metop/IASI. *IEEE Trans. Geosci. Remote Sens.* **2013**, *51*, 1160–1170. [CrossRef]
4. Chander, G.; Hewison, T.J.; Fox, N.; Wu, X.; Xiong, X.; Blackwell, W.J. Overview of intercalibration of satellite instruments. *IEEE Trans. Geosci. Remote Sens.* **2013**, *51*, 1056–1080. [CrossRef]
5. Iacovazzi, R.A., Jr.; Cao, C.Y. Reducing uncertainties of SNO-estimated inter-satellite AMSU-A brightness temperature biases for surface-sensitive channels. *J. Atmos. Ocean. Technol.* **2008**, *25*, 1048–1054. [CrossRef]
6. Yang, S.; Weng, F.; Yan, B.; Sun, N.; Goldberg, M. Special Sensor Microwave Imager (SSM/I) intersonsor calibration using s simultaneous conical overpass technique. *J. Appl. Meteorol. Climatol.* **2011**, *50*, 77–95. [CrossRef]
7. Ruf, C.S. Detection of calibration drifts in spaceborne microwave radiometers using a vicarious cold reference. *IEEE Trans. Geosci. Remote Sens.* **2000**, *38*, 44–52. [CrossRef]
8. Brown, S.T.; Ruf, C.S. Determination of an Amazon hot reference target for the on-orbit calibration of microwave radiometers. *J. Atmos. Ocean. Technol.* **2005**, *22*, 1340–1352. [CrossRef]
9. Wilheit, T.; Berg, W.; Jones, L.; Kroodsma, R.; McKague, D.; Ruf, C.; Sapiano, M. A consensus calibration based on TMI and WindSat. In Proceedings of the 2011 IEEE International Geoscience and Remote Sensing Symposium, Vancouver, BC, Canada, 24–29 July 2011; pp. 2641–2644.
10. Sapiano, M.R.P.; Berg, W.K.; McKague, D.S.; Kummerow, C.D. Toward an intercalibrated fundamental climate data record of the SSM/I sensors. *IEEE Trans. Geosci. Remote Sens.* **2013**, *51*, 1492–1503. [CrossRef]
11. Cardinali, C. Data Assimilation: Observation Impact on the Short Range Forecast. ECMWF Lecture Notes. Available online: https://software.ecmwf.int/wiki/download/attachments/31064618/ObservationImpactForecast.pdf (accessed on 9 September 2020).
12. Ferraro, R.R.; Weng, F.; Grody, N.; Zhao, L.; Meng, H.; Kongoli, C.; Pellegrino, P.; Qiu, S.; Dean, C. NOAA operational hydrological products derived from the AMSU. *IEEE Trans. Geosci. Remote Sens.* **2005**, *43*, 1036–1049. [CrossRef]
13. Boukabara, S.A.; Garrett, K.; Chen, W.; Iturbide-Sanchez, F.; Grassotti, C.; Kongoli, C.; Chen, R.; Liu, Q.; Yan, B.; Weng, F.; et al. MiRS: An all-weather 1DVAR satellite data assimilation and retrieval system. *IEEE Trans. Geosci. Remote Sens.* **2011**, *49*, 3249–3272. [CrossRef]
14. Saha, S.; Moorthi, S.; Pan, H.L.; Wu, X.; Wang, J.; Nadiga, S.; Tripp, P.; Kistler, R.; Woollen, J.; Behringer, D.; et al. The NCEP climate forecast system reanalysis. *Bull. Amer. Meteorol. Soc.* **2010**, *91*, 1015–1058. [CrossRef]
15. Rienecker, M.M.; Suarez, M.J.; Gelaro, R.; Todling, R.; Bacmeister, J.; Liu, E.; Bosilovich, M.G.; Schubert, S.D.; Takacs, L.; Kim, G.K.; et al. MERRA: NASA's modern-era retrospective analysis for research and applications. *J. Clim.* **2011**, *24*, 3624–3648. [CrossRef]
16. Dee, D.P.; Uppala, S.M.; Simmons, A.J.; Berrisford, P.; Poli, P.; Kobayashi, S.; Andrae, U.; Balmaseda, M.A.; Balsamo, G.; Bauer, D.P.; et al. The ERA-Interim reanalysis: Configuration and performance of the data assimilation system. *Q. J. R. Meteorol. Soc.* **2011**, *137*, 553–597. [CrossRef]
17. Moradi, I.; Meng, H.; Ferraro, R.R.; Bilanow, S. Correcting geolocation errors for microwave instruments aboard NOAA satellites. *IEEE Trans. Geosci. Remote Sens.* **2013**, *51*, 3625–3637. [CrossRef]
18. Yang, W.; Meng, H.; Ferraro, R.R.; Moradi, I.; Devaraj, C. Cross scan asymmetry of AMSU-A window channels: Characterization, correction and verification. *IEEE Trans. Geosci. Remote Sens.* **2013**, *51*, 1514–1530. [CrossRef]
19. Zou, C.Z.; Wang, W. Inter-satellite calibration of AMSU-A observations for weather and climate applications. *J. Geophys. Res.* **2011**, *116*, D23113. [CrossRef]
20. Zou, C.Z.; Goldberg, M.; Cheng, Z.; Grody, N.; Sullivan, J.; Cao, C.; Tarpley, D. Recalibration of microwave sounding unit for climate studies using simultaneous nadir overpasses. *J. Geophys. Res.* **2006**, *111*, D19114. [CrossRef]

21. Christy, J.R.; Spencer, R.W.; Norris, W.B.; Braswell, W.D.; Parker, D.E. Error estimates of version 5.0 of MSU-AMSU bulk atmospheric temperature. *J. Atmos. Ocean. Technol.* **2003**, *20*, 613–629. [CrossRef]

22. Mears, C.A.; Wentz, F.J. Construction of the Remote Sensing Systems V3.2 atmospheric temperature records from the MSU and AMSU microwave sounders. *J. Atmos. Ocean. Technol.* **2009**, *26*, 1040–1056. [CrossRef]

23. Robertson, F.R.; Bosilovich, M.G.; Chen, J.; Miller, T.L. The effect of satellite observing system changes on MERRA water and energy fluxes. *J. Clim.* **2011**, *24*, 5197–5217. [CrossRef]

24. Cao, C.; Weinreb, M.; Xu, H. Predicting simultaneous nadir overpasses among polar-orbiting meteorological satellites for the intersatellite calibration of radiometers. *J. Atmos. Ocean. Technol.* **2004**, *21*, 537–542. [CrossRef]

25. Cao, C.; Xu, H.; Sullivan, J.; McMillin, L.; Ciren, P.; Hou, Y.T. Inter-satellite radiance biases for the High-Resolution infrared Radiation Sounders (HIRS) on board NOAA-15, -16, and -17 from simultaneous nadir observations. *J. Atmos. Ocean. Technol.* **2005**, *22*, 381–395. [CrossRef]

26. Wang, L.; Goldberg, M.; Wu, X.; Cao, C.; Iacovazzi, R.; Yu, F.; Li, Y. Consistency assessment of Atmospheric Infrared Sounder and Infrared Atmospheric Sounding Interferometer radiances: Double differences versus simultaneous nadir overpasses. *J. Geophys. Res.* **2011**, *116*, D11. [CrossRef]

27. POSE Operational Status–POSE Status–OSPO. Available online: www.ospo.noaa.gov/Operations/POES/status.html (accessed on 9 September 2020).

28. John, V.O.; Holl, G.; Buehler, S.A.; Candy, B.; Saunders, R.W.; Parker, D.E. Understanding intersatellite biases of microwave humidity sounders using global simultaneous nadir overpasses. *J. Geophys. Res.* **2012**, *117*, D02305. [CrossRef]

29. Mo, T. Prelaunch calibration of the Advanced Microwave Sounding Unit-A for NOAA-K. *IEEE Trans. Microw. Theory Tech.* **1996**, *44*, 1460–1469. [CrossRef]

30. Mo, T.; Kigawa, S. A study of lunar contamination and on-orbit performance of the NOAA-18 Advanced Microwave Sounding Unit-A. *J. Geophys. Res.* **2007**, *112*, D20124. [CrossRef]

31. Mo, T. AMSU-A antenna pattern corrections. *IEEE Trans. Geosci. Remote Sens.* **1999**, *37*, 103–112.

32. Zou, C.Z.; Gao, M.; Goldberg, M. Error structure and atmospheric temperature trends in observations from the Microwave sounding Unit. *J. Clim.* **2009**, *22*, 1661–1681. [CrossRef]

33. Zou, C.Z.; Wang, W. Stability of the MSU-derived atmospheric temperature trend. *J. Atmos. Ocean. Technol.* **2010**, *27*, 1960–1971. [CrossRef]

34. Lu, Q.; Bell, W. Characterizing Channel Center Frequencies in AMSU-A and MSU Microwave Sounding Instruments. *J. Atmos. Ocean. Technol.* **2014**, *31*, 1713–1732. [CrossRef]

35. Han, Y.; Van Delst, P.; Liu, Q.; Weng, F.; Yan, B.; Treadon, R.; Derber, J. *JCSDA Community Radiative Transfer Model (CRTM)—Version 1*; NOAA Technical Report NESDIS 122; U.S. DOC NOAA: Washington, DC, USA, 2006.

36. Meissner, T.; Wentz, F.J. The emissivity of the ocean surface between 6 and 90 GHz over a large range of wind speeds and earth incidence angles. *IEEE Trans. Geosci. Remote Sens.* **2012**, *50*, 3004–3026. [CrossRef]

37. Norouzi, H.; Rossow, W.; Temimi, M.; Prigent, C.; Azarderakhsh, M.; Boukabara, S.; Khanbilvardi, R. Using microwave brightness temperature diurnal cycle to improve emissivity retrievals over land. *Remote Sens. Environ.* **2012**, *123*, 470–482. [CrossRef]

38. Tian, M.; Zou, X.; Weng, F. Use of Allan Deviation for characterizing satellite microwave sounder noise equivalent differential temperature (NEDT). *IEEE Geosci. Remote Sens. Lett.* **2015**, *12*, 2477–2480. [CrossRef]

39. Ferraro, R.R.; Nelson, B.R.; Smith, T.; Prat, O.P. The AMSU-based hydrological bundle climate data record—Description and comparison with other data sets. *Remote Sens.* **2018**, *10*, 1640. [CrossRef]

40. Yang, W.; John, V.O.; Zhao, X.; Lu, H.; Knapp, K.R. Satellite Climate Data Records: Development, Applications, and Societal Benefits. *Remote Sens.* **2016**, *8*, 331. [CrossRef]

 *remote sensing*

Article

The Reprocessed Suomi NPP Satellite Observations

Cheng-Zhi Zou [1,*], **Lihang Zhou** [2], **Lin Lin** [3], **Ninghai Sun** [4], **Yong Chen** [4], **Lawrence E. Flynn** [1], **Bin Zhang** [3], **Changyong Cao** [1], **Flavio Iturbide-Sanchez** [1], **Trevor Beck** [1], **Banghua Yan** [1], **Satya Kalluri** [1], **Yan Bai** [3], **Slawomir Blonski** [4], **Taeyoung Choi** [4], **Murty Divakarla** [5], **Yalong Gu** [4], **Xianjun Hao** [6], **Wei Li** [3], **Ding Liang** [4], **Jianguo Niu** [5], **Xi Shao** [3], **Larrabee Strow** [7], **David C. Tobin** [8], **Denis Tremblay** [4], **Sirish Uprety** [3], **Wenhui Wang** [3], **Hui Xu** [3], **Hu Yang** [3] and **Mitchell D. Goldberg** [2]

[1] Center for Satellite Applications and Research, NOAA/NESDIS, College Park, MD 20740, USA; Lawrence.E.Flynn@noaa.gov (L.E.F.); Changyong.Cao@noaa.gov (C.C.); Flavio.Iturbide@noaa.gov (F.I.-S.); Trevor.Beck@noaa.gov (T.B.); Banghua.Yan@noaa.gov (B.Y.); Satya.Kalluri@noaa.gov (S.K.)

[2] Joint Polar Satellite System, NOAA/NESDIS, Lanham, MD 20706, USA; Lihang.Zhou@noaa.gov (L.Z.); Mitch.Goldberg@noaa.gov (M.D.G.)

[3] ESSIC/CISESS, University of Maryland, College Park, MD 20740, USA; Lin.Lin@noaa.gov (L.L.); Bin.Zhang@noaa.gov (B.Z.); Yan.Bai@noaa.gov (Y.B.); wli12346@umd.edu (W.L.); Xi.Shao@noaa.gov (X.S.); Sirish.Uprety@noaa.gov (S.U.); Wenhui.Wang@noaa.gov (W.W.); huixu@umd.edu (H.X.); huyang@umd.edu (H.Y.)

[4] Global Science and Technology, College Park, MD 20740, USA; Ninghai.Sun@noaa.gov (N.S.); Yong.Chen@noaa.gov (Y.C.); Slawomir.Blonski@noaa.gov (S.B.); Taeyoung.Choi@noaa.gov (T.C.); Yalong.Gu@noaa.gov (Y.G.); Ding.Liang@noaa.gov (D.L.); Denis.Tremblay@noaa.gov (D.T.)

[5] I. M. Systems Group, Inc., College Park, MD 20740, USA; Murty.Divakarla@noaa.gov (M.D.); Jianguo.Niu@noaa.gov (J.N.)

[6] Global Environment and Natural Resources Institute/Environmental Science and Technology Center, George Mason University, Fairfax, VA 22030, USA; Xianjun.Hao@noaa.gov

[7] Department of Physics, University of Maryland Baltimore County, Baltimore, MD 21250, USA; strow@umbc.edu

[8] Space Science and Engineering Center, University of Wisconsin-Madison, Madison, WI 53715, USA; Dave.Tobin@ssec.wisc.edu

* Correspondence: Cheng-Zhi.Zou@noaa.gov

Received: 31 July 2020; Accepted: 3 September 2020; Published: 6 September 2020

Abstract: The launch of the National Oceanic and Atmospheric Administration (NOAA)/ National Aeronautics and Space Administration (NASA) Suomi National Polar-orbiting Partnership (S-NPP) and its follow-on NOAA Joint Polar Satellite Systems (JPSS) satellites marks the beginning of a new era of operational satellite observations of the Earth and atmosphere for environmental applications with high spatial resolution and sampling rate. The S-NPP and JPSS are equipped with five instruments, each with advanced design in Earth sampling, including the Advanced Technology Microwave Sounder (ATMS), the Cross-track Infrared Sounder (CrIS), the Ozone Mapping and Profiler Suite (OMPS), the Visible Infrared Imaging Radiometer Suite (VIIRS), and the Clouds and the Earth's Radiant Energy System (CERES). Among them, the ATMS is the new generation of microwave sounder measuring temperature profiles from the surface to the upper stratosphere and moisture profiles from the surface to the upper troposphere, while CrIS is the first of a series of advanced operational hyperspectral sounders providing more accurate atmospheric and moisture sounding observations with higher vertical resolution for weather and climate applications. The OMPS instrument measures solar backscattered ultraviolet to provide information on the concentrations of ozone in the Earth's atmosphere, and VIIRS provides global observations of a variety of essential environmental variables over the land, atmosphere, cryosphere, and ocean with visible and infrared imagery. The CERES instrument measures the solar energy reflected by the Earth, the longwave radiative emission from the Earth, and the role of cloud processes in the Earth's energy balance. Presently, observations from several instruments on S-NPP and JPSS-1 (re-named NOAA-20 after launch) provide near real-time monitoring of the environmental changes and improve weather forecasting by assimilation into numerical weather prediction models. Envisioning the need for consistencies in satellite retrievals, improving climate reanalyses, development of climate data records, and improving numerical weather forecasting, the NOAA/Center for Satellite Applications and Research (STAR) has been reprocessing the S-NPP observations for ATMS, CrIS, OMPS, and VIIRS through their life cycle. This article provides a summary of the instrument observing principles, data characteristics, reprocessing approaches, calibration algorithms, and validation results of the reprocessed sensor data records. The reprocessing generated consistent Level-1 sensor data records using unified and consistent calibration algorithms for each instrument that removed artificial jumps in data owing to operational changes, instrument anomalies, contaminations by anomaly views of the environment or spacecraft, and other causes. The reprocessed sensor data records were compared with and validated against other observations for a consistency check whenever such data were available. The reprocessed data will be archived in the NOAA data center with the same format as the operational data and technical support for data requests. Such a reprocessing is expected to improve the efficiency of the use of the S-NPP and JPSS satellite data and the accuracy of the observed essential environmental variables through either consistent satellite retrievals or use of the reprocessed data in numerical data assimilations.

Keywords: satellite reprocessing; satellite recalibration; suomi NPP and JPSS satellite instruments; fundamental climate data records; climate change monitoring

---

1. Introduction

Satellite observations have been playing a vital role in improving numerical weather prediction (NWP) during the past few decades. Direct use of the satellite radiance data in NWPs was demonstrated to considerably improve NWP forecasting skills in earlier data assimilation experiments [1,2]. After the initial success, the assimilation of satellite data in NWP models has been one of the major drivers in the continued improvement of the NWP forecasting skills over the last twenty years [3,4]. Satellite data assimilation has now become a standard practice in NWP forecasting in nearly all advanced weather forecasting operational centers in the world. Currently, more than 90% of the assimilated data for NWP are derived from satellite observations.

Due to its long-term continuity, satellite measurements have also been widely used to investigate global climate changes during the last four decades. Satellite climate data records (CDRs) have been the primary source for determining global climate trends in many essential climate variables. Such climate variables include, but are not limited to, global atmospheric temperatures from the mid-troposphere [5–11] to the upper stratosphere [12–17], sea ice concentration [18,19], and sea-level rise [20–25], etc. In addition, satellite observations are key input data sources in the development of climate reanalysis systems [26–30], which are broadly used for investigation of both the global climate change attribution and evolution of synoptic weather systems.

Current applications of satellite data in both NWP forecasting and climate reanalyses often include those produced based on operational calibration. An operational calibration generates satellite radiance data in level-1b swath format with necessary quality control procedure and provides all other necessary instrument telemetry information needed for instrument calibration. Such telemetry information includes, but is not limited to, raw counts data corresponding to the earth view, cold space view, and warm target blackbody view; instrument temperatures and blackbody temperatures, Earth view angles, spacecraft orbital parameters, and geophysical locations of views, etc. This calibration is, in general, conducted at satellite operational agencies, and can change frequently due to operational requirements on data latency, improvements and optimization of calibration algorithms, changes and improvements of quality control procedures, etc. As an example, the National Oceanic and Atmospheric Administration (NOAA) operational calibration algorithms for generating instrument sensor data records (SDRs or radiances) for the Suomi National Polar-orbiting Partnership (S-NPP) satellite typically proceed through three stages: beta, provisional, and validated maturity, with many updates of algorithms and calibration coefficients taking place between them [31–33]. Such changes can cause inconsistency in the level-1b satellite radiance data records and lead to radiance or brightness temperature (BT) bias jumps and drifts over time when compared with other observations or numerical model simulations.

Life-cycle satellite reprocessing can minimize or remove the inconsistency and improve the calibration accuracy by using unified and consistent calibration algorithms across the reprocessing period [34]. Unlike operational calibration and processing, reprocessing does not have a strict latency requirement; as a result, more efforts can be devoted to improving calibration algorithms and data consistency and accuracy. Such efforts are particularly beneficial to user applications that require long-term consistency of data products for reliable determination of climate trends observed with the data products. Reprocessing and recalibration result in consistent radiance fundamental climate data records (FCDRs) that are building blocks for CDRs. It is envisioned that reprocessing using the best science and most matured calibration algorithms provides the best FCDR inputs for essential climate variables and the development of critical CDRs [35]. In an end-to-end approach, FCDRs and CDRs can be generated simultaneously in a data production stream [9,10]. Examples of reprocessing included those for instrument observations onboard historical NOAA, National Aeronautics and Space Administration (NASA), the European Organization for the Exploitation of Meteorological Satellites (EUMETSAT), and the Defense Meteorological Satellite Program (DMSP) satellite series for the development of CDRs. The historical NOAA Polar-orbiting Operational Environmental Satellites (POES) carried the Advanced Very High-Resolution Radiometer (AVHRR), the Television Infrared Observation Satellite (TIROS) Operational Vertical Sounder (TOVS) suite of instruments, the Advanced Microwave Sounding Unit-A (AMSU-A), as well as other instruments. The TOVS suite of instruments consisted of three sensors: the High Resolution Infrared Radiation Sounder (HIRS), the Microwave Sounding Unit (MSU), and the Stratospheric Sounding Unit (SSU). The EUMETSAT MetOp series also carry AMSU-A and AVHRR. Together with AMSU-A, the TOVS instruments had been providing critical atmospheric temperature and moisture sounding observations globally for over four decades. Reprocessing and recalibration of these instruments had resulted in consistent FCDRs which were further used to develop the tropospheric and lower-stratospheric temperature CDRs from the MSU and AMSU-A observations [9,10], the high-quality middle to upper stratospheric temperature time series from the SSU observations [14], and the cloud parameter CDRs from the HIRS observations [36,37]. Reprocessing of AVHRR helped the generation of several long-term essential climate variables for climate change monitoring, including the total cloud amount, the Earth's radiation budget at the top of the atmosphere, the outgoing longwave radiation, and the absorbed solar radiation by the Earth–Atmosphere system, etc. [38,39]. As a long-lasting effort, the reprocessing for the DMSP Special Sensor Microwave Imager (SSM/I) was conducted many times over a period spanning two decades [40]. This effort was the basis for the development of a bundle of long-term essential climate variables consisting of surface wind speed over the ocean, rain rate, clouds, and total precipitable water, etc. [41,42].

In addition to CDR development, reprocessing contributes to the improved reanalysis products that merge data from many different observations through data assimilation to attain global climate analyses. As an example, recalibrated and consistent MSU observations developed by Zou et al. [9] were assimilated into NOAA Climate Forecast System Reanalysis (CFSR) and NASA's Modern-Era Retrospective Analysis for Research and Applications (MERRA) [26,27]. This assimilation improved the temporal consistency in bias correction patterns [27,43] and may have helped MERRA to produce a more realistic stratospheric temperature response following the eruption of Mount Pinatubo [44].

The goal of this article is to describe the reprocessing of instrument observations from S-NPP during the period from near its launch time to 8 March 2017 and to provide a perspective on reprocessing of its follow-on NOAA Joint Polar Satellite Systems (JPSS) satellites. The JPSS is the U.S. new generation polar-orbiting operational environmental satellite series making Earth and atmosphere observations for weather and climate applications with high spatial resolution and sampling rate. The S-NPP and JPSS-1 (renamed as NOAA-20 after launch) satellites were launched on 28 October 2011 and 18 November 2017, respectively, into afternoon orbits with a local time ascending node (LTAN) at 1:30 p.m. S-NPP is JPSS's experimental program which became NOAA's prime afternoon satellite since 1 May 2014. Three additional JPSS satellites are planned to be launched onto the same afternoon orbit every five years starting from 2017. These satellites constitute a continuous observation of the global environment for the next two decades and are a heritage of the past instrument observations.

The S-NPP and JPSS are equipped with five instruments, each with advanced design in Earth samplings, including the Advanced Technology Microwave Sounder (ATMS), Cross-track Infrared Sounder (CrIS), Ozone Mapping and Profiler Suite (OMPS), Visible Infrared Imaging Radiometer Suite (VIIRS), and Clouds and the Earth's Radiant Energy System (CERES). Among them, the ATMS is the new generation of microwave sounder measuring temperature and moisture profiles in the atmosphere, while CrIS is an advanced hyperspectral sounder providing more accurate and detailed atmospheric and moisture sounding observations for weather and climate applications. The OMPS instrument measures solar backscattered ultraviolet to provide information on the concentrations of ozone in the Earth's atmosphere, and the VIIRS provides global observations of a variety of essential environmental variables over the land, atmosphere, cryosphere, and ocean with visible and infrared imagery. Finally, the CERES measures the solar energy reflected by the Earth and the longwave radiation from the Earth, providing necessary information to understand the role of cloud processes in the Earth's energy balance.

Currently, observations from these instruments on S-NPP and NOAA-20 provide near real-time monitoring of the environmental changes and improve weather forecasting by assimilation into numerical weather prediction models. To facilitate user applications in further improvement of weather prediction and particularly climate change investigation, the NOAA/Center for Satellite Applications and Research (STAR) has been reprocessing the ATMS, CrIS, OMPS, and VIIRS onboard S-NPP using unified calibration algorithms that are based on optimal operational calibration. As a first version, the reprocessing used the S-NPP NOAA operational calibration algorithms and coefficients baselined on 8 March 2017, when all planned S-NPP post-launch updates were completed and in operation. The reprocessing period was from a starting date close to the S-NPP launch time, depending on when the instruments were turned on, to 8 March 2017 for most instruments. The reprocessing generates improved and consistent sensor data records when validated against other types of instrument observations or numerical simulations. This article summarizes this reprocessing procedure and validation results for the S-NPP, focusing on consistency check and stability assessment. This summary is organized instrument by instrument in the following sections. The next section is dedicated to ATMS, followed by CrIS, OMPS, and VIIRS. Each section consists of a description of instrument observation and calibration principles, calibration algorithms, reprocessing procedure, and validation results. Section 6 provides a conclusion and perspective to the JPSS reprocessing.

2. ATMS Reprocessing

2.1. The Instrument and Calibration Principles

The ATMS is a total power cross-track radiometer with 22 channels, combining all the channels of the heritage sensors, including AMSU-A and AMSU-B/Microwave Humidity Sounder (MHS), into a single sensor that spans from 23 to 183 GHz (Table 1). Such a design offers significant advantage in the reduction in instrument weight and the use of power. Among the channels, 1–2 and 16–17 are the window channels providing information on the atmospheric clouds, total precipitable water, surface emissivity, and water vapor concentration near the surface. Channels 3–15, often called the oxygen channels, use atmospheric oxygen absorption bands for temperature soundings from the surface to the upper stratosphere at approximately 1 hPa. The remaining channels, 18–22, use water vapor absorption lines at 183 GHz for humidity soundings from the lower to the upper troposphere at about 200 hPa. The ATMS channel frequencies are the same as those of AMSU-A and AMSU-B/MHS for most channels except for the addition of temperature-sounding channel 4 (51.74 GHz) and two water vapor channels at 183 GHz (see Table 1).

Table 1. Basic characteristics for the Advanced Technology Microwave Sounder (ATMS) channels. The abbreviations QV, QH, and AMSU-A refer to quasi-vertical, quasi-horizontal, and Advanced Microwave Sounding Unit-A, respectively.

ATMS Channel	Center Frequency (MHz)	Polarization	Maximum Bandwidth (MHz)	Calibration Accuracy (K)	3-dB Bandwidth (deg)	Reference Channels
1	23,800	QV	270	1.0	5.2	AMSU-A Ch1
2	31,400	QV	180	1.0	5.2	AMSU-A Ch2
3	50,300	QH	180	0.75	2.2	AMSU-A Ch3
4	51,760	QH	400	0.75	2.2	
5	52,800	QH	400	0.75	2.2	AMSU-A Ch4
6	$53{,}596 \pm 115$	QH	170	0.75	2.2	AMSU-A Ch5
7	54,400	QH	400	0.75	2.2	AMSU-A Ch6
8	54,940	QH	400	0.75	2.2	AMSU-A Ch7
9	55,500	QH	330	0.75	2.2	AMSU-A Ch8
10	$57{,}290.344(f_0)$	QH	330	0.75	2.2	AMSU-A Ch9
11	$f_0 \pm 217$	QH	78	0.75	2.2	AMSU-A Ch10
12	$f_0 \pm 322.2 \pm 48$	QH	36	0.75	2.2	AMSU-A Ch11
13	$f_0 \pm 322.2 \pm 22$	QH	16	0.75	2.2	AMSU-A Ch12
14	$f_0 \pm 322.2 \pm 10$	QH	8	0.75	2.2	AMSU-A Ch13
15	$f_0 \pm 322.2 \pm 4.5$	QH	3	0.75	2.2	AMSU-A Ch14
16	88,200	QV	2000	1.0	2.2	AMSU-B Ch16
17	165,500	QH	3000	1.0	1.1	AMSU-B Ch17
18	$183{,}310 \pm 7000$	QH	2000	1.0	1.1	AMSU-B Ch20
19	$183{,}310 \pm 4500$	QH	2000	1.0	1.1	
20	$183{,}310 \pm 3000$	QH	1000	1.0	1.1	AMSU-B Ch19
21	$183{,}310 \pm 1800$	QH	1000	1.0	1.1	
22	$183{,}310 \pm 1000$	QH	500	1.0	1.1	AMSU-B Ch18

The ATMS has two receiving antennas—one serving channels 1–15 and the other serving channels 16–22. ATMS scans the Earth within the range of 52.725° on each side of the nadir direction with an angular sampling interval of 1.11°, providing 96 Earth observations in a scan line with a swath width about 2600 km. Each of the 96 Earth samples takes about 18 milliseconds integration time. The beam width of the scans is 5.2° for channels 1–2, 2.2° for channels 3–16, and 1.1° for channels 17–22. This gives a ground nadir field of view (FOV) resolution of 75 km for channels 1–2, 32 km for channels 3–16, and 16 km for channels 17–22 for the S-NPP satellite orbital height of 829 km above the Earth.

When scanning the Earth, the signals received by the antennas are processed by the instrument and output as digital counts. These digital counts are then converted to the Earth radiances or brightness temperatures through an in-flight calibration system and instrument transfer function. Similar to its predecessor, AMSU-A, ATMS calibration relies on two calibration targets as end-point references: a cosmic space cold target and an onboard blackbody warm target. The cold space has a temperature of 2.73 K, and the warm target temperature is measured by platinum resistance thermometers (PRTs) embedded in the blackbody target. In each scan cycle, the instrument looks at these two targets, as well as the Earth, and the signals from these looks are recorded as digital counts. For linear transfer function, the Earth scene brightness temperature is completely determined by the two reference points that have known temperatures in a linear interpolation between the two targets and the Earth views.

In reality, however, the transfer function is slightly nonlinear. This nonlinearity is often assumed to be quadratically related to the Earth scene counts. The magnitude of the quadratic nonlinearity is characterized by a so-called nonlinear calibration coefficient and it was determined using pre-launch thermal vacuum test data in operational calibration [45]. Such a calibration system allows most of the system losses and instrument defects to be removed, since the calibration target views involve the same optical and electrical signal paths as the Earth scene views. Of interest, the nonlinear calibration coefficient can also be obtained in post-launch recalibration efforts by using post-launch simultaneous nadir overpass (SNO) matchups. For instance, optimal nonlinear calibration coefficients were obtained for the MSU and AMSU-A instruments using SNOs which removed or minimized time-varying biases related to instrument temperature variations [9,10].

2.2. Consistency and Stability of Reprocessed ATMS Data

Operational calibration algorithm and procedure have been described in detail in Weng et al. [45] and relevant references within it. This calibration generates the operational level-1 swath radiances that are broadly used in NWP data assimilations to improve NWP weather forecast and for climate reanalyses. However, the operational calibration has gone through a series of updates that have caused inconsistencies in time series. These included the update of processing calibration coefficients on 19 April 2012 at a post-launch instrument evaluation time after their initial implementation at the S-NPP launch time on 28 October 2011; the update of lunar intrusion correction on 20 February 2014; a change of calibration algorithm on 8 March 2017. The last update involved calibration algorithm changes from using the transfer function in its brightness-temperature form before to using the same transfer function but in its radiance form after 8 March 2017, and a fix of a sign bug in expressing the nonlinear term in the calibration equation. The change from brightness temperature to radiance forms in using the transfer function resulted in a better handling of and more accurate bias corrections in the cold space views owing to antenna emission [46]. These calibration changes caused large bias jumps in the brightness temperature time series before and after 8 March 2017 (Figure 1). Other updates had negligible effects on the consistency of time series when compared to the reprocessed time series (Figure 1).

The ATMS life cycle reprocessing used a fixed calibration algorithm taken from the operational calibration after the update on 8 March 2017. The fact that the reprocessed data agree exactly with those of the operational calibration after 8 March 2017 in Figure 1 demonstrates that the former dataset has a calibration accuracy the same as the latter. To examine the consistency and stability of the reprocessed radiance data, the same approach as proposed in Zou et al. [47] is used here, in which the ATMS observations are compared to the AMSU-A observations onboard the NASA's Aqua satellite. AMSU-A has the same channel frequencies as the ATMS (Table 1) for most temperature sounding channels so they observe the same layers of the atmospheric temperature. The Aqua satellite was launched on 4 May 2002 and its orbit has been fixed at close to 1:30 pm for its LTAN throughout its operation. In assessing the stability of historical microwave sounders, changes in diurnal sampling over time and calibration drift have been the main source of uncertainties when satellite orbits drift [47]. However, the similar overpass timing for S-NPP and Aqua in stable orbits naturally removes most of the diurnal differences between them, offering a great advantage in assessing consistency and stability between comparing instruments. In Zou et al. [47], a direct comparison of temperature anomalies between the two instruments shows little or no relative calibration drift for most channels. By comparing with Aqua AMSU-A, Zou et al. [47] suggest that the reprocessed S-NPP/ATMS instruments have achieved absolute radiometric stability in the measured atmospheric temperatures within 0.004 K per year for the time period between 2012 and 2018 for all analyzed channels. A similar comparison is shown in Figure 2 for the S-NPP/ATMS channel 8 and Aqua/AMSU-A channel 7 during the period 2012–2019. Trend differences between the two example channels are within 0.003 K/Year, slightly below the required stability of 0.004 K/Year for the temperature soundings for climate trend studies [48]. Similar results are also obtained for other channels when the quality of Aqua data is good enough for the comparison.

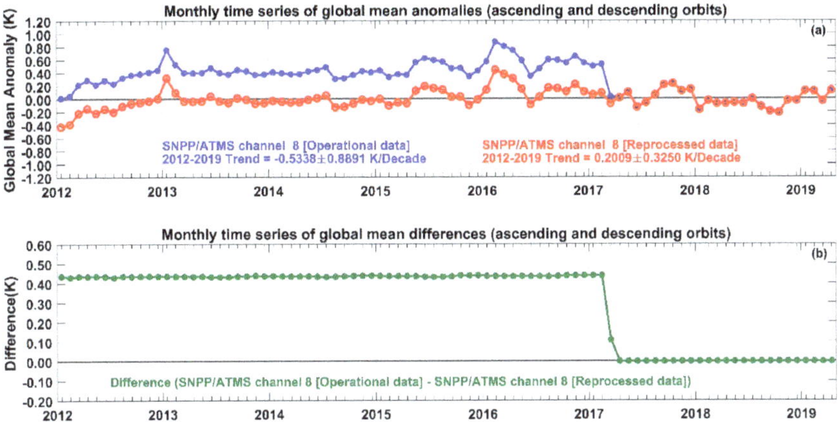

Figure 1. (**a**) Monthly global mean brightness temperature anomaly time series for ATMS channel 8 from operational calibrated (blue) and reprocessed (red) sensor data records, and (**b**) their differences (green). The global means are calculated using limb-adjusted scan positions from 29 to 68 for both operational calibrated and reprocessed datasets. The limb-adjustment and data processing details can be found in Zou et al. (2018). The bias jump between the operational calibrated and reprocessed data found in March 2017 was caused by the calibration update for the operational calibration on 8 March 2017. After that date, the two datasets are identical since they use the same calibration algorithm.

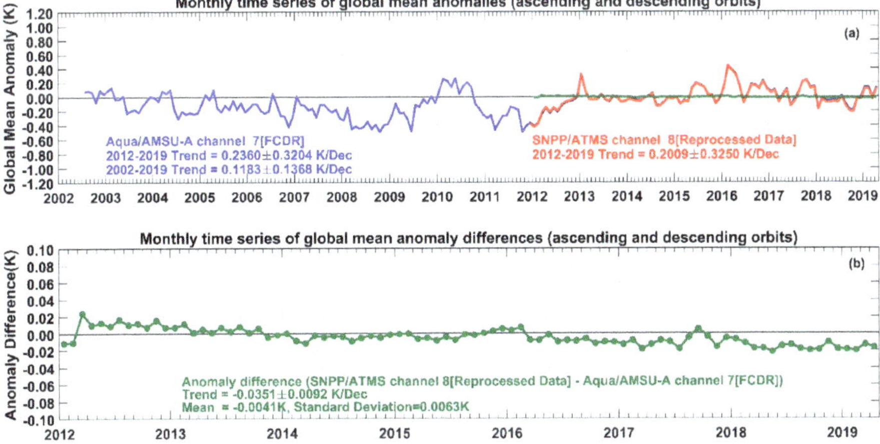

Figure 2. (**a**) Monthly global mean brightness temperature anomaly time series for AMSU-A channel 7 onboard Aqua (blue) versus ATMS channel 8 onboard Suomi National Polar-orbiting Partnership (S-NPP) (red), and (**b**) their difference time series (green). The AMSU-A and ATMS data are from June

2002 and December 2011 to April 2019, respectively. The AMSU-A anomaly time series are overlaid by ATMS during their overlapping period, with their differences shown as nearly a constant zero line in the same temperature scale. Amplified scale of temperature is used in (b) to show detailed features in the anomaly difference time series. The ATMS and AMSU-A data are from limb-adjusted scan positions of 29–68 and 8–23, respectively, and averaged over ascending and descending orbits. Uncertainties in trends represent 95% confidence intervals with autocorrelation adjustments.

The consistency and high radiometric stability in the reprocessed ATMS data have a broad impact on the climate trend observations from the satellite microwave sounders. Such features allow the climate trends to be inferred directly from the reprocessed ATMS observations with high confidence. With consistency and high radiometric stability, the reprocessed ATMS data could also be used as a reference when developing merged temperature time series from microwave sounders onboard multiple satellites [47]. Merged and harmonized satellite temperature products were developed by different research groups [5–10]. However, differences remain in the climate trend estimates between these research groups for the same satellite products owing to differences in bias correction algorithms applied in removing diurnal sampling drift or calibration drift. In this aspect, the stable ATMS observations can help in identifying potential drifts in the harmonized satellite temperature records and improve their accuracy by serving as a reference in developing algorithms for corrections of diurnal sampling and calibration-drifting errors.

The reprocessed ATMS data could also help resolve debates on observed differences in climate trends between different types of instruments and climate reanalyses. Radiosonde observations had been homogenized and extensively used for detecting atmospheric temperature climate trends [49–55]. Disagreement exists in climate trend estimates between the satellite and radiosonde observations [56,57]. By comparing with the stable ATMS observations, biases and their drifts over time in the radiosonde observations could be identified, which would, in turn, help in developing more accurate radiosonde data records for climate trend detection. Similarly, the Global Positioning System (GPS) Radio Occultation (RO) observations had been used for temperature trend investigations, but harmonized satellite data products of earlier versions showed large trend differences relative to the GPS-RO observations for the lower-stratospheric layer [58,59]. Comparisons between GPS-RO and the reprocessed stable ATMS observations could be helpful in identifying drift, if any, in the GPS-RO observations, or conversely, demonstrating their agreement in climate trend detection [60].

3. CrIS Reprocessing

3.1. The Instrument and Calibration Principles

The CrIS instrument is a Fourier transform spectrometer, providing double-sided interferogram measurements in which an interference pattern is produced when the incoming radiation passes through the interferometer. The optical and mechanical design of this instrument and the principles for interferogram data generation are described in detail in Han et al. [61] and the JPSS CrIS SDR Algorithm Theoretical Basis Document [62]. The CrIS interferometer includes a beamsplitter, a porch swing moving mirror, a stationary mirror with dynamic alignment, and a laser metrology system. The beamsplitter divides the incoming radiation into two beams that travel between different mirrors. An interference pattern is produced as the optical path difference between the two beams changes with the sweep of the moving mirrors. After passing through the interferometer, the radiation signal is transformed into time-varying interferogram data that are then output as digital counts from the analog-to-digital converter. To maintain a high signal-to-noise ratio in the interferogram measurements, CrIS is designed with a complex finite impulse response digital band-pass filter to reject out-band signals and to reduce noise in the interferogram data. After going through the filtering process, the interferogram counts are converted to calibrated radiance spectra first through a Fourier transform and then a radiometric calibration process on the ground using the internal calibration target and deep space views as calibration references. The calibrated and geolocated radiance data are referred to as the radiance spectra SDRs. They provide a total of 1305 apodized channels in the normal spectral resolution (NSR) operational mode, covering three spectral bands for sounding the atmosphere. These are the long-wave infrared (LWIR) band, from 650 to 1095 cm^{-1}, the mid-wave infrared (MWIR) band, from 1210 to 1750 cm^{-1}, and the short-wave infrared (SWIR) band, from 2155 to 2550 cm^{-1} (Table 2). The spectral resolutions for the CrIS SDR data at NSR are 0.625, 1.25, and 2.5 cm^{-1} for the LWIR, MWIR, and SWIR bands, respectively. The CrIS can also be operated in the full spectral resolution (FSR) mode, in which all the three bands have the same spectral resolution of 0.625 cm^{-1}, with a total of 2223 unapodized channels. Characterized by its high spectral resolution and wide spectral coverage, a large number of channels, as well as high signal-to-noise ratio, CrIS provides much improved vertical sounding resolution and accuracy in temperature and moisture information compared to the NOAA heritage HIRS.

CrIS contains nine detectors arranged on a 3×3 grid on a focal plane for each of the three spectral bands to receive the interferogram data. The interferometer optical axis is nominally centered in the middle of the fifth detector. The size and position of the detection field stop define the FOV for each detector, and the combined 3×3 FOVs define the field of regard (FOR). The nominal cross-track and in-track offset angles are 1.1° for each FOV. One typical CrIS scan sequence consists of 34 interferometer sweeps that comprise thirty FORs, or Earth scenes, two deep space observations and two internal calibration target measurements. CrIS scans the Earth within the scan angle range of 48.33° on each side of the nadir direction, and with an angular FOR sampling interval of 3.33°. The swath width of the CrIS scan is about 2200 km with a nadir footprint size of about 14 km for each FOV. Each scan takes about 8 s, where 0.2 s are required for each Earth scene, deep space or internal target interferogram measurement.

The Earth scene measurements are calibrated radiometrically for each channel independently using the instrument blackbody internal calibration target and the deep space views, whose radiance is negligible in the frequency range of CrIS measurements. The CrIS radiometric calibration relies on the proper radiometric nonlinearity correction [63]. The CrIS radiometric transfer function is linearly dominated and the small nonlinear response is characterized by a quadratic term multiplied by non-linear calibration coefficients. The CrIS SDRs also went through a complex spectral calibration aiming at removing instrument-design-related spectral self-apodization (shift in channel frequency or distortion in the spectral line shape) and ringing artifacts (spectral noise, see Strow et al. [64]). The spectral self-apodization is induced by the beam divergence that arose from the small angles between the incoming beam direction and the off-axis detectors in the 3×3 grid. The spectral noise was caused by the imaginary component out of the Fourier transform of the asymmetric interferogram, where the asymmetry arose from the phase delay between the two beams divided by the beamsplitter. Three main operations are included in the spectral calibration: application of a band-pass filter to suppress the noise signals in the guard bands that were amplified during the radiometric calibration, an instrument line shape correction to remove self-apodization effect, and spectral resampling to change the spectral resolution from the laser grid to the common user grid. All three operations are combined into a single matrix in the ground process, referred to as the correction matrix operator (CMO).

Table 2. Basic characteristics and requirements for the Cross-track Infrared Sounder (CrIS) normal spectral resolution (NSR) and full spectral resolution (FSR) sensor data records (SDRs).

Instrument	Frequency Band	Spectral Range (cm^{-1})	Number of In-Band Channels (Unapodized Channels)	Spectral Resolution (cm^{-1})	Effective Max. Path Difference (MPD) (cm)	Number of Channels with Guard Bands (N_b)	Decimation Factor (DF_b)	NEdN (mW/m^2/sr/cm^{-1})	Frequency Uncertainty (ppm)	Radiometric Uncertainty at 287 K BB (%)
NSR	LW	650 to 1095	713 * (717)	0.625	0.8	864	24	0.14	10	0.45
	MW	1210 to 1750	433 * (437)	1.25	0.4	528	20	0.06	10	0.58
	SW	2155 to 2550	159 * (163)	2.5	0.2	200	26	0.007	10	0.77
FSR	LW	650 to 1095	713 * (717)	0.625	0.8	874	24	0.14	10	0.45
	MW	1210 to 1750	865 * (869)	0.625	0.8	1052	20	0.084	10	0.58
	SW	2155 to 2550	633 * (637)	0.625	0.8	808	26	0.014	10	0.77

* Apodized channel.

3.2. Consistency and Stability of Reprocessed CrIS Data

The operational calibration algorithm, procedures for producing the S-NPP CrIS SDR data and the SDR validation and data quality have been described in detail in [61,63–67]. The operationally calibrated CrIS SDR data at the NSR mode are broadly used and assimilated at NWP centers to improve weather forecasting and climate reanalyses due to its high radiometric, spectral, and geometric accuracy, as well as excellent noise performance. However, the quality and calibration accuracy of the operational CrIS SDR data were continuously improved through a series of algorithm and software updates that have caused SDR inconsistencies, impacting its long-term stability. The key updates included: (i) update of the processing calibration coefficients on 11 April 2012; (ii) implementation of updates of non-linearity coefficients and instrument line shape parameters, as part of the operational processing system on 20 February 2014; (iii) transition to full spectral interferogram mode implemented in the Raw Data Record (RDR) on 4 December 2014; (iv) a change of the calibration algorithm on 8 March 2017 to include both NSR and FSR SDR data as part of the operational processing system; (v) separation of the CMO and engineering packet output; (vi) recalculating the resampling matrix using the latest metrology laser wavelength. Among these changes, the new calibration algorithm developed by Han and Chen [68] and implemented in March 2017 represented one of the major improvements in CrIS calibration. Instead of performing radiometric calibration first and spectral calibration second in the earlier calibration procedure, the new approach first applies the spectral calibration to the raw spectra after non-linear correction and the removal of the common phases from the radiance spectra, and then applies the radiometric calibration.

To generate the reprocessed SDR data product, a dedicated reprocessing system was developed based on the operational software and calibration algorithm updated on 8 March 2017. In the reprocessing system, the calibration coefficients, including the non-linearity coefficients, the instrument line shape parameters, and the geolocation mapping angles, were refined with the latest updates based on the work from CrIS SDR science team. Those calibration coefficients were included as part of the Engineering Packet in the RDR data stream. The reprocessing system takes advantage of the highly stable spectral emission line from a neon lightbulb to calibrate the metrology laser wavelength [64,65]. The resampling wavelength was updated based on the neon-calibrated metrology laser wavelength and it resulted in close to zero sampling errors in the spectral calibration. In the reprocessing system, all the S-NPP NSR SDRs were generated with the same calibration coefficients, resulting in improved consistency during the CrIS life-time mission. Figure 3a,b compare the FOV-to-FOV radiometric differences among the nine LWIR detectors before and after the reprocessing, respectively. The reprocessed SDR product shows consistent and smaller FOV-to-FOV radiometric differences throughout the reprocessing period (Figure 3b), in contrast to the performance observed for the operational SDR product, which shows a major discontinuity around February 2014. The FOV-to-FOV radiometric differences relative to the central detector (FOV5) are within 0.03 K for the reprocessed SDR product. This translates to a maximum FOV-to-FOV radiometric difference of 0.06 K, demonstrating the high radiometric accuracy and stability of the reprocessed SDR product. This uniformity of the reprocessed FOV-to-FOV radiometric performance is largely due to the improvement of the nonlinearity coefficients. This feature allows the NWP and reanalysis models to assimilate CrIS data from all of the FOVs without special treatment for different FOVs.

Figure 3. Time series of daily mean field of view (FOV)-to-FOV BT difference for the long-wave infrared (LWIR) band (17 channels averaged from 672 to 682 cm^{-1}) with respect to the center FOV 5 for (**a**) the operational and (**b**) the reprocessed SDRs, respectively. The time series are for clear sky over ocean. Scan-angle-corrected SDR data from all scan angles were used in calculating the daily means. The scan-angle correction was based on Community Radiative Transfer Model (CRTM) simulations with inputs from the European Centre for Medium Range Weather Forecasts (ECMWF) analyses. The gap during the period 05/08/2014 to 06/16/2014 in the plots was caused by missing ECMWF analyses at National Oceanic and Atmospheric Administration (NOAA)/ Center for Satellite Applications and Research (STAR).

Figure 4 shows the time series of the spectral errors of the reprocessed SDR product and compares its performance against the metrology laser wavelength, derived from the neon-calibration subsystem, and the operational SDR product performance. Figure 4 shows that the CrIS metrology laser wavelength varies within 4 ppm, as measured by the neon-lamp-calibration subsystem. Similarly, the spectral errors in the operational SDR product also vary within about 4 ppm. In contrast, the reprocessed SDR product has spectral errors less than 0.5 ppm, nearly an order of magnitude smaller than those spectral variations observed in the neon-calibrated metrology laser wavelength and the operational SDR product. The performance of the reprocessed SDR products demonstrates its remarkable spectral accuracy and stability.

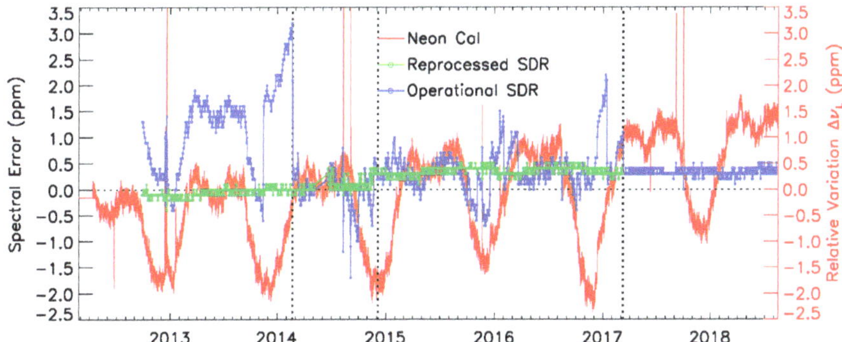

Figure 4. Long-term absolute spectral accuracy and stability for the LWIR band for the reprocessed CrIS SDRs (green line with open circle), compared to the operational SDRs (blue line with open circle) and neon-calibration subsystem (red line, indicated by "Neon Cal"). The absolute spectral error was obtained by simulating radiances at the top of the atmosphere using radiative transfer models under clear conditions and then finding the maximum correlations between the observed and simulated radiance by shifting the spectra in a certain range. The simulation was for the daily average of FOV5 at nadir direction (FORs 15 or 16), descending orbit over clear tropical ocean scenes. The three dashed vertical lines represent major algorithm update events as described in the main text.

4. OMPS Reprocessing

4.1. The Instruments and Calibration Principles

The S-NPP OMPS is composed of three sensors, the Nadir Mapper (NM), Nadir Profiler (NP), and Limb Profiler (LP), and only the first two instruments and their reprocessing are described here. Dittman et al. [69] and Seftor et al. [70] provided a detailed description of the optical design of the NM and NP instruments. The two instruments are nadir-viewing spectrometers that measure backscattered ultraviolet sunlight from the Earth's atmosphere and surface. The NM and NP share the same telescope with a dichroic beam splitter downstream redirecting the incoming radiation into either of the two spectrometers. The telescope has a 110° total across-track FOR, resulting in a 2800 km swath width at the Earth's surface. The dichroic beam splitter was optimized to reflect light for wavelengths shorter than 300 nm to the NP spectrometer and to transmit light for wavelengths longer than 310 nm to the NM spectrometer. There is a transition from reflection to transmission over the 300 nm to 310 nm interval. The telescope also includes a depolarizer to minimize the sensor linear polarization sensitivity [71].

After being split, the light from each spectrometer is dispersed via a diffraction grating onto corresponding dual charge-coupled devices (CCDs) with two-dimensional arrays that are located at each spectrometer's focal plane, comprising 740 individual spatial detector pixels and 340 spectral channels. The NM and NP, respectively, illuminate 196 and 147 of the 340 spectral samples. Given the total bandwidth for each of NM and NP, this results in a spectral resolution of ~0.41 nm for both instruments. In the spatial dimension, multiple pixels were summed together into a single "macro-pixel" to provide an instantaneous FOV (IFOV) much larger than the pixel size. Specifically, each IFOV is composed of 93 pixels for NP and of 20 pixels for all but the most extreme left and right off-nadir IFOVs for NM. The nadir-viewing resolution with such a composition is 50 km cross-track by 50 km along-track taking 7.6 s total integration time for NM and 250 km × 250 km taking 38 s total integration time for NP.

The OMPS nadir sensors retrieve estimates of ozone amounts by utilizing normalized radiances (NRs), defined as the ratio of measured Earth radiance to measured solar irradiance [72,73]. The NR is referred to as the albedo or reflectance at the top-of-the-atmosphere. Calibration of NR includes those of the Earth radiance and solar irradiance. Solar observations are made by putting one of the two reflective diffusers at the entrance aperture. They are rotated through seven different positions to obtain full coverage of the 110° FOV. One of them is the working diffuser providing solar irradiance measurement once every two weeks, and the other is a reference diffuser deployed every six months to monitor the stability of the working diffuser. The Earth radiance calibration assumes a close-to-linear relationship between the incoming light and the counts' values from the analog-to-digital converter output. Linearity is defined relative to the measured signals at two points: one at the bias level as the lower limit and one at the 75% of the prelaunch saturation point of the analog-to-digital converter as the upper limit. A non-linear correction is applied at the count level. Responses at individual pixels are converted from corrected counts using slope between the two points as well as other prelaunch knowledge. Details of the calculation procedure to create the Earth radiance can be found in Seftor et al. [70].

Wavelength calibration is conducted by utilizing the solar flux measurements and knowledge of the solar spectral line structure, that is, the solar Fraunhofer lines, as the reference. In this calibration, a reference wavelength registration is first performed in which laboratory reference spectral data were derived for each pixel on the CCD focal plane. They are then binned and averaged into the same data samples as the Earth view pixels to provide a reference for the Earth views. On-orbit wavelength calibration at each Earth view pixel is computed by periodically comparing the reference irradiance with the actually observed irradiance. The OMPS wavelength scales are subject to shifts due to thermal loading changes in various optical elements, referred to as the thermo-optical effect [74]. Among them, solar wavelength shifts up to 0.11 nm was observed for the NM due to a dramatic change in its operational temperature after the transition from the ground to on-orbit. Intra-orbit Earth wavelength shifts up to 0.05 nm are found for the NM in association with intra-orbit changes in housing temperature. In addition, both Earth and solar wavelengths drift in an annual cycle with a magnitude of 0.04 nm for NP and 0.02 nm for NM, respectively, due to seasonal variations of optical bench temperature associated with changes of the solar beta angle. Calibration algorithms were developed to correct these wavelength shifts [74,75]. The OMPS on-orbit wavelength calibration detects any of those shifts and then corrects the measured wavelength as a function of the spectral and spatial positions on the CCD focal plane using relevant calibration algorithms. The calibrated wavelength scales achieved an accuracy of better than 0.02 nm for NM and 0.01 nm for NP, respectively.

4.2. Consistency and Stability of Reprocessed OMPS Data

Both radiometric and wavelength calibrations for the operational NM and NP SDRs have been described in detail in Seftor et al. [70] and Pan et al. [74,75]. These calibration processes correct biases from several different error sources, including CCD dark current, electronic bias, nonlinearity, stray light, throughput degradation, and wavelength scales. Dark current is caused by electrons thermally excited into the CCD conduction band. Although the dark current is well characterized and stable before launch, it varies on orbit due to lattice damage caused by energetic solar wind or cosmic ray particles striking the CCD. Time-varying corrections are needed to remove the dark current effect. Figure 5 shows the time series of N-values, which are logs of NR, averaged over the tropical region for both operational and reprocessed SDRs for both NM and NP at selected channels. The OMPS operational SDRs were first released on 27 January 2012, three months after the S-NPP launch time, and then a Beta version was released shortly later on 13 March 2012. As a Beta version, the SDRs allow the users to get familiar with the data formats and parameters but are not appropriate for quantitative scientific studies and applications, as only initial radiometric and solar flux calibrations were applied. As such, a series of calibration updates were implemented after the release of the Beta version. Tables 3 and 4 provide the timelines for these calibration updates for NM and NP, respectively. For NM, weekly dark current calibration started on 21 December 2012, followed by a series of updates of the stray light calibration look up table (LUT). Among them, the updates on 10 July 2013, 21 November 2014, and 18 December 2014 caused large jumps in the operational NR time series for the 302 nm channel (Figure 5a). A major update in calibration algorithms was made on 9 September 2015, the release date of the validated maturity version (VMV) of SDRs, including updates of the solar LUT, wavelength calibration LUT, calibration constant, and stray light calibration LUT. The time series remained smooth for most of the period afterward, except for another jump on 9 July 2018 caused by the update of the stray light calibration LUT at that date.

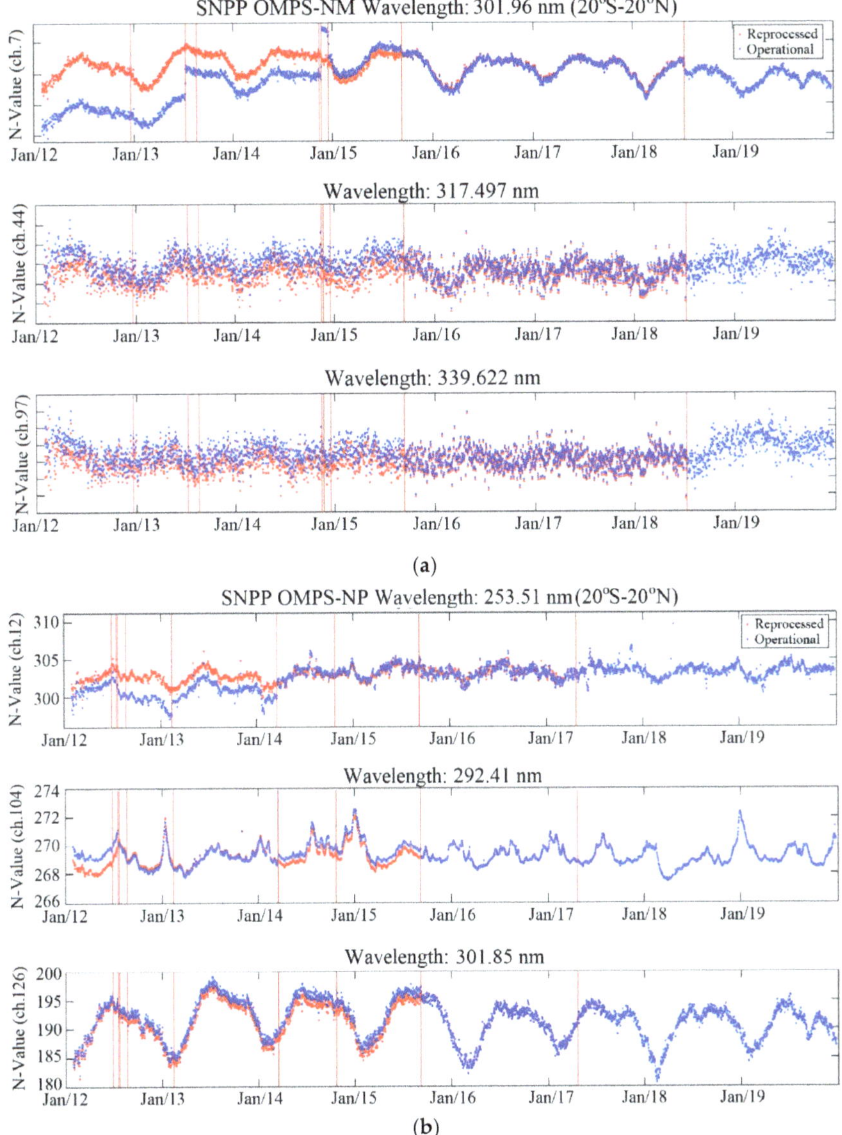

Figure 5. Daily N-value ($-100 \log_{10}(NR)$) time series for data over the tropical region (20° S–20° N) for (**a**) the Nadir Mapper (NM) and (**b**) the Nadir Profiler (NP) instruments at selected channels. Update events of calibration look up tables (LUTs) listed in Tables 3 and 4 are marked by the red vertical lines. Note that the channel wavelengths have errors as large as 0.1 nm for the operational SDRs due to improvement in wavelength scale calibration over time, and of less than 0.02 nm for the reprocessed SDRs.

Table 3. Timeline for the S-NPP/Ozone Mapping and Profiler (OMPS)/NM calibration LUTs updates.

Weekly Dark Current calibration started	12/21/2012
Stray light calibration LUT updates	07/10/2013, 08/20/2013, 11/21/2014, 12/18/2014, 09/09/2015, 07/09/2018
Observed Solar LUT update Wavelength calibration LUT update	11/13/2014
Solar LUT update Wavelength calibration LUT update Calibration constant update	09/09/2015

Table 4. Timeline for the S-NPP/OMPS/NP calibration LUTs updates.

Wavelength calibration LUT updates	6/26/2012, 7/22/2012, 8/19/2012, 10/23/2014, 09/09/2015
Solar LUT updates	07/17/2012, 09/09/2015, 04/20/2017
Weekly Dark Current calibration started	02/12/2013
Stray light calibration LUT update	03/18/2014
Wavelength calibration LUT biweekly update started	04/20/2017

For NP, major jumps in the operational NR time series occurred at four calibration updates: solar and wavelength LUT updates in the middle of July 2012, start of the weekly dark current calibration on 12 February 2013, update of stray light LUT on 18 March 2014, and solar and wavelength LUT updates on 9 September 2015 when the VMV was released.

In the reprocessing, consistent algorithms, tables and corrections were applied to both NM and NP throughout their reprocessing periods from 27 January 2012, the SDR releasing date, to 9 July 2018 for NM and 8 March 2017 for NP, respectively. The software codes and calibration LUTs that were used for reprocessing are the same as those used in operations on the end dates of reprocessing for both NM and NP, except that the operational processing used half-week delayed tables while the reprocessing used the current week tables for the dark current correction. The consistent calibration algorithms and tables in the reprocessing had effectively removed jumps in the operational processed NR time series, leading to consistent reprocessed time series for both NM and NP through their entire reprocessing periods (Figure 5a,b).

The operational and reprocessed SDRs are in better agreement during the later periods than the earlier periods of reprocessing in general, due to the use of the same calibration algorithms and LUTs applied at the end dates of reprocessing. The magnitudes of differences between the reprocessed and operational SDRs are different for different channels of the NM and NP instruments. For NM, large differences occurred for channels with shorter wavelengths, e.g., channel 302 nm before the VMV releasing date. On the other hand, the reprocessed and operational NR time series are almost the same for channels with a longer wavelength, i.e., channels 317.5 nm and 339.6 nm. For NP, significant differences between the reprocessed and operational N-values are observed for all selected channels before the VMV releasing date. There are two reasons for these phenomena. The first is that channels with shorter wavelengths have stronger ozone absorption leading to smaller radiances and are thus more sensitive to changes of calibration algorithms and tables. The other is that the dichroic beam splitter redirects 90% of the incoming sunlight to NP but only 10% to NM at 302 nm. The lower radiance levels for the channels received by NP caused noticeable sensitivity to changes of calibration algorithms for all of its channels. In contrast, only the NM channels with smaller signal levels are more sensitive to dark current and stray light changes since the incoming radiances are small.

Such differences in channel sensitivity to calibration changes help explain the slight differences between the operational and reprocessed SDRs during the period immediately before the end dates of reprocessing. For NM, calibration tables on the end date (9 July 2018) for reprocessing included the updated stray light correction LUTs on the same day and this was slightly different from those used in the operational processing on and after the VMV releasing date. This, together with differences in the dark current correction, caused slight differences between the reprocessed and operational NRs from 9 September 2015 to 9 July 2018 for all the selected NM channels (Figure 5a). On the other hand, LUTs used in the NP reprocessing are the same as those in operations on the VMV releasing date. This resulted in identical operational and reprocessed SDRs for NP channels 292 nm and 302 nm, which are not as sensitive to dark current corrections, from 9 September 2015 to 8 March 2017. However, due to the high sensitivity of channel 253 nm to the dark current correction, the reprocessed and operational SDRs for this channel show noticeable differences during the same period of time.

Changes in calibration constants and tables used in reprocessing also improved the consistency in the N-value measurement between the NM and NP. Figures 6 and 7 investigate the spectral dependence of the improvements by comparing the operational and reprocessed SDRs for a sample day in September 2012. Figure 6a shows that the main changes from the operational to reprocessed solar irradiances were in the overlap region from 300 nm to 310 nm for both the NM and NP. In contrast, the main changes to the Earth radiances were from 250 nm to 290 nm for the NP and from 300 nm to 310 nm for the NM (Figure 6b). These changes in solar irradiance and Earth radiance led to changes in the absolute N-values in the corresponding spectral regions (Figure 6c). Improvements in the stray light corrections listed in Tables 3 and 4 for the two instruments had the largest impact on the changes in these variables. Figure 7 shows the N-value differences for both the operational and reprocessed NM and NP relative to a smooth quadratic fit function that is used as a consistent reference to compare the NM and NP with different wavelength scales. Wavelike structures in the difference curves represent real signals from the solar and ozone absorption spectral features. The most striking feature in the plot is that the differences between the NM and NP for the reprocessed SDRs (differences between the dotted lines) are much smaller than those in the operational products (differences between the solid lines) in their overlap spectral region. This demonstrates the improved consistency between the NM and NP in the reprocessed SDRs. This improvement was largely due to the wavelength scale and irradiance calibration coefficient refinements over this interval as listed in Tables 3 and 4.

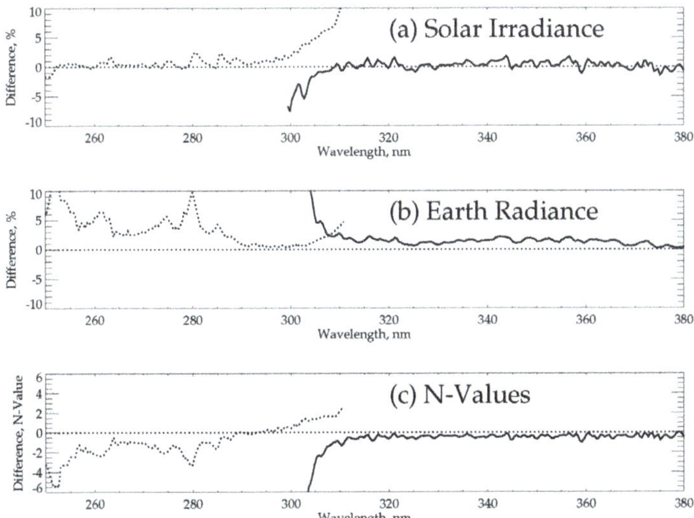

Figure 6. Comparison of the operational and reprocessed SDRs for a single day, 20 September 2012. (**a**) Daily average percentage differences of the solar irradiances versus wavelength between the operational and reprocessed SDRs; (**b**) similar differences but for the Earth radiances; (**c**) similar differences but for the N-values. The percentage differences are calculated with respect to the reprocessed data and the daily means are simple averages for 950 measured spectra with solar zenith angles less than 80 degrees.

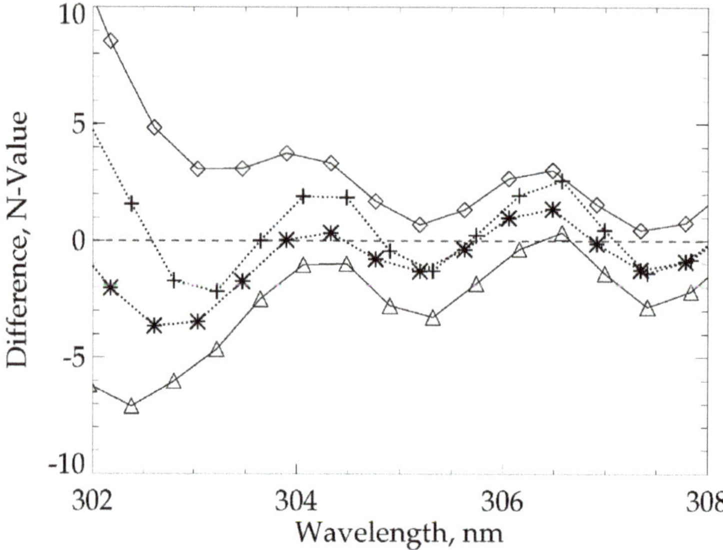

Figure 7. N-Values for NM (Δ and +) and NP (◇ and ✳) versus wavelength in the overlap spectral interval from 302 nm to 308 nm. The operational (solid lines) and reprocessed (dotted lines) SDRs are for the same day as used in Figure 6. The differences are calculated relative to fit quadratics which are smoothed functions of wavelength averaged over the operational or reprocessed NM and NP data.

Although reprocessing produced consistent time series, they are not considered as the final products as bias correction algorithms are still under improvement and will be implemented when they reach maturity. For the NM sensor, solar reference measurements show little throughput degradation and no time-dependent calibration or solar adjustments are made to the SDR products. The same Day 1 solar irradiance spectra and wavelength scales are used as the basis for the measurement-based adjustments for both reprocessed and operational products. The NM has an intra-orbital variation in the wavelength scale and this variation is estimated from the Earth-view radiances on a granule by granule basis. The new wavelength scales and the solar flux adjusted to these new wavelength scales have been reported in the SDR. Additionally, an error in the NM dark correction code has been identified and corrected code has been developed for the operational SDRs. However, reprocessing has not used this most updated code yet. The error is small for solar zenith angles less than 88°.

For the NP, solar reference measurements show wavelength-dependent degradation, however, no time-dependent calibration or solar adjustments for this degradation are made to the SDR products. This means that the NR values using the SDR information will have uncorrected calibration drifts from the Earth-view radiances. The NP also has an annual variation in the wavelength scale. The Day 1 solar irradiance spectra and wavelength scales are regularly updated by using the biweekly working diffuser solar measurements to account for this wavelength scale variation. Improved "Day 1" wavelength scales and unexpected behavior in night-side measurements are under investigation. Considerations are made to include degradation adjustments and solar activity in the biweekly table deliveries.

Once further improvements to account for these issues are made in the operational SDRs, reprocessing will be conducted again using consistent algorithms accounting for the improvement for an extended period of time longer than the reprocessing presented here. The initial set of reprocessed SDRs has already been used to generate total column ozone and ozone vertical profile CDRs, the Version 8 ozone datasets. The reprocessed SDRs were used as inputs to the Version 8 total ozone algorithm with a constant (non-time-varying) set of channel bias adjustments calculated to give agreement with the Earth Observing System (EOS) Aura Ozone Monitoring Instrument (OMI) Version 8 record [76].

The reprocessed SDRs were also used as inputs to the Version 8 ozone profile algorithm with a constant set of channel bias adjustments to give agreement with the NOAA-19 SBUV/2 Version 8 record in 2013 [77]. Daily updates to the solar irradiance were created to account for wavelength scale variations and throughput degradation that was not included in the current SDR reprocessing.

5. VIIRS Reprocessing

5.1. The Instrument and Calibration Principles

The VIIRS instrument is a whiskbroom scanning radiometer measuring reflected and emitted radiation from the Earth in the spectrum between 0.412 µm and 12.01 µm. A detailed description on the optical design and instrument characteristics is given in Cao et al. [78–80] and in the VIIRS algorithm theoretical basis document (ATBD) [81]. The VIIRS fore optics includes a rotating telescope assembly, and a rotating half angle mirror synchronized with the telescope. The aft optics consist of two dichroic beamsplitters with four-mirror anastigmat with all reflective design, three focal plane assemblies (FPA) with detector arrays designed to measure the visible/near infrared (VisNIR), the shortwave/midwave infrared (SW/MWIR), and the longwave infrared (LWIR) spectrum, with an additional day–night-band (DNB) FPA mounted adjacent to the VisNIR FPA. The onboard calibration system relies on a solar diffuser, solar diffuser stability monitor (SDSM), an onboard calibrator blackbody, and space view. The telescope scans the Earth between the angles of ±56.28° from nadir, resulting in a swath width of 3060 km at the nominal altitude of 829 km. Incoming radiation received by the telescope is reflected from the half-angle mirror into the aft-optics subsystem. The light is then spectrally and spatially divided by the beamsplitters and directed to the three FPA detector arrays with the DNB and VisNIR FPAs sharing the same optical path. The detector arrays are built in rectangular patterns, arranged with "bands" in the scan direction and detector numbers in the track direction [81]. The VIIRS instrument provides moderate resolution radiometric bands (M-bands) and fine resolution imaging bands (I-bands). Each M-band and I-band consists of 16 and 32 along-track detectors, respectively. VIIRS has a total of sixteen M-bands and five I-bands distributed among the three FPAs (Table 5). Among them, seven M-bands and two I-bands are in the VisNIR FPA, six M-bands and two I-bands in the SW/MWIR FPA, and three M-bands and one I-band in the LWIR FPA, respectively. Among the M-bands, eleven are reflective solar bands (RSBs) and five are the thermal emissive bands (TEBs). The I-bands include three RSBs and two TEBs. VIIRS uses a unique approach of pixel aggregation which controls the pixel growth towards the end of the scan. Such an aggregation helps a band maintain a nearly constant resolution over the entire scanning swath. At nadir, the FOV spatial resolutions are 750 m for the M-bands and 375 m for the I-bands.

The DNB module includes a CCD array and a focal plane interface electronics. The single DNB band contains 672 sub-detectors in the along-track direction by using multiple CCDs that provide multiple samples in the scan direction. The unique DNB detector technology allows the measurement of nightlights, reflected solar and/or moon lights with a large dynamic range, such as the reflected signals from as low as quarter moon illumination to the brightest daylight [80]. The DNB has a constant spatial resolution of 750 m across the scan, due to the advanced subpixel aggregation scheme with 32 aggregation zones from nadir to edge of the scan.

Table 5. Wavelength information for all Visible Infrared Imaging Radiometer Suite (VIIRS) bands and equivalent Moderate Resolution Imaging Spectroradiometer (MODIS) Thermal bands (from Cao et al. 2014). The abbreviations ViS/NIR, SW/MWIR, LWIR, and DNB stand for visible/near infrared, shortwave/midwave infrared, longwave infrared, and day–night-band, respectively.

	Band	M1	M2	M3	M4	I1	M5	M6	I2	M7
ViS/NIR	Wavelength (μm)	0.412	0.445	0.488	0.555	0.640	0.672	0.746	0.865	0.865
DNB	Band	DNB								
	Wavelength (μm)	0.5–0.9								
SW/MWIR	Band	M8	M	I3	M10	M11	I4	M12	M13	
	Wavelength (μm)	1.24	1.378	1.61	1.61	2.25	3.74	3.70	4.05	
LWIR	Band	M14	M15	I5	M16					
	Wavelength (μm)	8.55	10.763	11.450	12.013					
MODIS Thermal Bands	Band	B20	B22	B29	B31	B32				
	Wavelength (μm)	3.78	3.96	8.56	11.03	12.04				

Radiation received by the FPAs is converted to digital counts by the analog-to-digital converters as detector outputs, which are then converted to radiances using the on-board calibration system. VIIRS is a conventional differencing radiometer that uses the space view to determine zero radiance and observations of a known radiance source to determine the gain. For RSBs, the known radiance source is the on-board solar diffuser. The diffuser is fully illuminated once per orbit as the satellite passes from the dark side to the light side of the Earth near the South Pole and the reflected solar radiance from the diffuser is used as a reference to calibrate the Earth radiance and reflectance [80]. However, the solar diffuser degrades over time. The SDSM is therefore used to determine the degradation by directly measuring sunlight through an attenuation screen and comparing it with the reflected radiance from the solar diffuser.

For TEBs, the on-board blackbody serves as the calibration target. Six National Institute of Standards and Technology (NIST) traceable thermistors are embedded uniformly on the blackbody to measure its bulk temperature to ensure reliability and traceability of the measurements.

To measure radiances with seven orders of magnitude in dynamic range [80,82], the DNB is made of three sub focal plane arrays: the low-gain, mid-gain, and dual high-gain FPAs. The low-gain is used for high radiance or daytime observations, an intermediate-gain is used for mid-radiance, and a high-gain is used for low radiance or nighttime observations, with automatic switching between the gain stages to accommodate observed light sources. The low-gain stage calibration uses the same onboard solar diffuser and space view that are shared with the RSBs. For mid-gain and high-gain stages, however, measurements from the solar diffuser and space views are saturated, thus cannot be used for calibration directly. As a result, calibration of the mid-gain and high-gain stages rely on gain transfer from the low-gain stage through the gain ratio approach. In addition, because of an electronic timing difference between the calibrator view and Earth view, the dark ocean and blackbody views during a new moon are used to replace the space view and provide Earth view calibration offset for calibration of all three gain stages of DNB [81,83,84]. Detailed information on the DNB calibration is given in Uprety et al. [84].

Conversion from the Earth view digital counts to Earth view radiances is carried out through a quadratic calibration equation for all bands using their calibration target views as end-point references. Prelaunch calibration coefficients were determined in laboratory tests and used initially after launch. To account for onboard sensor degradation and other onboard calibration changes, a band-dependent time-varying scaling factor (also known as F-factor) is introduced in the calibration equation for the VIIRS instrument. Changes in scaling factors are parameterized as a function of time in the instrument calibration LUTs. Operational SDRs are generated using constantly updated LUTs with the algorithm and data processing software within a so-called Algorithm Development Library (ADL) framework. Details on the calibration equations for different bands were given in Cao et al. [80] and VIIRS ATBD [81].

Despite the comprehensive design of the RSB onboard calibration system with a solar diffuser and SDSM, residual degradation still exists which is not accounted for in the operational calibration using the onboard calibration system alone. This has an impact on applications that require extremely high long-term stability and accuracy. To mitigate this effect, rigorous monthly lunar calibration through spacecraft maneuvers has been operationalized to measure the moon irradiance at the same lunar phase angle [85]. In addition, vicarious calibration methods have also been used to account for the residual degradation, such as using the deep convective clouds, and twenty global calibration sites. The vicarious calibration has been used as feedback and correction for the operational calibration, as well as for reprocessing.

5.2. Consistency and Stability of Reprocessed VIIRS Data

The first VIIRS visible image was taken on 21 November 2011 when the VIIRS nadir door was opened, allowing Earth observations from the RSBs and DNB. Later on 18 January 2012, the cryo-cooler was opened for TEBs. VIIRS became thermally stable and functional two days later and has been continuously generating SDRs based on operational calibration algorithms since 19 January 2012 until present. Although assessments of the VIIRS calibration showed that it outperformed legacy sensors such as the AVHRR or Moderate Resolution Imaging Spectroradiometer (MODIS), there were many calibration and operational changes during the first 5 years of its mission for SDR improvement [80,86,87]. These changes caused time series of the VIIRS operational SDRs being inconsistent for climate change applications. To improve SDR consistency, reprocessing was conducted for the period from 2 January 2012 to 8 March 2017 for RSBs and DNB, and from 19 January 2012 to 8 March 2017 for TEBs using consistent calibration algorithms and calibration LUTs for all of the VIIRS bands. The reprocessing incorporated all improvements developed before the end date of reprocessing into the ADL framework, although different bands used different updated LUTs specific for the bands.

Uprety et al. [88] and Choi et al. [85] summarized the reprocessing for the RSBs. The biggest challenge in calibrating the RSBs is that the scaling factor derived from the onboard SDSM and solar diffuser has uncertainties due to errors in the measurement of the solar diffuser bi-directional reflectance function and screen vignetting function [89]. A series of updates was implemented in the operational calibration to improve the radiometric correction factor that included the update of the SDSM screen transmittance tables in early 2012, prelaunch calibration coefficients update in April 2014, optimized Robust Holt-Winters filter parameters for the characterization of solar diffuser degradation in May 2014, the transition from manual computation to automatic determination in software codes of the scaling factor, and solar vector error correction [87], etc. Although these updates largely improved radiometric accuracy in the RSBs, they also resulted in bias jumps in the operational SDR's time series (Figure 8). VIIRS reprocessing accommodated all of these changes. In addition, VIIRS reprocessing applied Kalman filter to determine gain coefficients for the M1–M7 and I1–I2 RSBs that help to remove long-term biases and improve data quality. Kalman filter allows calculation of the instrument degradation using multiple independent approaches and combining them for an optimal determination of gain values in calibration. Specifically, the Kalman filter combines calibration results from the latest solar diffuser-based calibration parameters with reduced seasonal oscillations, lunar, deep convective clouds, and extended SNO results. It also reconciles discrepancies between low-gain and high-gain calibrations.

An additional correction has also been applied for the M5 and M7 for the entire reprocessing period by comparing it with MODIS using SNO methods. Figure 8 shows the reflectance anomaly time series (reflectance minus an annual mean climatology) for the M5 band for the operational and reprocessed data over the Libya-4 desert area and their ratio. Both the operational and reprocessed data show similar temporal patterns in their monthly anomaly time series although there is a small bias between them. However, their ratio (reprocessed/operational) at the pixel level shows jumps associated with major calibration updates, suggesting inconsistency in the operational SDRs. This inconsistency has been removed in the reprocessed SDRs.

Uprety et al. [84] described in detail on the reprocessing of the DNB. The DNB is unique in wide relative spectral response function in the visible region designed especially for nighttime imaging. However, the nighttime observations are contaminated by stray light from the Sun when the satellite is in the twilight zone. Atmospheric airglow contamination impacted high-gain stage calibration using the dark ocean and resulted in low accuracy and substantial presence of negative radiance at views near the new moon. Algorithms to correct these biases along with other calibration updates were implemented at different times when SDRs were operationally generated, including an update of calibration coefficients using on-orbit data in late March 2012, stray-light correction since mid-2013, accommodation of changes in relative spectral response resulting from telescope throughput degradation in April 2013, removal of atmospheric airglow effect in early 2017, and minimizing strong striping in radiance for higher aggregation zones since January 2017. These calibration changes improved the quality of the operational SDRs over time, but also caused inconsistency in the SDR time series.

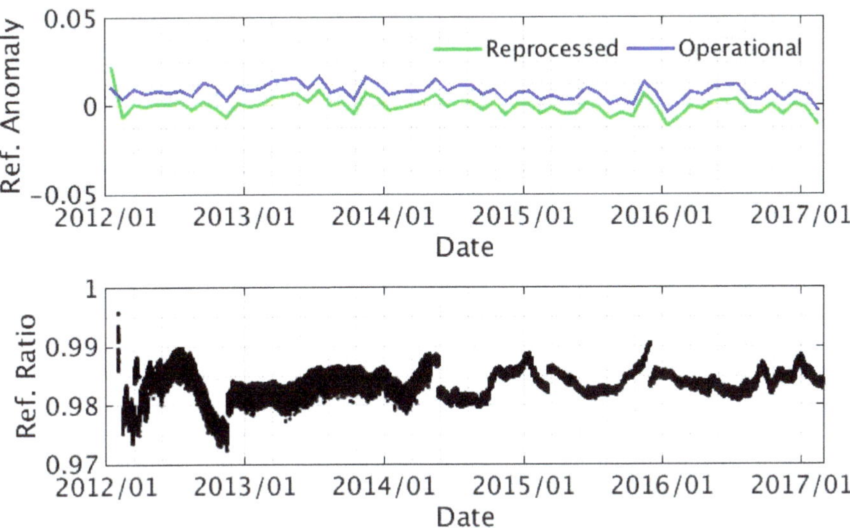

Figure 8. Upper panel: The reflectance monthly anomaly time series for the VIIRS reflective solar band (RSB) M5 for the reprocessed and operational SDRs over the Libya-4 desert calibration site, located in the Great Sand Sea. The anomaly values are computed as the reflectance minus an annual mean climatology derived from the reprocessed data from January 2012 to March 2017. Lower panel: The pixel-by-pixel reflectance ratio (reprocessed/operational) time series over the Libya-4 area. Jumps in time series correspond to events in algorithm changes or LUT updates. The thinner lines after mid-2014 show a better agreement in variability of individual pixels between the reprocessed and operational SDRs in comparison to those before mid-2014.

Reprocessing accommodated all calibration updates in the operational SDRs. Additionally, an improved stray-light correction algorithm was applied in the reprocessing to remove residual errors in the stray-light correction in the operational SDRs. Calibration improvements in the dark offset and gain ratios estimated every month during the new moon are updated periodically in the reprocessing. All of these improvements resulted in higher quality and consistent reprocessed DNB images [84].

The thermal band performance is very stable in general [86,90]. A main issue with the thermal band calibration is that a small bias on the order of 0.1 K was introduced in the brightness temperatures during the quarterly blackbody warm-up/cool-down (WUCD) periods—an operational procedure to assess the thermal band calibration nonlinearity [90]. This bias is caused by a calibration defect during the blackbody unsteady states when its temperature changes by nearly 50 K during a WUCD event. This bias is further amplified by up to 0.3 K in the sea surface temperatures through retrieval algorithms. Cao et al. [90] developed a diagnostic and correction method by introducing a compensatory term in the calibration equation to remove this bias. This correction algorithm was further analyzed and evaluated by Wang et al. [91] and has been implemented in the NOAA operational processing since 25 July 2019. In the reprocessing, the same correction algorithm for the WUCD effect was implemented for the entire reprocessing period from 19 January 2012 to 8 March 2017. Figure 9 compares brightness temperatures pixel by pixel at nadir between the reprocessed and operational SDRs over the tropical site (142° E, 2° N) separated into cloudy and clear sky conditions. As seen, the WUCD biases in the operational SDRs show up as regular spikes in the brightness temperature difference time series and they have been mostly mitigated in the reprocessing.

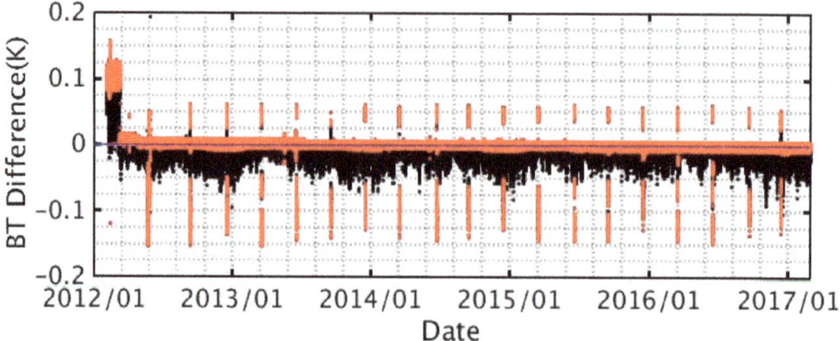

Figure 9. Brightness–temperature difference time series pixel by pixel at nadir for the reprocessed minus operational SDRs for the M15 thermal band at the tropical site (142° E, 2° N). The red (black) dots are for the clear (cloudy) sky pixels where the brightness temperatures are greater (smaller) than 260 K. The warm-up/cool-down (WUCD) biases in the operational SDRs show up as regular spikes in the brightness temperature difference time series and they have been mostly mitigated in the reprocessing. Note that the WUCD anomalies occurred in both the clear and cloudy difference time series, but those in the latter were overlaid by the former in the plot.

The VIIRS thermal bands are compared with the heritage sensor MODIS onboard Aqua for consistency and stability assessment. Some of the MODIS channels in the infrared region match quite well with the VIIRS thermal bands (Table 5), although their central wavelengths are slightly different. Nevertheless, the impact of these small wavelength differences is negligible in their stability assessment. The VIIRS band M15 (10.75 μm) and MODIS band 31 (11.03 μm) are selected here for comparison. These two bands have very close central wavelengths and bandwidths, and both are used for the sea surface temperature retrievals, respectively, in their own mission. In assessing the consistency and stability of similar sensors on different platforms, changes in diurnal sampling over time due to satellite orbital drifts pose a challenge in explaining comparison results. Fortunately, the similar overpass timing for S-NPP and Aqua in stable orbits naturally removes most of the diurnal differences between them, offering a great advantage in the stability assessment. Figure 10 shows the monthly global mean BT time series of near-nadir observations from 01/2012 to 03/2017 for both VIIRS and MODIS, ascending and descending nodes separately, and their differences. Seasonal variations dominate in both the VIIRS and MODIS global mean time series, but the ascending observations are a few Kelvin degrees larger than the descending data. This is because the former are daytime observations close to 1:30 pm local time while the latter are nighttime observations close to 1:30 am local time. In their difference time series, the VIIRS brightness temperatures are warmer by approximately 0.15 K than MODIS on average for both daytime and nighttime observations. This bias arises for the most part due to differences in the relative spectral response functions between the two sensors [92]. In addition, the S-NPP orbits drifted relative to Aqua in a zigzag pattern with a maximum of 10 min apart and a minimum near zero minute occurred in late 2014. This small orbital drift caused biases on the order of 0.03–0.08 K between the two instruments [93].

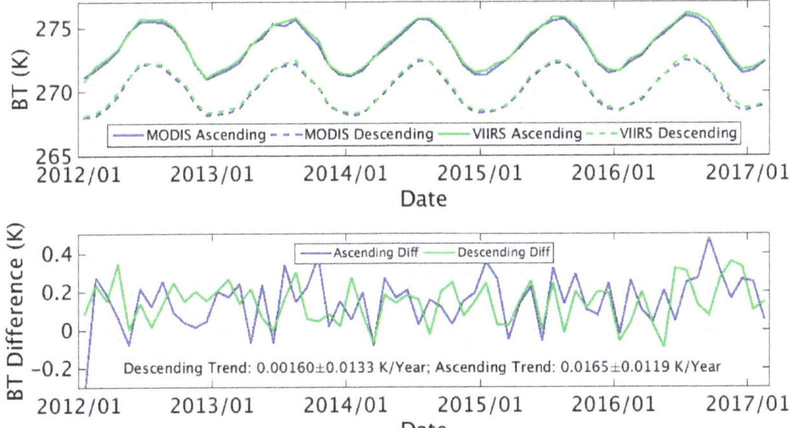

Figure 10. Monthly global mean time series of brightness temperatures for S-NPP VIIRS M15 and Aqua MODIS B31 bands (upper panel) and their difference (VIIRS-MODIS) time series (lower panel). Ascending and descending orbits are plotted separately. To minimize limb effects due to a large scan angle, only near-nadir pixels within a distance of 180 km were used for both MODIS and VIIRS. VIIRS data on 01/2012 were excluded due to insufficient observations.

Trends for the VIIRS minus MODIS time series are 0.0165 ± 0.0119 K/Year for daytime passes and 0.0016 ± 0.0133 K/Year for nighttime passes, respectively, during the five years from January 2012 to March 2017, with VIIRS being warmer. Uncertainties in trends represent 95% confidence intervals with autocorrelation adjustments. These uncertainties are relatively large that are mainly due to shorter observation length and larger magnitude of noise in the difference time series. These could be improved in future studies as the time series become longer and more observations from the larger scan angles, after scan-angle correction, are used in the global mean calculations to reduce noise [47]. Looking at the trend differences between VIIRS and MODIS, the trend value in daytime passes is an order of magnitude larger than that in the nighttime passes. This is most likely related to diurnal drift difference. Shao et al. [93] showed that the VIIRS M15 band was 0.084 K warmer than the MODIS B31 band in daytime passes in low latitudes, caused by the 10-min orbital drift of VIIRS relative to Aqua and a steeper temperature diurnal gradient near 1:30 pm local time. In contrast, orbital-drift related differences between the same two instrument bands were only of 0.028 K for nighttime passes due to a flat temperature diurnal gradient near 1:30 am local time [93], although orbital drift is also of 10-min magnitude between the two satellites. As such, the impact of diurnal changes on trend comparisons is negligible for the nighttime passes and the trend differences of the global nighttime time series best characterize the radiometric stability in the VIIRS and MODIS obviations. With the diurnal effect excluded, the small trend differences in nighttime passes suggest that there are little or nearly no relative drifting errors between the S-NPP/VIIRS and Aqua/MODIS observations for the comparing bands. As the two instruments were calibrated completely independently and it is unlikely that their biases are drifting to exactly the same direction to arrive at a near-zero relative drifting error, the most probable explanation is that both instruments have achieved an absolute radiometric stability within ±0.0016 K per year. Since these bands are mainly for retrieving sea surface temperatures, this stability satisfies the accuracy requirement for climate change measurement that allows reliable detection of the temperature climate trends at the surface [47,48].

Reprocessing also resulted in the improvement of geolocation accuracy. Wolfe et al. [94] described in detail the geometric calibration including prelaunch pointing and alignment measurements for all VIIRS bands. Prelaunch calibrated geolocation had biases up to −775 m in the track direction and 1118 m in the scan direction. Post-launch correction for these errors involves updates of a multi-parameter geolocation LUT containing instrument scan angle information, such as the satellite roll, yaw, and pitch angles, as well as other information. The first two updates, performed on 23 February 2012 and 18 April 2013, had reduced the geolocation error biases to within 2 m [94]. In addition, several other updates and refinement of scan angles in geolocation LUT were also performed to improve the geolocation accuracy in the NOAA operational processing during the period from three months after launch until 22 August 2013 [95]. The DNB operational SDR product did not have a terrain corrected geolocation until May 2014. These updates resulted in inconsistency in geolocation accuracy between the earlier and later periods of the VIIRS SDRs. In the reprocessing, optimal LUTs were applied to accommodate improved geolocation calibration algorithms and produced consistent geolocation accuracy throughout the reprocessing period for all the VIIRS band resolutions. Detailed description on the reprocessing improvements of the VIIRS geolocation can be found in Wang et al. [95].

6. Conclusions

In summary of the previous sections, the SDR data for the four instruments, ATMS, CrIS, OMPS, and VIIRS onboard S-NPP was reprocessed at NOAA/STAR for the period from near the launch time to 8 March 2017. The reprocessing was based on calibration algorithms and coefficients of the validated maturity version (the baseline version) for individual instruments. The reprocessing had effectively removed bias jumps in the operational SDR time series associated with changes in calibration algorithms and coefficients that were used to generate the SDRs. Preliminary assessments of the reprocessed SDR time series show much-improved consistency and stability over time compared to the operational SDRs as well as observations from legacy sensors onboard Aqua satellite. The reprocessed SDRs allow scientists to quantify their quality in the time dimension and open the opportunity for them to be used in a variety of environmental applications such as the development of climate data records, identifying NWP model errors, improving climate reanalyses as input datasets, and supporting satellite calibration and validation activities.

The reprocessed S-NPP SDR data are currently saved in a cluster computing system hosted by STAR and University of Maryland and accessible through the URL address: ftp://jlrdata.umd.edu/pub/SNPP_Reprocessing/. In addition, the transition of the reprocessed data to the NOAA/Comprehensive Large Array-data Stewardship System (CLASS) is being planned for archiving and distribution with operational support.

The reprocessing as described in this article was only the first attempt to develop consistent S-NPP SDRs. Since updates in calibration algorithms and coefficients are a continued activity in the operational generation of SDRs for improvement of SDR accuracy throughout the rest of the S-NPP mission, inconsistency could still occur beyond the end date of the first reprocessing period as presented here. To accommodate such changes, new versions of life-cycle reprocessing will be conducted now and then throughout the rest of the S-NPP mission whenever new changes and updates in the calibration algorithms and coefficients are significant enough to warrant a reprocessing. Similar to the baseline version, such reprocessing will use the latest calibration algorithms and coefficients to generate life-cycle consistent SDRs for a period that will be longer than the first reprocessing. The new reprocessing may be conducted individually for each instrument or together for all four instruments, and the results could replace the existing one for user applications or they could be used together for inter-comparison and mutual validation.

NOAA-20 has been successfully launched on 18 October 2017 into the same afternoon orbit as the S-NPP satellite. Similar reprocessing is being planned for NOAA-20 using consistent calibration algorithms and LUTs after its observations become a few years longer. Furthermore, such reprocessing could be implemented for future JPSS satellites, such as JPSS-2 and beyond, once these satellites are launched and operated for a few years. This will allow the JPSS instruments to achieve the lowest data uncertainties and long-term stability to augment the use of JPSS datasets in applications to evaluate and monitor societal impacts.

The reprocessing approach presented in this study emphasizes using the latest and consistent calibration algorithms taken from operations that have included all changes incurred during the operational calibration processes. This approach provides the reprocessed SDR products that not only have their own temporal consistencies but are also consistent with the latest operational products for user application support with required accuracies. Such an approach can be generalized and applied to satellite missions other than the JPSS series. This may include individual experimental missions such as those planned and launched by NASA, satellite mission series such as those from the NOAA Geostationary Operational Environmental Satellite (GOES) series and the second generation of the EUMETSAT MetOp (MetOp-SG) series, as well as other satellite missions. Reprocessing for individual instruments on a satellite mission, as for the historical heritage satellite instruments, was usually performed by individual research groups or investigators. These reprocessing approaches often used recalibration algorithms and coefficients derived by individual research groups. Although it had contributed a great deal to the reprocessing and CDR sciences, reprocessing with such approaches has become increasingly challenging as calibration processes become more and more complex with advancements in instrument designs. Given its importance and benefits to the user community, as well as for climate change monitoring, it would be most efficient if reprocessing is planned during the mission planning phases and then executed as a common best practice for science support by the satellite agencies responsible for the operational calibration activities. This would allow efficient transitions from operations to science and applications. In turn, feedback from users on the scientific quality of the reprocessed SDRs would help improve the next cycle of reprocessing, forming a transition process from research to operations.

Author Contributions: Conceptualization, C.-Z.Z., L.Z. and L.L.; Data curation, N.S., B.Z. and M.D.; Formal analysis, C.-Z.Z., Y.C., L.E.F., B.Z. and W.L.; Funding acquisition, M.D.G.; Methodology, L.S., D.C.T. and H.Y.; Project administration, C.-Z.Z.; Software, N.S., T.B., Y.B., S.B., T.C., Y.G., X.H., D.L., J.N., X.S., S.U., W.W. and H.X.; Supervision, S.K.; Visualization, Y.C., L.E.F., B.Z., X.H. and W.L.; Writing–original draft, C.-Z.Z., Y.C., L.E.F. and B.Z.; Writing–review and editing, L.Z., L.L., C.C., F.I.-S., B.Y., S.B., D.T., S.U. and W.W. All authors have read and agreed to the published version of the manuscript.

Funding: This research was funded by NOAA (Grants # NA14NES4320003 and NA19NES4320002) to the Cooperative Institute for Satellite Earth System Studies (CISESS) at the Earth System Science Interdisciplinary Center (ESSIC), University of Maryland.

Acknowledgments: The reprocessed S-NPP instrument sensor data records are available from the STAR/UMD web service: ftp://jlrdata.umd.edu/pub/SNPP_Reprocessing/. Detailed information on the S-NPP/JPSS instruments characteristics and their calibration performance can be found in the S-NPP/JPSS instrument documentation site https://www.star.nesdis.noaa.gov/jpss/Docs.php and calibration and validation site https://www.star.nesdis.noaa.gov/icvs/, respectively.

Conflicts of Interest: The authors declare no conflict of interest.

Disclaimer: The scientific results and conclusions, as well as any views or opinions expressed herein, are those of the author(s) and do not necessarily reflect those of NOAA or the Department of Commerce.

References

1. Derber, J.C.; Wu, W.-S. The use of TOVS cloud-cleared radiances in the NCEP SSI analysis system. *Mon. Weather Rev.* **1998**, *126*, 2287–2299. [CrossRef]
2. English, S.J.; Renshaw, R.J.; Dibben, P.C.; Smith, A.J.; Rayer, P.J.; Poulsen, C.; Saunders, F.W.; Eyre, J.R. A comparison of the impact of TOVS and ATOVS satellite sounding data on the accuracy of numerical weather forecasts. *Q. J. R. Meteorol. Soc.* **2000**, *126*, 2911–2931.
3. Thépaut, J.-N. Satellite data assimilation in numerical weather prediction: An overview. In Proceedings of the Annual Seminar on Recent Developments in Data Assimilation for Atmosphere and Ocean, ECMWF, Reading, UK, 8–12 September 2003; pp. 75–94.
4. Lorenc, A.C.; Marriott, R.T. Forecast sensitivity to observations in the Met Office Global NWP system. *Q. J. R. Meteorol. Soc.* **2014**, *140*, 209–224. [CrossRef]
5. Spencer, R.W.; Christy, J.R. Precision and radiosonde validation of satellite gridpoint temperature anomalies. Part I: MSU Channel 2. *J. Clim.* **1992**, *5*, 847–857. [CrossRef]

6. Christy, J.R.; Spencer, R.W.; Norris, W.B.; Braswell, W.D. Error estimates of version 5.0 of MSU-AMSU bulk atmospheric temperature. *J. Atmos. Oceanic Technol.* **2003**, *20*, 613–629. [CrossRef]

7. Mears, C.A.; Schabel, M.C.; Wentz, F.J. A reanalysis of the MSU channel 2 tropospheric temperature record. *J. Clim.* **2003**, *16*, 3650–3664. [CrossRef]

8. Mears, C.A.; Wentz, F.J. Sensitivity of satellite-derived tropospheric temperature trends to the diurnal cycle adjustment. *J. Clim.* **2016**, *29*, 3629–3646. [CrossRef]

9. Zou, C.-Z.; Goldberg, M.; Cheng, Z.; Grody, N.; Sullivan, J.; Cao, C.; Tarpley, D. Recalibration of microwave sounding unit for climate studies using simultaneous nadir overpasses. *J. Geophys. Res.* **2006**, *111*, 1–24. [CrossRef]

10. Zou, C.-Z.; Wang, W. Inter-satellite calibration of AMSU-A observations for weather and climate applications. *J. Geophys. Res. Atmos.* **2011**, *116*, 1–20. [CrossRef]

11. Santer, B.D.; Bonfils, C.J.W.; Fu, Q.; Fyfe, J.C.; Hegerl, G.C.; Mears, C.; Painter, J.F.; Po-Chedley, S.; Wentz, F.J.; Zelinka, M.D.; et al. Celebrating the anniversary of three key events. *Nat. Clim. Chang.* **2019**, *9*, 180–182. [CrossRef]

12. Wang, L.; Zou, C.-Z.; Qian, H. Construction of stratospheric temperature data records from Stratospheric Sounding Units. *J. Clim.* **2012**, *25*, 2931–2946. [CrossRef]

13. Thompson, D.W.J.; Seidel, D.J.; Randel, W.J.; Zou, C.-Z.; Butler, A.H.; Mears, C.; Osso, A.; Long, C.; Lin, R. The mystery of recent stratospheric temperature trends. *Nature* **2012**, *491*, 692–697. [CrossRef] [PubMed]

14. Zou, C.-Z.; Qian, H.; Wang, W.; Wang, L.; Long, C. Recalibration and merging of SSU observations for stratospheric temperature trend studies. *J. Geophys. Res. Atmos.* **2014**, *119*, 13,180–13,205. [CrossRef]

15. Zou, C.-Z.; Qian, H. Stratospheric temperature climate data record from merged SSU and AMSU-A observations. *J. Atmos. Ocean. Tech.* **2016**, *33*, 1967–1984. [CrossRef]

16. Randel, W.J.; Smith, A.K.; Wu, F.; Zou, C.-Z.; Qian, H. Stratospheric temperature trends over 1979–2015 derived from combined SSU, MLS and SABER satellite observations. *J. Clim.* **2016**, *29*, 4843–4859. [CrossRef]

17. Seidel, D.J.; Li, J.; Mears, C.; Moradi, I.; Nash, J.; Randel, W.J.; Saunders, R.; Thompson, D.W.J.; Zou, C.-Z. Stratospheric temperature changes during the satellite Era. *J. Geophys. Res.* **2016**, *121*, 664–681. [CrossRef]

18. Cavalieri, D.J.; Gloersen, P.; Parkinson, C.L.; Comiso, J.C.; Zwally, H.J. Observed hemispheric asymmetry in global sea ice changes. *Science* **1997**, *278*, 1104–1106. [CrossRef]

19. Comiso, J.C.; Meier, W.N.; Gersten, R. Variability and trends in the Arctic Sea ice cover: Results from different techniques. *J. Geophys. Res. Oceans* **2017**, *122*, 6883–6900. [CrossRef]

20. Leuliette, E.; Nerem, R.; Mitchum, G. Calibration of TOPEX/Poseidon and Jason altimeter data to construct a continuous record of mean sea level change. *Mar. Geodesy* **2004**, *27*, 79–94. [CrossRef]

21. Nerem, R.S.; Chambers, D.P.; Choe, C.; Mitchum, G.T. Estimating mean sea level change from the TOPEX and Jason altimeter missions. *Mar. Geodesy* **2010**, *33*, 435–446. [CrossRef]

22. Nerem, R.S.; Beckley, B.D.; Fasullo, J.T.; Hamlington, B.D.; Masters, D.; Mitchum, G.T. Climate-change–driven accelerated sea-level rise detected in the altimeter era. *Proc. Natl. Acad. Sci. USA* **2018**, *115*, 2022–2025. [CrossRef] [PubMed]

23. Ablain, M.; Legeais, J.F.; Prandi, P.; Marcos, M.; Fenoglio-Marc, L.; Dieng, H.B.; Benveniste, J.; Cazenave, A. Satellite altimetry-based sea level at global and regional scales. *Surv. Geophys.* **2017**, *38*, 9–33. [CrossRef]

24. Chen, X.; Zhang, X.; Church, J.A.; Watson, C.S.; King, M.A.; Monselesan, D.; Legresy, D.B.; Harig, C. The increasing rate of global mean sea-level rise during 1993–2014. *Nat. Clim. Chang.* **2017**, *7*, 492–495. [CrossRef]

25. Thompson, P.R.; Merrifield, M.A.; Leuliette, E.; Sweet, W.; Chambers, D.P.; Hamlington, B.D.; Jevrejeva, S.; Marra, J.J.; MItchum, G.T.; Nerem, R.S.; et al. Sea level variability and change [in *"State of the Climate in 2017"*]. *Bull. Am. Meteorol. Soc.* **2018**, *99*, S84–S87.

26. Saha, S.; Moorthi, S.; Pan, H.L.; Wu, X.; Wang, J.; Nadiga, S.; Tripp, P.; Kistler, R.; Woollen, J.; Behringer, D.; et al. The NCEP climate forecast system reanalysis. *Bull. Am. Meteorol. Soc.* **2010**, *91*, 1015–1058. [CrossRef]

27. Rienecker, M.M.; Suarez, M.J.; Gelaro, R.; Todling, R.; Bacmeister, J.; Liu, E.; Bosilovich, M.G.; Schubert, S.D.; Takacs, L.; Kim, G.K.; et al. MERRA—NASA's Modern-Era Retrospective Analysis for Research and Applications. *J. Clim.* **2011**, *24*, 3624–3648. [CrossRef]

28. Dee, D.P.; Uppala, S.M.; Simmons, A.J.; Berrisford, P.; Poli, P.; Kobayashi, S.; Andrae, U.; Balmaseda, M.A.; Balsamo, G.; Bauer, D.P.; et al. The ERA-Interim reanalysis: Configuration and performance of the data assimilation system. *Q. J. R. Meteorol. Soc.* **2011**, *137*, 553–597. [CrossRef]

29. Kobayashi, S.; Ota, Y.; Harada, Y.; Ebita, A.; Moriya, M.; Onoda, H.; Onogi, K.; Kamahori, H.; Kobayashi, C.; Endo, H.; et al. The JRA-55 Reanalysis: General specifications and basic characteristics. *J. Meteorol. Soc. Jpn.* **2015**, *93*, 5–48. [CrossRef]

30. Fujiwara, M.; Wright, J.S.; Manney, G.L.; Gray, L.J.; Anstey, J.; Birner, T.; Davis, S.; Gerber, E.P.; Harvey, V.L.; Hegglin, M.I.; et al. Introduction to the SPARC Reanalysis Intercomparison Project (S-RIP) and overview of the reanalysis systems. *Atmos. Chem. Phys.* **2017**, *17*, 1417–1452. [CrossRef]

31. Goldberg, M.D.; Kilcoyne, H.; Cikanek, H.; Mehta, A. Joint Polar Satellite System: The United States next generation civilian polar-orbiting environmental satellite system. *J. Geophys. Res. Atmos.* **2013**, *118*, 13463–13475. [CrossRef]

32. Zhou, L.; Divakarla, M.; Liu, X. An Overview of the Joint Polar Satellite System (JPSS) Science Data Product Calibration and Validation. *Remote Sens.* **2016**, *8*, 139. [CrossRef]

33. Zhou, L.; Divakarla, M.; Liu, X.; Layns, A.; Goldberg, M. An overview of the science performances and calibration. Validation of joint polar satellite system operational products. *Remote Sens.* **2019**, *11*, 698. [CrossRef]

34. NOAA/STAR. Workshop Report on JPSS Life-Cycle Data Reprocessing to Advance Weather and Climate Applications. Available online: https://www.star.nesdis.noaa.gov/star/documents/meetings/JPSS2016_LDRW/NPSSLifeCycleDataReprocessingWorkshopSummary.pdf (accessed on 28 July 2020).

35. National Research Council. *Climate Data Records from Environmental Satellites: Interim Report*; The National Academies Press: Washington, DC, USA, 2004; p. 135.

36. Chen, R.; Cao, C.; Menzel, P.W. Intersatellite calibration of NOAA HIRS CO2 channels for climate studies. *J. Geophys. Res.* **2013**, *118*, 5190–5203.

37. Menzel, W.P.; Frey, R.A.; Borbas, E.E.; Baum, B.A.; Cureton, G.; Bearson, N. Reprocessing of HIRS Satellite Measurements from 1980 to 2015: Development toward a Consistent Decadal Cloud Record. *J. Appl. Meteor. Climatol.* **2016**, *55*, 2397–2410. [CrossRef]

38. Stowe, L.L.; Jacobowitz, H.; Ohring, G.; Knapp, K.R.; Nalli, N.R. The Advanced Very High Resolution Radiometer (AVHRR) Pathfinder Atmosphere (PATMOS) Climate Dataset: Initial Analyses and Evaluations. *J. Clim.* **2002**, *15*, 1243–1260. [CrossRef]

39. Heidinger, A.K.; Foster, M.J.; Walther, A.; Zhao, X. The Pathfinder Atmospheres–Extended AVHRR Climate Dataset. *Bull. Am. Meteor. Soc.* **2014**, *95*, 909–922. [CrossRef]

40. Wentz, F.J. *SSM/I Version 7 Calibration Report*; RSS Technical Report 011012; Remote Sensing Systems: Santa Rosa, CA, USA, 2013; 46p, Available online: http://images.remss.com/papers/rsstech/2012_011012_Wentz_Version-7_SSMI_Calibration.pdf (accessed on 28 August 2020).

41. Mears, C.A.; Wang, J.; Smith, D.; Wentz, F.J. Intercomparison of total precipitable water measurements made by satellite-borne microwave radiometers and ground-based GPS instruments. *J. Geophys. Res. Atmos.* **2015**, *120*, 2492–2504. [CrossRef]

42. Mears, C.A.; Smith, D.K.; Ricciardulli, L.; Wang, J.; Huelsing, H.; Wentz, F.J. Construction and Uncertainty Estimation of a Satellite-Derived Total Precipitable Water Data Record Over the World's Oceans. *Earth Space Sci.* **2018**, *5*, 197–210. [CrossRef]

43. Zou, C.-Z. Atmospheric temperature climate data records from satellite microwave sounders. In *Satellite-Based Applications to Climate Change*; Qu, J.J., Powell, A., Sivakumar, M.V.K., Eds.; Springer: New York, NY, USA, 2013; pp. 107–125.

44. Simmons, A.J.; Poli, P.; Dee, D.P.; Berrisford, P.; Hersbach, H.; Kobayashi, S.; Peubey, C. Estimating low-frequency variability and trends in atmospheric temperature using ERA-Interim. *Q. J. R. Meteor. Soc.* **2014**, *140*, 329–353. [CrossRef]

45. Weng, F.; Zou, X.; Sun, N.; Yang, H.; Tian, M.; Blackwell, W.J.; Wang, X.; Lin, L.; Anderson, K. Calibration of Suomi national polar-orbiting partnership advanced technology microwave sounder. *J. Geophys. Res. Atmos.* **2013**, *118*, 11187–11200. [CrossRef]

46. Weng, F.; Yang, H. Validation of ATMS calibration accuracy using Suomi NPP pitch maneuver observations. *Remote Sens.* **2016**, *8*, 332. [CrossRef]

47. Zou, C.-Z.; Goldberg, M.; Hao, X. New generation of US microwave sounder achieves high radiometric stability performance for reliable climate change detection. *Sci. Adv.* **2018**, *4*, eaau0049. [CrossRef] [PubMed]

48. Ohring, G.; Wielicki, B.; Spencer, R.; Emery, B.; Datla, R. Satellite instrument calibration for measuring global climate change. *Bull. Am. Meteorol. Soc.* **2005**, *86*, 1303–1314. [CrossRef]

49. Christy, J.R.; Spencer, R.W.; Braswell, W.D. MSU tropospheric temperatures: Dataset construction and radiosonde comparisons. *J. Atmos. Oceanic Technol.* **2000**, *17*, 1153–1170. [CrossRef]
50. Parker, D.E.; Gordon, M.; Cullum, D.P.N.; Sexton, D.M.H.; Folland, C.K.; Rayner, N. A new gridded radiosonde temperature data base and recent temperature trends. *Geophys. Res. Lett.* **1997**, *24*, 1499–1502. [CrossRef]
51. Haimberger, L. Homogenization of radiosonde temperature time series using innovation statistics. *J. Clim.* **2007**, *20*, 1377–1403. [CrossRef]
52. Haimberger, L.; Tavolato, C.; Sperka, S. Towards the elimination of warm bias in historic radiosonde records—Some new results from a comprehensive intercomparison of upper air data. *J. Clim.* **2008**, *21*, 4587–4606. [CrossRef]
53. Haimberger, L.; Tavolato, C.; Sperka, S. Homogenization of the global radiosonde temperature dataset through combined comparison with reanalysis background series and neighboring stations. *J. Clim.* **2012**, *25*, 8108–8131. [CrossRef]
54. Sherwood, S.C.; Nishant, N. Atmospheric changes through 2012 as shown by iteratively homogenized radiosonde temperature and wind data (IUKv2). *Environ. Res. Lett.* **2015**, *10*, 054007. [CrossRef]
55. Christy, J.R.; Norris, W.B. Satellite and VIZ–radiosonde intercomparisons for diagnosis of nonclimatic influences. *J. Atmos. Ocean. Technol.* **2006**, *23*, 1181–1194. [CrossRef]
56. Thorne, P.W.; Lanzante, J.R.; Peterson, T.C.; Seidel, D.J.; Shine, K.P. Tropospheric temperature trends: History of an ongoing controversy. *WIREs Clim. Chang.* **2010**, *2*, 66–88. [CrossRef]
57. Seidel, D.J.; Gillett, N.P.; Lanzante, J.R.; Shine, K.P.; Thorne, P.W. Stratospheric temperature trends: Our evolving understanding. *WIREs Clim. Chang.* **2011**, *2*, 592–616. [CrossRef]
58. Ho, S.-P.; Kuo, Y.-H.; Zeng, Z.; Peterson, T.C. A comparison of lower stratosphere temperature from microwave measurements with CHAMP GPS RO data. *Geophys. Res. Lett.* **2007**, *34*, 1–5. [CrossRef]
59. Steiner, A.K.; Lackner, B.C.; Ladstädter, F.; Scherllin-Pirscher, B.; Foelsche, U.; Kirchengast, G. GPS radio occultation for climate monitoring and change detection. *Radio Sci.* **2011**, *46*, 1–17. [CrossRef]
60. Khaykin, S.M.; Funatsu, B.M.; Hauchecorne, A.; Godin-Beekmann, S.; Claud, C.; Keckhut, P.; Pazmino, A.; Gleisner, H.; Nielsen, J.K.; Syndergaard, S.; et al. Post-millennium changes in stratospheric temperature consistently resolved by GPS radio occultation and AMSU observations. *Geophys. Res. Lett.* **2017**, *44*, 7510–7518. [CrossRef]
61. Han, Y.; Revercomb, H.; Cromp, M.; Gu, D.; Johnson, D.; Mooney, D.; Scott, D.; Strow, L.; Bingham, G.; Borg, L.; et al. Suomi NPP CrIS measurements, sensor data record algorithm, calibration and validation activities, and record data quality. *J. Geophys. Res. Atmos.* **2013**, *118*, 12734–12748. [CrossRef]
62. JPSS Configuration Management Office. Joint Polar Satellite System (JPSS) Cross Track Infrared Sounder (CrIS) Sensor Data Records (SDR) Algorithm Theoretical Basis Document (ATBD) for Normal Spectral Resolution, JPSS Office, Document Code D0001-M01-S002. Available online: https://www.star.nesdis.noaa.gov/jpss/documents/ATBD/D0001-M01-S01-002_JPSS_ATBD_CRIS-SDR_nsr_20180614.pdf (accessed on 28 July 2020).
63. Tobin, D.; Revercomb, H.; Knuteson, R.; Taylor, J.; Best, F.; Borg, L.; DeSlover, D.; Martin, G.; Buijs, H.; Esplin, M.; et al. Suomi-NPP CrIS radiometric calibration uncertainty. *J. Geophys. Res. Atmos.* **2013**, *118*, 10,589–10,600. [CrossRef]
64. Strow, L.L.; Motteler, H.; Tobin, D.; Revercomb, H.; Hannon, S.; Buijs, H.; Predina, J.; Suwinski, L.; Glumb, R. Spectral calibration and validation of the Cross-track Infrared Sounder on the Suomi NPP satellite. *J. Geophys. Res. Atmos.* **2013**, *118*, 12486–12496. [CrossRef]
65. Chen, Y.; Han, Y.; Weng, F. Characterization of long-term stability of Suomi NPP Cross-Track Infrared Sounder spectral calibration. *IEEE Trans. Geosci. Remote Sens.* **2017**, *55*, 1147–1159. [CrossRef]
66. Zavyalov, V.; Esplin, M.; Scott, D.; Esplin, B.; Bingham, G.; Hoffman, E.; Lietzke, C.; Predina, J.; Frain, R.; Suwinski, L.; et al. Noise performance of the CrIS instrument. *J. Geophys. Res. Atmos.* **2013**, *118*, 13108–13120. [CrossRef]
67. Wang, L.; Zhang, B.; Tremblay, D.; Han, Y. Improved scheme for Cross-track Infrared Sounder geolocation assessment and optimization. *J. Geophys. Res. Atmos.* **2016**, *122*, 519–536. [CrossRef]
68. Han, Y.; Chen, Y. Calibration algorithm for Cross-Track Infrared Sounder full spectral resolution measurements. *IEEE Trans. Geosci. Remote Sens.* **2018**, *56*, 1008–1016. [CrossRef]

69. Dittman, M.G.; Ramberg, E.; Chrisp, M.; Rodriguez, J.V.; Sparks, A.L.; Zaun, N.H.; Hendershot, P.; Dixon, T.; Philbrick, R.H.; Wasinger, D. Nadir ultraviolet imaging spectrometer for the NPOESS Ozone Mapping and Profiler Suite (OMPS). In Proceedings of the International Symposium on Optical Science and Technology, SPIE Proceedings 4814, 2002, Earth Observing Systems VII. Seattle, WA, USA, 24 September 2002; William, L.B., Ed.; pp. 111–119.

70. Seftor, C.J.; Jaross, G.; Kowitt, M.; Haken, M.; Li, J.; Flynn, L.E. Postlaunch performance of the Suomi National Polar-orbiting Partnership Ozone Mapping and Profiler Suite (OMPS) nadir sensors. *J. Geophys. Res. Atmos.* **2014**, *119*, 4413–4428. [CrossRef]

71. McClain, S.C.; Maymon, P.W.; Chipman, R.A. Design and analysis of a depolarizer for the NASA Moderate Resolution Imaging Spectrometer—Tilt (MODIS-T). *Proc. SPIE* **1992**, *1746*, 375–385.

72. Bhartia, P.K.; McPeters, R.D.; Flynn, L.E.; Taylor, S.; Kramarova, N.A.; Frith, S.; Fisher, B.; DeLand, M. Solar Backscatter UV (SBUV) total ozone and profile algorithm. *Atmos. Meas. Tech.* **2013**, *6*, 2533–2548. [CrossRef]

73. Rodriguez, J.V.; Seftor, C.J.; Wellemeyer, C.G.; Chance, K. An overview of the nadir sensor and algorithms for the NPOESS ozone mapping and profiler suite (OMPS). *Proc. SPIE* **2003**, *4891*, 65–75.

74. Pan, C.; Weng, F.; Beck, T.; Flynn, L.; Ding, S. Recent improvements to Suomi NPP Ozone Mapper Profiler Suite nadir mapper sensor data records. *IEEE Trans. Geosci. Remote Sens.* **2017**, *99*, 1–7. [CrossRef]

75. Pan, C.; Weng, F.; Flynn, L. Spectral performance and calibration of the Suomi NPP OMPS Nadir Profiler sensor. *Earth Space Sci.* **2017**, *4*, 737–745. [CrossRef]

76. McPeters, R.D.; Frith, S.; Labow, G.J. OMI total column ozone: Extending the long-term data record. *Atmos. Meas. Tech.* **2015**, *8*, 4845–4850. [CrossRef]

77. McPeters, R.D.; Bhartia, P.K.; Haffner, D.; Labow, G.J.; Flynn, L. The version 8.6 SBUV ozone data record: An overview. *J. Geophys. Res. Atmos.* **2013**, *118*, 8032–8039. [CrossRef]

78. Cao, C.; Xiong, X.J.; Wolfe, R.; DeLuccia, F.; Liu, Q.M.; Blonski, S.; Lin, G.G.; Nishihama, M.; Pogorzala, D.; Oudrari, H.; et al. *Visible Infrared Imaging Radiometer Suite (VIIRS) Sensor Data Record (SDR) User's Guide, Version 1.2*; NOAA Technical Report NESDIS 142A; NESDIS/NOAA/Department of Commerce: Washington, DC, USA, 2013; 46p.

79. Cao, C.; Xiong, J.; Blonski, S.; Liu, Q.; Uprety, S.; Shao, X.; Bai, Y.; Weng, F. Suomi NPP VIIRS sensor data record verification, validation, and long-term performance monitoring. *J. Geophys. Res. Atmos.* **2013**, *118*, 11,664–11,678. [CrossRef]

80. Cao, C.; DeLuccia, F.; Xiong, X.; Wolfe, R.; Weng, F. Early on-orbit performance of the Visible Infrared Imaging Radiometer Suite (VIIRS) onboard the Suomi National Polar-orbiting Partnership (S-NPP) satellite. *IEEE Trans. Geosci. Remote Sens.* **2014**, *52*, 1142–1156. [CrossRef]

81. VIIRS ATBD, Joint Polar Satellite System (JPSS) Visible Infrared Imaging Radiometer Suite (VIIRS) Sensor Data Records (SDR) Algorithm Theoretical Basis Document (ATBD). 2013. Available online: https://ncc. nesdis.noaa.gov/documents/documentation/ATBD-VIIRS-RadiometricCal_20131212.pdf (accessed on 28 July 2020).

82. Cao, C.; Bai, Y. Quantitative analysis of VIIRS DNB nightlight point source for light power estimation and stability monitoring. *Remote Sens.* **2014**, *6*, 11915–11935. [CrossRef]

83. Cao, C.; Bai, Y.; Wang, W.; Choi, T. Radiometric inter-consistency of VIIRS DNB on Suomi NPP and NOAA-20 from observations of reflected lunar lights over deep convective clouds. *Remote Sens.* **2019**, *11*, 934. [CrossRef]

84. Uprety, S.; Cao, C.; Gu, Y.; Shao, X.; Blonski, S.; Zhang, B. Calibration improvements in S-NPP VIIRS DNB Sensor Data Record using version 2 reprocessing. *IEEE Trans. Geosci. Remote Sens.* **2019**, *57*, 9602–9611. [CrossRef]

85. Choi, T.; Shao, X.; Cao, C.; Weng, F. Radiometric Stability Monitoring of the Suomi NPP Visible Infrared Imaging Radiometer Suite (VIIRS) Reflective Solar Bands Using the Moon. *Remote Sens.* **2016**, *88*, 15. [CrossRef]

86. Xiong, X.; Bulter, J.; Chiang, K.; Efremova, B.; Fulbright, J.; Lei, N.; McIntire, J.; Oudrari, H.; Sun, J.; Wang, Z.; et al. VIIRS on-orbit calibration methodology and performance. *J. Geophys. Res. Atmos.* **2014**, *119*, 5065–5078. [CrossRef]

87. Blonski, S.; Cao, C. Suomi NPP VIIRS Reflective Solar Bands operational calibration reprocessing. *Remote Sens.* **2015**, *7*, 16131–16149. [CrossRef]

88. Uprety, S.; Cao, C.; Xiong, X.; Wang, W.; Zhang, B.; Choi, T.; Blonski, S.; Shao, X. Improving S-NPP VIIRS Reflective Solar Band (RSB) calibration accuracy through reprocessing. In *Global Space-Based Inter-Calibration System (GSICS) Quarterly Newsletter*; 2018; Volume 12, p. 15. Available online: https://repository.library.noaa. gov/view/noaa/19087 (accessed on 5 September 2020). [CrossRef]

89. Sun, J.; Wang, M. VIIRS reflective solar bands calibration progress and its impact on ocean color products. *Remote Sens.* **2016**, *8*, 194. [CrossRef]

90. Cao, C.; Wang, W.; Blonski, S.; Zhang, B. Radiometric traceability diagnosis and bias correction for the Suomi NPP VIIRS long-wave infrared channels during blackbody unsteady states. *J. Geophys. Res. Atmos.* **2017**, *122*, 5285–5297. [CrossRef]

91. Wang, W.; Cao, C.; Ignatov, A.; Liang, X.; Li, Z.; Wang, L.; Zhang, B.; Blonski, S.; Li, J. Improving the Calibration of Suomi NPP VIIRS Thermal Emissive Bands during Blackbody Warm-Up/Cool-Down. *IEEE Trans. Geosci. Remote Sens.* **2019**, *57*, 1977–1994. [CrossRef]

92. Li, Y.; Wu, A.; Xiong, X. Inter-comparison of S-NPP VIIRS and Aqua MODIS thermal emissive bands using hyperspectral infrared sounder measurements as a transfer reference. *Remote Sens.* **2016**, *8*, 72. [CrossRef]

93. Shao, X.; Cao, C.; Xiong, X.; Liu, T.; Zhang, B.; Uprety, S. Orbital variations and impacts on observations from S-NPP, NOAA 18-20, and AQUA sun-synchronous satellites. In Proceedings of the SPIE, 10764, 2018, Earth Observing Systems XXIII, 107641U. San Diego, CA, USA, 7 September 2018; 2018.

94. Wolfe, R.E.; Lin, G.; Nishihama, M.; Tewari, K.P.; Tilton, J.C.; Isaacman, A.R. Suomi NPP VIIRS prelaunch and on-orbit geometric calibration and characterization. *J. Geophys. Res. Atmos.* **2013**, *118*, 11508–11521. [CrossRef]

95. Wang, W.; Cao, C.; Bai, Y.; Blonski, S.; Schull, M.A. Assessment of the NOAA S-NPP VIIRS geolocation reprocessing improvements. *Remote Sens.* **2017**, *9*, 974. [CrossRef]

 *remote sensing*

Letter

Evaluating the Absolute Calibration Accuracy and Stability of AIRS Using the CMC SST

Hartmut H. Aumann *, Steven E. Broberg, Evan M. Manning , Thomas S. Pagano and Robert C. Wilson

Jet Propulsion Laboratory, California Institute of Technology, Pasadena, CA 91101, USA;
Steven.E.Broberg@jpl.nasa.gov (S.E.B.); Evan.M.Manning@jpl.nasa.gov (E.M.M.);
Thomas.S.Pagano@jpl.nasa.gov (T.S.P.); Robert.C.Wilson@jpl.nasa.gov (R.C.W.)
* Correspondence: Hartmut.Aumann@jpl.nasa.gov

Received: 26 July 2020; Accepted: 17 August 2020; Published: 25 August 2020

Abstract: We compare the daily mean and standard deviation of the difference between the sea surface skin temperature (SST) derived from clear sky Atmospheric InfraRed Sounder (AIRS) data from seven atmospheric window channels between 2002 and 2020 and collocated Canadian Meteorological Centre (CMC) SST data from the tropical oceans. After correcting the mean difference for cloud contamination and diurnal effects, the remaining bias relative to the CMC SST, is reasonably consistent with estimates of the AIRS absolute accuracy based on the uncertainty of the pre-launch calibration. The time series of the bias produces trends well below the 10 mK/yr level required for climate change evaluations. The trends are in the 2 mK/yr range for the five window channels between 790 and 1231 cm^{-1}, and +5 mK/yr for the shortwave channels. Between 2002 and 2020, the time series of the standard deviation of the difference between the AIRS SST and the CMC SST dropped fairly steadily to below 0.4 K in several AIRS window channels, a level previously only seen in gridded SST products relative to the Argo buoys.

Keywords: Infrared; hyperspectral; climate

1. Introduction

The absolute radiometric calibration accuracy of any sounder is a complicated function of its design, its on-orbit thermal environment, likely degradation on orbit, and the scene temperature. The new generation of hyperspectral infrared sounders was designed to produce very accurate and stable data, to meet the 100 mK absolute accuracy and 10 mK/yr stability required for "climate quality" [1]. The achieved absolute accuracy and stability of the sounders at this level is uncertain. Pagano, T.S. et al. [2] estimated the Atmospheric InfraRed Sounder (AIRS) absolute calibration uncertainty based on the SI traceability of the pre-launch calibration to be in the 50 to 200 mK range for 300K scenes. Lower bounds on the absolute accuracy can be established by noting that most sounders make measurements in channels which have identical or functionally identical spectral response functions. The relative brightness temperature differences measured with these channels should show no bias. IASI (Infrared Atmospheric Sounder Interferometer) [3] on the on the MetOp satellites has four detectors, the Crosstrack Interferometer Sounder, CrIS [4] on SNPP and JPSS1 satellites has nine detectors in each of three bands at nominally identical frequencies. The Atmospheric Infrared Sounder, AIRS [5], has several channels with nearly identical spectral response functions. Differences between these nominally identical channels and trends in the differences constitute lower bounds on the absolute accuracy and stability of the calibration. They are lower bounds because there are shared elements in the calibration, which cancel in the difference. Another method is to note that all hyperspectral sounders make measurements in many atmospheric window channels. Each of these window channels can be used to derive a surface temperature. The differences between the derived temperatures and a reliable surface truth can be used as a measure of the absolute radiometric accuracy and stability.

The general approach of comparing a sea surface skin temperature (SST) derived directly from AIRS clear sky Level 1B radiances to a gridded SST product was proposed before launch and was subsequently tested, most recently in Aumann et al., 2019, using the NOAA generated RTG (Real Time Global) SST [6]. The RTGSST became noisy in about 2017 [7], and was discontinued in 2019. Deriving an SST directly from individual window channels under clear sky conditions retains traceability to the calibration. Note that the Level 2 product, which uses the best 350 of 2378 AIRS channels to simultaneously derive a surface skin temperature and temperature and water vapor profiles from cloud-cleared and tuned level 1B radiances [8] loses the traceability of the calibration of individual channels.

For the ground truth we use the Canadian Meteorological Centre (CMC) SST, which has been produced on a 0.2 degree grid since 1991 (v2.0) and on a 0.1 degree grid since 2016 (v3.0). In the following, we refer to the CMC SST simply as CMC. The primary references for the CMC are in situ observations of the SST by buoys (excluding Argo) and ships from the International Comprehensive Ocean-Atmosphere Data Set (ICOADS) program. The grid is filled by optimally interpolating between the buoy measurements with surface temperatures deduced from space-borne sensors and ship reports. Sensors used in the production of the CMC include the AVHRR from NOAA-18 and 19, the European Meteorological Operational-A (METOP-A) and Operational-B (METOP-B), and data from the Advanced Microwave Scanning Radiometer 2 (AMSR2) onboard the GCOM-W satellite. No AIRS data are used for the CMC production. In a decadal average, the CMC agrees with the independent Argo buoys at the 10 mK level and has a trend of -1.9 ± 1.0 mK/yr (2 sigma) [9].

2. Data

We used the L1B v5 calibrated AIRS data available from the GSFC/DISC since September 2002. AIRS on the NASA's Earth Observing System (EOS) Aqua spacecraft is in a 1:30 AM ascending node polar orbit at 703 km altitude. We derived seven independent SSTs for seven independent window channels at 2615, 2508, 1231, 1128, 961, 901 and 790 cm^{-1}, representing seven of the fifteen AIRS focal plane detector modules under clear sky conditions for the 30S-30N oceans. Table 1 summarizes the associate module names and absolute calibration uncertainty for 290K scenes based on Pagano et al. 2020.

Table 1. The seven Atmospheric InfraRed Sounder (AIRS) window channels and associated module ID, atmospheric correction, and calibration uncertainty.

Channel [cm^{-1}]	Detector Module	Typical atm. Correction [K]	Calibration Uncertainty at 290K [K]
2615.3	M1a	1.0	0.11
2508.1	M2a	2.2	0.05
1231.3	M4d	2.9	0.08
1128.5	M5	2.7	0.22
961.4	M7	2.1	0.05
901.0	M8	2.7	0.12
790.3	M9	4.7	0.16

We use a clear sky filter based on a 3×3 footprint spatial coherence test (SCT [7]), which basically measures the absolute value of the difference between the brightness temperature of the center pixel at 1231 cm^{-1} and its four nearest neighbors. Samples where this difference is less than a 0.5 K threshold and where the calculated SST using the 1231 cm^{-1} channel differs from the CMC by less than 4 K are identified as SCT clear. The latter condition eliminates low stratus clouds, akin to a 10 sigma rejection test. The absolute calibration cancels in the SCT, since the same 1231 cm^{-1} detector is used to make the measurements of the nearest neighbors. Since the noise-equivalent delta temperature of this channel is 0.07 K, the impact of random noise on the SCT using a 0.5 K threshold is negligible. Our analysis uses the daily mean and standard deviation (stddev) of the observed brightness temperature (obs) and the brightness temperature calculated (calc). Although the selected channels are in atmospheric windows, atmospheric water vapor causes absorption ranging from 1 K at 2615 cm^{-1} to 4.7 K at 790 cm^{-1} (Table 1, column 3).

Defining bt$_{airs}$ as the observed brightness temperature in a window channel, tr as the combined corrections for atmospheric transmission and surface emissivity [10], then the relationship between obs, calc, CMC and SST$_{airs}$ is shown in Equation (1).

$$(\text{obs-calc}) = \text{bt}_{airs} - (\text{CMC-tr}) = (\text{bt}_{airs} + \text{tr})\text{-CMC} = \text{SST}_{airs}\text{-CMC}. \tag{1}$$

The difference between the 1231.3 and 1227.7 cm^{-1} channels is used to derive the atmospheric transmission correction due to water vapor [7] for all channels. Defining Q = bt1231 − bt1227, we can write

$$\text{SST}_{airs} = \text{bt}_{airs} + a_0 + a_1 * Q + a_2 * Q^2 + a_3/\cos(sza) \tag{2}$$

where sza is the satellite zenith angle (between −50 and +50 degrees). The coefficients were regression trained on 1403 open ocean profiles from the European Centre for Medium-range Weather Forecasting (ECMWF) as described in Aumann et al. 2019. The ocean profiles were converted to the AIRS spectra training set using SARTA, the Stand Alone Radiative Transfer Model (RTM) developed for AIRS [11]. SARTA is based on the 2008 version of HITRAN with a pre-release of version 3.2 of the MT_CKD water continuum.

For each day we matched the longitudes and latitudes of the clear ocean footprints within ±30 degrees of the equator to the nearest grid point of the CMC. We calculated SST$_{airs}$ for each clear footprint and evaluated the daily mean and standard deviation of SST$_{airs}$-CMC. The daily number of clear SST matchups, fluctuates daily and seasonally, but is typically about 10,000.

3. Results

Figures 1–5 show the time series of the mean and the standard deviation for the day and night overpasses for the midwave channels. The night mean has been shifted by 0.38 K to make the day and night plots approximately overlay. Figure 6 shows the results from night overpasses for the 2615 and 2508 cm^{-1} "shortwave" channels. Daytime results from the shortwave band cannot be used because of reflected solar light. Results from all channels are presented in Table 2.

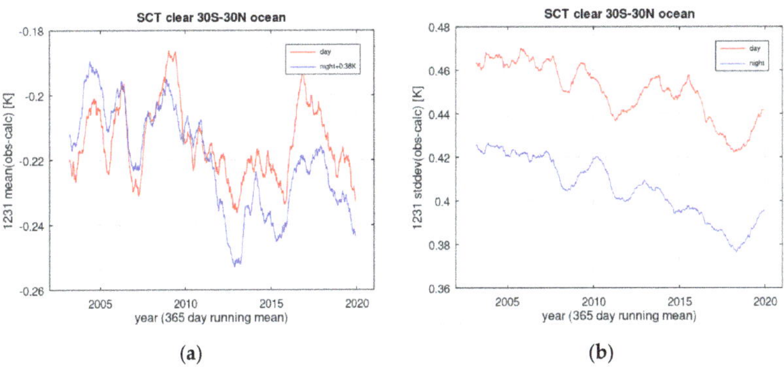

Figure 1. (**a**) Mean and (**b**) standard deviation at 1231 cm^{-1} with one-year smoothing.

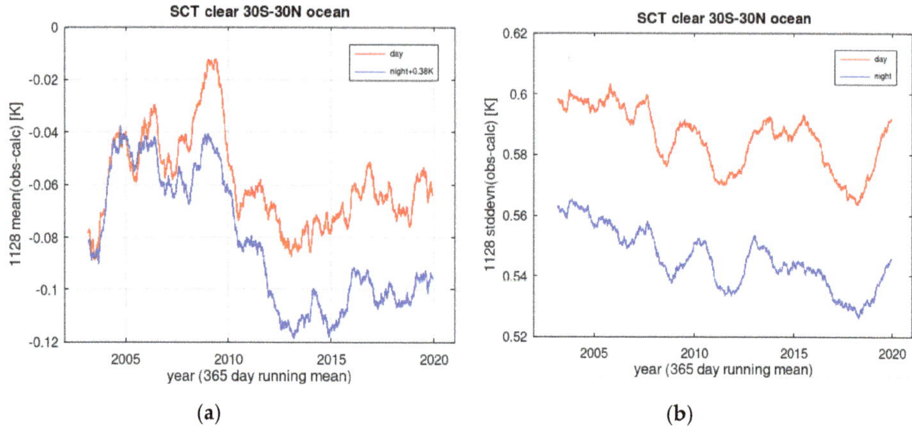

(**a**)   (**b**)

Figure 2. (**a**) Mean and (**b**) stddev at 1128 cm^{-1} with one-year smoothing.

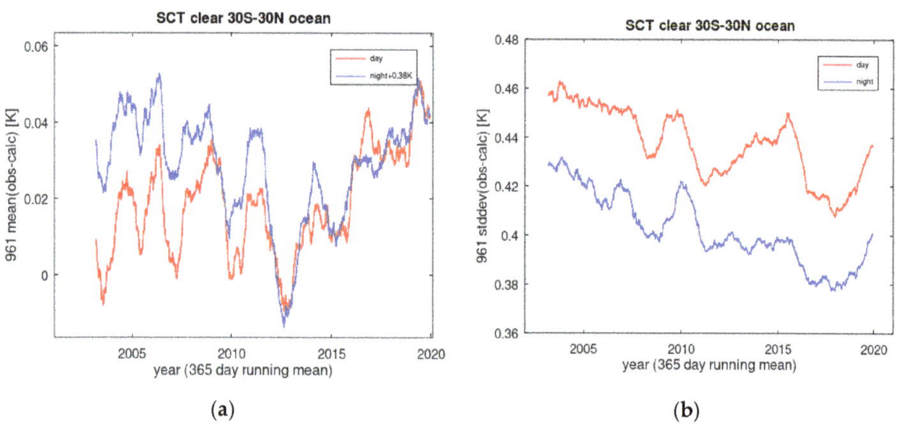

(**a**)   (**b**)

Figure 3. (**a**) Mean and (**b**) standard deviation at 961 cm^{-1} with one-year smoothing.

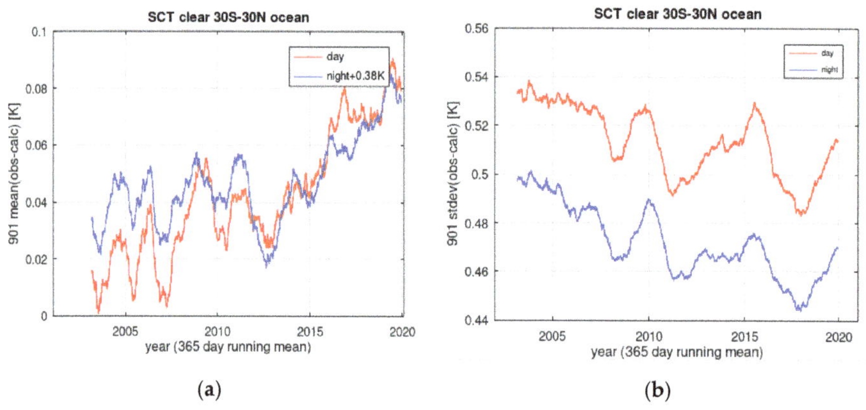

(**a**)   (**b**)

Figure 4. (**a**) Mean and (**b**) standard deviation at 901 cm^{-1} with one-year smoothing.

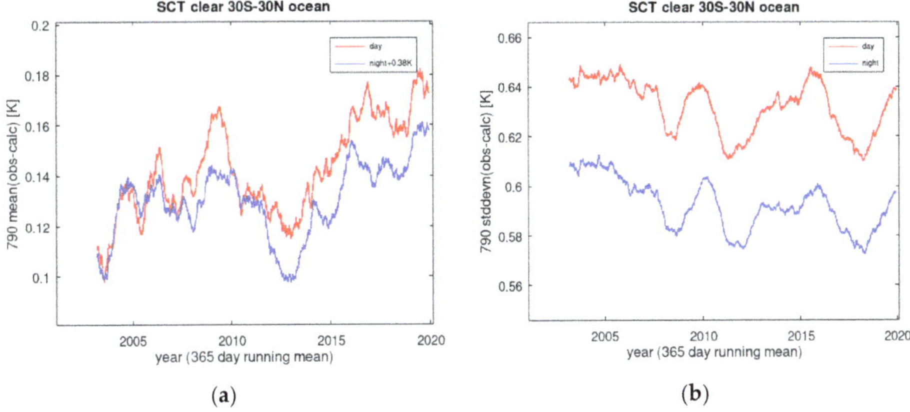

Figure 5. (**a**) Mean and (**b**) standard deviation at 790 cm^{-1} with one-year smoothing.

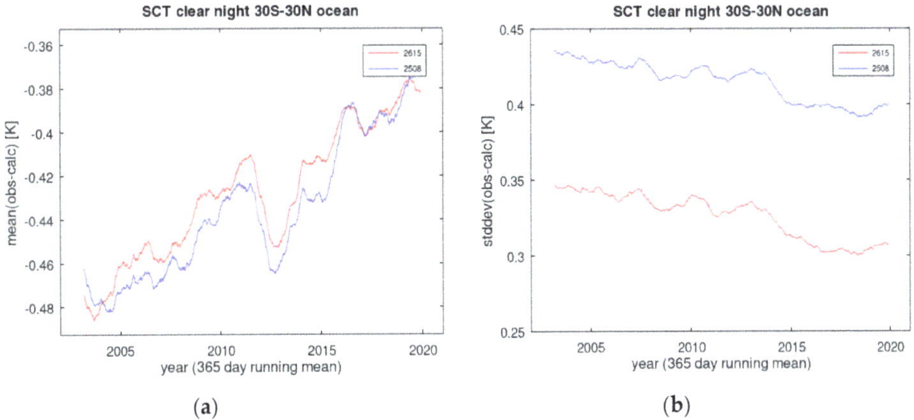

Figure 6. (**a**) Mean and (**b**) standard deviation at 2615 and 2508 cm^{-1} with one-year smoothing.

Table 2. Summary of results from night (a) and day (b) overpasses.

(a) Night

Channel [cm^{-1}]	Night Bias [K]	Stddev [K]	SCT = 0 Bias [K]	SCT Slope	0.38K Night Bias Corr. [K]	Trend ± 1σ [mK/yr]
2615.3	−0.59	0.326	−0.38	−0.22	−0	+5.6 ± 0.2
2508.1	−0.435	0.415	−0.38	−0.18	−0	+5.7 ± 0.2
1231.3	−0.59	0.405	−0.49	−0.22	−0.11	−2.2 ± 0.1
1128.5	−0.404	0.486	−0.32	−0.20	+0.18	−0.4 ± 0.2
961.4	−0.351	0.405	−0.22	−0.27	+0.11	−0.6 ± 0.1
901.0	−0.333	0.472	−0.21	−0.27	+0.11	+1.8 ± 0.2
790.3	−0.250	0.593	−0.14	−0.26	+0.12	+1.4 ± 0.2

Table 2. *Cont.*

(b) Day

Channel [cm^{-1}]	Day Bias [K]	stddev	SCT = 0 Bias [K]	SCT Slope	Day Bias Corrected	Trend± 1σ [mK/yr]
2615.3	3.83	4.14	3.82	+0.13	na	+10.3 ± 1.4
2508.2	1.92	2.18	1.96	−0.02	na	+9.1 ± 0.7
1231.3	−0.22	0.451	−0.15	−0.15	−0.15	−0.6 ± 0.2
1128.5	−0.02	0.516	+0.05	−0.13	+0.05	+1.7 ± 0.2
961.4	+0.02	0.438	+0.11	−0.20	+0.11	+1.2 ± 0.2
901.0	+0.04	0.514	+0.19	−0.19	+0.19	+3.8 ± 0.2
790.3	+0.14	0.633	+0.24	−0.18	+0.24	+2.7 ± 0.2

4. Discussion

We discuss our results in terms of the bias, standard deviation, and anomaly trend of (obs-calc).

4.1. The Bias in (Obs-Calc)

Fiedler et al. [9] used 7 years of matchups to find the mean CMC minus Argo floating buoys difference to be the −10 mK level. This is an order of magnitude smaller than the bias between AIRS and the CMC seen in Table 2. The major contributors to this bias are residual cloud contamination, corrections for the diurnal effect and skin effects, water vapor correction uncertainty, and the absolute calibration of AIRS. Uncertainties in the absolute calibration of AIRS are dominated by uncertainties in the scan mirror polarization, the effective (as opposed to the telemetered) temperature of the on-board blackbody calibrator (OBC), and the pre-launch determined non-linearity coefficients. Errors due to non-linearity should be minimal, because the typical brightness temperatures of the observations are close to the OBC temperature. The bias in the shortwave channels (2615 cm^{-1} and 2508 cm^{-1}) is higher than what is expected based on those uncertainties. There are indications of contamination of the scan mirror, causing an increase in scattering at these frequencies. This degradation could work its way into the radiometric calibration through either the OBC view, Space view, or the Earth view (via contribution from neighboring pixels) [12], resulting in the higher than expected bias relative to the CMC.

4.1.1. Cloud Contamination.

If 1% of a 300 K ocean footprint were to be covered by a 220K cloud, the effective mean temperature of the footprint at 1231 cm^{-1} would decrease by 0.5 K. With a SCT clear threshold of 0.5 K, this footprint and an adjacent totally clear footprint would be identified as SCT clear. The presence of some clouds in some footprints which are identified as SCT clear will thus create a cold bias. The magnitude of the cloud contamination can be estimated by changing the threshold of the SCT filter. Figure 7 shows the decrease in the bias as the SCT threshold is changed from 2K to 0.5K. A tighter threshold results in a steep decrease in the yield as well. The SCT = 0.5 K threshold represents a practical limit. Extrapolated to SCT = 0 (as shown in Figure 7), the AIRS 1231 cm^{-1} bias for day and night becomes −0.14 K and −0.49 K, respectively. The SCT = 0 extrapolated biases for all channels are listed in Table 2.

4.1.2. Diurnal and Skin Effects

The CMC represent the daily mean temperature at the buoy level, while SST.airs is the surface skin temperature at the time of the overpass. The skin is on average 0.2 K colder than the buoys [6], with a very small wind speed sensitivity [13]. Seasonally averaged between 20S and 20N, the buoy temperatures are 0.20 K warmer than the mean for the 1:30 PM overpasses and 0.11 K colder than the mean for the

1:30 AM overpasses [14]. Therefore, we expected SST.airs −CMCSST to be −0.2 K −0.11 K = −0.31 K at night, −0.20 K + 0.2 K = 0 K during the day. The observed (SCT = 0 extrapolated) day/night bias difference is 0.35 K. This 40 mK difference could be due to applying the buoy-mean difference from 20S–20N to the 30S–30N oceans.

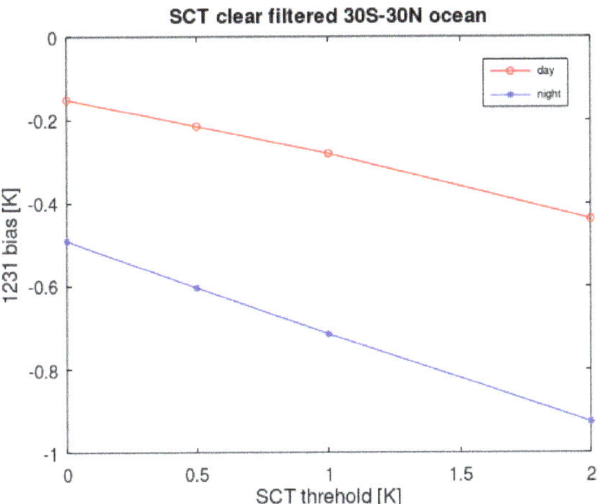

Figure 7. Cold bias as a function of the spatial coherence threshold (SCT).

4.1.3. Bias (Obs-Calc) Residuals

In the context of evaluating the absolute calibration, it is legitimate to correct the bias for cloud contamination and the diurnal effect. Table 2, column 6, lists the diurnal cycle and cloud contamination corrected bias. The residual bias ranges from −150 to + 240 mK.

The residual bias can at best only partially be attributed to an uncertainty of the transmission correction. The transmission correction ranges from 1K at 2616 cm^{-1} to 5K at 790 cm^{-1}. The transmission correction was based on the water continuum used by the RTM and the vertical water vapor distribution ECMWF profiles in the training set. These profiles were a mix of convective (more wet) and subsidence (more dry) case, but the transmission correction was applied to clear sky (subsidence) cases. If the transmission correction were too weak by 1%, the correction would be between 10 and 50 mK too small, depending on the channel. This is too little to shift the −150 mK residual bias to zero, and would make the +240 mK bias even larger. In addition, the corrected residuals show only a weak correlation with the magnitude of the transmission correction. This suggests that the residual bias is dominated by absolute calibration effects. The SI traceable absolute calibration uncertainty for the AIRS detector modules in this study (Table 1, 3rd column), are reasonably consistent with the estimated bias residuals (Table 2, column 6). The comparison with the CMC does not significantly improve the absolute calibration uncertainty estimates.

4.2. The Standard Deviation (stddev(Obs-Calc))

The time series of the daily stddev(SST.airs-CMC) (Figures 1–6, right panels) show a different temporal structure than the time series of the daily mean. This structure is likely related to changes in the CMC. The stddev decreases steadily in all channels, day and night. Since the AIRS instrument or calibration did not change, the decrease in the stddev indicates a steady improvement in the fidelity of the CMC, most likely related to an increasing skill in the ingest of more satellite data. At 2615, 1231 and 961 cm^{-1} the stddev is less than 0.4 K for the night data. This indicates a level of agreement between the CMC and our independent SST observations seen previously only relative to the Argo buoys in the 2000–2011 period [9].

The stddev for the day observations of the midwave channels is about 15% higher than that of the night observations at 1231 cm^{-1}. The most likely explanation for this is that the higher daytime yield in clear is associated with higher cloud contamination and associated additional noise. If the SCT threshold is raised from 0.5 K to 1 K, the yield of "clear" increases by a factor of 3 during the day, the cold bias increases from −0.21 K to −0.28 K, and the stddev increases from 0.44 K to 0.55 K.

The day and night results for the 2508 and 2616 cm^{-1} channels are included in Table 2 to illustrate an unexpected daytime effect. The SST.airs calculation for these channels did not account for solar reflected radiation. It can be seen that the solar reflected component is about 2 K at 2508 cm^{-1}, 4 K at 2616 cm^{-1}, and highly variable due to its sensitivity to wave angle, whitecaps and residual cloud contamination. This causes the stddev for the day observations with the shortwave channels to be almost an order of magnitude larger than at night.

4.3. Anomaly Trends in Mean(Obs-Calc)

The time series of the mean(obs-calc) has a seasonal component, which is removed in the time series of the anomaly of mean(obs-calc). The anomaly trends of the mean(obs-calc), referred to as just "trends", are not consistent for the different channels. The trends in the midwave channels are day/night inconsistent based on the 1σ error bars, with trends ranging from −2.2 to +3.4 mK/yr. If we were to ignore the error bars and treat the 10 trends from the midwave channels as 10 independent samples of a distribution, we would state the trend as +0.9 mK/yr with 1.2 mK/yr 2σ confidence. This does not tell if AIRS is warming, or an artifact in the CMC causes the CMC to get colder.

Inspection of Figures 1–5 show what appears to be a change in the state of the instrument after 2012. The change appears to have a different effect on different detector modules, and differs between day and night, This eliminates the CMC as a potential reason. This effect is currently under investigation by the AIRS calibration team. Also visible in the time series of the bias is a 40 mK dip in the 1231 cm^{-1} channel, seen day and night and centered on 2013, which is seen even more clearly in the 2615 and 2508 cm^{-1} channels (Figure 6). This pattern is common to all channels and suggests an artifact in the CMC. The same pattern is seen in the comparison of CrIS SNPP data and the CMC SST [15].

The character of the anomaly trend of (obs-calc) of the two shortwave channels (Figure 6) is very different from that of the midwave channels. We see a 5 mK/yr warming trend at night and almost twice that during the day. This trend was briefly interrupted in 2013 by a 40 mK dip mentioned in the last paragraph, but then the trend continues. We interpret this trend to be an artifact of the L1B V5 calibration which is currently under investigation by the AIRS calibration team [12].

The EOS Aqua spacecraft with AIRS was launched into its 1:30 PM orbit in 2002, also known as the A-train. It is expected to exit the A-train in January 2022 and slowly drift to an increasingly later ascending node and lower altitude. The AIRS L1B v5 calibration is SI traceable and has been unchanged since launch. With 18 years of data, artifacts at the 100 mK absolute level, and trends well below 10 mK/yr, become visible in the AIRS data and in the reference truth data. After the exit from the A-train the AIRS calibration team will analyze the available data and correct those artifacts which can be physically related to events, voltages or temperatures on the spacecraft. This provides the opportunity for final refinements of the AIRS L1B calibration.

The AIRS instrument was designed to measure climate change. This task was facilitated by the actively maintained 1:30 PM ascending node of the EOS AQUA spacecraft orbit. The AIRS design life was 5 years, but by 2022 AIRS will have provided a continuous 20 year data record for climate change studies, the longest continuous data record to date from any temperature sounder.

5. Conclusions

We compare the daily mean and standard deviation of the difference between SST derived from clear AIRS data and collocated CMC data from the tropical oceans. After correcting the mean for cloud contamination and diurnal effects, the remaining bias relative to the CMC at the 100 mK level is reasonably consistent with estimates of the AIRS absolute accuracy based on the uncertainty of the pre-launch calibration. The anomaly time series of the bias has channel-dependent trends, but all well below the 10 mK/yr level required for climate change evaluations. The trends are in the 2 mK/yr range for the midwave channels, but +5 mK/yr for the shortwave channels. The trend in the shortwave channels is likely due to a scan mirror degradation. The time series of the standard deviation of the difference between the AIRS SST and the CMC dropped steadily to below 0.4 K in several AIRS window channels, a level previously only seen in the CMC relative to the Argo buoys.

Author Contributions: Conceptualization, H.H.A.; Methodology, H.H.A.; Software, H.H.A., E.M.M. and R.C.W.; writing–original draft, H.H.A.; writing–review and editing, H.H.A., S.E.B., E.M.M., T.S.P., R.C.W.; Project Administration, T.S.P. All authors have read and agreed to the published version of the manuscript.

Funding: This research was carried out at the Jet Propulsion Laboratory, California Institute of Technology, under contract with NASA.

Acknowledgments: Jorge Vasquez, JPL, suggested the use of the CMC SST. The daily AIRS Calibration Data Subset (ACDS) is available free of charge from https://disc.gsfc.nasa.gov/datasets/AIRXBCAL_005/summary?keywords=AIRXBCAL. The CMC is freely available from https://podaac.jpl.nasa.gov/dataset/CMC0.2deg-CMC-L4-GLOB-v2.0?ids=&values=&search=CMC.

References

1. Ohring, G.; Wielicki, B.; Spencer, R.; Datla, R. Satellite Instrument Calibration for Measuring Global Climate Change: Report of a Workshop. *Bull. Am. Meteor. Soc.* **2005**, *86*, 1303–1314. [CrossRef]

2. Pagano, T.S.; Aumann, H.H.; Broberg, S.E.; Canas, C.; Manning, E.M.; Overoye, K.O.; Wilson, R.C. SI-Traceability and Measurement Uncertainty of the Atmospheric Infrared Sounder Version 5 Level 1B Radiances. *Remote Sens.* **2020**, *12*, 1338. [CrossRef]

3. Blumstein, D.; Chalon, G.; Carlier, T.; Buil, C.; Hebert, P.; Maciaszek, T.; Ponce, G.; Phulpin, T.; Tournier, B.; Simeoni, D.; et al. IASI Instrument: Technical Overview and Measured Performances. In *Infrared Spaceborne Remote Sensing XII, Proceedings of the Optical Science and Technology, the SPIE 49th Annual Meeting, Denver, CO, USA, 2–6 August 2004*; Strojnik, M., Ed.; SPIE: Bellingham, DC, USA, 2008; Volume 5543, pp. 196–207.

4. Glumb, R.J.; Williams, F.L.; Funk, N.; Chateauneuf, F.; Roney, A.; Allard, R. Cross-Track Infrared Sounder (CrIS) Development Status. In *Infrared Spaceborne Remote Sensing XI, Proceedings of the Optical Science and Technology, SPIE'S 48th Annual Meeting, San Diego, CA, USA, 3–8 August 2003*; SPIE: Bellingham, DC, USA, 2003; Volume 5152.

5. Aumann, H.H.; Chahine, M.T.; Gautier, C.; Goldberg, M.; Kalnay, E.; McMillin, L.; Revercomb, H.; Rosenkranz, P.W.; Smith, W.L.; Staelin, D.H.; et al. AIRS/AMSU/HSB on the Aqua Mission: Design, Science Objectives, Data Products and Processing Systems. *IEEE Trans. Geosci. Remote Sens.* **2003**, *41*, 253–264. [CrossRef]

6. Donlon, C.J.; Minnett, P.J.; Gentemann, C.; Nightingale, T.J.; Barton, I.J.; Ward, B.; Murray, M.J. Toward Improved Validation of Satellite Sea Surface Skin Temperature Measurements for Climate Research. *J. Clim.* **2002**, *15*, 353–369. [CrossRef]

7. Aumann, H.H.; Broberg, S.; Manning, E.; Pagano, T. Radiometric Stability Validation of 17 Years of AIRS Data Using Sea Surface Temperatures. *Geophys. Res. Lett.* **2019**, *46*, 12504–12510. [CrossRef]

8. Susskind, J.; Schmidt, G.A.; Lee, J.N.; Iredell, L. Recent global warming as confirmed by AIRS. *Environ. Res. Lett.* **2019**, *14*, 044030. [CrossRef]

9. Fiedler, E.K.; McLaren, A.; Banzon, V.; Brasnett, B.; Ishizaki, S.; Kennedy, J.; Rayner, N.; Roberts_Jones, J.; Corlett, G.; Merchant, C.; et al. Intercomparison of long-term sea surface temperature analysis using the GHRSST Multi-Product Ensemble (GMPE) system. *Remote Sens. Environ.* **2019**, *222*, 18–23. [CrossRef]

10. Masuda, K.; Takahima, T.; Takayama, Y. Emissivity of pure and sea water from the model sea surface in the infrared window regions. *Remote Sens. Environ.* **1988**, *24*, 313–329. [CrossRef]

11. Strow, L.L.; Hannon, S.E.; DeSouza-Machado, S.; Mottler, H.E.; Tobin, D.C. Validation of the atmospheric infrared sounder radiative transfer algorithm. *JGR* **2006**, *111*, D09S06. [CrossRef]

12. Wilson, R.C.; Manning, E.; Pagano, T.; Aumann, H.; Broberg, S. Examining the possible effect of scan mirror contamination in the AIRS Instrument. In *Infrared Spaceborne Remote Sensing, Proceedings of the Optical Science and Technology, SPIE San Diego, CA, USA, 23–27 August 2020*; SPIE: Bellingham, DC, USA, 2020.

13. Gentemann, C.L.; Minnett, P.J. Radiometric measurements of ocean surface thermal variability. *J. Geophys. Res.* **2008**, *113*, C08017. [CrossRef]
14. Kennedy, J.J.; Brohan, P.; Tett, S.F.B. A global climatology of the diurnal variations in sea-surface temperature and implications for MSU temperature trends. *GRL* **2007**, *34*, L05712. [CrossRef]
15. Aumann, H.H.; Manning, E.M.; Wilson, R.C.; Vasquez, J. Evaluation of bias and trends in AIRS and CrIS SST measurements relative to globally gridded SST products. *TGRS* **2020**. under review.

Article

Spatio-Temporal Variability of Aerosol Optical Depth, Total Ozone and NO_2 Over East Asia: Strategy for the Validation to the GEMS Scientific Products

Sang Seo Park [1],*, Sang-Woo Kim [2], Chang-Keun Song [1], Jong-Uk Park [2] and Kang-Ho Bae [1]

[1] School of Urban & Environmental Engineering, Ulsan National Institute of Science and Technology, Ulsan 44919, Korea; cksong@unist.ac.kr (C.-K.S.); baegh1223@unist.ac.kr (K.-H.B.)
[2] School of Earth and Environmental Sciences, Seoul National University, Seoul 08826, Korea; sangwookim@snu.ac.kr (S.-W.K.); jonguk7628@snu.ac.kr (J.-U.P.)
* Correspondence: sangseopark@unist.ac.kr; Tel.: +82-52-217-2895

Received: 26 May 2020; Accepted: 13 July 2020; Published: 14 July 2020

Abstract: In this study, the spatio-temporal variability of aerosol optical depth (AOD), total column ozone (TCO), and total column NO_2 (TCN) was identified over East Asia using long-term datasets from ground-based and satellite observations. Based on the statistical results, optimized spatio-temporal ranges for the validation study were determined with respect to the target materials. To determine both spatial and temporal ranges for the validation study, we confirmed that the observed datasets can be statistically considered as the same quantity within the ranges. Based on the thresholds of $R^2>0.95$ (temporal) and $R>0.95$ (spatial), the basic ranges for spatial and temporal scales for AOD validation was within 30 km and 30 min, respectively. Furthermore, the spatial scales for AOD validation showed seasonal variation, which expanded the range to 40 km in summer and autumn. Because of the seasonal change of latitudinal gradient of the TCO, the seasonal variation of the north-south range is a considerable point. For the TCO validation, the north-south range is varied from 0.87° in spring to 1.05° in summer. The spatio-temporal range for TCN validation was 20 min (temporal) and 20–50 km (spatial). However, the nearest value of satellite data was used in the validation because the spatio-temporal variation of TCN is large in summer and autumn. Estimation of the spatio-temporal variability for respective pollutants may contribute to improving the validation of satellite products.

Keywords: AOD; total ozone; NO_2; Validation; GEMS

1. Introduction

Air quality is affected by pollutants on both a regional and global scale. However, air quality studies have traditionally been based on ground-based in-situ network measurements with intensive field experiments. To understand emission and long-range transport patterns, remote sensing techniques with space-borne observations are essential. Monitoring air quality from satellites is a key method in the regional and global monitoring of air pollutions with temporally continuous datasets.

Satellite observation is used to assess air quality, provide information on the amount of pollutants [1,2] and transport patterns of pollutants [3]. Furthermore, the spatio-temporal variation of emissions of specific pollutants can also be identified based on the estimation processes from the satellite-based observation dataset [4–8]. Environmental monitoring satellites have been launched mainly for the observation of the total column amount of ozone and specific pollutants, such as tropospheric ozone [9], nitrogen dioxide (NO_2) [10–12], sulfur dioxide (SO_2) [13,14], and aerosols [15–17]. However, observation data from environmental satellite have a retrieval uncertainty relating to the data accuracy [18,19]. For this reason, inter-comparison and validation processes are essential studies in the satellite observation projects.

For the validation of satellite observation data, ground-based measurements with optical instruments have been widely adopted as reference data [15,19]. Due to the differences in spatial and temporal scales between ground and satellite observations, the spatio-temporal collocation range is essential to assume the mean value calculations of satellite observation datasets. The closest pixel from the ground-based site is assumed to be the reference dataset for inter-comparisons between ground and satellite observations. This comparison method is able to neglect the spatial variation of pollutant concentrations. However, the closest pixel method has a problem relating to the representability of data [20,21]. Another reason for the differences between ground-based and satellite-based observations is known to be the colocation mismatch uncertainty (CMU) [20]. CMU is caused by radiance uncertainty, which is due to cloud and surface reflectance and differences in viewing geometry.

Otherwise, the averaging method near the ground-based observation site is one of the most widely used methods in data validation studies [22–24]. The averaging method is less affected by the measurement noise, and it is suitable for spatio-temporally homogeneous species. However, especially in pollutant observations, the spatial scale for high concentrations of pollutants is presented on a city-scaled range, which is similar to or smaller than the spatial resolution of the satellite observation. Although recently developed environmental satellites (Tropospheric Monitoring Instrument (TROPOMI) [25], Geostationary Environmental Monitoring Spectrometer (GEMS) [26], Tropospheric Emissions: Monitoring of Pollution (TEMPO) [27], and Sentinel-4 [28]) designed to the advanced spatial resolution (less than 10 km), satellite observation still inaccurately captured the high concentration of city-scaled pollutant emissions due to the assumptions during the retrieval processes, including horizontal smoothing and small sensitivity of pollutants near the surface [19,29–31].

The Korean geostationary environmental satellite (GEMS) was launched in February 2020 to monitor the air quality over Asia. GEMS is one of the global constellation instruments that observes air quality [26]. Over East Asia, several pollutants are simultaneously mixed and transported on a regional scale [32]. Furthermore, intensive anthropogenic pollutions can affect the spatio-temporal variations in the amounts for atmospheric pollutants. However, the number of ground-based observation networks is limited and cannot cover all major emission source regions. In addition, the characteristics of pollutant emissions in Asia are very complex. Therefore, the strategy for the validation plans in GEMS products, based on the ground-based and other satellite measurements, are more important than other regions.

In this study, we identified the best validation strategies for the various GEMS scientific products—total ozone, aerosol optical depth (AOD), and NO_2—which are relatively well-established observation networks, such as the Aerosol Robotic Network (AERONET) [23,33–35] or Pandora observation network [19], in East Asia. To identify the spatio-temporal range for the validation study, the inter-comparison was executed between identical products from satellite and ground-based measurements. In Section 2, we introduce the overall explanation of the datasets used. Section 3 shows the overall method for this study. Section 4 shows the temporal and spatial range for validation with respect to the scientific products, and Section 5 suggests the strategy of GEMS validation. Section 6 shows the conclusion and summary of this paper.

2. Instruments

2.1. Ozone Monitoring Instrument (OMI)

The Ozone Monitoring Instrument (OMI) is an environmental purposed optical instrument onboard the Aura satellite, launched in 2004. The main purpose of this sensor is the monitoring of the total column amount of ozone and trace gases for air quality and climate studies by using the hyperspectral UV-visible radiance (270–500 nm). The horizontal resolution is 13×24 km^2 at nadir [9,36]. The overpass time of the Aura satellite is approximately 13:30 local time, thus the trace gas amount from the OMI satellite sensor has limitations to monitor its diurnal variation. In this study, the scientific products for total column amount of NO_2 (TCN) and ozone (TCO) from OMI are used to identify the spatial variation of trace gases.

For TCO, OMI has two different algorithm products: a TOMS-based algorithm and an algorithm based on differential optical absorption spectroscopy (DOAS). The TOMS-based algorithm (OMTO3) was developed for TCO observation using two UV wavelength data (331.2 and 317.5 nm) [37,38], and estimates the TCO by comparing the observed and simulated radiance. The DOAS-based algorithm (OMDOAO3) uses the spectral radiance fitting method to estimate the slant column amount, and finally converts it to the vertical column amount after dividing the airmass factor (AMF) [38,39]. Because the operational algorithm for GEMS TCO is based on the TOMS-based algorithm [25], the TCO with version 3 of the TOMS-based algorithm (i.e., OMTO3) was used as a reference OMI TCO dataset for this study.

For the TCN, the OMI standard product (OMNO2 hereafter) was used as the DOAS spectral fitting method with the spectral radiance data from 402 to 465 nm. The OMNO2 was recently improved for the vertical profile variations on a regional scale. By considering the regional variation of vertical profiles, the AMF also precisely considered the spatio-temporal variation of NO_2 [12,40]. Although OMNO2 provides the total, stratospheric and tropospheric column amount of NO_2, we only used the total column amount data for the study of spatial variability.

Because the spatial variability of TCN is basically considered within the regional scale, it is important to consider the issue of horizontal pixel resolution before analyzing the scientific products. As mentioned above, the spatial resolution of OMI is 13×24 km^2 at nadir, but the pixel size for the East-West direction changes considerably depending on the viewing angle. Particularly in the off-nadir position, the east-west pixel size effectively reaches up to 100 km [41]. Due to the coarse spatial resolution, the retrieved amount of TCN has a large amount of uncertainty by the sub-pixel cloud existence. To avoid the data pixels for off-nadir position, we only used the TCN data with the Xtrack position range of 5–49 in this study. In addition, several kinds of data quality flags were also considered before the data analysis. For the TCN, the algorithm quality flags relating to AMF, the algorithm process, and vertical column conversion were considered. Because of the accuracy problem of TCN in cloudy pixels, the pixels less than 0.25 for cloud radiative fraction were only selected as the analysis dataset for the TCN.

2.2. Pandora Spectrophotometer

The Pandora spectrophotometer (Pandora hereafter) is a ground-based UV-visible hyperspectral sensor that uses the sun as a light source. The spectral resolution and sampling are 0.6 nm (full width at half maximum; FWHM) and 0.23 nm, respectively [42,43]. As the Pandora retrieves the TCN and TCO with two-minute resolution, the retrieved data has been widely used in inter-comparison and validation studies for several ground-based [44,45], air-borne [31], and satellite observations [19,29,42–50].

Pandora in South Korea was first installed at Yonsei University (Latitude: 37.564 °N, Longitude: 126.934 °E) and Pusan National University (Latitude: 35.235 °N, Longitude: 129.083 °E) in 2012 [45,51]. The observation sites in Asia make up the Pandora Asia Network (PAN). In this study, we used the observation data from two sites in South Korea. To consider the data quality of the retrieved products, the level 3 TCN and TCO observation data were used after considering the threshold value in the normalized root mean square error (RMSE) of the spectral fitting residual, and the uncertainties in the total column amount during the retrieval. Detailed criteria for total ozone and NO_2 are summarized in Table 1, which is based on the previous studies [45,51,52].

Table 1. Criteria of Data selection for Pandora.

Species	Criteria
Total Ozone	Normalized RMSE < 0.05 SZA < 75 ° Uncertainty < 2 DU
NO_2	Normalized RMSE < 0.05 Uncertainty < 0.05 DU SZA < 70 ° Wavelength shift < 0.01 nm

2.3. CIMEL Sunphotometer

The CIMEL sunphotometer is the main instrument of the Aerosol Robotic Network (AERONET) [33]. This multi-band sunphotometer is composed of 8 shortwave channels (340, 380, 440, 500, 670, 870, 940, and 1020 nm), and measures the physical and optical properties of aerosol, such as aerosol optical depth (AOD), single scattering albedo (SSA), and size information by using direct sun and sky scanning methods [34].

In this study, the Level 1.5 all-point dataset are mainly selected to the instantaneous value of AOD at 500 nm. Although the optically thin cloud-screening issue remains in Level 1.5 datasets, the real-time cloud masking and instrument quality check are adopted during the process to the Level 1.5 datasets. In South Korea, the long-term observed AERONET sites are located in Seoul (Yonsei University; YSU; Latitude: 37.564 °N, Longitude: 126.934 °E), Anmyeon (Latitude: 36.539 °N, Longitude: 126.330 °E), and Gwangju (Gwangju Institute of Science and Technology; GIST; Latitude: 35.228°N, Longitude: 126.843 °E). Recently, the version 3 of the AERONET aerosol observation dataset was published [35]. Because the version 3 datasets considers and corrects the diurnal dependence of AOD by the instrument [35], the temporal variability from the original observation data from direct sun measurement can be regarded as the physical difference of AOD in the real atmosphere.

2.4. MODIS

The moderate resolution imaging spectroradiometer (MODIS) is onboard the polar orbit satellites, Terra (Equatorial Crossing time: 10:30 AM) and Aqua (Equatorial Crossing time: 1:30 PM). To coincide with the overpass time of the Aura satellite, this study only used the MODIS product in the Aqua satellite sensor. The MODIS scientific products show the atmospheric parameters, including aerosol information. The daily level 2 aerosol product (MYD04) was used in this study. The MYD04 is composed of two different aerosol retrieval algorithms: dark target [53–56] and deep-blue algorithms [57,58].

Basically, the MODIS aerosol product has a horizontal resolution of 10×10 km^2. Contrary to the OMI spatial pixel resolution, the spatial resolution of MODIS is almost the same in all pixels. Therefore, the constraint of the spatial pixel position was ignored during the analysis. From the previous studies, the accuracy of products from the representative algorithm was shown to be $\pm\,0.05 \pm 0.15 \times$ AOD over land and $\pm\,0.03 \pm 0.05 \times$ AOD over ocean (dark target), and $\pm\,0.05 \pm 20$ % over arid and semi-arid areas (deep blue) [57,58].

3. Methods

This study used ground-based and satellite-based datasets to investigate the effective temporal and spatial variability, respectively. By using ground-based datasets with high-temporal resolution, the lag-correlation studies on the time lag of 10–60 min were adopted to estimate the temporal variabilities of AOD, TCO, and TCN. For the selection of specific data considering time lag, the observed data with closest data of selected time lag were selected in this study. Figure 1 shows the number of observation data from the AERONET ground-based measurement for the temporal variability analysis since 2012. Because of the cloud contamination or data quality problems during the observation, the total number of observation data per month varies greatly during the observation periods. Although there is seasonal variability in the number of data, observation datasets from the instruments can be used in the temporal variability studies. Similar to the AERONET ground-based measurements, the TCO and TCN data have been used by the ground-based Pandora observation datasets in South Korea since 2012.

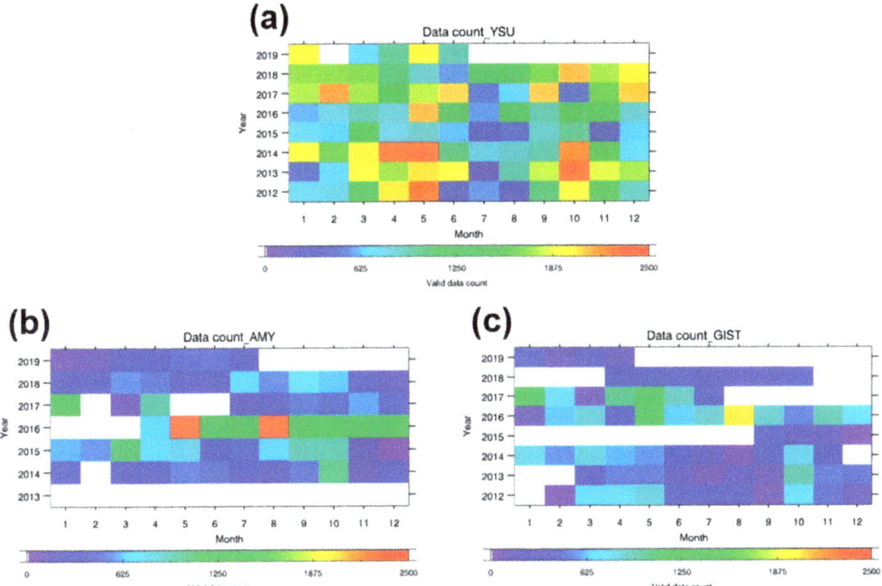

Figure 1. Monthly variation of number of observation data at (**a**) Yonsei University (YSU), (**b**) Anmyeon, and (**c**) Gwangju Institute of Science and Technology (GIST).

For the spatial variability study, the pixels within the spatial range were selected and these valid datasets were totally averaged for comparisons with data from the center of pixels. In addition, the correlation between the compared datasets and the scientific data at the center of spatial range was calculated. To focus on the Asia region, satellite data within 20–50 °N and 90–150 °E were selected for the latitudinal and longitudinal spatial ranges, respectively. Figure 2 shows the schematic figures for the spatial variability estimation of AOD from MODIS. Because the variation in the spatial resolution of MODIS AOD is negligible, the 5-pixels were selected if the spatial range was in the 10 km range, as shown in the red pixels. Furthermore, the 13-pixels (red and blue) were also used in the average calculation if the spatial range was 20 km. To identify the spatial variability with respect to the spatial range, the spatial range was assumed to change from 10 to 100 km radius with every 10 km interval. This study focuses on the quantitative consistency with spatio-temporal changes in the year 2016. Therefore, the spatio-temporal variabilities for specific products were estimated by the correlation coefficient, slope, and intercept from the linear regression analysis.

For the TCO, the spatio-temporal variability is slightly different compared to those of TCN and AOD. The global distribution of TCO has seasonal dependence with several natural effects, such as solar cycle, Quasi-Biennial Oscillation (QBO), El Nino/Southern Oscillation (ENSO), and stratospheric aerosols [59]. In addition, the regional TCO has spatio-temporal variation by the size of the jet-stream and vortex strength. They are related to the Brewer-Dobson circulation [60,61]. The spatio-temporal variation of TCO is a scale with latitude and longitude of several degrees. For this reason, the Level 3 gridded datasets from OMI were used for the spatial variation of TCO. By estimating the TCO variation, the difference in the TCO between adjacent grids was used in the north-south and east-west direction, after considering the data quality.

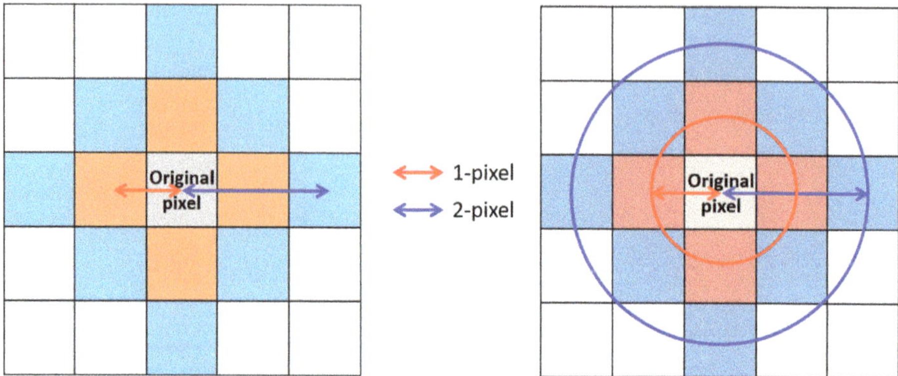

Figure 2. Pixel selection for spatial variability estimation of aerosol optical depth (AOD) from the moderate resolution imaging spectroradiometer (MODIS).

4. Results and Discussion

4.1. AOD

Figure 3 and Table 2 show the temporal variability of AOD at 500 nm by using data from Sunphotometer in three ground-based stations in Korea. To identify the temporal variability of AOD, the correlation coefficient of determination (R^2), mean bias (MB), and root mean square error (RMSE) were used for the statistical score. In time lags larger than 30 min, the temporal variability of AOD is extremely enhanced, especially for an AOD larger than 1.0. In addition, the RMSE by the temporal variation is significantly larger than the long-term uncertainty of the instrument itself [62]. As shown in Table 2, the spread of data increases linearly with the increasing time lag. Because the large AOD cases are caused by the transport of aerosol layers, including dust and anthropogenic pollutants, the spatio-temporal variability of aerosol conditions can change drastically. As the time lag increased, the R^2 values were slightly decreased with increasing absolute values of MB and RMSE (Table 2). On average, the R^2 difference with 10 min lag increase ranges from −0.0095 to −0.0164, and those differences are more sensitive the larger the time lag is.

The RMSE between the original and time-lagged datasets also increased when the time lag increased. Compared to the expected error range of the satellite algorithm over land, the RMSE that was smaller than the absolute expected error range of satellite algorithm (0.05) only satisfied the case shorter than 20 min of time lag (10 min in YSU). Furthermore, the RMSE difference with 10 min time lag increase is 0.007 to 0.012, and the absolute increasing tendency is largest in YSU. Because the mean value of AOD in Seoul is larger than the other two observation sites, the RMSE in the same time lag is also larger. From the RMSE and R^2, the temporal variability is largest in Seoul, and those in the other two observation sites are almost the same. If we assume R^2>0.95 and RMSE<0.05 for the same propensity for temporal range, the temporal range of data with the same propensity is 20 min in South Korea.

Table 2. Statistical results for temporal variability of AOD with time lag change at (**a**) Yonsei University, (**b**) Anmyeon, and (**c**) GIST.

Time Lag (minute)	R^2	Mean Bias	RMSE
(a) Yonsei University			
10	0.989	−0.00007	0.037
20	0.978	−0.00002	0.054
30	0.968	−0.00005	0.066
40	0.958	−0.00026	0.074
50	0.947	−0.00062	0.083
60	0.936	−0.00068	0.092
(b) Anmyeon			
10	0.988	0.00010	0.032
20	0.977	0.00129	0.044
30	0.966	0.00226	0.054
40	0.952	0.00267	0.063
50	0.940	0.00326	0.073
60	0.929	0.00265	0.080
(c) GIST			
10	0.983	0.00038	0.032
20	0.973	0.00113	0.039
30	0.957	0.00044	0.053
40	0.946	0.00232	0.056
50	0.927	0.00166	0.069
60	0.909	0.00141	0.079

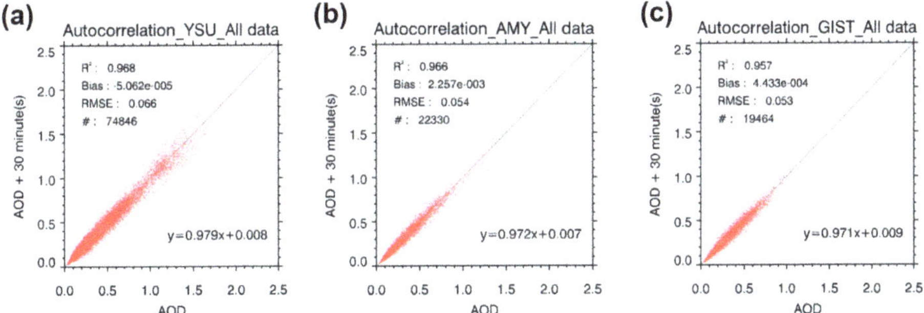

Figure 3. Temporal variability of AOD for 30 min time lag from Sunphotometer at (**a**) Yonsei University, (**b**) Anmyeon, and (**c**) GIST.

For the spatial variability of AOD from the satellite measurements, the distribution of the correlation coefficient (R_d) was used. For the R_d, each R value was estimated in a single granule in 5-min intervals. Based on each R value, the R_d value was finally calculated to the mean and standard deviation of R values with respect to the colocation range. Figure 4 shows the R_d value as a function of the colocation range. Focusing on each correlation coefficient value with its respective granule, the R value largely varies by up to ~0.5. The large variation of correlation coefficient in each granule was partially caused by the low AOD values in the background areas, such as the ocean areas. However, statistically estimated mean R_d was higher than the respective R value. The mean R_d ranged from 0.981 in 10 km colocation range to 0.877 in 100 km colocation range. As the colocation range increased, the mean value of R_d slightly decreased as the standard deviation of R_d increased. In addition, as the colocation range increased, the tendency to decrease the mean of R_d and increase the standard deviation of R_d slowed together.

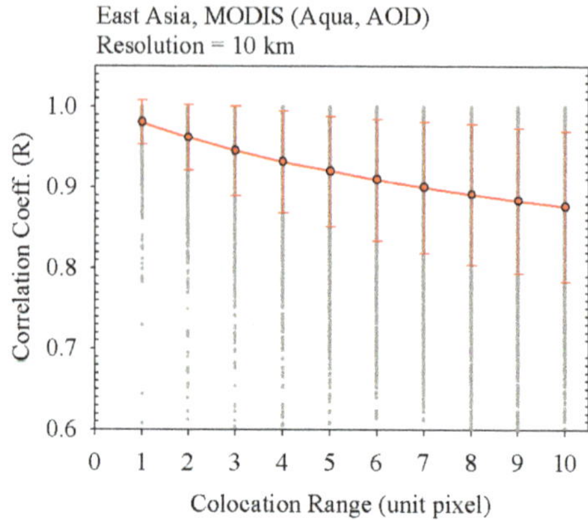

Figure 4. Distribution of correlation coefficient (R_d) for AOD in East Asia with respect to the colocation range.

For the seasonal variation, the decreasing tendency of R_d as increasing colocation range has slight seasonal dependency as shown in Figure 5. While the mean value of R_d has weak seasonal dependence, the standard deviation of R_d is larger in winter and spring than those in summer and autumn. In East Asia, regional transport of aerosol frequently occurs during the late autumn to early spring season through the winter season by the occurrence of dust transport [63–65] and haze by anthropogenic emissions [66,67]. On the contrary, the trans-boundary transport of aerosol weakens due to decreasing wind speed in the summer and autumn season [68]. For the seasonal difference of aerosol transport patterns, the regional inhomogeneity of AOD increases in winter and spring, thus, the variation of correlation coefficient with changing colocation range is more sensitive.

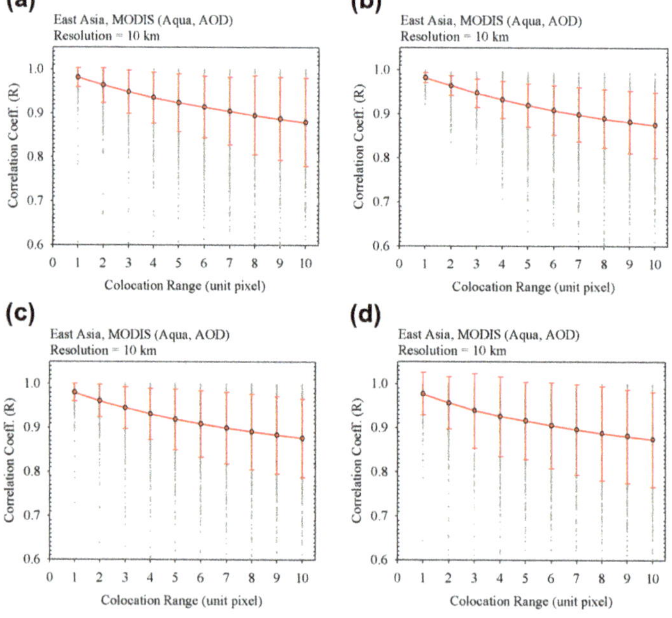

Figure 5. Distribution of correlation coefficient (R_d) for AOD in East Asia with respect to the colocation range in (**a**) spring, (**b**) summer, (**c**) autumn, and (**d**) winter.

4.2. Total Ozone

Table 3 show the temporal variability of TCO in Seoul and Busan. From the statistical analysis, the R^2 and RMSE are 0.966–0.992 and 3.37–7.00 DU in Seoul, and 0.987–0.996 and 2.17–3.98 DU in Busan. In a similar way to the other species, the R^2 has a decreasing tendency and the RMSE has an increasing tendency as the time lag increases. However, the variation of R^2 as a changing time lag is relatively small because the temporal variation of TCO largely depends on the stratospheric ozone variability [69]. Focusing on the RMSE value, the variability up to 3% of TCO is considerable in the analysis [38,70]. In Seoul and Busan, the annual mean value of TCO ranges from 300–340 DU [44,71]. For this reason, the temporal variability of TCO is negligible from the Pandora observation.

Table 3. Statistical results for temporal variability of total column amount of ozone (TCO) with time lag change in (**a**) Seoul and (**b**) Busan.

Time Lag (minute)	R^2	Mean Bias (DU)	RMSE (DU)
(a) Seoul			
10	0.992	0.009	3.370
20	0.986	0.018	4.492
30	0.980	0.030	5.276
40	0.975	0.022	5.938
50	0.970	0.026	6.512
60	0.966	0.029	6.999
(b) Busan			
10	0.996	−0.023	2.175
20	0.994	−0.047	2.616
30	0.993	−0.068	2.974
40	0.991	−0.084	3.313
50	0.989	−0.100	3.647
60	0.987	−0.116	3.981

However, several previous studies have shown that the TCO suddenly changes during winter and spring seasons due to the enhancement of ozone concentration at the upper troposphere/lower stratosphere (UT/LS) in the jet-stream outflow regions [71–73]. In East Asia, the UT/LS ozone enhancement frequently occurs, and sudden changes of TCO have frequently been observed based on the daily observation datasets from ground-based measurements [71]. This phenomenon also affects the spatio-temporal variability of TCO. For this reason, the spatio-temporal variability, considering seasonal change, is also executed to consider the TCO variation due to the UT/LS ozone enhancement.

Figure 6 shows the spatial variability of the TCO using Level 3 gridded datasets from OMI in 2016. Because the latitudinal change of TCO is significantly larger than the longitudinal change, the average of the latitude range for 1 DU change was adopted for the estimation of spatial variability of TCO in this study. For the daily statistical results, the spatial variability ranged from 0.237 (12 March) to 0.396 degree/DU (16 October). Categorized by the seasons, as shown in Figure 6b, the spatial variability is largest in spring (0.29 degree/DU), and smallest in autumn (0.35 degree/DU). Compared to the spring and autumn, there is a 20% difference in the spatial variability. In a similar way to the temporal variability of TCO, large spatial variability in springtime is caused by the sudden increase in the ozone concentration in UT/LS. In addition, the latitudinal gradient of TCO is also related to the Brewer-Dobson circulation, and the intensity of the Brewer-Dobson circulation is strong in wintertime. For this reason, the latitudinal range has to be adjusted according to the season.

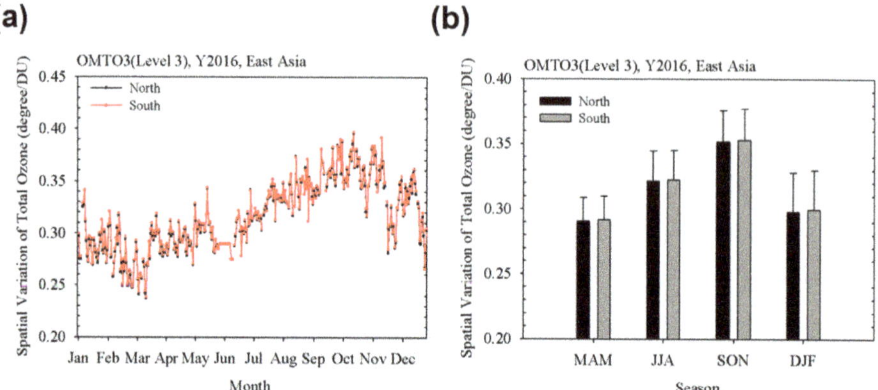

Figure 6. (**a**) Daily change and (**b**) seasonal change for latitudinal variability of the TCO by using Level 3 gridded datasets from the Ozone Monitoring Instrument (OMI) in Year 2016.

4.3. NO_2

Table 4 shows the temporal variability for TCN from the Pandora measurements. In a similar way to the AOD analysis, the temporal variability for TCN also used the R^2 and RMSE in their statistical analysis. Over the two ground-based observation sites, the R^2 ranged from 0.880 to 0.565 in Busan and from 0.897 to 0.638 in Seoul. In addition, the RMSE ranged from 0.29 to 0.56 DU, and from 0.15 to 0.31 DU in Seoul and Busan, respectively. Compared to the AOD and TCO results, the R^2 is too low. Because the diurnal variation of the NO_2 was clearly shown in the urban region due to the short lifetime of NO_2 [51], temporal variability of TCN is larger compared to those of AOD and TCO. The RMSE in Seoul is two times larger than that in Busan. This RMSE difference is caused by the absolute value difference of TCN in Busan and Seoul. From the previous study, the mean value of TCN in Seoul is two times larger than that in Busan during the MAPS-Seoul campaign [51]. Therefore, the relative value of RMSE to the mean value of TCN is similar in both two observation sites.

Table 4. Statistical results for temporal variability of total column amount of NO_2 (TCN) with time lag change in (**a**) Seoul and (**b**) Busan.

Time Lag (minute)	R^2	Mean Bias (DU)	RMSE (DU)
(a) Seoul			
10	0.897	−0.004	0.285
20	0.829	−0.008	0.371
30	0.774	−0.011	0.431
40	0.727	−0.015	0.478
50	0.681	−0.018	0.522
60	0.638	−0.020	0.561
(b) Busan			
10	0.880	−0.006	0.153
20	0.802	−0.013	0.199
30	0.733	−0.019	0.234
40	0.670	−0.025	0.264
50	0.615	−0.031	0.289
60	0.565	−0.038	0.311

As the time lag increases, the decrease of R^2 per 10-min time lag shows -0.069 (10 min) to -0.043 (60 min) in Seoul and -0.078 (10 min) to -0.050 (60 min) in Busan. Based on the sensitivity of R^2 for time lag, temporal variability is larger in Busan than in Seoul. The large temporal variability in Busan is also identified by the MB and relative value of RMSE. The main reason is the difference in the TCN level in the two regions. Because of the uncertainty from the instrument, the TCN includes the systematic variability during the observation. Without considering the instrument uncertainty, diurnal variation of TCN, due to the emission and photochemical reaction, can also affect the temporal variability in specific points. The temporal variability is similar in the two sites, although the correlation is slightly weaker in Busan than in Seoul. Based on the R^2 and RMSE values in the 10-min time lag, if the change in R^2 decreased by 0.1 and the RMSE increased by 50%, the temporal range with the data of the same propensity is 30 min in South Korea.

The R_d value for the TCN as a function of colocation range is shown in Figure 7. Because of the pixel size of OMI, there were no cases within 10 km for the colocation range. In a similar way to the R_d distribution for AOD, a significant tendency to decrease was found with an increasing colocation range. However, the absolute value of R was always smaller than those of the AOD cases. For all cases of the colocation range, the R value did not exceed 0.81. For this reason, it is difficult to apply the threshold of the spatial colocation range for aerosol to those for NO_2. Focusing on the seasonal variation, the R_d value has strong seasonal variability, as shown in Figure 8. Because the photo-chemical reaction vigorously activates in summer [74], the spatial variability of NO_2 is enhanced during the summer season. In addition, the size of areas in high concentration of NO_2, near the downtown or industrial region, may be reduced due to the photochemical reaction change. For this reason, the R value is up to 0.569 (colocation range: 20 km). Otherwise, the R value is from 0.666 (colocation range: 100 km) to 0.925 (colocation range: 20 km) in wintertime. Therefore, we have to consider the spatial variability of NO_2 because of the strong seasonal variations and inhomogeneity near the emission source regions.

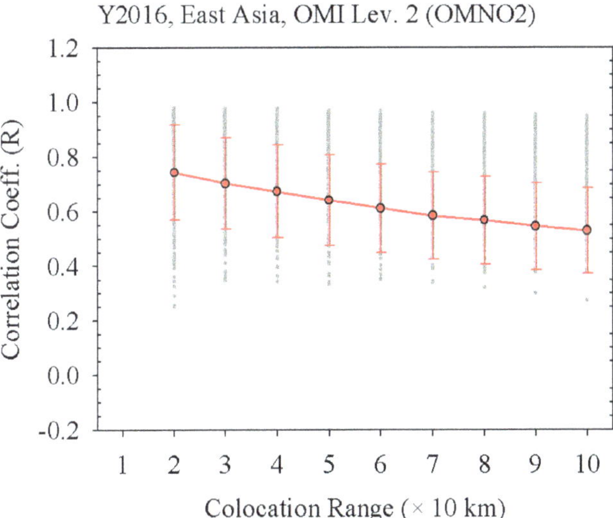

Figure 7. Distribution of correlation coefficient (R_d) for TCN in East Asia with respect to the colocation range.

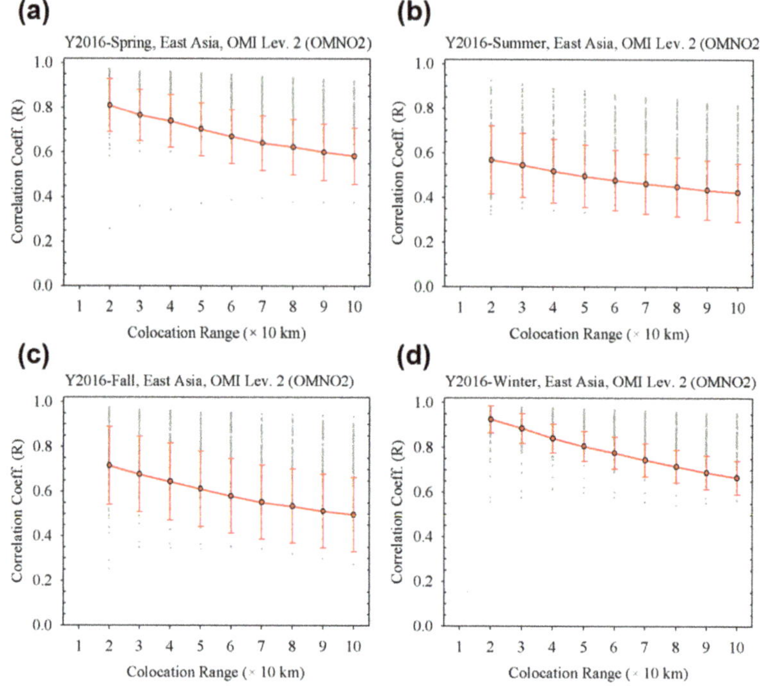

Figure 8. Distribution of correlation coefficient (R_d) for TCN in East Asia with respect to the colocation range in (**a**) spring, (**b**) summer, (**c**) autumn, and (**d**) winter.

5. Validation Strategy for GEMS Scientific Products

Information of spatio-temporal variability for the amount of species is necessary to identify the temporal and spatial ranges. To determine the spatial and temporal ranges, the datasets observed within the range should be considered statistically as a same quantity. In addition, it is also important to ensure a consistently sufficient amount of data over time and space. Based on the above spatio-temporal variability in Section 4, we also listed the strategy of the validation for GEMS products in Asia.

Table 5 shows the summary of strategy of the validation plans for the GEMS scientific products, considering spatio-temporal ranges. To determine the spatio-temporal range for the validation, the spatial range is determined by the satellite observation and temporal range is considered by the temporal variation of datasets from ground-based observations. Basically, the validation uses the averaging methods within valid ranges, because the trace gas observations include a large amount of errors during the retrieval process.

Table 5. Strategy of the validation plans for the GEMS scientific products in East Asia.

Species	Spatial	Temporal
Aerosol	30 km (Whole Season) 40 km (Summer~Autumn)	30 min (Whole Season)
TCO	0.87 ° (Spring) ~1.05 ° (Summer)	Negligible for sub-daily scale
TCN	20 km (Spring) ~50 km (Winter)	20 min

For the AOD, the spatio-temporal range for validation is based on $R^2 > 0.95$ (temporal) and $R > 0.95$ (spatial). From these thresholds, the spatial and temporal validation range is within 30 km and 30 min in all seasons. Especially considering the seasonal variation, the spatial range expands to 40 km in summer and autumn. For the TCO, the temporal variation is negligible due to the time scale of the stratospheric ozone variation. However, the spatial scale is considerable for the latitudinal range. Because of strict target accuracy of TCO, the latitudinal range for TCO is 0.87° in spring to 1.05° in summer, if the threshold of spatial variation of TCO is assumed to be 3 DU. For the TCN, it is difficult to establish rigorous criteria for temporal and spatial agreement. For this reason, the R^2 for temporal colocation is assumed to be 0.8. In addition, the rapid increase of RMSE—more than a 50% increase by a 10-min range increase—is also a considerable threshold for temporal colocation. Considering these two criteria, 20 min is suitable for the threshold of temporal range for validation. In the spatial scale, the seasonal considerations are essential. For this reason, the spatial range is 20 km and 50 km for spring and winter, respectively. However, the criteria of R are not satisfied during summer and autumn. In those seasons, the nearest value of satellite data is prioritized for the validation.

6. Summary and Conclusions

As the aerosol and trace gases have large spatial and temporal variabilities due to the complex emission sources and various chemical process, the colocation methods of the dataset are an important consideration factor during intercomparison and validation. From the ground-based and satellite observation datasets, the spatio-temporal variability of AOD, TCO, and TCN were identified over East Asia to define the spatio-temporal range for the GEMS scientific products. Based on the statistical results (R^2, RMSE, and MB), optimized spatio-temporal ranges were determined with respect to the target materials. R_d was also used for the satellite observation dataset to define the mean status of the spatial variability of aerosol and trace gases.

For the AOD, the temporal range, as data with the same propensity, was 20 min in South Korea, assuming $R^2 > 0.95$ and RMSE<0.05 for the temporal scales. From the R_d distribution, the spatial range within 20 km was able to be considered the same value for the validation studies. The R_d distribution also has weak seasonal dependence due to the trans-boundary transport change. For the TCO, the R^2 and RMSE was estimated to be from 0.966–0.996 and 2.17–7.00 DU for temporal variation, and 0.237–0.396 degree/DU for spatial variability. The spatio-temporal variability of TCO is affected by the latitudinal gradient change related to the Brewer-Dobson circulation. For the TCN, the R^2 and RMSE for temporal variation ranged from 0.565–0.897 and 0.15–0.56 DU, respectively. For all cases of the spatial colocation range, however, the R_d was always smaller than 0.81, which makes it difficult to assume the same value in all spatial ranges. Because of the photochemical reaction change, the spatial variability of TCN has strong seasonal dependence.

Based on the spatio-temporal variability from the observation data, we suggest the basic strategy of the GEMS scientific products. Based on the thresholds of $R^2 > 0.95$ (temporal) and $R > 0.95$ (spatial), the basic ranges for spatial and temporal scales for AOD validation were found to be within 30 km and 30 min, respectively. In addition, the spatial range was expanded to 40 km in summer and autumn for the consideration of seasonal characteristics. For the TCO validation, the north-south range was the only considerable factor for the validation study due to the seasonal change in the latitudinal gradient. The latitudinal range was set to be from 0.87° in spring to 1.05° in summer. However, the validation criteria for the TCN was difficult because the spatio-temporal variation of TCN was large. Thus, the nearest value in the satellite data was used as the representative data for the validation in summer and autumn. Furthermore, the spatio-temporal range was 20 min and 20–50 km in other seasons.

Although the spatio-temporal variation studies were done using several observation sites, the spatial coverage of ground-based observations had trouble identifying the temporal variability of pollutants. Particularly for formaldehyde (HCHO), it was difficult for the ground-based dataset to observe the total column amount based on the remote sensing instrument. In the future, a geostationary orbit satellite will observe the diurnal variation of atmospheric pollutions, thus the diurnal variation of the spatio-temporal range will be studied.

Author Contributions: Conceptualization, C.-K.S.; formal analysis, S.S.P., J.-U.P. and K.-H.B.; methodology, S.S. Park and C.-K.S.; project administration, S.-W.K. and C.K.S; resources, S.-W.K.; validation, S.S.P. and J.-U.P.; visualization, J.U.P and K.-H.B.; writing—original draft, S.S.P; writing—review & editing, S.-W.K. and C.-K.S. All authors have read and agreed to the published version of the manuscript.

Funding: This research was funded by the Korea Ministry of Environment (MOE) as Public Technology Program based on Environmental Policy (2017000160001), and also supported by the National Strategic Project-Fine Particle of the National Research Foundation of Korea (NRF) funded by the Ministry of Science and ICT (MSIT), the Ministry of Environment (ME), and the Ministry of Health and Welfare (MOHW) (NRF-2017M3D8A1092021).

Acknowledgments: The authors would like to thank the editor and the reviewers' feedback and suggestions.

Conflicts of Interest: The authors declare no conflicts of interest.

References

1. Van Donkelaar, A.; Martin, R.V.; Park, R.J. Estimating ground-level PM2.5 using aerosol optical depth determined from satellite remote sensing. *J. Geophys. Res.* **2006**, *111*, D21201. [CrossRef]

2. Zhang, X.; Kondragunta, S.; Schmidt, C.; Kogan, F. Near real time monitoring of biomass burning particulate emissions (PM2.5) across contiguous United States using multiple satellite instruments. *Atmos. Environ.* **2008**, *42*, 6959–6972. [CrossRef]

3. Duncan, B.N.; Lamsal, L.N.; Thompson, A.M.; Yoshida, Y.; Lu, Z.; Streets, D.G.; Hurwitz, M.M.; Pickering, K.E. A space based, high resolution view of notable changes in urban NOx pollution around the world (2005–2014). *J. Geophys. Res. Atmos.* **2016**, *121*, 976–996. [CrossRef]

4. Curci, G.; Palmer, P.I.; Kurosu, T.P.; Chance, K.; Visconti, G. Estimating European volatile organic compound emissions using satellite observations of formaldehyde from the Ozone Monitoring Instrument. *Atmos. Chem. Phys.* **2010**, *10*, 11501–11517. [CrossRef]

5. Fioletov, V.E.; McLinden, C.A.; Krotkov, N.; Moran, M.D.; Yang, K. Estimation of SO$_2$ emissions using OMI retrievals. *Geophys. Res. Lett.* **2011**, *38*, L21811. [CrossRef]

6. Kim, S.-W.; Heckel, A.; McKeen, S.A.; Frost, G.J.; Hsie, E.-Y.; Trainer, M.K.; Richter, A.; Burrows, J.P.; Peckham, S.E.; Grell, G.A. Satellite-observed U.S. power plant NOx emissions reductions and their impact on air quality. *Geophys. Res. Lett.* **2006**, *33*, L22812. [CrossRef]

7. Turner, A.J.; Jacob, D.J.; Wecht, K.J.; Maasakkers, J.D.; Lundgren, E.; Andrews, A.E.; Biraud, S.C.; Boesch, H.; Bowman, K.W.; Deutscher, N.M.; et al. Estimating global and North American methane emissions with high spatial resolution using GOSAT satellite data. *Atmos. Chem. Phys.* **2015**, *15*, 7049–7069. [CrossRef]

8. Wiedinmyer, C.; Quayle, B.; Geron, C.; Belote, A.; McKenzie, D.; Zhang, X.; O'Neill, S.; Wynne, K.K. Estimating emissions from fires in North America for air quality modeling. *Atmos. Environ.* **2006**, *40*, 3419–3432. [CrossRef]

9. Levelt, P.F.; Van den Oord, G.H.J.; Dobber, M.R.; Malkki, A.; Visser, H.; de Vries, J.; Stammes, P.; Lundell, J.O.V.; Saari, H. The ozone monitoring instrument. *IEEE Trans. Geosci. Remote Sens.* **2006**, *44*, 1093–1101. [CrossRef]

10. Boersma, K.F.; Eskes, H.J.; Veefkind, J.P.; Brinksma, E.J.; van der, A.R.J.; Sneep, M.; van den Oord, G.H.J.; Levelt, P.F.; Stammes, P.; Gleason, J.F.; et al. Near-real time retrieval of tropospheric NO$_2$ from OMI. *Atmos. Chem. Phys.* **2007**, *7*, 2103–2118. [CrossRef]

11. Boersma, K.F.; Eskes, H.J.; Dirksen, R.J.; van der, A.R.J.; Veefkind, J.P.; Stammes, P.; Huijnen, V.; Kleipool, Q.L.; Sneep, M.; Claas, J.; et al. An improved tropospheric NO$_2$ column retrieval algorithm for the Ozone Monitoring Instrument. *Atmos. Meas. Tech.* **2011**, *4*, 1905–1928. [CrossRef]

12. Bucsela, E.J.; Krotkov, N.A.; Celarier, E.A.; Lamsal, L.N.; Swartz, W.H.; Bhartia, P.K.; Boersma, K.F.; Veefkind, J.P.; Gleason, J.F.; Pickering, K.E. A new stratospheric and tropospheric NO$_2$ retrieval algorithm for nadir-viewing satellite instruments: Applications to OMI. *Atmos. Meas. Tech.* **2013**, *6*, 2607–2626. [CrossRef]

13. Lee, C.; Richter, A.; Weber, M.; Burrows, J.P. SO$_2$ retrieval from SCIAMACHY using the Weighting Function DOAS (WFDOAS) technique: Comparison with Standard DOAS retrieval. *Atmos. Chem. Phys.* **2008**, *8*, 6137–6145. [CrossRef]

14. Li, C.; Joiner, J.; Krotkov, N.A.; Bhartia, P.K. A fast and sensitive new satellite SO$_2$ retrieval algorithm based on principal component analysis: Application to the ozone monitroring instrument. *Geophys. Res. Lett.* **2013**, *40*, 6314–6318. [CrossRef]

15. Ahn, C.; Torres, O.; Jethva, H. Assessment of OMI near-UV aerosol optical depth over land. *J. Geophys. Res. Atmos.* **2014**, *119*, 2457–2473. [CrossRef]

16. Torres, O.; Decae, R.; Veefkind, P.; de Leeuw, G. *OMI Aerosol Retrieval Algorithm, OMI Algorithm Theoretical Basis Document, Clouds, Aerosols and Surface UV Irradiance*; NASA-KNMI ATBD-OMI-03; Harvard Library: Cambridge, MA, USA, 2002; Volume III, pp. 47–71.

17. Veihelmann, B.; Levelt, P.F.; Stammes, P.; Veefkind, J.P. Simulation study of the aerosol information content in OMI spectral reflectance measurements. *Atmos. Chem. Phys.* **2007**, *7*, 3115–3127. [CrossRef]

18. Griffin, D.; Zhao, X.; McLinden, C.A.; Boersma, F.; Bourassa, A.; Dammer, E.; Degenstein, D.; Eskes, H.; Fehr, L.; Fioletov, V.; et al. High-Resolution Mapping of Nitrogen Dioxide with TROPOMI: First Results and Validation over the Canadian Oil Sands. *Geophys. Res. Lett.* **2018**, *46*, 1049–1060. [CrossRef]

19. Herman, J.; Abuhassan, N.; Kim, J.; Kim, J.; Dubey, M.; Raponi, M.; Tzortziou, M. Underestimation of column NO_2 amounts from the OMI satellite compared to diurnally varying ground-based retrievals from multiple PANDORA spectrometer instruments. *Atmos. Meas. Tech.* **2019**, *12*, 5593–5612. [CrossRef]

20. Virtanen, T.H.; Kolmonen, P.; Sogacheva, L.; Rodriguez, E.; Saponaro, G.; de Leeuw, G. Collocation mismatch uncertainties in satellite aerosol retrieval validation. *Atmos. Meas. Tech.* **2018**, *11*, 925–938. [CrossRef]

21. Ichoku, C.; Chu, D.A.; Mattoo, S.; Kaufman, Y.J.; Remer, L.A.; Tanre, D.; Slutsker, I.; Holben, B.N. A spatio-temporal approach for global validation and analysis of MODIS aerosol products. *Geophys. Res. Lett.* **2002**, *29*, 1616–1619. [CrossRef]

22. Cheng, M.M.; Jiang, H.; Guo, Z. Evaluation of long-term tropospheric NO_2 columns and the effect of different ecosystem in Yangtze River Delta. *Proced. Environ. Sci.* **2012**, *13*, 1045–1056. [CrossRef]

23. Mishchenko, M.I.; Liu, L.; Geogdzhayev, I.V.; Travis, L.D.; Cairns, B.; Lacis, A.A. Toward unified satellite climatology of aerosol properties. 3. MODIS versus MISR versus AERONET. *J. Quant. Spectrosc. Radiat. Transf.* **2010**, *111*, 540–552. [CrossRef]

24. Su, C.-H.; Ryu, D.; Young, R.I.; Western, A.W.; Wagner, W. Inter-comparison of microwave satellite soil moisture retrievals over the Murrumbidgee Basin, southeast Australia. *Remote Sens. Environ.* **2013**, *134*, 1–11. [CrossRef]

25. De Vries, J.; Voors, R.; Ording, B.; Dingjan, J.; Veefkind, P.; Antje, L.; Kleiipool, Q.; Hoogeveen, R.; Aben, I. TROPOMI on ESA's Sentinel 5p ready for launch and use. In Proceedings of the 4th International Conference on Remote Sensing and Geoinformation of the Environment (SPIE 9688), Cyprus, Greece, 4–8 April 2016, 96880B. [CrossRef]

26. Kim, J.; Jeong, U.; Ahn, M.-H.; Kim, J.H.; Park, R.J.; Lee, H.; Song, C.H.; Choi, Y.-S.; Lee, K.-H.; Yoo, J.-M.; et al. New Era of Air Quality Monitoring from Space: Geostationary Environment Monitoring Spectrometer (GEMS). *Bull. Am. Meteorol. Soc.* **2020**, *101*, E1–E22. [CrossRef]

27. Zoogman, P.; Liu, X.; Suleiman, R.M.; Pennington, W.F.; Flittner, D.E.; Al-Saadi, J.A.; Hilton, B.B.; Nicks, D.K.; Newchurch, M.J.; Carr, J.L.; et al. Tropospheric emissions: Monitoring of pollution (TEMPO). *J. Quant. Spectrosc. Radiat. Transf.* **2017**, *186*, 17–39. [CrossRef]

28. Ingmann, P.; Veihelmann, B.; Langen, J.; Lamarre, D.; Stark, H.; Courrèges-Lacoste, G.B. Requirements for the GMES Atmosphere Service and ESA's implementation concept: Sentinels-4/-5 and -5p. *Remote Sens. Environ.* **2012**, *120*, 58–69. [CrossRef]

29. Goldberg, D.L.; Lamsal, L.N.; Loughner, C.P.; Swartz, W.H.; Lu, Z.; Streets, D.G. A high-resolution and observationally constrained OMI NO_2 satellite retrieval. *Atmos. Chem. Phys.* **2017**, *17*, 11403–11421. [CrossRef]

30. Liu, M.; Lin, J.; Boersma, K.F.; Pinardi, G.; Wang, Y.; Chimot, J.; Wagner, T.; Xie, P.; Eskes, H.; Van Roozendael, M.; et al. Improved aerosol correction for OMI tropospheric NO_2 retrieval over East Asia: Constraint from CALIOP aerosol vertical profile. *Atmos. Meas. Tech.* **2019**, *12*, 1–21. [CrossRef]

31. Judd, L.M.; Al-Saadi, J.A.; Janz, S.J.; Kowalewski, M.G.; Pierce, R.B.; Szykman, J.J.; Valin, L.C.; Swap, R.; Cede, A.; Mueller, M.; et al. Evaluating the impact of spatial resolution on tropospheric NO_2 column comparisons within urban areas using high-resolution airborne data. *Atmos. Meas. Tech.* **2019**, *12*, 6091–6111. [CrossRef]

32. Pan, X.; Uno, I.; Wang, Z.; Nishizawa, T.; Sugimoto, N.; Yamamoto, S.; Kobayashi, H.; Sun, Y.; Fu, P.; Tang, X.; et al. Real-time observational evidence of changing Asian dust morphology with the mixing of heavy anthropogenic pollution. *Sci. Rep.* **2017**, *7*, 335. [CrossRef]

33. Holben, B.N.; Eck, T.F.; Slutsker, I.; Tanre, D.; Buis, J.P.; Setzer, A.; Vermote, E.; Reagan, J.A.; Kaufman, Y.J.; Nakajima, T.; et al. AERONET-A Federated Instrument Network and Data Archive for Aerosol Characterization. *Remote Sens. Environ.* **1998**, *66*, 1–16. [CrossRef]

34. Dubovik, O.; Smirnov, A.; Holben, B.N.; King, M.D.; Kaufman, Y.J.; Eck, T.F.; Slutsker, I. Accuracy assessments of aerosol optical properties retrieved from Aerosol Robotic Network (AERONET) Sun and sky radiance measurements. *J. Geophys. Res. Atmos.* **2000**, *105*, 9791–9806. [CrossRef]

35. Giles, D.M.; Sinyuk, A.; Sorokin, M.G.; Schafer, J.S.; Smirnov, A.; Slutsker, I.; Eck, T.F.; Holben, B.N.; Lewis, J.R.; Campbell, J.R.; et al. Advancements in the Aerosol Robotic Network (AERONET) Version 3 database—Automated near-real-time quality control algorithm with improved cloud screening for Sun photometer aerosol optical depth (AOD) measurements. *Atmos. Meas. Tech.* **2019**, *12*, 169–209. [CrossRef]

36. Buchard, V.; Brogniez, C.; Auriol, F.; Bonnel, B.; Lenoble, J.; Tanskanen, A.; Bojkov, B.; Veefkind, P. Comparison of OMI ozone and UV irradiance data with ground-based measurements at two French sites. *Atmos. Chem. Phys.* **2008**, *8*, 4517–4528. [CrossRef]

37. Bhartla, P.K.; Wellemeyer, C.W. *TOMS-V8 Total O3 Algorithm in OMI Algorithm Theoretical Basis Document*; Bhartia, P.K., Ed.; NASA Goddard Space Flight Center: Greenbelt, MD, USA, 2002; Volume II, pp. 15–32.

38. McPeters, R.; Kroon, M.; Labow, G.; Brinksma, E.; Balis, D.; Petropavlovskikh, I.; Veefkind, J.P.; Bhartia, P.K.; Levelt, P.F. Validation of the Aura Ozone Monitoring Instrument total column ozone product. *J. Geophys. Res. Atmos.* **2008**, *113*, D15S14. [CrossRef]

39. Veefkind, J.P.; de Haan, J.F.; Brinksma, E.J.; Kroon, M.; Levelt, P.F. Total Ozone from the Ozone Monitoring Instrument (OMI) using the DOAS technique. *IEEE Trans. Geosci. Remote Sens.* **2006**, *44*, 1239–1244. [CrossRef]

40. Krotkov, N.A.; Lamsal, L.N.; Celarier, E.A.; Swartz, W.H.; Marchenko, S.V.; Bucsela, E.J.; Chan, K.L.; Wenig, M.; Zara, M. The version 3 OMI NO_2 standard product. *Atmos. Meas. Tech.* **2017**, *10*, 3133–3149. [CrossRef]

41. De Graaf, M.; Sihler, H.; Tilstra, L.G.; Stammes, P. How big is an OMI pixel? *Atmos. Meas. Tech.* **2016**, *9*, 3607–3618. [CrossRef]

42. Herman, J.; Cede, A.; Spinei, E.; Mount, G.; Tzortziou, M.; Abuhassan, N. NO_2 column amounts from groundbased Pandora and MFDOAS spectrometers using the direct-sun DOAS technique: Intercomparisons and application to OMI validation. *J. Geophys. Res.* **2009**, *114*, D13307. [CrossRef]

43. Tzortziou, M.; Herman, J.R.; Cede, A.; Abuhassan, N. High precision, absolute total column ozone measurements from the Pandora spectrometer system: Comparisons with data from a Brewer double monochromator and Aura OMI: Pandora total column ozone retrieval. *J. Geophys. Res. Atmos.* **2012**, *117*, D16303. [CrossRef]

44. Baek, K.; Kim, J.H.; Herman, J.R.; Haffner, D.P.; Kim, J. Validation of Brewer and Pandora measurements using OMI total ozone. *Atmos. Environ.* **2017**, *160*, 165–175. [CrossRef]

45. Kim, J.; Kim, J.; Cho, H.-K.; Herman, J.; Park, S.S.; Lim, H.K.; Kim, J.-H.; Miyagawa, K.; Lee, Y.G. Intercomparison of total column ozone data from the Pandora spectrophotometer with Dobson, Brewer, and OMI measurements over Seoul, Korea. *Atmos. Meas. Tech.* **2017**, *10*, 3661–3676. [CrossRef]

46. Flynn, C.M.; Pickering, K.E.; Crawford, J.H.; Lamsal, L.; Krotkov, N.; Herman, J.; Weinheimer, A.; Chen, G.; Liu, X.; Szykman, J. Relationship between column-density and surface mixing ratio: Statistical analysis of O3 and NO_2 data from the July 2011 Maryland DISCOVER-AQ mission. *Atmos. Environ.* **2014**, *92*, 429–441. [CrossRef]

47. Ialongo, I.; Herman, J.; Krotkov, N.; Lamsal, L.; Boersma, K.F.; Hovila, J.; Tamminen, J. Comparison of OMI NO_2 observations and their seasonal and weekly cycles with ground-based measurements in Helsinki. *Atmos. Meas. Techn.* **2016**, *9*, 5203–5212. [CrossRef]

48. Lamsal, L.N.; Krotkov, N.A.; Celarier, E.A.; Swartz, W.H.; Pickering, K.E.; Bucsela, E.J.; Gleason, J.F.; Martin, R.V.; Philip, S.; Irie, H. Evaluation of OMI operational standard NO_2 column retrievals using in situ and surface-based NO_2 observations. *Atmos. Chem. Phys.* **2014**, *14*, 11587–11609. [CrossRef]

49. Lamsal, L.N.; Janz, S.J.; Krotkov, N.A.; Pickering, K.E.; Spurr, R.J.D.; Kowalewski, M.G.; Loughner, C.P.; Crawford, J.H.; Swartz, W.H.; Herman, J.R. High-resolution NO_2 observations from the Airborne Compact Atmospheric Mapper: Retrieval and validation. *J. Geophys. Res. Atmos.* **2017**, *122*, 1953–1970. [CrossRef]

50. Nowlan, C.R.; Liu, X.; Leitch, J.W.; Chance, K.; Gonzalez Abad, G.; Liu, C.; Zoogman, P.; Cole, J.; Delker, T.; Good, W.; et al. Nitrogen dioxide observations from the Geostationary Trace gas and Aerosol Sensor Optimization (GeoTASO) airborne instrument: Retrieval algorithm and measurements during DISCOVER-AQ Texas 2013. *Atmos. Meas. Techn.* **2016**, *9*, 2647–2668. [CrossRef]

51. Chong, H.; Lee, H.; Koo, J.-H.; Kim, J.; Jeong, U.; Kim, W.; Kim, S.-W.; Herman, J.R.; Abuhassan, N.K.; Ahn, J.-Y.; et al. Regional Characteristics of NO_2 column Densities from Pandora Observations during the MAPS-Seoul Campaign. *Aerosol Air Qual. Res.* **2018**, *18*, 2207–2219. [CrossRef]

52. Tzortziou, M.; Herman, J.R.; Cede, A.; Loughner, C.P.; Abuhassan, N.; Naik, S. Spatial and temporal variability of ozone and nitrogen dioxide over a major urban estuarine ecosystem. *J. Atmos. Chem.* **2015**, *72*, 287–309. [CrossRef]

53. Remer, L.A.; Tanre, D.; Kaufman, Y.J.; Ichoku, C.; Mattoo, S.; Levy, R.; Chu, D.A.; Holben, B.; Dubovik, O.; Smirnov, A.; et al. Validation of MODIS aerosol retrieval over ocean. *Geophys. Res. Lett.* **2002**, *29*, 1618. [CrossRef]

54. Remer, L.; Kaufman, Y.; Tanre, D.; Mattoo, S.; Chu, D.; Martins, J.V.; Li, R.-R.; Ichoku, C.; Levy, R.C.; Kelidman, R.G.; et al. The MODIS aerosol algorithm, products, and validation. *J. Atmos. Sci.* **2005**, *62*, 947–973. [CrossRef]

55. Tanre, D.; Kaufman, Y.J.; Herman, M.; Mattoo, S. Remote sensing of aerosol properties over oceans using the MODIS/EOS spectral radiances. *J. Geophys. Res.* **1997**, *102*, 16971–16988. [CrossRef]

56. Levy, R.C.; Remer, L.A.; Kleidman, R.G.; Mattoo, S.; Ichoku, C.; Kahn, R.; Eck, T.F. Global evaluation of the Collection 5 MODIS dark-target aerosol products over land. *Atmos. Chem. Phys.* **2010**, *10*, 10399–10420. [CrossRef]

57. Hsu, N.C.; Tsay, S.-C.; King, M.D. Deep Blue Retrievals of Asian Aerosol Properties during ACE-Asia. *IEEE Trans. Geosci. Remote Sens.* **2006**, *44*, 3180–3195. [CrossRef]

58. Hsu, N.C.; Jeong, M.-J.; Betternhausen, C.; Sayer, A.M.; Hansell, R.; Seftor, C.S.; Huang, J.; Tsay, S.-C. Enhanced Deep Blue aerosol retrieval algorithm: The second generation. *J. Geophys. Res. Atmos.* **2013**, *118*, 9296–9315. [CrossRef]

59. Steinbrecht, W.; Hassler, B.; Claude, H.; Winkler, P.; Stolarski, R.S. Global distribution of total ozone and lower stratospheric temperature variations. *Atmos. Chem. Phys.* **2003**, *3*, 1421–1438. [CrossRef]

60. Perlwitz, J.; Graf, H.-F. The Statistical Connection between Tropospheric and Stratospheric Circulation of the Northern Hemisphere in Winter. *J. Clim.* **1995**, *8*, 2281–2295. [CrossRef]

61. Salby, M.L.; Callaghan, P.F. Interannual Changes of the Stratospheric Circulation: Relationship to Ozone and Tropospheric Structure. *J. Clim.* **2002**, *15*, 3673–3685. [CrossRef]

62. Kim, S.-W.; Yoon, S.-C.; Dutton, E.G.; Kim, J.; Wehrli, C.; Holben, B.N. Global Surface-Based Sun Photometer Network for Long-Term Observations of Column Aerosol Optical Properties: Intercomparison of Aerosol Optical Depth. *Aerosol Sci. Technol.* **2008**, *42*, 1–9. [CrossRef]

63. Chun, Y.; Boo, K.-O.; Kim, J.; Park, S.-U.; Lee, M. Synopsis, transport, and physical characteristics of Asian dust in Korea. *J. Geophys. Res.* **2001**, *106*, 18461–18469. [CrossRef]

64. Kim, S.-W.; Choi, I.-J.; Yoon, S.C. A multi-year analysis of clear-sky aerosol optical properties and direct radiative forcing at Gosan, Korea (2001–2008). *Atmos. Res.* **2010**, *95*, 279–287. [CrossRef]

65. Kurosaki, Y.; Mikami, Y. Recent frequent dust events and their relation to surface wind in East Asia. *Geophys. Res. Lett.* **2003**, *30*, 1736. [CrossRef]

66. Cheng, X.; Zhao, T.; Gong, S.; Xu, X.; Han, Y.; Yin, Y.; Tang, L.; He, H.; He, J. Implications of East Asian summer and winter monsoons for interannual aerosol variations over central-eastern China. *Atmos. Environ.* **2016**, *129*, 218–228. [CrossRef]

67. Jeong, J.I.; Park, R.J. Winter monsoon variability and its impact on aerosol concentrations in East Asia. *Environ. Pollut.* **2017**, *221*, 285–292. [CrossRef]

68. Kim, S.-W.; Yoon, S.-C.; Kim, J.; Kim, S.-Y. Seasonal and monthly variationso f columnar aerosol optical properties over east Asia determined from multi-year MODIS, LIDAR, and AERONET Sun/sky radiometer measurements. *Atmos. Environ.* **2007**, *41*, 1634–1651. [CrossRef]

69. Fioletov, V.E.; Bodeker, G.E.; Miller, A.J.; McPeters, R.D.; Stolarski, R. Global and zonal total ozone variations estimated from ground-based and satellite measurements: 1964–2000. *J. Geophys. Res.* **2002**, *107*, 4647. [CrossRef]

70. Balis, D.; Kroon, M.; Koukouli, M.E.; Brinksma, E.J.; Labow, G.; Veefkind, J.P.; McPeters, R.D. Validation of Ozone Monitoring Instrument total ozone column measurements using Brewer and Dobson spectrophotometer ground-based observations. *J. Geophys. Res.* **2007**, *112*, D24S46. [CrossRef]

71. Park, S.S.; Kim, J.; Cho, H.K.; Lee, H.; Lee, Y.; Miyagawa, K. Sudden increase in the total ozone density due to secondary ozone peaks and its effect on total ozone trends over Korea. *Atmos. Environ.* **2012**, *47*, 226–235. [CrossRef]

72. Hwang, S.-H.; Kim, J.; Cho, G.-R. Observation of secondary ozone peaks near the tropopause over the Korean peninsula associated with stratosphere-troposphere exchange. *J. Geophys. Res.* **2007**, *112*, D16305. [CrossRef]

73. Lemoine, R. Secondary maxima in ozone profiles. *Atmos. Chem.Phys.* **2004**, *4*, 1085–1096. [CrossRef]
74. Shah, V.; Jacob, D.J.; Li, K.; Silvern, R.F.; Zhai, S.; Liu, M.; Lin, J.; Zhang, Q. Effect of changing NOx lifetime on thhe seasonality and long-term trends of satellite-observed tropospheric NO$_2$ columns over China. *Atmos. Chem. Phys.* **2020**, *20*, 1483–1495. [CrossRef]

 *remote sensing*

Article

Evaluation of the ERA5 Sea Surface Skin Temperature with Remotely-Sensed Shipborne Marine-Atmospheric Emitted Radiance Interferometer Data

Bingkun Luo * and **Peter J. Minnett**

Rosenstiel School of Marine and Atmospheric Science, University of Miami, 4600 Rickenbacker Causeway, Miami, FL 33149, USA; pminnett@rsmas.miami.edu
* Correspondence: lbk@rsmas.miami.edu

Received: 15 May 2020; Accepted: 8 June 2020; Published: 9 June 2020

Abstract: Sea surface temperature is very important in weather and ocean forecasting, and studying the ocean, atmosphere and climate system. Measuring the sea surface skin temperature (SST_{skin}) with infrared radiometers onboard earth observation satellites and shipboard instruments is a mature subject spanning several decades. Reanalysis model output SST_{skin}, such as from the newly released ERA5, is very widely used and has been applied for monitoring climate change, weather prediction research, and other commercial applications. The ERA5 output SST_{skin} data must be rigorously evaluated to meet the stringent accuracy requirements for climate research. This study aims to estimate the accuracy of the ERA5 SST_{skin} fields and provide an associated error estimate by using measurements from accurate shipboard infrared radiometers: the Marine-Atmosphere Emitted Radiance Interferometers (M-AERIs). Overall, the ERA5 SST_{skin} has high correlation with ship-based radiometric measurements, with an average difference of~0.2 K with a Pearson correlation coefficient (R) of 0.993. Parts of the discrepancies are related to dust aerosols and variability in air-sea temperature differences. The downward radiative flux due to dust aerosols leads to significant SST_{skin} differences for ERA5. The SST_{skin} differences are greater with the large, positive air–sea temperature differences. This study provides suggestions for the applicability of ERA5 SST_{skin} fields in a selection of research applications.

Keywords: ERA5; evaluation; sea surface skin temperature; M-AERI

1. Introduction

Sea-surface temperature (SST) has been declared to be an Essential Climate Variable (ECV; [1]) by the Global Climate Observing System (GCOS). SST data are essential in many areas of research, such as climate change and weather forecasting [2–4].

SST observations are unevenly distributed in terms of space and time. The retrieval of the sea surface skin temperature (SST_{skin}) both by radiometers on earth observation satellites [3,4] and shipboard instruments [5,6] has been developed over many years and is a mature subject. Climate change research usually needs consistent SST data, which may be acquired by long series of measurements. However, weather and ocean forecasting typically require the best estimate data, collected by as many observations as possible within a specific period of time, and available within a short interval after the measurements are taken. Reanalysis datasets usually strike a balance between these two requirements, trying to generate long-term, consistent, high-quality data [7]. Over the past few decades, a number of reanalyses, such as the European Centre for Medium-Range Weather Forecasts (ECMWF) re-analyses, ERA-Interim [8] and ERA5 [9,10]; the National Centers for Environmental Predictions (NCEP)—National

Center for Atmospheric Research Climate Forecast System Reanalysis [11]; the NASA Modern Era Retrospective-Analysis for Research and Applications (MERRA) [12] and MERRA-2 [13,14]; the Japanese global atmospheric reanalysis JRA-55 [15], have drawn a lot of attention. These reanalysis products have created long-term global SST fields, from 1979 to present. This study focuses on evaluating the latest generation of high-resolution SST_{skin} from ERA5.

Several previous researchers have evaluated the performance of ERA5 using observations from field campaigns and meteorological stations. Graham, et al. [16] used radiosondes which have not been assimilated into any reanalyses to validate the ERA5 wind speed, humidity and air temperature data in the Arctic Fram Strait relative to MERRA-2, JRA-55 and ERA-Interim; the newly released ERA5 has a higher correlation with the independent radiosonde data than the other reanalyses, and with less bias. Hirahara, et al. [7] validated the high-resolution SSTs used in ECMWF, specifically, the HadISST [17] and OSTIA [18]; their optimal usage for ERA5 and performance is well described: these two products are in good agreement in the global SST fields: the spread of the global mean SST is about 0.02K, but with locally larger biases in eddy-active regions. Nogueira [19] presented a comprehensive inter-comparison of the rainfall over the last 40 years between the Global Precipitation Climatology Project (GPCP) and ERA5 reanalysis; the convective rainfall and moisture convergence patterns are better represented in ERA5 than ERA-Interim. The significant rainfall underestimation over the mid-latitude oceans in ERA-Interim has been significantly improved in ERA5. Mahto and Mishra [20] evaluated ERA5 hydrologic application data such as precipitation, runoff, soil moisture and surface temperatures against the observations from India Meteorological Department, revealing that ERA5 products perform better than other reanalysis data.

The performance of ERA5 SST_{skin} has not been evaluated. A key limitation is the paucity of surface-based SST_{skin}-related field campaigns or stations. In general, a popular SST validation source is the drifting buoy array, with thermometers mounted 10–20 cm below the sea surface, but the temperature differences between that depth and the surface [4,21,22] may introduce errors in the validation.

Independent SST_{skin} derived from the Marine-Atmosphere Emitted Radiance Interferometers (M-AERI; [6]) are used in this study to perform an assessment of ERA5 SST_{skin} and evaluate the potential inaccuracies associated with dust aerosols and sensitivity to air–sea temperature differences. Data from a series of NOAA Aerosols and Ocean Science Expeditions (AEROSE; [23]) and Royal Caribbean International (RCI) cruises are used in this study. In addition, in many research cruises where radiometric SST_{skin} were made, atmospheric temperature and humidity profiles were also measured. The datasets have not been submitted to any assimilation schemes, so the M-AERI data used here are independent of the ERA5 fields.

We organize this paper as follows: The M-AERI-retrieved SST_{skin} data, ERA5-derived SST_{skin} data, and other MERRA-2 inputs are introduced in Section 2. Details of the cruises are also introduced in Section 2. In Section 3, we present the overall statistics of the comparisons. The results of the error analysis are discussed in Section 4 with day/night differences, air–sea temperature difference effects, and dust aerosol effects.

2. Materials and Methods

2.1. ERA5 SST_{skin} Data

The ERA5 reanalysis model output was generated using the four-dimensional variational (4D-VAR) analysis systems [9,10]. The ERA5 is the improved version of ERA-Interim [8], and is available from the ECMWF archive (https://cds.climate.copernicus.eu/#!/search?text=ERA5&type=dataset). ERA5 is available on a regular latitude-longitude grid at a spatial resolution of 31 km (0.25° × 0.25°) [9,10].

The ERA5 SST_{skin} product is based on a model simulation with data from satellite-derived SSTs. The temperature of the depth where there is no diurnal signal is the foundation temperature; the foundation temperature for ERA5 is taken from the Operational Sea Surface Temperature and Sea Ice Analysis (OSTIA) analysis [18], which is a blended product from various satellite-retrieved SST and in situ data. As to near-surface effects, ocean temperature variability is represented by three physical processes: the thermal skin cool layer during both day and night, the diurnal heating warm layer during the day, and the salinity saturation effect near the surface [10].

The cool skin effect originates from the heat loss to the atmosphere, the temperature difference between the skin layer (T_{skin}) and at the foundation depth (T_{fnd}) can be expressed as [24,25]:

$$T_{skin} - T_{fnd} = \frac{\delta}{\rho_w c_w k_w}(Q + R_s f_s) \tag{1}$$

where R_s is the net solar radiation at the surface, f_s is the fraction of the surface-absorbed solar radiation, ρ_w is the water density, c_w is the volumetric heat capacity, k_w is the molecular thermal conductivity of water and δ is the skin layer thickness. Q is the net heat flux in this cool layer:

$$Q = H + E + LW \tag{2}$$

where H, E, and LW denote the surface sensible heat flux, latent heat flux and net long wave radiation at the surface, respectively.

The f_s can be given as:

$$f_s = 0.065 + 11\delta - \frac{6.6 \times 10^{-5}}{\delta}(1 - e^{-\delta/0.0008}) \tag{3}$$

The diurnal warming layer [25,26] is due to the solar absorption during the daytime; the diurnal warming effect may be affected by surface wind, by cloud amount and type, by free convection, or by internal waves [27]. The ERA5 diurnal warming calculations are based on Takaya, et al. [28] and can be expressed as:

$$\frac{\partial\left(T_{-\delta} - T_{fnd}\right)}{\partial t} = \frac{Q + R_s - R(-d)}{d\rho_w c_w v/(v+1)} - \frac{(v+1)ku_{*w}}{d\phi_t(d/L)}\left(T_{-\delta} - T_{fnd}\right) \tag{4}$$

where $T_{-\delta}$ is the temperature below the cool skin layer, d is the depth of the diurnal warm layer, which is set as 3 m, v is the profile shape and it is set as 0.3, u_{*w} is the water friction velocity, $\phi_t(d/L)$ is the stability function and L is the Obukhov length; $R(-d)$ is the solar radiation absorbed at depth -d, which is

$$R(-d) = R_s \times 0.28e^{-71.5d} + R_s \times 0.27e^{-2.8d} + R_s \times 0.45e^{-0.06d}\phi_t(d/L) \tag{5}$$

The nondimensional shear stability function, $\phi_t(d/L)$, is

$$\phi_h(\zeta) = \begin{cases} 1 + 5\dfrac{-z}{L}, & \dfrac{-z}{L} > 0 \\[2mm] \left(1 - 16\dfrac{-z}{L}\right)^{-\frac{1}{2}}, & \dfrac{-z}{L} < 0 \end{cases} \tag{6}$$

The Obukhov length L is

$$L = \rho_w c_w u_{*w}^3/(kF_d) \tag{7}$$

Equation (4) has been integrated in time to derive the warm layer effect; during daytime, the warm layer effect $\left(T_{-\delta} - T_{fnd}\right)$ from Equation (4) and the cool layer effect $(T_{skin} - T_{fnd})$ from Equation (1) have been added together to derive T_{skin}.

Different reanalysis schemes use different choices of these parameter settings: for example, according to Akella, et al. [29] and Gentemann and Akella [21], the NASA MERRA-2 temperature profile uses 2 m and 0.2 for the diurnal warm layer depth, d, and the diurnal profile shape, v, respectively, but ERA5 uses 3 m and 0.3 [25]. It is essential to evaluate the newly updated ERA5 data.

2.2. M-AERI Data

Self-calibrating, ship-based radiometers provide SST_{skin} that is more directly comparable to the ERA5 SST_{skin} than the temperatures at the depth of the drifting buoy measurements. This study utilizes the M-AERI [6,30], a ship-based spectro-radiometer mounted a few meters above the sea surface on the ships, as shown in Figure 1, to validate the ERA5 SST_{skin}.

Figure 1. Installations of Marine-Atmosphere Emitted Radiance Interferometers (M-AERI)s on cruise ships (**a,b**). The instruments are inside hermetically sealed aluminum enclosures, with the fore optics on the aft, sheltered sides of the enclosures. The smaller boxes contain air-conditioning units to limit temperature and humidity variations in the instrument enclosures. (**c**): An M-AERI installed on the bridge wing of the R/V Alliance. (**d**): The M-AERI is calibrated in the laboratory before and after each deployment using an external validation procedure.

The internal calibration of the M-AERIs is checked in the laboratory using an SI-traceable water-bath blackbody calibration target [31–33] (Figure 1). The M-AERI viewing geometry is shown in Figure 2; each M-AERI contains two internal blackbodies, one at ambient temperature and the other heated, that provide a two-point calibration before and after each measurement of the sea-surface and sky infrared emissions. The sky emission measurement is used for correcting the sky radiance. After the interferometer sequentially measures the sea and sky emissions over a specified time interval, the scan mirror rotates to the apertures of the two blackbody cavities to provide a real-time two-point calibration of the measured emission spectra.

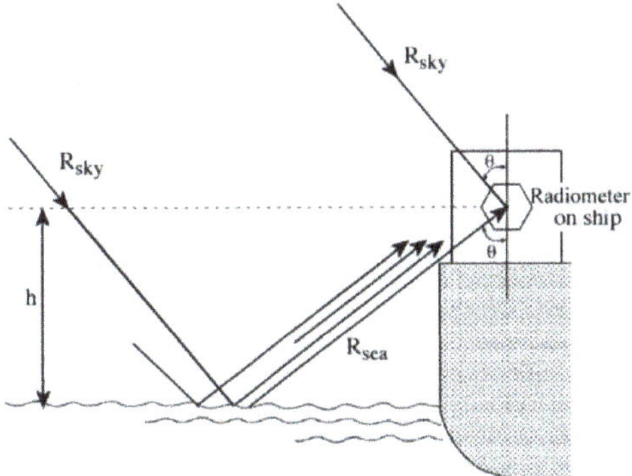

Figure 2. M-AERI view geometry (from [30]). R_{sky}, and R_{sea} are the spectral infrared radiances measured in the direction of sky and sea surface. The R_{sky} provides a correction for the sky radiation that is reflected at the sea surface.

The SST_{skin} derived from M-AERI instruments can be expressed as:

$$SST_{skin} = B^{-1}\left(\frac{R_{water}(\lambda,\theta) - (1 - \varepsilon(\lambda,\theta))R_{sky}(\lambda,\theta) - R_h(\lambda,\theta)}{\varepsilon(\lambda,\theta)}\right) \tag{8}$$

where B is the Planck function; R_{water}, R_{sky}, and R_h are the spectral radiance measured in the direction of the sea surface, emitted by the atmosphere above the instrument, and below the instrument (both directly into the measured beam and reflected at the sea surface). λ is the wavelength of the radiance, θ is the angle from vertical of the measurement, and ε is the surface emissivity at λ and θ. The detailed technical description, including the atmospheric correction, is given by Minnett, et al. [6]. The SST_{skin} derived from the M-AERI spectra has an uncertainty ~ 0.04 K. M-AERI deployments are monitored from the laboratory via a satellite Internet link.

M-AERI spectral measurements are also used to derive a near-surface air temperature [34]. Thus, the M-AERI spectral measurements can provide better air temperature than by conventional contact thermometers. Accurate M-AERI-derived air temperature have also been used in this study to characterize the conditions in the lower atmosphere in the comparisons with ERA5 data.

Until the recent suspension of cruises in response to the Covid-19 pandemic, there were four M-AERIs operational—three on ships of Royal Caribbean International (RCI): Celebrity Equinox (May 2014–March 2020), Allure of the Seas (June 2014–March 2020), and Adventure of the Seas (January, 2018–March 2020). The fourth is usually deployed on research vessels, such as on the NOAA ship Ronald H Brown (RHB) for a circumnavigation from March to October 2018. Figure 3 shows the tracks of deployments on several research vessels that have provided matchups in a wide range of environmental conditions. The RCI ships have provided a rich source of measurements in the western North Atlantic Ocean, Caribbean Sea, and the Mediterranean Sea.

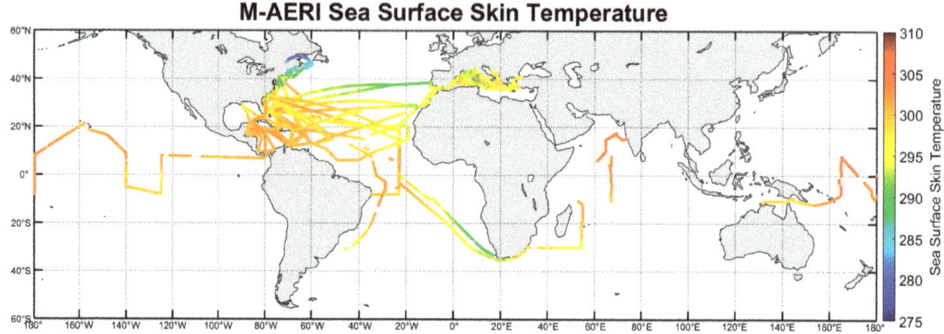

Figure 3. Tracks of ships with M-AERIs installed that provided data for this study. Most of the cruise ship tracks are repeated many times. The colors indicate the M-AERI SST_{skin}.

Data from nine campaigns from 2004 to 2019 were taken during the AEROSE project [23] on the NOAA Ship Ronald H. Brown and the R/V Alliance. AEROSE comprises Atlantic field campaigns to conduct in situ measurements of the effects of Saharan dust aerosol on the tropical and subtropical Atlantic Ocean. The dust effects on satellite-derived SST_{skin} have been quantified using AEROSE data [35,36]. The SST_{skin} provided by AEROSE is valuable to validate ERA5 SST_{skin} data under the dust-polluted air layers.

Table 1. Summarizes the times and regions of M-AERIs deployed on RCI ships; Table 2 summarizes the same information, but for AEROSE cruises.

Table 1. Details of the Royal Caribbean International (RCI) cruises used in this study.

CRUISES	AREA	START	END	DAYS OF DATA
2014 ALLURE	Caribbean Sea	2014-08-24	2014-12-31	130
2014 EQUINOX	Caribbean Sea	2014-11-16	2014-12-31	46
2015 ALLURE	Caribbean Sea, North Atlantic Ocean, and Mediterranean Sea	2015-01-01	2015-12-26	360
2016 EQUINOX	Caribbean Sea, North Atlantic Ocean, and Mediterranean Sea	2016-01-02	2016-12-31	365
2017 EQUINOX	Caribbean Sea	2017-01-01	2017-12-31	365
2017 ALLURE	Caribbean Sea	2017-10-02	2017-11-26	56
2018 EQUINOX	Caribbean Sea	2018-01-11	2018-09-23	255
2018 ADVENTURE	Caribbean Sea and US East Coast	2018-02-12	2018-12-31	322
2018 ALLURE	Caribbean Sea	2018-02-18	2018-10-14	238
2019 ADVENTURE	Caribbean Sea and US East Coast	2019-01-01	2019-10-30	302
TOTAL	–	**2014-08-24**	**2019-10-30**	**2439**

Table 2. Details of the AEROSE and other cruises used in this study.

CRUISES	AREA	START	END	DAYS OF DATA
2004 RHB		2004-02-13	2004-04-13	61
2006 RHB		2006-05-27	2006-07-14	49
2007 RHB		2007-05-07	2007-05-28	22
2008 RHB	North Atlantic Ocean, South Atlantic, Indian and Pacific Oceans	2008-04-29	2008-05-19	21
2011 RHB		2011-07-21	2011-08-20	31
2013 RHB		2013-11-11	2013-12-08	28
2015 ALLIANCE		2015-11-17	2015-12-14	28
2018 RHB		2018-03-07	2018-10-23	231
2019 RHB		2019-02-24	2019-03-29	34
TOTAL	–	**2004-02-13**	**2019-03-29**	**505**

2.3. MERRA-2

Dust effects on satellite derived SST_{skin} have been discussed by Luo, et al. [35]; high concentrations of dust aerosol are also a problem for reanalyses [37], and dust appears to degrade the quality of MERRA-2 SST_{skin} [37]. MERRA-2 aerosol dust fields are used to quantify the effect of Sahara aerosol dusts on the ERA5-derived SST_{skin}.

NASA's Goddard Earth Sciences MERRA-2 dataset provides atmospheric and surface fields [13,14], some of which are useful for this study. The data were downloaded from http://disc.sci.gsfc.nasa.gov/mdisc/. The MERRA-2 aerosol analysis system [14,38] provides the assimilated aerosol-related radiation output and dust scattering aerosol optical thickness (AOT) for this study.

The MERRA-2 AOT profile is taken from the variable labelled tavg1_2d_aer_Nx, which is a 1-hourly time-averaged aerosol diagnostic product. The surface net downward longwave flux due to aerosols is taken from the variable tavg1_2d_rad_Nx, which is a 1-hourly time-averaged radiation product and contains the surface-absorbed shortwave and longwave radiation, top of atmosphere incoming shortwave flux, cloud fraction, surface albedo, etc. The surface net downward longwave flux due to aerosol used is calculated as:

$$\text{LW\_aer\_rad} = \text{LW}\downarrow_{\text{with\_aerosol}} - \text{LW}\downarrow_{\text{clear}} \tag{9}$$

where $\text{LW}\downarrow_{\text{with\_aerosol}}$ is the MERRA-2 LWGNTCLR product, meaning surface net downward longwave flux assuming clear sky (cloud-free), and $\text{LW}\downarrow_{\text{clear}}$ is the MERRA-2 LWGNTCLRCLN product, meaning surface net downward longwave flux assuming clear sky and no aerosol.

The MERRA-2 dataset has a spatial resolution of 0.625° (longitude) and 0.5° (latitude), being different from ERA5 which has 0.25° × 0.25° resolution. Therefore, MERRA-2 aerosol and radiation data are bi-linearly interpolated to the ERA5 positions in this study.

3. Results

In a skin-to-skin temperature comparison, SST_{skin} values from ERA5 are directly compared with M-AERI SST_{skin}. The comparison of ERA5 SST_{skin} with M-AERI SST_{skin} values can be made by populating a matchup data base (MUDB). Each MUDB record includes the ERA5 SST_{skin} corresponding to a set of times and locations of a M-AERI measurement. The data vector also contains the M-AERI near-surface air temperature, MERRA-2 AOT, MERRA-2 radiation profile and other instrumental variables. The ERA5 SST_{skin} were temporally and spatially bi-linearly interpolated to the ship positions and times. Moreover, because RCI cruises are often near coasts and ERA5 has a horizontal resolution of 31 km, we calculate the distance to the land of each ship-board measurement and apply a filter to exclude the matchup points which are less than 32 km to land. In addition, some oceanic features, such as upwelling and freshwater input, are stronger near coasts; the corresponding SST_{skin} variations within 31 km cannot be determined from ERA5 data. For these reasons, the filter has been used to avoid significant errors due to the ERA5 spatial resolution.

3.1. Statistics of SST_{skin} Comparisons

The scatter plot in Figure 4 shows that there are a few matchups with significant bias, but that there is good quantitative agreement between ERA5 and M-AERI data. The histogram of the differences of ERA5 SST_{skin} minus M-AERI SST_{skin} are shown in Figure 4 (right) with a well-defined histogram peak; most of the differences fall into the range of −1 K to 1 K.

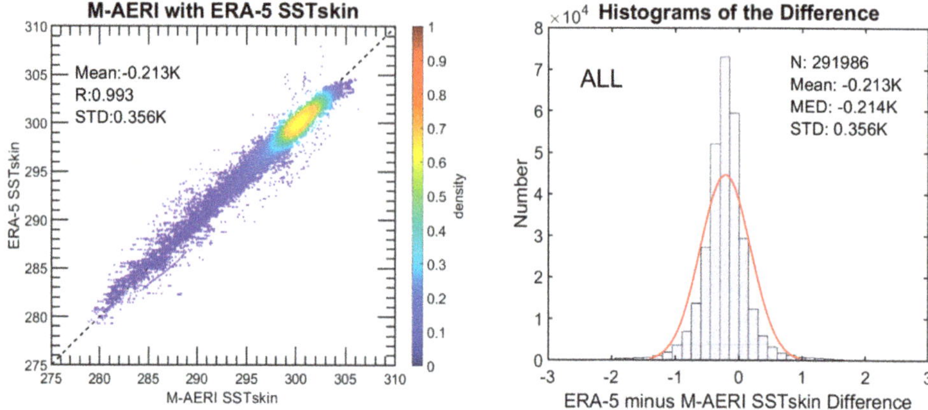

Figure 4. Left: Scatter plot of M-AERI SST$_{skin}$ with ERA5 SST$_{skin}$. Right: Histogram of the SST$_{skin}$ difference. All of the units are K.

Table 3 shows the statistics of the ERA5 SST$_{skin}$ minus M-AERI SST$_{skin}$ differences during AEROSE cruises and Table 4 shows the same statistics for the RCI cruises. The mean differences are −0.190 K for AEROSE cruises and −0.220 K for RCI cruises. The overall standard deviations (STD) are 0.348 K and 0.358 K. Robust standard deviations (RSD) are less sensitive to outliers and are a better representation of the ERA5 SST$_{skin}$ algorithm performance [39]. The robust statistics of the difference are the best assessment of the ERA5 SST$_{skin}$ performances, which are between 0.239 K and 0.247 K, similar for both cruises and smaller than the STD. Table 5 summarizes the statistics of the SST$_{skin}$ differences for all of the cruises, comprising a total of 291,986 match-up pairs. ERA5 SST$_{skin}$ values are generally in good agreement with the corresponding M-AERI data, with a median difference of -0.214 K and an RSD of 0.356 K.

Table 3. Statistics of ERA5 SST$_{skin}$ minus M-AERI SST$_{skin}$ for each AEROSE cruise. The unit is K.

CRUISES	N*	MEAN	MED	STD	RMS	RSD	R	E
2004 RHB	5805	−0.212	−0.165	0.460	0.507	0.342	0.979	0.949
2006 RHB	3908	−0.152	−0.124	0.383	0.413	0.357	0.976	0.944
2007 RHB	1257	0.024	−0.029	0.441	0.442	0.415	0.971	0.942
2008 RHB	1592	0.020	−0.012	0.482	0.483	0.366	0.968	0.935
2011 RHB	2264	−0.038	−0.005	0.327	0.329	0.308	0.996	0.993
2013 RHB	7099	−0.201	−0.193	0.230	0.305	0.180	0.981	0.927
2015 ALLIANCE	5547	−0.299	−0.318	0.242	0.385	0.228	0.991	0.952
2018 RHB	38,108	−0.167	−0.148	0.282	0.328	0.206	0.994	0.984
2019 RHB	8378	−0.329	−0.299	0.502	0.601	0.380	0.963	0.895
TOTAL	**73,958**	**−0.190**	**−0.170**	**0.348**	**0.396**	**0.247**	**0.991**	**0.978**

Note: N* means number of valid match-up points. Med: median; STD: standard deviation; RMS: root mean square; RSD: robust standard deviation. R: Pearson correlation coefficient. E: Nash–Sutcliffe efficiency coefficient.

Table 4. Statistics of ERA5 SST$_{skin}$ minus M-AERI SST$_{skin}$ for each RCI cruise. The unit is K.

CRUISES	N*	MEAN	MED	STD	RMS	RSD	R	E
2014 ALLURE	9811	−0.196	−0.199	0.262	0.327	0.233	0.972	0.914
2014 EQUINOX	5421	−0.293	−0.288	0.247	0.383	0.219	0.953	0.780
2015 ALLURE	34,658	−0.208	−0.231	0.367	0.422	0.265	0.991	0.975
2016 EQUINOX	28,673	−0.188	−0.205	0.371	0.416	0.272	0.995	0.987
2017 EQUINOX	41,945	−0.244	−0.238	0.270	0.364	0.211	0.983	0.938
2017 ALLURE	5031	−0.145	−0.133	0.218	0.262	0.206	0.959	0.884
2018 EQUINOX	29,779	−0.266	−0.240	0.291	0.395	0.213	0.981	0.928
2018 ADVENTURE	7266	−0.170	−0.182	0.480	0.509	0.213	0.992	0.977
2018 ALLURE	27,215	−0.257	−0.252	0.274	0.376	0.238	0.982	0.933
2019 ADVENTURE	28,229	−0.169	−0.218	0.548	0.574	0.272	0.994	0.986
TOTAL	**218,028**	**−0.220**	**−0.228**	**0.358**	**0.420**	**0.239**	**0.993**	**0.981**

Note: N* means number of valid match-up points. Med: median; STD: standard deviation; RMS: root mean square; RSD: robust standard deviation. R: Pearson correlation coefficient. E: Nash–Sutcliffe efficiency coefficient.

Table 5. Statistics of ERA5 SST$_{skin}$ minus M-AERI SST$_{skin}$. The unit is K.

CRUISES	N*	MEAN	MED	STD	RMS	RSD	R	E
AEROSE	73,958	−0.190	−0.170	0.348	0.396	0.247	0.991	0.978
RCI	218,028	−0.220	−0.228	0.358	0.420	0.239	0.993	0.981
TOTAL	**291,986**	**−0.213**	**−0.214**	**0.356**	**0.415**	**0.243**	**0.993**	**0.980**

3.2. SST$_{skin}$ Bias Distribution

Figure 5 shows the SST$_{skin}$ differences (ERA5 minus M-AERI) distribution. The map shows the locations from the matchup database. Although the figure does not include the matchup points within 32 km of the coast, the differences are still sometimes greater towards coasts, such as in the Mediterranean Sea and Northwest Atlantic Ocean.

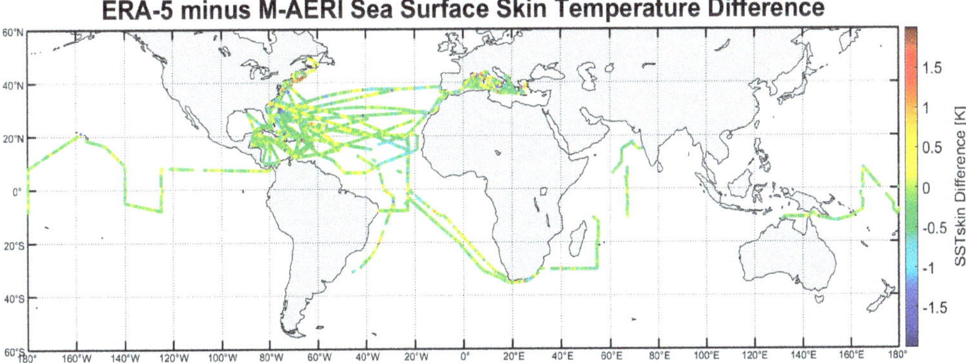

Figure 5. ERA5 SST$_{skin}$ minus M-AERI SST$_{skin}$ along the ship tracks. The operations of M-AERIs are suspended during rain or when sea spray reaches the instrument, thus causing some gaps. Other gaps are the result of instrument failure. Comparisons in and close to ports and coasts are not used in the analyses presented here.

The map is representative of the whole data set. A cool skin effect is present all of the time, and the diurnal heating is present during the daytime when wind speeds are low. To compare the performance of ERA5 SST$_{skin}$ derivation algorithms during the daytime and nighttime, the SST$_{skin}$ difference has been separated as 7 AM–5 PM as daytime, and 7 PM–5 AM as nighttime. The histograms of the results are presented in Figure 6. There are 88,955 matchups during the daytime, and 166,849 matchups during the nighttime.

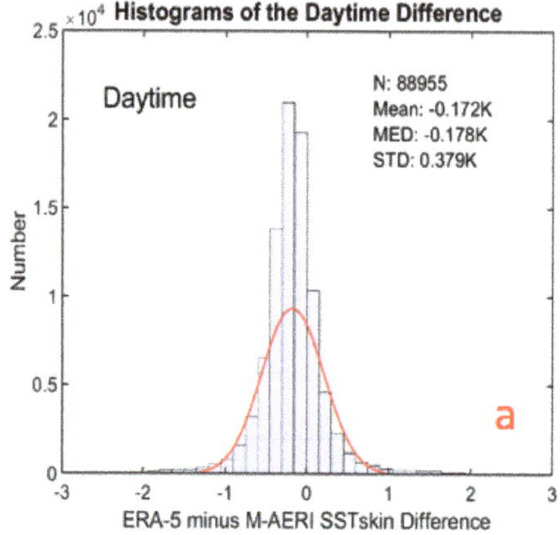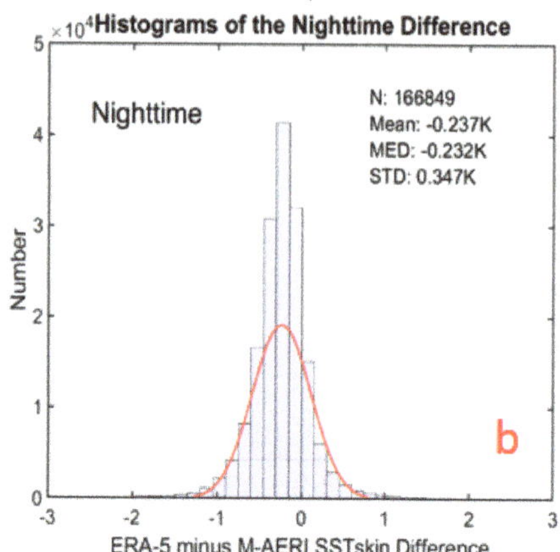

Figure 6. (**a**): Histograms of the daytime SST_{skin} differences between ERA5 and M-AERI for all cruises. (**b**): Corresponding nighttime SST_{skin} differences.

The comparison, based on 88,955 daytime matchup pairs, showed that ERA5 had an average SST_{skin} difference of −0.172 K; the nighttime had an average SST_{skin} difference of −0.237 K, with an average STD of 0.347 K. A statistical two-sample t-test rejects the null hypothesis and the means between day and night should therefore be considered as dissimilar. The effects of diurnal heating in the upper ocean is expected to be small during the nighttime and the SST_{skin} variation should be less than during the daytime. However, the nighttime SST_{skin} had larger discrepancies with the M-AERI than the daytime by an average of 0.065 K. One possible reason for the larger nighttime difference may be due to the variations in the air–sea temperature difference, which will be discussed in the Section 4.2.

4. Discussion

This study is intended to provide better knowledge of the characteristics of the errors. Discussion in this section about the accuracy of the ERA5 fields is split into two parts: air–sea temperature differences, and aerosol dust effects.

4.1. Air–Sea Difference Effect

Accurate air temperatures derived from M-AERI spectra [34] are part of the matchup records. Figure 7 shows the M-AERI air temperature minus M-AERI SST_{skin} along the cruise tracks between 60°W and 90°W. Advection of the air over strong SST gradients, such as in the Gulf Stream area, could lead to anomalous air–sea temperature differences, where anomalous means different from the usual open-ocean distribution. To investigate the possible consequence of air–sea temperature differences, we focus an analysis from 0°N to 50°N, and 50°W to 100°W in the Atlantic region. The corresponding ERA5 minus M-AERI SST_{skin} differences are displayed in Figure 8.

The ERA5 minus M-AERI SST_{skin} difference is related to the air temperature minus SST_{skin}. Renfrew, et al. [40] compared the R/V Knorr surface meteorological measurements with ECMWF and NCEP reanalysis over the Labrador Sea during February to March of 1997. Since the sensible heat flux is directly related to the air–sea temperature difference when the air–sea temperature difference is large, the sensible heat flux is high. Smith, et al. [41] also highlighted the shortcomings of the surface heat flux parameterization, finding that the latent heat fluxes contain significant systematic errors dependent on dry stability (SST minus air temperature).

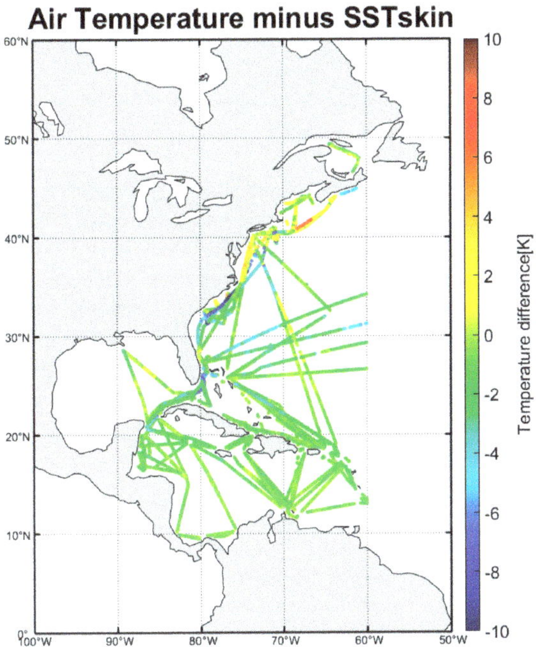

Figure 7. M-AERI-derived air temperature minus M-AERI SST$_{skin}$ between 60°W and 90°W in the Atlantic area. The color indicates the difference according to the scale at the right.

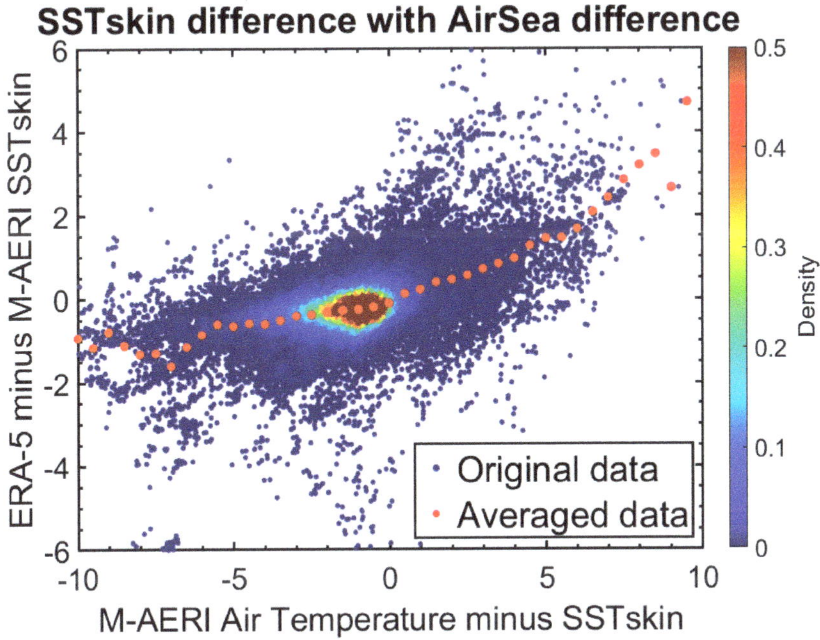

Figure 8. ERA5–M-AERI SST$_{skin}$ differences with air–sea temperature difference for the cruise data in Figure 7. The SST$_{skin}$ differences are large with large air–sea temperature differences. Temperature differences are in K. Averaging is over 0.5 K bins.

Figure 8, using the data shown in Figure 6, compares the ERA5–M-AERI SST_{skin} differences during the daytime and the nighttime. The air temperature is usually warmer during the daytime, and, for the daytime SST_{skin} difference statistics shown in the histograms of Figure 6a, it is less negative than nighttime. According to Equations (1) and (2), the cool skin effect is strongly dependent on the heat flux parameterizations employed in the ERA5 SST_{skin} scheme.

4.2. Dust Aerosol Effects

The Saharan Air Layer and the associated dust outflow can flow over the Atlantic Ocean [42]. The radiative impact of mineral dust is one of the major contributors to the satellite-retrieved SST_{skin} inaccuracies in this region [35]. The Saharan dust layer has also been a problem for the reanalysis of SST_{skin} fields [37] and the numerical weather prediction [43]. The dust aerosols, transported across the Atlantic Ocean within the Saharan Air Layer, contribute to formation of shallow stratocumulus clouds under the base of the Saharan Air Layer [44,45]; satellite measurements frequently showed dust within the SAL layer between 1 km and 5 km altitude, and the presence of narrow stratocumulus clouds below the dust layer [46].

The SST_{skin} data collected during the cruises provide an opportunity to investigate the accuracies of the ERA5 SST_{skin} values near the regions susceptible to strong Saharan dust outbreaks in the tropical and subtropical Atlantic Ocean. Figure 9 shows the ERA5–M-AERI SST_{skin} differences along cruise tracks from 2004 to 2019, indicating that there are strong negative SST_{skin} biases near the Saharan dust region. Plots of the corresponding MERRA-2 AOT data are given in Figure 10. ERA5–M-AERI SST_{skin} differences increase with strong aerosol dust outflow.

The cloud influence on errors in ERA5 downwelling longwave radiation at the surface has been discussed by Silber, et al. [47]; however, the dust aerosol influence on the surface downwelling longwave radiation has not been studied. Numerical weather prediction models are usually under the effects of the longwave radiation and other model errors related to aerosol indirect effects [47]. The Saharan dust layer induces a vertical dipole effect [43,48], which warms within the dust layer and introduces a cooling of the surface below. The thermal dipole effect can lead to increased atmospheric stability during the daytime and decreased stability during the nighttime; the diurnal cycle of precipitation and wind speed is affected [49]. The dust layer radiative effect has been included in the NASA MERRA-2 reanalysis product. To derive the surface net downward longwave flux due to aerosols along the cruise tracks, we have matched the MERRA-2 radiation to the times and locations of the M-AERI measurements, then computed the surface net downward longwave flux due to aerosol according to Equation (9). Figure 11 shows the aerosol downwelling longwave radiation at the sea surface. Figure 12 shows the M-AERI and ERA5 SST_{skin} scatterplot with surface net downward longwave flux due to dust aerosols, and Figure 13 gives the relation with ERA5 SST_{skin} bias. It can be seen that the intense downward longwave flux leads to substantially significant SST_{skin} differences for ERA5; the averaged SST_{skin} difference can be as large as 1 K when the aerosol radiative flux is above 10 Wm^{-2}.

The atmospheric thermal structure change due to aerosol radiative effect will introduce changes in reanalysis models. Interactive-aerosol, which is a feature implemented in NASA GEOS-5 Global Forecasting System, was studied by Reale, et al. [48]; the consideration of the interactive aerosols radiative effects can increase the accuracy of the African easterly jet representation. Similarly, the ERA5 SST_{skin} scheme's improvements in accuracy would be expected if these aerosol effects were taken into account.

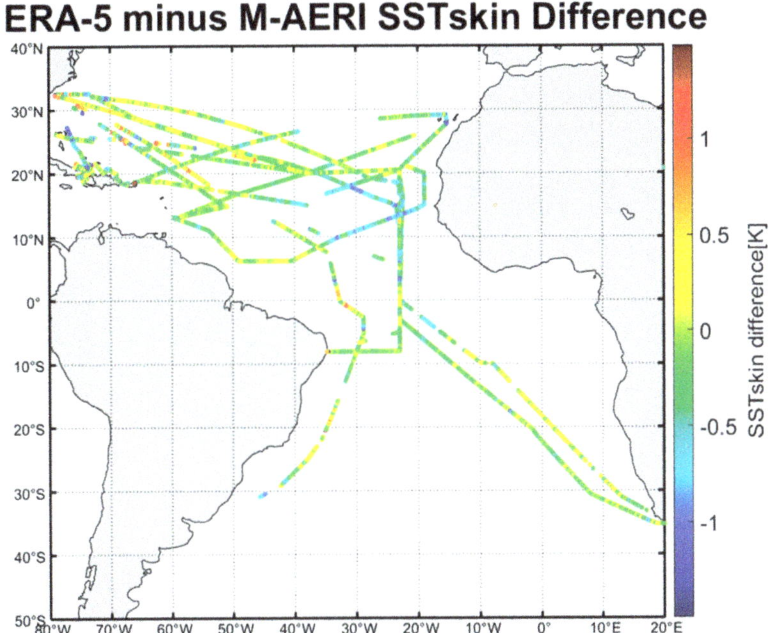

Figure 9. ERA5–M-AERI SST$_{skin}$ differences along the AEROSE cruise tracks and Adventure of the Seas 2018 and 2019 cruises. The color indicates the difference in K according to the color scale at the right.

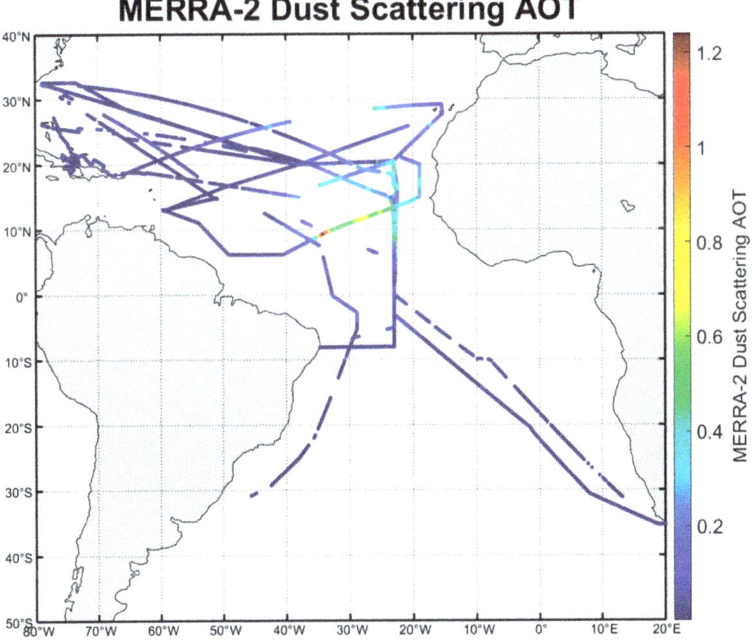

Figure 10. As Figure 9, but for MERRA-2 dust scattering aerosol optical thickness (AOT).

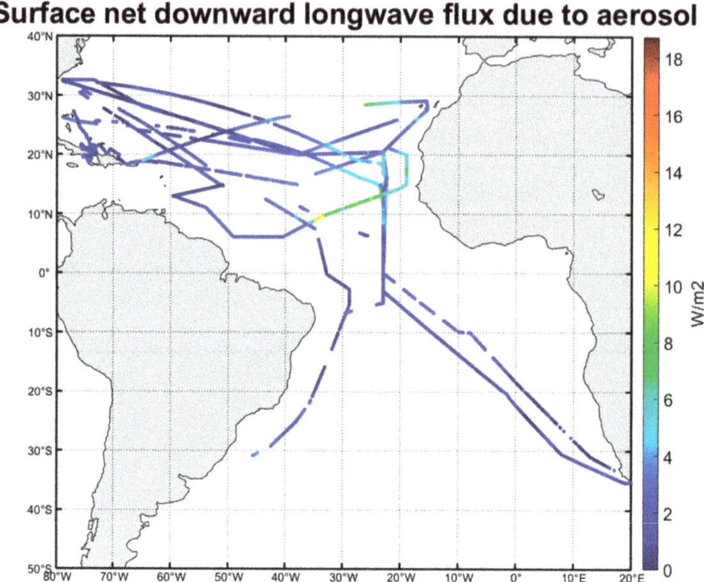

Figure 11. As Figure 9, but for surface net downward longwave flux due to aerosols. The color indicates the longwave flux in W/m^2 according to the color scale at the right.

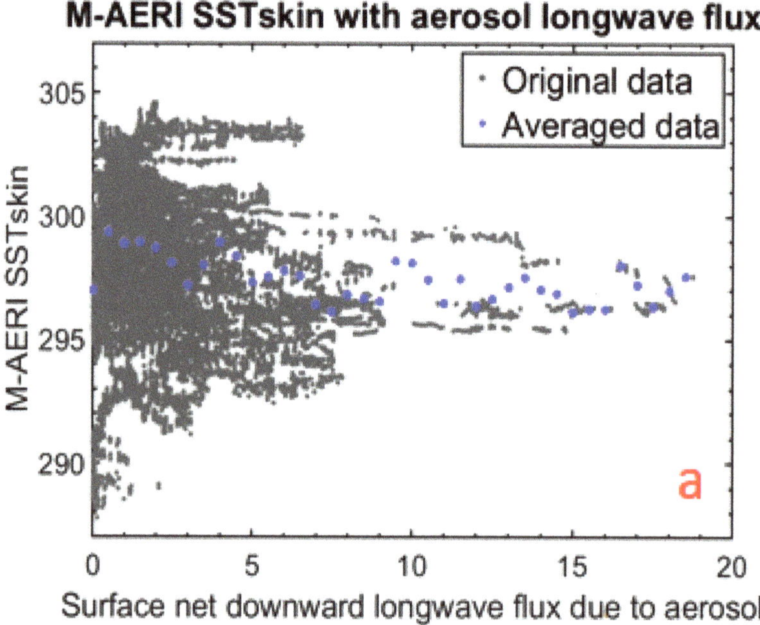

Figure 12. *Cont.*

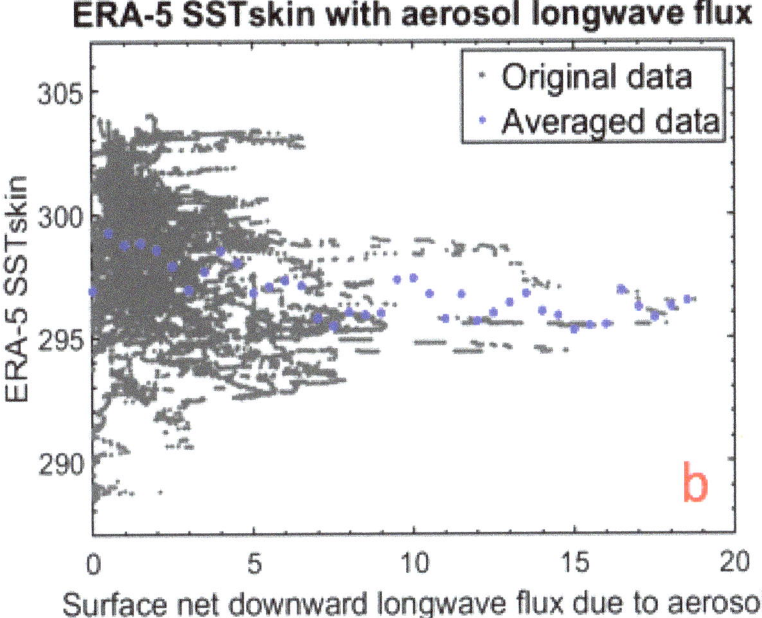

Figure 12. M-AERI SST_{skin} (**a**) and ERA5 SST_{skin} (**b**) with surface net downward longwave flux due to aerosols along cruise tracks, the averaged SST_{skin} differences are shown in blue dots.

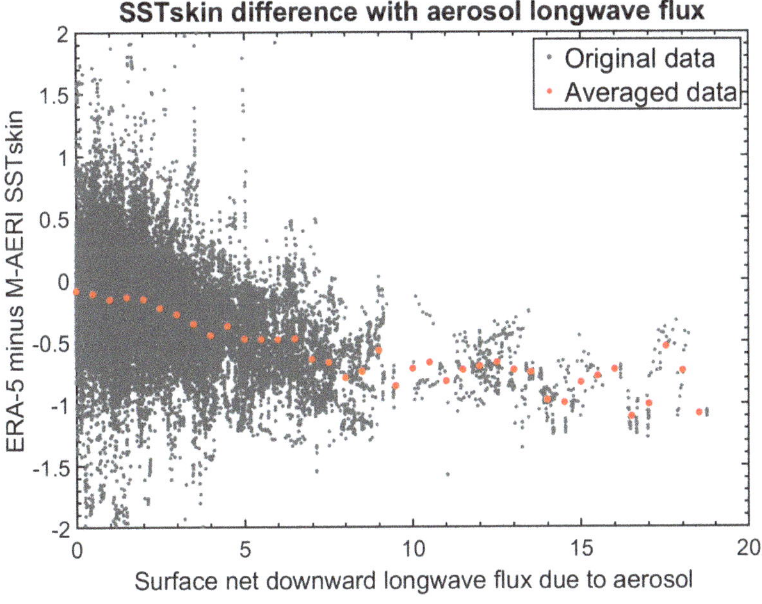

Figure 13. ERA5–M-AERI SST_{skin} differences with surface net downward longwave flux due to aerosols along cruise tracks, the averaged SST_{skin} differences (red dots) are large above 5 W/m².

5. Conclusions

SST is an important parameter in the global climate system. In recent years, it has become increasingly apparent that those involved in the fields of climate change studies and weather prediction require highly accurate estimates of the errors and uncertainties of the reanalysis data. By assessing the accuracy of the ERA5-derived SST_{skin}, this study was aimed at improving the understanding of the strengths and weaknesses of ERA5 data. The use of high-accuracy shipboard radiometers with calibration traceability to SI-standards permitted the determination of the accuracies of ERA5 SST_{skin}.

The independent SST_{skin} observations from research vessels and RCI cruise ships provide a valuable way to validate ERA5 SST_{skin} values, including in areas influenced by Saharan dust aerosol. This study developed a matchup technique by using a subset of ERA5 data that coincide with the shipboard M-AERI measurements deployed for the validation of satellite-derived SST_{skin} [50,51]. The statistics in this study are considered as skin-to-skin temperature comparisons, which avoid the subsurface temperature variability inherent in comparisons with in situ sea temperature measurements. The results indicate good performance of the ERA5 SST_{skin} algorithm, with an average bias of −0.213 K, RSD of 0.243 K and STD of 0.356 K. The accuracy of the ERA5 SST_{skin} during the daytime is generally better than during the nighttime. The overall Pearson correlation coefficient (R) is 0.993 and the Nash–Sutcliffe efficiency coefficient (E) is 0.980; ERA5 and M-AERI have a very strong correlation with each other. The contributions of the atmospheric temperature effects should be paid attention to, as the ERA5 SST_{skin} bias appears to be straightforwardly related to the air–sea temperature differences. The ERA5 SST_{skin} difference with respect to the M-AERI measurements in the Saharan dust outflow regions, with aerosol distributions taken from the MERRA-2 AOT, indicates that the SST_{skin} derived by ERA5 is affected by the downward aerosol longwave flux. The averaged difference can be as large as 1 K when the aerosol downward longwave flux is above 10 W/m^2.

However, more work is needed to evaluate the ERA5 SST_{skin} dependence on other factors, such as wind speed, water vapor, smoke, sea salt aerosol, and clouds. It is difficult to draw any firm conclusions concerning the accuracy of ERA5 SST_{skin} at the global level, due to the quite limited geographical area in this research. We anticipate that further comparison studies will be extended to wider geographic areas in the future. Moreover, further research will include the important dust effect on SST.

Author Contributions: Conceptualization, methodology, software, validation, data curation, writing—original draft preparation: B.L.; writing—review and editing, visualization, supervision, project administration, funding acquisition: P.J.M. All authors have read and agreed to the published version of the manuscript.

Funding: This research was funded by RSMAS Mary Roche endowed Fellowship, NASA Physical Oceanography program (Grant # NNX14AK18G), and Future Investigators in NASA Earth and Space Science and Technology (FINESST) Program (Grant # 80NSSC19K1326), NASA Earth Science Senior Review 2017 (Grant # 80NSSC18K0534).

Acknowledgments: This work has benefited from discussions with colleagues at RSMAS. The at-sea support of the Officers, crew, and colleagues of the NOAA Ship Ronald H Brown and R/V Alliance is appreciated. Royal Caribbean International (RCI) is thanked for hosting our M-AERIs on the Allure of the Seas, Celebrity Equinox and Adventure of the Seas, including providing the instrument installations and internet connectivity.

Conflicts of Interest: The authors declare no conflict of interest.

References

1. Bojinski, S.; Verstraete, M.; Peterson, T.C.; Richter, C.; Simmons, A.; Zemp, M. The Concept of Essential Climate Variables in Support of Climate Research, Applications, and Policy. *Bull. Am. Meteorol. Soc.* **2014**, *95*, 1431–1443. [CrossRef]

2. Luo, B.; Minnett, P.J.; Szczodrak, M.; Kilpatrick, K.; Izaguirre, M. Validation of Sentinel-3A SLSTR derived Sea-Surface Skin Temperatures with those of the shipborne M-AERI. *Remote Sens. Environ.* **2020**, *244*, 111826. [CrossRef]

3. Minnett, P.J.; Alvera-Azcárate, A.; Chin, T.M.; Corlett, G.K.; Gentemann, C.L.; Karagali, I.; Li, X.; Marsouin, A.; Marullo, S.; Maturi, E.; et al. Half a century of satellite remote sensing of sea-surface temperature. *Remote Sens. Environ.* **2019**, *233*, 111366. [CrossRef]

4. Donlon, C.J.; Robinson, I.; Casey, K.S.; Vazquez-Cuervo, J.; Armstrong, E.; Arino, O.; Gentemann, C.; May, D.; LeBorgne, P.; Piollé, J.; et al. The Global Ocean Data Assimilation Experiment High-resolution Sea Surface Temperature Pilot Project. *Bull. Am. Meteorol. Soc.* **2007**, *88*, 1197–1213. [CrossRef]

5. Wimmer, W.; Robinson, I.S.; Donlon, C.J. Long-term validation of AATSR SST data products using shipborne radiometry in the Bay of Biscay and English Channel. *Remote Sens. Environ.* **2012**, *116*, 17–31. [CrossRef]

6. Minnett, P.J.; Knuteson, R.O.; Best, F.A.; Osborne, B.J.; Hanafin, J.A.; Brown, O.B. The Marine-Atmospheric Emitted Radiance Interferometer (M-AERI), a high-accuracy, sea-going infrared spectroradiometer. *J. Atmos. Ocean. Technol.* **2001**, *18*, 994–1013. [CrossRef]

7. Hirahara, S.; Balmaseda, M.A.; Boisseson, E.D.; Hersbach, H. *Sea Surface Temperature and Sea Ice Concentration for ERA5*; ECMWF: Reading, UK, 2016.

8. Dee, D.P.; Uppala, S.M.; Simmons, A.J.; Berrisford, P.; Poli, P.; Kobayashi, S.; Andrae, U.; Balmaseda, M.A.; Balsamo, G.; Bauer, P.; et al. The ERA-Interim reanalysis: Configuration and performance of the data assimilation system. *Q. J. R. Meteorol. Soc.* **2011**, *137*, 553–597. [CrossRef]

9. (CDS), C.C.C.S.C.D.S. Copernicus Climate Change Service (C3S) (2017): ERA5: Fifth Generation of ECMWF Atmospheric Reanalyses of the Global Climate. Available online: https://cds.climate.copernicus.eu/cdsapp#!/home (accessed on 26 February 2020).

10. Hennermann, K.; Berrisford, P. ERA5 Data Documentation. Available online: https://confluence.ecmwf.int/display/CKB/ERA5%3A+data+documentation (accessed on 26 February 2020).

11. Saha, S.; Moorthi, S.; Wu, X.; Wang, J.; Nadiga, S.; Tripp, P.; Behringer, D.; Hou, Y.-T.; Chuang, H.-Y.; Iredell, M.; et al. The NCEP Climate Forecast System Version 2. *J. Clim.* **2014**, *27*, 2185–2208. [CrossRef]

12. Rienecker, M.M.; Suarez, M.J.; Gelaro, R.; Todling, R.; Bacmeister, J.; Liu, E.; Bosilovich, M.G.; Schubert, S.D.; Takacs, L.; Kim, G.-K.; et al. MERRA: NASA's Modern-Era Retrospective Analysis for Research and Applications. *J. Clim.* **2011**, *24*, 3624–3648. [CrossRef]

13. Gelaro, R.; McCarty, W.; Suárez, M.J.; Todling, R.; Molod, A.; Takacs, L.; Randles, C.A.; Darmenov, A.; Bosilovich, M.G.; Reichle, R.; et al. The Modern-Era Retrospective Analysis for Research and Applications, Version 2 (MERRA-2). *J. Clim.* **2017**, *30*, 5419–5454. [CrossRef]

14. Randles, C.A.; Da Silva, A.M.; Buchard, V.; Colarco, P.R.; Darmenov, A.; Govindaraju, R.; Smirnov, A.; Holben, B.; Ferrare, R.; Hair, J.; et al. The MERRA-2 Aerosol Reanalysis, 1980—Onward, Part I: System Description and Data Assimilation Evaluation. *J. Clim.* **2017**, *30*, 6823–6850. [CrossRef]

15. Kobayashi, S.; Ota, Y.; Harada, Y.; Ebita, A.; Moriya, M.; Onoda, H.; Onogi, K.; Kamahori, H.; Kobayashi, C.; Endo, H.; et al. The JRA-55 Reanalysis: General Specifications and Basic Characteristics. *J. Meteorol. Soc. Jpn. Ser. II* **2015**, *93*, 5–48. [CrossRef]

16. Graham, R.M.; Hudson, S.R.; Maturilli, M. Improved performance of ERA5 in Arctic gateway relative to four global atmospheric reanalyses. *Geophys. Res. Lett.* **2019**, *46*, 6138–6147. [CrossRef]

17. Rayner, N.A. Global analyses of sea surface temperature, sea ice, and night marine air temperature since the late nineteenth century. *J. Geophys. Res.* **2003**, *108*. [CrossRef]

18. Donlon, C.J.; Martin, M.; Stark, J.; Roberts-Jones, J.; Fiedler, E.; Wimmer, W. The Operational Sea Surface Temperature and Sea Ice Analysis (OSTIA) system. *Remote Sens. Environ.* **2012**, *116*, 140–158. [CrossRef]

19. Nogueira, M. Inter-comparison of ERA-5, ERA-interim and GPCP rainfall over the last 40 years: Process-based analysis of systematic and random differences. *J. Hydrol.* **2020**, *583*, 124632. [CrossRef]

20. Mahto, S.S.; Mishra, V. Does ERA-5 Outperform Other Reanalysis Products for Hydrologic Applications in India? *J. Geophys. Res. Atmos.* **2019**, *124*, 9423–9441. [CrossRef]

21. Gentemann, C.L.; Akella, S. Evaluation of NASA GEOS-ADAS Modeled Diurnal Warming Through Comparisons to SEVIRI and AMSR2 SST Observations. *J. Geophys. Res. Ocean.* **2018**, *123*, 1364–1375. [CrossRef]

22. Minnett, P.J.; Smith, M.; Ward, B. Measurements of the oceanic thermal skin effect. *Deep Sea Res. Part. II Top. Stud. Oceanogr.* **2011**, *58*, 861–868. [CrossRef]

23. Nalli, N.R.; Joseph, E.; Morris, V.R.; Barnet, C.D.; Wolf, W.W.; Wolfe, D.; Minnett, P.J.; Szczodrak, M.; Izaguirre, M.A.; Lumpkin, R.; et al. Multiyear Observations of the Tropical Atlantic Atmosphere: Multidisciplinary Applications of the NOAA Aerosols and Ocean Science Expeditions. *Bull. Am. Meteorol. Soc.* **2011**, *92*, 765–789. [CrossRef]

24. Fairall, C.; Bradley, E.; Godfrey, J.; Wick, G.; Edson, J.; Young, G. Cool-skin and warm-layer effects on sea surface temperature. *J. Geophys. Res.* **1996**, *101*, 1295–1308. [CrossRef]

25. ECMWF. Part IV: Physical Processes. In *IFS Documentation CY43R1*; ECMWF: Reading, UK, 2016.

26. Zeng, X.; Beljaars, A. A prognostic scheme of sea surface skin temperature for modeling and data assimilation. *Geophys. Res. Lett.* **2005**, *32*, L14605. [CrossRef]

27. Gentemann, C.L.; Donlon, C.J.; Stuart-Menteth, A.; Wentz, F.J. Diurnal signals in satellite sea surface temperature measurements. *Geophys. Res. Lett.* **2003**, *30*, 1140–1143. [CrossRef]

28. Takaya, Y.; Bidlot, J.-R.; Beljaars, A.C.M.; Janssen, P.A.E.M. Refinements to a prognostic scheme of skin sea surface temperature. *J. Geophys. Res. Ocean.* **2010**, *115*. [CrossRef]

29. Akella, S.; Todling, R.; Suarez, M. Assimilation for skin SST in the NASA GEOS atmospheric data assimilation system. *Q. J. R. Meteorol. Soc.* **2017**, *143*, 1032–1046. [CrossRef]

30. Minnett, P.J. The Validation of Sea Surface Temperature Retrievals from Spaceborne Infrared Radiometers. In *Oceanography from Space, Revisited*; Barale, V., Gower, J.F.R., Alberotanza, L., Eds.; Springer Science+Business Media B.V.: Berlin, Germany, 2010; pp. 273–295.

31. Fowler, J.B. A third generation water bath based blackbody source. *J. Res. Natl. Inst. Stand. Technol.* **1995**, *100*, 591–599. [CrossRef]

32. Rice, J.P.; Butler, J.J.; Johnson, B.C.; Minnett, P.J.; Maillet, K.A.; Nightingale, T.J.; Hook, S.J.; Abtahi, A.; Donlon, C.J.; Barton, I.J. The Miami2001 Infrared Radiometer Calibration and Intercomparison: 1. Laboratory Characterization of Blackbody Targets. *J. Atmos. Ocean. Technol.* **2004**, *21*, 258–267. [CrossRef]

33. Theocharous, E.; Fox, N.; Barker-Snook, I.; Niclòs, R.; Santos, V.G.; Minnett, P.J.; Göttsche, F.; Poutier, L.; Morgan, N.; Nightingale, T. The 2016 CEOS infrared radiometer comparison: Part II: Laboratory comparison of radiation thermometers. *J. Atmos. Ocean. Technol.* **2019**, *36*, 1079–1092. [CrossRef]

34. Minnett, P.J.; Maillet, K.; Hanafin, J.; Osborne, B. Infrared interferometric measurements of the near-surface air temperature over the oceans. *J. Atmos. Ocean. Technol.* **2005**, *22*, 1019–1032. [CrossRef]

35. Luo, B.; Minnett, P.J.; Gentemann, C.; Szczodrak, G. Improving satellite retrieved night-time infrared sea surface temperatures in aerosol contaminated regions. *Remote Sens. Environ.* **2019**, *223*, 8–20. [CrossRef]

36. Nalli, N.R.; Clemente-Colón, P.; Minnett, P.J.; Szczodrak, M.; Jessup, A.; Branch, R.; Morris, V.; Goldberg, M.D.; Barnet, C.; Wolf, W.W.; et al. Ship-based measurements for infrared sensor validation during AEROSE 2004. *J. Geophys. Res.* **2006**, *111*, D09S04. [CrossRef]

37. Luo, B.; Minnett, P.J.; Szczodrak, M.; Nalli, N.R.; Morris, V.R. Accuracy assessment of MERRA-2 and ERA-Interim sea-surface temperature, air temperature and humidity profiles over the Atlantic Ocean using AEROSE measurements. *J. Clim.* **2020**. [CrossRef]

38. Buchard, V.; Randles, C.A.; da Silva, A.M.; Darmenov, A.; Colarco, P.R.; Govindaraju, R.; Ferrare, R.; Hair, J.; Beyersdorf, A.J.; Ziemba, L.D.; et al. The MERRA-2 Aerosol Reanalysis, 1980 Onward. Part II: Evaluation and Case Studies. *J. Clim.* **2017**, *30*, 6851–6872. [CrossRef]

39. Merchant, C.J.; Harris, A.R. Toward the elimination of bias in satellite retrievals of skin sea surface temperature. 2: Comparison with in situ measurements. *J. Geophys. Res.* **1999**, *104*, 23579–23590. [CrossRef]

40. Renfrew, I.A.; Moore, G.W.K.; Guest, P.S.; Bumke, K. A Comparison of Surface Layer and Surface Turbulent Flux Observations over the Labrador Sea with ECMWF Analyses and NCEP Reanalyses. *J. Phys. Oceanogr.* **2002**, *32*, 383–400. [CrossRef]

41. Smith, S.R.; Legler, D.M.; Verzone, K.V. Quantifying Uncertainties in NCEP Reanalyses Using High-Quality Research Vessel Observations. *J. Clim.* **2001**, *14*, 4062–4072. [CrossRef]

42. Foltz, G.R.; McPhaden, M.J. Trends in Saharan dust and tropical Atlantic climate during 1980–2006. *Geophys. Res. Lett.* **2008**, *35*. [CrossRef]

43. Mulcahy, J.; Walters, D.; Bellouin, N.; Milton, S. Impacts of increasing the aerosol complexity in the Met Office global numerical weather prediction model. *Atmos. Chem. Phys.* **2014**, *14*, 4749–4778. [CrossRef]

44. Kishcha, P.; da Silva, A.; Starobinets, B.; Long, C.; Kalashnikova, O.; Alpert, P. Saharan dust as a causal factor of hemispheric asymmetry in aerosols and cloud cover over the tropical Atlantic Ocean. *Int. J. Remote Sens.* **2015**, *36*, 3423–3445. [CrossRef]

45. Amiri-Farahani, A.; Allen, R.J.; Neubauer, D.; Lohmann, U. Impact of Saharan dust on North Atlantic marine stratocumulus clouds: Importance of the semidirect effect. *Atmos. Chem. Phys.* **2017**, *17*, 6305–6322. [CrossRef]

46. Adams, A.M.; Prospero, J.M.; Zhang, C. CALIPSO-Derived Three-Dimensional Structure of Aerosol over the Atlantic Basin and Adjacent Continents. *J. Clim.* **2012**, *25*, 6862–6879. [CrossRef]

47. Silber, I.; Verlinde, J.; Wang, S.-H.; Bromwich, D.H.; Fridlind, A.M.; Cadeddu, M.; Eloranta, E.W.; Flynn, C.J. Cloud Influence on ERA5 and AMPS Surface Downwelling Longwave Radiation Biases in West Antarctica. *J. Clim.* **2019**, *32*, 7935–7949. [CrossRef]
48. Reale, O.; Lau, K.; da Silva, A. Impact of interactive aerosol on the African Easterly Jet in the NASA GEOS-5 global forecasting system. *Weather Forecast.* **2011**, *26*, 504–519. [CrossRef]
49. Heinold, B.; Tegen, I.; Schepanski, K.; Hellmuth, O. Dust radiative feedback on Saharan boundary layer dynamics and dust mobilization. *Geophys. Res. Lett.* **2008**, *35*. [CrossRef]
50. Minnett, P.J.; Corlett, G.K. A pathway to generating Climate Data Records of sea-surface temperature from satellite measurements. *Deep Sea Res. Part. II Top. Stud. Oceanogr.* **2012**, *77–80*, 44–51. [CrossRef]
51. Donlon, C.J.; Minnett, P.J.; Fox, N.; Wimmer, W. Strategies for the Laboratory and Field Deployment of Ship-Borne Fiducial Reference Thermal Infrared Radiometers in Support of Satellite-Derived Sea Surface Temperature Climate Data Records. In *Experimental Methods in the Physical Sciences, Vol 47, Optical Radiometry for Ocean Climate Measurements*; Zibordi, G., Donlon, C.J., Parr, A.C., Eds.; Academic Press: Cambridge, MA, USA, 2014; Volume 47, pp. 557–603.

Letter

Evaluation of the Diurnal Variation of Upper Tropospheric Humidity in Reanalysis Using Homogenized Observed Radiances from International Geostationary Weather Satellites

Yunheng Xue [1,2,3] , Jun Li [2,*], Zhenglong Li [2], Mathew M. Gunshor [2] and Timothy J. Schmit [4]

[1]   Institute of Atmospheric Physics, Chinese Academy of Sciences, Beijing 100029, China; yxue44@wisc.edu
[2]   Cooperative Institute for Meteorological Satellite Studies, University of Wisconsin-Madison, Madison, WI 53705, USA; zli@ssec.wisc.edu (Z.L.); matg@ssec.wisc.edu (M.M.G.)
[3]   University of Chinese Academy of Sciences, Beijing 100029, China
[4]   Center for Satellite Applications and Research, CoRP/ASPB, Madison, WI 53705, USA; tim.j.schmit@noaa.gov
*   Correspondence: jun.li@ssec.wisc.edu

Received: 16 April 2020; Accepted: 18 May 2020; Published: 19 May 2020

Abstract: A near global dataset of homogenized clear-sky 6.5-μm brightness temperatures (BTs) from international geostationary (GEO) weather satellites has recently been generated and validated. In this study, these radiance measurements are used to construct the diurnal variation of upper tropospheric humidity (UTH) and to evaluate these diurnal variations simulated by five reanalysis datasets over the 45° N–45° S region. The features of the diurnal variation described by the new dataset are comparable with previous observational studies that a land–sea contrast in the diurnal variation of UTH is exhibited. Distinct diurnal variations are observed over the deep convective regions where high UTH exists. The evaluation of reanalysis datasets indicates that reanalysis systems still have considerable difficulties in capturing the observed features of the diurnal variation of UTH. All five reanalysis datasets present the largest wet biases in the afternoon when the observed UTH experiences a diurnal minimum. Reanalysis can roughly reproduce the day–night contrast of UTH but with much weaker amplitudes and later peak time over both land and ocean. Comparison of the geographical distribution of the diurnal variation shows that both ERA5 and MERRA-2 could capture the larger diurnal variations over convective regions. However, the diurnal amplitudes are widely underestimated, especially over convective land regions, while the phase biases are relatively larger over open oceans. These results suggest that some deficiencies may exist in convection and cloud parameterization schemes in reanalysis models.

Keywords: diurnal variation; upper tropospheric humidity; homogenized radiances; GEO weather satellites; evaluation of reanalysis

---

1. Introduction

Atmospheric water vapor (WV) is one of the major absorption gases of the outgoing longwave radiation. Although the WV content decreases with altitude rapidly, the outgoing longwave radiation (OLR) at the top of the atmosphere (TOA) is more sensitive to the upper tropospheric WV, and even small variations in upper tropospheric humidity (UTH) may lead to a significant impact on the radiation energy budget and climate feedback [1–4]. However, UTH is one of the least well-monitored atmospheric variables due to its high variability in both space and time and the lack of accurate conventional observations in the upper troposphere [5–7].

The reanalysis datasets are widely used to monitor and project the variability of key climate variables and must be continuously assessed to understand their strengths and weaknesses [8–10]. Previous studies have indicated that reanalysis has difficulties to accurately simulate the WV above the tropopause [9–11]. Valid observations for assimilation in the upper troposphere are sparse, and thus reanalysis data heavily rely on the "first guess" from their host models [12]. Given the importance of the UTH for radiative forcing, the uncertainties of UTH may lead to misrepresentations in radiative and dynamical processes in reanalysis. The diurnal variation is one of the most fundamental modes in the climate system. Simulations of diurnal variations of atmospheric variables are one important check of the reliability of a reanalysis system [13]. Some studies [14–16] have suggested that the deficiencies in simulating diurnal variation of UTH can help further identify the potential problems in convection and cloud parameterization in reanalysis systems. Therefore, it is important to assess how well reanalysis data capture the observed diurnal variability of UTH.

Geostationary (GEO) weather satellites monitor infrared (IR) radiation at the WV absorption bands with high spatiotemporal resolution and large spatial coverage. It was shown that the clear sky near 6.7-μm WV brightness temperature at nadir view is linearly related to the natural logarithm of UTH [17] based on simplified radiative theory and certain assumptions of atmospheric profiles:

$$\ln UTH = a + bT_{6.7clr},\qquad(1)$$

where a (~31.5) and b (~−0.1) can be treated as nearly constant values for interpretation purposes. This simplified equation demonstrates that, to a reasonable degree of accuracy, the near 6.7-μm WV absorption spectral region radiances can be interpreted in terms of a more familiar water vapor measurements, i.e., UTH. The UTH is defined as the mean relative humidity averaged over a broad layer between approximately 200 and 500 hPa, indicated by the moisture Jacobian function of the near 6.7-μm WV band [18,19]. As a result, the WV radiances from GEO weather satellites have been widely used for studying the diurnal variation of UTH [11,14,15,17]. Studies have found that the diurnal cycle of UTH has a land–sea contrast with larger amplitude and later peak time over land than over the oceans [14–16]. The diurnal variation of UTH usually lags deep convection and high clouds, indicating the importance of deep convection in moistening the upper troposphere through the evaporation/sublimation of the clouds [15]. Scientists have also tried to validate the diurnal characteristics of UTH in climate models and reanalysis datasets. For example, the simulation of UTH in two climate models are compared with microwave and IR measurements from polar orbiting satellites over selected convective regions [20]. The diurnal variation of UTH in five reanalysis datasets over the convectively active regions of Africa and the Atlantic Ocean has been evaluated with 6.7-μm WV band radiances from Meteosat-5 [11]. However, these studies have usually been limited to polar orbiting satellites that have large temporal gaps or limited to a single GEO satellite due to the spectral differences among international GEO weather satellites.

Recently, a new homogenized IR 6.5-μm WV absorption band radiance dataset (referred to as homogenized WV radiances data hereafter) from multiple international GEO weather satellites has been successfully generated [19]. This homogenized WV radiances dataset maintains the high spatial and temporal resolution of GEO satellites and has near global coverage of the tropics and mid-latitudes. It provides a great opportunity to perform a near global assessment of diurnal variation of UTH in reanalysis datasets.

The purpose of this study is to construct the diurnal variation of UTH with the homogenized WV radiance data on a near global scale. This dataset will then be further used to evaluate the capability of five reanalysis datasets to capture these observed diurnal variations. The paper is organized as follows: the homogenized WV radiance data, reanalysis simulations, and diurnal analysis method are described in Section 2. The main results of the observed diurnal variation of UTH from GEO weather satellites and the evaluation of the five different reanalysis datasets are given in Section 3. Section 4 presents the discussion, while the conclusions are in Section 5.

2. Data and Methodologies

2.1. Homogenized WV Radiances from International GEO Weather Satellites

The homogenized WV radiances (expressed as equivalent brightness temperatures, or BTs) data were generated by homogenizing seven international GEO weather satellite imagers' WV radiances (see Table 1 in Li et al. [19]) to the nadir view radiances of GOES-15 Imager 6.5-μm WV band. The homogenization process accounts for both spectral differences and the limb (angle) effect between other GEO imagers and GOES-15 Imager. The cloud detection process is based on a simple cloud mask scheme in post-processing to remove the contaminations by high clouds. The accuracies of the homogenized clear sky WV radiances data have been validated with the independent hyperspectral sounder [21] Infrared Atmospheric Sounding Interferometer (IASI) radiances from both Metop-A and Metop-B. This 3-hourly observation archive covers the years from 2015 to 2017 with the spatial coverage of 45° N–45° S and all longitudes. The reader is referred to Li et al. [19] for detailed technical approaches on this dataset.

2.2. Reanalysis Datasets

Recently, the quality of modern reanalysis systems has been much improved due to the great efforts made in forecast models and data assimilation (DA) systems [22–24]. In this study, five reanalysis datasets were evaluated. They were the European Centre for Medium-Range Weather Forecasts' (ECMWF) newly released fifth generation reanalysis (ERA5, [25]), the ECMWF Interim Reanalysis (ERA-Interim, [22]), the National Centers for Environmental Prediction's (NCEP) Climate Forecast System reanalysis, version 2 (CFSv2, [26], which is also referred to as CFSR), the Modern-Era Retrospective Analysis for Research and Applications version 2 (MERRA-2, [24]), and the 55-year modern Japanese Reanalysis Projects (JRA55, [23]). Some of the basic information of the five reanalysis datasets is listed in Table 1. It should be noted that the temporal interval is 6 h for ERA-Interim, CFSR and JRA-55, while ERA5 and MERRA-2 can provide 3-hourly meteorological variables, which is consistent with the homogenized WV radiances.

Table 1. Basic characteristics of reanalysis datasets evaluated. IFS: Integrated Forecasting System. GEOS: Goddard Earth Observing System Model developed by the National Aeronautics and Space Administration (NASA)'s Global Modeling and Assimilation Office (GMAO). CFS: The National Centers for Environmental Prediction's (NCEP) Climate Forecast System. GSM: Global Spectral Model of the Japan Meteorological Agency (JMA).

Reanalysis	ERA5	ERA-Interim	CFSR	MERRA-2	JRA-55
Source	ECMWF	ECMWF	NCEP	NASA GMAO	JMA
Forecast Model	IFS Cycle 41r2	IFS Cycle 31r2	CFS	GEOS 5.12.4	JMA GSM
Assimilation Scheme	4D-VAR	4D-VAR	3D-VAR	3D-VAR	4D-VAR
Vertical Resolution (Pressure Level)	37	37	37	42	37; 27 for WV profiles
Horizontal Resolution	$0.25° \times 0.25°$	$0.75° \times 0.75°$	$0.5° \times 0.5°$	$0.5° \times 0.625°$	$1.25° \times 1.25°$
Temporal Resolution	1 hourly	6 hourly	6 hourly	3 hourly	6 hourly

2.3. Methodologies

The evaluation in this study was based on the WV radiances rather than the WV retrievals. A profile-to-radiance approach [11,12,27] was adopted. Atmospheric profiles of temperature and humidity from reanalysis datasets were input with surface information into the Community Radiative Transfer Model (CRTM) v2.1.3 [28] using Optical Depth in Pressure Space (ODPS) coefficients to simulate the clear-sky GOES-15 Imager 6.5-μm WV band BTs at nadir view. Many studies [29–31] have validated the CRTM capability of simulating IR radiances with the rigorous line-by-line radiative transfer model (LBLRTM) [32]. The LBLRTM provides spectral radiance calculations with high accuracies and is widely regarded as a standard benchmark for RTM model evaluations. It was shown that both the bias and root-mean-square error of CRTM are mostly below 0.15 K, indicating that the CRTM is quite accurate for clear sky IR radiance simulations. Therefore, the large differences between observed radiances and simulated radiances could be mainly attributed to the deficiencies in reanalysis datasets, which can then be interpreted to the UTH uncertainties based on Equation (1). The cloud mask from observations was used for reanalysis to exclude the grids that may contain the cloud contamination in reanalysis and to ensure the reanalysis simulations have the same sampling gaps as observations. All the data, including the observations and the reanalysis simulations, were then re-gridded to a $0.5° \times 0.5°$ horizontal grid format using an inverse distance squared weighted interpolation method.

To characterize the main signal of the periodically repeated diurnal signal and reduce the weather noise, a "composite day" was necessary to be first prepared by averaging the BT fields at each time step for a given period at each grid. The Fourier decomposition of 3-hourly diurnal cycle composites are widely used in diurnal analysis [2,15,16,33–36]. A first-order Fourier series was fitted to the daily composite to estimate the amplitude and phase of the BT diurnal variation for each $0.5° \times 0.5°$ grid box:

$$\mathrm{BT(t)} = \overline{\overline{\mathrm{BT}}} + \mathrm{A}\cos\left[\frac{2\pi}{24}(\mathrm{t} - \mathrm{P})\right] + residual, \tag{2}$$

where $\overline{\overline{\mathrm{BT}}}$ is the diurnal mean, t is the local solar time (LST) in hours, A represents the amplitude of the BTs, and P represents the diurnal phase of BTs. It should be noted that the diurnal phase of BTs corresponds to the LST showing maximum value of BTs. Since there is a strong negative correlation (-0.968) between the BT and the corresponding value of the ln *UTH* [17], the diurnal variation of BT can be easily interpreted to the diurnal variation of UTH. A large amplitude of BTs represents a large diurnal amplitude of UTH, and the LST for maximum UTH (referred to as the phase of UTH hereafter) corresponds to the LST showing the minimum value of BT, which is denoted by P + 12 in Equation (2). The diurnal amplitude of BTs and the phase of UTH will then be displayed in vector maps (Figures 2 and 5) to highlight the geographical distribution of the diurnal variation of UTH.

The Fisher statistical significance test (F-test) was used to determine the statistical significance [33,37] of the Fourier first harmonic fit. The results are only shown where the first harmonic fit is statistically significant at the 90% confidence level [37] in all vector figures.

3. Results

3.1. Characteristics of Observed Diurnal Variation

3.1.1. Observed Diurnal Anomaly

The diurnal anomaly is calculated by the daily composite at a given time minus the daily mean. The diurnal anomalies of the area-weighted average of observed 6.5-μm WV band BTs over the near global area (45° N–45° S) for the entire 3 years, as well as that over the northern (0°–45° N) and southern (0°–45° S) hemisphere for two different seasons, December-January-February (DJF) and June-July-August (JJA), are shown in Figure 1. The weights were calculated by the cosine of the latitude [12,38,39] and the area were separated for (a) land and (b) ocean to display the land–sea contrast. Overall, the amplitude of the diurnal anomaly of 3-year mean BTs is larger over land than over ocean. The BTs reach the maximum and minimum values over land at 15:00–18:00 LST and 3:00–6:00 LST, respectively. In contrast, the BTs over ocean show maxima and minima at 14:00–16:00 LST and 0:00–3:00 LST, respectively. This indicates that the upper troposphere tends to be more humid in the nighttime (0:00–6:00 LST) and drier in the midafternoon to early evening (14:00–20:00 LST). These results are comparable with previous observational studies [14–16,40]. The diurnal anomalies over the two hemispheres for two different seasons show that the diurnal variation of BTs (UTH) is strong in the summer hemisphere, suggesting its relationship with the active convections. In the study period, observations for two DJF seasons (2015/16, 2016/17) and three JJA seasons (2015, 2016, and 2017) are available and have been used in the following analysis to present the strong diurnal signals in different seasons. It should be noted that the observations from Meteosat-8, one of the international GEO satellites used in the homogenized WV radiances data, are only available since November 2016. Therefore, the regions near 50° E–80° E measured by Meteosat-8 just have observations for one DJF and one JJA season.

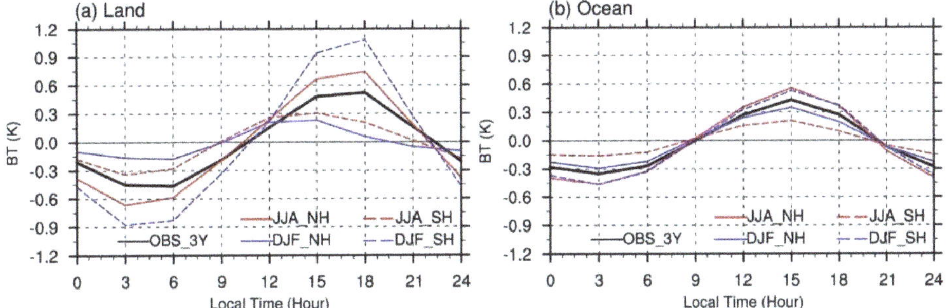

Figure 1. (**a**) Diurnal anomalies of the observed clear-sky 6.5-μm water vapor (WV) band brightness temperature (BT) averaged over (**a**) land regions and (**b**) ocean regions over global area (45° N–45° S) for the entire period of 2015–2017 (black solid line), and over the northern hemisphere (NH; 0°–45° N), and southern hemisphere (SH; 0°–45° S) for two seasons (December-January-February, DJF and June-July-August, JJA).

3.1.2. Observed Global Distribution

Figure 2 shows the geographical distributions of seasonal mean BTs for boreal winter (DJF) and summer (JJA) in 2015–2017. Also shown are the two important components of the diurnal variation of the UTH: the amplitude and the phase derived using first order (24-h) Fourier analysis mentioned in Section 2.3. The results are spatially smoothed to a 2.5° × 2.5° grid resolution to reduce the influence of mesoscale and microscale disturbance and make the results more clarified. The distributions of observed BTs are continuous throughout the coverage of different satellites, which further indicates that this homogenized WV radiances dataset has good performance in homogeneity and consistency. As indicated by Equation (1), the distribution of seasonal mean BTs is highly related to the distributions of seasonal mean UTH, with colder temperature corresponding to higher WV content in the upper troposphere. The convective regions are thus clearly indicated by the cold BT fields in Figure 2a,b, such as South Africa, the "Marine Continent" of the western Pacific, South America in DJF, and the Central Africa and India monsoon regions in JJA. The movement of the cold BT (high UTH) fields between DJF and JJA is consistent with the seasonal transition of the intertropical convergence zone (ITCZ) and the movement of deep convection centers.

The land–sea contrast for the diurnal variation of UTH is clearly revealed in Figure 2c,d. Larger diurnal amplitudes of UTH are observed over the deep convective regions (high UTH area in Figure 2a,b), especially over convective land regions. In contrast, the diurnal amplitudes are much weaker over non-convective subtropical regions where the warm BT bands are dominant. In general, the UTH peaks earlier over ocean in early night (0:00–3:00 LST) than over land at around late night to early morning (3:00–6:00 LST). The seasonal differences of the diurnal phase of UTH over these convective regions appear to be small over both land and ocean.

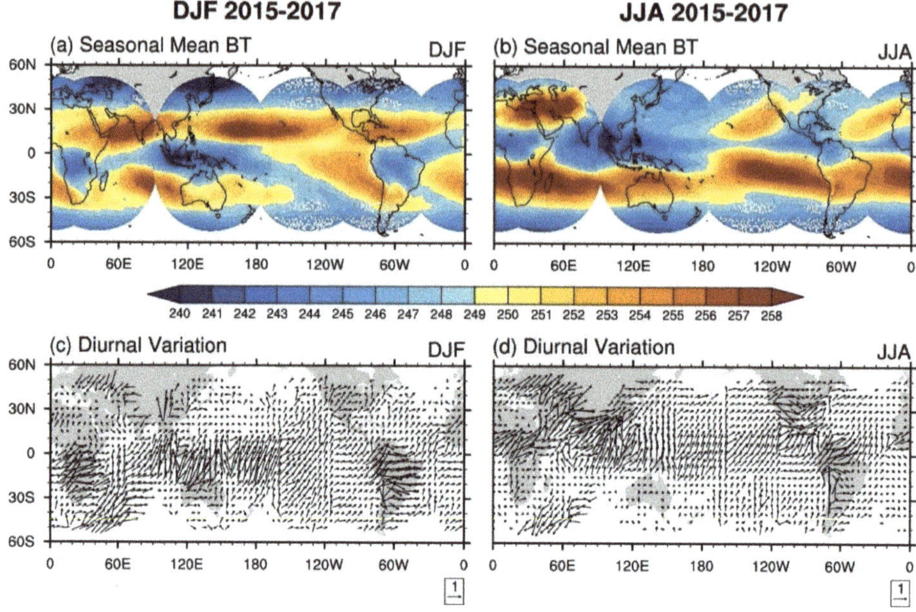

Figure 2. Geographical distributions of (**upper panel**) observed seasonal mean BTs (unit: K) and (**lower panel**) observed diurnal amplitudes of BTs and diurnal phases of UTH for (**left panel**) DJF and (**right panel**) JJA in 2015–2017. The length of the arrow in vector figures denotes the diurnal amplitude (unit: K). The orientation of the arrow with respect to a 24-h clock depicts the diurnal phase (local standard time, LST). For example, arrows pointing upward indicate the UTH peaks at 00:00 LST (midnight). For clarity, results are only shown in every other grid point.

3.2. Comparisons between Observed and Simulated Diurnal Variation

3.2.1. Diurnal Variation of Brightness Temperature Differences

The BT differences (BTDs) are defined as the simulated BTs from reanalysis datasets minus the observed BTs. As inferred from Equation (1), a negative BTD corresponds to a wet UTH bias while a positive BTD indicates a dry UTH bias. The diurnal variation of near global mean BTDs over land and ocean for DJF and JJA are shown in Figure 3. The BTDs are negative at each time step in one composite day, which indicates an overall wet bias in the upper troposphere in all reanalysis datasets. The BTDs typically reach the maximum absolute value near 15:00–18:00 LST over land and 12:00–15:00 LST over ocean, when observed BTs experience a diurnal maximum (Figure 1). This indicates that the reanalysis datasets tend to have larger UTH biases in a drier upper troposphere. Previous studies [41,42] have shown that a dry upper troposphere is usually related to the descending branches of large-scale circulations. Therefore, this may suggest that the large-scale circulation, especially the descending branches, is not well simulated in reanalysis systems.

The BT bias is largest in MERRA-2, about 1 K more negative than other datasets, indicating the wettest upper troposphere in MERRA-2 when compared with the other four reanalysis datasets. Additionally, the BT bias in MERRA-2 is larger in JJA than in DJF, which is not obvious in other reanalysis datasets. In contrast, the JRA55 has a comparable simulation of UTH to observations with the smallest mean BTD of −1.56 K among all reanalysis datasets. According to Equation (1), a BTD of 1 K corresponds to an uncertainty in $\frac{UTH}{UTH}$ of approximately −0.1 [17,43]. Therefore, if the UTH is 50%, then this BTD in JRA55 can be roughly estimated as a UTH wet bias of 7.8%. Figure 3 also shows that when compared with the ERA-Interim, the overall UTH simulations in ERA5 have been improved with smaller BTDs in both DJF and JJA seasons and over both land and ocean.

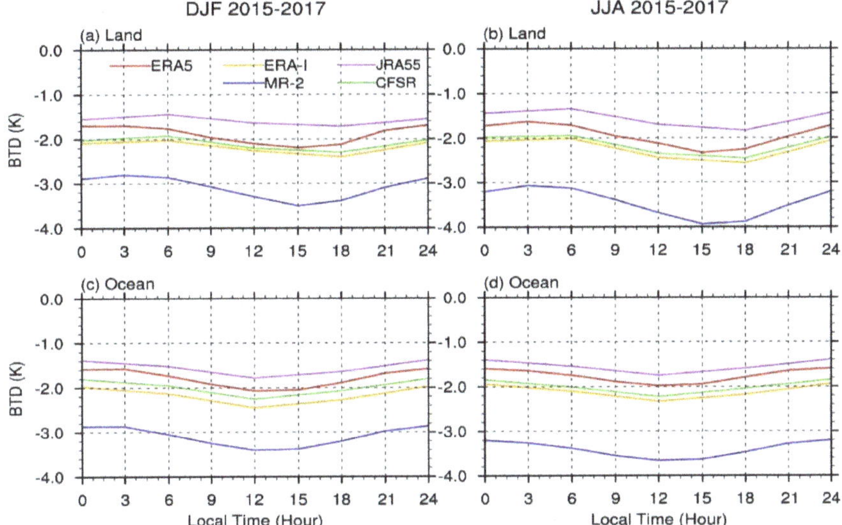

Figure 3. The diurnal variation of the BT differences (BTDs), which are calculated by the simulated BTs in reanalysis datasets minus the observed BTs, over (**a,b**) land and (**c,d**) ocean between 45° N and 45° S for (left panel) DJF and (right panel) JJA in 2015–2017.

3.2.2. Diurnal Anomaly

The diurnal anomalies of simulated BTs for area-weighted average over land and ocean for DJF and JJA in 2015–2017 are shown in Figure 4 along with the observations. Overall, all reanalyses can roughly reproduce the day–night contrast of UTH, with the maximum BTs (minimum UTH) found in the late afternoon to early evening and the minimum BTs (maximum UTH) in the nighttime to the early morning. However, differences are clearly shown in both the diurnal amplitude and phase when compared with observed diurnal anomalies. The amplitudes of the diurnal variation are significantly weaker in reanalysis datasets over both land and ocean. In other words, the reanalysis datasets tend to overestimate (underestimate) the moisture content when the upper troposphere has dry (wet) anomalies indicated by the warm (cold) BT anomalies. Furthermore, the LST for the minimum BTs (maximum UTH) in reanalysis datasets lags about 3 h behind the observations, especially over ocean. As a result, the observed global land–sea contrast in the phase of UTH is not well represented in reanalysis datasets.

The discrepancies in different reanalysis datasets are distinct over land. In particular, the diurnal amplitude in MERRA-2 is the smallest compared with other reanalysis datasets over land. The observed diurnal amplitude is larger over global land than over ocean, especially in boreal summer in the study period, while the land–sea contrast of the diurnal amplitude is not obvious in reanalysis datasets. In addition, the observed diurnal amplitude is slightly larger in JJA than in DJF over land in the study period, which is not obvious in the reanalysis datasets either. The seasonal differences of the diurnal variation in both observations and reanalysis are small over ocean. In general, the reanalysis datasets cannot simulate the main characteristic (amplitude and phase) of the diurnal anomaly of near global mean UTH well.

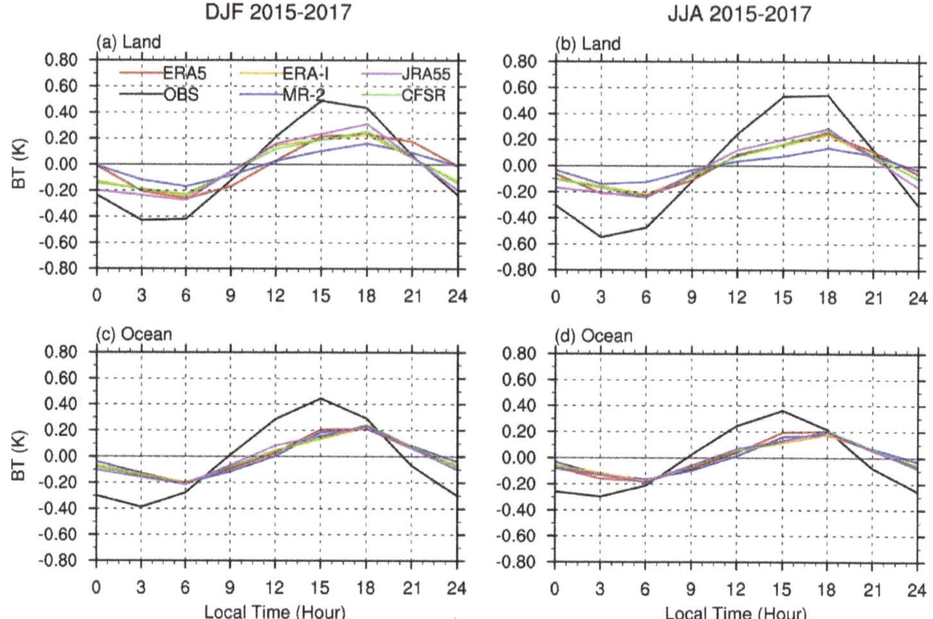

Figure 4. Diurnal anomalies of BTs simulated from ERA5, MERRA-2, ERA-Interim, CFSR, and JRA55 averaged over (**a,b**) land and (**c,d**) ocean regions between 45° N and 45° S for (left panel) DJF and (right panel) JJA in 2015–2017. The black lines denote the diurnal anomalies of observed BTs for the same period. The MR-2 is short for MERRA-2.

3.2.3. Global Distribution

The geographical distribution of the diurnal amplitude of BTs and the phase of UTH are also constructed from reanalysis datasets for comparison to satellite observations. Since only ERA5 and MERRA-2 can provide 3-hourly simulated BTs for the daily composite that could be accurately decomposed using Fourier analysis, the following evaluations are mainly focused on these two reanalysis datasets. In addition, the main deficiencies in reanalysis for DJF and JJA are similar, thus only the results for the JJA season are presented.

It is clearly shown in Figure 5 that the diurnal variations in both ERA5 and MERRA-2 are much weaker than the observations (Figure 2c,d) on the global scale. First, the number of grids which have significant diurnal (24-h) signal are smaller in reanalysis datasets, indicated by the lower density of vectors. Second, the diurnal amplitudes of BTs are much smaller when compared with observations. Nevertheless, the two reanalysis datasets could still roughly capture the spatial distribution of the observed diurnal variation of UTH, that is, larger diurnal variation in deep convective regions.

To illustrate the diurnal phase, the histograms of the diurnal phase summarized over land and ocean grids for observations and reanalysis datasets are displayed in the lower panel of Figure 5. Only those grids with statistically significant diurnal (24-h) components were counted. The results clearly show a phase shift in reanalysis data with respect to observations over ocean. The observed maximum UTH over ocean shows a broad peak time from 22:00 LST and mostly occurs around 02:00–03:00 LST, whereas the UTH in ERA5 and MERRA-2 usually experiences a maximum around 04:00–05:00 LST over most of the oceans, which is about 2 h later than observations, and does not show the late evening (~22:00) peak well. The phase differences between observations and ERA5 are relatively small. However, the diurnal phase of UTH over land in MERRA-2 is not well defined, exhibiting a broad range of peak time, from 23:00–12:00, with no obvious dominant single peak.

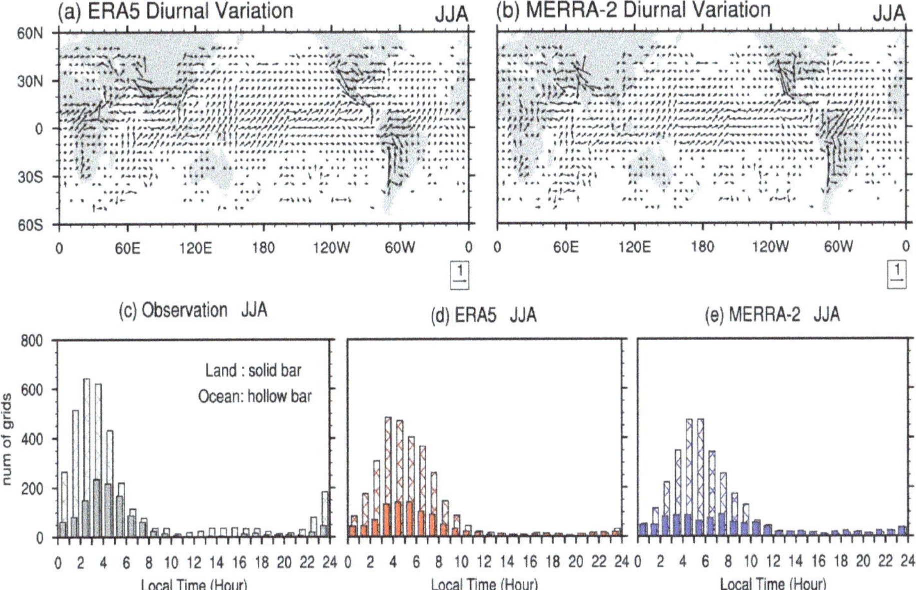

Figure 5. (upper panel) Geographical distributions of diurnal amplitudes of BTs and diurnal phases (peak time) of UTH in (**a**) ERA5 and (**b**) MERRA-2 for JJA of 2015–2017. The length of the arrow denotes the diurnal amplitude (unit: K). The orientation of the arrow with respect to a 24-h clock depicts the diurnal phase (local standard time, LST). For example, arrows pointing upward indicate the UTH peaks at 00:00 LST (midnight). For clarity, results are only shown in every other grid point. (lower panel). Histograms of the diurnal phase of UTH for land (solid bar) and ocean (hollow bar) grids from (**c**) satellite observations, (**d**) ERA5, and (**e**) MERRA-2. Only those grids with significant diurnal (24-h) signal were counted.

The quantitative differences of amplitude and phase between the simulated and observed diurnal variations of UTH (simulations minus observations) are provided in Figure 6. For the diurnal amplitude, the ERA5 and MERRA-2 tend to underestimate the diurnal amplitude of BTs by about −0.3 K over much of the oceans. In contrast, the amplitude biases are much larger over convective regions, especially over continental convective regions such as Central Africa, South America, and India. A few studies [14–16] have revealed that the diurnal variation of UTH is highly regulated by deep convections. For example, Chung et al. [14] investigated the relationship between UTH and convective activities over tropical Africa with the Meteosat-8 measurements. They found that deep convections could transport the cloud condensates and WV upward into the upper troposphere during the developing period. These condensates will be detrained into the surrounding environment in the decay period through the cirrus anvil clouds spreading, and then increase the moisture content in the upper troposphere by evaporation/sublimation. As a result, the larger amplitude biases over convective regions suggest that there might be some deficiencies in convection and cloud parameterization schemes in reanalysis models.

For the diurnal phase, a substantial time lag is widely seen, especially over ocean. The peak time of UTH in ERA5 mostly lags observations by about 1.5–2 h, in comparison to 2.5–3 h in MERRA-2 over ocean. This result is similar to the result of Tian et al. [15], who found that the maximum of UTH in Geophysical Fluid Dynamics Laboratory (GFDL) global atmosphere and land model (AM2/LM2) is usually 3 h later than observed. Unlike the amplitude biases, the phase biases are much larger over ocean than over land. Studies [15,44] have argued that the diurnally varying sea surface temperature (SST) is important to the oceanic diurnal phase. Therefore, the SST boundary condition in reanalysis models might be one possible reason for the larger phase lagging over ocean. Although the UTH in reanalysis datasets generally peaks later than that in observations, there are some exceptions where the UTH peaks earlier than that in observations, such as the north of Arabia, the southern Indian Ocean around 30 °S, and north of South America in MERRA-2 in boreal summer. These regional exceptions might be indicative of some local-scale disturbance that needs more investigation. Furthermore, the limited observations over those regions covered only by Meteosat-8 (between about 50° E–80° E) might also introduce some uncertainties. More investigations will be conducted over these regions once more observations are available.

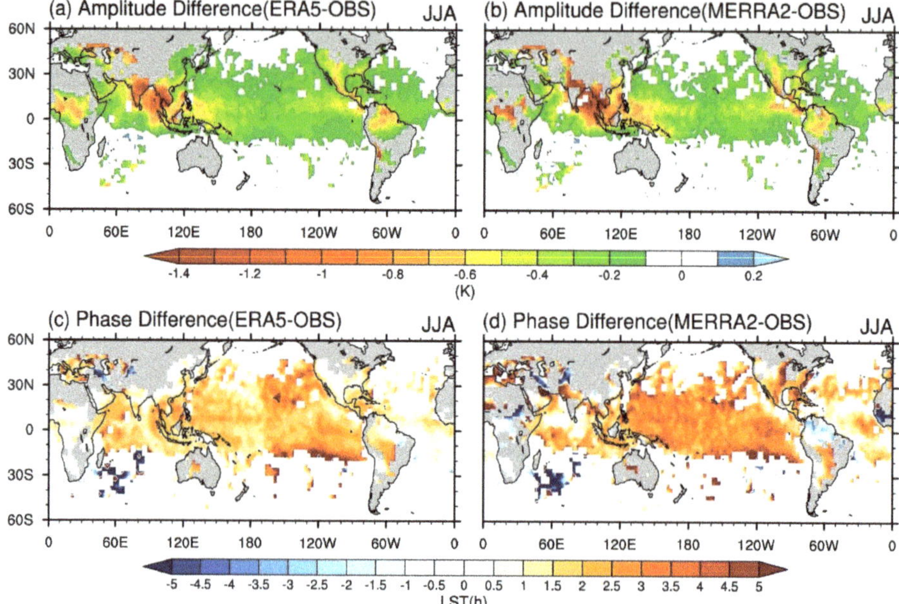

Figure 6. Geographical distributions of differences in (upper panel) diurnal amplitudes of BTs and (lower panel) diurnal phases of UTH between (**a,c**) ERA5, (**b,d**) MERRA-2 and observations (reanalysis minus observations) in JJA of 2015–2017. For clarity, results are only shown where the amplitude bias is more than 0.1 K and the phase bias is more than 0.5 h.

4. Discussion

As the observations are clear sky BTs, this study mainly represents the diurnal variation of UTH under clear sky conditions rather than all sky conditions. Although the clear sky sampling might lead to an underestimation of the real moisture environment in the upper troposphere and may also introduce a small bias in the phase of UTH [11,20], the comparison between simulated and observed BT is generally based on the same condition to ensure the differences are not critically affected by the clear-sky sampling bias.

The results from the comparisons are generally consistent over the global scale, such that the reanalysis datasets tend to underestimate the diurnal amplitude and have a later peak time of UTH than observed. However, some geographic differences are seen. For example, a notable exception is over the southern Indian Ocean, where the observations show an afternoon maximum of UTH in boreal summer, whereas ERA5 and MERRA-2 simulations show a morning peak, which is earlier than the observations. These regional differences might be related to local scale convections, topography and limited observations. Longer-time measurements from the Meteosat-8 satellite over the Indian Ocean may help to further examine these differences in the near future.

There are some possible reasons that reanalysis datasets do not capture the major characteristics of the diurnal variation of UTH well. For example, there might be some deficiencies in the convective and cloud parameterization in reanalysis models. The valid observations over convective regions might not be effectively used in the DA systems. Inspired by Kim et al. [36], the simulations of diurnal variations might be improved by increasing the horizontal resolution of models to better represent the local-scale circulations. Although the specific causes are not identified, these evaluation results can provide useful information and important feedback to the model and DA communities for further improving the performance of reanalysis systems.

Recently, with measurements from advanced imagers onboard the new generation of international geostationary weather satellites such as Himawari-8/-9 [45], GOES-16/-17 [46], FengYun-4A [47], and GEO-KOMPSAT-2A, tropospheric moisture information can be obtained with high temporal and spatial resolutions, the WV radiances from the new generation of international GEO weather satellites, once homogenized, can be used to evaluate the diurnal characteristics of both middle tropospheric and upper tropospheric humidity in models and reanalysis systems.

5. Conclusions

This study uses a new homogenized 6.5-μm WV radiances data from international GEO weather satellites to construct the diurnal variation of upper tropospheric moisture and to evaluate the diurnal variations from five modern reanalysis datasets globally between 45° N and 45° S for DJF and JJA of 2015–2017. The main results show that:

1. The diurnal variation of UTH constructed from the new homogenized WV radiances data generally agrees with previous observational studies which were mostly limited to regional scale or short time period [2,11,14–16,20,48]. Larger diurnal variations are observed over the deep convective regions where the mean UTH is high. The land–sea contrast for the diurnal variation of UTH is clearly revealed; that is, the diurnal amplitude is relatively larger over land and the UTH usually peaks earlier over ocean.

2. All five reanalysis datasets show wet biases in the upper troposphere, with the largest bias found in MERRA-2 and smallest bias in JRA55. The wet biases tend to reach the maximum at the time when the UTH reaches the minimum, indicating that reanalysis datasets have slightly more difficulties in simulating the moisture in a drier upper troposphere.

3. Accurately depicting the characteristics in the diurnal variation of UTH is still a challenging task for current reanalysis systems. The diurnal amplitudes of global mean BTs are much weaker in the five reanalysis datasets, and the LST for the minimum BTs (maximum UTH) usually lags about 3 h behind the observations.

4. Both ERA5 and MERRA-2 could roughly capture the larger diurnal variations over deep convective regions. However, the diurnal amplitudes are widely underestimated, especially over convective land regions, which possibly suggests some deficiencies in convection parameterization schemes in reanalysis models. In contrast, the phase biases are relatively larger over the ocean.

Author Contributions: Conceptualization, J.L.; methodology, Y.X., and J.L.; software, Y.X.; validation, Y.X., Z.L., and J.L.; formal analysis, Y.X., Z.L., and J.L.; investigation, Y.X., Z.L., M.M.G., and T.J.S.; resources, J.L.; data curation, Z.L.; writing-original draft preparation, Y.X.; writing-review and editing, J.L., Z.L., M.M.G., T.J.S., and Y.X.; visualization, Y.X.; supervision, J.L.; project administration, Z.L.; funding acquisition, Z.L. All authors have read and agreed to the published version of the manuscript.

Funding: This work is partly supported by NSFC (41775045) and China Scholarship Council (CSC) (Yunheng Xue). It is also partially supported by NOAA's GOES-R series science program NA15NES4320001 (Zhenglong Li, Jun Li, and Mathew M. Gunshor).

Acknowledgments: The new homogenized WV radiance data from international GEO imagers can be obtained by applying the regression coefficients with the Matlab reader (ftp://ftp.ssec.wisc.edu/ABS/zli/GEO_WV_Homogenization_Coeffs/regression_coef_supplemental_materials.tar) to the radiance measurements available from the Data Center at the Space Science and Engineering Center at University of Wisconsin-Madison (https://www.ssec.wisc.edu/datacenter/). The ERA5 and ERA-Interim data are obtained from ECMWF at https://www.ecmwf.int/en/forecasts/datasets/browse-reanalysis-datasets. The CFSR and MERRA-2 reanalysis are obtained from https://rda.ucar.edu and https://disc.gsfc.nasa.gov/, respectively. The JRA55 reanalysis project carried out by JMA is available at https://rda.ucar.edu/. The views, opinions and findings contained in this report are those of the authors and should not be construed as an official NOAA or U.S. government position, policy, or decision.

Conflicts of Interest: The authors declare no conflict of interest.

References

1. Held, I.M.; Soden, B.J. Water vapor feedback and global warming. *Annu. Rev. Energy Environ.* **2000**, *25*, 441–475. [CrossRef]
2. Chung, E.-S.; Sohn, B.-J.; Schmetz, J. Diurnal variation of outgoing longwave radiation associated with high cloud and UTH changes from Meteosat-5 measurements. *Meteorol. Atmos. Phys.* **2009**, *105*, 109–119. [CrossRef]
3. Allan, R.P. The role of water vapour in Earth's energy flows. *Surv. Geophys.* **2012**, *33*, 557–564. [CrossRef]
4. Houghton, J.T.; Ding, Y.; Griggs, D.J.; Noguer, M.; Van der Linden, P.J.; Dai, X.; Maskell, K.; Johnson, C.A. *Climate Change 2001: The Scientific Basis*; The Press Syndicate of the University of Cambridge: Cambridge, UK, 2001.
5. Elliott, W.P.; Gaffen, D.J. On the utility of radiosonde humidity archives for climate studies. *Bull. Am. Meteorol. Soc.* **1991**, *72*, 1507–1520. [CrossRef]
6. Miloshevich, L.M.; Vömel, H.; Whiteman, D.N.; Lesht, B.M.; Schmidlin, F.J.; Russo, F. Absolute accuracy of water vapor measurements from six operational radiosonde types launched during AWEX-G and implications for AIRS validation. *J. Geophys. Res. Atmos.* **2006**, *111*. [CrossRef]
7. Wang, J.; Carlson, D.J.; Parsons, D.B.; Hock, T.F.; Lauritsen, D.; Cole, H.L.; Beierle, K.; Chamberlain, E. Performance of operational radiosonde humidity sensors in direct comparison with a chilled mirror dew-point hygrometer and its climate implication. *Geophys. Res. Lett.* **2003**, *30*. [CrossRef]
8. Davis, S.M.; Hegglin, M.I.; Fujiwara, M.; Dragani, R.; Harada, Y.; Kobayashi, C.; Long, C.; Manney, G.L.; Nash, E.R.; Potter, G.L.; et al. Assessment of upper tropospheric and stratospheric water vapor and ozone in reanalyses as part of S-RIP. *Atmospheric Chem. Phys.* **2017**, *17*, 12743–12778. [CrossRef]
9. Jiang, J.H.; Su, H.; Zhai, C.; Wu, L.; Minschwaner, K.; Molod, A.M.; Tompkins, A.M. An assessment of upper troposphere and lower stratosphere water vapor in MERRA, MERRA2, and ECMWF reanalyses using Aura MLS observations. *J. Geophys. Res. Atmos.* **2015**, *120*, 11468–11485. [CrossRef]
10. Dessler, A.E.; Davis, S.M. Trends in tropospheric humidity from reanalysis systems. *J. Geophys. Res. Atmos.* **2010**, *115*. [CrossRef]
11. Chung, E.-S.; Soden, B.J.; Sohn, B.J.; Schmetz, J. An assessment of the diurnal variation of upper tropospheric humidity in reanalysis data sets. *J. Geophys. Res. Atmos.* **2013**, *118*, 3425–3430. [CrossRef]
12. Iacono, M.J.; Delamere, J.S.; Mlawer, E.J.; Clough, S.A. Evaluation of upper tropospheric water vapor in the NCAR Community Climate Model (CCM3) using modeled and observed HIRS radiances. *J. Geophys. Res. Atmos.* **2003**, *108*, ACL 1-1–ACL 1-19. [CrossRef]
13. Yang, G.-Y.; Slingo, J. The Diurnal Cycle in the Tropics. *Mon. Weather Rev.* **2001**, *129*, 784–801. [CrossRef]
14. Chung, E.S.; Sohn, B.J.; Schmetz, J.; Koenig, M. Diurnal variation of upper tropospheric humidity and its relations to convective activities over tropical Africa. *Atmos. Chem. Phys.* **2007**, *7*, 2489–2502. [CrossRef]
15. Tian, B.; Soden, B.J.; Wu, X. Diurnal cycle of convection, clouds, and water vapor in the tropical upper troposphere: Satellites versus a general circulation model. *J. Geophys. Res. Atmos/* **2004**, *109*. [CrossRef]
16. Soden, B.J. The diurnal cycle of convection, clouds, and water vapor in the tropical upper troposphere. *Geophys. Res. Lett.* **2000**, *27*, 2173–2176. [CrossRef]
17. Soden, B.J.; Bretherton, F.P. Upper tropospheric relative humidity from the GOES 6.7 μm channel: Method and climatology for July 1987. *J. Geophys. Res.* **1993**, *98*, 16669–16688. [CrossRef]
18. Di, D.; Ai, Y.; Li, J.; Shi, W.; Lu, N. Geostationary satellite-based 6.7 μm band best water vapor information layer analysis over the Tibetan Plateau. *J. Geophys. Res. Atmos.* **2016**, *121*, 4600–4613. [CrossRef]
19. Li, Z.; Li, J.; Gunshor, M.; Moeller, S.-C.; Schmit, T.J.; Yu, F.; McCarty, W. Homogenized Water Vapor Absorption Band Radiances From International Geostationary Satellites. *Geophys. Res. Lett.* **2019**, *46*, 10599–10608. [CrossRef]

20. Kottayil, A.; John, V.O.; Buehler, S.A.; Mohanakumar, K. Evaluating the Diurnal Cycle of Upper Tropospheric Humidity in Two Different Climate Models Using Satellite Observations. *Remote Sens.* **2016**, *8*, 325. [CrossRef]

21. Menzel, W.P.; Schmit, T.J.; Zhang, P.; Li, J. Satellite-based atmospheric infrared sounder development and applications. *Bull. Am. Meteorol. Soc.* **2018**, *99*, 583–603. [CrossRef]

22. Dee, D.P.; Uppala, S.M.; Simmons, A.J.; Berrisford, P.; Poli, P.; Kobayashi, S.; Andrae, U.; Balmaseda, M.A.; Balsamo, G.; Bauer, P. The ERA-Interim reanalysis: Configuration and performance of the data assimilation system. *Q. J. R. Meteorol. Soc.* **2011**, *137*, 553–597. [CrossRef]

23. Kobayashi, S.; Ota, Y.; Harada, Y.; Ebita, A.; Moriya, M.; Onoda, H.; Onogi, K.; Kamahori, H.; Kobayashi, C.; Endo, H. The JRA-55 reanalysis: General specifications and basic characteristics. *J. Meteorol. Soc. Jpn. Ser. II* **2015**, *93*, 5–48. [CrossRef]

24. Gelaro, R.; McCarty, W.; Suárez, M.J.; Todling, R.; Molod, A.; Takacs, L.; Randles, C.A.; Darmenov, A.; Bosilovich, M.G.; Reichle, R. The modern-era retrospective analysis for research and applications, version 2 (MERRA-2). *J. Clim.* **2017**, *30*, 5419–5454. [CrossRef] [PubMed]

25. Hersbach, H.; Dee, D. ERA5 reanalysis is in production. *ECMWF Newsl.* **2016**, *147*, 5–6.

26. Saha, S.; Moorthi, S.; Wu, X.; Wang, J.; Nadiga, S.; Tripp, P.; Behringer, D.; Hou, Y.-T.; Chuang, H.; Iredell, M.; et al. The NCEP Climate Forecast System Version 2. *J. Clim.* **2013**, *27*, 2185–2208. [CrossRef]

27. Jiang, X.; Li, J.; Li, Z.; Xue, Y.; Di, D.; Wang, P.; Li, J. Evaluation of Environmental Moisture from NWP Models with Measurements from Advanced Geostationary Satellite Imager—A Case Study. *Remote Sens.* **2020**, *12*, 670. [CrossRef]

28. Han, Y. *JCSDA Community Radiative Transfer Model (CRTM): Version 1*; National Oceanic and Atmospheric Administration: Washington, DC, USA, 2006.

29. Di, D.; Li, J.; Han, W.; Bai, W.; Wu, C.; Menzel, W.P. Enhancing the Fast Radiative Transfer Model for FengYun-4 GIIRS by Using Local Training Profiles. *J. Geophys. Res. Atmos.* **2018**, *123*, 12583–12596. [CrossRef]

30. Ding, S.; Yang, P.; Weng, F.; Liu, Q.; Han, Y.; Van Delst, P.; Li, J.; Baum, B. Validation of the community radiative transfer model. *J. Quant. Spectrosc. Radiat. Transf.* **2011**, *112*, 1050–1064. [CrossRef]

31. Saunders, R.; Rayer, P.; Brunel, P.; Von Engeln, A.; Bormann, N.; Strow, L.; Hannon, S.; Heilliette, S.; Liu, X.; Miskolczi, F.; et al. A comparison of radiative transfer models for simulating Atmospheric Infrared Sounder (AIRS) radiances. *J. Geophys. Res. Atmos.* **2007**, *112*. [CrossRef]

32. Clough, S.A.; Iacono, M.J.; Moncet, J.-L. Line-by-line calculations of atmospheric fluxes and cooling rates: Application to water vapor. *J. Geophys. Res. Atmos.* **1992**, *97*, 15761–15785. [CrossRef]

33. Taylor, P.C. Tropical Outgoing Longwave Radiation and Longwave Cloud Forcing Diurnal Cycles from CERES. *J. Atmos. Sci.* **2012**, *69*, 3652–3669. [CrossRef]

34. Lindfors, A.V.; Mackenzie, I.A.; Tett, S.F.B.; Shi, L. Climatological Diurnal Cycles in Clear-Sky Brightness Temperatures from the High-Resolution Infrared Radiation Sounder (HIRS). *J. Atmo. Ocean. Technol.* **2011**, *28*, 1199–1205. [CrossRef]

35. Lee, M.-I.; Schubert, S.D.; Suarez, M.J.; Bell, T.L.; Kim, K.-M. Diurnal cycle of precipitation in the NASA Seasonal to Interannual Prediction Project atmospheric general circulation model. *J. Geophys. Res. Atmos.* **2007**, *112*. [CrossRef]

36. Kim, H.; Lee, M.-I.; Cha, D.-H.; Lim, Y.-K.; Putman, W.M. Improved representation of the diurnal variation of warm season precipitation by an atmospheric general circulation model at a 10 km horizontal resolution. *Clim. Dyn.* **2019**, *53*, 6523–6542. [CrossRef]

37. Itterly, K.F.; Taylor, P.C. Evaluation of the Tropical TOA Flux Diurnal Cycle in MERRA and ERA-Interim Retrospective Analyses. *J. Clim.* **2014**, *27*, 4781–4796. [CrossRef]

38. Lavers, D.A.; Ralph, F.M.; Waliser, D.E.; Gershunov, A.; Dettinger, M.D. Climate change intensification of horizontal water vapor transport in CMIP5. *Geophys. Res. Lett.* **2015**, *42*, 5617–5625. [CrossRef]

39. Xue, Y.; Li, J.; Menzel, W.P.; Borbas, E.; Ho, S.-P.; Li, Z.; Li, J. Characteristics of Satellite Sampling Errors in Total Precipitable Water from SSMIS, HIRS, and COSMIC Observations. *J. Geophys. Res. Atmos.* **2019**, *124*, 6966–6981. [CrossRef]

40. Udelhofen, P.M.; Hartmann, D.L. Influence of tropical cloud systems on the relative humidity in the upper troposphere. *J. Geophys. Res. Atmos.* **1995**, *100*, 7423–7440. [CrossRef]

41. Van de Berg, L.; Pyomjamsri, A.; Schmetz, J. Monthly mean upper tropospheric humidities in cloud-free areas from meteosat observations. *Int. J. Climatol.* **1991**, *11*, 819–826. [CrossRef]

42. Shi, L.; Schreck, C.J.; Schröder, M. Assessing the Pattern Differences between Satellite-Observed Upper Tropospheric Humidity and Total Column Water Vapor during Major El Niño Events. *Remote Sens.* **2018**, *10*, 1188. [CrossRef]

43. Soden, B.J.; Bretherton, F.P. Interpretation of TOVS water vapor radiances in terms of layer-average relative humidities: Method and climatology for the upper, middle, and lower troposphere. *J. Geophys. Res. Atmos.* **1996**, *101*, 9333–9343. [CrossRef]

44. Chen, S.S.; Houze, R.A., Jr. Diurnal variation and life-cycle of deep convective systems over the tropical Pacific warm pool. *Q. J. R. Meteorol. Soc.* **1997**, *123*, 357–388. [CrossRef]

45. Bessho, K.; Date, K.; Hayashi, M.; Ikeda, A.; Imai, T.; Inoue, H.; Kumagai, Y.; Miyakawa, T.; Murata, H.; Ohno, T.; et al. An Introduction to Himawari-8/9—Japan's New-Generation Geostationary Meteorological Satellites. *J. Meteorol. Soc. Jpn. Ser. II* **2016**, *94*, 151–183. [CrossRef]

46. Schmit, T.J.; Gunshor, M.M.; Menzel, W.P.; Gurka, J.J.; Li, J.; Bachmeier, A.S. Introducing the next-generation Advanced Baseline Imager on GOES-R. *Bull. Am. Meteorol. Soc.* **2005**, *86*, 1079–1096. [CrossRef]

47. Yang, J.; Zhang, Z.; Wei, C.; Lu, F.; Guo, Q. Introducing the New Generation of Chinese Geostationary Weather Satellites, Fengyun-4. *Bull. Am. Meteorol. Soc.* **2016**, *98*, 1637–1658. [CrossRef]

48. Zhang, Y.; Klein, S.A.; Liu, C.; Tian, B.; Marchand, R.T.; Haynes, J.M.; McCoy, R.B.; Zhang, Y.; Ackerman, T.P. On the diurnal cycle of deep convection, high-level cloud, and upper troposphere water vapor in the Multiscale Modeling Framework. *J. Geophys. Res. Atmos.* **2008**, *113*. [CrossRef]

 *remote sensing*

Article

Characterization of the High-Resolution Infrared Radiation Sounder Using Lunar Observations

Martin J. Burgdorf [1,*] , **Thomas G. Müller** [2], **Stefan A. Buehler** [1] and **Marc Prange** [1] and **Manfred Brath** [1]

[1] Meteorological Institute, Department of Earth Sciences, Faculty of Mathematics, Informatics and Natural Sciences, Universität Hamburg, Bundesstraße 55, 20146 Hamburg, Germany; stefan.buehler@uni-hamburg.de (S.A.B.); marc.prange@uni-hamburg.de (M.P.); manfred.brath@uni-hamburg.de (M.B.)

[2] Independent Researcher, 85748 Garching, Germany; Thomas.Mueller@mnet-online.de

[*] Correspondence: martin.burgdorf@uni-hamburg.de; Tel.: +49-4042-838-8121

Received: 1 April 2020; Accepted: 3 May 2020; Published: 7 May 2020

Abstract: The High-Resolution Infrared Radiation Sounder (HIRS) has been operational since 1975 on different satellites. In spite of this long utilization period, the available information about some of its basic properties is incomplete or contradictory. We have approached this problem by analyzing intrusions of the Moon in the deep space view of HIRS/2 through HIRS/4. With this method we found: (1) The diameters of the field of view of HIRS/2, HIRS/3, and HIRS/4 have the relative proportions of 1.4° to 1.3° to 0.7° with all channels; (2) the co-registration differs by up to 0.031° among the long-wave and by up to 0.015° among the shortwave spectral channels in the along-track direction; (3) the photometric calibration is consistent within 0.7% or less for channels 2–7 (1.2% for HIRS/2), similar values were found for channels 13–16; (4) the non-linearity of the short-wavelength channels is negligible; and (5) the contribution of reflected sunlight to the flux in the short-wavelength channels can be determined in good approximation, if the emissivity of the surface is known.

Keywords: infrared sounder; calibration; moon; surface

1. Introduction

The High-resolution Infra-Red Radiation Sounder (HIRS) performs temperature/humidity sounding on satellites in sun-synchronous orbit since the seventies. The first HIRS instrument has been operated on Nimbus-6 from 1975 through 1983. Starting with its first evolution, HIRS/2, built by the Optical Division of ITT Aerospace, it is equipped with 19 infrared channels and forms part of the TOVS sounding instrument suite (TIROS [Television Infrared Observation Satellite] Operational Vertical Sounder) on NOAA-6 to -19. The channel frequencies of each instrument can be found at [1]. It flies as well on TIROS-N, Metop-A, and Metop-B. In all these years the instrument evolved to HIRS/4 and has accumulated a large set of observations relevant to the study of long-term variations of temperature and upper tropospheric humidity (UTH), amongst other things. A trend analysis over three decades of HIRS channel 12 measurements, trying to find changes in tropical UTH, is described for example in [2]. Their investigation, however, was hampered by significant inter-satellite biases, the reason for which was not easily identified. Different spectral response functions, the on board black body calibration system or non-linear response of the detectors could all be at fault. The situation is similar with other channels [3]. A full understanding of the various effects contributing to the systematic errors of HIRS that manifest themselves as biases is further complicated by the fact that one encounters contradictory or incomplete information in the literature even about basic properties of this instrument. Examples are:

- According to the first performance report from ITT Aerospace, the FoV (field of view) of HIRS/2 has a diameter of 1.22° [4]. This value increased to 1.25° in various books, for example [5], and reached a temporary height on the OSCAR web page with some 1.4° [6]. OSCAR gives the resolution in km at s.s.p., which introduces an uncertainty, because the altitude of a satellite on a sun-synchronous orbit can vary by almost 50 km during the mission, and its mean value is not the same for different satellites. The discrepancies get even larger for HIRS/4, where most web pages and documents give a value of some 0.7°, except for the ESA Metop performance page with its extravagant claim of 1.4° for the shortwave channels 13–19 and 1.3° for the long-wave channels 1–12 [7].
- Experts on HIRS cannot agree, either, as to how good the spectral channels co-registration is. This property of the instrument can only be determined in flight for window channels, where it is possible to identify characteristic features on the surface of the Earth. As HIRS has a beamsplitter and two completely different optical paths for long-wave and shortwave channels, systematic pointing differences between these groups of channels are to be expected. Investigations into this matter on ground are not necessarily representative for the conditions in flight, because the strong vibrations during the launch phase can affect the optical path of the instruments.
- The central wavelengths of several channels of HIRS are very similar, but spectral uncertainties remain [8]. This concerns in particular channels 13–16, which lie between 4.4 and 4.6 µm, and channels 1–7, which lie between 13.3 and 15 µm, i.e., close to the frequency of maximum spectral radiance for an object with a brightness temperature of 300 K. These wavelengths cover absorption bands of nitrous oxide and carbon dioxide in the atmosphere and produce therefore quite different flux values for Earth scenes, which makes it difficult to directly determine the inter-channel homogeneity of the sounding channels in flight.
- Estimates of the non-linearity varied by more than a factor of two. A non-linear effect was first detected in flight with HIRS/2 on NOAA-10, where it was maximum in the ninth and tenth channel and lowest in the shortwave channels [9]. When [10] determined later the non-linearity terms for the long-wave channels 4 and 6 of HIRS/4 on Metop-B in flight, they found them exactly 3.333 times smaller than the pre-launch values. Their method worked by identifying those non-linearity terms that produced the smallest orbital mean bias against IASI (Infrared Atmospheric Sounding Interferometer) on the same satellite. The underlying assumption is here that non-linearity is the only reason for bias. An independent confirmation of this claim is highly desirable.
- The impact of reflected solar radiance and the low signal-to-noise ratio at low temperatures adversely affect the accuracy and precision of radiance measurements with the short-wavelength channels, according to [11]. The exact value of the reflected solar radiance depends on the scan position, solar zenith angle, etc. [12], which makes its calculation difficult. It can, however, be very much simplified by assuming diffuse reflection—a concept that can be tested with the *geometric albedo* of the Moon.

The aim of our investigation is to shed some light on these and other issues concerning the performance of HIRS/2–HIRS/4 in flight, or at least to propose new ways of addressing the open questions. This is not only of interest to meteorologists trying to understand biases and other peculiarities in the data from HIRS, but provides as well helpful suggestions for verifying the compliance of future infrared sounders with requirements. Here we make use of intrusions of the Moon in the deep space view (DSV) of HIRS during routine operations. As the Moon has no atmosphere, its infrared spectrum has no narrow, variable features. The hemisphere it presented to the weather satellites was more or less the same for all observations: The sub-observer longitude varied between 351.2° and 6.8°, and the sub-observer latitude, i.e., the apparent planetodetic longitude and latitude of the nearest point of the target seen by HIRS, varied between −5.6° and +6.8°. We could not detect any significant correlation between either coordinate and the measured brightness temperatures. All of these things are strong evidence in favour of a basic assumption relevant for thermo-physical modeling, namely that the disk-integrated properties of the Moon were the same in all observations.

Its temperature, however, varies with the illumination by the Sun, and these variations are much larger than those of the Earth's upper troposphere [13]. Hence observations of Moon intrusions make it possible to monitor the performance of an instrument over a very large range of flux values. They provide new insights into the effects causing systematic uncertainties in the measurements. Such deeper understanding is essential to produce fundamental climate data records from HIRS with consistent calibration. Such a data set could for example form the basis for new climate data records of upper tropospheric humidity as a complement to the corresponding microwave data set [14].

In the next section we describe the method used for identifying suitable observations of the Moon with HIRS, and how we derived brightness temperatures from the raw data. The results are presented in Section 3, and we show for each of the five items mentioned above, how the Moon can prove itself useful by providing new insights. In Section 4 we assess the relevance of these results by comparing them with expectations. Finally, we draw conclusions in the last section on the future use of the Moon in calibration and validation of instruments on weather satellites.

2. Materials and Methods

The first step in our efforts to take advantage of the intrusions of the Moon in the DSV of HIRS was to identify such events in the raw data (level 1b), which are supplied by the NOAA Comprehensive Large Array-data Stewardship System and which were read and procesed by us using Typhon [15]. The HIRS lunar contamination status can be found in plots that are available on the web page of the Center for Satellite Applications and Research Integrated Calibration/Validation System Long-Term Monitoring [16]. This web page, however, does not include monitoring data of HIRS/2, and it does not show the lunar contamination status before 2016. Another, but much more laborious method, is searching the raw data for anomalies in the counts from the DSV during periods of time when the Moon was close to its pointing direction. In Figure 1 we show an example of the signal from deep space for all calibration lines of one orbit. The radiometric calibrations with deep space happen every 256 s, i.e., one gets $6100/256 \approx 24$ such calibration lines per orbit. In between there are Earth scans. Each calibration line contains at least 46 useful measurements, and they are shown in Figure 1 as 24×46 counts. The plot betrays immediately the scan that was affected by the presence of the Moon, because the Moon adds flux to the otherwise empty space. As can be seen in the figure, HIRS produces a lower number of counts for stronger incoming flux. The signal is already reverse after processing in the amplifier chains and before analog to digital conversion, see Section 2.1 in [17]. In a second step we checked with the aid of the infrared light curve of the intrusion of the Moon in the DSV, whether the Moon was fully included in the FoV of HIRS or not. As the number of counts is inversely proportional to the fraction of the lunar disk that falls inside the FoV, this number decreases while the Moon is approaching the center of the FoV and increases while the Moon is leaving the FoV. It is constant at a high level, as long as the Moon is completely outside the FoV, and at a low level, as long as the Moon is completely inside the FoV, apart from random fluctuations of the counts. When during an orbit the Moon is at the orange position in Figure 2, then only a fraction of the lunar disk can be present in the FoV of HIRS, and a falling number of counts is followed immediately by a rise in the number of counts. Whenever it seemed like HIRS got infrared radiation from the whole lunar disk, we looked for additional evidence by comparing the counts with the signal from another intrusion of the Moon with similar phase angle. We usually carried out our search for Moon intrusions with light curves from channel 8, which means that the Moon might not have been fully included in the FoV of the short-wavelength channels, due to systematic pointing differences between long- and shortwave channels, see Section 3.1.2.

The blue circle in Figure 2 is close to the celestial equator and its center coincides with the direction of the orbital axis of the satellite carrying HIRS. When this direction is sufficiently close to the orbit of the Moon, i.e., the orange line, then the Moon will appear every month in the DSV. At other times of the year, however, such an alignment cannot happen. It is also possible, that only Earth scenes were observed, when the Moon crossed the direction of the DSV, or that only part of the Moon fell into the FoV. In consequence, a whole year might pass without a single, useful Moon intrusion.

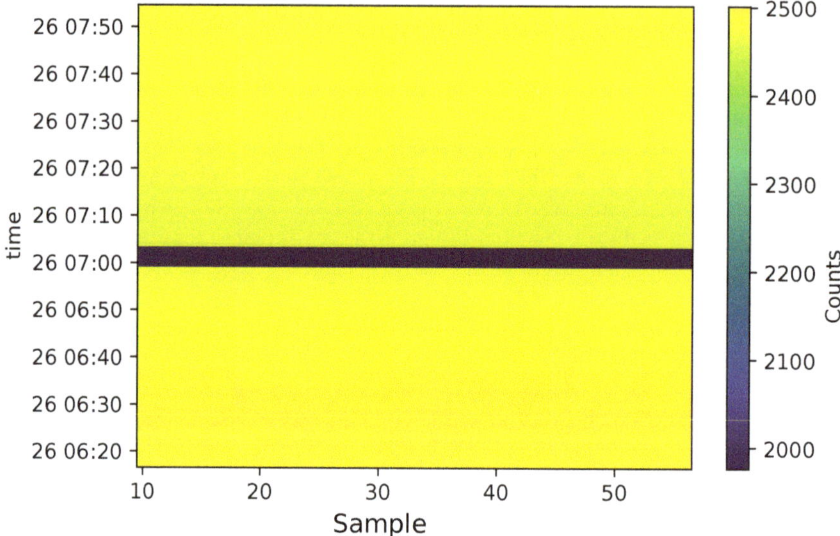

Figure 1. Plot of the numeric counts corresponding to space radiance with HIRS/3 on NOAA-17 during one orbit on 26 September 2002. The first samples were excluded, because the radiometer scanning mirror needs some time to reach its final pointing. The Moon was present in the FoV at 7:01 UTC and provided additional flux. Each scan takes 6.4 s.

We identified a total of 20 suitable intrusions of the Moon in the DSV of seven satellites. This number is large enough for a representative subset of all Moon intrusions and for demonstrating the methods we have developed to learn more about HIRS. It is a far cry, however, from a complete inventory of Moon intrusions, because it is heavily biased towards recent years and the latest satellites. We chose this approach in order to get a balanced set of Moon intrusions from different versions of HIRS, although HIRS/2 flew on more satellites than all other versions combined. Depending on the specific instrumental effect under investigation, we chose those Moon intrusions among our set of 20 that were particularly well suited. Examples are instances of different instruments observing the Moon at the same phase angle or observations, where the Moon appeared in all channels in spite of their small misalignment.

The observations of the Moon are not part of the standard processing of the raw data from HIRS, so we had to calibrate them ourselves. This was done by calculating the average counts $\overline{X}_{sp}$ from the previous and the following DSV calibration line, i.e., 256 s before and 256 s after the Moon intrusion, and using the counts $\overline{X}_{bb}$ from the black body (bb) calibration line that is closest in time to the Moon intrusion. The space radiance R_{sp} is zero for all channels of HIRS, the black body radiance R_{bb} is calculated from the temperature T_{bb} of the black body. The reference counts are the average from some 47 samples, because the first eight to ten samples were taken while the scan mirror was still in motion. This average of 47 samples enters the equation of calibration, without non-linearity, as used by AAPP (ATOVS [Advanced TIROS Operational Vertical Sounder] and AVHRR [Advanced Very High Resolution Radiometer] Pre-processing Package) [18]. The counts obtained, when the Moon is in the FoV, are the average of less than 47 samples, because usually the whole disk of the Moon does not remain in the FoV during the entire duration of the calibration line (6.4 s). The radiant flux density received by HIRS from the lunar surface is calculated according to

$$R = G \cdot \overline{X}_{Moon} + I \tag{1}$$

where $\overline{X}_{Moon}$ is the average counts from the space target. G is defined as

$$G = \frac{R_{bb} - R_{sp}}{\overline{X}_{bb} - \overline{X}_{sp}} \tag{2}$$

where the radiance of the black body is calculated according to Planck's law

$$R_{bb} = \frac{c_1}{\lambda^5 \cdot (e^{\frac{c_2}{\lambda \cdot T_{bb}^*}} - 1)} \tag{3}$$

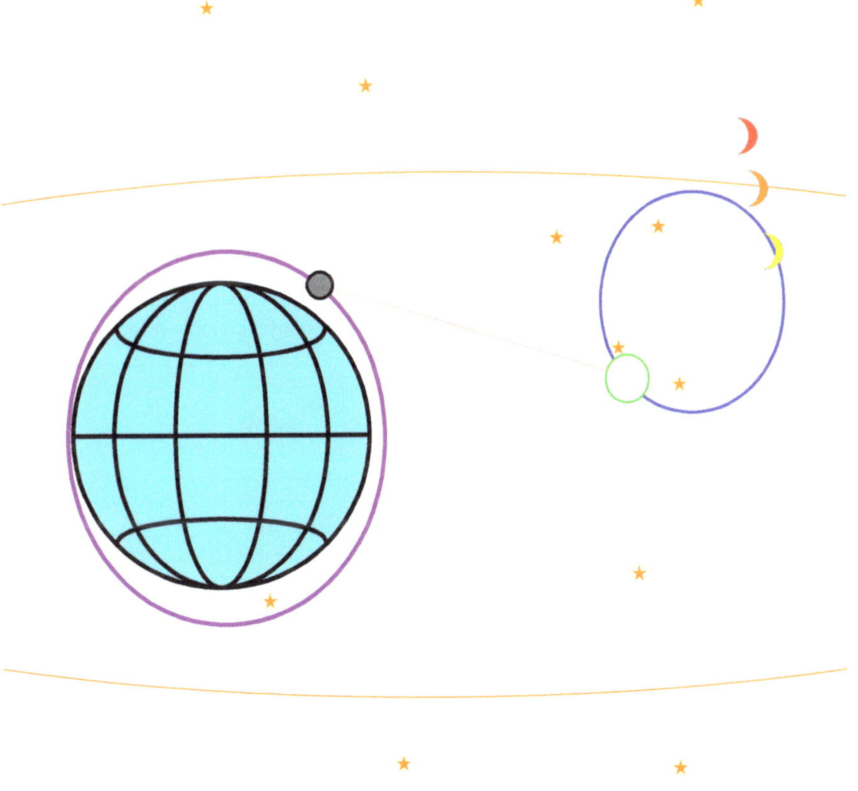

Figure 2. The pointing direction of the DSV describes a circle (blue) in the sky during one orbit of the satellite. When the position of the Moon on its orbit around the Earth (orange) coincides with this circle, it will appear for a short while in the FoV of HIRS. The orbit of the Moon is tilted against the celestial equator, so its minimum distance from the line of sight (straight red line) of the DSV of an instrument on a polar, sun-synchronous orbit (violet) is different each month. The orange position corresponds to a case, where part of the Moon stays outside the FoV, the yellow position corresponds to a case, where HIRS gets the radiation of the complete lunar disk. During that time the signal remains constant.

The temperature of the black body needs a band correction with channel specific constants b and c, which are given in AAPP.

$$T_{bb}^* = b + c \cdot T_{bb} \tag{4}$$

The temperature of the black body is measured with n calibrated platinum sensors, where $n = 4$, except for for HIRS/4, where it is 5.

$$T_{bb} = \frac{\sum_{i=1}^{n} T_i}{n} \tag{5}$$

Every platinum sensor has its own resistance/temperature relationship.

$$T_i = \sum_{j=0}^{5} a_{ij} \cdot \overline{PRT}_i^{j} \tag{6}$$

The measured radiance is zero, when the instrument points at empty space.

$$I = -G \cdot \overline{X}_{sp} \tag{7}$$

with

a_{ij} = conversion coefficient (numeric counts to temperature)
$\overline{PRT}_i$ = mean numeric counts associated to PRT (platinum resistance thermometer) number i
$c_1 = 3.74 \cdot 10^{-16}$ W m^2
$c_2 = 1.44 \cdot 10^{-2}$ K m
λ = wavelength

The Moon has a smaller apparent diameter than the FoV of HIRS, but the calibration targets and also the Earth scenes are extended. For objects that do not fill the FoV, one has to divide R by the fraction of the FoV they cover and the included energy, i.e., the fraction of the flux that actually originates within the FoV as opposed to the contribution from stray light, like this:

$$R_{Moon} = R \cdot (d_{FOV} / d_{Moon})^2 / \eta \tag{8}$$

with:

R_{Moon} = radiance of the lunar disk
d_{FOV} = diameter of the optical field of view
d_{Moon} = diameter of the Moon as seen from the position of the satellite
η = total energy contained within a circle of $1.8°$ ($1 - \eta$ is the fraction of the flux reaching the detector from outside the field of view. Such flux is present with extended sources, e.g., the black body, but not with the Moon. Numerical values can be found in [17,19]).

For the diameter of the FOV we assumed $1.4°$ for HIRS/2, $1.3°$ for HIRS/3, and $0.7°$ for HIRS/4. The included energy is 0.97 for HIRS/2 [4] and 0.98 for HIRS/3 and HIRS/4 [19]. As no uncertainties were reported for these values, it is not clear whether this difference between HIRS/2 and the following versions of the instrument is significant.

We did not correct for changes in temperature of the instrument with a self emission model, although they are known to affect the calibration measurements of the deep space view [20]. This means for our investigation that the self emission could be slightly different at the time of the intrusion of the Moon in the DSV than with the deep space calibration lines before and after. We mitigate this problem by using for our calibration the average of the counts from the deep space calibration lines before and after, but this method removes only the effects of a linear drift of the temperature. By comparing the counts from many sets of three consecutive deep space calibration lines when the Moon was nowhere to be seen, we concluded that the absence of a self emission model adds an uncertainty of one or two counts to the cold calibration reference, but that it does not introduce a systematic error. This does not rule out the possibility of long term effects on the photometric calibration, but they are something altogether different. We did not include a non-linearity term in our equation of calibration, either, because there is no consensus on the correct values for this term, and as a consequence it is set to zero in AAPP. Furthermore we would compromise our aim of deriving upper limits on the non-linearity, if we applied a non-linearity correction already in our processing of the data.

3. Results

As the Moon has got no atmosphere, its spectral energy distribution is close to that of a grey body, with gradual, small variations of its brightness temperature. For a comparison of the disk-integrated flux density of the Moon from HIRS with a thermo-physical model, see [21]. None of the spectral lines familiar from Earth's atmosphere are present on the Moon, and this special quality allows checks of the performance of HIRS in flight that are much more difficult or even impossible in the framework of the routine calibration procedure. An example is checking the coregistration of sounding channels, because one cannot identify surface features on Earth with them. We give in the following several illustrations of how lunar intrusions in the deep space view can serve as diagnostic tool for an infrared sounder. The absolute photometric calibration is not among them, because a sufficiently accurate model of the lunar radiance in all channels of HIRS is not available yet [21].

3.1. System Characteristics

3.1.1. Optical Field of View

According to Equation (8) the measured radiance of the Moon is proportional to the solid angle of the FoV. Hence, it is possible to determine the ratio of the FoVs of different instruments with high accuracy, provided that they observed the Moon at very similar phase angles so that they got more or less the same radiance from the Moon. In this case they must measure the same value, if the assumed FoVs are correct. Variations of the included energy η are negligible, because this value is almost the same for all instruments, viz. close to one, and not controversial in the literature. With other words, the intrusions of the Moon in the DSV put us in a position to find out, which numbers in Table 1 are correct.

Table 1. Diameter of the field of view of different versions of HIRS according to different sources. The values from the NASA and OSCAR web pages are approximate, because they needed to be converted from km at s.s.p. to degrees.

Source	HIRS/2	HIRS/3	HIRS/4
	Degrees	Degrees	Degrees
[4]	1.22	-	-
[5]	1.25	-	-
[18]	1.25	1.25	0.72
[19]	-	1.4 (SW), 1.3 (LW)	0.7
[6]	≈1.38	≈1.27	≈0.67
[7]	1.25	1.25	1.4 (SW), 1.3 (LW)
[22]	-	≈1.32	≈0.66
Moon intrusions in DSV	1.4 ± 0.03	1.3 ± 0.03	0.7 ± 0.01

Table 2 is a collection of nine pairs of observations of the Moon at similar phase angle, but different times. In most cases different versions of HIRS are involved. There were for example intrusions of the Moon in the DSV of HIRS/2 on NOAA-14 on 1996-05-28 and with a very similar phase angle in the DSV of HIRS/3 on NOAA-17 on 2002-09-26. The average brightness temperature of the Moon for all twelve long-wave channels was calculated in order to reduce the uncertainty. We used the shifted, central wavelengths provided by ECMWF for our calculations. In some cases we could do the same calculation also for most shortwave channels; these values are given in Table 3. Because of the poor alignment between long- and shortwave channels, and because there are less shortwave channels to begin with, the calculated brightness temperatures at short wavelengths are averages of much fewer values. As we have chosen channel 8 to identify the Moon intrusions in the deep space view, other long-wave channels are more likely to provide useful data than the shortwave channels.

Table 2. Ratio of the average brightness temperature T^{br} of the Moon as measured with the long-wave channels 1-12 of HIRS on different satellites. This ratio would be one for perfect instruments. The uncertainties reflect the random scatter of the ratios among the different channels. The first column gives the absolute value of the phase angles of the Moon; the pairs were chosen just so these angles are almost the same for either measurement. The value in bold face refers to the only pair, where both measurements were made with the same instrument on the same satellite, but at different times. It is also the only pair, where the measurements were made close to minimum and maximum distance between the Sun and the Moon, which explains the large ratio.

|Phase Angle|	$T^{br}_{HIRS/2}/T^{br}_{HIRS/3}$	$T^{br}_{HIRS/2}/T^{br}_{HIRS/4}$
34.6°/34.8°		0.996 ± 0.001 (NOAA-11/Metop-B)
46.3°/47.5°		1.003 ± 0.002 (NOAA-14/NOAA-18)
51.1°/50.5°	1.007 ± 0.005 (NOAA-14/NOAA-17)	
70.8°/70.6°		1.011 ± 0.002 (NOAA-11/Metop-A)

|Phase Angle|	$T^{br}_{HIRS/3}/T^{br}_{HIRS/4}$	$T^{br}_{HIRS/time1}/T^{br}_{HIRS/time2}$
24.8°/23.8°	1.000 ± 0.002 (NOAA-15/NOAA-18)	
40.1°/40.9°		0.998 ± 0.002 (NOAA-11/NOAA-14)
47.5°/48.5°		1.006 ± 0.002 (NOAA-18/Metop-B)
69.3°/68.0°		**1.014 ± 0.002 (NOAA-14/NOAA-14)**
73.8°/73.1°	0.991 ± 0.002 (NOAA-17/NOAA-18)	

Table 3. Ratio of the average brightness temperature T^{br} of the Moon as measured with channels 13-17 (no channel but 17 was used for the calculation of the pair 24.8°/23.8°) of HIRS on different satellites. As the brightness temperatures were derived from less than six measurements, we did not calculate uncertainties based on their scatter. The value in bold face refers to the only pair, where both measurements were made with the same instrument on the same satellite, but at different times. It is also the only pair, where the measurements were made close to minimum and maximum distance between the Sun and the Moon, which explains the large ratio.

|Phase Angle|	$T^{br}_{HIRS/2}/T^{br}_{HIRS/4}$	$T^{br}_{HIRS/3}/T^{br}_{HIRS/4}$	$T^{br}_{HIRS/time1}/T^{br}_{HIRS/time2}$
24.8°/23.8°		1.007 (NOAA-15/NOAA-18)	
40.1°/40.9°		0.997 (NOAA-11/NOAA-14)	
46.3°/48.5°	0.998 (NOAA-14/NOAA-18)		
69.3°/68.0°			1.011 (NOAA-14/NOAA-14)
73.8°/73.1°		0.996 (NOAA-17/NOAA-18)	

Long-Wave Channels

The calculations were carried out assuming a FoV of 1.4° for HIRS/2, 1.3° for HIRS/3, and 0.7° for HIRS/4. There is at least one comparison for each possible combination of versions of HIRS. Besides, we found three pairs of observations of the Moon with very similar phase angle that were performed with the same version of HIRS. When comparing observations of the Moon with different versions of HIRS, one has to take the slightly different central wavelengths of each channel into account. This inconsistency is particularly pronounced with channel 12, where the central wavelength is 6.7 μm with HIRS/2 and 6.5 μm with HIRS/3 and 4, according to the numbers given by ECMWF. Adding to the confusion in the literature about the HIRS system characteristics, ref. [23] claimed a central wavelength of 6.5 μm also for HIRS/2, but only on two satellites [2], however, demonstrated convincingly that the central wavelength was shifted from 6.7 μm to 6.5 μm only with the launch of HIRS/3 on NOAA-15 in 1998. This shift resulted in a BT difference of 8 K [24] for Earth scenes, because the absorption caused by water vapour in the atmosphere varies strongly between these two wavelengths. On the Moon, however, the emissivity and with it the brightness temperature remain almost constant between 6.5 and 6.7 μm [25]. This is also true for the other channels. In the example of the pair HIRS/2 and HIRS/3 mentioned above we find a value of 1.031 for the ratio of the average radiance of the Moon measured with HIRS/2 and HIRS/3, but 1.007 for the ratio of the corresponding brightness temperatures. This difference is caused in part by the different wavelengths of the channels in version 2 and 3 of HIRS. Hence we determined the ratios of the *brightness temperature* rather than the ratios of the radiance in each pair for all twelve long-wavelength channels and calculated their average and standard deviation of the mean for Tables 2 and 3. We note that the biggest difference is found with two measurements made with the same instrument, viz. HIRS/2 on NOAA-14. Surprisingly the smaller flux was measured here for the smaller phase angle, i.e., closer to full Moon. The explanation for this unusual ratio is that this is also the pair with the largest ratio in the Sun->Moon distance: It amounts to $\frac{1.521 \cdot 10^{13} \text{ cm}}{1.475 \cdot 10^{13} \text{ cm}} = 1.031$. The different brightness temperatures measured on the Moon reflect therefore in this special case the fact that the solar irradiance at perihelion is 106% of the value at aphelion. In all other cases we find differences in measured brightness temperature among the various instruments below 1.1%. This corresponds to less than 4% difference in flux density, which gives an upper limit of the random uncertainty of the diameter of the FoV of about 2%.

Shortwave Channels

Our selection criterion for the Moon intrusions was based on the long-wave channels, but unfortunately there is a systematic misalignment between long- and shortwave channels, because their optical paths are separated by a beamsplitter [4]. Hence we have only five pairs of observations at the same phase angle with the shortwave channels. A direct comparison between measurements with HIRS/2 and HIRS/3 is not among them, but the excellent agreement between HIRS/2 and HIRS/4 and between HIRS/3 and HIRS/4 suggests that the FoVs we assumed are correct also with the shortwave channels. In particular we have proven wrong the occasional claims of different FoVs for different channels of HIRS/3 or HIRS/4 in documents, e.g., [19], or web pages [7] dedicated to HIRS. We note that also the shortwave channels produce the highest ratio of brightness temperatures for the pair with the largest difference in the Sun->Moon distance.

Our measurements suggest that the diameter of the FoV is 1.4° for HIRS/2, 1.3° for HIRS/3, and 0.7° for HIRS/4 with all channels. These values are relevant for the comparison of HIRS data with those from other instruments, when they observed simultaneously the same Earth scene.

3.1.2. Spectral Channels Co-Registration

In a few, rare cases, the light curve of the Moon intrusion shows both decreasing and increasing counts (see Figure 3). The lack of constant signal means that the Moon was never fully included in the FoV, because HIRS would receive the radiation from the complete disk as long as this is the case. At least, however, the moment of its closest approach to the pointing direction of the DSV happened during the calibration procedure. In this case it is possible to determine exactly the time of this closest approach for each channel and to derive from this information the HIRS spectral channels coregistration in the along-track direction. We did that by fitting a second order polynomial to the light curve of channel 8, shown in Figure 3, and the light curves of all other channels. Then we determined the number of the sample, where the second order polynomial reached its minimum. The uncertainty of the position of the minimum was calculated from the uncertainties of the parameters produced in the polynomial fit. Then we converted the number of the sample to an angular displacement. For this last step we followed the method described by [26]. The whole procedure allows us to find out, whether the different channels point in the same direction. This assumption is often taken for granted by meteorologists when working with data from HIRS.

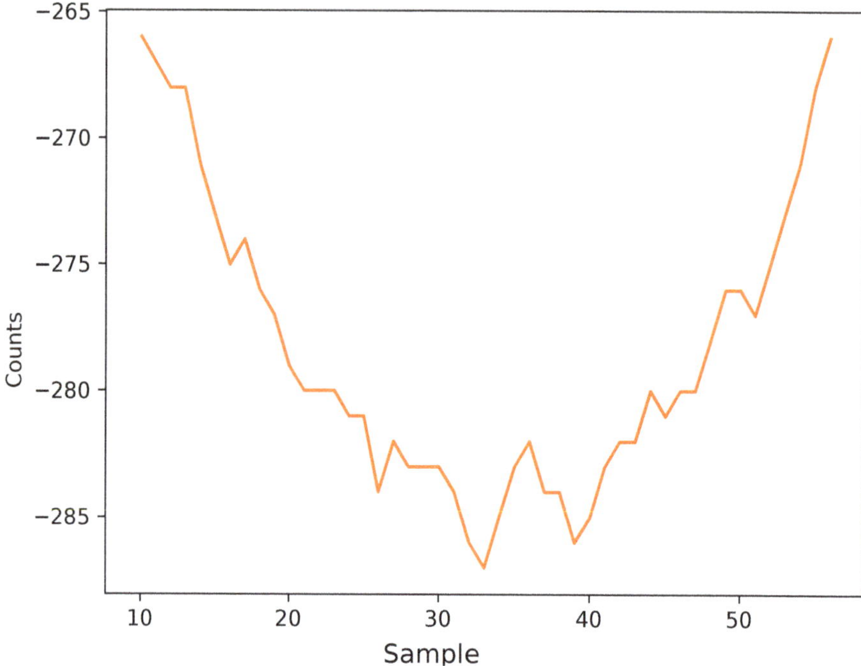

Figure 3. Signal from all samples taken during the calibration line performed by HIRS/4 on NOAA-19 on 2012-03-04 between 5:07:0.8 and 5:07:7.2 UTC. The first positions were excluded, because the radiometer scanning mirror needs some time to reach its final pointing. The Moon came closest to the center of the FoV of channel 8—the one plotted here—for scanposition 34.

Table 4 lists the sample number of this closest approach for each channel, except for number 1. Channel 1 was excluded because of its poor signal-to-noise ratio. Its SNR is small, because the difference in counts between low and high fluxes is smaller with channel 1 than with the other channels. There is a clear trend in the sense that this sample number decreases along the rows of the table, but there is a discontinuity between channels 12 and 13, i.e., between SW and LW. This suggests the presence of chromatic aberration. As the long-wave and shortwave optical paths have no lenses in common [17], their variations in refractive index are different. On the other hand the correlation between pointing direction and wavelength is in case of LW significantly higher than the correlation between pointing direction and channel number, because channel 10 does not fit the sequence of decreasing wavelength. This fact demonstrates that it is not some tilt of the filter wheel that matters for the misalignment of the different channels, but rather a property of the lenses.

During the calibration procedure, the instrument stays at the same scan position, which is 68 for space view, but its pointing direction in the sky changes because of the movement of the satellite on its orbit around the Earth. The angular distance between the pointing directions of two consecutive samples of the space calibration line is:

$$\Delta\phi = t \cdot sin\theta \cdot 360° / P = 0.0019° \tag{9}$$

with:

t = dwell time = 100 msec

θ = space view position relative to the orbital axis = 161.1° (pointing away from the Sun)

P = orbital period = 101.5 min for NOAA-19.

Table 4. Sample number of the smallest distance between the position of the Moon and the center of the DSV for the space calibration line from 2012-03-04 at 5:07 UTC with HIRS/4 on NOAA-19. The differences in sample number correspond to differences in the pointing direction of the different channels in the along-track direction. The numbers given in the fourth and fifth row are relative to the position of the Moon in channel 19, and they are positive, if the pointing direction of a channel is displaced in the flight direction.

Channel Number	2	3	4	5	6
Wavelength/μm	14.685	14.526	14.232	13.973	13.635
Sample Number	39.8 ± 10.3	38.1 ± 4.6	36.9 ± 3.9	36.4 ± 5.4	37.8 ± 3.0
Displacement/°	0.0444 ± 0.0197	0.0412 ± 0.0088	0.0389 ± 0.0075	0.0379 ± 0.0103	0.0406 ± 0.0057
Displacement/km	0.67	0.63	0.59	0.58	0.62

Channel Number	7	8	9	10	11
Wavelength/μm	13.347	11.124	9.729	12.456	7.352
Sample Number	39.7 ± 4.2	33.9 ± 1.4	28.3 ± 2.9	41.6 ± 4.0	26.8 ± 4.1
Displacement/°	0.0442 ± 0.008	0.0331 ± 0.0027	0.0224 ± 0.0056	0.0479 ± 0.0077	0.0195 ± 0.0079
Displacement/km	0.67	0.5	0.34	0.73	0.3

Channel Number	12	13	14	15	16
Wavelength/μm	6.529	4.577	4.517	4.479	4.451
Sample Number	21.5 ± 7.5	25.2 ± 1.1	23.5 ± 0.7	23.0 ± 0.9	22.2 ± 0.9
Displacement/°	0.0094 ± 0.0144	0.0165 ± 0.0021	0.0132 ± 0.0013	0.0122 ± 0.044	0.0107 ± 0.0425
Displacement/km	0.14	0.25	0.2	0.19	0.16

Channel Number	17	18	19
Wavelength/μm	4.131	3.971	3.757
Sample Number	20.3 ± 0.9	17.5 ± 1.0	16.6 ± 0.4
Displacement/°	0.0071 ± 0.0389	0.0017 ± 0.0335	0 ± 0.0008
Displacement/km	0.11	0.003	0

The fourth and fifth row (*Displacement*) of Table 4 give the differences between the positions of the Moon in the along-track direction found with channel 19 and the other channels. These values are plotted in Figure 4. There is a systematic shift in the position of the Moon as seen through the different filters, and the slope of position as a function of wavelength is larger for the SW channels than for the LW channels. This misalignment must be taken into consideration for estimating the overall uncertainty of the result, when flux densities measured in different channels are combined to calculate climate variables.

3.1.3. Inter-Channel Uniformity

The sounding channels of HIRS measure flux densities at several different wavelengths in order to characterise the exact shape of a spectral feature. In order to obtain meaningful results, the different channels do not only have to point in the same direction, they also must have a consistent flux calibration. These preconditions are usually not questioned, when the measurements are used to retrieve atmospheric variables. It is desirable to check the validity of these assumptions, and therefore we want to derive now upper limits for the systematic discrepancies between channels.

As the Moon has got no atmosphere, the N_2O (SW) and CO_2 (LW) sounding channels should always give almost the same lunar brightness temperature. In Table 5 we give the values for the brightness temperatures and their standard deviations based on channels 2–7 for different versions of HIRS. According to the last column of the table it is typically 1.0 K for HIRS/2 and 0.6 K for HIRS/4, corresponding to about 1.2% or 0.7%, respectively, in radiance. These figures, however, are only an upper limit of the inter-channel bias, because the brightness temperature decreases slightly with increasing wavelength for the channels considered here, and this systematic trend inflates the calculated standard deviations. This finding was expected in the light of the properties of the lunar soil [25] between 9 and 11 μm, because it means that also the radiance and as a consequence the emissivity of the lunar soil decreases by about 1%. We note the fact that HIRS observed the disk-integrated radiance at non-zero phase angles of the Moon, i.e., the measured spectral energy distribution is the average over quite different angles of incidence and reflection. Hence, any systematic trends of brightness temperature with wavelength seen by HIRS could differ from those shown in the plots by [25].

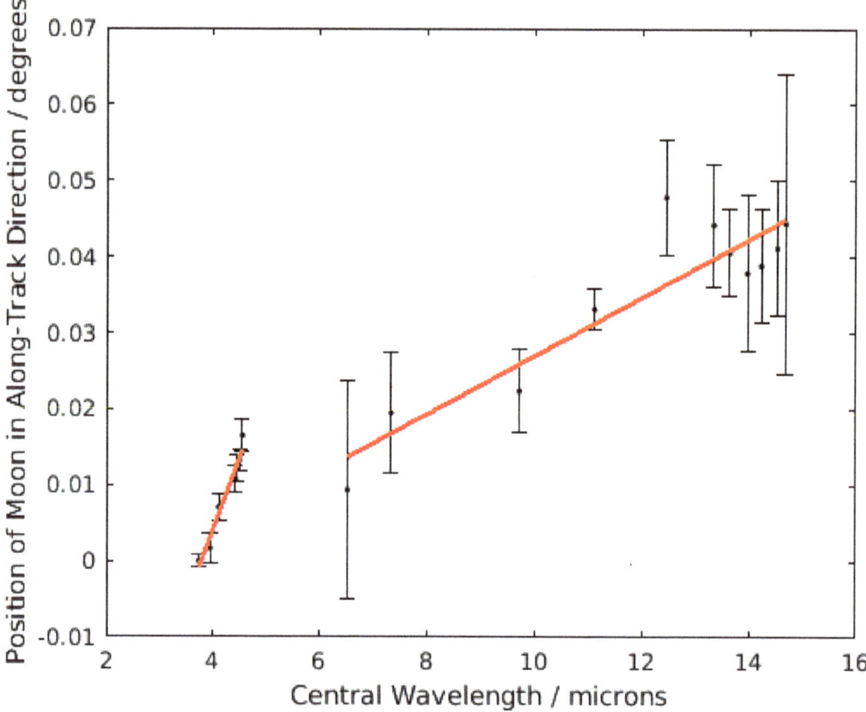

Figure 4. Position of the Moon in the along-track direction according to each infrared channel of HIRS/4 on NOAA-19 during its intrusion in the DSV on 2012-03-04 at 5:07 UTC. The values are relative to the position of the Moon in channel 19, and they are positive, if the pointing direction of a channel is displaced in the flight direction. The relationship between the central wavelength of the channels and the displacement is plotted as two different red lines for the SW and the LW channels. The channel numbers are, from left to right: 19, 18, 17, 16, 15, 14, 13, 12, 11, 9, 8, 10, 7, 6, 5, 4, 3, 2.

Table 5. Average brightness temperatures and their standard deviations for channels 2–7 at different phase angles of the Moon. The central wavelengths of these long-wave CO_2 channels lie between 13.3 and 14.8 μm for HIRS/2, 3, and 4; their calibration has been studied in detail by [8].

Instrument	Satellite	Phase Angle	$<T_B^{Ch2-7}>$	$\sigma(T_B^{Ch2-7})$
		Degrees	**K**	**K**
HIRS/2	NOAA-11	−34.6	343.3	0.9
HIRS/2	NOAA-11	−40.1	335.3	0.9
HIRS/2	NOAA-11	−57.5	306.6	1.2
HIRS/2	NOAA-11	−70.8	282.5	1.0
HIRS/2	NOAA-14	−46.3	323.7	1.1
HIRS/2	NOAA-14	−51.1	310.8	1.1
HIRS/3	NOAA-15	+24.8	349.1	0.6
HIRS/4	NOAA-18	+23.8	348.1	0.8
HIRS/4	NOAA-19	−15.6	359.1	0.6
HIRS/4	Metop-B	+34.8	343.8	0.5
HIRS/4	Metop-B	+48.5	320.3	0.7

The SW channels show an even stronger trend of increasing brightness temperature towards smaller wavelengths than the LW channels. There are two reasons for this:

- At longer wavelengths one sees a temperature, which is close to the disk-average temperature of the Moon, but at shorter wavelengths the radiance is dominated by the hottest (sub-solar) surface areas on the Moon.
- At shorter wavelengths the share of reflected sunlight becomes larger. It should be subtracted from the flux density that HIRS receives from the Moon, before one can analyze the inter-shortwave-channel uniformity (see Section 3.2).

Our measurements prove that the inter-channel uniformity of the carbon dioxide sounding channels has no systematic component larger than the random scatter of the measurements. The measurements with HIRS are trustworthy.

3.1.4. Non-Linearity

The correct equation of calibration is needed for calculating the correct radiance and its uncertainty, and therefore it is important to know, whether the relationship between counts and radiance is linear or not. The HIRS operational calibration algorithm [27] sets all non-linearity coefficients to zero, because their effect is supposed to be negligible. A detailed investigation of this question, however, was only carried out for a few LW channels [10]. We use the observations of the Moon to derive an upper limit on the non-linearity coefficient of most SW channels. In doing so we take advantage of the fact that the sub-solar region of the Moon reaches temperatures of almost 400 K, i.e., more than 100 K above the temperatures of the black body and typical Earth scenes. The shortwave channels have central wavelengths between 3.7 and 4.6 μm, where the radiance grows exponentially with temperature according to Planck's law for short wavelengths. This means that, when the DSV is pointed at full Moon, the SW channels receive flux densities that are several times higher than those the black body can provide. As the non-linearity term in the measurement equation increases with the square of the counts, it must feature in observations of the Moon, if it is there at all.

For the calculation of the non-linearity term we follow the definition of [10]:

$$R^{nl}_{Moon} = ((\frac{R_{bb}}{X_{bb} - X_{sp}} - q \cdot (X_{bb} - X_{sp})) \cdot (X_{Moon} - X_{sp}) + q \cdot (X_{Moon} - X_{sp})^2) \cdot (d_{FOV}/d_{Moon})^2/\eta \qquad (10)$$

with:

q = non-linearity
R^{nl}_{Moon} = radiance of the Moon after correction for non-linearity.

The non-linearity correction makes a difference d_{nl} (in percent of the lunar radiance) of

$$d_{nl} = \frac{R^{nl}_{Moon} - R_{Moon}}{R_{Moon}} \cdot 100 = \frac{q \cdot (X_{Moon} - X_{bb}) \cdot (X_{Moon} - X_{sp}) \cdot (d_{FOV}/d_{Moon})^2}{R_{Moon} \cdot \eta} \cdot 100 \qquad (11)$$

On 6/15, 1997, there was an intrusion of the Moon in the DSV of HIRS/2 on NOAA-14, and on 7/21, 2019, in the DSV of HIRS/4 of Metop-B. The phase of the Moon was in either case some 47°, therefore R_{Moon} was almost the same. The FoV is different with the different versions—HIRS/2, HIRS/3, and HIRS/4—but in each case big enough to fully include the Moon, and therefore big enough to receive the flux from the whole disk. The situation is different, however, with the black body, because this calibration reference has a larger diameter than any FoV of HIRS. Therefore the flux received from the black body is proportional to the radius of the FoV squared. With other words, because the DSV of HIRS/4 has only half the radius of the one of HIRS/2, the flux density obtained from the black body S_{bb} of HIRS/4 is four times smaller than S_{bb} of HIRS/2. In the case we consider here, where the Moon is observed at a phase angle of 47°, it provides a similar flux density as the black body of HIRS/2 does. This means, however, that the term $X_{Moon} - X_{bb}$ is close to zero, whereas the same term is much larger with the small FoV of HIRS/4, and d_{nl} of all SW channels is more than ten times higher for HIRS/4 than for HIRS/2—in case of channel 17 the ratio even amounts to 236. Hence for an estimate of the non-linearity coefficient one can assume that the non-linearity is negligible with the Moon intrusion of HIRS/2, and we take the fluxes measured with this instrument as reference. They agree, however, within 1.5% with the flux values obtained with HIRS/4. Hence we conclude that the non-linearity, if uncorrected, causes at most an error of this size with HIRS/4. The corresponding upper limits for the values of q are given in Table 6.

Table 6. Upper limit of the non-linearity and the ratio of flux densities received from the Moon and the black body for channels 13-17 of HIRS/4 on Metop-B.

Channel	Central Wavelength	Maximum Non-linearity Term	Dynamic Range ($\frac{S_{Moon}}{S_{bb}}$)
	μm	10^{-8} mW m^{-2} cm sterad^{-1} counts^{-2}	
13	4.575	1.1	3.0
14	4.532	1.1	3.1
15	4.476	0.8	3.1
16	4.458	0.7	3.2
17	4.130	0.6	4.0

The non-linearity terms of channels 13–17 are at least a factor ten smaller than the pre-launch values for the LW channels [10]. The shorter the wavelength, the larger the flux difference between Moon and black body, and the tighter the constraint on the non-linearity coefficient. Our data are compatible with $q = 0$ for all SW channels and lend support to the equation of calibration used in AAPP. It is planned to extend the search for non-linearity effects to observations of the Moon at a variety of phase angles in [21].

3.2. Reflected Solar Radiance

As the STD of the shortwave channels is so large that in the Antarctic June the observed radiance for example in channel 19 is at the level of instrument noise [11], they are best used at daytime. This rule applies also to observations of the Moon: When it is full, the reflected sunlight alone gives already a satisfactory signal-to-noise ratio. In order to calculate its flux density, we assume that the Sun is a black body with the temperature [28]:

$$T_{blackbody\odot} = T_{eff\odot} = (L_\odot/\sigma)^{1/4} = 5778K \tag{12}$$

with:

$T_{eff\odot}$ = solar effective temperature

$L_\odot$ = solar absolute luminosity

σ = Stefan-Boltzmann constant.

This approximation is good enough for our purpose of getting an estimate of the contribution of reflected sunlight to Earth or Moon scenes at wavelengths around 4 μm [29]. We want to demonstrate that the thermal emission of the Moon in the shortwave channels can be determined accurately enough by a correction of the measured flux density that only requires the reflectance of the scene and its distance from the Sun. For this we assume that the Moon reflects 20% of the incoming radiation at 4 μm [25]. Table 7 gives the brightness temperatures of the Moon for three different phase angles from the SW sounding channels of HIRS/4 on three different satellites. The reflected sunlight was subtracted, taking the distance between the Sun and the Moon at the time of its intrusion in the DSV into account. None of the measured brightness temperatures differs by more than 0.4 K from the average value of channels 13–16, suggesting an even better inter-channel uniformity than with the long-wave channels.

Table 7. Brightness temperature of the Moon in the four channels with central wavelengths around 4.5 μm for three different phase angles. T_B is the brightness temperature of the Moon after subtraction of the reflected sunlight.

Phase angle	Satellite	Distance Sun-Moon	T_B^{Ch13}	T_B^{Ch14}	T_B^{Ch15}	T_B^{Ch16}	$T_B^{average}$
Degrees		10^{13} cm	K	K	K	K	K
48.5	Metop-B	1.5228	338.45	337.89	337.91	338.02	338.07
70.6	Metop-A	1.4806	318.63	319.03	318.84	318.85	318.84
−73.1	NOAA-18	1.4883	312.92	312.66	312.44	312.51	312.58

The reflected sunlight is always less then 8% of the overall flux density received from the Moon in the examples of Table 7, and the average temperatures given in the last column of this table are our best estimate for the brightness temperature of the Moon in the shortwave range of HIRS. The Sun's share in the measured flux densities increases, however, towards shorter wavelengths, because the Sun is on the Rayleigh-Jeans branch of the Planck function, and the Moon is on the Wien branch. Hence we can now use the average brightness temperatures from Table 7 to calculate the radiance from the Moon at the central wavelengths of channels 17–19, if there was no reflected sunlight present, and then calculate the albedo of the Moon at these wavelengths from the difference between the actually measured flux density and what we would get from the thermal radiation of the Moon alone. Here we assume that no thermal radiation of the Moon is emitted by its night side, because it is very cold and on the Wien branch of the Planck function. The results are listed in Table 8—all values are compatible with the emissivity determined by [25] and do not vary much among the three channels 17–19. This consistency, especially at the smallest phase angle, proves that the values for the albedo are close to the truth, else they would change towards shorter wavelengths, where the ratio between emitted and reflected infrared light shifts quickly in favour of the latter. A value for the reflectivity of the Moon at 4 μm that is significantly larger than 20%, as for example proposed by [30], would cause larger inconsistencies among the values in Table 8 and is therefore off the mark.

This method can be applied also the other way round, when the emissivity of an Earth scene is known, for example when the satellite flies over the Sahara. In this case the unwanted contribution from reflected sunlight to the overall signal can be subtracted, and the shortwave window channels can supply trustworthy measurements.

Table 8. Contribution by reflected sunlight to the radiance obtained from the Moon ΔR and geometric albedo p of the Moon at the central wavelengths of the three SW window channels with HIRS/4. T_B is the brightness temperature of the Moon after subtraction of the reflected Sun light.

Phase Angle	T_B	ΔR^{Ch17}	p^{Ch17}	ΔR^{Ch18}	p^{Ch18}	ΔR^{Ch19}	p^{Ch19}
Degrees	K	10^6 MJy sterad^{-1}		10^6 MJy sterad^{-1}		10^6 MJy sterad^{-1}	
48.5	338.07	3.31	0.23	3.54	0.23	3.85	0.23
70.6	318.84	3.04	0.20	2.92	0.18	3.08	0.17
−73.1	312.58	2.91	0.20	2.99	0.19	2.92	0.17

4. Discussion

The random uncertainty of our determination of the diameter of the FoV amounts to 2%, but strictly speaking we have only proven that the relative proportions of the FoVs of the different versions of HIRS are 1.4:1.3:0.7. Absolute values for the FoVs can only be calculated, if absolute values for the radiance of the Moon are known with high accuracy. Existing models of the brightness temperature of the Moon, however, have typical uncertainties of 5 K [31], and only few observations of the Moon with other infrared sounders than HIRS have been analyzed and published. As the DIVINER Lunar Radiometer Experiment on the Lunar Reconnaissance Orbiter did neither cover the wavelength range between 3.0 and 7.5 μm nor from 8.6 to 12.5 μm, the measurements we present from HIRS must be considered a unique source of information about disk-integrated brightness temperatures of the

Moon. Given the fact that values between 0.69° and 0.7° for the FoV of HIRS/4 are well established in the literature, we are confident that our assumptions about the size of the FoV are correct.

The ratio of the lunar radiance from our observations close to perihelion and aphelion was 1.056 according to the measurements in channel 4 (14.2 μm). This value is very close to the 6% seasonal variation in the Moon's thermal emission found with CERES (Clouds and the Earth's Radiant Energy System) [32]. The solar flux that the Moon absorbs and also its emitted thermal flux are inversely proportional to the square of its distance from the Sun. $\sqrt{1.056} = 1.028$, which is quite close to the ratio of the distances: 1.031. Detecting the effect of the eccentricity of the Earth's orbit on the temperature of the Moon's surface with only two observations is an impressive demonstration of the performance of HIRS, and we conclude that the brightness temperature of the Moon in the thermal infrared is 1%–2% higher at perihelion than at aphelion for a phase angle of some 70°.

According to [19] the channel to channel registration is less than 0.01° for LW and less than 0.007° for SW. These values do not agree with our findings: The pointing direction in the along-track direction alone differs already by up to 0.031° among the long-wave channels of HIRS/4 on NOAA-19, and by up to 0.015° among the shortwave channels, based on accurate measurements of when the Moon came closest to the center of the FoV of each channel. The channel to channel registration in flight is hence at least two times worse than claimed in the KLM User's Guide for SW and three times for LW. Our results, however, are very similar to actual measurements of the centroid location of the SW channels of HIRS for NIMBUS F before launch [33]. The average pointing direction of all SW channels differs by 0.026° from the direction of the LW channels in the along-track direction, therefore there is a significant misalignment between the two groups of channels. As a consequence of this and a possible misalignment in the along-scan direction as well, the Moon was never fully included in the sounding shortwave channels in a third of the intrusions we found with the long-wave channels. This problem is worst for HIRS/4, because of its small FoV. We attribute the misalignment to different chromatic aberration of the lenses in the long-wave and shortwave optical paths and recommend to take this defect into account in the design of similar instruments in the future.

The small, but significant differences among the brightness temperatures measured by the various N_2O and CO_2 sounding channels contain information about the wavelength dependence of the emission properties of the bulk material on the Moon. As the disk-integrated fluxes, however, are the sum of areas with quite different distances from the "sub-Sun" and "sub-HIRS" point, a thermo-physical model is needed to interpret our findings—a task that goes beyond the scope of this article.

Our method is not able to reproduce the biases between different satellites detected by other authors in the past [3,11], because we do not have observations of the Moon at exactly the same phase angle with the satellite pairs they used. Besides, the uncertainty of a single observation of the Moon in a given channel would have to be a small fraction of a Kelvin, and we cannot achieve that without a self emission model for HIRS. We conclude, however, on the basis of our investigation into non-linearity that any biases that may be present in the shortwave channels are rather caused by errors in the HIRS spectral response function, as stated by [11].

Finally we mention the fact that a quite simple method for subtracting the contribution from sunlight in the shortwave channels produced surprisingly good results: Only the reflectance of the scene at the central wavelength of the channel and its distance from the Sun are needed. Hence we believe this technique could easily be applied to HIRS Earth-viewing measurements where both variables are readily available, such as from the study of surface emissivity and reflectance of northern Africa at 11.1 μm, 8.3 μm, and 4 μm, which was carried out by [34]. Although the determination of the reflected sunlight gets less reliable, when the Sun-scene-HIRS angle (phase angle when the scene is the Moon) is close to 90°, it should be good enough for most of the swath of HIRS, which extends from −49.5° to +49.5° around nadir. The Metop satellites have a local equator crossing time of 9:30, i.e., seen from the nadir pixel on Earth, the Sun has an hour angle of 37.5° when the satellite crosses the equator - again a value much smaller than 90°. This means that over tropical regions with known reflectance it should be possible to eliminate the reflected sunlight without major impact on the overall uncertainty of the measurements.

5. Conclusions

The Moon has been observed with HIRS on many different satellites. We identified a few of these observations that offered particularly illuminating information about basic properties of the instrument. A basic calibration of the raw data was sufficient to characterize various effects with an impact on the performance of the instrument. In some cases this concerned properties that have never been determined in flight before. We have described the methods employed and given examples of the accuracy that can be achieved. The accuracy of the measured brightness temperature of the Moon might be further improved by correcting for the HIRS instrument self-emission [20].

Although it was not our intention to present a thorough study of all observations of the Moon from all satellites that carried HIRS, we were able to fill some gaps in the knowledge about this instrument. We ended the confusion about the size of the field of view, characterised how the co-registration of channels depends on their central wavelength in flight, and supplied upper limits on the non-linearity of the shortwave channels. All of these things are essential for a proper estimate of the uncertainties of the data from HIRS and as well for judging its compliance with the requirements.

The Moon has also been observed with other infrared sounders, e.g., CERES [35] or IASI, and therefore it offers unique possibilities for cross calibration. This includes comparisons with future instruments like IASI—New Generation or the Meteorological Imager on Metop Second Generation.

The (disk-integrated) Moon data, obtained with different versions of HIRS in different wavelength channels, are very consistent. Hence, they are well suited to verify/benchmark thermo-physical model (TPM) techniques, which are widely used for (disk-integrated) thermal IR measurements of other airless bodies (like asteroids, satellites, trans-Neptunian objects, or inactive comets). The benchmarked TPM of the Moon would then also help to calibrate thermal IR instruments of other satellites, e.g., interplanetary missions like the Origins Spectral Interpretation Resource Identification Security—Regolith Explorer or Hayabusa2. Both of them have looked at the Moon during swing-by maneuvers with their IR instruments to obtain an in-flight calibration [36,37]. We aim for a thermophysical model of the Moon using the available global properties and also a well-established directional hemispherical emissivity. This model will take the true observing and illumination geometries (as seen from the satellites) into account. Eventually we intend to establish the Moon as a calibration reference with empirical uncertainties for infrared instruments to evaluate their calibration accuracy and to assess their long-term calibration stability. Similar efforts are already underway with microwave instruments [38] and optical sensors in the framework of inter-agencies collaborations, for example at ESA [39] and EUMETSAT [40].

Author Contributions: Conceptualization, M.J.B., T.G.M., S.A.B. and M.B.; software, M.J.B.; validation, T.G.M.; formal analysis, M.J.B. and M.P.; investigation, M.J.B. and T.G.M.; writing–original draft preparation, M.J.B.; writing–review and editing, T.G.M., S.A.B. and M.B.; supervision, S.A.B.; funding acquisition, S.A.B. All authors have read and agreed to the published version of the manuscript.

Funding: This research was funded by Deutsche Forschungsgemeinschaft, project number 421761264. With this work we contribute to the Cluster of Excellence "CLICCS—Climate, Climatic Change, and Society" funded by the Deutsche Forschungsgemeinschaft DFG (EXC 2037, Project Number 390683824), and to the Center for Earth System Research and Sustainability (CEN) of Universität Hamburg. TM received funding from the European Union's Horizon 2020 Research and Innovation Programme, under Grant Agreement no 687378, as part of the project "Small Bodies Near and Far" (SBNAF).

Acknowledgments: The authors would like to thank Hu Yang for his comments on the first version of the manuscript. The level 1b files of HIRS were read and processed using Typhon [15].

Conflicts of Interest: The authors declare no conflict of interest.

Abbreviations

The following abbreviations and mathematical symbols are used in this manuscript:

AAPP	ATOVS and AVHRR Pre-processing Package
ATOVS	Advanced TIROS Operational Vertical Sounder
AVHRR	Advanced Very High Resolution Radiometer
BB	Black Body
BT	Brightness Temperature
CERES	Clouds and the Earth's Radiant Energy System
DSV	Deep Space View
ESA	European Space Agency
FoV	Field of View
HIRS	High-resolution Infrared Radiation Sounder
IASI	Infrared Atmospheric Sounding Interferometer
IR	InfraRed
ITT	International Telephone and Telegraph
LW	Long-Wave
Metop	Meteorological operational satellite
NOAA	National Oceanic and Atmospheric Administration
OSCAR	Observing Systems Capability Analysis and Review tool
PRT	Platinum Resistance Thermometer
s.s.p.	sub-satellite point
SNR	signal-to-noise ratio
sp	space
STD	STandard Deviation
SW	ShortWave
TIROS	Television InfraRed Observation Satellite
TOVS	TIROS Operational Vertical Sounder
TPM	Thermo-Physical Model
UTH	Upper Tropospheric Humidity
ΔR	contribution by reflected sunlight to the radiance from the Moon
η	total energy contained within a circle of $1.8°$, i.e., fraction of flux received that is not straylight
λ	shifted, central wavelength of a channel
σ	Stefan-Boltzmann constant
θ	space view position relative to the orbital axis = $161.1°$ (pointing away from the Sun)
a_{ij}	conversion coefficient (numeric counts to temperature)
b	channel specific band-correction coefficient in the file `calcoef.dat` in AAPP
c	channel specific band-correction coefficient in the file `calcoef.dat` in AAPP
c_1	$3.74 \cdot 10^{-16}$ W m^2
c_2	$1.44 \cdot 10^{-2}$ K m
d_{FOV}	diameter of the optical field of view
d_{Moon}	diameter of the Moon as seen from the position of the satellite
d_{nl}	non-linearity correction as percentage of the measured radiance
G	gain
I	offset term in measurement equation
$L_{\odot}$	solar absolute luminosity
n	number of platinum sensors on the black body
p	geometric albedo
P	orbital period = 101.5 min for NOAA-19
$\overline{PRT}_i$	mean numeric counts associated to PRT number i
q	non-linearity
R	radiant flux density received by HIRS from the lunar surface
R_{bb}	radiance of the black body
R_{Moon}	radiance of the lunar disk
R_{Moon}^{nl}	radiance of the Moon after correction for non-linearity
R_{sp}	radiance of the space target, zero for all channels
S_{bb}	flux density obtained from the black body
S_{Moon}	flux density obtained from the Moon
t	dwell time = 100 msec
T_B	brightness temperature of the Moon without reflected sunlight
T_{bb}	temperature of the black body as measured with PRTs

$T_{eff\odot}$	solar effective temperature
T_i	temperature of the black body as measured with PRT number i
T_{bb}^*	temperature of the black body after band correction
T^{br}	brightness temperature
$\overline{X}_{bb}$	average counts from the black body
$\overline{X}_{Moon}$	average counts from the space target, Moon in FoV
$\overline{X}_{sp}$	average counts from the space target, no Moon in FoV

References

1. EUMETSAT NWP SAF. Visible/IR Spectral Response Functions. 2017. Available online: https://www.nwpsaf.eu/site/software/rttov/download/coefficients/spectral-response-functions/ (accessed on 4 May 2020).
2. Shi, L.; Bates, J.J. Three decades of intersatellite-calibrated High-Resolution Infrared Radiation Sounder upper tropospheric water vapor. *J. Geophys. Res.* **2011**, *116*, D04108. [CrossRef]
3. Shi, L. Intersatellite Differences of HIRS Longwave Channels Between NOAA-14 and NOAA-15 and Between NOAA-17 and METOP-A. *IEEE Trans. Geosci. Remote* **2013**, *51*, 1414–1424. [CrossRef]
4. Koenig, E.W. Performance of the HIRS/2 Instrument on TIROS-N. In *Remote Sensing of Atmospheres and Oceans*; Deepak, A., Ed.; Academic Press: New York, NY, USA, 1980; pp. 67–94.
5. Cracknell, A.P. *The Advanced Very High Resolution Radiometer*; Taylor & Francis: London, UK, 1997; p. 31.
6. WMO OSCAR. Space-Based Capabilities—Instruments. 2016. Available online: http://www.wmo-sat.info/oscar/instruments (accessed on 4 May 2020).
7. ESA. HIRS/4 Performance. Undated. Available online: https://www.esa.int/Applications/Observing$_$the$_$Earth/\Meteorological$_$missions/MetOp/Performance8 (accessed on 4 May 2020).
8. Chen, R.; Cao, C.; Menzel, W.P. Intersatellite calibration of NOAA HIRS CO_2 channels for climate studies. *J. Geophys. Res.-Atmos.* **2013**, *118*, 5190–5203. [CrossRef]
9. Dong, C.; Liu, Q.; Li, G.; Zhang, F. The study of in-orbit calibration accuracy of NOAA satellite infrared sounder and its effect on temperature profile retrievals. *Adv. Atmos. Sci.* **1990**, *7*, 211–219. [CrossRef]
10. Chen, R.; Cao, C. Physical analysis and recalibration of MetOp HIRS using IASI for cloud studies. *J. Geophys. Res.* **2012**, *117*, D03103. [CrossRef]
11. Cao, C. Intersatellite Radiance Biases for the High-Resolution Infrared Radiation Sounders (HIRS) on board NOAA-15, -16, and -17 from Simultaneous Nadir Observations. *J. Atmos. Ocean Tech.* **2005**, *22*, 381–395. [CrossRef]
12. Norman, M.N.; Becker, F. Terminology in thermal infrared remote sensing of natural surfaces. *Agric. Forest. Meteorol.* **1995**, *77*, 153–166. [CrossRef]
13. Maghrabi, A.H. On the measurements of the moon's infrared temperature and its relation to the phase angle. *Adv. Space Res.* **2014**, *53*, 329–347. [CrossRef]
14. Lang, T.; Buehler, S.A.; Burgdorf, M.J.; Hans, I.; John, V. MW UTH CDR. Submitted to SCIENTIFIC DATA 2020. Available online: http://cedadocs.ceda.ac.uk/1411/1/Attachment_0-13.pdf (accessed on 31 March 2020).
15. Lemke, O.; Kluft, L.; Mrziglod, J.; Pfreundschuh, S.; Holl, G.; Larsson, R.; Yamada, T.; Mieslinger, T.; Doerr, J. Atmtools/Typhon. Zenodo 2020. Available online: https://zenodo.org/record/3626449 (accessed on 31 March 2020).
16. NOAA STAR ICVS. Long-Term Monitoring. 2019 Available online: https://www.star.nesdis.noaa.gov/icvs/ (accessed on 4 May 2020).
17. Koenig, E.W. *Final Report on the High Resolution Infrared Radiation Sounder/Mod 2*; Goddard Space Flight Center: Greenbelt, MD, USA, 1979; N80-33055.
18. Labrot, T.; Lavanant, L.; Whyte, K.; Atkinson, N.; Brunel, P. AAPP Documentation Scientific Description. In *Satellite Application Facility for Numerical Weather Prediction*; NWPSAF-MF-UD-001; 2019. Available online: https://nwpsaf.eu/site/download/documentation/aapp/NWPSAF-MF-UD-001_Science.pdf (accessed on 31 March 2020).
19. Robel, J.; Graumann, A. *NOAA KLM User's Guide*; National Oceanic and Atmospheric Administration: Silver Spring, MD, USA, 2014.

20. Chang, T.; Cao, C. Analysis of MetOp/HIRS instrument self-emission and its impact on on-orbit calibration. In *Sensors, Systems, and Next-Generation Satellites XV*; International Society for Optics and Photonics: Bellingham, WA, USA, 2011.

21. Müller, T.G.; Burgdorf, M.J.; Lagoa, V.A. The Moon in the thermal IR. In preparation. Will be made available online: http://home.mnet-online.de/tmueller/ (accessed on 7 May 2020)

22. NASA. Polar Operational Environmental Satellites. Undated. Available online: https://www.star.nesdis.noaa.gov/icvs/ (accessed on 4 May 2020).

23. Holl, G.; Mittaz, J.P.D.; Merchant, C.J. Error Correlations in High-Resolution Infrared Radiation Sounder (HIRS) Radiances. *Remote Sens.* **2019**, *11*, 1337. [CrossRef]

24. Chung, E.-S.; Soden, B.J.; Huang, X.; Shi, L.; John, V.O. An assessment of the consistency between satellite measurements of upper tropospheric water vapor. *J. Geophys. Res.-Atmos.* **2016**, *121*, 2874–2887. [CrossRef]

25. Salisbury, J.W.; Basu, A.; Fischer, E.M. Thermal Infrared Spectra of Lunar Soils. *Icarus* **1997**, *130*, 125–139. [CrossRef]

26. Bonsignori, R. In-orbit verification of microwave humidity sounder spectral channels coregistration using the moon. *J. Appl. Remote Sens.* **2018**, *12*, 025013. [CrossRef]

27. Cao, C.; Jarva, K.; Ciren, P. An improved algorithm for the operational Calibration of the High-Resolution Infrared Radiation Sounder. *J. Atmos. Ocean Tech.* **2007**, *24*, 169–181. [CrossRef]

28. Lodders, K.; Fegley, B. *The Planetary Scientist's Companion*; Oxford University Press: New York, NY, USA, 1998; p. 95.

29. Rieke, G.H.; Blaylock, M.; Decin, L.; Engelbracht, C.; Ogle, P.; Avrett, E.; Carpenter, J.; Cutri, R.M.; Armus, L.; Gordon, K.; et al. Absolute Physical Calibration in the Infrared, *Astron. J.* **2008**, *135*, 2245–2263. [CrossRef]

30. Shaw, J.A. Modeling infrared lunar radiance. *Opt. Eng.* **1998**, *38*, 1763–1764. [CrossRef]

31. Keihm, S.J.; Peters, K.; Langseth, M.G.; Chute, J.L. Apollo 15 measurement of lunar surface brightness temperatures thermal conductivity of the upper 1 1/2 meters of regolith. *Earth Planet. Sci. Lett.* **1973**, *19*, 337–351. [CrossRef]

32. Matthews, G. Celestial body irradiance determination from an underfilled satellite radiometer. *Appl. Opt.* **2008**, *47*, 4981–4993. [CrossRef] [PubMed]

33. Koenig, E.W. *High Resolution Infrared Radiation Sounder for the Nimbus F Spacecraft*; Goddard Space Flight Center: Greenbelt, MD, USA, 1975.

34. Chédin, A.; Péquignot, E.; Serrar, S.; Scott, N. A. Simultaneous determination of continental surface emissivity and temperature from NOAA 10/HIRS observations. *J. Geophys. Res.-Atmos.* **2004**, *109*, D20110. [CrossRef]

35. Daniels, J.; Smith, G.L.; Priestley, K.J.; Thomas, S. Using lunar observations to validate pointing accuracy and geolocation, detector sensitivity stability and static point response of the CERES instruments. In *Remote Sensing of Clouds and the Atmosphere XIX; and Optics in Atmospheric Propagation and Adaptive Systems XVII*; International Society for Optics and Photonics: Bellingham, WA, USA, 2014.

36. Simon, A.A.; Reuter, D.C.; Emery, J.; Cosentino, R.G.; Gorius, N.; Lunsford, A.; Lauretta, D.S. OSIRIS-REx Earth Flyby. In Proceedings of the 50th Lunar and Planetary Science Conference, The Woodlands, TX, USA, 18–22 March 2019.

37. Okada, T.; Fukuhara, T.; Tanaka, S.; Taguchi, M.; Arai, T.; Senshu, H.; Demura, H.; Ogawa, Y.; Kouyama, T.; Sakatani, N.; et al. Earth and moon observations by thermal infrared imager on Hayabusa2 and the application to detectability of asteroid 162173 Ryugu. *Planet. Space Sci.* **2018**, *158*, 46–52. [CrossRef]

38. Yang, H.; Burgdorf, M.J. A Study of Lunar Microwave Radiation Based on Satellite Observations. *Remote Sens.* **2020**, *12*, 1129. [CrossRef]

39. ESA. Enabling & Support. 2018. Available online: https://www.esa.int/Enabling$_$Support/Preparing$_$for$_$the$_$Future/\Discovery$_$and$_$Preparation/Moon$_$holds$_$key$_$to$_$improving$_$satellite$_$views$_$of$_$Earth (accessed on 4 May 2020).

40. EUMETSAT. Science Studies. Undated. Available online: https://www.eumetsat.int/website/home/Data/ScienceActivities/\ScienceStudies/Moonasreferenceforlongtermstabilityassessmentofopticalsensors/index.html (accessed on 4 May 2020).

 *remote sensing*

Article

Surface Diffuse Solar Radiation Determined by Reanalysis and Satellite over East Asia: Evaluation and Comparison

Hou Jiang [1,2], Yaping Yang [1,3,*], Hongzhi Wang [4], Yongqing Bai [1,2] and Yan Bai [1,3]

[1] State Key Laboratory of Resources and Environmental Information System, Institute of Geographic Sciences and Natural Resources Research, Chinese Academy of Sciences, Beijing 100101, China; jiangh.18b@igsnrr.ac.cn (H.J.); baiy@lreis.ac.cn (Y.B.)
[2] College of Resources and Environment, University of Chinese Academy of Sciences, Beijing 100049, China
[3] Jiangsu Center for Collaborative Innovation in Geographical Information Resource Development and Application, Nanjing 210023, China; baiy@igsnrr.ac.cn
[4] College of Environment and Planning, Henan University, Kaifeng 475004, China; wanghz@lreis.ac.cn
[*] Correspondence: yangyp@igsnrr.ac.cn; Tel.: +86-10-6488-9452

Received: 6 April 2020; Accepted: 27 April 2020; Published: 28 April 2020

Abstract: Recently, surface diffuse solar radiation (R_{dif}) has been attracting a growing interest in view of its function in improving plant productivity, thus promoting global carbon uptake, and its impacts on solar energy utilization. To date, very few radiation products provide estimates of R_{dif}, and systematic validation and evaluation are even more scare. In this study, R_{dif} estimates from Reanalysis Fifth Generation (ERA5) of European Center for Medium-Range Weather Forecasts and satellite-based retrieval (called JiEA) are evaluated over East Asia using ground measurements at 39 stations from World Radiation Data Center (WRDC) and China Meteorological Administration (CMA). The results show that JiEA agrees well with measurements, while ERA5 underestimates R_{dif} significantly. Both datasets perform better at monthly mean scale than at daily mean and hourly scale. The mean bias error and root-mean-square error of daily mean estimates are -1.21 W/m^2 and 20.06 W/m^2 for JiEA and -17.18 W/m^2 and 32.42 W/m^2 for ERA5, respectively. Regardless of over- or underestimation, correlations of estimated time series of ERA5 and JiEA show high similarity. JiEA reveals a slight decreasing trend at regional scale, but ERA5 shows no significant trend, and neither of them reproduces temporal variability of ground measurements. Data accuracy of ERA5 is more robust than JiEA in time but less in space. Latitudinal dependency is noted for ERA5 while not for JiEA. In addition, spatial distributions of R_{dif} from ERA5 and JiEA show pronounced discrepancy. Neglect of adjacency effects caused by horizontal photon transport is the main cause for R_{dif} underestimation of ERA5. Spatial analysis calls for improvements to the representation of clouds, aerosols and water vapor for reproducing fine spatial distribution and seasonal variations of R_{dif}.

Keywords: surface diffuse solar radiation; temporal trend; spatial pattern; atmospheric factor

1. Introduction

Surface solar radiation (R_s) drives the global energy, water and carbon cycles of by affecting sensible and latent heat fluxes, longwave emission, water vapor and circulations in the atmosphere and the ocean [1–3]. Determining the variations of R_s is essential for understanding global climate changes, particularly the rate of global warming and its effects on glacial melt and sea level rise [4,5]. R_s data with different spatiotemporal resolutions are urgently required in diverse application fields, such as global numerical weather prediction, agricultural meteorology, climate monitoring and solar electricity.

Moreover, the accuracy of R_s data greatly influences simulations of runoff, evapotranspiration, gross primary productivity, growth and yield of crops [6–9].

In addition to R_s, surface diffuse solar radiation (R_{dif}) takes on greater importance in monitoring and modeling ecosystem carbon uptake [10]. R_{dif} tends to increase plant productivity, as it enhances light use efficiency of plants by penetrating more radiation into deeper canopies, thus improving photosynthesis in shaded leaves [11–13]. It was reported that changes of R_{dif} affect the global land carbon sink [11], according to a high-quality R_{dif} dataset for quantifying its effects on carbon dynamics of terrestrial ecosystems. The fraction of R_{dif} is also a necessary input to agricultural models, such as the Soil Water Atmosphere Plant (SWAP), Forest Biomass, Assimilation, Allocation, and Respiration (FöBAAR) and Yale Interactive Terrestrial Biosphere (YIB), for early assessment of crop yield [14,15] or radiation-use efficiency of forests [16]. Besides, the spatially continuous high-resolution hourly ratio of R_s and R_{dif} is required for a comprehensive assessment of the potential of rooftop solar photovoltaics and policy-making regarding the renewable energy sector [17].

Currently, R_s products are available from four common sources, namely direct measurements of surface radiation networks [3], simulations based on radiation transfer models [18,19], estimates from reanalysis systems [18,20,21] and retrievals from satellite observations [10,22–25]. Direct surface measurements are regarded as a reliable reference for data validation from simulations, reanalysis and satellite retrievals [26–29]. However, R_{dif} is very rare among these products, for example, R_s measurements are attainable at 119 stations in China while only 17 of them measure R_{dif} [26,30]; the Global Land Surface Satellite (GLASS) provides global 5-km resolution, 3-h interval R_s but lacks an R_{dif} map [25]. Nonetheless, many algorithms have tried to determine the fraction of R_{dif}, [10,31,32]. Greuell et al. [31] retrieved global, direct and diffuse irradiance (3 km, 15 min) from Spinning Enhanced Visible and Infrared Imager (SEVIRI) observations through a physics-based and empirically adjusted algorithm. Ryu et al. [10] produced incident shortwave radiation (SW), photosynthetically active radiation (PAR) and diffuse PAR datasets (5 km, 4 day) by combining an atmospheric radiative transfer model with an artificial neural network (ANN) based on Moderate Resolution Imaging Spectroradiometer (MODIS) atmosphere and land products. To date, mature kilometer-scale hourly radiation datasets (including R_{dif}) with global and multiyear coverage are still rare [26,33]. Most products are generated over specific regions like Europe, North America and China. To the best of our knowledge, only two products provide multiyear hourly R_{dif} over East Asia, i.e., Reanalysis Fifth Generation (ERA5) provided by the European Center for Medium-Range Weather Forecasts (ECMWF) and satellite-based products produced by Jiang et al. [22] (hereafter called JiEA for short).

However, these radiation products generally contain large uncertainties. The reported root-mean-square error (RMSE) of instantaneous R_s retrievals under all-sky conditions range from 60 to 140 W/m^2 (~15%–30%) depending on local cloud climatology [33]. In addition, multisource products usually show inconsistent temporal trends and spatial distributions [29,34], which could hamper their applicability for assessing global brightening or dimming and local climate responses to radiation changes [5,35]. Therefore, it is necessary and important to compare different products and understand their discrepancies. Zhang et al. [29] compared four satellite products of R_s using comprehensive ground measurements at stations around the world and found that satellite estimates capture the seasonal variations of R_s well and have acceptable data accuracy at the monthly time scale, with an overestimation of approximately 10 W/m^2. Zhang et al. [36] evaluated two R_s estimates of global reanalyses using homogenized surface measurements in China and pointed out the pronounced overestimation of the reanalyses. The significant spatiotemporal difference of data accuracy mainly results from atmospheric factors, including cloud coverage, aerosol optical depth and water vapor content. There are large numbers of references that concentrate on the data accuracy of R_s, but to date very few studies have been devoted to the evaluation of R_{dif}.

The purpose of this study is to evaluate and compare R_{dif} estimates from ERA5 and JiEA using surface in situ measurements and to investigate the spatial pattern and seasonal variations of R_{dif}

over East Asia. The reliability of different data is discussed at both the site level and the regional scale in combination with the spatial distribution of atmospheric factors that mostly relate to the retrievals of solar radiation. This study provides a reference for rational use of these data and opens new perspectives for improving R_{dif} estimation.

This paper is organized as follows. The ground measurements and diffuse radiation products used are briefly described in Section 2. Section 3 explains various validation metrics and the method for comparative analysis. Section 4 presents the results of site-level validation and analysis of spatiotemporal deviations at different time scales, followed by a discussion on the reliability of these data, especially concerning their spatial pattern, in Section 5. A conclusion is finally given in Section 6.

2. Data

2.1. Ground Measurements

The ground measurements used to evaluate R_{dif} estimates are obtained from two data centers: the World Meteorological Organization's (WMO) World Radiation Data Center (WRDC) (22 stations) and that of the China Meteorological Administration (CMA) (17 stations). Figure 1 shows the geographical distribution of the selected 39 stations from WRDC and CMA, with detailed information provided in Table S1.

The WRDC is one of the recognized World Data Centers sponsored by the WMO, which centrally collects and archives radiometric data from the world to ensure the availability of these data for research by the international scientific community. Daily and monthly totals of surface energy components such as global radiation (i.e., R_s), diffuse radiation (i.e., R_{dif}) and radiation balance are available from the official website (http://wrdc.mgo.rssi.ru/) after a simple registration process. Daily totals of global and diffuse radiation are determined where ground measurements for all time intervals of the daytime are available, along with an auxiliary procedure to avoid undue losses due to the gaps in the data for sunrise and sunset hours. Monthly totals are the sum of the entire daily totals of the month. If less than ten days with missing records exist, a monthly mean of the available daily records is calculated, then a monthly total is calculated by multiplying the monthly mean by the number of days in the calendar month. A monthly value is not provided if missing records within the month exceed ten days. A subset of 22 WRDC stations (red circles in Figure 1), which provide at least one-year monthly series of diffuse radiation within the period from 2007 to 2014, was selected for this study.

The CMA Meteorological Information Center have released daily and monthly meteorological data at 122 routine weather stations. Radiation-related elements include net radiation, downward shortwave radiation (i.e., R_s), reflected shortwave radiation and diffuse radiation (i.e., R_{dif}). R_{dif} measurements are conducted at 17 stations (blue triangles in Figure 1). The procedure to calculate daily and monthly totals is the same as that adopted by WRDC. Additional quality control measures before release include a spatial and temporal consistency check and manual inspection and correction. Furthermore, hourly measurements of these stations are attainable from National Meteorological Science Data Center (http://data.cma.cn/) on reasonable request. Herein, hourly measurements in 2007 and 2008 and daily/monthly measurements from 2007 to 2014 of diffuse radiation at these 17 CMA stations are available for evaluation.

R_s and R_{dif} at stations are widely measured through thermoelectric pyranometer, which has a spectral response of 0.3–3.0 μm, a thermal effect of less than 5% and an annual stability of about 5%. Pyranometers are exposed to the sun to measure R_s. For measuring R_{dif}, pyranometric sensors are shaded by an additional component (e.g., shadow-ball or rotating shadow band) to prevent direct solar radiation from reaching the sensor. The shading mechanism hides the minimum of sky outside the small solid angle of the sensor to receive the maximum R_{dif} from the whole sky dome.

Previous studies point out that systematic errors are very common in radiation measurements due to equipment failure and operational problems [30] and that it is necessary to examine measured values carefully before subsequent utilization [28,29,37]. In this study, we first applied the physical threshold test [38] and then the method based on reconstructed data [29] to the measurements of R_s associated with the selected records of R_{dif} for further quality control. If the measured value of R_s failed to pass the quality check, the corresponding R_{dif} was eliminated. In addition, R_{dif} should not be larger than R_s. The numbers of valid records from each station at hourly, daily and monthly scales are listed in Table S1.

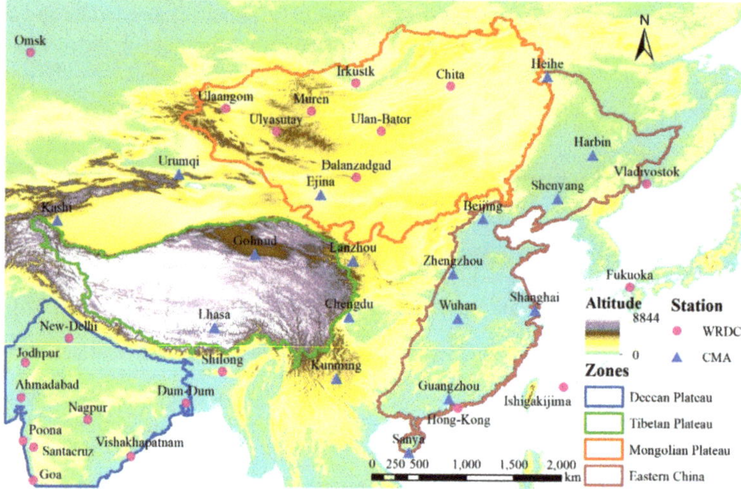

Figure 1. Locations of used radiation stations and zone boundary for statistical analysis. Hourly, daily and monthly measurements of diffuse radiation are available for 17 stations (blue triangles) from China Meteorological Administration (CMA). Daily and monthly measurements for other stations (red circles) are obtained from World Radiation Data Center (WRDC). Detailed information of all stations can be found in Table S1. Four polygons define the boundary of Deccan Plateau, Tibetan Plateau, Mongolian Plateau and Eastern China for regional analysis in this study.

2.2. Diffuse Radiation Products

Two datasets that provide estimates of R_{dif} over East Asia are evaluated and compared in our study, i.e., the latest global climate reanalysis provided by ECMWF [21] and satellite-based products produced by Jiang et al. [22].

ECMWF Reanalysis Fifth Generation (EAR5) is the fifth-generation ECMWF atmospheric reanalysis of the global climate to replace the old ERA-Interim. It is produced using a 4D-Var assimilation system of ECMWF's Integrated Forecast System (IFS), namely IFS Cycle 41r2, which guarantees significant increase in forecast accuracy and computational efficiency. The advanced system is also combined with vast amounts of historical observations to generate globally consistent time series of multiple climate variables. ERA5 provides hourly estimates of many atmospheric, land-surface and sea-state parameters together with their uncertainties at reduced spatial and temporal resolutions. The parameters used in this study involve "surface solar radiation downwards" and "total sky direct solar radiation at surface", which represent the amount of shortwave radiation (surface direct and diffuse solar radiation) and the amount of direct radiation reaching the surface of the Earth, respectively. Estimates of R_{dif} can be derived by subtracting total sky direct solar radiation from surface solar radiation downwards. To date, these hourly data are available in the Climate Data Store (https://climate.copernicus.eu/climate-reanalysis) on regular latitude–longitude grids at 0.25° × 0.25° resolution from 1979 to present.

Satellite-based diffuse radiation products (hereafter, called JiEA for short) are from the work of Jiang et al. [22], where a deep learning algorithm was developed to retrieve R_s from Multifunctional Transport Satellites (MTSAT) data. They concentrate on overcoming the negative impact of spatial adjacency effects on R_{dif} estimation through convolutional neural networks (CNNs). Spatial adjacency effects refer to the phenomena that some photons out of the field of view are reflected by the surface then scattered by the atmosphere, thus finally entering into the field of view to change the amount of solar radiation within the field of view. CNN is used to handle this effect by gaining knowledge of the spatial distribution of clouds/aerosols from satellite image blocks. This algorithm is originally designed for estimates of R_s and further extended through a transfer learning approach for estimates of R_{dif}. Currently, a dataset from 2007 to 2018 is freely available from Pangaea at https://doi.pangaea.de/10.1594/PANGAEA.904136 [39]. This dataset provides gridded estimates of R_s and R_{dif} at $0.05° \times 0.05°$ resolution within $71°$–$141°$E and $15°$–$60°$N, mainly covering East Asia. In view of the difference of spatial coverage from ERA5, East Asia in this study is referred to as the maximum overlapped extent of the two datasets.

3. Methods

3.1. Validation Metrics

Ground measurements are regarded as the reference for evaluation of R_{dif} from different datasets. To quantify the accuracy of R_{dif} estimates, a set of metrics including Pearson correlation coefficient (R), (relative) mean bias error (MBE, rMBE), (relative) mean absolute bias error (MABE, rMABE), (relative) root-mean-square error (RMSE, rRMSE), bias and absolute percentage bias (APE) are used. These metrics are defined as follows:

$$R = \frac{\sum_{i=1}^{n}\left(\hat{y}_i - \overline{\hat{y}}\right)(y_i - \overline{y})}{\sqrt{\sum_{i=1}^{n}\left(\hat{y}_i - \overline{\hat{y}}\right)^2}\sqrt{\sum_{i=1}^{n}(y_i - \overline{y})^2}} \tag{1}$$

$$MBE = \frac{1}{n}\sum_{i=1}^{n}(\hat{y}_i - y_i), \ rMBE = MBE/\overline{y} \tag{2}$$

$$MABE = \frac{1}{n}\sum_{i=1}^{n}|\hat{y}_i - y_i|, rMABE = MABE/\overline{y} \tag{3}$$

$$RMES = \sqrt{\frac{1}{n}\sum_{i=1}^{n}(\hat{y}_i - y_i)^2}, rRMSE = RMSE/\overline{y} \tag{4}$$

$$bias = \hat{y}_i - y_i, i = 1, 2, \ldots, n \tag{5}$$

$$APE = \left|\frac{\hat{y}_i - y_i}{y_i}\right|, i = 1, 2, \ldots, n \tag{6}$$

where n is the number of data samples, y means ground-measured R_{dif} values whose mean value is $\overline{y}$ and $\hat{y}_i$ represents corresponding estimated values whose mean value is $\overline{\hat{y}}$. R measures the strength and direction of a linear relationship between $\hat{y}_i$ and y_i. R ranges from -1 to 1, and a closer value to 1 indicates a strong positive linear relationship between estimated and measured R_{dif}. MBE is the mean difference between compared variables, representing the systematic error of R_{dif} products to under- or overestimate. MABE is the mean of absolute differences between $\hat{y}_i$ and y_i and gives the average magnitude of under- or overestimation of R_{dif} compared to ground measurements. RMSE represents the standard deviation of the differences between $\hat{y}_i$ and y_i. Compared to MABE, RMSE is more sensitive to outliers. To eliminate the scale-dependency (i.e., influence from numbers of samples) of these metrics, their relative values are also available through dividing the original values by the mean of the reference measurements. For temporal and spatial evaluation and comparison of R_{dif}, these metrics were calculated according to different grouping strategies, i.e., 12 months, 8 years, and 39 stations.

In addition, we demonstrated the probability density functions (PDFs) of bias and cumulative distribution functions (CDFs) of APE for comparison of data accuracy within different tolerance ranges of deviations. The bias indicates the under- or overestimation of each estimated R_{dif} value. PDF is a statistical expression that defines the probability distribution of a random variable. When PDF is graphically portrayed, the total area of an interval (expressed as bin width during statistical process) under the curve equals the probability of the random variable occurring. Herein, PDF determines the likelihood of calculated bias falling into a specific range. APE expresses the deviation of each estimate in percentage, and the associated CDF gives the proportion of APE with values less than a certain threshold.

3.2. Time Series Decomposition

Time series decomposition is a common way to identify the change of different components of interest [35,40], and it involves separating a time series into several distinct components. Three components are typically of interest, i.e., the trend, seasonal periodicity and stochastic irregular anomalies. The additive functional form has been widely used to observe the bias and errors of R_s [41,42]. It assumes that a monthly R_{dif} time series, $\mathbf{R}(t)$, can be decomposed into the low-frequency climatological contributions, consisting of the long-term trends $\bar{\mathbf{R}}(t)$, the climatological seasonal cycles $\tilde{\mathbf{R}}(t)$ and high-frequency deviations $\mathbf{R}'(t)$:

$$\mathbf{R}(t) = \bar{\mathbf{R}}(t) + \tilde{\mathbf{R}}(t) + \mathbf{R}'(t) \tag{7}$$

where t defines the length of R_{dif} time series. The trends $\bar{\mathbf{R}}(t)$ describe the gradual variations and can be estimated by using moving averages or parametric regression models [3,43]. The seasonal cycles $\tilde{\mathbf{R}}(t)$ capture level shifts that repeat systematically within the same period between successive years. The anomalies $\mathbf{R}'(t)$ exhibit autocorrelation and cycles of unpredictable duration. For identifiability from $\bar{\mathbf{R}}(t)$, $\tilde{\mathbf{R}}(t)$ and $\mathbf{R}'(t)$ are assumed to fluctuate around zero.

Since the periodicity of R_{dif} data is monthly, a 13-term moving window is used for estimating the long-term trend by setting weight 1/24 for the first and last terms and weight 1/12 for the interior terms. Then $\bar{\mathbf{R}}(t)$ is removed from the original series to obtain the detrended time series. Assuming a stable seasonal component that has constant amplitude across the series, $\tilde{\mathbf{R}}(t)$ can be determined by averaging detrended time series for each month over the whole period, i.e., by averaging all of the January values, then all of the February values and so on for the remaining months. Finally, $\mathbf{R}'(t)$ is determined by removing $\bar{\mathbf{R}}(t)$ and $\tilde{\mathbf{R}}(t)$ from the original time series. If only $\tilde{\mathbf{R}}(t)$ is removed, the rest is called a deseasonalized time series $\mathbf{R}'_d(t)$:

$$\mathbf{R}'_d(t) = \bar{\mathbf{R}}(t) + \mathbf{R}'(t) = \mathbf{R}(t) - \tilde{\mathbf{R}}(t) \tag{8}$$

In this study, the similarity of two time series from different datasets was measured by the Pearson correlation coefficient (Equation (1)) of their corresponding $\mathbf{R}'_d(t)$. The significance at the 95% confidence level was obtained through an F-test on the linear regression model of the two time series. In particular, we considered the increasing/decreasing trend of different components over time. We fit a linear regression model between the components and associated time index, and the slope coefficient was regarded as the indicator of increasing/decreasing trend versus time. In addition, the 95% confidence bounds of the slope coefficient were given by the F-test on the regression model.

4. Results

4.1. Evaluation Against Ground Measurements

The hourly R_{dif} estimates of ERA5 and JiEA are compared with the quality-controlled ground measurements from 17 CMA stations. It is stressed that such comparisons are conducted at their original spatial resolutions. ERA5 has an overall correlation coefficient R of 0.71, a negative bias of

29.69 W/m^2, an MABE of 63.83 W/m^2 and an RMSE of 92.29 W/m^2, whereas these values are 0.85, 8.54 W/m^2, 50.43 W/m^2, and 66.36 W/m^2 for JiEA. It is apparent that R_{dif} estimates of JiEA correlate better than ERA5 with in situ measurements at the selected stations. Evidence comes from their density scatterplots; Figure 2a shows that more points are concentrated on the lower side of 1:1 line, while in Figure 2b almost all data pairs are symmetrically distributed around the 1:1 line. This is also the reason why ERA5 exhibits a relatively serious underestimation of R_{dif}, with an rMBE of 18.4%. For low-radiation estimation, ERA5 performs better than JiEA, as high-density (red) scatters are on both sides of 1:1 line in Figure 2a, while they are obviously inclined to the upper side in Figure 2b. The PDF of JiEA resembles the Gaussian distribution with a mean slightly larger than zero, coinciding with the observed overestimation and the density scatterplots. Although the peak of ERA5's PDF nears zero, the curve is significantly asymmetric, revealing a high probability of underestimation. The performance of JiEA is superior to ERA5 when setting the tolerance of absolute percentage bias lower than 0.51, while few estimates of ERA5 would exceed one (Figure 2d). At hourly scale, the time systems of R_{dif} estimates and ground measurements deserve attention [44]. For example, measurements at some stations might be recorded according to the local time and then converted to universal time (usually the time system of satellite acquisition and climate reanalysis) when stored into a standard database. There is consequently a change of original values due to a resampling of data series in time; in any case, returning to the original values is impossible due to the asystematic shift of a fraction of an hour before and after conversion. That change would have negative impacts on evaluation results at hourly scale. It is pointed out that this impact does not hold if we deal with daily, monthly or yearly averages or sums of solar radiation data.

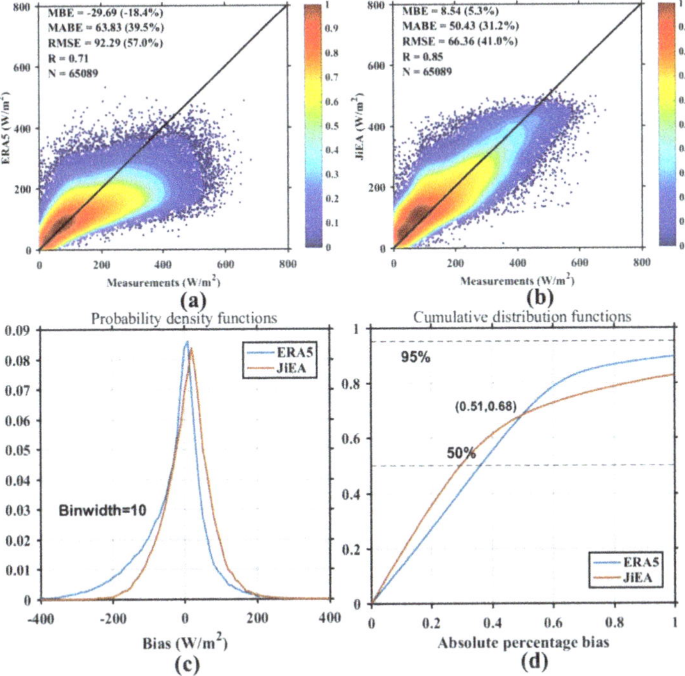

Figure 2. Evaluation results of hourly R_{dif} estimates. (**a**) Density scatterplots between ERA5 estimates and CMA measurements. (**b**) Scatterplots for estimates of JiEA. At the upper left corner shows the values of validation metrics with their relative values in the brackets. Black lines represent the 1:1 lines. (**c**) probability distribution functions (PDFs) of bias for ERA5 (blue line) and JiEA (orange line); (**d**) The related cumulative distribution functions (CDFs) of absolute percentage bias.

Daily mean R_{dif} estimates are evaluated using measurements at all 39 stations from WRDC and CMA. As indicated by various metrics, the overall accuracy of JiEA exceeds that of ERA5. Similar differences to hourly-scale evaluation are observed between ERA5 and JiEA from Figure 3a–d. The PDF of JiEA is symmetrically distributed with a zero mean, while that of ERA5 indicates a high probability of a negative bias. The proportion of JiEA samples whose accuracy is higher than ERA5 reaches up to 84% (Figure 3d). However, some apparently questionable estimates exist for the results of JiEA (e.g., scatters at the lower right corner of Figure 3b). Failure of these estimates might result from the difference between ground and satellite measurements, in that ground measurements represent an average state over the sample time interval whereas only instantaneous state is manifested by satellite images [45]. For instance, when coming across fast-moving clouds, a satellite sensor may scan a cloudy sky, but ground stations are covered by cloud shadows only within a momentary period (less than sample time interval). In this case, ground measurements would be greater than satellite-based estimates. The same evaluation is conducted in four typical regions, i.e., Eastern China, Mongolian Plateau, Tibetan Plateau and Deccan Plateau, whose boundaries are defined in Figure 1. Data accuracy of JiEA is always better than ERA5 except for the Mongolian Plateau (Figure S1). Both ERA5 and JiEA achieve more accurate estimates of R_{dif} over the Mongolian Plateau and Tibetan Plateau than over other regions. Particularly, ERA5 seriously underestimates R_{dif} over the Deccan Plateau and shows a large difference compared to JiEA. This is probably due to their inappropriate representation or modeling of aerosols, clouds and their interactions with solar radiation in the atmosphere [2,46,47] for Eastern China and India where rapid economic development and high-speed urbanization have caused heavy pollution [27,48]. Besides, frequent cloudy and rainy weather in India and South China also leads to the difficulty in estimating R_{dif} [49,50].

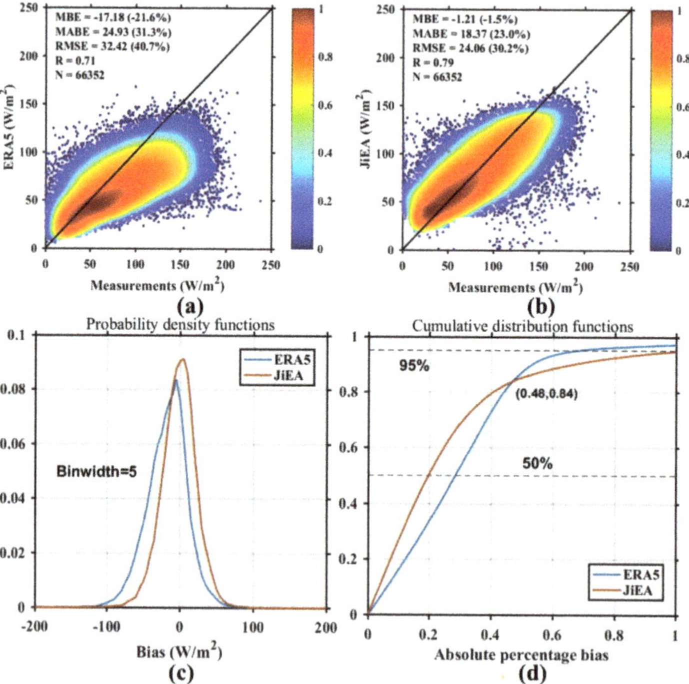

Figure 3. Evaluation results of daily mean R_{dif} estimates. (**a–d**) Analogous to Figure 2 but at daily mean scale using measurements from WRDC and CMA.

The differences in data accuracy between JiEA and ERA5 are more obvious at monthly mean scale (Figure S2). JiEA almost achieves zero deviation on average (a negative MBE of 0.92 W/m^2 and zero-centered PDF). ERA5 underestimates most parts of the selected samples, and the largest underestimation is greater than 50 W/m^2. The accuracy of 94% of samples exceeds ERA5 with absolute percentage bias lower than 0.39. We also depict the PDF and CDF of JiEA after upscaling the original monthly data to 0.25° grids (dotted black lines in Figure S2c,d) and observe no significant change comparing to the original ones, suggesting that the above comparisons are hardly affected by the different spatial resolutions of the two datasets. As pointed out by previous studies [51–53], the evaluation results are likely affected by the spatial representativeness of ground measurements. The comparison of Figures S2b and S3 indicates that ground R_{dif} measurements at the selected stations are more representative for 0.05° × 0.05° spatial grids than 0.25° × 0.25°. In this regard, the deviations in comparison to ground measurements are not completely attributed to the performance of models or algorithms [33,44,54].

The monthly maximum (minimum) of R_{dif} appears in June/July (December) and approximates to 110 (47), 90 (36) and 107 (44) W/m^2 for measurements, ERA5 and JiEA, respectively. It is clear that results from JiEA are closer to the measured values than those of ERA5. At the selected 39 stations, the measured yearly R_{dif} is 79.78 W/m^2 on average, and the ratio of R_{dif} to R_s (173.97 W/m^2) equals 45.86%. The estimates of JiEA (R_{dif}: 78.41 W/m^2, R_s: 171.95 W/m^2, R_{dif} ratio: 45.60%) are basically consistent with the measurements; on the contrary, ERA5 seems to underestimate R_{dif} as well as its fraction (R_{dif}: 63.26 W/m^2, R_s: 190.10 W/m^2, R_{dif} ratio: 33.28%). For the whole East Asia region, JiEA provides a mean R_{dif} of 71.89 W/m^2, accounting for 41.84% of R_s (171.81 W/m^2), while ERA5's estimate of R_{dif} (63.40 W/m^2) only accounts for 34.78% of R_s (182.28 W/m^2).

4.2. Temporal Difference of Data Accuracy

The temporal stability of data accuracy is critical for detection of the long-term trend of time series products [55,56]. One of the advantages of reanalysis products is their potential to provide geographically and physically consistent estimates of regional climate changes [57–59]. We illustrate the average seasonal (Figure 4a) and interannual (Figure 4b) variations of different metrics to examine the temporal consistency. Considering that the absolute amount of R_{dif} varies greatly among months and years, relative errors (rMABE and rRMSE) are discussed. Although the overall accuracy of ERA5 is inferior to JiEA, ERA5 shows a good robustness in time. The change of R is less than 0.1 and those of rMABE and rRMSE are less than 5% for ERA5, while the maximum disparity is doubled for JiEA. Snow/ice cover is the factor most likely to be responsible for the worse accuracy of satellite-based estimates in winter. The similarity of spectral and physical properties of cloud and surface snow covers hampers the identification of clouds and retrievals of cloud optical depths over snow/ice surface [60,61], subsequently resulting in a lower accuracy in satellite estimation of solar radiation [18,24,46]. Due to the lack of a physical basis, machine learning based methods always suffer from their dependence on the representativeness of training samples, and consequently their generalizability is limited [33,62,63]. As shown in Figure 4b, although a perfect performance is achieved in 2008, when the training set for the deep network behind JiEA is constructed, data accuracy of other years becomes much worse, with a maximum disparity of 8% with respect to rRMSE. On the contrary, the accuracy of ERA5 is relatively stable over time.

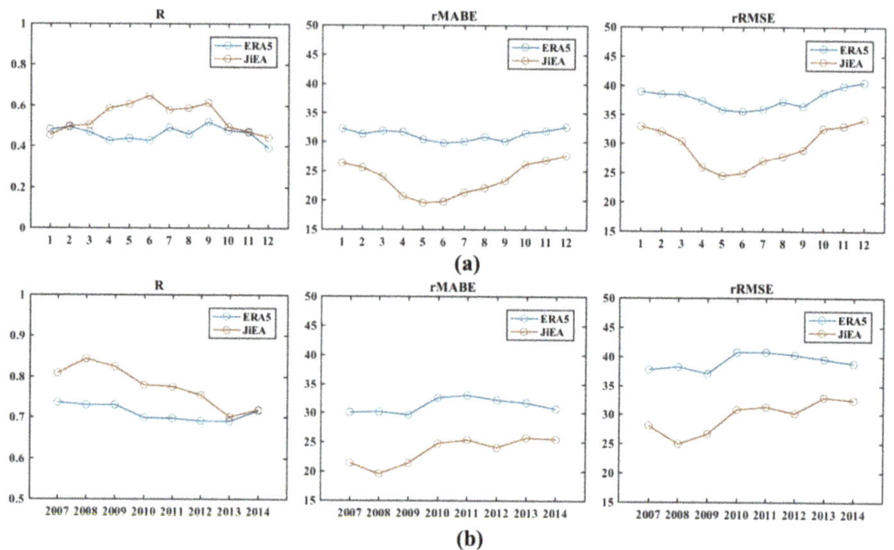

Figure 4. Temporal variation of R, rMABE and rRMSE: (**a**) results among different months; (**b**) results among different years. We illustrate the mean value of all stations at daily mean scale.

To examine whether the two datasets can capture the changing trend and seasonal cycles of R_{dif}, we pick out 24 stations that provide relatively complete monthly measurements for time series analysis. Very few missing values are substituted by the average of existing records of the same month. The results of time series decomposition are presented in Figure S4. ERA5 and JiEA reflect roughly similar trends that are consistent with the measured ones at most stations. However, there are issues with significantly different and even contradictory trends, such as for Ulan-Bator, Harbin, Lhasa and Urumqi. We speculate that this may be due to the combined effects of local pollution and climate change. For example, the increase of particles and aerosols in the atmospheric layer near the surface caused by air pollution actually leads to greater measured R_{dif} at stations, but reanalysis and satellite-based estimates do not respond to such pollution because of information loss. With respect to time series anomalies, ERA5 tends to level off, while JiEA and measurements exhibit stronger fluctuations. Specifically, the observed increasing/decreasing trends from estimates are identical with measurements at 14 and 12 stations for ERA5 and JiEA, respectively, but none of them passes the significance test at the 95% confidence level. This confirms the conclusion that neither satellite retrievals nor reanalysis can accurately reproduce the decadal variability and trend revealed by combining homogenized measurements and sunshine-duration-derived R_s [29,64]. The discrepant trends between estimates and measurements during the validation period might be attributed to inappropriate responses of models to undulated aerosols over these regions [28]. In view of aerosol's dominant contribution to the decadal trends in R_s [65], an inclusion of aerosol variability in the reanalysis and satellite retrieval is necessary for an accurate detection of changes of R_{dif}, which result from the scattered solar radiation on particles in the atmosphere (aerosols).

At regional levels, the trends of ERA5 remain highly constant for all regions except the Deccan Plateau, while JiEA shows slightly decreasing trends with slopes versus times ranging from 0.03 to 0.09 W m^{-2} yr^{-1} (Figure S5). The results of JiEA are in line with the reported insignificant trend (slope = −0.03, $p > 0.1$) of R_s over China between 2001 and 2016 [10]. Regardless of the large difference with the measured series, the deseasonalized time series of ERA5 and JiEA correlate for most parts of East Asia (Figure S5f). The phenomenon of estimates being weakly correlated with measurements reflects the difficulty in reproducing temporal variations at fine spatial resolutions and implies that the constructed variations of R_{dif} from gridded products are reasonable at coarse scales.

4.3. Spatial Distribution of Biases

The data accuracy shows notable spatial differentiation at the selected stations (Figure 5). For ERA5 data, latitude holds a positive correlation with R and a negative correlation with rMABE and rRMSE (Figure 6a), with correlation coefficients of 0.70, −0.75 and −0.55, respectively. This latitudinal dependency is prevalent among radiation datasets, such as GEWEX-SRB [66], ISCCP-FD [29] and UMD-SRB [20]. Serious underestimation occurs at stations on the Deccan Plateau, followed by the Eastern China (Figure S6a), which might be attributed to the inappropriate aerosol representations [29]. Local air pollution has caused high aerosol concentrations in these regions [30,67], but representation of aerosol absorption under a cloud layer is not included in current algorithms [20,29]. Although dimming of R_s is observed in Eastern China, absorption and scattering of solar radiation by aerosols and clouds increase the fraction of diffuse radiation [30,68].

With respect to JiEA data, the latitudinal dependency is not as pronounced as for ERA5 (Figures 5 and 6). R shows a positive correlation with latitude (correlation coefficient equals 0.64), along with similar spatial distribution to ERA5. However, rMABE and rRMSE are positively correlated with latitude. The linear relationship is greatly weakened when only considering CMA stations (brown dots in Figure 6a). Moreover, the spatial difference of rMBE is almost negligible in China (Figure S6b). These results support that this deep learning based method results in high robustness in space [22]. As pointed out by previous studies [22–24], the ability of machine learning methods depends on the representativeness of training samples; therefore, some large deviations appear at stations outside China, such as Irkutsk, Omsk, Fukuoka and Ishigakijima. It is surprising that JiEA provides satisfying R_{dif} estimates at stations in India; this might be due to the similarity in atmospheric scattering mechanisms with South China.

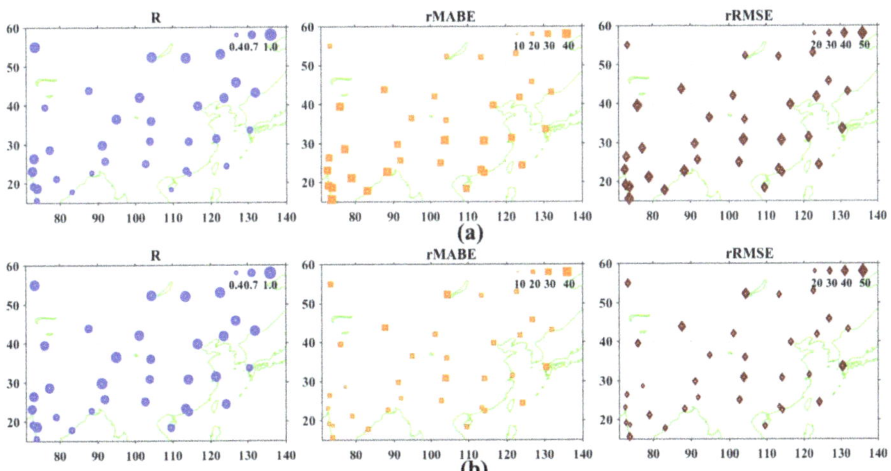

Figure 5. Spatial mapping of R, rMABE and rRMSE: (**a**) results for ERA5; (**b**) results for estimates of JiEA. Values are calculated from valid records of each station at daily means.

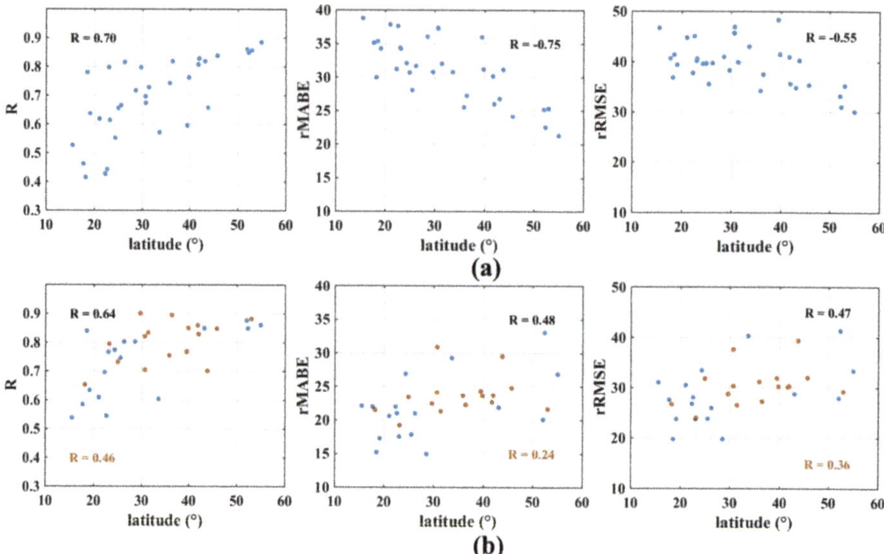

Figure 6. Latitudinal dependency of data accuracy: (**a**) results for ERA5; (**b**) results for estimates of JiEA. Spatial distribution corresponds to Figure 5. The correlation coefficients (R) indicate a linear relationship. For (**b**), brown dots represent the selected CMA stations.

5. Discussion

The data accuracy of ERA5 and JiEA is evaluated using ground measurements, and the results show that both datasets provide acceptable estimates of R_{dif} at the selected stations. For research on global climate change, homogeneous data with global coverage including focal hotspot regions like the Arctic, the Antarctic, the Tibet Plateau and others are always required [69]. In the field of solar energy applications, finer spatial resolution and wider ranges of temporal resolution are usually emphasized [17,62]. The WMO Observing System Capability Analysis and Review Tool (OSCAR) collects user-defined quantitative requirements with respect to the spatial resolution, timescale, coverage and quality for downward short-wave irradiance at the Earth's surface (Table S2). Although it is reported that the overall accuracy of R_s has entered the gate of intermediate level requirements (*Break.* in Table S2) [22,26,33,62], R_{dif} estimates from ERA5 and JiEA can only meet the minimum requirement (*Thres.* in Table S2) at monthly mean scale according to above evaluation results.

Last but not least, we concentrate on the spatial distribution of R_{dif} over East Asia. We show the annual average from 2007 to 2014 of R_{dif} estimates and its fraction (relative to R_s) from ERA5 and JiEA at 0.25° grids (Figure 7). It is apparent that the two datasets illustrate significantly different spatial patterns, with the largest differences on the Tibetan Plateau, Deccan Plateau and Taklimakan Desert (Figure 7e). Both the amount and ratio of JiEA are in line with the application level products of SolarGIS [70] over all of East Asia (Figure S7a,b). In contrast, the R_{dif} distribution of ERA5 is in agreement with the diffuse photosynthetically active radiation (PAR) from Breathing Earth System Simulator (BESS) [10], except at low latitudes (Figure S7c). In the absence of densely distributed in situ measurements, it is difficult to judge which pattern is reliable. However, subjective judgements can be made according to common sense in combination with the spatial mappings of atmospheric factors mostly related to the estimate of R_{dif} (Figure 8). Previous studies show that cloud parameters (cloud coverage and optical thickness) and aerosols are two of the most important factors for R_s estimation [2,71–74]. The amount of water vapor plays a vital role in radiation scattering and leads to altitudinal disparity [75–77]. Herein, we take MODIS-derived parameters as references, including cloud fraction (CF), cloud optical thickness (COT), aerosol optical depth (AOD) and water vapor.

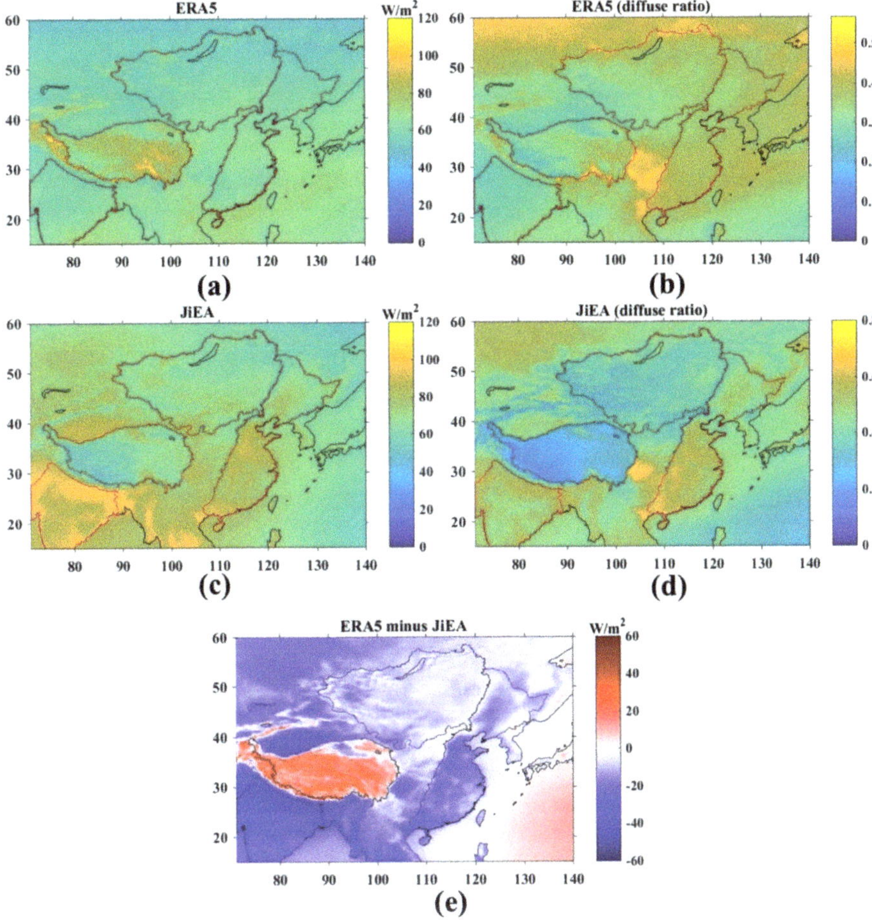

Figure 7. Spatial distribution of R_{dif} over East Asia: (**a**,**b**) annual average (2007–2014) and its fraction in relation to R_s estimated by ERA5 datasets; (**c**,**d**) analogous to (a,b) after upscaling estimates of JiEA to 0.25°; (**e**) the difference between ERA5 and JiEA (ERA5 minus JiEA).

Both BESS and SolarGIS confirm that ERA5 seriously underestimates R_{dif} at low latitudes. The large amount of R_{dif} is the combined result of high-density downward radiation and strong scattering effects of water vapor. An additional contribution comes from aerosols for the southern Himalayas and from clouds (high CF and middle COT) for South China. The amount of R_{dif} from JiEA is approximately equal to that from SolarGIS, but their diffuse ratios are discrepant, implying an underestimation of R_s by JiEA. In Sichuan Basin and the middle and lower reaches of the Yangtze River, COT and water vapor account for a large amount of R_{dif}. In North China, AOD occupies the dominant role in affecting the estimate of R_{dif}. Due to the low R_s caused by high CF, the ratio of R_{dif} to R_s can reach around 0.7 in these regions (Figure 7d). As indicated by site-level evaluation (Beijing, Chengdu, Wuhan and Shanghai in Figure S4), the underestimation of ERA5 seems certain. Another area of concern is the Taklimakan Desert, where both CF and COT are low but aerosols and atmospheric water vapor are high. Therefore, we believe that the high amount (~90 W/m^2) and middle ratio (~0.5) of R_{dif} is possible. In addition, the regional average is very close to the measured values at Kashi station (Figure S4). Based on the above analysis, we are confident in the reliability of JiEA and believe that underestimation indeed occurs for ERA5 in related regions. Global simulations of surface solar radiation like ERA5 use a one-dimensional atmospheric radiative transfer model for computation efficiency. As a result, radiation retrievals are unable to tackle the adjacency effects caused by photons which are reflected by the surface out of the field of view and then scattered into the field of view by the atmosphere [78]. That effect directly results in the increase of R_{dif}. Neglect of adjacency effects can account for up to 5% underestimation in incident shortwave radiation on the land surface [10]. In particular, multiple reflections and scattering events off the sides of clouds lead to stronger adjacency effects and consequently to worse underestimation [33,54]. JiEA relies on a CNN-based module to capture the spatial pattern of clouds to deal with adjacency effects [22] and avoids underestimation of R_{dif} radiation to some degree.

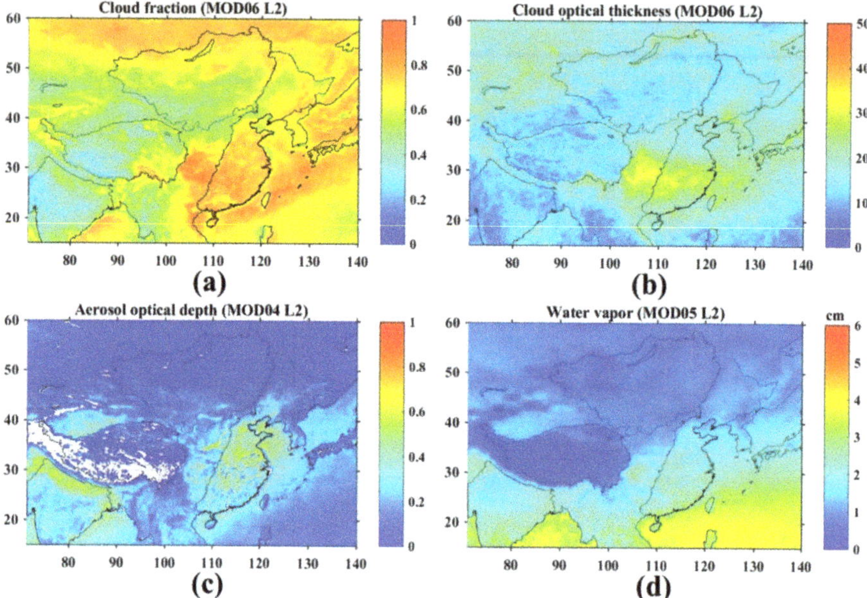

Figure 8. Spatial distribution of atmospheric parameters most relevant to R_{dif} estimation: (**a**) cloud fraction; (**b**) cloud optical thickness; (**c**) aerosol optical depth; (**d**) atmospheric water vapor. We show the averages of monthly results in 2010 of MODIS derived parameters (https://neo.sci.gsfc.nasa.gov/).

On the Tibetan Plateau, ERA5 provides the highest estimates of R_{dif}, significantly greater than those of JiEA. An exception appears in the Tarim Basin. Regardless of overestimation or underestimation, the inner spatial distribution of R_{dif} estimated by ERA5 and JiEA is highly similar (Figure S5f) and agrees well with relevant atmospheric factors (Figure 8). Measurements at Golmud station that is located in the Tarim Basin support the results of JiEA, while the high similarity between observed time series at Lhasa station and ERA5's estimates confirms the potential underestimation of JiEA on the Tibetan Plateau (Figure S4c). One cause of JiEA's underestimation might be the excessive constraint that assumes an idealized state without diffuse radiation at the top of Mt. Everest [22]. The underestimation might also result from misidentification between ice clouds and liquid water clouds, whose radiative effects are significantly different [79,80]. The high probability of ice clouds on the Tibetan Plateau [80] tends to cause more R_{dif} than equivalent liquid water clouds. Previous studies demonstrate that cloud parameters (liquid/ice cloud types are inclusive) are critical in determining R_s [46,81]. This reminds us that we cannot accurately retrieve surface radiation from passive satellite signals alone, and even the best model needs to integrate atmospheric parameters. Therefore, integration of radiation transfer models and deep learning might be the next research focus.

Surface conditions may also influence the estimate of surface radiation. The most frequently mentioned one relates to snow/ice cover, which is often mistaken for clouds. In particular, retrievals of cloud optical depths over such surfaces are accompanied by large uncertainties [61]. It is even more challenging over short-lived snow or ice [33,60]. The high-level R_{dif} of ERA5 on the Tibetan Plateau and the Pamir Plateau is likely affected by snow/ice because observed seasonal variations of R_{dif} (Figure S8) are not consistent with variations of atmospheric factors (Figure S9) but show high similarity to snow/ice cover (Figure S10). Except for this specific issue, both datasets conform to common sense on how atmospheric factors influence R_{dif} in seasonal cycles, proving their strong ability to capture seasonal variation of R_{dif} at regional scale (Figure S5f).

6. Conclusions

Although R_s estimates are widely available from many radiation products, only ERA5 reanalysis and satellite-based JiEA provide estimates of R_{dif} over East Asia. Comprehensive evaluation and comparison are of great importance for rational use of these data and in-depth understanding of temporal trends and spatial differences of R_{dif}. In this study, estimates of R_{dif} at the surface are evaluated by comparing to quality-controlled measurements from WRDC and CMA and are mutually compared with respect to temporal variations and spatial distributions by referring to the spatial pattern of related atmospheric factors.

Hourly R_{dif} estimates of JiEA agree well with CMA measurements with an R of 0.85, MBE of 8.54 W/m^2, MABE of 50.43 W/m^2, and RMSE of 66.36 W/m^2, while ERA5 performs a little worse with an R of 0.71, negative MBE of 29.69 W/m^2, MABE of 63.83 W/m^2 and RMSE of 92.29 W/m^2. The performance of ERA5 is better than JiEA for low-radiation estimates. The overall accuracy of JiEA also exceeds ERA5 at daily means, with 84% of winning samples. Some problematic estimates occur for JiEA, likely due to the failure to handle extreme cases. Their performances are different in different regions. Particularly, ERA5 seriously underestimates R_{dif} on the Deccan Plateau. At monthly means, the RMSE of R_{dif} estimates decreases to 12.92 and 21.13 W/m^2 for JiEA and ERA5, respectively. These comparisons are hardly affected by their different spatial resolution, but the evaluation results are dependent on the spatial representativeness of ground measurement.

Data accuracy of ERA5 shows strong temporal consistency and latitudinal dependency. On the contrary, the accuracy of JiEA fluctuates in time and is robust in space. Therefore, we would like to recommend using ERA5 reanalysis data for trend detection and satellite-based JiEA for regional comparisons. Deseasonalized monthly time series of ERA5 and JiEA are highly correlated with each other but differ from the ground-observed series, indicating that gridded products are unable to reproduce temporal variability at site level. At the regional scale, we observe a slight decreasing trend of R_{dif} from JiEA and no trend from ERA5 within the validation period. Both time series analysis at stations and seasonal variations of spatial distribution show that ERA5 and JiEA are capable of capturing the seasonal cycle of R_{dif} effectively, although deviations still exist.

Notable differences of spatial distribution of R_{dif} from the two datasets appear on the Tibetan Plateau, where the underestimation of JiEA might be due to the misidentification between ice clouds and liquid water clouds, while the overestimation of ERA5 seems related to surface snow/ice cover. References to the spatial distribution of atmospheric factors support R_{dif} estimates of JiEA and confirm the general underestimation of ERA5 over East Asia. Neglect of adjacency effects caused by photon transport is regarded as the main cause for ERA5's underestimation. Our analysis calls for the integration of physical models and new technologies (e.g., deep learning) to obtain accurate estimates of R_{dif}.

Supplementary Materials: The following are available online at http://www.mdpi.com/2072-4292/12/9/1387/s1, Figure S1: Evaluation results of R_{dif} in different regions at daily mean scale. Figure S2: Evaluation results of monthly mean R_{dif} estimates. Figure S3: The effects of spatial resolution on evaluation results. Figure S4: Results of time series decomposition. Figure S5: Results of time series decomposition in different regions. Figure S6. Spatial distribution of rMBE. Figure S7. Spatial distribution of reference data. Figure S8. Seasonal spatial distribution of two datasets. Figure S9. Seasonal spatial distribution of atmospheric parameters most relating to R_{dif} estimation.

Figure S10. Seasonal snow/ice cover. Table S1. Basic information of surface radiation stations involved in this study. Table S2. Requirements defined for downward short-wave irradiance at Earth surface.

Author Contributions: Conceptualization, H.J. and Y.Y.; methodology, H.J.; software, H.W.; validation, H.W., Y.B. (Yongqing Bai) and Y.B. (Yan Bai); formal analysis, H.J.; investigation, H.W.; resources, Y.Y.; data curation, Y.B. (Yan Bai); writing—original draft preparation, H.J.; writing—review and editing, H.J.; visualization, Y.B. (Yongqing Bai); supervision, Y.B. (Yongqing Bai); project administration, H.W.; funding acquisition, Y.Y. All authors have read and agreed to the published version of the manuscript.

Funding: This research was funded by Strategic Priority Research Program (A) of the Chinese Academy of Sciences (XDA19020304), Multidisciplinary Joint Expedition for China–Mongolia–Russia Economic Corridor (2017FY101300), Branch Center Project of Geography, Resources and Ecology of Knowledge Center for Chinese Engineering Sciences and Technology (CKCEST-2019-1-4), National Earth System Science Data Sharing Infrastructure (2005DKA32300).

Acknowledgments: We would like to thank all people fighting with COVID-19, the Copernicus Climate Change Service for the online available ECMWF ERA5 datasets, and NASA Earth Observations for the public atmospheric parameters.

Conflicts of Interest: The authors declare no conflict of interest. The founding sponsors had no role in the design of the study; in the collection, analyses, or interpretation of data; in the writing of the manuscript, and in the decision to publish the results.

References

1. Wang, K.; Dickinson, R.; Wild, M.; Liang, S. Evidence for decadal variation in global terrestrial evapotranspiration between 1982 and 2002: 1. Model development. *J. Geophys. Res.* **2010**, *115*, D20112.

2. Wild, M.; Folini, D.; Schär, C.; Loeb, N.; Dutton, E.; König-Langlo, G. The global energy balance from a surface perspective. *Clim. Dyn.* **2012**, *40*, 3107–3134.

3. Wild, M.; Ohmura, A.; Schär, C.; Müller, G.; Folini, D.; Schwarz, M.; Hakuba, M.Z.; Sanchez-Lorenzo, A. The global energy balance archive (GEBA) version 2017: A database for worldwide measured surface energy fluxes. *Earth Syst. Sci. Data* **2017**, *9*, 601–613.

4. Yang, W.; Guo, X.; Yao, T.; Yang, K.; Zhao, L.; Shenghai, L.; Zhu, M. Summertime surface energy budget and ablation modeling in the ablation zone of a martime tibetan glacier. *J. Geophys. Res.* **2011**, *116*, D14116.

5. Wild, M.; Ohmura, A.; Makowski, K. Impact of global dimming and brightening on global warming. *Geophys. Res. Lett.* **2007**, *34*, L04702.

6. Jia, B.; Xie, Z.; Dai, A.; Shi, C.; Chen, F. Evaluation of satellite and reanalysis products of downward surface solar radiation over east asia: Spatial and seasonal variations. *J. Geophys. Res. (Atmos.)* **2013**, *118*, 3431–3446.

7. Mokhtari, A.; Noory, H.; Vazifedoust, M. Improving crop yield estimation by assimilating LAI and inputting satellite-based surface incoming solar radiation into SWAP model. *Agric. For. Meteorol.* **2018**, *250*, 159–170.

8. Rap, A.; Scott, C.E.; Reddington, C.; Mercado, L.; Ellis, R.; Garraway, S.; Evans, M.; Beerling, D.J.; Mackenzie, A.; Hewitt, C.N.; et al. Enhanced global primary production by biogenic aerosol via diffuse radiation fertilization. *Nat. Geosci.* **2018**, *11*, 640–644.

9. Zhang, M.; Yu, G.-R.; Zhuang, J.; Gentry, R.; Fu, Y.-L.; Sun, X.; Zhang, L.-M.; Wen, X.; Wang, Q.-F.; Han, S.-J.; et al. Effects of cloudiness change on net ecosystem exchange, light use efficiency, and water use efficiency in typical ecosystems of China. *Agric. For. Meteorol.* **2011**, *151*, 803–816.

10. Ryu, Y.; Jiang, C.; Kobayashi, H.; Detto, M. Modis-derived global land products of shortwave radiation and diffuse and total photosynthetically active radiation at 5 km resolution from 2000. *Remote Sens. Environ.* **2017**, *204*, 812–825.

11. Mercado, L.; Bellouin, N.; Sitch, S.; Boucher, O.; Huntingford, C.; Wild, M.; Cox, P. Impact of changes in diffuse radiation on the global land carbon sink. *Nature* **2009**, *458*, 1014–1017. [PubMed]

12. Alton, P.; North, P.R.J.; Los, S. The impact of diffuse sunlight on canopy light-use efficiency, gross photosynthetic product and net ecosystem exchange in three forest biomes. *Glob. Chang. Biol.* **2007**, *13*, 776–787.

13. Kanniah, K.; Beringer, J.; North, P.R.J.; Hutley, L. Control of atmospheric particles on diffuse radiation and terrestrial plant productivity: A review. *Prog. Phys. Geogr.* **2012**, *36*, 210–238.

14. Choudhury, B. A sensitivity analysis of the radiation use efficiency for gross photosynthesis and net carbon accumulation by wheat. *Agric. For. Meteorol.* **2000**, *101*, 217–234.

15. Holzman, M.E.; Carmona, F.; Rivas, R.; Niclòs, R. Early assessment of crop yield from remotely sensed water stress and solar radiation data. *ISPRS J. Photogramm. Remote Sens.* **2018**, *145*, 297–308.

16. Lee, M.; Hollinger, D.; Keenan, T.; Ouimette, A.; Ollinger, S.; Richardson, A. Model-based analysis of the impact of diffuse radiation on CO_2 exchange in a temperate deciduous forest. *Agric. For. Meteorol.* **2017**, *249*, 377–389.

17. Bódis, K.; Kougias, I.; Jäger-Waldau, A.; Taylor, N.; Szabó, S. A high-resolution geospatial assessment of the rooftop solar photovoltaic potential in the european union. *Renew. Sustain. Energy Rev.* **2019**, *114*, 109309.

18. Qin, J.; Tang, W.; Yang, K.; Lu, N.; Niu, X.; Liang, S. An efficient physically based parameterization to derive surface solar irradiance based on satellite atmospheric products. *J. Geophys. Res. Atmos.* **2015**, *120*, 4975–4988.

19. Kobayashi, H.; Iwabuchi, H. A coupled 1-D atmosphere and 3-D canopy radiative transfer model for canopy reflectance, light environment, and photosynthesis simulation in a heterogeneous landscape. *Remote Sens. Environ.* **2008**, *112*, 173–185.

20. Xia, X.; Wang, P.; Chen, H.; Liang, F. Analyis of downwelling surface solar radiation in China from national centers for environmental prediction reanalysis, satellite estimates, and surface observations. *J. Geophys. Res.* **2006**, *111*, D09103.

21. Copernicus Climate Change Service (C3S). ERA5: Fifth generation of ECMWF atmospheric reanalyses of the global climate. Climate Data Store (CDS) 2017. Available online: https://cds.climate.copernicus.eu/cdsapp#!/home (accessed on 27 April 2020).

22. Jiang, H.; Lu, N.; Qin, J.; Tang, W.; Yao, L. A deep learning algorithm to estimate hourly global solar radiation from geostationary satellite data. *Renew. Sustain. Energy Rev.* **2019**, *114*, 109327.

23. Linares-Rodriguez, A.; Ruiz-Arias, J.; Pozo-Vazquez, D.; Tovar-Pescador, J. An artificial neural network ensemble model for estimating global solar radiation from meteosat satellite images. *Energy* **2013**, *61*, 636–645.

24. Lu, N.; Qin, J.; Yang, K.; Sun, J. A simple and efficient algorithm to estimate daily global solar radiation from geostationary satellite data. *Energy* **2011**, *36*, 3179–3188.

25. Zhang, X.; Liang, S.; Zhou, G.; Wu, H.; Zhao, X. Generating global land surface satellite incident shortwave radiation and photosynthetically active radiation products from multiple satellite data. *Remote Sens. Environ.* **2014**, *152*, 318–332.

26. Jiang, H.; Yang, Y.; Bai, Y.; Wang, H. Evaluation of the total, direct, and diffuse solar radiations from the ERA5 reanalysis data in China. *IEEE Geosci. Remote Sens. Lett.* **2020**, *17*, 47–51.

27. Soni, V.; Pandithurai, G.; Pai, D. Evaluation of long-term changes of solar radiation in india. *Int. J. Climatol.* **2012**, *32*, 540–551.

28. Wang, Y.; Trentmann, J.; Yuan, W.; Wild, M. Validation of CM SAF CLARA-A2 and SARAH-E surface solar radiation datasets over China. *Remote Sens.* **2018**, *10*, 1977.

29. Zhang, X.; Liang, S.; Wild, M.; Jiang, B. Analysis of surface incident shortwave radiation from four satellite products. *Remote Sens. Environ.* **2015**, *165*, 186–202.

30. Yang, S.; Wang, X.; Wild, M. Homogenization and trend analysis of the 1958-2016 in-situ surface solar radiation records in China. *J. Clim.* **2018**, *31*, 4529–4541.

31. Greuell, W.; Meirink, J.F.; Wang, P. Retrieval and validation of global, direct, and diffuse irradiance derived from SEVIRI satellite observations. *J. Geophys. Res. (Atmos.)* **2013**, *118*, 2340–2361.

32. Ridley, B.; Boland, J.; Lauret, P. Modeling of diffuse solar fraction with multiple predictors. *Renew. Energy* **2010**, *35*, 478–483.

33. Huang, G.; Li, Z.; Li, X.; Liang, S.; Yang, K.; Wang, D.; Zhang, Y. Estimating surface solar irradiance from satellites: Past, present, and future perspectives. *Remote Sens. Environ.* **2019**, *233*, 111371.

34. Wild, M. Global dimming and brightening: A review. *J. Geophys. Res.* **2009**, *114*, D00D16.

35. Sanchez-Lorenzo, A.; Enriquez-Alonso, A.; Wild, M.; Trentmann, J.; Vicente-Serrano, S.M.; Sanchez-Romero, A.; Posselt, R.; Hakuba, M.Z. Trends in downward surface solar radiation from satellites and ground observations over europe during 1983–2010. *Remote Sens. Environ.* **2017**, *189*, 108–117.

36. Zhang, X.; Lu, N.; Jiang, H.; Yao, L. Evaluation of reanalysis surface incident solar radiation data in China. *Sci. Rep.* **2020**, *10*, 3494.

37. Shi, G.-Y.; Hayasaka, T.; Ohmura, A.; Chen, Z.-H.; Wang, B.; Zhao, J.-Q.; Che, H.; Xu, L. Data quality assessment and the long-term trend of ground solar radiation in China. *J. Appl. Meteorol. Climatol.* **2007**, *47*, 13319–13337.

38. Roebeling, R.; Putten, E.; Genovese, G.; Rosema, A. Application of Meteosat derived meteorological information for crop yield predictions in Europe. *Int. J. Remote Sens.* **2004**, *25*, 5389–5401.

39. Jiang, H.; Lu, N. High-resolution surface global solar radiation and the diffuse component dataset over China. *PANGAEA* **2019**. [CrossRef]

40. Jing, W.; Zhang, P.; Zhao, X. A comparison of different grace solutions in terrestrial water storage trend estimation over tibetan plateau. *Sci. Rep.* **2019**, *9*, 1765.

41. Schwarz, M.; Folini, D.; Hakuba, M.; Wild, M. From point to area: Worldwide assessment of the representativeness of monthly surface solar radiation records. *J. Geophys. Res. Atmos.* **2018**, *123*, 13857–13874.

42. Hakuba, M.; Folini, D.; Sanchez-Lorenzo, A.; Wild, M. Spatial representativeness of ground-based solar radiation measurements—Extension to the full Meteosat disk. *J. Geophys. Res.* **2014**, *119*, 11760–11771.

43. Sanchez-Lorenzo, A.; Wild, M.; Trentmann, J. Validation and stability assessment of the monthly mean cm saf surface solar radiation dataset over Europe against a homogenized surface dataset (1983–2005). *Remote Sens. Environ.* **2013**, *134*, 355–366.

44. Espinar, B.; Blanc, P. Satellite images applied to surface solar radiation estimation. In *Solar Energy at Urban Scale*, 1st ed.; Beckers, B., Ed.; ISTE Ltd: London, UK, 2012; pp. 57–98.

45. Tang, W.; Li, J.; Yang, K.; Qin, J.; Zhang, G.; Wang, Y. Dependence of remote sensing accuracy of global horizontal irradiance at different scales on satellite sampling frequency. *Sol. Energy* **2019**, *193*, 597–603.

46. Tang, W.; Qin, J.; Yang, K.; Shaomin, L.; Lu, N.; Niu, X. Retrieving high-resolution surface solar radiation with cloud parameters derived by combining modis and mtsat data. *Atmos. Chem. Phys.* **2016**, *16*, 2543–2557.

47. Li, Z.; Zhao, X.; Kahn, R.; Mishchenko, M.; Remer, L.; Lee, K.-H.; Wang, M.; Laszlo, I.; Nakajima, T.; Maring, H. Uncertainties in satellite remote sensing of aerosols and impact on monitoring its long-term trend: A review and perspective. *Ann. Geophys.* **2009**, *27*, 2755–2770.

48. Wang, Y.; Wild, M.; Sanchez-Lorenzo, A.; Manara, V. Urbanization effect on trends in sunshine duration in China. *Ann. Geophys.* **2017**, *35*, 839–851.

49. Qin, J.; Chen, Z.; Yang, K.; Liang, S.; Tang, W. Estimation of monthly-mean daily global solar radiation based on modis and trmm products. *Appl. Energy* **2011**, *88*, 2480–2489.

50. Tang, W.; Yang, K.; He, J.; Qin, J. Quality control and estimation of global solar radiation in China. *Sol. Energy* **2010**, *84*, 466–475.

51. Huang, G.; Li, X.; Huang, C.; Shaomin, L.; Ma, Y.; Chen, H. Representativeness errors of point-scale ground-based solar radiation measurements in the validation of remote sensing products. *Remote Sens. Environ.* **2016**, *181*, 198–206.

52. Hakuba, M.; Folini, D.; Sanchez-Lorenzo, A.; Wild, M. Spatial representativeness of ground-based solar radiation measurements. *J. Geophys. Res.* **2013**, *118*, 10362.

53. Schwarz, M.; Folini, D.; Hakuba, M.; Wild, M. Spatial representativeness of surface-measured variations of downward solar radiation. *J. Geophys. Res. Atmos.* **2017**, *122*, 13319–13337.

54. Wyser, K.; O'Hirok, W.; Gautier, C. A simple method for removing 3-D radiative effects in satellite retrievals of surface irradiance. *Remote Sens. Environ.* **2005**, *94*, 335–342.

55. Wang, K.; Ma, Q.; Wang, X.; Wild, M. Urban impacts on mean and trend of surface incident solar radiation. *Geophys. Res. Lett.* **2014**, *41*, 4664–4668.

56. Imamovic, A.; Tanaka, K.; Folini, D.; Wild, M. Global dimming and urbanization: Did stronger negative ssr trends collocate with regions of population growth? *Atmos. Chem. Phys.* **2016**, *16*, 2719–2725.

57. Yang, S.-K.; Hou, Y.-T.; Miller, A.; Campana, K. Evaluation of the earth radiation budget in NCEP-NCAR reanalysis with ERBE. *J. Clim.* **1999**, *12*, 477–493.

58. Zhang, Q.; Qian, Y.F. Monthly mean surface albedo estimated from NCEP/NCAR reanalysis radiation data. *Acta Geogr. Sin.* **1999**, *54*, 309–317.

59. Hu, G.; Zhao, L.; Wu, X.; Li, R.; Wu, T.; Su, Y.; Hao, J. Evaluation of reanalysis air temperature products in permafrost regions on the Qinghai-Tibetan Plateau. *Theor. Appl. Climatol.* **2019**, *138*, 1457–1470.

60. Pinker, R.; Li, X.; Meng, W.; Yegorova, E. Toward improved satellite estimates of short-wave radiative fluxes—Focus on cloud detection over snow: 2. Results. *J. Geophys. Res.* **2007**, *112*, D09204.

61. Platnick, S.; Meyer, K.; King, M.; Wind, G.; Amarasinghe, N.; Marchant, B.; Arnold, G.; Zhang, Z.; Hubanks, P.; Holz, R.; et al. The MODIS cloud optical and microphysical products: Collection 6 updates and examples from Terra and Aqua. *IEEE Trans. Geosci. Remote Sens.* **2017**, *55*, 502–525.

62. Zhang, J.; Zhao, L.; Deng, S.; Xu, W.; Zhang, Y. A critical review of the models used to estimate solar radiation. *Renew. Sustain. Energy Rev.* **2017**, *70*, 314–329.

63. Voyant, C.; Notton, G.; Kalogirou, S.; Nivet, M.-L.; Paoli, C.; Motte, F.; Fouilloy, A. Machine learning methods for solar radiation forecasting: A review. *Renew. Energy* **2017**, *105*, 569–582.

64. Wang, K.; Ma, Q.; Li, Z.; Wang, J. Decadal variability of surface incident solar radiation over China: Observations, satellite retrievals, and reanalyses. *J. Geophys. Res. Atmos.* **2015**, *120*, 6500–6514.

65. Wang, Y.; Yang, Y. China's dimming and brightening: Evidence, causes and hydrological implications. *Ann. Geophys.* **2013**, *32*, 41–55.

66. Wu, F.; Fu, C. Assessment of GEWEX/SRB version 3.0 monthly global radiation dataset over China. *Meteorol. Atmos. Phys.* **2012**, *112*, 155–166.

67. Norris, J.; Wild, M. Trends in aerosol radiative effects over China and eapan inferred from observed cloud cover, solar "dimming," and solar "brightening". *J. Geophys. Res.* **2009**, *114*, D10.

68. Sweerts, B.; Pfenninger, S.; Yang, S.; Folini, D.; van der Zwaan, B.; Wild, M. Estimation of losses in solar energy production from air pollution in China since 1960 using surface radiation data. *Nat. Energy* **2019**, *4*, 657–663.

69. Hollmann, R.; Müller, R.; Gratzki, A. CM-SAF surface radiation budget: First results with AVHRR data. *Adv. Space Res.* **2006**, *37*, 2166–2171.

70. Perez, R.; Ineichen, P.; Moore, K.; Kmiecik, M.; Chain, C.; George, R.; Vignola, F. A new operational satellite-to-irradiance model. *Sol. Energy* **2002**, *73*, 307–317.

71. Wild, M. Enlightening global dimming and brightening. *Bull. Am. Meteorol. Soc.* **2012**, *93*, 27–37.

72. Wang, K.; Dickinson, R.; Ma, Q.; Augustine, J.; Wild, M. Measurement methods affect the observed global dimming and brightening. *J. Clim.* **2013**, *26*, 4112–4120.

73. Wang, K.; Dickinson, R.; Wild, M.; Liang, S. Atmospheric impacts on climatic variability of surface incident solar radiation. *Atmos. Chem. Phys. Discuss.* **2012**, *12*, 9581–9592.

74. Geogdzhayev, I.; Rossow, W.; Cairns, B.; Carlson, B.; Lacis, A.; Travis, L. Long-term satellite record reveals likely recent aerosol trend. *Science* **2007**, *315*, 1543.

75. Zhang, Y.; Rossow, W.; Lacis, A.; Oinas, V. Calculation of radiative fluxes from the surface to top of atmosphere based on ISCCP and other global data sets: Refinements of the radiative transfer model and the input data. *J. Geophys. Res.* **2004**, *109*, D19105.

76. Ma, Y.; Pinker, R. Modeling shortwave radiative fluxes from satellites. *J. Geophys. Res. (Atmos.)* **2012**, *117*, 23202.

77. Hatzianastassiou, N.; Matsoukas, C.; Fotiadi, A.; Pavlakis, K.; Drakakis, E.; Hatzidimitriou, D.; Vardavas, I. Global distribution of earth's surface shortwave radiation budget. *Atmos. Chem. Phys.* **2005**, *5*, 2847–2867.

78. Kaufman, Y. Atmospheric effect on spatial resolution of surface imagery. *Appl. Opt.* **1984**, *23*, 3400.

79. Yi, B.; Rapp, A.; Yang, P.; Baum, B.; King, M. A comparison of Aqua MODIS ice and liquid water cloud physical and optical properties between Collection 6 and Collection 5.1: Pixel-to-pixel comparisons. *J. Geophys. Res. Atmos.* **2017**, *122*, 4528–4549.

80. Yi, B.; Rapp, A.; Yang, P.; Baum, B.; King, M. A comparison of Aqua MODIS ice and liquid water cloud physical and optical properties between Collection 6 and Collection 5.1: Cloud radiative effects. *J. Geophys. Res. Atmos.* **2017**, *122*, 4550–4564.

81. Tang, W.; Yang, K.; Qin, J.; Li, X.; Niu, X. A 16-year dataset (2000–2015) of high-resolution (3 hour, 10 km) global surface solar radiation. *Earth Syst. Sci. Data* **2019**, *11*, 1905–1915.

 *remote sensing*

Article

NOAA Operational Microwave Sounding Radiometer Data Quality Monitoring and Anomaly Assessment Using COSMIC GNSS Radio-Occultation Soundings

Robbie Iacovazzi [1,*], Lin Lin [2], Ninghai Sun [1] and Quanhua Liu [3]

[1] GST Incorporated, 7855 Walker Drive, Suite 200, Greenbelt, MD 20770, USA; Ninghai.Sun@noaa.gov
[2] UMD/ESSIC, 5825 University Research Court, Suite #4001, College Park, MD 20740, USA; Lin.Lin@noaa.gov
[3] NOAA/NESDIS/STAR, 5830 University Research Court, College Park, MD 20740, USA; Quanhua.Liu@noaa.gov
[*] Correspondence: Robert.Iacovazzi@noaa.gov; Tel.: +1-301-683-3553

Received: 28 January 2020; Accepted: 29 February 2020; Published: 4 March 2020

Abstract: National Oceanographic and Atmospheric Administration (NOAA) operational Advanced Technology Microwave Sounder (ATMS) and Advanced Microwave Sounding Unit-A (AMSU-A) data used in numerical weather prediction and climate analysis are essential to protect life and property and maintain safe and efficient commerce. Routine data quality monitoring and anomaly assessment is important to sustain data effectiveness. One valuable parameter used to monitor microwave sounder data quality is the antenna temperature (Ta) difference (O-B) computed between direct instrument Ta measurements and forward radiative transfer model (RTM) brightness temperature (Tb) simulations. This requires microwave radiometer data to be collocated with atmospheric temperature and moisture sounding profiles, so that representative boundary conditions are used to produce the RTM-simulated Tb values. In this study, Constellation Observing System for Meteorology, Ionosphere, and Climate/Formosa Satellite Mission 3 (COSMIC) Global Navigation Satellite System (GNSS) Radio Occultation (RO) soundings over the ocean and equatorward of 60° latitude are used as input to the Community RTM (CRTM) to generate simulated NOAA-18, NOAA-19, Metop-A, and Metop-B AMSU-A and S-NPP and NOAA-20 ATMS Tb values. These simulated Tb values, together with observed Ta values that are nearly simultaneous in space and time, are used to compute Ta O-B statistics on monthly time scales for each instrument. In addition, the CRTM-simulated Tb values based on the COSMIC GNSS RO soundings can be used as a transfer standard to inter-compare Ta values from different microwave radiometer makes and models that have the same bands. For example, monthly Ta O-B statistics for NOAA-18 AMSU-A Channels 4–12 and NOAA-20 ATMS Channels 5–13 can be differenced to estimate the "double-difference" Ta biases between these two instruments for the corresponding frequency bands. This study reveals that the GNSS RO soundings are critical to monitoring and trending individual instrument O-B Ta biases and inter-instrument "double-difference" Ta biases and also to estimate impacts of some sensor anomalies on instrument Ta values.

Keywords: remote sensing; joint polar satellite system; advanced technology microwave sounder; COSMIC-1; GNSS radio occultation; satellite instrument performance monitoring and anomaly detection; data quality tracking

---

1. Introduction

The Advanced Technology Microwave Sounder (ATMS) and Advanced Microwave Sounding Unit (AMSU)-A satellite instruments have been critical in improving numerical weather prediction

(NWP) [1,2] and extending the long-term mid-tropospheric temperature climate time series of the Microwave Sounding Unit (MSU) [3–6], the predecessor instrument of AMSU-A and ATMS. Such projects have revealed that inherent calibration-related antenna temperature (Ta) biases and bias trends within and between operating AMSU-A and/or ATMS instruments must be detected and corrected in order to utilize these satellite data in NWP and climate analyses without the risk of significant errors. Thus, monitoring the quality of National Oceanographic and Atmospheric Administration (NOAA) operational microwave radiometer data is critical to ensuring that NOAA meets its mission of protecting life and property, and maintaining safe and efficient commerce.

The history of AMSU-A and ATMS radiometers used operationally by NOAA are listed in Table 1. The table shows that the AMSU-A radiometer has been manifested on NOAA Polar Operational Environmental Satellite (POES) and EUMETSAT Polar System (EPS) Metop satellite platforms. Meanwhile, ATMS has been manifested on the NOAA Joint Polar Satellite System (JPSS) and Suomi National Polar Partnership (S-NPP) satellite platforms. Given the importance of microwave sounding satellite instruments to weather forecasting and climate analysis, it is imperative to monitor their data quality.

Table 1. NOAA operational microwave sounder instrument name, satellite platform, and operational onset date and August 2019 status.

Instrument	Satellite Platform	Operational Onset Date	August 2019 Operational Status
AMSU-A	POES NOAA-15	1998DEC15	(Operating) Channel 11 and 14 not functional
AMSU-A	POES NOAA-16	2001MAR20	(Decommissioned 2014JUN09)
AMSU-A	POES NOAA-17	2002JUN24	(Decommissioned 2003OCT)
AMSU-A	POES NOAA-18	2005AUG30	(Operating)
AMSU-A	POES NOAA-19	2009JUN02	(Operating)
AMSU-A	EPS Metop-A	2007MAY15	(Operating) Channel 7 and 8 not functional
AMSU-A	EPS Metop-B	2013JAN29	(Operating)
AMSU-A	EPS Metop-C	2019MAR21	(Operating)
ATMS	S-NPP	2012MAR06	(Operating)
ATMS	JPSS NOAA-20	2018MAY30	(Operating)

Microwave radiometer Ta measurement monitoring, important for diagnosing instrument performance degradation of the sensors listed in Table 1, has predominately been carried out using the Simultaneous Nadir Overpass (SNO) and the Radiative Transfer Model (RTM) Background Simulation (BS) methods, summarized briefly below.

The SNO method is based on the fact that many polar-orbiting satellites revolve around the Earth at slightly different periods, which causes them to occasionally view the same nadir location at nearly the same time. Ideally, identical radiometers flown on different satellites that simultaneously view the same exact Earth target should produce redundant observations. Any deviation from these results would be attributable to relative calibration differences between the "identical" radiometers. Taking advantage of this concept, the SNO method was developed to estimate and track relative calibration-related measurement biases between visible/infrared radiometers flown on-board different polar-orbiting satellites [7–9]. For a given pair of polar-orbiting satellites, the SNO method analysis is performed for near-nadir observations close to satellite orbital intersections that have a relatively small time difference (~30 s). The SNO method was extended to microwave radiometers by Iacovazzi and Cao [10,11].

The RTM-BS method entails simple comparison of radiance, Ta or brightness temperature (Tb) values observed by an instrument with respect to simulated radiance or Tb by an RTM. In order to implement this method in clear-sky regions, atmospheric sounding data that includes pressure, temperature, water vapor mixing ratio, and ozone mixing ratio are needed to establish representative boundary conditions for the RTM. There are diverse atmospheric sounding sources that can be used to support this method. These include soundings generated from traditional radiosondes, NWP model output, and Global Navigation Satellite System (GNSS) Radio-Occultation (RO). Once simulated background (B) Tb values are computed, they can be subtracted from observed (O) Ta values to yield O-B Ta biases. Application of this method to the S-NPP ATMS using National Center for Environmental Prediction (NCEP) Global Forecast System (GFS) 6-hr forecast outputs can be found in a paper by Weng et al. [12].

In this study, we implement the RTM-BS method by harnessing the Constellation Observing System for Meteorology, Ionosphere, and Climate/Formosa Satellite Mission 3 (COSMIC-1/FORMOSAT3, hereafter referred to as COSMIC for brevity) GNSS-RO sounding data to establish boundary conditions for the Community Radiative Transfer Model (CRTM). Simulated Tb from the CRTM is used to estimate monthly-average O-B Ta bias for each operational AMSU-A and ATMS instrument commissioned after 1 January 2000 except for Metop-C AMSU-A—i.e., NOAA-18, NOAA-19, Metop-A and Metop-B AMSU-A and S-NPP and NOAA-20 ATMS. Furthermore, using the CRTM-simulated Tb values as a transfer standard, the monthly-average O-B Ta bias values for each instrument can be the foundation to inter-compare Ta observations from different microwave radiometer makes and models. Note that the Metop-C AMSU-A is not analyzed in this study because the Ta product was undergoing validation, while the bulk of this study was being performed.

In the next section, we briefly describe the AMSU-A and ATMS instruments and their data, as well as COSMIC GNSS RO observations. Section 3 discusses the method of microwave radiometer and GNSS RO data collocation, as well as providing a brief discussion of the CRTM. Section 4 provides monthly-average O-B Ta bias trends, and double-difference Ta bias trends based on monthly-average O-B Ta bias values from each pair of operational microwave radiometers. Section 4 also provides examples of the power of this method to support operational microwave radiometer anomaly assessment. Finally, in Section 5 a summary is provided.

2. Instruments and Observations

Section 2.1 offers overviews of the AMSU/ATMS instruments, and the common sounding channels between them that are selected for this study. Section 2.2 describes the nature of the COSMIC GNSS RO observations, which form the basis of most of the atmospheric sounding inputs needed by the CRTM.

2.1. AMSU and ATMS Microwave Radiometer Instrument Overviews

The Advance Microwave Sounding Unit (AMSU) is composed of two units, AMSU-A and AMSU-B. Since the AMSU-A instrument contains atmospheric sounding channels that produce data that can be easily compared with the CRTM-simulated Tb values generated using soundings based on the COSMIC GNSS RO data, the AMSU-A will only be described here. The 15-channel AMSU-A satellite radiometer was designed to replace the 4-channel Microwave Sounding Unit (MSU), which has flown on several POES missions since 1979. In Table 2, the specifications for central frequency, Noise Equivalent Delta Temperature (NEDT) and accuracy, polarization, and nominal beam width and field of view size at nadir of each AMSU-A radiometer channel is listed [13]. The physics behind the choice of these channels can be visualized with the aid of Figure 1, which is actually a combination of figures taken from [14,15].

Table 2. Specifications for AMSU-A channel central frequency, Noise Equivalent Delta Temperature (NEDT) and accuracy, polarization, and nominal beam width and field of view size at nadir.

Channel Number	Central Frequency (MHz)	Specified NEDT/Accuracy (K)	Polarization	3dB Beam width and FOV Nadir Size (deg & km)
1	23,800	0.30/1.0	V	3.3 & 50
2	31,400	0.30/1.0	V	3.3 & 50
3	50,300	0.40/1.0	V	3.3 & 50
4	52,800	0.25/1.0	V	3.3 & 50
5	53,596 ± 115	0.25/1.0	H	3.3 & 50
6	54,400	0.25/1.0	H	3.3 & 50
7	54,940	0.25/1.0	V	3.3 & 50
8	55,500	0.25/1.0	H	3.3 & 50
9	f0 = 57,290.344	0.25/1.0	H	3.3 & 50
10	f0 ± 217	0.40/1.0	H	3.3 & 50
11	f0 ± 322.2 ± 48	0.40/1.0	H	3.3 & 50
12	f0 ± 322.2 ± 22	0.60/1.0	H	3.3 & 50
13	f0 ± 322.2 ± 10	0.80/1.0	H	3.3 & 50
14	f0 ± 322.2 ± 4.5	1.20/1.0	H	3.3 & 50
15	89,000	0.50/1.0	V	3.3 & 50

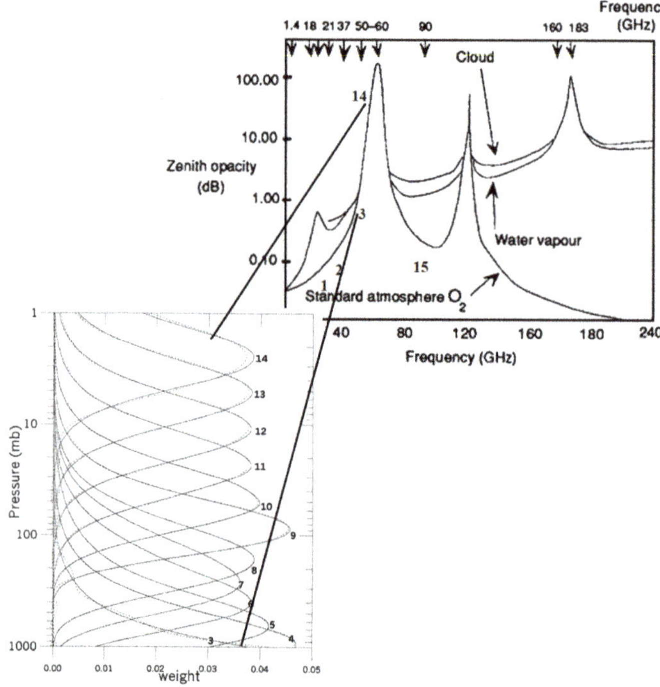

Figure 1. Atmospheric zenith opacity as a function of microwave frequency (top—[15]), where AMSU-A Channels (Chs) 1–3 and 14–15 are provided for reference. Additionally given are the weighting functions for AMSU-A sounding Chs 3–14 (bottom—[14] © Copyright 2000 AMS) in the oxygen absorption band.

In this figure, Channels (Chs) 1, 2, and 15 are considered surface channels in the absence of clouds and water vapor because of their relatively low atmospheric zenith opacity. The Ch 1 and 2 Ta values are directly proportional to surface emissivity, which has values of approximately 1.0 and 0.5 over land and ocean, respectively. Over ocean, relatively cool Ch 1 and 2 Ta values at a given location increase with cloud liquid water and precipitation amount, which leads to estimates of these physical quantities over ocean. Sea ice is also detected in these channels using similar reasoning. All of these properties of Chs 1 and 2 can be found in Ch 15, except that the atmosphere becomes increasingly opaque in this channel with increasing water vapor. Furthermore, Ch 15 is sensitive to ice particle scattering, which lends it to be used to estimate convective precipitation. The AMSU-A Chs 3 to 14 are sounding channels that utilize the strongly increasing opacity of the atmosphere as frequency approaches the center of the 60 GHz oxygen absorption band. Note that sounding Chs 3 and 4 are strongly influenced by the surface, so Chs 4 to 12 are the channels used in conjunction with the COSMIC GNSS RO soundings.

The nominal beam width and NEDT specified in Table 2 are a source of variability in the Ta bias statistics. Nominal beam width determines the size of the region where 50 % of the microwave energy is coming from in a given scene. A nominal beam width of 3.3 degrees leads to a nadir footprint of nearly 50 km. The AMSU-A instrument scans ±48.33° from nadir to complete a total of 30 field of view (FOV) measurements along scan lines, which leads to a swath width of 2243 km.

On-board calibration of total power microwave sounding radiometers, such as AMSU-A and ATMS, is achieved by observing cold space and a well-characterized internal blackbody target during each revolution of the scan reflector antenna. Each AMSU-A eight-second full scan revolution begins by the scan antenna sampling earth scenes. After observing the earth, the reflector rotates such that cold space is measured when the antenna moves to a position that points to an unobstructed view of space—i.e., between the Earth's limb and the spacecraft horizon. After the cold space observations, the internal blackbody is viewed when the antenna rotates to the instrument anti-nadir direction, where the blackbody is located. After these observations, the scan reflector rotates back to an earth view, and then continues the next scan cycle. The calibration measurements are used to accurately determine the so-called radiometer transfer function that relates the measured digitized output (i.e., counts) to a radiance, which then can be expressed as radiometric Ta through the Planck function. More about the calibration of AMSU-A can be found in the NOAA KLM User's Guide [16].

The ATMS is a cross-track scanning microwave radiometer that is manifested on the S-NPP and JPSS satellites. In Table 3, the specifications of spectral value, NEDT and accuracy, polarization and nominal beam width in the cross- and along-track directions and the field of view size at nadir of each ATMS radiometer channel is listed [17].

Table 3. Specifications for ATMS channel central frequency, NEDT and accuracy, polarization, and nominal beam width and field of view size at nadir.

Channel Number	Central Frequency (MHz)	Specified NEDT/Accuracy (K)	Polarization	3dB Beam width and FOV Nadir Size (deg & km)
1	23,800	0.70/1.0	V	6.3/5.2 & 75
2	31,400	0.80/1.0	V	6.3/5.2 & 75
3	50,300	0.9/0.75	H	3.3/2.2 & 32
4	51,760	0.7/0.75	H	3.3/2.2 & 32
5	52,800	0.7/0.75	H	3.3/2.2 & 32
6	$53,596 \pm 115$	0.7/0.75	H	3.3/2.2 & 32
7	54,400	0.7/0.75	H	3.3/2.2 & 32
8	54,940	0.7/0.75	H	3.3/2.2 & 32
9	55,500	0.7/0.75	H	3.3/2.2 & 32
10	$f0 = 57,290.344$	0.75/0.75	H	3.3/2.2 & 32
11	$f0 \pm 217$	1.2/0.75	H	3.3/2.2 & 32
12	$f0 \pm 322.2 \pm 48$	1.2/0.75	H	3.3/2.2 & 32
13	$f0 \pm 322.2 \pm 22$	1.5/0.75	H	3.3/2.2 & 32
14	$f0 \pm 322.2 \pm 10$	2.4/0.75	H	3.3/2.2 & 32
15	$f0 \pm 322.2 \pm 4.5$	3.6/0.75	H	3.3/2.2 & 32
16	88,200	0.5/1.0	V	3.3/2.2 & 32
17	165,500	0.6/1.0	H	2.2/1.1 & 16
18	$183,310 \pm 7000$	0.8/1.0	H	2.2/1.1 & 16
19	$183,310 \pm 4500$	0.8/1.0	H	2.2/1.1 & 16
20	$183,310 \pm 3000$	0.8/1.0	H	2.2/1.1 & 16
21	$183,310 \pm 1800$	0.8/1.0	H	2.2/1.1 & 16
22	$183,310 \pm 1000$	0.9/1.0	H	2.2/1.1 & 16

The ATMS instrument scans ±52.725° from nadir to complete a total of 96 field of view (FOV) measurements along scan lines. ATMS has a swath width of 2700 km, which leaves almost no data gap even near the equator. The instrument is configured with a total of 22 channels at microwave frequencies ranging from 23 to 183 GHz. Calibration of ATMS is similar to that described in the previous subsection for AMSU-A, except some important radiometric corrections have been made to the data to improve accuracy. More information about ATMS calibration can be found in articles by Weng et al. [12], Han et al. [18], and Weng et al. [17].

When comparing the ATMS and AMSU-A instrument channels, ATMS includes a channel at 51.76 GHz that is not present in AMSU-A. Additionally, the 89.0 GHz channel of AMSU-A is not equivalent to the 88.2 GHz channel of ATMS. On the other hand, ATMS shares the same central frequencies with 14 AMSU-A channels. Not all of these shared channels are included in the present analysis though. The reason for this is that the COSMIC GNSS RO soundings report atmospheric pressure, temperature and water vapor between the surface and 40 km at 100 m intervals. For this study, measurements of surface parameters, and sounding parameters above 3 mb, are assumed to be climatological values. Therefore, the CRTM simulations performed using the synthesis of COSMIC GNSS RO sounding measurements, and climatological surface and upper atmosphere sounding estimates, leads to the highest skill in comparing CRTM simulated and AMSU-A and ATMS observed Ta values for common AMSU-A and ATMS channels that have weighting functions contained almost entirely between the surface and 40 km. This represents AMSU-A Chs 4–12 and ATMS Chs 5–13.

Another important architectural change between ATMS and AMSU-A is that the polarization of AMSU-A Ch 4 and 7 (V-pol) and ATMS Ch 5 and 8 (H-pol) are different. Since AMSU-A Ch 4 and ATMS Ch 5 receive a large portion of energy from the surface, this will lead to quite different absolute values for them over ocean. On the other hand, the double difference with respect to CRTM-simulated values should eliminate this large absolute temperature bias. For AMSU-A Ch 7 and ATMS Ch 8, there should not be such an absolute difference, since radiation originating from the atmosphere in the microwave is considered unpolarized, and the amount of energy from the surface for this sounding channel is much smaller.

2.2. COSMIC GNSS RO Observations Overview

The GNSS RO sounding method is a limb-sounding technique that makes use of radio signals emitted from GNSS satellites for determining the temperature and water vapor profiles of the Earth's atmosphere. Assuming spherical symmetry of the atmospheric refractive index, vertical profiles of bending angle and refractivity can be derived from the raw RO measurements of the excess Doppler shift of the radio signals transmitted by GNSS satellites [19]. Since temperature and water vapor variations in the atmosphere can elicit small variations of this altitude-dependent refractivity, profiles of refractivity and their subtle variability can then be used to generate profiles of the temperature and water vapor retrieval using a one-dimensional variational data assimilation (1D-Var) algorithm [20,21]. The COSMIC Data Analysis and Archive Center (CDAAC) wet (wetPrf) retrieval is used for this software tool. A brief description of a 1D-Var algorithm for GNSS RO retrieval is provided at https://cdaac-www.cosmic.ucar.edu/cdaac/doc/overview.html.

The COSMIC satellite system consists of a constellation of six low-Earth-orbit (LEO) micro-satellites, and was launched on April 15, 2006. Each LEO follows a circular orbit 512 km above the Earth surface, with an inclination angle of 72°. Currently, since only one COSMIC instrument is functioning, there are only up to about 250 soundings daily, and this number can be much lower for some days. The vertical resolution is 0.1 km from surface to 39.9 km, and each GNSS-RO measurement quantifies an integrated atmospheric refraction effect over a few hundred kilometers along a ray path centered at the tangent point. The global mean differences between COSMIC and high-quality reanalysis within the height range between 8 and 30 km are estimated to be about 0.65 K [22], while a more recent study of soundings over the Arctic estimated a structural uncertainty (due to different data processing approaches) of about 0.72 K [23]. The precision of COSMIC GNSS RO soundings, estimated by comparison of closely collocated Challenging Minisatellite Payload (CHAMP) and Scientific Application Satellite-C (SAC-C) GNSS-RO soundings, is approximately 0.05 K in the upper troposphere and lower stratosphere [24]. In the water vapor abundant region in the lower troposphere (e.g., when temperature is greater than

270 K), the precision reduces to about 0.1 K. In the ionosphere regions, GNSS profiles become less accurate due to residual ionospheric effect. The estimated precision of COSMIC GNSS RO soundings is approximately 0.2 K in the ionosphere.

3. Method

In this section, a summary of the method to monitor and trend AMSU-A and ATMS Temperature Data Record (TDR) with RTM-BS output is described. The first subsection describes the methods to screen the AMSU-A and ATMS data, and the COSMIC GNSS RO soundings, and to collocate these data. The second subsection offers a brief description of the CRTM, and the final subsection presents a list of the statistical analysis of the observed minus background (O-B) Ta values computed from AMSU-A and ATMS observations and associated CRTM simulations.

3.1. Microwave Radiometer and COSMIC GNSS RO Observation Screening and Collocation Method Subsection

In this inter-comparison study, the data associated with the COSMIC GNSS RO soundings and the microwave radiometer measurements must be screened and then efficiently collocated. Screening criteria applied to both data sets include limiting them to ocean regions equatorward of 60° latitude. Since the physical properties of Ta for the sounding channels are affected by clouds, a cloud detection algorithm similar to [25] is applied to separate the data in clear sky conditions over ocean from the total microwave radiometer measurements [12]. As the GNSS radio signal passes through the atmosphere, its ray path is bent over due to variations of atmospheric refraction. Therefore, the geolocation of the perigee point (also called tangent point) of a single GNSS RO profile varies with altitude. Therefore, soundings are rejected where the location variation of sounding measurement geolocation versus altitude is more than 150 km, because the GNSS RO resolution is about 300 km. This eliminates 5% or less of soundings that may have relatively large vertical variations of geolocation with altitude. When the sounding geolocation variability with altitude is constrained, it allows spatial constraints to be placed on the collocation process that makes it much more efficient.

The collocation criteria are set by a time difference of no more than three hours and a horizontal spatial separation of less than 50 km. If there are multiple microwave radiometer pixel measurements satisfying these collocation criteria, the one that is closest to the related COSMIC sounding is chosen and others are discarded. As an efficiency, the initial spatial bounding circle to screen collocations is established to ensure that if no matchup is found at the first viable sounding level, then it is impossible that there with be a matchup at any higher sounding level. For a given GNSS RO sounding, this spatial bounding circle is defined by a sum of (1) the maximum distance between the lowest sounding level location and all the other sounding level locations above, (2) the 50 km distance threshold, and (3) a 25 km distance padding to be conservative on the side of finding matchups. If no matchups in an ATMS granule or AMSU-A file are found at the lowest level with this screen, then the program goes to the next granule or file without trying to find matchups for the higher sounding levels.

As mentioned above, the geolocation of the tangent point of a single GNSS RO profile varies with altitude. On the other hand, a satellite measurement at a specific frequency represents a weighted average of radiation emitted from different layers of the atmosphere. The magnitude of such a weighting is determined by a channel-dependent weighting function (WF). The measured radiation is most sensitive to the atmospheric temperature at the altitude where the WF reaches a maximum. The WF also varies with scan angle [26]. For each channel, the altitude of the peak WF is the lowest at nadir and increases with the scan angle. Considering the geolocation change of the perigee point of a GNSS RO profile with altitude, the geolocation of a given GNSS RO sounding at the altitude where the WF for each collocated microwave radiometer field of view (FOV) of a particular sounding channel reaches the maximum is used for implementing the spatial collocation criteria of less than 50 km. The altitude of the maximum WF is determined by inputting the U.S. standard atmosphere into the CRTM.

3.2. Summary of the Community Radiative Transfer Model

The CRTM is developed and distributed by the US Joint Center for Satellite Data Assimilation (JCSDA). The model is publicly available and may be downloaded from ftp://ftp.emc.ncep.noaa.gov/jcsda/CRTM/REL-2.1.3/. The CRTM is a sensor-channel-based radiative transfer model [27–30] and is widely used for microwave and infrared satellite data assimilation and remote sensing applications. It includes modules that compute the satellite-measured thermal radiation from gaseous absorption, absorption and scattering of radiation by aerosols and clouds, and emission and reflection of radiation by the Earth surface. The input to the CRTM includes atmospheric state variables—e.g., temperature, water vapor, pressure, and ozone concentration at user defined layers, and optionally, liquid water content and mean particle size profiles for up to six cloud types and surface state variables and parameters including the emissivity, skin temperature, and wind. In addition to CRTM (i.e., the forward model), the corresponding tangent-linear, adjoint, and K-matrix models have also been included in the CRTM package.

As outlined in Section 2.1, in this study the vertical profiles of temperature, water vapor and pressure are obtained as a hybrid of COSMIC GNSS RO sounding retrievals and climatological values of surface parameters and sounding parameters above 3 mb (~40 km). The mixing ratio profile of ozone is set to be equal to the U.S. standard atmospheric state. For simplicity, no cloud or aerosols are considered in the radiative transfer simulation. The emissivity is derived from the CRTM oceanic surface model at microwave frequencies.

3.3. AMSU-A and ATMS O-B Antenna Temperature Statistics

The mission-life time series and statistical results presented in the subsequent section are based on NOAA operational AMSU-A and ATMS observed minus CRTM-simulated background (O-B) monthly-mean Ta bias values. For a given on-orbit AMSU-A or ATMS unit, these values are computed from the set of O-B Ta biases determined from all individual collocated radiometer and COSMIC GNSS RO data accumulated over each month. The mission-life standard deviation of these monthly-mean O-B Ta bias data provides an estimate of their uncertainty and can be used to approximate the mission-life population standard deviation.

For temporal periods of overlap between sensors, the monthly-mean Ta biases between sensors can be determined by assigning the CRTM-simulated values as a calibration transfer standard. In this case, simply subtracting the monthly-mean O-B Ta bias values for different instruments establishes these sensor-to-sensor "double-difference" Ta bias statistics. For the purpose of this study, only operational satellites launched after January 1, 2000 will be considered, which includes NOAA-18, NOAA-19, Metop-A and Metop-B AMSU-A and S-NPP and NOAA-20 (JPSS-1) ATMS. Furthermore, the data records here are only processed after October 2012 for AMSU-A data, January 2015 for S-NPP ATMS and December 2017 for JPSS-1 ATMS.

It is important here to stress that the output of CRTM microwave sounding radiometer simulations is radiance or Tb. Meanwhile, this analysis uses AMSU-A and ATMS Ta measurements, because AMSU-A and ATMS both have Ta products, while only ATMS has an official operational Tb product. Although CRTM Ta estimates are available for direct comparison with AMSU-A Ta measurements, we choose to generate CRTM AMSU-A and ATMS Tb simulated values. This makes inter-satellite comparisons between AMSU-A and ATMS more "apples-to-apples." Using the CRTM generated Tb for both AMSU-A and ATMS allows the CRTM to be used as a transfer standard to compute O-B Ta double difference. The key to trending O-B Ta for each instrument is being able to visualize changes, which is in no way compromised by using CRTM simulated Tb.

4. Results and Discussion

In the next three subsections, the AMSU-A and ATMS O-B Ta bias analysis results for all collocated radiometer and COSMIC GNSS RO data that pass the imposed spatial and cloud screening criteria are presented. Section 4.1 focuses on fine-resolution analysis represented by one-day and one-month global O-B Ta bias maps, to provide an example of the data distribution and population at these time scales. Coarser resolution analyses are depicted in the last two subsections. In Section 4.2, global monthly-mean O-B Ta bias time series plots are given, as well as mission-life mean and standard deviation plots computed from these time series. Finally, in Section 4.3 statistical analysis related to the differences between monthly-mean O-B Ta bias results from instrument pairs—i.e., the sensor-to-sensor "double-difference" Ta bias values—are presented in a similar manner as Section 4.2. Sections 4.2 and 4.3 also highlight microwave sounder anomaly detection and investigation examples to reveal how these monitoring parameters are used in "day-in-the-life" instrument calibration maintenance and sustainment.

4.1. One-Day and –Month O-B Ta Bias Maps

According to information about the COSMIC-1/FORMOSAT3 Program captured by the Earth Observation (EO) Portal (https://directory.eoportal.org/web/eoportal/satellite-missions), five of the six COSMIC constellation spacecraft were operating in October 2012, which is the beginning of the time range of this analysis. This configuration led to about 1500 to 2000 GNSS RO soundings per day, uniformly distributed around the globe. By mid-2019, near the end of this analysis, number of COSMIC GNSS RO soundings per day had dropped to about 400–500 due to either satellite instrument decommission or operational instabilities. In this study, we use the COSMIC-1/FORMOSAT3 acquired between October 2012 and August 2019 to perform our analysis.

The data analysis screening process presented in Section 3.1 limits the GNSS RO soundings to clear-sky ocean regions equatorward of 60° latitude, and the collocation process with the operational satellite microwave radiometer data limits the sample even more. By the end of these processes, the number of collocated radiometer and COSMIC GNSS RO data points on a given day in 2012 could be in the hundreds, while near the end of this analysis in August 2019 that number drops into the tens. The mission-life average of the number of collocated radiometer and COSMIC GNSS RO data matchups accumulated over the course of a month for each channel is given in Table 4.

Table 4. The mission-life average number of collocated radiometer and COSMIC GNSS RO data matchups accumulated over the course of a month for NOAA-18, NOAA-19, Metop-A, and Metop-B AMSU-A Channels 4–12, and for Suomi-NPP and JPSS-1 ATMS Channels 5–13.

	AMSU-A					ATMS	
Ch #	NOAA-18	NOAA-19	Metop-A	Metop-B	Ch #	S-NPP	NOAA-20
4	1990	2205	1881	1864	5	1089	415
5	1541	1860	1627	1633	6	1571	718
6	2380	2619	2219	2207	7	1594	725
7	2403	2651	0	2229	8	1602	728
8	2408	2660	3570	2233	9	1609	728
9	6589	6554	5794	5687	10	4355	1852
10	6590	6557	5794	5688	11	4365	1868
11	6591	6558	5795	5686	12	4369	1862
12	6596	6562	5800	5688	13	4381	1868

Examples of daily global O-B Ta bias maps for NOAA-18 AMSU-A Ch 4 and Ch 12 are provided in Figure 2A,B, respectively. In these figures, each point represents the location of a COSMIC GNSS RO sounding and collocated NOAA-18 AMSU-A observation that was acquired on 31 October 2012. These figures show clearly that upper-air sounding channels such as AMSU-A Ch 12 have a great deal more points than surface-influenced channels like AMSU-A Ch 4. This is due to the fact that data associated with GNSS RO sounding levels closest to the surface are more likely to be missing or have the bad data quality flag set. The color shade of each point represents an O-B Ta bias range, to crudely quantify its value. As expected, the data distribution is fairly uniform over the global oceans equatorward of 60° latitude. Daily O-B Ta bias values at GNSS RO sounding locations are accumulated over a month to create datasets for monthly O-B Ta bias statistics. The monthly O-B Ta bias maps for NOAA-18 AMSU-A Ch 4 and Ch 12 are provided in Figure 3A,B so the reader can gain an appreciation of the large number of observations that are the foundation of monthly statistics.

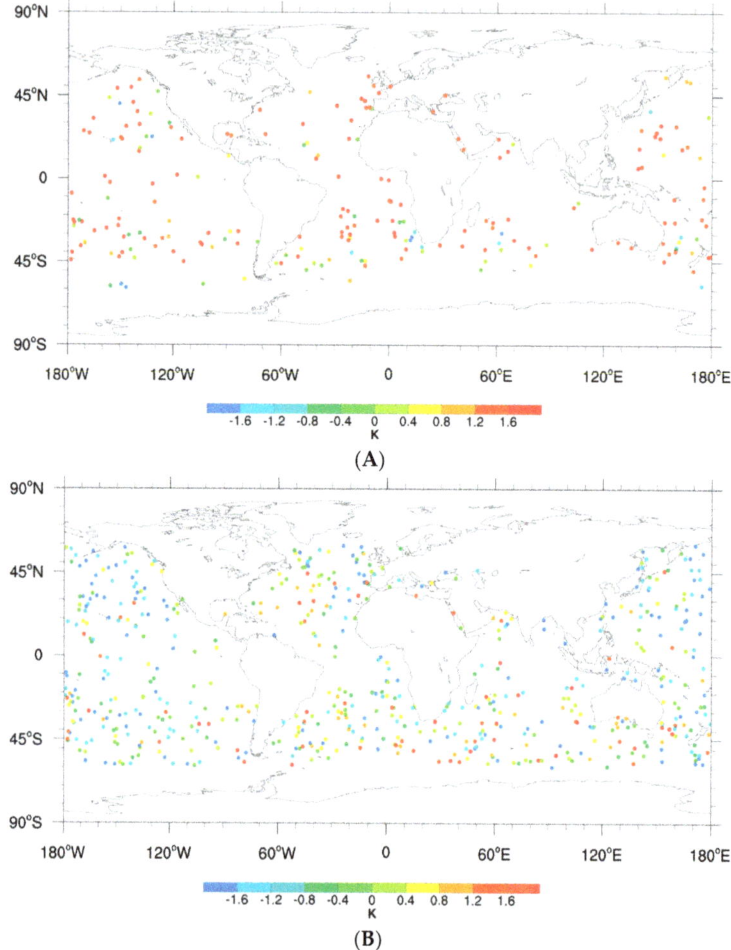

Figure 2. Daily global O-B Ta bias maps for NOAA-18 AMSU Ch 4 (**A**) and Ch 12 (**B**) for 31 October 2012.

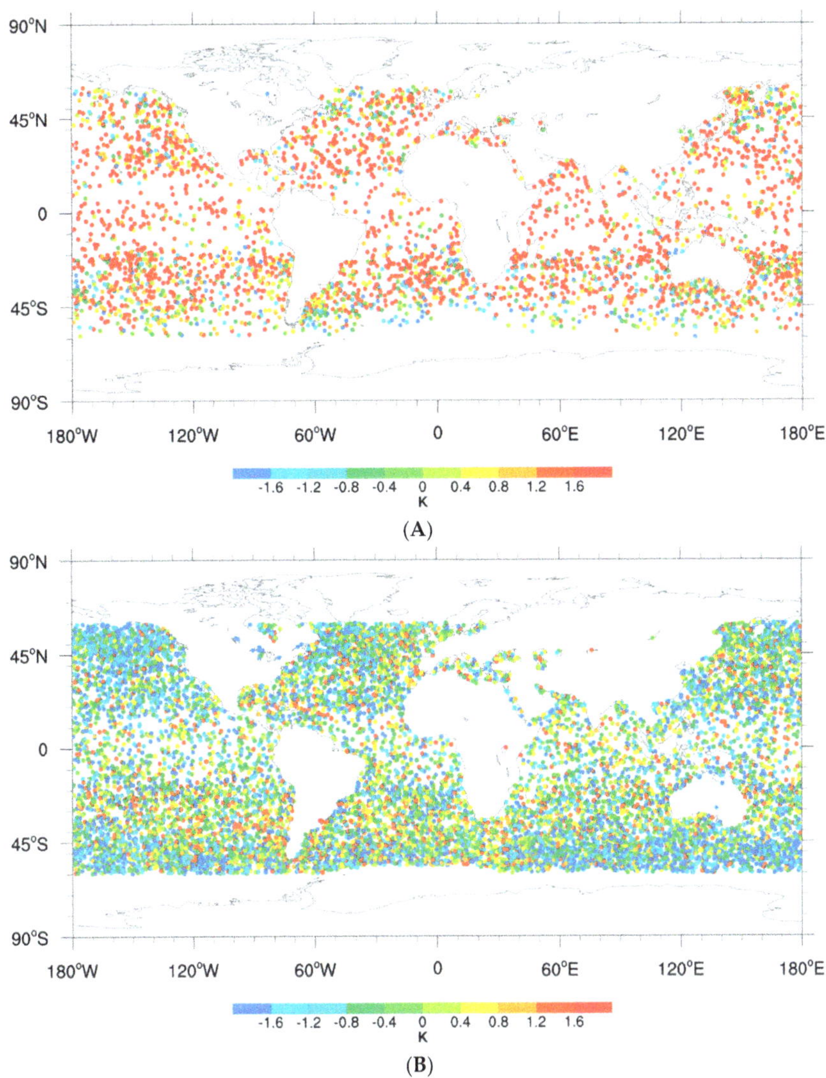

Figure 3. Monthly global O-B Ta bias maps for NOAA-18 AMSU Ch 4 (**A**) and Ch 12 (**B**) for October 2012.

As mentioned above, the number of functioning COSMIC instruments changed dramatically by the end of the record, which resulted in much smaller sample sizes. On the other hand, these reduced sample sizes remain globally well distributed and are adequate to be statistically robust. This is shown in Figure 4A,B with the monthly global O-B Ta bias maps for NOAA-18 AMSU-A Ch 4 and Ch 12 for August 2019.

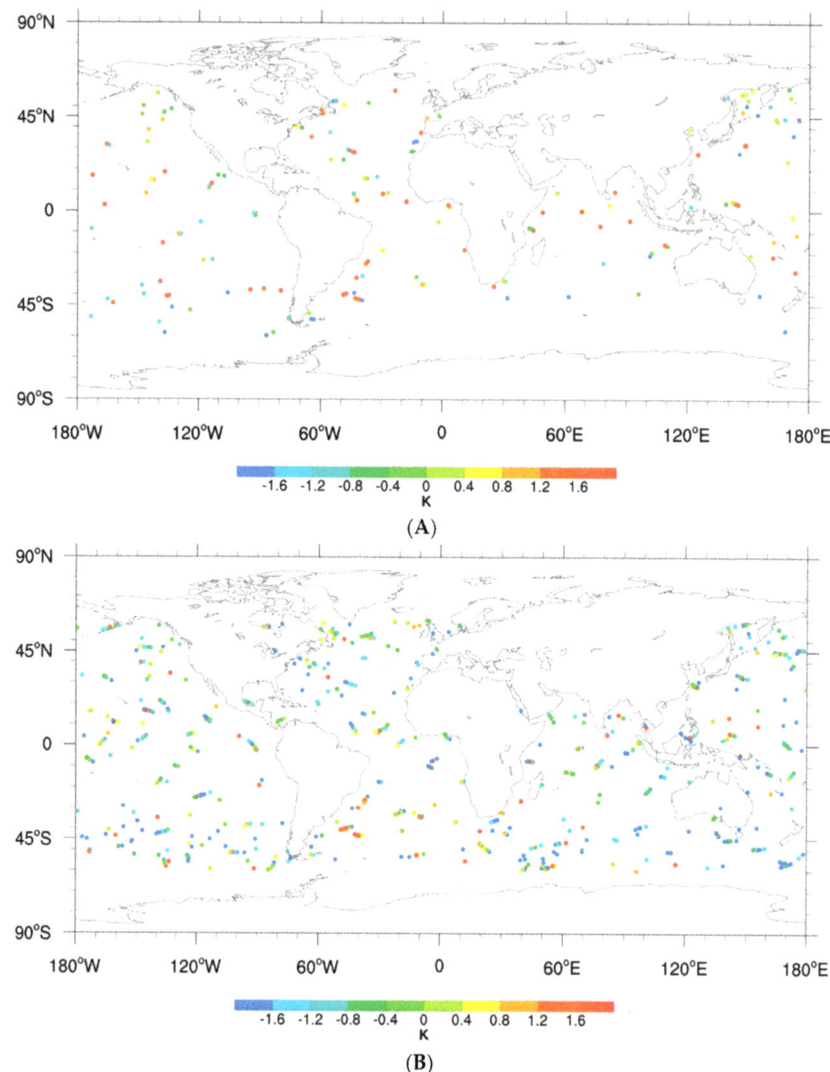

Figure 4. Monthly global O-B Ta bias maps for NOAA-18 AMSU Ch 4 (**A**) and Ch 12 (**B**) for August 2019.

4.2. Monthly and Mission-Life O-B Ta Bias Statistical Results

A goal of this paper is to help NOAA operational microwave radiometer data users gain insight into the stability of radiometer observations over instrument mission-life time periods. For this purpose, in this section monthly and mission-life O-B Ta bias statistical results are given for NOAA-18, NOAA-19, Metop-A, and Metop-B AMSU-A, and S-NPP and JPSS-1 ATMS. These statistics include the monthly-mean O-B Ta bias values, as well as mission-life averages of these monthly-mean values, which is called the "mission-life mean" in this study. Mission-life statistics also include the standard deviation of the monthly-mean values. This "mission-life standard deviation" is used to assign an uncertainty to the method.

In Figure 5A, the mission-life mean O-B Ta bias values are plotted as a function of radiometer sounding channel. For ATMS Ch 5 (AMSU-A Ch 4) these O-B Ta bias values are about 1 K, while they vary between about −1.5 K to 0 K for ATMS Ch 6–13 (AMSU-A Ch 5–12). These salient results are reminiscent of temperature biases found in the COSMIC wetPrf data relative to radiosondes by Wang et al. [31]. In their study, comparisons of COSMIC wetPrf and radiosonde temperatures revealed a positive wetPrf bias that increased from 0.0 K to 0.5 K as the atmospheric layer pressure increased from 700 hPa to 925 hPa. For layers with pressure less than 700 hPa, theses biases between COSMIC wetPrf and radiosonde temperatures were about −0.2 K to −0.3 K. Figure 5A also shows a prominent NOAA-19 Ch 8 O-B Ta bias outlier of about 0.5 K. This bias outlier differs substantially from the Ta bias value of about −1.25 K for the other instrument makes and models for this 55.5 GHz frequency channel. This figure further reveals that except for the one outlier, the mission-life mean O-B Ta bias values cluster within about 0.5 K of each other for channels with identical frequencies.

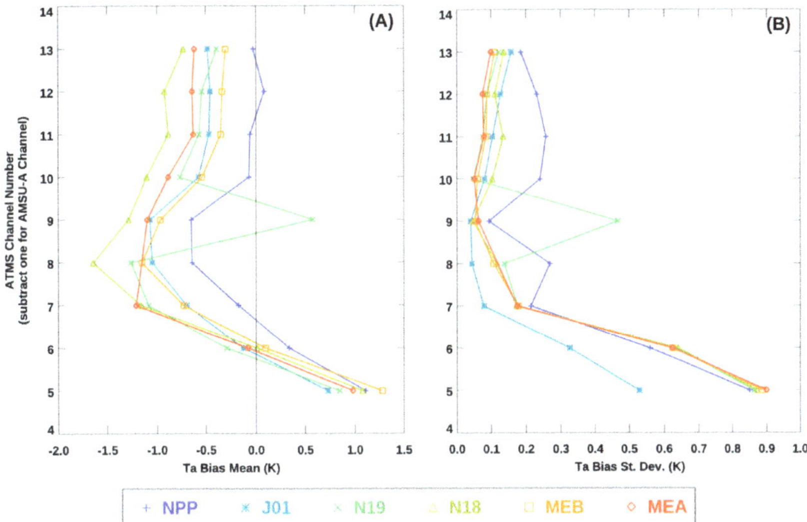

Figure 5. Mission-life mean (**A**) and standard deviation (**B**) O-B Ta bias values for NOAA-18 (N18), NOAA-19 (N19), Metop-A (MEA), and Metop-B (MEB) AMSU-A, and S-NPP (NPP) and JPSS-1 (J01) ATMS are plotted for each ATMS channel. The legend below the figures denotes the satellite identifier associated with each line plot. Note in this figure that the corresponding AMSU-A channel is the ATMS channel number minus one.

Figure 5B displays the mission-life standard deviation O-B Ta bias values, which represent a measure of the method uncertainty. This figure shows that these values are typically less than 0.2 K, except for NOAA-19 AMSU-A Ch 8 and ATMS Ch 5–6 (AMSU-A Ch 4–5). The ATMS Ch 5–6 (AMSU-A Ch 4–5) have values greater than 0.6 K for all instruments other than JPSS-1 ATMS, which has standard deviation values greater than 0.3 K. The Wang et al. study [31] mentioned above revealed that the standard deviation of COSMIC wetPrf and radiosonde temperature biases in the layers from 700 hPa to 150 hPa were 30% smaller than those between 700 hPa and 925 hPa. The even greater relative standard deviation found in Figure 5B for ATMS Ch 5–6 and AMSU Ch 4–5 may be explained by the surface contamination that plagues these radiometer channels. This is not an issue for the comparisons with radiosondes. Meanwhile, Figure 5B also shows the NOAA-19 AMSU-A Ch 8 mission-life standard deviation O-B Ta bias value of about 0.5 K is an outlier in comparison to the values of about 0.1 K for the other instrument makes and models at the 55.5 GHz frequency. Mission-life time series of monthly-mean O-B Ta bias values are able to provide some clarity into the stability of the AMSU and ATMS instruments, and these are found in Figure 6 for all relevant channels.

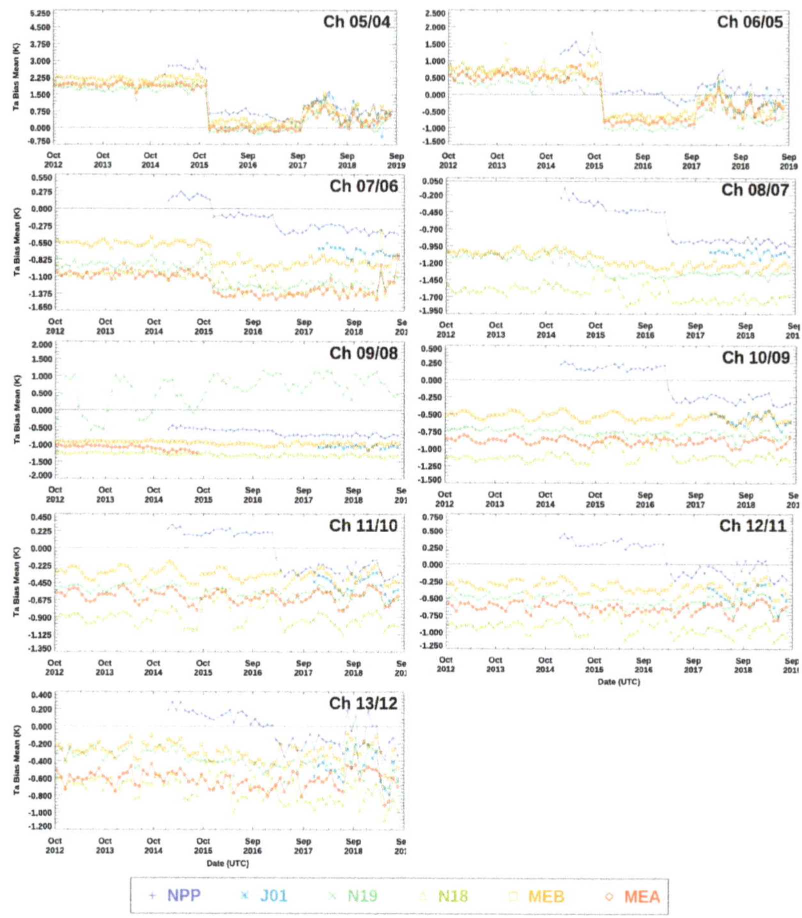

Figure 6. Time series of monthly-mean O-B Ta bias for NOAA-18 (N18), NOAA-19 (N19), Metop-A (MEA), and Metop-B (MEB) AMSU-A, and S-NPP (NPP) and JPSS-1 (J01) ATMS for ATMS/AMSU Ch 5–13/4–12. The legend below the figures denotes the satellite identifier associated with each line plot.

There are many noteworthy features found in these figures. There is a relatively large dip (−2.0 K to −0.5 K) in the O-B Ta bias for ATMS/AMSU-A Chs 5–7/4–6 after October 2015, which is due to an update of the GNSS RO "ROAM" [32]. The ROAM is the program name given to the original FORTRAN-77 software that inverts RO signals into physical parameters. These parameters include L1, L2 and ionosphere free bending angles, impact parameter, neutral atmospheric refractivity and "dry" pressure and temperature, height over mean sea-level, latitude and longitude of the estimated ray tangent point (in the Earth fixed reference frame), and azimuth of the occultation plane [33].

Also evident in the figures is a decrease of about −0.5 K for S-NPP ATMS Chs 7–13 in March 2017. According to JPSS Center for Satellite Applications and Research (STAR) Program reports, a Block 2.0 data processing "Build" Transition to Operations (TTO) occurred on 8 March 2017. After this transition, S-NPP ATMS Ta decreases were reported during the post-release software validation activity. These decreases resulted from thermal vacuum testing coefficient updates (version 003) in the ATMS parameter coefficient table. Meanwhile, a rise and fall of O-B Ta bias greater than 0.5 K in ATMS/AMSU-A Chs 5–6/4–5 between September 2017 and September 2018 is apparent that does not have a clear origin. Finally, the outlier in NOAA-19 Ch 8 can be clearly seen.

The origins of the NOAA-19 Ch 8 outlier can be visualized with the aid of the STAR Integrated Cal/Val System (ICVS) (https://www.star.nesdis.noaa.gov/icvs/index.php) instrument engineering and housekeeping data plots. Figure 7 represents the NOAA-19 AMSU-A Ch 8 mission-life trend of NEDT distributed by the STAR ICVS. This figure shows significant increases in NOAA-19 AMSU-A Ch 8 NEDT. The following report was logged by the NOAA Office of Satellite and Product Operations on 22 December 2009: "The NOAA-19 AMSU-A Channel 8 NEdT/Gain began experiencing noise on 21 December 2009 (JDAY 355). Only Channel 8 of NOAA-19's AMSU-A seems to be experiencing increased noise levels in the NEdT." Simultaneously, there was a large decrease in instrument gain for this channel, after which it became relatively unstable. These unstable gain variabilities are also reflected in the monthly-mean O-B Ta time series (see Figure 8). In Figure 8, NOAA-19 AMSU-A Ch 8 gain and monthly-mean O-B Ta bias variations are shown along with four numbered circles depicting the October 1 date for each year from 2012 to 2015. This shows that phases of the gain and monthly-mean O-B Ta variations are related and simply opposite in sign of each other. This is expected as the measured radiance variation is a function of the inverse gain value.

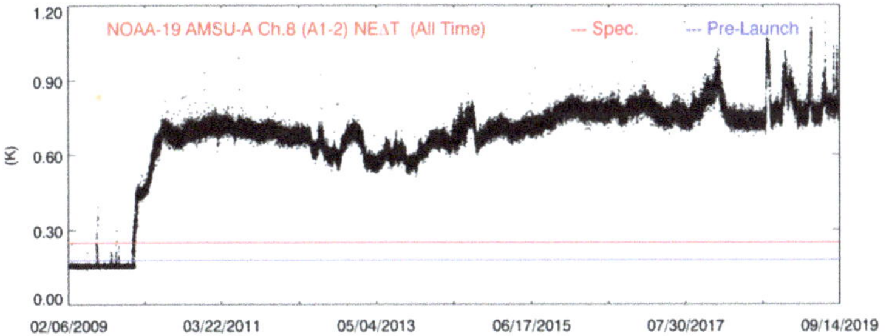

Figure 7. NEDT for NOAA-19 AMSU Channel 8, as shown by the STAR ICVS.

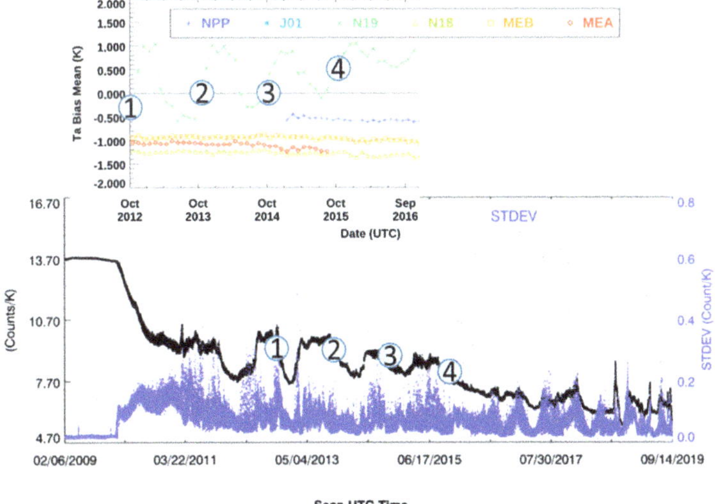

Figure 8. NOAA-19 AMSU Ch 8 Gain (Bottom) and monthly-mean O-B Ta bias (Top). The four circled numbers on these figures represent the October 1 date for years 2012–2015.

The ATMS (AMSU-A) Ch 10–13 (9–12) are found to have subtle but noticeable annual cycles, as shown in Figure 6. The amplitude of these variations is on the scale of 0.1 K to 0.4 K for ATMS (AMSU-A) Ch 10 (9) and Ch 13 (12), respectively. This is also reflected in the general tendency for the mission-life standard deviation shown in Figure 5B to increase from about 0.05 K to over 0.1 K over this channel range. These channels have peak microwave sounder weighting functions between 80 hPa and 20 hPa, as shown in Figure 1. Two plausible explanations for these annual cycles are the seasonal 2 K uncertainty of COSMIC GNSS RO sounding temperatures at levels above 30 km (about 10 hPa), as shown by [34], as well as the influence of using the time-independent US Standard Atmosphere temperature and water vapor sounding above 40 km.

4.3. Monthly and Mission-Life Double-Difference Inter-Sensor Ta Bias Statistical Results

As mentioned in Section 3.3, the difference between monthly-mean O-B Ta biases of two operational AMSU-A and/or ATMS microwave instruments—i.e., the double difference—computed during their overlap periods provides an indirect estimate of the monthly-mean Ta biases between those instruments. For NOAA-18, NOAA-19, Metop-A and Metop-B AMSU-A and S-NPP and JPSS-1 ATMS microwave instruments, there are six AMSU-A to AMSU-A, eight ATMS to AMSU-A, and one ATMS to ATMS instrument pair(s) possible, where each pair is capable of producing "double-difference" monthly-mean Ta biases. The character of the double-difference Ta biases is revealed in Figure 9. This figure represents the mean of AMSU-A to AMSU-A, ATMS to AMSU-A and ATMS to ATMS monthly minimum and maximum double-difference Ta biases computed over the period of on-orbit operational instrument overlaps. This simplification of presentation is chosen to avoid having 15 plots on one graph.

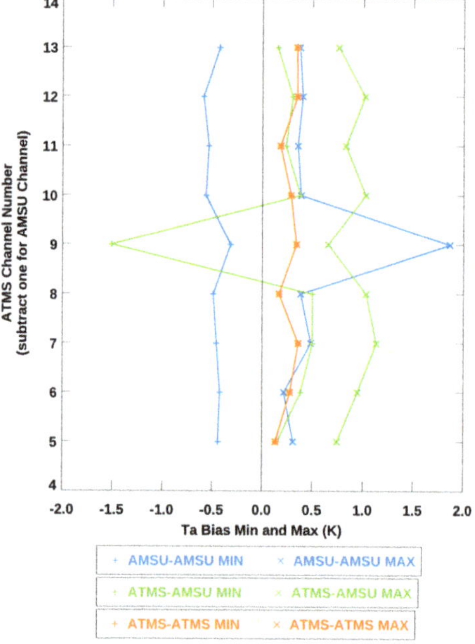

Figure 9. Operational overlap period means of the AMSU-A to AMSU-A, ATMS to AMSU-A, and ATMS to ATMS monthly minimum and maximum double-difference Ta biases. The legends below the figure denote the instrument pair type and statistic associated with each line plot. Note in this figure that the corresponding AMSU-A channel is the ATMS channel number minus one.

Figure 9 shows that the minimum (maximum) double-difference Ta biases between overlapping AMSU-A instrument pairs is on average about −0.5 K (0.5 K), respectively. The exception to this is the double-difference Ta biases associated with NOAA-19 AMSU-A Ch 8. These biases can be as high as 2 K, as a result of the large noise and gain anomalies in this channel discussed in the previous section. The results of overlapping ATMS and AMSU-A instrument pair double-difference Ta biases reveal average minimum (maximum) values of approximately 0.25 K (0.75 K) when the results from double differences with respect to NOAA-19 AMSU-A Ch 8 are neglected. These results reveal that the AMSU-A to AMSU-A (ATMS to AMSU-A) instrument pairs have measurements that are on average no greater that about 1 K (0.5 K) of each other. There is only one overlapping ATMS instrument pair—i.e., between S-NPP and JPSS-1—that has average double-difference Ta biases of approximately 0.3 K. It is important to note that nothing can be said about the absolute accuracies of any of these instruments.

In order to gain a more detailed understanding of these results, in Figure 10 time series of the minimum and maximum monthly-mean Ta bias for the AMSU-A to AMSU-A, ATMS to AMSU-A, and ATMS to ATMS instrument pair(s) are given for all relevant channels.

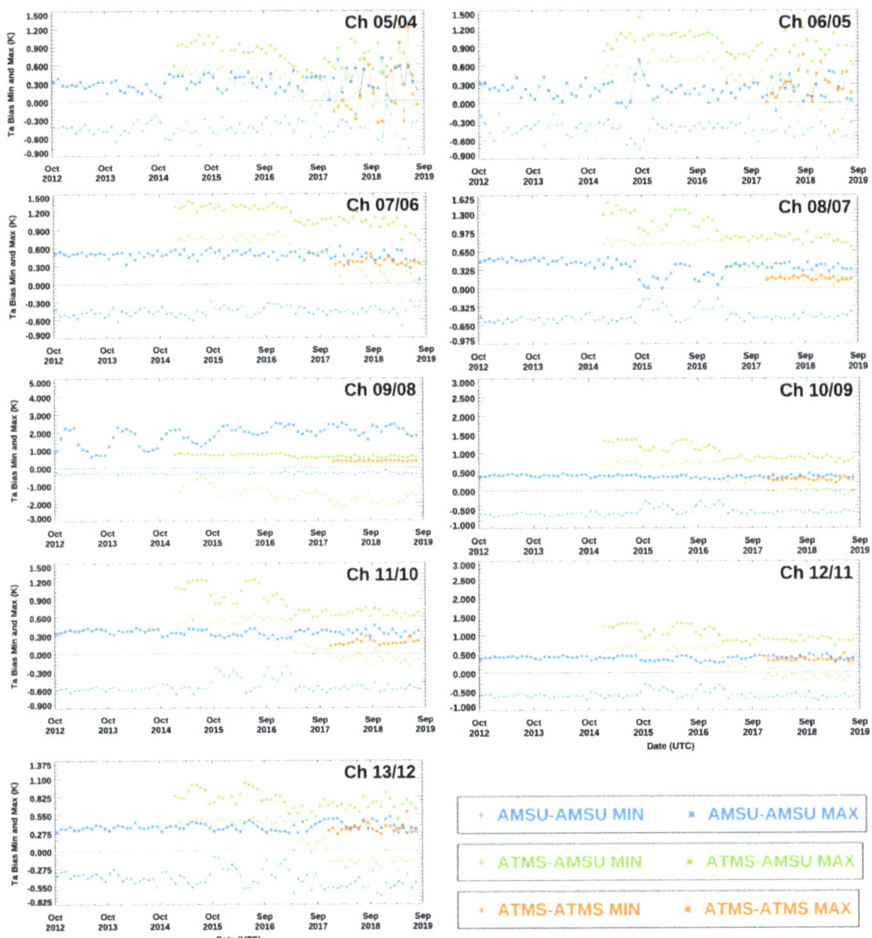

Figure 10. Time series of monthly minimum and maximum AMSU-A to AMSU-A, ATMS to AMSU-A, and ATMS to ATMS double differences Ta biases computed using available monthly-mean O-B Ta biases values for ATMS/AMSU-A Chs 5–13/4–12. The legend in the bottom right corner denotes the pair of instruments and the statistic for a given time series.

The new information that is accessible in these figures is that the annual cycles that were clearly present in the monthly-mean O-B Ta bias values of Figure 6 for ATMS (AMSU-A) Ch 10–13 (9–12) have largely disappeared in the double-difference Ta time series. A clear exception to this is the elevated values of the AMSU-A to AMSU-A double difference minimum values from June 2015 to June 2017. These are associated with excursions of NOAA-18 AMSU-A O-B Ta bias that are unusually large over this period, which can be seen clearly in Figure 6 for AMSU-A Ch 7 and Chs 9–11.

In the previous section, NOAA-19 AMSU-A Ch 8 monthly-mean O-B Ta bias changes were linked to significant instrument noise and gain changes displayed in the STAR ICVS. Thus, we turn to the STAR ICVS to also investigate a potential NOAA-18 AMSU-A anomaly as well. The NOAA 18 AMSU-A Radio Frequency (RF) Multiplexer (MUX) temperature, and Ch 10 cold space counts, warm space counts, and gain from the STAR ICVS are given in Figure 11A–D, respectively. Note that the RF MUX temperature is considered to be an indicator of the instrument temperature. In Figure 11D, the ATMS Ch 11 and AMSU-A Ch 10 monthly-mean O-B Ta biases are provided to facilitate comparison. During the June 2015 to June 2017 period, these plots show instrument temperature excursions of over 10 K that are at least 3 times larger than temperature ranges recorded for dates outside this period. These large instrument temperature changes are reflected in Ch 10 cold space and warm target counts, which are used to compute the instrument gain. The resulting changes of instrument gain can be clearly seen to have a similar annual cycle signature compared to the NOAA-18 AMSU-A Ch 10 monthly-mean O-B Tb biases (see Figure 11D).

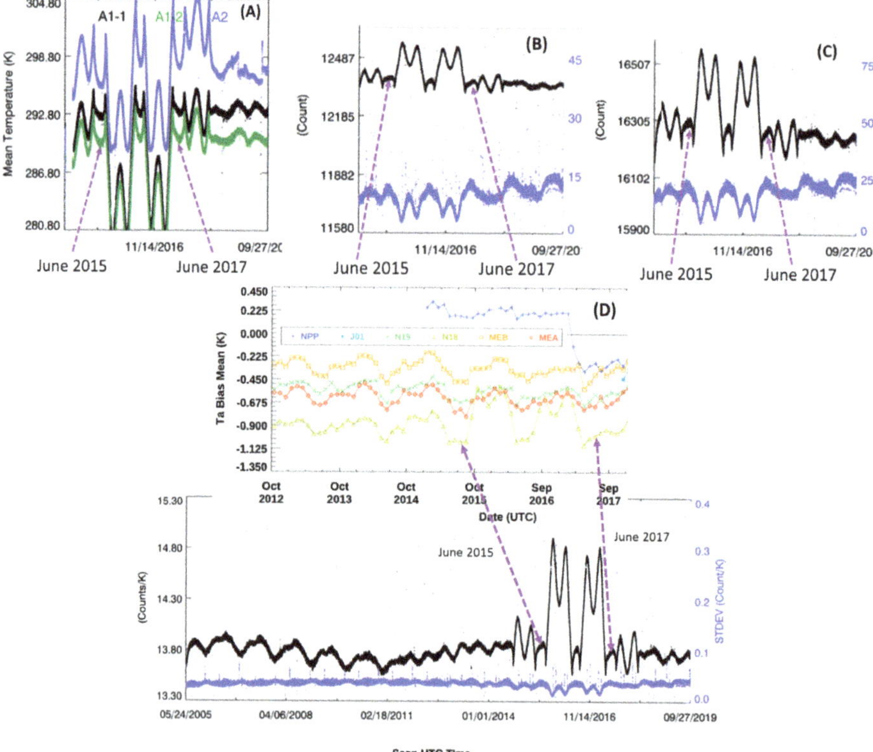

Figure 11. NOAA-18 AMSU-A RF MUX temperature (**A**) and Ch 10 cold space (**B**) and blackbody (**C**) temperatures. The final plot (**D**) includes ATMS Ch 11/AMSU-A Ch 10 monthly-mean O-B Ta bias (upper) and NOAA-18 AMSU-A Ch 10 Gain (lower). The dashed purple arrows delineate the June 2015 and June 2017 in each plot.

An explanation for such large changes in NOAA-18 AMSU-A thermal characteristics during this time may be due to the uncontrolled drift of the NOAA-18 satellite orbit local equator crossing time (LECT). Figure 12 shows that the orbits of all NOAA polar-orbiting satellites before NOAA-20 were allowed to have dramatic LECT changes over their lifetimes. In this figure, it is clear that NOAA-18 has passed through a "terminator" orbit, where its orbital plane has LECT nodes near 0600 and 1800 and is perpendicular to incident solar radiation. The Earth Terminator is defined as the circle that divides its daylight side from its night side. This satellite polar-orbiting configuration could cause solar radiation shining from the side of, or slightly underneath, the satellite to directly heat AMSU-A, or allow it to be shadowed by other instruments, for much of its orbit.

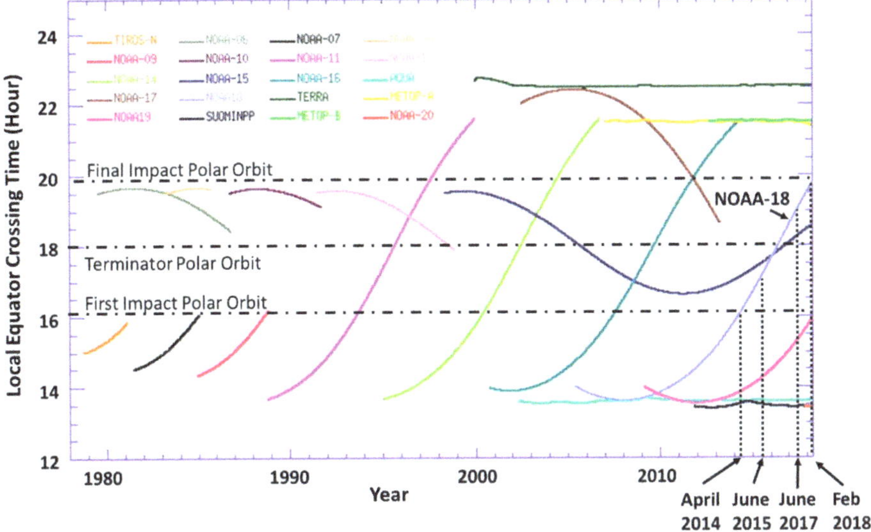

Figure 12. Graph of LECT versus Year for NOAA, Earth Observing System (EOS), and Suomi-NPP polar orbiting satellites. Additionally included is a line representing the 0600/1800 "Terminator Polar Orbit" LECT, and lines representing first and final impact LECT for polar orbiting satellites. The figure is courtesy from the STAR JPSS web site at https://www.star.nesdis.noaa.gov/jpss/images/orbit-drift.jpg.

This affect can be visualized with the aid of the illustration in Figure 13. In this illustration, the location a polar-orbiting satellite crossing the equator where the underside of the satellite is first subject to direct solar illumination is show. At an orbital height of about 848 km, the longitudinal arc angle between the nearest Earth Terminator point and the satellite LECT point is about 28 degrees, which takes about 1 h and 52 min to subtend at the earth's rotation rate. Thus, if the tangent point is 0600/1800, then orbital LECT nodes starting at 0408/1608 could begin seeing impacts of solar radiation to instrument thermal characteristics, which would end when satellite LECT drifts to 0752/1952. At this point, instrument thermal behavior would completely go back to normal.

Figures 11 and 12 can be used to roughly test these predictions. According to Figure 12 and these predictions, direct solar radiation should begin affecting the NOAA-18 AMSU-A instrument around April 2014 and end in February 2018. The bottom panel of Figure 11 shows that the AMSU-A instrument gain was clearly anomalous from about August 2014 to April 2018, which is within reasonable agreement for such a back-of-the-envelope theoretical treatment of the phenomenon. The top panel of Figure 11 reflects the large monthly O-B Ta bias anomalies between June 2015 and June 2017 and little or no response to the smaller gain anomalies found at the beginning and end of this direct solar radiation period.

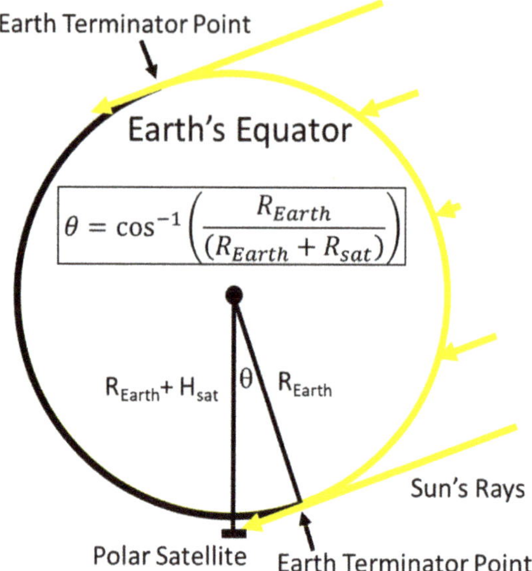

Figure 13. An illustration depicting a polar-orbiting satellite with an altitude H_{Sat} above the Earth's equator at an LECT that allows Sun's rays to begin to illuminate the nadir pointing side of the satellite platform. The angle θ depicts the longitudinal angle between satellite nadir and the Earth Terminator Point where a solar ray striking the satellite meets the earth horizon.

4.4. Lessons Learned in Using COSMIC RO Soundings to Track and Trend Operational Microwave Sounder Data

Users of GNSS RO soundings most importantly need to assess their fitness-for-purpose related to the task they want to accomplish. For this study, the purpose of using COSMIC wetPrf soundings is to monitor temporal changes in microwave sounder radiometer O-B Ta biases, and to establish estimates of inter-satellite Ta biases. The microwave sounder radiometer mission-life mean and standard deviation O-B Ta biases clearly show similar character to documented temperature biases between COSMIC wetPrf and radiosonde soundings [31]. Absolute accuracy of O-B Ta biases is not essential to operational microwave sounding radiometer data monitoring, so it is not necessary to account for these COSMIC wetPrf sounding artifacts. On the other hand, radiometric tracking and trending does depend on understanding GNSS RO wetPrf sounding quality stability over space and time. In this case, research by Fan et al. [34] revealed seasonality in COSMIC temperature soundings. Knowledge of this allows us to anticipate and screen out these signals when detecting anomalies in the microwave sounder instrument data. In addition, computing monthly mean O-B Ta bias "double difference" values remove COSMIC wetPrf sounding bias effects when establishing inter-satellite microwave sounder Ta biases. These COSMIC wetPrf sounding effects are considered the same for any two co-orbiting satellites, so they cancel out in the "double difference."

Examples from Sections 4.2 and 4.3 reveal how the O-B Ta bias statistics—computed with support of the CRTM and GNSS RO soundings—can be used to monitor operational microwave radiometer Ta products. Additionally, when these bias statistics are compared to instrument engineering and calibration data, they act together as integral parts of a holistic Integrated Calibration/Validation System that can discern instrument change impacts on these Ta products. This represents a success story related to the use of GNSS RO generated soundings. On the other hand, it was important for the researchers of this study to understand the requirements for, and the strengths and weaknesses of, the GNSS RO generated soundings.

5. Conclusions

The O-B Ta bias parameter computed from operational satellite microwave sounding radiometer Ta observations and collocated forward RTM Tb simulations has been found to be key to monitoring data quality and performing initial instrument anomaly investigations. In this study, COSMIC GNSS RO atmospheric temperature and moisture (wetPrf) sounding profiles over ocean and equatorward of 60° latitude are used as input to the CRTM to generate simulated NOAA-18, NOAA-19, Metop-A, and Metop-B AMSU-A and S-NPP and JPSS-1 ATMS Tb values. These simulated Tb values, together with observed Ta values that are nearly simultaneous in space and time, are used to compute O-B Ta bias statistics on monthly time scales for each instrument. In addition, the CRTM-simulated Tb values based on the COSMIC GNSS RO soundings can be used as a transfer standard to inter-compare Ta values from different microwave radiometer makes and models that have the same bands. This is accomplished by computing the "double difference" between monthly-mean Ta bias O-B values from pairs of co-orbiting operational microwave sounding instruments for the corresponding frequency bands.

The collocated radiometer and COSMIC GNSS RO data points available for this research were found to be geographically well distributed and statistically robust, even though the number of samples on a given day dropped from the hundreds to the tens from October 2012 to August 2019. It is discovered that the upper-air sounding channels such as AMSU-A Ch 12 have a great deal more points than surface-influenced channels like AMSU-A Ch 4. This is due to the fact that data associated with GNSS RO sounding levels closest to the surface are more likely to be missing or have the bad data quality flag set.

Mission-life mean O-B Ta bias values plotted as a function of radiometer sounding channel reveal that ATMS Ch 5 (AMSU-A Ch 4) O-B Ta bias values are about 1 K, while they vary between about −1.5 K to 0 K for ATMS Ch 6–13 (AMSU-A Ch 5–12). This behavior is similar to comparisons of COSMIC wetPrf and radiosonde temperature profiles [31]. There is a prominent NOAA-19 Ch 8 O-B Ta bias outlier of about 0.5 K, which differs substantially from the Ta bias value of about −1.25 K for the other instrument makes and models for this 55.5 GHz frequency channel. The mission-life mean O-B Ta bias values cluster within about 0.5 K of each other for channels with identical frequencies. Meanwhile, the mission-life standard deviation O-B Ta bias values are typically less than 0.2 K, except for NOAA-19 AMSU-A Ch 8 and ATMS Ch 5–6 (AMSU-A Ch 4–5). The ATMS Ch 5–6 (AMSU-A Ch 4–5) have values greater than 0.6 K for all instruments other than JPSS-1 ATMS, which has a standard deviation value greater than 0.3 K. This upper and lower sounding channel standard deviation disparity is large with respect to COSMIC wetPrf and radiosonde temperature sounding comparisons for levels with pressure less than and greater than 700 hPa [31]. Surface influence of the radiometer channels, which is absent in the radiosonde data, could explain this. Meanwhile, the NOAA-19 AMSU-A Ch 8 mission-life standard deviation O-B Ta bias value of about 0.5 K is an outlier in comparison to the values of about 0.1 K for the other instrument makes and models at the 55.5 GHz frequency.

Minimum (maximum) double-difference Ta biases between overlapping AMSU-A instrument pairs is on average about −0.5 K (0.5 K), respectively. The exception to this is the double-difference Ta biases associated with NOAA-19 AMSU-A Ch 8. These biases can be as high as 2 K as a result of the large noise and gain anomalies. The results of overlapping ATMS and AMSU-A instrument pair double-difference Ta biases reveal average minimum (maximum) values of approximately 0.25 K (0.75 K) when the results from double differences with respect to NOAA-19 AMSU-A Ch 8 are neglected. These results reveal that the AMSU-A to AMSU-A (ATMS to AMSU-A) instrument pairs have measurements that are on average no greater than about 1 K (0.5 K) of each other. There is only one overlapping ATMS instrument pair—i.e., between S-NPP and JPSS-1—that has average double-difference Ta biases of approximately 0.3 K. It is important to note that nothing can be said about the absolute accuracies of any of these instruments. One obvious anomaly in the double difference Ta bias values manifested as elevated AMSU-A to AMSU-A minimum values from June 2015 to June 2017. These were shown to be associated with excursions of NOAA-18 AMSU-A O-B Ta bias that are unusually large over this period.

A significant finding of this study is that efforts to gain insight into mission-life mean and standard deviation O-B Ta bias statistics outliers can be supported with time series of monthly-mean O-B, and "double-difference", Ta bias values, along with relevant instrument engineering and housekeeping data plots from the STAR ICVS. This was exemplified by investigations into the NOAA-18 AMSU-A O-B Ta bias excursions between June 2015 and June 2017 and the long-term ongoing NOAA-19 AMSU-A Ch 8 anomaly. NOAA operational ATMS and AMSU-A data used in numerical weather prediction and climate analysis are essential to protect life and property and maintain safe and efficient commerce. Routine data quality monitoring and anomaly assessment, such as that provided by statistics and time series of individual instrument O-B Ta biases and inter-instrument "double-difference" Ta biases computed with the aid of GNSS RO sounding profiles, is an important tool to sustain data effectiveness. The study also reveals that it is important for users of the GNSS RO sounding profiles to understand the requirements, and strengths and weaknesses, of these data.

In the Introduction, several references are given regarding (1) the Simultaneous Nadir Overpass method to detect inter-satellite Ta biases, as well as (2) the use of NWP output parameters coupled with the CRTM to generate O-B Ta statistics. In future work, research will be performed to compare and contrast results from these legacy methods and the method provided in this paper to highlight their effectiveness for operational microwave sounding instrument data integrity monitoring. In addition, as COSMIC-2 data are now readily available as part of their post-launch check-out, they will be studied for their ability to track and trend microwave sounding instrument data as well.

Author Contributions: Conceptualization, methodology, and software, L.L. and R.I.; formal analysis and investigation, R.I.; resources, Q.L., N.S., and Q.L.; data curation, R.I., N.S., and L.L.; writing—original draft preparation, R.I.; writing—review and editing, R.I., Q.L., and N.S.; supervision, Q.L. and N.S.; project administration and funding acquisition, Q.L. All authors have read and agreed to the published version of the manuscript.

Funding: This research was funded by the Joint Polar Satellite System Program.

Acknowledgments: The author would like to acknowledge all STAR colleagues for their input during the process of writing this manuscript.

Conflicts of Interest: The authors declare no conflicts of interest.

References

1. English, S.J.; Renshaw, R.J.; Dibben, P.C.; Smith, A.J.; Rayer, P.J.; Poulsen, C.; Saunders, F.W.; Eyre, J.R. A comparison of the impact of TOVS and ATOVS satellite sounding data on the accuracy of numerical weather forecasts. *Q. J. Royal Met. Soc.* **2000**, *126*, 2911–2931.
2. McNally, A.P.; Derber, J.C.; Wu, W.-S.; Katz, B.B. The use of TOVS level-1 radiances in the NCEP SSI analysis system. *Q. J. Royal Met. Soc.* **2000**, *129*, 689–724. [CrossRef]
3. Zou, C.-Z.; Goldberg, M.D.; Cheng, Z.; Grody, N.C.; Sullivan, J.T.; Cao, C.; Tarpley, D. Recalibration of microwave sounding unit for climate studies using simultaneous nadir overpasses. *J. Geophys. Res.* **2006**, *111*, D19114. [CrossRef]
4. Spencer, R.W.; Christy, J.R.; Braswell, W.D.; Norris, W.B. Estimation of tropospheric temperature trends from MSU Channels 2 and 4. *J. Atmos. Ocean. Technol.* **2006**, *23*, 417–423. [CrossRef]
5. Mears, C.; Schabel, M.; Wentz, F. A reanalysis of the MSU Channel 2 tropospheric temperature record. *J. Clim.* **2003**, *16*, 3650–3664. [CrossRef]
6. Prabhakara, C.; Iacovazzi, R., Jr.; Yoo, J.-M.; Dalu, G. Global warming: Evidence from satellite observations. *Geophys. Res. Lett.* **2000**, *27*, 3517–3520. [CrossRef]
7. Cao, C.; Heidinger, A.K. Intercomparison of the longwave infrared channels of MODIS and AVHRR/NOAA-16 using simultaneous nadir observations at orbit intersections. In *Earth Observing Systems VII*; William, L.B., Ed.; International Society for Optics and Photonics: Bellingham, WA, USA, 2002; Volume 4814, pp. 306–316.
8. Cao, C.; Weinreb, M.; Xu, H. Predicting simultaneous nadir overpasses among polar-orbiting meteorological satellites for the intersatellite calibration of radiometers. *J. Atmos. Ocean. Technol.* **2004**, *21*, 537–542. [CrossRef]

9. Cao, C.; Xu, H.; Sullivan, J.; McMillin, L.; Ciren, P.; Hou, Y. Intersatellite radiance biases for the High Resolution Infrared Radiation Sounders (HIRS) on-board NOAA-15, -16, and -17 from simultaneous nadir observations. *J. Atmos. Ocean. Technol.* **2005**, *22*, 381–395. [CrossRef]

10. Iacovazzi, R., Jr.; Cao, C. Quantifying EOS Aqua and NOAA POES AMSU-A brightness temperature biases for weather and climate applications utilizing the SNO method. *J. Atmos. Ocean. Technol.* **2007**, *24*, 1895–1909. [CrossRef]

11. Iacovazzi, R., Jr.; Cao, C. Reducing uncertainties of SNO-estimated intersatellite AMSU-A brightness temperature biases for surface-sensitive channels. *J. Atmos. Ocean. Technol.* **2008**, *25*, 1048–1054. [CrossRef]

12. Weng, F.; Zou, X.; Wang, X.; Yang, S.; Goldberg, M.D. Introduction to Suomi national polar-orbiting partnership advanced technology microwave sounder for numerical weather prediction and tropical cyclone applications. *J. Geophys. Res.* **2012**, *117*, D19112. [CrossRef]

13. Mo, T. *Calibration of the Advanced Microwave Sounding Unit-A Radiometers for NOAA-L and NOAA-M*; NOAA Technical Report; NESDIS: Silver Spring, MD, USA, 1999; Volume 92, 62p.

14. Goldberg, M.D.; Crosby, D.S.; Zhou, L. The limb adjustment of AMSU-A observations: Methodology and validation. *J. Applied Met.* **2000**, *40*, 70–83. [CrossRef]

15. Murphy, R.; Le Vine, D.M.; Barath, F.; Barrett, E.; Bernstein, R.L.; Clark, C.A.; Dozier, J.; Kakar, R.; Njoku, E.; Runge, E.; et al. *Earth Observing System Volume IIe: HMRR High-Resolution Multifrequency Microwave Radiometer*; NASA: Washington, DC, USA; Goddard Space Flight Center: Greenbelt, Maryland, USA, 1987; Volume 20771, 59p. Available online: https://babel.hathitrust.org/cgi/pt?id=umn.31951d014844120&view=1up&seq=4 (accessed on 3 March 2020).

16. Goodrum, G.; Kidwell, K.B.; Winston, W.; Aleman, R. *NOAA KLM User's Guide*; Indiana University Bloomington Virtual CD ROM/Floppy Disk Library, 1999. Available online: http://webapp1.dlib.indiana.edu/virtual_disk_library/index.cgi/2790181/FID3711/klm/index.htm. (accessed on 7 January 2020).

17. Weng, F.; Zou, X.; Sun, N.; Yang, H.; Tian, M.; Blackwell, W.J.; Wang, X.; Lin, L.; Anderson, K. Calibration of Suomi national polar-orbiting partnership advanced technology microwave sounder. *J. Geophys. Res. Atmos.* **2013**, *118*, 11–187. [CrossRef]

18. Han, Y.; Weng, F.; Zou, X.; Yang, H.; Scott, D. Characterization of geolocation accuracy of Suomi NPP Advanced Technology Microwave Sounder measurements. *J. Geophys. Res. Atmos.* **2016**, *121*, 4933–4950. [CrossRef]

19. Zou, X.; Vandenberghe, F.; Wang, B.; Gorbunov, M.E.; Kuo, Y.-H.; Sokolovskiy, S.; Chang, J.C.; Sela, J.G.; Anthes, R.A. A ray-tracing operator and its adjoint for the use of GPS/MET refraction angle measurements. *J. Geophys. Res.* **1999**, *104*, 22301–22318. [CrossRef]

20. Healy, S.; Eyre, J. Retrieving temperature, water vapor and surface pressure information from refractivity-index profiles derived by radio occultation: A simulation study. *Q. J. Royal Meteorol. Soc.* **2000**, *126*, 1661–1683. [CrossRef]

21. Palmer, P.I.; Barnett, J.; Eyre, J.; Healy, S. A non-linear optimal estimation inverse method for radio occultation measurements of temperature, humidity, and surface pressure. *J. Geophys. Res.* **2000**, *105*, 17513–17526. [CrossRef]

22. Kishore, P.; Namboothiri, S.P.; Jiang, J.H.; Sivakumar, V.; Igarashi, K. Global temperature estimates in the troposphere and stratosphere: A validation study of COSMIC/FORMOSAT-3 measurements. *Atmos. Chem. Phys. Discuss.* **2008**, *8*, 8327–8355. [CrossRef]

23. Yu, X.; Xie, F.; Ao, C.O. Evaluating the lower-tropospheric COSMIC GPS radio occultation sounding quality over the Arctic. *Atmos. Meas. Tech.* **2018**, *11*, 2051–2066. [CrossRef]

24. Anthes, R.A.; Bernhardt, P.A.; Chen, Y.; Cucurull, L.; Dymond, K.F.; Ector, D.; Healy, S.B.; Ho, S.-P.; Hunt, D.C.; Kuo, Y.-H.; et al. The COSMIC/FORMOSAT-3 mission: Early results. *Bull. Am. Meteorol. Soc.* **2008**, *89*, 313–333. [CrossRef]

25. Weng, F.; Zhao, L.; Ferraro, R.; Poe, G.; Li, X.; Grody, N. Advanced microwave sounding unit cloud and precipitation algorithms. *Radio Sci.* **2003**, *38*, 8086–8096. [CrossRef]

26. Zou, X.; Lin, L.; Weng, F. Absolute calibration of ATMS upper level temperature sounding channels using GPS RO observations. *IEEE Trans. Geosci. Remote Sens.* **2014**, *52*, 1397–1406. [CrossRef]

27. Han, Y.; van Delst, P.; Liu, Q.; Weng, F.; Yan, B.; Treadon, R.; Derber, J. *JCSDA Community Radiative Transfer Model (CRTM)—Version 1*; NOAA Technical Report; NESDIS: Silver Spring, MD, USA, 2006; Volume 122, 40p.

28. Weng, F. Advances in radiative transfer modeling in support of satellite data assimilation. *J. Atmos. Sci.* **2007**, *64*, 3803–3811. [CrossRef]

29. Han, Y.; van Delst, P.; Weng, F.; Liu, Q.; Groff, D.; Yan, B.; Chen, Y.; Vogel, R. 2010: Current status of the JCSDA community radiative transfer model (CRTM). In Proceedings of the 17th International ATOVS Study Conference, Monterey, CA, USA, 14–20 April 2010; World Meteorological Organization: Geneva, Switzerland, 2010.
30. Shouguo, D.; Yang, P.; Weng, F.; Liu, Q.; Han, Y.; van Delst, P.; Li, J.; Baum, B. Validation of the community radiative transfer model. *J. Quant. Spectrosc. Radiat. Transf.* **2011**, *112*, 1050–1064.
31. Wang, B.-R.; Liu, X.-Y.; Wang, J.-K. Assessment of COSMIC radio occultation retrieval product using global radiosonde data. *Atmos. Meas. Tech.* **2013**, *6*, 1073–1083. [CrossRef]
32. Sokolovskiy, S.; CDAAC Team. Improvements, Modifications, and Alternate Approaches in the Processing of GPS RO Data. ECMWF/EUMETSAT Radio Occultation Meteorology (ROM) Satellite Applications Facility (SAF): Reading, UK. Available online: https://cdaac-www.cosmic.ucar.edu/cdaac/doc/documents/Sokolovskiy_newroam.pdf (accessed on 8 January 2020).
33. Sokolovskiy, S. Algorithms for Inverting Radio Occultation Signals in the Neutral Atmosphere. COSMIC Data Analysis and Archive Center (CDAAC). Available online: https://cdaac-www.cosmic.ucar.edu/cdaac/doc/documents/roam05.doc (accessed on 8 January 2020).
34. Fan, Z.Q.; Sheng, Z.; Shi, H.Q.; Yi, X.; Jiang, Y.; Zhu, E.Z. Comparative assessment of COSMIC Radio occultation data and TIMED/SABER satellite data over china. *J. Appl. Meteorol. Clim.* **2015**, *54*, 1931–1943. [CrossRef]

Article

The New Potential of Deep Convective Clouds as a Calibration Target for a Geostationary UV/VIS Hyperspectral Spectrometer

Yeeun Lee [1], Myoung-Hwan Ahn [1,*]  **and Mina Kang [2]**

[1]   Department of Climate and Energy Systems Engineering, Ewha Womans University, 52 Ewhayeodae-gil, Seodaemun-gu, Seoul 03760, Korea; dungia@ewhain.net

[2]   Department of Atmospheric Science and Engineering, Ewha Womans University, 52 Ewhayeodae-gil, Seodaemun-gu, Seoul 03760, Korea; mina@ewhain.net

*   Correspondence: terryahn65@ewha.ac.kr; Tel.: +82-3277-4462

Received: 17 December 2019; Accepted: 30 January 2020; Published: 1 February 2020

Abstract: As one of geostationary earth orbit constellation for environmental monitoring over the next decade, the Geostationary Environment Monitoring Spectrometer (GEMS) has been designed to observe the Asia-Pacific region to provide information on atmospheric chemicals, aerosols, and cloud properties. In order to continuously monitor sensor performance after its launch in early 2020, we suggest in this paper deep convective clouds (DCCs) as a possible target for the vicarious calibration of the GEMS, the first ultraviolet and visible hyperspectral sensor onboard a geostationary satellite. The Tropospheric Monitoring Instrument (TROPOMI) and the Ozone Monitoring Instrument (OMI) are used as a proxy for GEMS, and a conventional DCC-detection approach applying a thermal threshold test is used for DCC detection based on collocations with the Advanced Himawari Imager (AHI) onboard the Himawari-8 geostationary satellite. DCCs are frequently detected over the GEMS observation area at an average of over 200 pixels within a single observation scene. Considering the spatial resolution of the GEMS (3.5×8 km^2), which is similar to the TROPOMI and its temporal resolution (eight times a day), the availability of DCCs is expected to be sufficient for the vicarious calibration of the GEMS. Inspection of the DCC reflectivity spectra estimated from OMI and TROPOMI data also shows promising results. The estimated DCC spectra are in good agreement within a known uncertainty range with comparable spectral features even with different observation geometries and sensor characteristics. When DCC detection is improved further by applying both visible and infrared tests, the variability of DCC reflectivity from TROPOMI data is reduced from 10% to 5%. This is mainly due to the efficient screening out of cold, thin cirrus clouds in the visible test and of bright, warm clouds in the infrared test. Precise DCC detection is also expected to contribute to the accurate characterization of cloud reflectivity, which will be investigated further in future research.

Keywords: GEMS; UV; VIS; hyperspectral data; deep convective cloud; vicarious calibration; OMI; TROPOMI

---

1. Introduction

With the global transport of anthropogenic chemicals in the atmosphere becoming a controversial issue over recent years, satellites have been considered a key tool for keeping track of chemicals given their wide spatial coverage. In the Asia-Pacific region, Geostationary Korea Multi-Purpose Satellite-2B (GEO-KOMPSAT-2B, GK2B) is expected to perform this role following its planned launch in February 2020 using an ultraviolet (UV) and visible (VIS) hyperspectral sensor called the Geostationary Environment Monitoring Spectrometer (GEMS). The GEMS has been designed to observe the Asia-Pacific region including the Korea Peninsula and surrounding areas and continuously monitor atmospheric conditions by measuring the concentration of atmospheric chemicals and tracking aerosol properties [1,2]. To ensure the consistency of these measurements, the onboard calibration with solar diffusers and light-emitting diode (LED) is deployed in the GEMS calibration system, which converts light from a scene into calibrated spectral data (Level 1B). However, it has been frequently reported by previous satellite programs that residual errors in Level-1B data introduce some level of uncertainty to higher-level products [3–8]. It is also highly probable for a sensor's characteristics to change over time due to both internal and external factors, and this makes it necessary for the sensor to be continuously monitored and calibrated.

Vicarious calibration is a well-known approach for monitoring and improving sensor performance by periodically comparing it with reference targets. To successfully perform the calibration, it is important to select a suitable target that is stable enough to be repeatedly observed and well-characterized under different observation conditions. Because of these requirements, particular observation targets have been used for calibration, such as snow and ice over polar regions, bright clouds, deserts, and artificial sites [9–14]. However, geostationary earth orbit (GEO) sensors are limited in selecting the target because each sensor only covers a particular spatial region, while low-earth orbit (LEO) sensors cover the entire surface of the Earth. Considering that the GEMS only measures UV and VIS radiance reflected by the atmosphere and the Earth's surface, the variation in the measurements also imposes limitations on the selection of a stable target.

Deep convective clouds (DCCs), in this respect, are an excellent candidate as a calibration target for the GEMS considering their physical and radiative properties. DCCs are frequently observed over the tropical western Pacific (TWP) region with their tops reaching up to or over the tropopause due to the strong vertical convection [15–18]. This means that the backscattered radiation from these clouds is less affected by the Earth's surface and the troposphere, where most atmospheric components reside. The reflective properties of the cloud tops have also been fairly well-characterized due to their spatially uniform and less penetrative features, especially in the VIS and infrared (IR) spectral regions [19–22]. With these characteristics, DCCs have been widely used as a reference target for the monitoring of VIS and IR satellite sensors [23–31]. However, little attention has been paid to the applicability of DCCs as reference targets in the UV spectral region because there are not many UV sensors in operation, especially onboard GEO satellites. In this study, we aim to explore the applicability of DCCs as a reference target for the GEMS. Some of the advantages of using DCCs as a target are still valid even at shorter wavelengths, such as lower dependence on atmospheric conditions, the distinct brightness of the clouds, and the low spectral dependence in the reflected radiance from the clouds [32].

To select only spatially homogeneous clouds, we apply a DCC-detection routine with the IR brightness temperature (TB) threshold suggested by Doelling et al. [27] and an adaption of the UV–VIS threshold. Combining thermal and reflective signals is expected to facilitate the selection of suitable DCCs because each radiative property provides different types of information on the clouds [33,34]. In Section 2, to evaluate the applicability of DCC calibration, we firstly check whether DCCs occur over the TWP region in high enough numbers to provide reliable statistical parameters. Because the GEMS does not cover the IR region, we use TB and reflectivity data from the Advanced Himawari Imager (AHI) onboard a geostationary weather satellite (Himawari-8) to derive simple climatology for DCCs over the TWP region. After confirming that there are a sufficient number of DCCs over the TWP region, UV–VIS hyperspectral data from the Ozone Monitoring Instrument (OMI) onboard Aura and the Tropospheric Monitoring Instrument (TROPOMI) onboard Sentinel-5 Precursor (S5P) are used as a proxy for the GEMS for the spectral analysis of DCCs. In Section 3, we compare OMI and TROPOMI DCCs to confirm whether the detected DCCs reflect a sufficiently stable and bright signal to reduce the different sensor characteristics as having homogeneous spectral features. Based on these results, DCC detection thresholds are tested to optimize detection for further characterization of cloud reflectivity. In Section 4, we verify the effectiveness of the optimized DCC detection using TROPOMI observations and cloud properties from TROPOMI Level-2 data products. Preliminary results and the limitations of our proposed method are also presented in this section. In Section 5, conclusions are presented with the remarks on the future research.

2. Data and Methods

2.1. UV–VIS Hyperspectral Sensor

2.1.1. GEMS

The GEMS covers the Asia-Pacific region (5°S–45°N and 75°E–145°E), observing the Earth in an east-west direction with a fixed north–south field of view (FOV) of 7.73° [2]. For the retrieval of the concentrations of atmospheric gases (O_3, NO_2, SO_2, and HCHO) and aerosol properties, the GEMS has been designed to provide a continuous spectrum from 300 to 500 nm, with a spectral resolution of better than 0.6 nm every 0.2 nm. As the first hyperspectral UV–VIS sensor onboard a GEO satellite, the GEMS is expected to provide critical information for the monitoring of the regional transport of atmospheric chemicals at hourly intervals during the daytime as part of the GEO constellation [35].

Prior to the launch of the satellite, on-ground sensor characterization and calibration have been conducted during the preparatory phase for the GEMS. While in orbit, the GEMS relies on the onboard calibration consisting of solar diffusers and LED to evaluate and maintain calibration quality. As part of the onboard calibration system, the LED serves as a stable light source to monitor the non-linearity of the electronic response and the aliveness of each pixel at the detector level. Solar measurements have also been designed to monitor and calibrate changes in the sensor response with two transmissive diffusers: a working and reference diffuser. The working diffuser has been designed to observe the sun on a daily basis which makes it to gradually degrade over the course of the mission. Thus, a reference diffuser identical to the working diffuser but observing the sun only once every six months has been included in the calibration system. However, because most components of the sensor are expected to degrade over time, it is important to isolate the degradation of each component of the sensor and accurately calibrate the changes. Because onboard calibration has been incorporated into the calibration system, an independent method for evaluating the overall performance of the calibration system would be useful for maintaining the data quality of the GEMS in the long-term as a back-up calibration strategy.

2.1.2. OMI and TROPOMI

The OMI and TROPOMI are hyperspectral sensors that encompass both the spectral range and the observation area of the GEMS. Operating in a sun-synchronous polar orbit, both sensors take radiance measurements in the ascending node of the satellites at around the local solar time (LST) of 13:30. The top-level specifications for the GEMS, OMI, and TROPOMI are summarized in Table 1. Launched in October 2017, the TROPOMI has stricter data quality requirements compared to other sensors. Because the spatial and spectral resolution of the GEMS is quite similar to the resolution of the TROPOMI, the GEMS and TROPOMI are strongly expected to be reciprocal candidates for inter-calibration once the GEMS goes into operation.

Table 1. Sensor specifications for the GEMS, OMI, and TROPOMI.

Sensor	GEMS	OMI		TROPOMI	
Orbit type	Geosynchronous (nadir at 128°E)	Sun-synchronous mean LST – 13:45)		Sun-synchronous (mean LST – 13:35)	
Spectral range	300–500 nm	UV-2 VIS	307–383 nm 349–504 nm	Band 3 Band 4	320–405 nm 405–500 nm
Spectral resolution	< 0.60 nm	UV-2 VIS	0.42 nm 0.63 nm	Band 3 Band 4	0.55 nm
Spectral sampling	< 0.20 nm/pixel	UV-2 VIS	0.14 nm/pixel 0.21 nm/pixel	Band 3 Band 4	0.20 nm/pixel
Spatial resolution	3.5×8 km^2 (at Seoul)	13×24 km^2 (along × across track)		5.5×3.5 km^2 (along × across track)	

* The spatial resolution of TROPOMI Band 3-4 has been updated from 7 to 5.5 km along track since 6 August 2019 [36]. UV-2 and VIS indicate the Level 1B products of OMI while Band 3 and Band 4 indicate the products of TROPOMI.

2.2. DCC Climatology

To check whether there are sufficient DCCs available within the GEMS field of regards (FOR), especially over the TWP region, we apply a conventional DCC-detection approach using threshold tests for TB and the uniformity of the clouds [26]. The threshold values used for each test and the constraints for the observation angles and spatial coverage are summarized in Table 2. For the TB test, we use an 11-μm window channel with a threshold of 205 K, which is set considering the trade-off between the precision of DCC detection and the number of detected DCCs as presented by previous studies [27,28]. In addition, for the uniformity test, a relaxed threshold value (from 1 K to 2 K) is used to account for the lower spatial resolution of the GEMS. The relaxation of the threshold could broaden the range of available data with little change to the effectiveness of the DCC detection [28]. The maximum solar and viewing zenith angle is also limited to 40° because DCC reflectivity changes considerably when the solar and viewing angles are too large [21].

Table 2. DCC detection thresholds.

Condition	Threshold
Infrared brightness temperature (TB_{IR})	$TB_{IR} < 205$ K
Spatial uniformity (TB_{IR})	Standard deviation of $TB_{IR} < 2$ K
Spatial uniformity (R_{VIS})	Standard deviation of $R_{VIS} < 0.03$
Solar and viewing zenith angle (θ_0 and θ)	$\theta_0 < 40°, \theta < 40°$
GEMS observation area	5°S–45°N, 75°E–145°E

2.2.1. AHI Data Processing

AHI measurements are used because this imager onboard a GEO satellite provides VIS ($R_{0.47}$) and IR ($TB_{10.4}$) channels at a higher temporal resolution while fully covering the TWP region with its full-disk observation (see Table 3). Because $R_{0.47}$ has a higher spatial resolution than $TB_{10.4}$, spatially averaged $R_{0.47}$ is employed. To test the availability of DCCs under GEMS observation conditions, the spatial resolution of the GEMS is simulated using 4×4 pixels for each of the VIS and IR channels, while the mean of $TB_{10.4}$ and the standard deviations of $R_{0.47}$ and $TB_{10.4}$ are used for DCC detection.

Table 3. AHI VIS and IR channels for DCC detection.

AHI	Ch01 ($R_{0.47}$)	Ch13 ($TB_{10.4}$)
Channel	VIS	IR (window channel)
Wavelength	0.47 μm	10.4 μm
Spatial resolution	1×1 km^2	2×2 km^2
Observation interval	Every 10 min	
Spatial coverage	Full-disk scan (nadir at 140.7°E)	

2.2.2. Frequency Distribution

DCC climatology data from the AHI for July 2016 to June 2017 with a spatial grid and sampling frequency that matches that of the GEMS are collected. In the Asia-Pacific region, most DCCs are observed near the tropics and are distributed quite evenly over the GEMS observation area, as reported in previous studies [37–39]. In Figure 1, the spatial distribution of the frequency of DCCs exhibits a unique arc-shaped boundary, which is attributed to the limitations imposed by the current study (i.e., the viewing zenith angle (θ) should be smaller than 40°). Given that the viewing zenith angle is a fixed value over time for each pixel of a GEO sensor, the spatial distribution may be limited to those pixels that satisfy the angle condition.

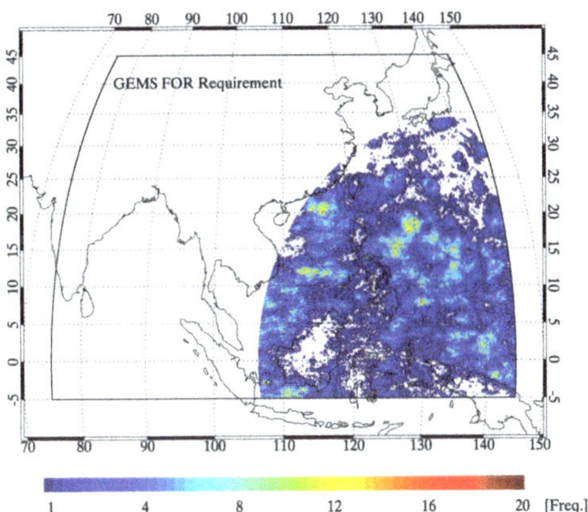

Figure 1. Frequency distribution of DCCs matched to the GEMS FOV over the GEMS observation area from AHI data taken at three-day intervals for the period July 2016–June 2017. Here, the frequency is calculated as the number of DCCs occurring over the year at three-day intervals at each AHI grid point binned to 8×8 km^2.

Figure 2 shows the temporal variation in the number of DCCs observed in a single observation scene. On a specific day, most DCCs are detected at noon (02:00–04:00 UTC) when the sun directly passes over the target area. This can be attributed to the constraint on the solar zenith angle for DCC detection because the solar zenith angle is also limited in the same way as the viewing zenith angle. The constraint along with the seasonal deep convection in the Northern Hemisphere might also cause the seasonal variation in the frequency of DCCs. As shown in Figure 2b, DCCs mostly occur from late summer to early autumn over the TWP region because atmospheric convection is strongly dependent on high moisture levels and the latent heat that accumulates during summer [38,40–42].

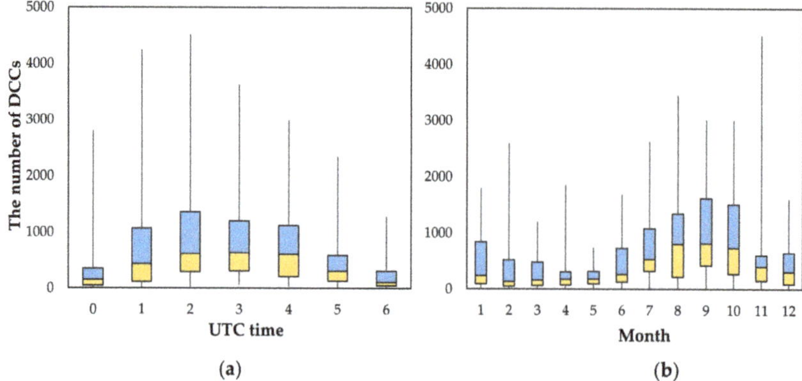

Figure 2. (**a**) Hourly and (**b**) monthly distribution of the number of DCCs observed in a single scene over the GEMS observation area corresponding to the GEMS FOV from AHI data taken at three-day intervals for the period July 2016–June 2017. The yellow and blue boxes represent the lower and upper quartile to the median, respectively.

Even with the limitation imposed by the viewing angular geometry and the seasonality, the average number of DCCs in a single observation scene is still larger than 200 pixels even in the month with the minimum frequency. Because the GEMS observes the Earth eight times a day, at least 50,000 DCCs can be detected a month from the GEMS when using the conventional DCC-detection approach with collocated AHI data. This number could be higher if the Advanced Meteorological Imager (AMI) onboard GEO-KOMPSAT-2A (GK2A) is used, which is stationed over 128.2°E as with the GEMS; thus, the coverage could be expanded further to the west.

2.3. DCC Reflectivity Spectrum

After confirming the availability of DCCs over the GEMS coverage area, DCC reflectivity spectra obtained from the OMI and TROPOMI are compared to confirm that the DCC measurements show similar spectral features and a sufficiently stable signal to be compared across different sensor characteristics and optical paths. In the UV–VIS spectral region, the reflected radiation from ice clouds is significantly affected by the angle condition [43], and this means it is important to precisely detect DCCs for the accurate characterization of cloud reflectivity.

2.3.1. Collocation Process

Because the OMI and TROPOMI only cover the UV–VIS and UV–SWIR spectral regions, respectively, DCC detection for both sensors could be performed with the collocated VIS and IR channels of the AHI. For the collocation between GEO and LEO sensors, we apply the collocation criteria suggested by the Global Space-based Inter-Calibration System (GSICS) community [44]. Because collocated LEO and GEO data are not directly compared in this study, the viewing angle does not match between the sensors. As shown in Figure 3, hyperspectral data satisfying the spatial and angle conditions (see Table 2) are collected first, and then the AHI VIS and IR channels matching the temporal collocation criteria are called. With the collected data, AHI pixels observed at nearly the same time ($\Delta t < 5$ min) as the OMI and TROPOMI pixels are collocated when the pixels simultaneously satisfy the spatial threshold (located within a half of shorter FOV of a LEO sensor). With the collocated AHI pixels, the average $TB_{10.4}$ and the standard deviations of $TB_{10.4}$ and $R_{0.47}$ are employed for DCC detection.

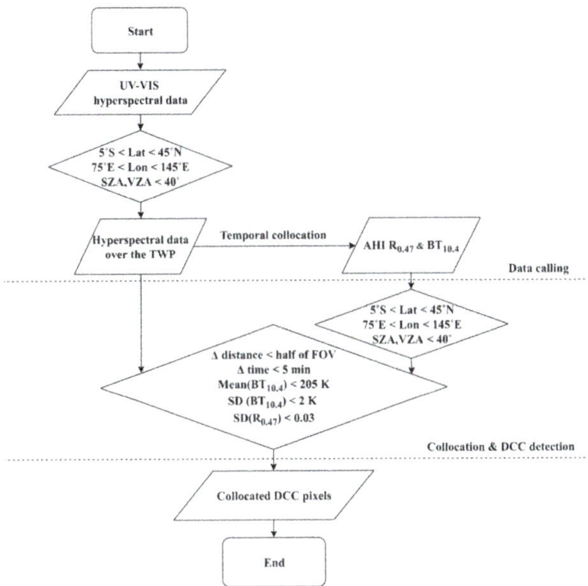

Figure 3. Flow chart of the collocation process between a UV–VIS hyperspectral sensor and a VIS–IR imager for the generation of DCC data. Mean and standard deviation (SD) are calculated with AHI pixels satisfying the spatiotemporal collocation criteria.

2.3.2. Apparent Reflectivity of DCCs

The GEMS, OMI, and TROPOMI provide the spectral radiance data that are used as input for the retrieval of geophysical information from the atmosphere. Because the uncertainty in the measured radiance due to the optical path of the instrument could be mitigated by using irradiance, which has the same optical depth [45], here, we use reflectivity for the spectral analysis. Because the OMI and TROPOMI provide solar observations once a day, timely matched irradiance with radiance is used to calculate the reflectivity. The radiance measured over the DCCs can be written as:

$$I_\lambda(\theta_0, \theta, \varphi) = R_\lambda(\theta_0, \theta, \varphi)\frac{F_\lambda}{\pi}e^{-(\frac{\mu+\mu_0}{\mu\mu_0})\tau_\lambda(z)} \tag{1}$$

where $I_\lambda(\theta_0, \theta, \varphi)$ is the measured upwelling radiance at wavelength λ with solar zenith angle θ_0, satellite zenith angle θ, and relative azimuth angle φ. The measured radiance is strongly affected by cloud reflectivity $R_\lambda(\theta_0, \theta, \varphi)$ and incoming solar irradiance F_λ at the top of the atmosphere (TOA). The equation also includes the attenuation caused by atmospheric extinction occurring when the incoming and outgoing radiation passes through the atmosphere. The atmospheric optical depth $\tau_\lambda(z)$ from the cloud top altitude z to the TOA is determined by both absorption and scattering. Here, we consider only Rayleigh scattering to simplify the problem and neglect the backscattered radiation from the atmosphere above the DCCs. The angle component μ is the cosine of the zenith angle. Thus, cloud reflectivity using the measured radiance and irradiance can be given as:

$$R_\lambda(\theta_0, \theta, \varphi) = \frac{\pi I_\lambda(\theta_0, \theta, \varphi)}{F_\lambda}e^{(\frac{\mu+\mu_0}{\mu\mu_0})\tau_\lambda(z)} \tag{2}$$

Here, the optical depth $\tau_\lambda(z)$ is estimated using the approximation suggested by Bodhaine et al. [46] that considers the altitude and Rayleigh scattering in the atmosphere. The cloud altitude is set to approximately 16 km because the cloud top of DCCs nearly reaches the tropopause in the equatorial region [31,47]. This means that the optical depth of the atmosphere above the clouds is within the range of 0.0005–0.0025 from 300 to 500 nm. Because Mie scattering and atmospheric absorption in the upper troposphere are not included in the calculation, the reflectivity is the apparent reflectivity of the DCCs, though it is referred to as simply DCC reflectivity in this study.

3. Results

3.1. DCCs Detected Using the OMI and TROPOMI

Figure 4 shows the DCCs identified using OMI and TROPOMI data for a particular cloudy scene (3 July 2018 06:10 UTC and 3 July 2018 06:40 UTC, respectively). The TROPOMI observes the Earth about 30 min earlier than the OMI, thus, the cloud distributions are slightly different. As shown in the figure, the number of DCCs obtained from the OMI–AHI collocations is appreciably smaller than that from the TROPOMI–AHI collocations. One of the main reasons for this difference is the lower spatial resolution of the OMI. An OMI pixel is about 15 times larger than a TROPOMI pixel, thus, many of the small-scale DCCs detected as DCCs using TROPOMI data are missed by the OMI because of the threshold and uniformity tests. The lower spatial resolution of the OMI for small-scale DCCs increases not only the TB but also the spatial variability in the IR and VIS channels, leading the pixel to be labeled as a non-DCC.

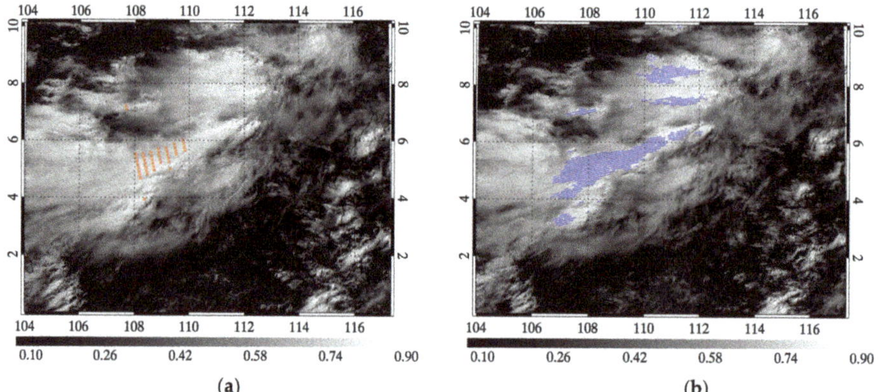

Figure 4. DCCs plotted on an AHI $R_{0.47}$ image: (**a**) OMI DCCs (orange dots) on 3 July 2018 06:10 UTC and (**b**) TROPOMI DCCs (blue dots) on 3 July 2018 06:40 UTC.

Data quality issues that arise during the long-term operation of the OMI also affect the availability of OMI observations. For instance, the row anomaly (RA) effect [32] renders nearly a quarter of all OMI pixels (especially those close to the nadir observations) unavailable for analysis. Figure 5 shows the measured reflectivity of the OMI as a function of the position (i.e., row) of the detector (i.e., the charge-coupled device, CCD) and the reflectivity spectrum affected by the RA effect. The measurements in rows 24–41 contaminated by the RA effect are eliminated during data processing. However, as shown in Figure 5a, the reflectivity for the row numbers close to the nadir port is also significantly lower, even though the rows are not flagged as RA-affected pixels. When these observations are detected together as DCCs, the reflectivity spectrum is significantly lower compared to the DCC reflectivity of the TROPOMI. Thus, in this study, the pixels in rows 41–48 are also eliminated, which are possibly affected by the RA effect but which are not flagged as such (https://projects.knmi.nl/omi/research/product/rowanomaly-messages.php). As shown in Figure 5b, with the elimination of the measurements in the CCD rows close to the nadir (41–48), the mean reflectivity becomes much closer to the DCC reflectivity of the TROPOMI. Because the rows close to the nadir port generally have a low viewing zenith angle, which satisfies the angle condition for DCC detection, the RA effect significantly influences the availability of DCC observations from the OMI.

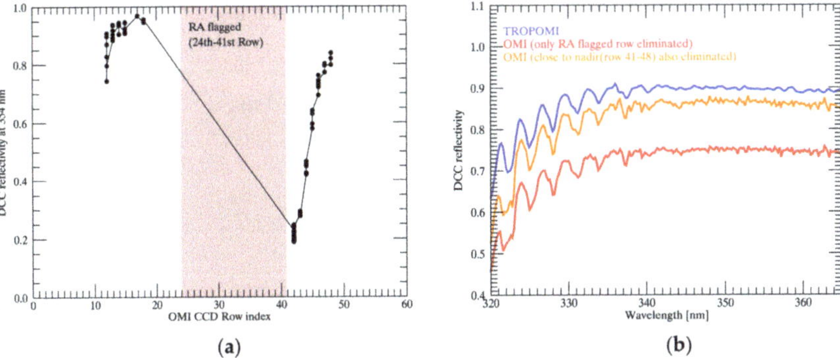

Figure 5. (**a**) DCC $R_{0.354}$ binned depending on the position of the detector in the OMI. The shaded red box indicates the RA flagged rows in the northern region of the orbit. (**b**) The DCC mean reflectivity spectrum of the OMI (the UV-2 product) containing the RA-affected observations in comparison with TROPOMI DCCs

3.2. DCC Reflectivity Spectrum

The DCC spectra of the OMI and TROPOMI observed over a year at 10-day intervals are presented in Figure 6, which shows the mean and standard deviation of the radiance and reflectivity at each wavelength. Solar measurements observed on the same day and the scan angle position (i.e., the position on the detector) of each DCC measurement are also displayed together. Because of the previously mentioned data quality issues, the number of DCCs from OMI observations over the year is only 3% of that from TROPOMI observations. However, even with this considerable difference in the number of measurements, the mean reflectivity of the OMI and TROPOMI is very similar at about 0.90 and 0.85, respectively, with nearly invariant spectral features except for both ends of the wavelength range. The spectral features at both ends are attributed to ozone absorption (300–345 nm) and the pixel saturation of the TROPOMI (450–500 nm) [36]. The results indicate that the DCCs observed by the satellite sensor reflect a mostly stable signal even with differences in sensor characteristics, the number of measurements, and the observation angle geometry.

However, some differences are also observed between the spectra of the OMI and TROPOMI. At shorter wavelengths, TROPOMI reflectivity is slightly higher than that of the OMI; as the wavelength increases, this difference becomes much smaller. This might be caused by the degradation of the diffuser in the solar measurements of the TROPOMI because this degradation is more significant at shorter wavelengths. This degradation is to be addressed in future updates of TROPOMI L1B data in early 2020, as announced in the S5P validation report [36]. TROPOMI DCCs also have less spectral noise because the OMI solar measurements have more spectral noise across all wavelengths, as shown in Figure 6a. There are also sharp peaks at 393 and 397 nm corresponding to the Ca II K and Ca II H Fraunhofer lines, respectively, which are caused by the beam-filling effect of the atmosphere above the clouds. Because rotational Raman scattering occurs in the atmosphere, scattered radiation is added to the upwelling radiation from the clouds [48]. However, OMI reflectivity exhibits negative peaks, which appears unrealistic considering that the beam-filling effect predominantly occurs with radiance. These peaks are caused by missing data at particular wavelengths for OMI irradiance. When calculating reflectivity, missing data are approximated by linear interpolation, which may not accurately reproduce the spectral features, especially for higher peaks.

These results indicate that the TROPOMI still requires further minor updates but that DCCs are a promising target given the theoretically well-matched spectral features and lower spectral noise. The abundance of data is also an advantage of using the measurements in further research. However, even with the well-explained spectral features, DCC measurements still exhibit large systematic differences, as indicated by the standard deviations in Figure 6c, reaching nearly 10% and 12% for the OMI and TROPOMI, respectively. Because this systematic difference increases as the number of measurements increases, the TROPOMI has a higher systematic difference than the OMI. This indicates that, as the observation period becomes longer, the difference among the DCCs could increase considerably. The difference might be too large to regard the DCC detection properly done and this also make the characterization of the cloud reflectivity complicated without knowing the reason of the difference. Thus, in Section 3.3, the thresholds for conventional DCC detection are tested to reduce the systematic differences in DCC measurements and improve the accuracy of DCC detection. Because the OMI has some data-quality issues, we use only TROPOMI and AHI observations for this analysis.

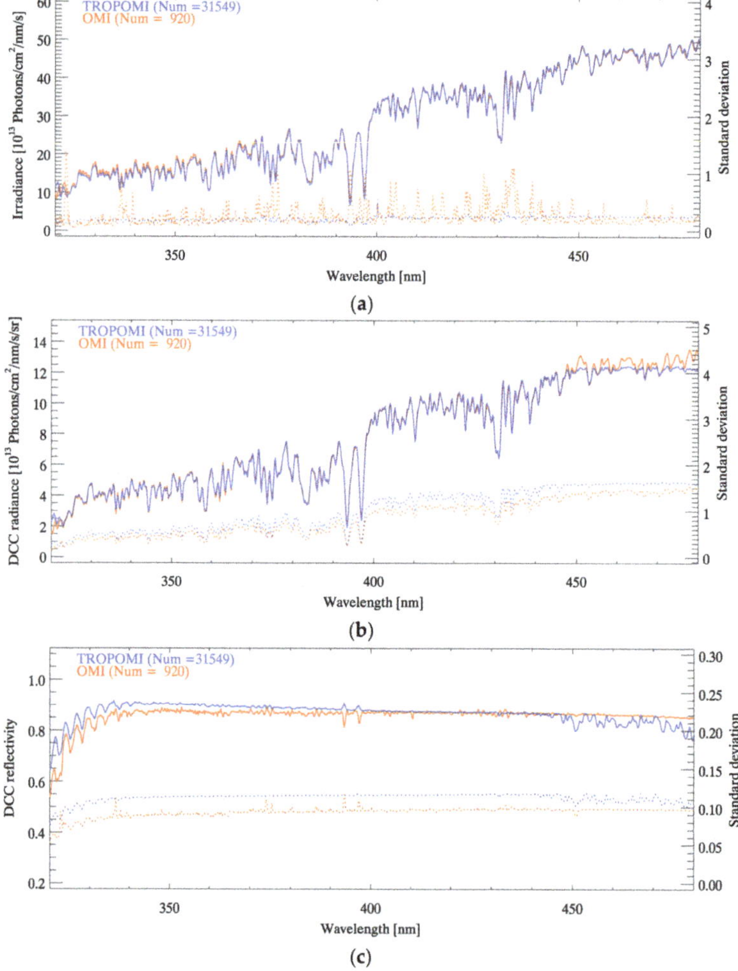

Figure 6. Mean and standard deviation of the (**a**) irradiance, (**b**) radiance, and (**c**) reflectivity spectra of OMI and TROPOMI DCCs observed for the period July 2018–June 2019 at 10-day intervals. The solid and dashed lines represent the mean and standard deviation at each wavelength, respectively.

3.3. Improvement in DCC Detection

3.3.1. Comparison of VIS and IR Radiation

Figures 7 and 8 present the characteristics of the DCCs detected by the VIS and IR channels. This comparison provides insights into whether DCC detection is accurate when detecting only the colder and brighter cloud cores. Figure 7 shows the horizontal distribution of DCCs found over Typhoon Chaba in October 2016. For a one-to-one comparison, an AHI $R_{0.47}$ image is binned to match the spatial resolution of $TB_{10.4}$. As shown in Figure 7a, DCCs (identified as blue dots) are mainly found over the typhoon center, which has a cold $TB_{10.4}$, with a symmetrical distribution around the center. However, Figure 7b, which presents the DCCs detected over the $R_{0.47}$ image, is interesting in that the blue dots over the right side of the typhoon center have a lower $R_{0.47}$ of about 0.7. These are thin cirrus clouds that have spread out from the typhoon center following strong upper air outflow. Because these cirrus clouds have colder cloud tops composed of ice particles, the clouds are detected as DCCs using the conventional detection method even though their reflectivity is much lower than genuine DCCs.

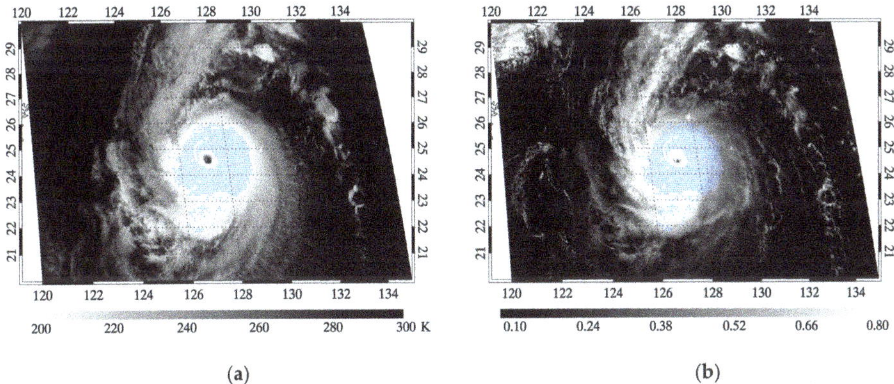

(a)   (b)

Figure 7. AHI DCCs plotted as blue dots on (**a**) 2-km AHI $TB_{10.4}$ and (**b**) 2-km $R_{0.47}$ images of Typhoon Chaba (3 October 2016 03:30 UTC)

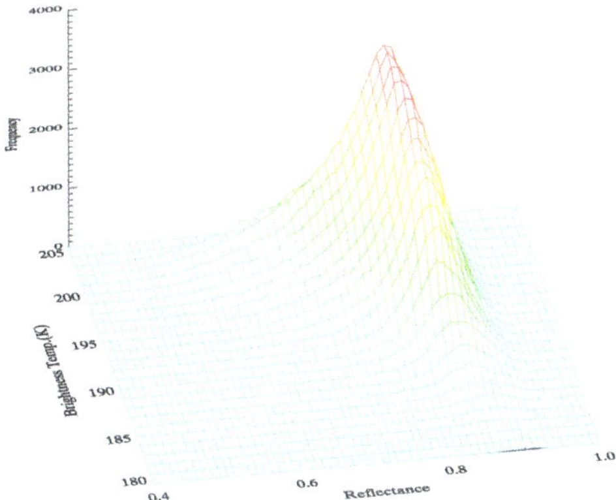

Figure 8. Two-dimensional histogram of DCCs detected using AHI $R_{0.47}$ and $TB_{10.4}$ over the GEMS observation area with AHI data taken at three-day intervals for the period July 2016–June 2017.

The difference in the radiative properties of the DCCs is also demonstrated in Figure 8, which presents a two-dimensional histogram of $R_{0.47}$ and $TB_{10.4}$ of the DCCs. Based on the histogram, it can be inferred that an increase in $TB_{10.4}$ also increases the skewness of the distribution of $R_{0.47}$. This may be due to the increase in the proportion of detected DCCs with a lower reflectivity. This also indicates that DCCs usually have colder cloud tops, a higher reflectivity, and an optically thicker vertical structure, while cloud edges and cirrus clouds have similar colder but darker cloud tops. Consequently, these results show that $TB_{10.4}$ might be less effective as a DCC detection threshold, especially for UV–VIS measurements.

One of the few attempts to use DCCs for the monitoring of a UV–VIS hyperspectral sensor used only the UV reflectivity threshold for DCC detection [32]. In that study, OMI pixels with a higher reflectivity at 354 nm ($R_{0.354} > 0.9$) were identified as DCCs and then used for the monitoring of the temporal stability of the radiometric calibration of the OMI. At 354 nm, ozone absorption becomes weaker while Rayleigh scattering becomes stronger, and these interactions with the atmosphere reduce the proportion of the directly transmitted light from the clouds which shows higher angle dependence even though the dependence is not very significant over bright clouds [34]. Even with this simple form of detection, the average cloud reflectivity was fairly constant regardless of the wavelength (which is a characteristic of DCC reflectivity), and thus, they used DCC reflectivity for the long-term monitoring of spectral dependence in sensor performance. However, DCC reflectivity still exhibited seasonal and inter-annual variation, which was attributed to differences in cloudiness, angle dependence, and residual atmospheric effects (refer to Figure 32 in [32]). Although it is not easy to quantify, it is highly possible that these attributions could be increased when the detected DCCs are bright but low-lying warm clouds. For example, the optical path for warm clouds is much longer than that for DCCs, causing increased variability in the measured reflectivity due to the increased contribution from tropospheric air. By the same token, the angular variation of the measured reflectivity also increases with increasing optical depth.

To further clarify the issues associated with warm clouds, Figures 9 and 10 show the spatial distribution of $TB_{10.4}$ and $R_{0.47}$ in warm clouds and the spectral reflectivity of clouds with different $TB_{10.4}$ values, respectively. This demonstrates the importance of the IR threshold for the accurate detection of DCCs, especially in relation to high-altitude clouds with minimal influence from the troposphere. The blue dots in Figure 9 show DCCs with high reflectivity (TROPOMI $R_{0.354} > 0.9$) and warm IR temperatures (AHI $TB_{10.4} > 260$ K). In this case, most of the blue dots are located over the cloud edges with bright reflectivity, although their temperatures are much warmer than the nearby convection core. Thus, if we use the UV–VIS radiation threshold only, it would be difficult to screen out bright but warm clouds that are close to the cloud core.

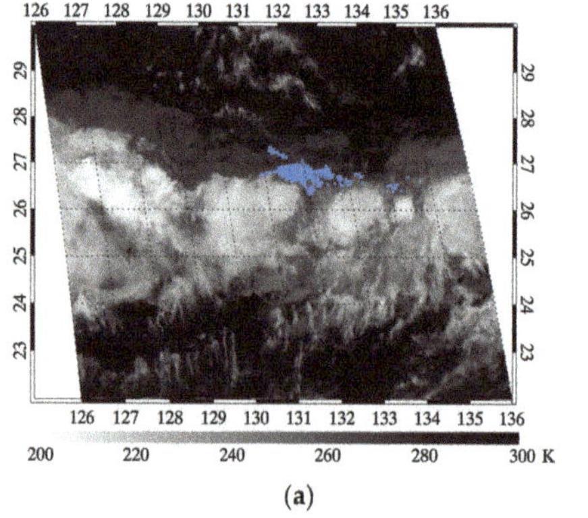

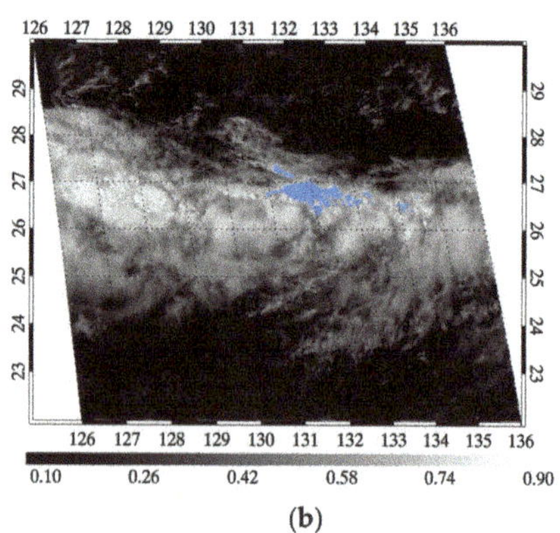

(**a**)   (**b**)

Figure 9. TROPOMI cloud pixels (AHI $TB_{10.4} > 260$ K, TROPOMI $R_{0.354} > 0.9$) plotted as blue dots on (**a**) 2-km AHI $TB_{10.4}$ and (**b**) 2-km $R_{0.47}$ images (20 June 2019 04:30 UTC).

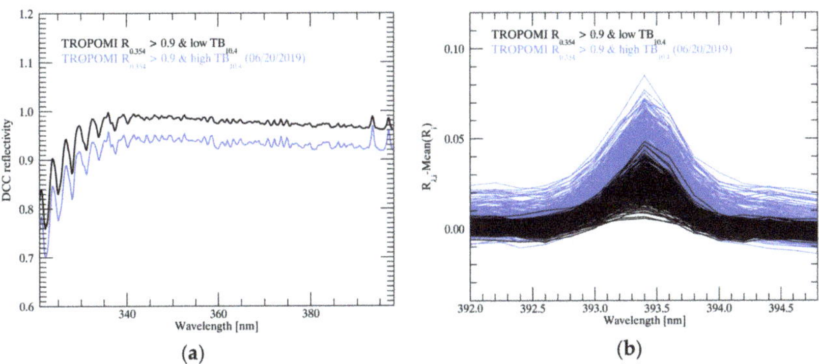

Figure 10. (**a**) Mean reflectivity spectrum of TROPOMI DCCs detected using UV reflectivity (TROPOMI $R_{0.354} > 0.9$) and (**b**) spectral anomaly spectra (i: each DCC pixel, λ: wavelength). The black and blue lines represent a cold and warm IR temperature, respectively.

Figure 10 shows the reflectivity spectrum of bright DCCs with different brightness temperatures. The blue line in Figure 10a represents the average reflectivity spectrum of the blue dots in Figure 9, while the black line represents that of the clouds satisfying the conventional DCC detection thresholds. The reflectivity spectrum including bright but warm clouds (blue line) clearly has a smaller reflectivity compared to the bright and cold clouds, which is attributed to the radiative interaction with the tropospheric atmosphere. The tropospheric effects in the measured reflectivity is also presented in Figure 10b because the beam-filling effect increases cloud reflectivity with greater rotational Raman scattering from the tropospheric atmosphere [48].

The results in Figure 8, Figure 9, Figure 10 make it clear that using VIS and IR information together could effectively screen out cirrus clouds and cloud edges as well as ensure the detection of only colder cloud tops for the better utilization of DCC reflectivity.

3.3.2. DCC Detection with Additional VIS Reflectivity

Based on the previous analysis, we develop an updated DCC detection approach utilizing both reflectivity and TB. In order to adapt the reflectivity test, it is important to set an appropriate threshold for reflectivity; a stricter threshold (e.g., 0.9) could produce more stable statistics but reduce the availability of the data, while a more relaxed threshold (e.g., 0.6) could increase the number of data points but increase the variability. Thus, the optimal reflectivity threshold for DCC detection needs to be set by weighing both sides (i.e., data availability and the stability of the reflectivity distribution). Here, we choose an optimal value by analyzing the variation in statistical parameters as a function of different threshold values.

Figure 11a,b presents the DCC frequency distribution for TROPOMI $R_{0.354}$ with the addition of the AHI $R_{0.47}$ threshold and the uniformity threshold for AHI $R_{0.47}$, respectively. The use of $R_{0.354}$ is based on a previous implementation with the OMI [32]. As shown in Figure 11a, applying the AHI $R_{0.47}$ test reduces the spread of the TROPOMI $R_{0.354}$ distribution and generates a distribution that closely follows a normal distribution. However, some low-reflectivity data remains because of the atmospheric effects and collocation uncertainty between the AHI and TROPOMI measurements. Figure 11b also shows that cloud pixels with higher spatial inhomogeneity account for a large proportion of the center of the distribution. This means that the overshooting tops near the cloud core may have a lower spatial uniformity, which cannot be eliminated by the reflectivity threshold.

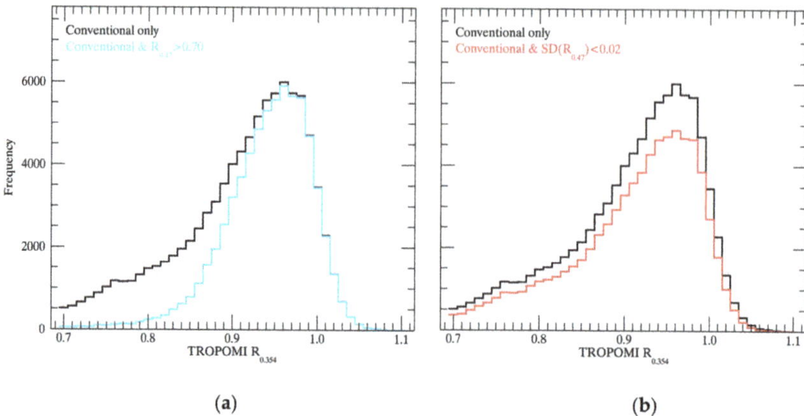

(**a**)   (**b**)

Figure 11. Frequency distribution of TROPOMI R_{354} with an additional AHI $R_{0.47}$ restriction for data from July 2018–June 2019 taken at five-day intervals.

Table 4 presents the statistics for TROPOMI DCC $R_{0.354}$ with the application of different AHI $R_{0.47}$ thresholds to determine the optimal threshold that produces a fairly normal distribution without eliminating too many observations. TROPOMI $R_{0.354}$ is also applied together as the detection threshold to reduce collocation uncertainty by restricting the tail of the distribution (TROPOMI $R_{0.354} > 0.7$). The results show that, as the AHI $R_{0.47}$ threshold increases, the distribution becomes very close to normal even though the number of detected DCCs decreases exponentially. The standard deviation of the reflectivity decreases linearly and the VIS threshold increases when the kurtosis increases exponentially. Interestingly, only skewness converges at a particular AHI $R_{0.47}$ threshold (0.64). Because TROPOMI $R_{0.354}$ reflectivity is skewed to the left due to the darker cirrus clouds with a lower reflectivity, the skewness of the distribution has a negative value regardless of the AHI $R_{0.47}$ threshold.

Table 4. Statistics for TROPOMI DCC $R_{0.354}$ depending on the addition of an AHI $R_{0.47}$ threshold for DCC detection compared to the conventional DCC detection (w/o column) using DCC measurements for July 2018–June 2019 taken at five-day intervals.

AHI $R_{0.47}$	w/o	0.60	0.62	0.64	0.66	0.68	0.70	0.72	0.74	0.76
Count	91630	90752	89861	88286	86138	83569	80475	76696	71469	64857
Mean	0.916	0.917	0.919	0.922	0.925	0.929	0.933	0.938	0.943	0.949
Median	0.932	0.933	0.934	0.936	0.938	0.940	0.943	0.946	0.951	0.956
Mode*	0.960	0.960	0.960	0.960	0.960	0.960	0.960	0.960	0.960	0.960
SD*	0.076	0.074	0.072	0.070	0.067	0.063	0.060	0.057	0.053	0.050
Skewness	−0.779	−0.769	−0.765	−0.761	−0.767	−0.780	−0.803	−0.848	−0.917	−1.021
Kurtosis	-0.027	-0.008	0.029	0.110	0.255	0.448	0.706	1.056	1.570	2.225

* The bin size used to calculate the mode is set to 0.01. SD indicates the standard deviation.

Table 5 presents the statistics for TROPOMI DCC $R_{0.354}$ with the application of different thresholds for the uniformity test for AHI $R_{0.47}$. As shown in Table 5, the central value and the spread of the distribution changes only slightly with the different thresholds for the uniformity test for $R_{0.47}$. The kurtosis and skewness also change as the uniformity increases, though they do not change dramatically, as with the reflectivity threshold.

Table 5. Statistics for TROPOMI DCC $R_{0.354}$ depending on the uniformity threshold for AHI $R_{0.47}$ for DCC detection compared to the conventional DCC detection (w/o column) using DCC measurements for July 2018–June 2019 taken at five-day intervals.

SD* of AHI $R_{0.47}$	w/o	0.025	0.024	0.023	0.022	0.021	0.020	0.019	0.018	0.017
Count	91630	84159	82454	80629	78626	76557	74241	71737	69135	66243
Mean	0.916	0.916	0.916	0.916	0.916	0.916	0.916	0.916	0.916	0.916
Median	0.932	0.933	0.933	0.933	0.933	0.933	0.933	0.932	0.932	0.932
Mode*	0.960	0.960	0.960	0.960	0.960	0.960	0.960	0.960	0.960	0.960
SD*	0.076	0.075	0.075	0.075	0.074	0.074	0.074	0.074	0.074	0.074
Skewness	−0.779	−0.788	−0.789	−0.792	−0.793	−0.795	−0.797	−0.796	−0.795	−0.797
Kurtosis	−0.027	−0.002	0.006	0.012	0.017	0.019	0.026	0.025	0.023	0.026

* The bin size used to calculate the mode is set to 0.01. SD indicates the standard deviation.

In summary, the skewness of distribution of TROPOMI $R_{0.354}$ might become close to 0 with a brighter AHI $R_{0.47}$ threshold until the number of DCCs is significantly lower. When it comes to spatial inhomogeneity, some DCCs with a relatively low uniformity are eliminated with the stricter uniformity test mostly at the center of the distribution. Because the AHI $R_{0.47}$ threshold and the uniformity test might simultaneously affect the statistics for the reflectivity distribution of TROPOMI $R_{0.354}$, the optimal threshold value for DCC detection needs to be set considering both effects. Figure 12a,b show the number of available DCCs and the skewness of the distribution, respectively, as a function of the detection thresholds. Considering the distribution of each variable, the optimal thresholds for AHI $R_{0.47}$ and the uniformity test for DCC detection are set at 0.70 and 0.018, respectively, because at that point, the available number of DCCs is still high even with a relatively low skewness of −0.70.

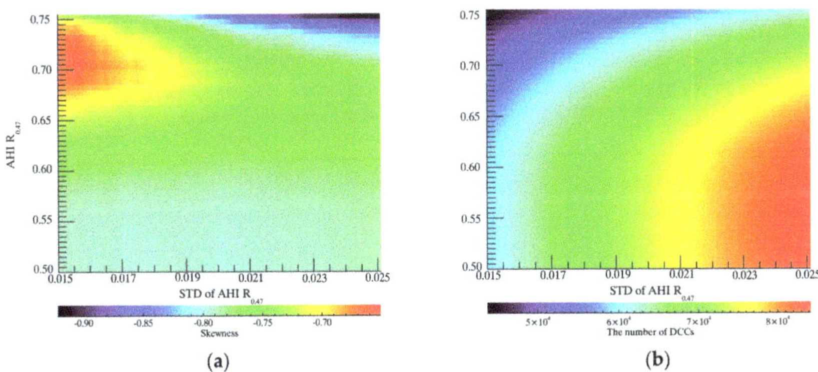

Figure 12. (a) Skewness and (b) the number of DCCs as a function of the AHI $R_{0.47}$ and uniformity test for DCC detection using DCC measurements for July 2018–June 2019 taken at 5-day intervals.

4. Discussion

4.1. Verification of the Updated DCC Detection Method

The results in Section 3 show that DCCs have different radiative properties depending on the way to detect the DCCs. For the thermal threshold test, it would be most effective to screen out the low-altitude clouds, in this case warm clouds having longer optical path lengths. VIS reflectivity can also be a useful indicator for detecting only optically thick clouds that are bright enough to reflect most of the incoming radiation. Using both radiative properties, DCC detection can be improved further to detect only optically thick and high-altitude cloud targets that exhibit homogeneous spectral features and higher reflectivity with lower variation.

4.1.1. Spectral Analysis of DCC Reflectivity

Figure 13 highlights the advantages of applying the updated DCC detection method with the threshold values suggested in Section 3.3.2. The DCC mean reflectivity spectra at the Fraunhofer lines are presented to compare the spectral features of the DCCs detected by different detection methods, including the UV threshold test (TROPOMI $R_{0.354} > 0.9$) and the IR threshold test (AHI $BT_{10.4} < 205$ K). In Figure 13a, the mean reflectivity spectra show similar spectral features but differences in reflectivity as the DCCs detected using the UV threshold test show the highest values. However, in Figure 13b, the spectral features of anomaly spectra exhibit more variance when only the UV detection threshold is used. In the figure, the DCCs detected using the updated DCC detection method have lower peaks at the Fraunhofer lines, which indicates that the atmosphere above the clouds might be much thinner when the DCCs are detected using the thermal radiation threshold.

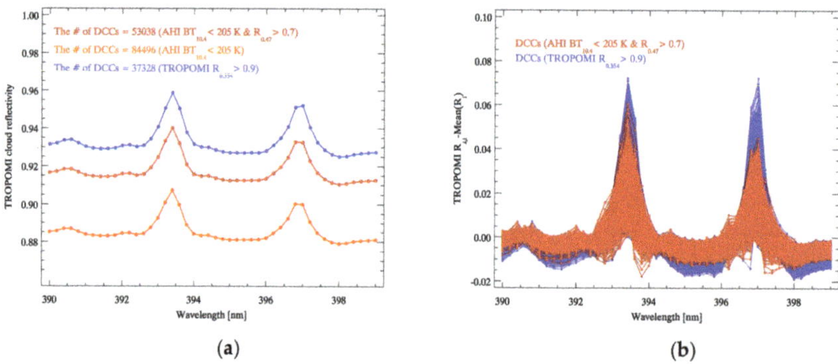

Figure 13. (**a**) Mean reflectivity and (**b**) anomaly spectra (i: each DCC pixel; λ: wavelength) of DCCs detected using different DCC detection threshold tests. The blue, red, orange lines represent the UV threshold test only, the updated DCC detection method, and the IR threshold test only, respectively. DCC measurements are from July 2018–June 2019 taken at five-day intervals.

4.1.2. Cloud Properties of DCCs

The cloud properties obtained from TROPOMI Level 2 cloud products are presented in this section in order to identify the practical range of cloud properties for the DCCs detected using different DCC detection threshold tests. Cloud optical thickness and cloud top height are used for this analysis because these properties represent the optical and physical features of the clouds, respectively. The cloud properties are retrieved from the O_2 A-band at 760 nm, while the clouds are treated as scattering layers [49]. In Figure 14a, the optical thickness of the DCCs detected using the IR detection threshold is lower than that of the DCCs detected using the UV detection threshold. However, as shown in Figure 14b, the cloud top height is much higher when the IR threshold test is used for DCC detection. These results indicate that the UV and IR DCC detection thresholds complement each other in limiting various cloud properties while accurately detecting only those DCCs with homogeneous cloud properties. These results closely correspond with the analysis in Section 3.3.

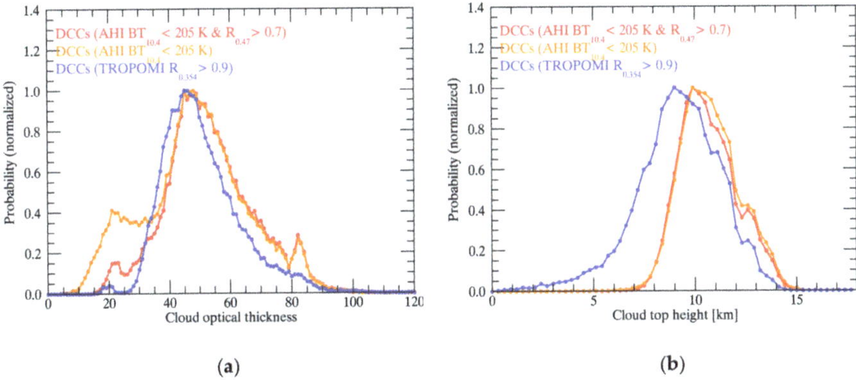

Figure 14. Same as Figure 13 for histograms of (**a**) cloud optical thickness and (**b**) cloud top height from the TROPOMI Level -2 cloud product for DCCs.

4.2. Feasibility and Limitations

In this section, we present the feasibility of using DCC calibration for a UV and VIS hyperspectral sensor based on our updated DCC detection method. As mentioned in Section 1, DCC calibration has been generally used with meteorological sensors to update radiometric calibration coefficients, which typically change over the course of the operation period. A meteorological sensor can be calibrated with the well-calibrated sensor after the normalization of various observation conditions, such as the angle dependence, spectral response functions, and different center wavelengths. DCC calibration for environmental sensors still has a long way to go in terms of normalization, but in this study, we present preliminary results for the temporal variability in the TROPOMI DCC observations.

Figure 15a presents the seasonal distribution of TROPOMI DCCs for data collected over the period of a year with probability density functions (PDFs). Even though the number of DCCs is not sufficient to calculate a representative PDF for the observations, the PDFs have similar distribution patterns regardless of the number of DCCs in each season. However, given that distribution modes are generally used to monitor the calibration accuracy of meteorological sensors, the PDF modes are too variable since the bidirectional reflectivity of the DCCs and the disparity in the cloud optical properties have not been sufficiently accounted for so far. However, the temporal variability caused by these uncertainties could cancel each other out as the reflectivity ratio between two different wavelengths represents in Figure 15b. The ratio of DCC reflectivity at 354 and 397 nm is used because reflectivity at 397 nm (Ca II H line) is affected both by scattered and directly transmitted light. Even with the highly expected variability, the ratio of the mean reflectivity at both wavelengths appears relatively stable within 0.99–1.01.

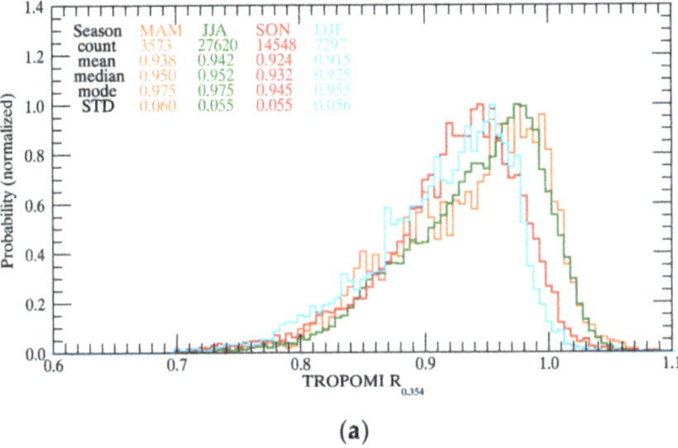

(**a**)

Figure 15. *Cont.*

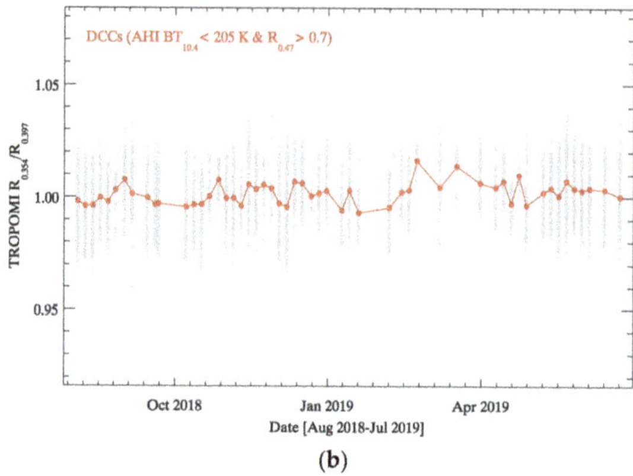

Figure 15. (**a**) Probability density function for TROPOMI R$_{0.354}$ over time (MAM: March to May; JJA: June to August; SON: September to November; DJF: December to February) and (**b**) time series of mean reflectivity ratio of R$_{0.354}$ and R$_{0.397}$ (grey diamonds are individual values) for the DCCs detected using the updated DCC detection method.

5. Conclusions

As the first UV–VIS hyperspectral sensor onboard a GEO satellite, the GEMS covers the Asia-Pacific region, including the TWP region. To develop a vicarious calibration approach based on the current availability of calibration targets, the present study tests DCCs to determine whether optically thick clouds provide a sufficiently stable and bright signal to allow the radiometric calibration of sensors with different hardware characteristics and observation conditions especially in the UV–VIS spectral region. For feasibility testing, the VIS and IR channels of the AHI are used with UV–VIS hyperspectral data from the OMI and TROPOMI, as a surrogate for the GEMS. To mitigate the calibration uncertainty caused by degradation and high-frequency perturbations of the instrument optical paths, reflectivity (i.e., the ratio between radiance and irradiance) is used. The cloud reflectivity is calculated by taking account of the solar zenith angle, the satellite zenith angle, and Rayleigh scattering above the clouds.

To ensure a sufficient number of DCCs over the GEMS observation area, AHI data from a year-long period that match the spatial and temporal resolution of the GEMS are analyzed. The DCCs detected using the conventional approach (i.e., thermal temperature tests and uniformity tests) have a clear seasonality, with a maximum in September and a minimum in April. Spatially, the viewing zenith angle also limits the number of DCCs because the AHI observes the target area with a higher viewing zenith angle compared to the GEMS. This limitation of the satellite zenith angle is expected to be improved with the AMI onboard GK2A, which has the potential to be collocated with the GEMS as stationed nearby at 128.2°E. Even with these limitations, DCCs occur in more than 200 pixels on average in a single observation scene, which appears to be sufficient for the proposed statistical approach considering the observation frequency and the spatial resolution of the GEMS.

Although the number of DCCs detected by the OMI and TROPOMI is significantly different, mainly due to the poor spatial resolution and degraded quality of OMI data, a comparison between the estimated spectral reflectivity of the DCCs shows comparable results even with clear differences in sensor characteristics, viewing geometry, and the number of data points. Given that more accurate calibration is essential for achieving the final goal of the mission, the results look promising in terms of applying the proposed method to various UV and VIS environmental sensors for inter-calibration. However, a closer inspection of the reflectivity spectra shows that there is high variability in the standard deviation (up to 10%), which is mainly due to the false classification of thin cirrus clouds as DCCs, which have a cold cloud temperature with a low optical depth. Furthermore, inspection of an alternative approach using only reflectivity tests for DCC detection leads to the false detection of warm clouds having a high reflectivity and a lower cloud top altitude. Thus, we devise an updated DCC detection approach using both thermal and reflectivity tests to screen out cold, thin cirrus clouds and bright, warm clouds. Based on the variation in the statistical parameters of DCC reflectivity with different reflectivity threshold values, the threshold value for the reflectivity test is determined to be 0.7, which produces a distribution close to normal with the location values of the distribution converging and retains as many observations as possible. However, certain issues remain that lead to a spread in reflectivity caused by the variation in cloud properties and angle dependence, including the bidirectional reflectivity distribution of DCCs. The long-term variability in DCC reflectivity based on the updated detection method needs to be analyzed, with the results used to minimize such variation and to demonstrate the applicability of the new approach for hyperspectral UV–VIS sensors. Additionally, since the updated DCC detection can still be dependent on the calibration accuracy of the meteorological sensor such as AHI (further AMI), it must also be investigated hereafter to properly perform the DCC calibration for the environmental sensors.

Author Contributions: M.-H.A. designed and supervised the study; Y.L. performed the experiments, analyzed the data, and prepared the manuscript; M.K. contributed to the analysis of results. All authors contributed to the edition of the manuscript. All authors have read and agreed to the published version of the manuscript.

Funding: This research is supported by the Basic Science Research Program through the National Research Foundation of Korea (NRF), funded by the Ministry of Education (2018R1A6A1A08025520). Also, this research was supported by the Korea Ministry of Environment (MOE) as "Public Technology Program based Environmental Policy (2017000160002).

Acknowledgments: We would like to thank the National Meteorological Satellite Center (NMSC) of the Korea Meteorological Administration (KMA) for providing the AHI Level 1B data. The valuable comments from the anonymous reviewers are also greatly helpful to improve the manuscript.

Conflicts of Interest: The authors declare no conflict of interest.

References

1. Kim, H.-O.; Kim, H.-S.; Lim, H.-S.; Choi, H.-J. Space-Based Earth Observation Activities in South Korea [Space Agencies]. *IEEE Geosci. Remote Sens. Mag.* **2015**, *3*, 34–39. [CrossRef]
2. Choi, W.J.; Moon, K.-J.; Yoon, J.; Cho, A.; Kim, S.; Lee, S.; Ko, D.H.; Kim, J.; Ahn, M.H.; Kim, D.-R.; et al. Introducing the geostationary environment monitoring spectrometer. *J. Appl. Remote Sens.* **2019**, *13*, 1. [CrossRef]
3. Veefkind, J.P.; De Leeuw, G.; Stammes, P.; Koelemeijer, R.B.A. Regional distribution of aerosol over land, derived from ATSR-2 and GOME. *Remote Sens. Environ.* **2000**, *74*, 377–386. [CrossRef]
4. Wan, Z.; Zhang, Y.; Zhang, Q.; Li, Z.L. Validation of the land-surface temperature products retrieved from terra moderate resolution imaging spectroradiometer data. *Remote Sens. Environ.* **2002**, *83*, 163–180. [CrossRef]
5. Boersma, K.F.; Eskes, H.J.; Veefkind, J.P.; Brinksma, E.J.; Van Der A, R.J.; Sneep, M.; Van Den Oord, G.H.J.; Levelt, P.F.; Stammes, P.; Gleason, J.F.; et al. Atmospheric Chemistry and Physics Near-real time retrieval of tropospheric NO_2 from OMI. *Atmos. Chem. Phys* **2007**, *7*, 2103–2118. [CrossRef]
6. Gloudemans, A.M.S.; Schrijver, H.; Hasekamp, O.P.; Aben, I. Error analysis for CO and CH_4 total column retrievals from SCIAMACHY 2.3 µm spectra. *Atmos. Chem. Phys.* **2008**, *8*, 3999–4017. [CrossRef]
7. Loyola, D.G.; Koukouli, M.E.; Valks, P.; Balis, D.S.; Hao, N.; Van Roozendael, M.; Spurr, R.J.D.; Zimmer, W.; Kiemle, S.; Lerot, C.; et al. The GOME-2 total column ozone product: Retrieval algorithm and ground-based validation. *J. Geophys. Res. Atmos.* **2011**, *116*, 1–11. [CrossRef]
8. Yoshida, Y.; Ota, Y.; Eguchi, N.; Kikuchi, N.; Nobuta, K.; Tran, H.; Morino, I.; Yokota, T. Retrieval algorithm for CO_2 and CH_4 column abundances from short-wavelength infrared spectral observations by the Greenhouse gases observing satellite. *Atmos. Meas. Tech.* **2011**, *4*, 717–734. [CrossRef]
9. Dinguirard, M.; Slater, P.N. Calibration of space-multispectral imaging sensors: A review. *Remote Sens. Environ.* **1999**, *68*, 194–205. [CrossRef]

10. Kowalewski, M.G.; Jaross, G.; Cebula, R.P.; Taylor, S.L.; van den Oord, G.H.J.; Dobber, M.R.; Dirksen, R. Evaluation of the Ozone Monitoring Instrument's pre-launch radiometric calibration using in-flight data. In Proceedings of the Optics and Photonics 2005, San Diego, CA, USA, 22 August 2005.

11. Jaross, G.; Warner, J. Use of Antarctica for validating reflected solar radiation measured by satellite sensors. *J. Geophys. Res. Atmos.* **2008**, *113*, 1–13. [CrossRef]

12. Brook, A.; Dor, E. Ben Supervised vicarious calibration (SVC) of hyperspectral remote-sensing data. *Remote Sens. Environ.* **2011**, *115*, 1543–1555. [CrossRef]

13. Sterckx, S.; Livens, S.; Adriaensen, S. Rayleigh, Deep Convective Clouds, and Cross-Sensor Desert Vicarious Calibration Validation for the PROBA-V Mission. *IEEE Trans. Geosci. Remote Sens.* **2013**, *51*, 1437–1452. [CrossRef]

14. Bhatt, R.; Doelling, D.R.; Wu, A.; Xiong, X.; Scarino, B.R.; Haney, C.O.; Gopalan, A. Initial stability assessment of S-NPP VIIRS reflective solar band calibration using invariant desert and deep convective cloud targets. *Remote Sens.* **2014**, *6*, 2809–2826. [CrossRef]

15. Di Giuseppe, F.; Tompkins, A.M. Three-dimensional radiative transfer in tropical deep convective clouds. *J. Geophys. Res. D Atmos.* **2003**, *108*, 4741. [CrossRef]

16. Luo, Z.; Liu, G.Y.; Stephens, G.L. CloudSat adding new insight into tropical penetrating convection. *Geophys. Res. Lett.* **2008**, *35*, 2–6. [CrossRef]

17. Setvák, M.; Lindsey, D.T.; Rabin, R.M.; Wang, P.K.; Demeterová, A. Indication of water vapor transport into the lower stratosphere above midlatitude convective storms: Meteosat Second Generation satellite observations and radiative transfer model simulations. *Atmos. Res.* **2008**, *89*, 170–180. [CrossRef]

18. Fan, J.; Leung, L.R.; Rosenfeld, D.; Chen, Q.; Li, Z.; Zhang, J.; Yan, H. Microphysical effects determine macrophysical response for aerosol impacts on deep convective clouds. *Proc. Natl. Acad. Sci. USA* **2013**, *110*, E4581–E4590. [CrossRef]

19. Loeb, N.G.; Manalo-Smith, N.; Kato, S.; Miller, W.F.; Gupta, S.K.; Minnis, P.; Wielicki, B.A. Angular Distribution Models for Top-of-Atmosphere Radiative Flux Estimation from the Clouds and the Earth's Radiant Energy System Instrument on the Tropical Rainfall Measuring Mission Satellite. Part I: Methodology. *J. Appl. Meteorol.* **2003**, *42*, 240–265. [CrossRef]

20. Yongxiang, H.; Wielicki, B.A.; Ping, Y.; Stackhouse, P.W.; Lin, B.; Young, D.F. Application of deep convective cloud albedo observation to satellite-based study of the terrestrial atmosphere: Monitoring the stability of spaceborne measurements and assessing absorption anomaly. *IEEE Trans. Geosci. Remote Sens.* **2004**, *42*, 2594–2599. [CrossRef]

21. Sohn, B.-J.; Ham, S.-H.; Yang, P. Possibility of the Visible-Channel Calibration Using Deep Convective Clouds Overshooting the TTL. *J. Appl. Meteorol. Climatol.* **2009**, *48*, 2271–2283. [CrossRef]

22. Bhatt, R.; Doelling, D.; Scarino, B.; Haney, C.; Gopalan, A. Development of Seasonal BRDF Models to Extend the Use of Deep Convective Clouds as Invariant Targets for Satellite SWIR-Band Calibration. *Remote Sens.* **2017**, *9*, 1061. [CrossRef]

23. Schmetz, J.; Tjemkes, S.A.; Gube, M.; Van De Berg, L. Monitoring deep convection and convective overshooting with METEOSAT. *Adv. Sp. Res.* **1997**, *19*, 433–441. [CrossRef]

24. Doelling, D.R.; Nguyen, L.; Minnis, P. On the use of deep convective clouds to calibrate AVHRR data. In Proceedings of the Optical Science and Technology, the SPIE 49th Annual Meeting, Denver, CO, USA, 26 October 2004.

25. Minnis, P.; Doelling, D.R.; Nguyen, L.; Miller, W.F.; Chakrapani, V. Assessment of the Visible Channel Calibrations of the VIRS on TRMM and MODIS on Aqua and Terra. *J. Atmos. Ocean. Technol.* **2008**, *25*, 385–400. [CrossRef]

26. Doelling, D.; Morstad, D.; Bhatt, R.; Scarino, B. Algorithm Theoretical Basis Document (ATBD) for Deep Convective Cloud (DCC) technique of calibrating GEO sensors with Aqua-MODIS for GSICS. Available online: http://gsics.atmos.umd.edu/pub/Development/AtbdCentral/GSICS_ATBD_DCC_NASA_2011_09.pdf (accessed on 31 January 2020).

27. Doelling, D.R.; Morstad, D.; Scarino, B.R.; Bhatt, R.; Gopalan, A. The Characterization of Deep Convective Clouds as an Invariant Calibration Target and as a Visible Calibration Technique. *IEEE Trans. Geosci. Remote Sens.* **2013**, *51*, 1147–1159. [CrossRef]

28. Wang, W.; Cao, C. DCC Radiometric Sensitivity to Spatial Resolution, Cluster Size, and LWIR Calibration Bias Based on VIIRS Observations. *J. Atmos. Ocean. Technol.* **2015**, *32*, 48–60. [CrossRef]

29. Wang, W.; Cao, C. Monitoring the NOAA operational VIIRS RSB and DNB calibration stability using monthly and semi-monthly deep convective clouds time series. *Remote Sens.* **2016**, *8*, 1–19. [CrossRef]

30. Yu, F.; Wu, X. Radiometric inter-calibration between Himawari-8 AHI and S-NPP viirs for the solar reflective bands. *Remote Sens.* **2016**, *8*, 1–16. [CrossRef]
31. Ai, Y.; Li, J.; Shi, W.; Schmit, T.J.; Cao, C.; Li, W. Deep convective cloud characterizations from both broadband imager and hyperspectral infrared sounder measurements. *J. Geophys. Res.* **2017**, *122*, 1700–1712. [CrossRef]
32. Schenkeveld, V.M.E.; Jaross, G.; Marchenko, S.; Haffner, D.; Kleipool, Q.L.; Rozemeijer, N.C.; Veefkind, J.P.; Levelt, P.F. In-flight performance of the Ozone Monitoring Instrument. *Atmos. Meas. Tech.* **2017**, *10*, 1957–1986. [CrossRef]
33. Sneep, M.; de Haan, J.F.; Stammes, P.; Wang, P.; Vanbauce, C.; Joiner, J.; Vasilkov, A.P.; Levelt, P.F. Three-way comparison between OMI and PARASOL cloud pressure products. *J. Geophys. Res.* **2008**, *113*, 1–11. [CrossRef]
34. Vasilkov, A.; Joiner, J.; Spurr, R.; Bhartia, P.K.; Levelt, P.; Stephens, G. Evaluation of the OMI cloud pressures derived from rotational Raman scattering by comparisons with other satellite data and radiative transfer simulations. *J. Geophys. Res.* **2008**, *113*, D15S19. [CrossRef]
35. Kim, J.; Jeong, U.; Ahn, M.-H.; Kim, J.H.; Park, R.J.; Lee, H.; Song, C.H.; Choi, Y.-S.; Lee, K.-H.; Yoo, J.-M.; et al. New Era of Air Quality Monitoring from Space: Geostationary Environment Monitoring Spectrometer (GEMS). *Bull. Am. Meteorol. Soc.* **2019**, *84*, 00. [CrossRef]
36. Lambert, J.-C.; Keppens, A.; Hubert, D.; Langerock, B.; Eichmann, K.-U.; Kleipool, Q.; Sneep, M.; Verhoelst, T.; Wagner, T.; Weber, M.; et al. *Quarterly Validation Report of the Copernicus Sentinel-5 Precursor Operational Data Products #04: April 2018–August 2019*; Tropomi: De Bilt, The Netherlands, 20 April; pp. 1–125.
37. Hong, G.; Heygster, G.; Miao, J.; Kunzi, K. Detection of tropical deep convective clouds from AMSU-B water vapor channels measurements. *J. Geophys. Res. D Atmos.* **2005**, *110*, 1–15. [CrossRef]
38. Liu, C.; Zipser, E.J.; Nesbitt, S.W. Global distribution of tropical deep convection: Different perspectives from TRMM infrared and radar data. *J. Clim.* **2007**, *20*, 489–503. [CrossRef]
39. Sassen, K.; Wang, Z.; Liu, D. Cirrus clouds and deep convection in the tropics: Insights from CALIPSO and CloudSat. *J. Geophys. Res.* **2009**, *114*, D00H06. [CrossRef]
40. Alcala, C.M.; Dessler, A.E. Observations of deep convection in the tropics using the Tropical Rainfall Measuring Mission (TRMM) precipitation radar. *J. Geophys. Res. Atmos.* **2002**, *107*, 4792. [CrossRef]
41. Jiang, J.H.; Wang, B.; Goya, K.; Hocke, K.; Eckermann, S.D.; Ma, J.; Wu, D.L.; Read, W.G. Geographical distribution and interseasonal variability of tropical deep convection: UARS MLS observations and analyses. *J. Geophys. Res. Atmos.* **2004**, *109*. [CrossRef]
42. Stubenrauch, C.J.; Rossow, W.B.; Kinne, S.; Ackerman, S.; Cesana, G.; Chepfer, H.; Di Girolamo, L.; Getzewich, B.; Guignard, A.; Heidinger, A.; et al. Assessment of Global Cloud Datasets from Satellites: Project and Database Initiated by the GEWEX Radiation Panel. *Bull. Am. Meteorol. Soc.* **2013**, *94*, 1031–1049. [CrossRef]
43. Ahmad, Z.; Bhartia, P.K.; Krotkov, N. Spectral properties of backscattered UV radiation in cloudy atmospheres. *J. Geophys. Res. D Atmos.* **2004**, *109*, D01201. [CrossRef]
44. Hewison, T.J.; Wu, X.; Yu, F.; Tahara, Y.; Hu, X.; Kim, D.; Koenig, M. GSICS inter-calibration of infrared channels of geostationary imagers using metop/IASI. *IEEE Trans. Geosci. Remote Sens.* **2013**, *51*, 1160–1170. [CrossRef]
45. Dobber, M.; Kleipool, Q.; Dirksen, R.; Levelt, P.; Jaross, G.; Taylor, S.; Kelly, T.; Flynn, L.; Leppelmeier, G.; Rozemeijer, N. Validation of Ozone Monitoring Instrument level 1b data products. *J. Geophys. Res.* **2008**, *113*, 1–12. [CrossRef]
46. Bodhaine, B.A.; Wood, N.B.; Dutton, E.G.; Slusser, J.R. On Rayleigh optical depth calculations. *J. Atmos. Ocean. Technol.* **1999**, *16*, 1854–1861. [CrossRef]
47. Pinto da Silva Neto, C.; Alves Barbosa, H.; Assis Beneti, C.A. A method for convective storm detection using satellite data. *Atmósfera* **2016**, *29*, 343–358. [CrossRef]
48. Joiner, J.; Bhartia, P.K.; Cebula, R.P.; Hilsenrath, E.; McPeters, R.D.; Park, H. Rotational Raman scattering (Ring effect) in satellite backscatter ultraviolet measurements. *Appl. Opt.* **1995**, *34*, 4513. [CrossRef] [PubMed]
49. Loyola, D.G.; Gimeno García, S.; Lutz, R.; Argyrouli, A.; Romahn, F.; Spurr, R.J.D.; Pedergnana, M.; Doicu, A.; Molina García, V.; Schüssler, O. The operational cloud retrieval algorithms from TROPOMI on board Sentinel-5 Precursor. *Atmos. Meas. Tech* **2018**, *11*, 409–427. [CrossRef]

MDPI
St. Alban-Anlage 66
4052 Basel
Switzerland
Tel. +41 61 683 77 34
Fax +41 61 302 89 18
www.mdpi.com

Remote Sensing Editorial Office
E-mail: remotesensing@mdpi.com
www.mdpi.com/journal/remotesensing